Handbook of Petrochemical Processes

Chemical Industries

Founding Editor:
Heinz Heinemann

Series Editor:
James G. Speight

The **Chemical Industries** Series offers in-depth texts related to all aspects of the chemical industries from experts and leaders in academia and industry. The titles explore recent developments and best practices that facilitate successful process control and commercialization of industrial processes and products to help meet changing market demands and match the stringent emission standards. The series focuses on technologies, process development improvements, and new applications to ensure proper performance in industrial units and evaluation of novel process designs that will result in production of valuable products from efficient and economical processes.

Modeling of Processes and Reactors for Upgrading of Heavy Petroleum
Jorge Ancheyta

Synthetics, Mineral Oils, and Bio-Based Lubricants: Chemistry and Technology,
Second Edition
Leslie R. Rudnick

Transport Phenomena Fundamentals, Third Edition
Joel Plawsky

The Chemistry and Technology of Petroleum, Fifth Edition
James G. Speight

Refining Used Lubricating Oils
James Speight and Douglas I. Exall

Petroleum and Gas Field Processing, Second Edition
Hussein K. Abdel-Aal, Mohamed A. Aggour, and Mohamed A. Fahim

Handbook of Refinery Desulfurization
Nour Shafik El-Gendy and James G. Speight

Handbook of Petroleum Refining
James G. Speight

Advances in Refining Catalysis
Deniz Uner

Lubricant Additives: Chemistry and Applications, Third Edition
Leslie R. Rudnick

Handbook of Petrochemical Processes
James G. Speight

For more information about this series, please visit: https://www.crcpress.com/Chemical-Industries/book-series/CRCCHEMINDUS

Handbook of Petrochemical Processes

James G. Speight

CRC Press
Taylor & Francis Group
Boca Raton London New York

CRC Press is an imprint of the
Taylor & Francis Group, an **informa** business

CRC Press
Taylor & Francis Group
6000 Broken Sound Parkway NW, Suite 300
Boca Raton, FL 33487-2742

First issued in paperback 2021

© 2019 by Taylor & Francis Group, LLC
CRC Press is an imprint of Taylor & Francis Group, an Informa business

No claim to original U.S. Government works

ISBN 13: 978-1-03-223623-0 (pbk)
ISBN 13: 978-1-4987-2970-3 (hbk)

DOI: 10.1201/9780429155611

Library of Congress Cataloging-in-Publication Data

Names: Speight, James G., author.
Title: Handbook of petrochemical processes / James G. Speight.
Description: Boca Raton, FL : CRC Press/Taylor & Francis Group, [2019] |
Series: Chemical industries
Identifiers: LCCN 2019003675 | ISBN 9781498729703 (hardback : acid-free paper)
Subjects: LCSH: Petroleum—Refining. | Petroleum chemicals. | Chemical processes.
Classification: LCC TP690.3 .S64 2019 | DDC 665.5/3—dc23
LC record available at https://lccn.loc.gov/2019003675

Visit the Taylor & Francis Website at
http://www.taylorandfrancis.com

and the CRC Press Web site at
http://www.crcpress.com

Contents

Preface

The petrochemical industry had its modern origins in the later years of the 19th century. However, the production of products from naturally occurring bitumen is a much older industry. There is evidence that the ancient Bronze Age towns of Tuttul (Syria) and Hit (also spelled Heet, Iraq) used bitumen from seepages as a caulking material and mastic. Also, Arabian scientists knew that attempts to distill the bitumen caused it to decompose into a variety of products.

By the time that the 19th century had dawned, it was known that kerosene, a fuel for heating and cooking, was the primary product of the petroleum industry in the 1800s. Rockefeller and other refinery owners considered gasoline a useless by-product of the distillation process. But all of that changed around 1900 when electric lights began to replace kerosene lamps, and automobiles came in the scene. New petroleum fuels were also needed to power the ships and airplanes used in World War I. After the war, an increasing number of farmers began to operate tractors and other equipment powered by oil. The growing demand for petrochemicals and the availability of petroleum and natural gas caused the industry to quickly expand in the 1920s and 1930s. During World War II, vast amounts of oil were produced and made into fuels and lubricants. The United States supplied more than 80% of the aviation gasoline used by the allies during the war. American oil refineries also manufactured synthetic rubber, toluene (an ingredient in TNT), medicinal oils, and other key military supplies.

The term *petrochemicals* represents a large group of chemicals manufactured from petroleum and natural gas as distinct from fuels and other products that are also derived from petroleum and natural gas by a variety of processes and used for a variety of commercial purposes. Petrochemical products include such items as plastics, soaps and detergents, solvents, drugs, fertilizers, pesticides, explosives, synthetic fibers and rubbers, paints, epoxy resins, and flooring and insulating materials. Petrochemicals are found in products as diverse as aspirin, luggage, boats, automobiles, aircraft, polyester clothes, and recording discs and tapes. It is the changes in product demand that have been largely responsible for the evolution of the petroleum industry from the demand for asphalt mastic used in ancient times to the current high demand for gasoline, other liquid fuels an products as well increasing demand for as a wide variety of petrochemicals.

As a result, the petrochemical industry is a huge field that encompasses many commercial chemicals and polymers. The organic chemicals produced in the largest volumes are methanol, ethylene, propylene, butadiene, benzene, toluene, and xylenes. Ethylene, propylene, and butadiene, along with butylenes, are collectively called olefins, which belong to a class of unsaturated aliphatic hydrocarbons having the general formula C_nH_{2n}. Olefins contain one or more double bonds, which make them chemically reactive. Benzene, toluene, and xylenes, commonly referred to as aromatics, are unsaturated cyclic hydrocarbons containing one or more rings. Olefins, aromatics, and methanol are precursors to a variety of chemical products and are generally referred to as primary petrochemicals. Given the number of organic chemicals and the variety and multitude of ways by which they are converted to consumer and industrial products, this report focuses primarily on these seven petrochemicals, their feedstock sources, and their end uses.

Furthermore, because ethylene and propylene are the major building blocks for petrochemicals, alternative ways for their production have always been sought. The main route for producing ethylene and propylene is steam cracking, which is an energy-extensive process. Fluid catalytic cracking (FCC) is also used to supplement the demand for these low molecular weight olefins.

Basic chemicals and plastics are the key building blocks for manufacture of a wide variety of durable and nondurable consumer goods. Considering the items we encounter every day—the clothes we wear, construction materials used to build our homes and offices, a variety of household appliances and electronic equipment, food and beverage packaging, and many products used in various modes of transportation—chemical and plastic materials provide the fundamental building blocks that enable the manufacture of the vast majority of these goods. Demand for chemicals and plastics is driven by global economic conditions, which are directly linked to demand for consumer goods.

The search for alternative ways to produce monomers and chemicals from sources other than crude oil. In fact, Fisher–Tropsch technology, which produces in addition to fuels, low molecular weight olefins, could enable nonpetroleum feedstocks (such as extra heavy oil, tar sand bitumen, coal, oil shale, and biomass) to be used as feedstocks for petrochemicals.

In the book, the reactions and processes involved in transforming petroleum-based hydrocarbons into the chemicals that form the basis of the multi-billion dollar petrochemical industry are reviewed and described. In addition, the book includes information on new process developments for the production of raw materials and intermediates for petrochemicals. This book will provide the readers with a valuable source of information containing insights into petrochemical reactions and products, process technology, and polymer synthesis. The book will also provide the reader with descriptions of role of nonpetroleum sources in the production of chemicals and present to the reader alternate routes to chemicals.

Dr. James G. Speight
CD&W Inc.,
Laramie,
Wyoming 82070, USA

About the Author

 **Dr. James G. Speight** has doctorate degrees in Chemistry, Geological Sciences, and Petroleum Engineering and is the author of more than 80 books in petroleum science, petroleum engineering, and environmental sciences.

Dr. Speight has more than 50 years of experience in areas associated with (i) the properties, recovery, and refining of reservoir fluids, conventional petroleum, heavy oil, and tar sand bitumen; (ii) the properties and refining of natural gas, gaseous fuels; (iii) the production and properties of petrochemicals; (iv) the properties and refining of biomass, biofuels, biogas, and the generation of bioenergy; and (v) the environmental and toxicological effects of fuels. His work has also focused on safety issues, environmental effects, remediation, and safety issues as well as reactors associated with the production and use of fuels and biofuels. He is the author of more than 70 books in petroleum science, petroleum engineering, biomass and biofuels, and environmental sciences.

Although he has always worked in private industry which focused on contract-based work, he has served as a Visiting Professor in the College of Science, University of Mosul, Iraq and has also been a Visiting Professor in Chemical Engineering at the following universities: University of Missouri-Columbia, the Technical University of Denmark, and the University of Trinidad and Tobago.

In 1996, Dr. Speight was elected to the Russian Academy of Sciences and awarded the Gold Medal of Honor that same year for outstanding contributions to the field of petroleum sciences. In 2001, he received the Scientists without Borders Medal of Honor of the Russian Academy of Sciences and was also awarded the Einstein Medal for outstanding contributions and service in the field of Geological Sciences. In 2005, the Academy awarded Dr. Speight the Gold Medal—Scientists without Frontiers, Russian Academy of Sciences, in recognition of Continuous Encouragement of Scientists to Work Together Across International Borders. In 2007, Dr. Speight received the Methanex Distinguished Professor award at the University of Trinidad and Tobago in recognition of excellence in research.

1 The Petrochemical Industry

1.1 INTRODUCTION

The constant demand for hydrocarbon products such as liquid fuels is one of the major driving forces behind the petroleum industry. However, the other driving force is a major group of hydrocarbon products (petrochemicals) that are the basis of a major industry. There is a myriad of products that have evolved through the short life of the petroleum industry, either as bulk fractions or as single hydrocarbon products (Tables 1.1 and 1.2). And the complexities of product composition have matched the evolution of the products. In fact, it is the complexity of product composition that has served the industry well and, at the same time, had an adverse effect on product use.

A *petrochemical* is a chemical product developed from petroleum that has become an essential part of the modern chemical industry (Table 1.3) (Speight, 1987; Parkash, 2003; Gary et al., 2007; Speight, 2014; Hsu and Robinson, 2017; Speight, 2017). The *chemical industry* is, in fact, the *chemical process industry* by which a variety of chemicals are manufactured. The chemical process industry is, in fact, subdivided into other categories that are: (i) the chemicals and allied product industries in which chemicals are manufactured from a variety of feedstocks and may then be put to further use, (ii) the rubber and miscellaneous product industries which focus on the manufacture of rubber and plastic materials, and (iii) petroleum refining and related industries which, on the basis of the following chapters in this text, is now self-explanatory. Thus, the petrochemical industry falls under the subcategory of *petroleum and related industries*.

In the context of this book, the definition of petrochemicals excludes fuel products, lubricants, asphalt, and petroleum coke but does include chemicals produced from other feedstocks such as coal, oil shale, and biomass, which could well be the sources of chemicals in the future. Thus, petrochemicals are, in the strictest sense, different to petroleum products insofar, as the petrochemicals are the basic building blocks of the chemical industry. Petrochemicals are found in products as diverse as plastics, polymers, synthetic rubber, synthetic fibers, detergents, industrial chemicals, and fertilizers (Table 1.3). Petrochemicals are used for production of several feedstocks and monomers and monomer precursors. The monomers after polymerization process create several polymers, which ultimately are used to produce gels, lubricants, elastomers, plastics, and fibers.

By way of definition and clarification as it applies to the petrochemical and chemical industry, primary raw materials are naturally occurring substances that have not been subjected to chemical

TABLE 1.1
The Various Distillation Fractions of Petroleum

Product	Lower Carbon Number[a]	Upper Carbon Number[a]	Lower b.p. °C[a]	Upper b.p. °C[a]	Lower b.p. °F[a]	Upper b.p. °F[a]
Liquefied petroleum gas	C3	C4	−42	−1	−44	31
Naphtha	C5	C17	36	302	97	575
Kerosene	C8	C18	126	258	302	575
Light gas oil	C12	>C20	216	421	>345	>650
Heavy gas oil	>C20		>345		>650	
Residuum	>C20		>345		>660	

[a] The carbon number and boiling point difficult to assess accurately because of variations in production parameters from refinery-to-refinery and are inserted for illustrative purposes only.

TABLE 1.2

Properties of Hydrocarbon Products from Petroleum

	Molecular Weight	Specific Gravity	Boiling Point °F	Ignition Temperature °F	Flash Point °F	Flammability Limits in Air % v/v
Benzene	78.1	0.879	176.2	1,040	12	1.35–6.65
n-Butane	58.1	0.601	31.1	761	−76	1.86–8.41
isobutane	58.1		10.9	864	−117	1.80–8.44
n-Butene	56.1	0.595	21.2	829	Gas	1.98–9.65
isobutene	56.1		19.6	869	Gas	1.8–9.0
Diesel fuel	170–198	0.875			100–130	
Ethane	30.1	0.572	−127.5	959	Gas	3.0–12.5
Ethylene	28.0		−154.7	914	Gas	2.8–28.6
Fuel oil No. 1		0.875	304–574	410	100–162	0.7–5.0
Fuel oil No. 2		0.920		494	126–204	
Fuel oil No. 4	198.0	0.959		505	142–240	
Fuel oil No. 5		0.960			156–336	
Fuel oil No. 6		0.960			150	
Gasoline	113.0	0.720	100–400	536	−45	1.4–7.6
n-Hexane	86.2	0.659	155.7	437	−7	1.25–7.0
n-Heptane	100.2	0.668	419.0	419	25	1.00–6.00
Kerosene	154.0	0.800	304–574	410	100–162	0.7–5.0
Methane	16.0	0.553	−258.7	900–1,170	Gas	5.0–15.0
Naphthalene	128.2		424.4	959	174	0.90–5.90
Neohexane	86.2	0.649	121.5	797	−54	1.19–7.58
Neopentane	72.1		49.1	841	Gas	1.38–7.11
n-Octane	114.2	0.707	258.3	428	56	0.95–3.2
isooctane	114.2	0.702	243.9	837	10	0.79–5.94
n-Pentane	72.1	0.626	97.0	500	−40	1.40–7.80
isopentane	72.1	0.621	82.2	788	−60	1.31–9.16
n-Pentene	70.1	0.641	86.0	569	—	1.65–7.70
Propane	44.1		−43.8	842	Gas	2.1–10.1
Propylene	42.1		−53.9	856	Gas	2.00–11.1
Toluene	92.1	0.867	321.1	992	40	1.27–6.75
Xylene	106.2	0.861	281.1	867	63	1.00–6.00

changes after being recovered. Currently, through a variety of intermediates petroleum and natural gas are the main sources of the raw materials because they are the least expensive, most readily available, and can be processed most easily into the primary petrochemicals. An aromatic petrochemical is also an organic chemical compound but one that contains, or is derived from, the basic benzene ring system.

Primary petrochemicals include: (i) olefin derivatives such as ethylene, propylene, and butadiene; (ii) aromatic derivatives such as benzene, toluene, and the isomers of xylene (BTX); and (iii) methanol. However, although petroleum contains different types of hydrocarbon derivatives, not all hydrocarbon derivatives are used in producing petrochemicals. Petrochemical analysis has made it possible to identify some major hydrocarbon derivatives used in producing petrochemicals (Speight, 2015). From the multitude of hydrocarbon derivatives, those hydrocarbon derivatives serving as major raw materials used by petrochemical industries in the production of petrochemicals are: (i) the raw materials obtained from natural gas processing such as methane, ethane, propane, and

---

TABLE 1.3

Examples of Products from the Petrochemical Industry

Group	Areas of Use
Plastics and polymers	Agricultural water management
	Packaging
	Automobiles
	Telecommunications
	Health and hygiene
	Transportation
Synthetic rubber	Transportation industry
	Electronics
	Adhesives
	Sealants
	Coatings
Synthetic fibers	Textile
	Transportation
	Industrial fabrics
Detergents	Health and hygiene
Industrial chemicals	Pharmaceuticals
	Pesticides
	Explosives
	Surface coating
	Dyes
	Lubricating oil additives
	Adhesives
	Oil field chemicals
	Antioxidants
	Printing ink
	Paints
	Corrosion inhibitors
	Solvents
	Perfumes
	Food additives
Fertilizers	Agriculture

---

butane; (ii) the raw materials obtained from petroleum refineries such as naphtha and gas oil; and (iii) the raw materials such as benzene, toluene and the xylene isomers obtained when extracted from reformate (the product of reforming processes through catalysts called catalytic reformers in petroleum refineries (Parkash, 2003; Gary et al., 2007; Speight, 2008, 2014; Hsu and Robinson, 2017; Speight, 2017).

Thus, petrochemicals are chemicals derived from petroleum and natural gas and, for convenience of identification, petrochemicals can be divided into two groups: (i) primary petrochemicals and (ii) intermediates and derivatives (Figure 1.1).

Primary petrochemicals include: olefins (ethylene, propylene, and butadiene), aromatics (benzene, toluene, and xylenes), and methanol. Petrochemical intermediates are generally produced by chemical conversion of primary petrochemicals to form more complicated derivative products. Petrochemical derivatives can be made in a variety of ways: (i) directly from primary petrochemicals; (ii) through intermediate products which still contain only carbon and hydrogen; and (iii) through intermediates which incorporate chlorine, nitrogen, or oxygen in the

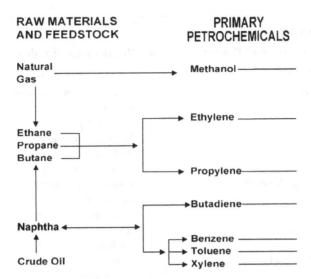

FIGURE 1.1   Raw materials and primary petrochemicals. (Speight, J.G. 2007. *The Chemistry and Technology of Petroleum.* 4th Edition. CRC Press, Boca Raton, FL. Figure 27.3, p. 784.)

finished derivative. In some cases, they are finished products; in others, more steps are needed to be arrived at the desired composition.

Moreover, petrochemical feedstocks can be classified into several general groups: olefins, aromatics, and methanol; a fourth group includes inorganic compounds and synthesis gas (mixtures of carbon monoxide and hydrogen). In many instances, a specific chemical included among the petrochemicals may also be obtained from other sources, such as coal, coke, or vegetable products. For example, materials such as benzene and naphthalene can be made from either petroleum or coal, while ethyl alcohol may be of petrochemical or vegetable origin.

Thus, primary petrochemicals are not end products, but are the chemical building blocks for a wide range of chemical and manufactured materials. For example, *petrochemical intermediates* are generally produced by chemical conversion of primary petrochemicals to form more complicated derivative products (Parkash, 2003; Gary et al., 2007; Speight, 2008, 2014; Hsu and Robinson, 2017; Speight, 2017). Petrochemical derivative products can be made in a variety of ways: (i) directly from primary petrochemicals; (ii) through intermediate products which still contain only carbon and hydrogen; and (iii) through intermediates which incorporate chlorine, nitrogen, or oxygen in the finished derivative. In some cases, they are finished products; in others, more steps are needed to arrive at the desired composition. Some typical petrochemical intermediates are: (i) vinyl acetate ($CH_3CO_2CH=CH_2$) for paint, paper, and textile coatings; (ii) vinyl chloride ($CH_2=CHCl$) for polyvinyl chloride (PVC); (iii) ethylene glycol ($HOCH_2CH_2OH$) for polyester textile fibers; and (iv) styrene ($C_6H_5CH=CH_2$), which is important in rubber and plastic manufacturing. Of all the processes used, one of the most important is polymerization (Chapter 11). It is used in the production of plastics, fibers, and synthetic rubber, the main finished petrochemical derivatives.

Following from this, secondary raw materials, or intermediate chemicals (Chapters 5 and 6), are obtained from a primary raw material through a variety of different processing schemes. The intermediate chemicals may be low-boiling hydrocarbon compounds such as methane, ethane, propane, and butane or higher-boiling hydrocarbon mixtures such as naphtha, kerosene, or gas oil. In the latter cases (naphtha, kerosene, and gas oil), these fractions are used (in addition to the production of fuels) as feedstocks for cracking processes to produce a variety of petrochemical products (e.g., ethylene, propylene, benzene, toluene, and the xylene isomers), which are identified by the relative placement of the two methyl groups on the aromatic ring:

1,2-dimethylbenzene   1,3-dimethylbenzene   1,4-dimethylbenzene
(*ortho*-xylene)   (*meta*-xylene)   (*para*-xylene)

Also, by way of definition, petrochemistry is a branch of chemistry in which the transformation of petroleum (crude oil) and natural gas into useful products or feedstock for other process is studied. A petrochemical plant is a plant that uses chemicals from petroleum as a raw material (the feedstock) are usually located adjacent to (or within the precinct of) a petroleum refinery in order to minimize the need for transportation of the feedstocks produced by the refinery (Figure 1.2). On the other hand, specialty chemical plants and fine chemical plants are usually much smaller than a petrochemical plant and are not as sensitive to location.

Furthermore, a paraffinic petrochemical is an organic chemical compound, but one that does not contain any ring systems such as a cycloalkane (naphthene) ring or an aromatic ring. A naphthenic petrochemical is an organic chemical compound that contains one or more cycloalkane ring systems. An aromatic petrochemical is also an organic chemical compound but one that contains, or is derived from, the basic benzene ring system.

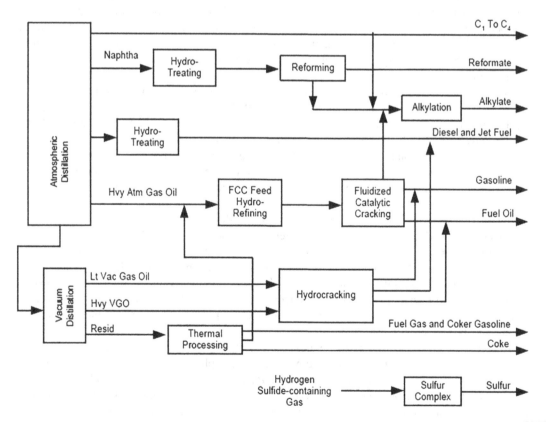

FIGURE 1.2 Schematic diagram of a refinery showing the production of products during the distillation and during thermal processing (e.g., visbreaking, coking, and catalytic cracking).

Petroleum products (in contrast to *petrochemicals*) are those hydrocarbon fractions that are derived from petroleum and have commercial value as a bulk product (Tables 1.1 and 1.2) (Parkash, 2003; Gary et al., 2007; Speight, 2014; Hsu and Robinson, 2017; Speight, 2017). These products are generally not accounted for in petrochemical production or used in statistics. Thus, in the context of this definition of petrochemicals, this book focuses on chemicals that are produced from petroleum as distinct from *petroleum products*, which are organic compounds (typically hydrocarbon compounds) that are burned as a fuel. In the strictest sense of the definition, a petrochemical is any chemical that is manufactured from petroleum and natural gas as distinct from fuels and other products, which are derived from petroleum and natural gas by a variety of processes and used for a variety of commercial purposes (Chenier, 2002; Meyers 2005; Naderpour, 2008; Speight, 2014). Petrochemical products include such items as plastics, soaps and detergents, solvents, drugs, fertilizers, pesticides, explosives, synthetic fibers and rubbers, paints, epoxy resins, and flooring and insulating materials.

Moreover, the classification of materials as petrochemicals is used to indicate the source of the chemical compounds, but it should be remembered that many common petrochemicals can be made from other sources, and the terminology is therefore a matter of source identification. However, in the setting of modern industry, the term petrochemicals, is often used in an expanded form to include chemicals produced from other fossil fuels such as coal or natural gas, oil shale, and renewable sources such as corn or sugar cane as well as other forms of biomass (Chapter 3). It is in the expanded form of the definition that the term petrochemical is used in this book.

In fact, in the early days of the chemical industry, coal was the major source of chemicals (it was not then called the *petrochemical industry*) and it was only after the discovery of petroleum and the recognition that petroleum could produce a variety of products other than fuels that the petrochemical industry came into being (Spitz, 1988; Speight, 2013, 2014, 2017). For several decades both coal and petroleum served as the primary raw materials for the manufacture of chemicals. Then during the time of World War II, petroleum began to outpace coal as a source of chemicals—the exception being the manufacture of synthetic fuels from coal because of the lack of access to petroleum by German industry.

To complete this series of definitions and to reduce the potential for any confusion that might occur later in this text, specialty chemicals (also called *specialties* or *effect chemicals*) are particular chemical products, which provide a wide variety of effects on which many other industry sectors rely. Specialty chemicals are materials used on the basis of their performance or function. Consequently, in addition to *effect chemicals* they are sometimes referred to as *performance chemicals* or *formulation chemicals*. The physical and chemical characteristics of the single molecules or the formulated mixtures of molecules and the composition of the mixtures influence the performance of the end product.

On the other hand, the term *fine chemicals* is used in distinction to *heavy chemicals*, which are produced and handled in large lots and are often in a crude state. Since their inception in the late 1970s, fine chemicals have become an important part of the chemical industry. Fine chemicals are typically single, but often complex pure chemical substances, produced in limited quantities in multipurpose plants by multistep batch chemical or biotechnological processes and are described by specifications to which the chemical producers must strictly adhere. Fine chemicals are used as starting materials for specialty chemicals, particularly pharmaceutical chemicals, biopharmaceutical chemicals, and agricultural chemicals.

To return to the subject of petrochemicals, a petroleum refinery converts raw crude oil into useful products such as liquefied petroleum gas (LPG), naphtha (from which gasoline is manufactured), kerosene from which diesel fuel is manufactured, and a variety of gas oil fractions—of particular interest is the production of naphtha that serves as a feedstock for several processes that produce petrochemical feedstocks (Table 1.4). However, each refinery has its own specific arrangement and combination of refining processes largely determined by the market demand (Parkash, 2003; Gary et al., 2007; Speight, 2014; Hsu and Robinson, 2017; Speight, 2017). The most common

TABLE 1.4
Naphtha Production

Primary Process	Primary Product	Secondary Process	Secondary Product
Atmospheric distillation	Naphtha		Light naphtha
			Heavy naphtha
	Gas oil	Catalytic cracking	Naphtha
	Gas oil	Hydrocracking	Naphtha
Vacuum distillation	Gas oil	Catalytic cracking	Naphtha
		Hydrocracking	Naphtha
	Residuum	Coking	Naphtha
		Hydrocracking	Naphtha

petrochemical precursors are various hydrocarbon derivatives, olefin derivatives, aromatic derivatives (including benzene, toluene and xylene isomers), and synthesis gas (also called syngas—a mixture of carbon monoxide and hydrogen).

A typical crude oil refinery produces a variety of hydrocarbon derivatives, olefin derivatives, and aromatic derivatives by processes such as coking and fluid catalytic cracking of various feedstocks. Chemical plants produce olefin derivatives by steam cracking natural gas liquids, such as ethane (CH_3CH_3) and propane ($CH_3CH_2CH_3$) to produce ethylene ($CH_2=CH_2$) and propylene ($CH_3CH=CH_2$), respectively. A steam cracking unit (Figure 1.3) is, in theory, one of the simplest operations in a refinery—essentially a hot reactor into which steam and the feedstock are introduced, but in reality the steam cracking is one of the most technically complex and energy-intensive plants in the refining industry and in the petrochemical industry. The equipment typically operates over the range 175°C–1125°C (345°F–2055°F) and from a near vacuum, i.e., <14.7 psi to high pressure (1,500 psi). While the fundamentals of the process have not changed in recent decades, improvements continue to be made to the energy efficiency of the furnace, ensuring that the cost of production is continually reduced.

In more general terms, steam cracking units use a variety of feedstocks, for example, (i) ethane, propane, and butane from natural gas; (ii) naphtha, a mixture of C_5–C_8 or C_5–C_{10} hydrocarbon

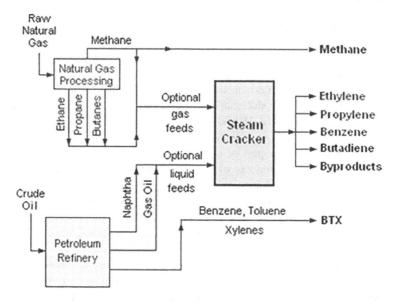

FIGURE 1.3 Representation of a steam cracking operations.

derivatives from the distillation of crude oil; (iii) gas oil; and (iv) resids—also called residue or residua—from the primary distillation of crude oil. In the steam cracking process, a gaseous or liquid hydrocarbon feedstock is diluted with steam and then briefly heated in a furnace, obviously without the presence of oxygen. Typically, the reaction temperature is high (up to 1125°C (2055°F)), but the reaction is only allowed to take place very briefly (short residence time). The residence time is even reduced to milliseconds (resulting in gas velocities reaching speeds beyond the speed of sound) in order to improve the yield of desired products. After the cracking temperature has been reached, the gas is quickly quenched to stop the reaction in a transfer line exchanger. The product-type and product yield produced in the cracking unit depend on (i) the composition of the feed, (ii) the hydrocarbon to steam ratio, (iii) the cracking temperature, and (iv) the residence time of the feedstock in the hot zone.

The advantages of steam cracking are that the process reduced the need for repeated product distillation that produces a wider range of products. However, the disadvantage is that the process may not produce the product that is needed in high enough yield. In fact, aromatic derivatives, such as benzene (C_6H_6), toluene ($C_6H_5CH_3$), and the xylenes (ortho-, meta-, and para-isomers, $H_3CC_6H_4CH_3$) are produced by reforming naphtha, which is a low-boiling liquid product obtained by distillation from crude oil (Tables 1.1 and 1.4) (Parkash, 2003; Gary et al., 2007; Speight, 2014; Hsu and Robinson, 2017; Speight, 2017). With higher molecular weight (higher-boiling) feedstocks, such as gas oil, it is important to ensure that the feedstock does not crack to form carbon, which is normally formed at this temperature. This is avoided by passing the gaseous feedstock very quickly and at very low pressure through the pipes which run through the furnace.

On the basis of chemical structure, petrochemicals are categorized into a variety of petrochemical products, which are named according to the chemical character of the constituents:

- Paraffin derivatives: such as methane (CH_4), ethane (C_2H_6), propane (C_3H_8), the butane isomers (C_4H_{10}), and higher molecular weight hydrocarbon derivatives up to and including low-boiling mixtures such as naphtha.
- Olefin derivatives: such as ethylene ($CH_2=CH_2$) and propylene ($CH_3CH=CH_2$), which are important sources of industrial chemicals and plastics; the diolefin derivative butadiene ($CH_2=CHCH=CH_2$) is used in making synthetic rubber.
- Aromatic derivatives: such as benzene (C_6H_6), toluene *($C_6H_5CH_3$), and the xylene isomers ($CH_3C_6H_4CH_3$) which are identified by the relative placement of the two methyl groups on the aromatic ring (1,2-$CH_3C_6H_4CH_3$, 1,3-$CH_3C_6H_4CH_3$, and 1,4-$CH_3C_6H_4CH_3$) and which have a variety of uses—benzene is a raw material for dyes and synthetic detergents, and benzene and toluene for isocyanates while the xylene isomers are used in the manufacture of plastics and synthetic fibers.
- Synthesis gas: a mixture of carbon monoxide (CO) and hydrogen (H_2) that is sent to a Fischer–Tropsch reactor to produce naphtha-range and kerosene-range hydrocarbon derivative as well as methanol (CH_3OH) and dimethyl ether (CH_3OCH_3).

Ethylene and propylene, the major part of olefins are the basic source in preparation of several industrial chemicals and plastic products whereas butadiene is used to prepare synthetic rubber. Benzene, toluene and the xylene isomers are major components of aromatic chemicals. These aromatic petrochemicals are used in the manufacturing of secondary products like synthetic detergents, polyurethanes, plastic, and synthetic fibers. Synthesis gas comprises carbon monoxide and hydrogen which are basically used to produce ammonia and methanol, which are further used to produce other chemical and synthetic substances.

Low-boiling olefins (light olefins)—ethylene and propylene—are the most important intermediates in the production of plastics and other chemical products. The current end use of ethylene worldwide is for (i) the manufacture of polyethylene, which is used in plastics; (ii) the manufacture of ethylene oxide/glycol, which is used in fibers and plastic; (iii) the manufacture of ethylene dichloride, which is used in polyvinyl chloride polymers; and (iv) the manufacture of ethylbenzene, which is used in styrene polymers. The current end use of propylene worldwide is in (i) the manufacture of polypropylene, which is used in plastics; (ii) the manufacture of acrylonitrile; (iii) the manufacture of cumene, which is used in phenolic resin; (iv) the manufacture of propylene oxide; and (v) the manufacture of 8% oxo-alcohol derivatives.

Also, non-olefin petrochemicals are typically aromatic derivatives (benzene, toluene and the xylene isomers or simply BTX). Most of the benzene is used to make (i) styrene, which is used in the manufacture of polymers and plastics; (ii) phenol, which is used in the manufacture of resins and adhesives by way of cumene; and (iii) cyclohexane, which is used in the manufacture of nylon (Chapters 8 and 11). Benzene is also used in the manufacture of rubber, lubricants, dyes, detergents, drugs, explosives, and pesticides. Toluene is used as a solvent (for making paint, rubber, and adhesives), a gasoline additive or for making toluene isocyanate (toluene isocyanate is used for making polyurethane foam), phenol, and trinitrotoluene (generally known as TNT). Xylene is used as a solvent and as an additive for making fuels, rubber, leather and terephthalic acid [14-HO$_2$CC$_6$H$_4$CO$_2$H, also written as 1,4-CO$_2$HC$_6$H$_4$CO$_2$H or 1,4-C$_6$H$_4$(CO$_2$H)$_2$]), which is used in the manufacture of polymers.

Terephthalic acid

The primary focus of this book is on chemical products derived from petroleum and natural gas but *chemicals* from sources such as coal and biomass are also included as alternate feedstocks (Tables 1.5 and 1.6). However, petrochemicals in the strictest sense are chemical products derived from petroleum, although many of the same chemical compounds are also obtained from other

TABLE 1.5
Alternative Feedstocks for the Production of Petrochemicals

Chemicals	Petroleum-Natural Gas Feedstock	Alternate Feedstock
Methane	Natural gas	Coal, as byproduct of separation of coke gases
	Refinery gas	Coal hydrogenation
Ammonia	Methane	From coal via water gas
Methyl alcohol	Methane	From coal via water-gas reaction
Ethylene	Pyrolysis of low-boiling hydrocarbon derivatives	Dehydration of ethyl alcohol
Acetylene	Ethylene	Calcium carbide
Ethylene glycol	Ethylene	From coal via carbon monoxide and formaldehyde
Acetaldehyde	Paraffin gas oxidation	Fermentation of ethyl alcohol
	Oxidation of ethylene	Acetylene
Acetone	Propylene	Destructive distillation of wood
		Pyrolysis of acetic acid
		Acetylene-steam reaction
Glycerol	Propylene	Byproduct of soap manufacture
Butadiene	1- and 2-Butenes	Ethyl alcohol; acetaldehyde via 1:3-butanediol
	Butane	Acetylene and formaldehyde from coal
Aromatic hydrocarbons	Aromatic-rich fractions by catalytic reforming	Byproducts of coal tar distillation
	Naphthene-rich fractions by catalytic reforming	

TABLE 1.6
Illustration of the Production of Petrochemical
Starting Materials from Petroleum and Natural Gas

Feedstock	Process	Product
Petroleum	Distillation	Light ends
		Methane
		Ethane
		Propane
		Butane
	Catalytic cracking	Ethylene
		Propylene
		Butylenes
		Higher olefins
	Catalytic reforming	Benzene
		Toluene
		Xylenes
	Coking	Ethylene
		Propylene
		Butylenes
		Higher olefins
Natural gas	Refining	Methane
		Ethane
		Propane
		Butane

fossil fuels such as coal and natural gas or from renewable sources such as corn, sugar cane, and other types of biomass (Matar and Hatch, 2001; Meyers, 2005; Speight, 2008, 2013, 2014; Clark and Deswarte, 2015). But first there is the need to understand the origins of the industry and, above all, the continuing need for the petrochemical industry.

1.2 HISTORICAL ASPECTS AND OVERVIEW

When coal came to prominence as a fuel during the Industrial Revolution, there was a parallel development relating to the use of coal for the production of chemicals. Byproduct liquids and gases from coal carbonization processes became the basic raw materials for the organic chemical industry and the production of metallurgical coke from coal was essential to the development of steel manufacture (Speight, 2013). Coal tar constituents were used for the industrial syntheses of dyes, perfumes, explosives, flavorings, and medicines. Processes were also developed for the conversion of coal to fuel gas and to liquid fuels.

By the time that the decade of the 1930s had dawned, the direct and indirect liquefaction technologies became available for the substantial conversion of coals to liquid fuels and chemicals. Subsequently, the advent of readily available petroleum and natural gas, and the decline of the steel industry, reduced dependence on coal as a resource for the production of chemicals and materials. For the last several decades as the 20th century came to a close and the 21st century dawned, the availability of coal tar chemicals has depended on the production of metallurgical coke which is, in turn, tied to the fortunes and future of the steel industry. The petroleum era was ushered in by the discovery of petroleum at Titusville, Pennsylvania in 1859. Although the petroleum era was ushered in by the discovery of petroleum at Titusville, Pennsylvania in 1859, the production of chemicals from natural gas and petroleum has been a recognized industry only since the early 20th century. Nevertheless, the petrochemical industry has made quantum leaps in the production of a wide variety of chemicals (Chenier, 2002), which being based on starting feedstocks from petroleum is termed *petrochemicals*.

Following from this, the production of chemicals from natural gas and petroleum has been a recognized industry since the early decades of the 20th century. Nevertheless, the lead up and onset of World War II led to the development and expansion of the petrochemical industry which, since that time, has made quantum leaps in terms of the production of a wide variety of chemicals (Chenier, 2002; Meyers, 2005; Naderpour, 2008; Speight, 2014; EPCA, 2016; Hsu and Robinson, 2017). At this time, coal alone could no longer satisfy the demand for basic chemicals that had increased by the demands of World War II and the production of chemicals from coal tar or some agricultural products was not sufficient and led to the major development of chemicals production from petroleum. During the 1950s and 1960s, the increased demand for liquid fuels increased phenomenally and, paralleling the demand for fuels, the onset of the *age of plastics* (which also included demand for rubber, fibers, surfactants, pesticides, fertilizers, pharmaceuticals, dyes, solvents, lubricating oils, and food additives) caused an increase in the demand for chemicals from petroleum and natural gas.

This trend has continued until the present decade and demand for the manufacture of chemicals will continue for the foreseeable future.

1.3   THE PETROCHEMICAL INDUSTRY

The petrochemical industry, as the name implies, is based upon the production of chemicals from petroleum. However, there is more to the industry than just petroleum products. The petrochemical industry also deals with chemicals manufactured from the byproducts of petroleum refining, such as natural gas, natural gas liquids, and (in the context of this book) other feedstocks such as coal, oil shale, and biomass. The structure of the industry is extremely complex, involving thousands of chemicals and processes and there are many inter-relationships within the industry with products of one process being the feedstocks of many others. For most chemicals, the production route from feedstock to final products is not unique, but includes many possible alternatives. As complicated as it may seem; however, this structure is comprehensible, at least in general form.

At the beginning of the production chain are the raw feedstocks: petroleum, natural gas, and alternate carbonaceous feedstocks tar. From these are produced a relatively small number of important building blocks which include primarily, but not exclusively, the lower-boiling olefins and aromatic derivatives, such as ethylene, propylene, butylenes, butadiene, benzene, toluene, and the xylene isomers. These building blocks are then converted into a complex array of thousands of intermediate chemicals. Some of these intermediates have commercial value in and of themselves, and others are purely intermediate compounds in the production chains. The final products of the petrochemical industry are generally not consumed directly by the public, but are used by other industries to manufacture consumer goods.

Thus, on a scientific basis, as might be expected, the petrochemical industry is concerned with the production and trade of petrochemicals that have a wide influence on lifestyles through the production of commodity chemicals and specialty chemicals that have a marked influence on lifestyles:

Petroleum/natural gas → bulk chemicals (commodity chemicals) → specialty chemicals

The basis of the petrochemical industry and, therefore, petrochemicals production consists of two steps: (i) feedstock production from primary energy sources to feedstocks and (ii) and petrochemicals production from feedstocks.

Petroleum/natural gas → feedstock production → petrochemical products

This simplified equation encompasses the multitude of production routes available for most chemicals. In the actual industry, many chemicals are products of more than one method, depending upon local conditions, corporate polices, and desired byproducts. There are also additional methods available, which have either become obsolete and are no longer used, or which have never been used commercially but could become important as technology, supplies, and other factors change. Such versatility, adaptability, and dynamic nature are three of the important features of the modern petrochemical industry.

Thus, the petrochemical industry began as suitable byproducts became available through improvements in the refining processes. As the decades of the 1920s and 1930s closed, the industry developed in parallel with the crude oil industry and has continued to expand rapidly since the 1940s as the crude oil refining industry was able to provide relatively cheap and plentiful raw materials (Speight, 2002; Gary et al., 2007; Lee et al., 2007; Speight, 2011, 2014; Hsu and Robinson, 2017; Speight, 2017). The supply–demand scenario as well as the introduction of many innovations has resulted in basic chemicals and plastics becoming the key building blocks for manufacture of a wide variety of durable and nondurable consumer goods. Chemicals and plastic materials provide the fundamental building blocks that enable the manufacture of the vast majority of consumer goods. Moreover, the demand for chemicals and plastics is driven by global economic conditions, which are directly linked to demand for consumer goods.

At the start of the production chain is the selection and preparation of the feedstock from which the petrochemicals will be produced. Typically, the feedstock is a primary energy source (such as crude oil, natural gas, coal, and biomass) are extracted and then converted into feedstocks (such as naphtha, gas oil, and/or methanol). In the production of petrochemicals, the feedstocks are converted into basic petrochemicals, such as ethylene ($CH_2=CH_2$) and aromatic derivatives, which are then separated from each other. Thus, petrochemicals or products derived from these feedstocks, along with other raw materials, are converted to a wide range of products (Table 1.3).

Therefore, the history of the industry has always been strongly influenced by the supply of primary energy sources and feedstocks. Thus, the petrochemical industry directly interfaces with the petroleum industry and the natural gas industry, which proves the feedstocks (Chapter 2), and especially the downstream sector, as well as the potential for the introduction and use of nonconventional feedstocks (Chapter 3). A major part of the petrochemical industry is made up of the polymer (plastics) industry (Chapter 11).

The petrochemical industry is currently the biggest of the industrial chemical sectors and petrochemicals represent the majority of all chemicals shipped between the continents of the world (EPCA, 2016). Petrochemicals have a history that began in the 19th century that has experienced many changes. However, from the beginning there have been underlying trends, which shaped the evolution of the industry to modern times. From the start, it was an industry that was destined to become a global sector because of the contribution the product makes to raise the standards of living of much of the population of the world. These same influences have also shaped the rate and nature of the expansion and the structure of the industry as it exist in the 21st century.

In the petrochemical industry, the organic chemicals produced in the largest volumes are methanol (methyl alcohol, CH_3OH), ethylene ($CH_2=CH_2$), propylene ($CH_3CH=CH_2$), butadiene $CH_2=CHCH=CH_2$), benzene (C_6H_6), toluene ($C_6H_5CH_3$), and the xylene isomers ($H_3CC_6H_4CH_3$). Ethylene, propylene, and butadiene, along with butylenes, are collectively called olefins, which belong to a class of unsaturated aliphatic hydrocarbon derivatives having the general formula C_nH_{2n}. Olefin derivatives contain one or more double bonds (>C=C<), which make them chemically reactive and, hence, the starting materials for many products. Benzene, toluene, and xylenes, commonly referred to as aromatics, are unsaturated cyclic hydrocarbon derivatives containing one or more rings.

As stated above, some of the chemicals and compounds produced in a refinery are destined for further processing and as raw material feedstocks for the fast growing petrochemical industry. Such nonfuel uses of crude oil products are sometimes referred to as its non-energy uses. Petroleum products and natural gas provide two of the basic starting points for this industry: methane (Figure 1.4), naphtha, including benzene, toluene, and the xylene isomers (Figure 1.5) and refinery gases which

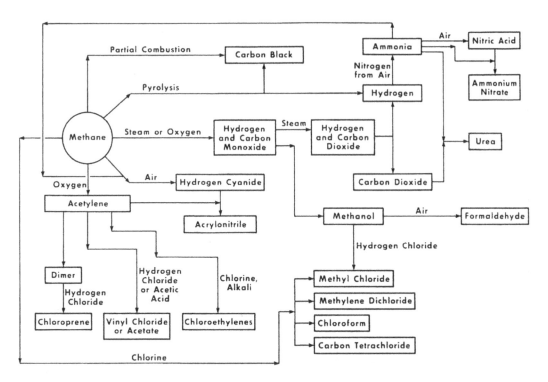

FIGURE 1.4 Chemicals from methane. (Speight, J.G. 2007. *The Chemistry and Technology of Petroleum.* 4th Edition. CRC Press, Boca Raton, FL. Figure 27.8, p. 800.)

contain olefin derivatives such as ethylene (Figure 1.6), propylene (Figure 1.7), and, potentially, all of the butylene isomers, (Figures 1.7 and 1.8) (Parkash, 2003; Gary et al., 2007; Speight, 2014; Hsu and Robinson, 2017; Speight, 2017).

Petrochemical intermediates are generally produced by chemical conversion of primary petrochemicals to form more complicated derivative products. Petrochemical derivative products can be made in a variety of ways: directly from primary petrochemicals; through intermediate products which still contain only carbon and hydrogen; and, through intermediates which incorporate chlorine, nitrogen, or oxygen in the finished derivative. In some cases, they are finished products; in others, more steps are needed to arrive at the desired composition.

The end products number in the thousands, some going on as inputs into the chemical industry for further processing. The more common products made from petrochemicals include adhesives, plastics, soaps, detergents, solvents, paints, drugs, fertilizers, pesticides, insecticides, explosives, synthetic fibers, synthetic rubber, and flooring and insulating materials.

Petrochemical products include such items as plastics, soaps and detergents, solvents, drugs, fertilizers, pesticides, explosives, synthetic fibers and rubbers, paints, epoxy resins, and flooring and insulating materials. Petrochemicals are found in products as diverse as aspirin, luggage, boats, automobiles, aircraft, polyester clothes, and recording discs and tapes.

The petrochemical industry has grown with the petroleum industry (Goldstein, 1949; Steiner, 1961; Hahn, 1970) and is considered by some to be a mature industry. However, as is the case with the latest trends in changing crude oil types, it must also evolve to meet changing technological needs. The manufacture of chemicals or chemical intermediates from a variety of raw materials is well established (Wittcoff and Reuben, 1996). And the use of petroleum and natural gas is an excellent example of the conversion of such raw materials to more valuable products. The individual chemicals made from petroleum and natural gas are numerous and include industrial chemicals,

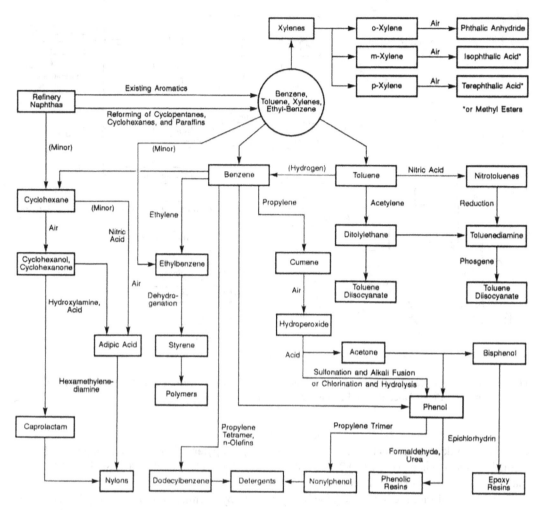

FIGURE 1.5 Chemicals from benzene, toluene, and the xylene isomers. (Speight, J.G. 2007. *The Chemistry and Technology of Petroleum*. 4th Edition. CRC Press, Boca Raton, FL. Figure 27.7, p. 798.)

household chemicals, fertilizers, and paints, as well as intermediates for the manufacture of products, such as synthetic rubber and plastics.

Petrochemicals are generally considered chemical compounds derived from petroleum either by direct manufacture or indirect manufacture as byproducts from the variety of processes that are used during the refining of petroleum. Gasoline, kerosene, fuel oil, lubricating oil, wax, asphalt, and the like are excluded from the definition of petrochemicals, since they are not, in the true sense, chemical compounds but are in fact intimate mixtures of hydrocarbon derivatives.

The classification of materials as petrochemicals is used to indicate the source of the chemical compounds, but it should be remembered that many common petrochemicals can be made from other sources, and the terminology is therefore a matter of source identification.

The starting materials for the petrochemical industry are obtained from crude petroleum in one of two general ways. They may be present in the raw crude oil and, as such, are isolated by physical methods, such as distillation or solvent extraction (Parkash, 2003; Gary et al., 2007; Speight, 2014; Hsu and Robinson, 2017; Speight, 2017). On the other hand, they may be present, if at all, in trace amounts and are synthesized during the refining operations. In fact, unsaturated (olefin) hydrocarbon derivatives, which are not usually present in crude oil, are nearly always manufactured as

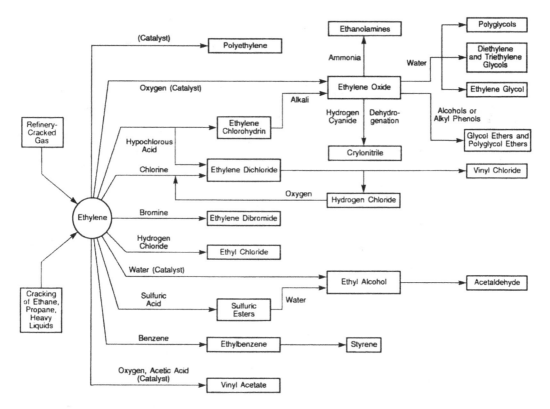

FIGURE 1.6 Chemicals from ethylene. (Speight, J.G. 2007. *The Chemistry and Technology of Petroleum.* 4th Edition. CRC Press, Boca Raton, FL. Figure 27.5, p. 788.)

intermediates during the various refining sequences (Parkash, 2003; Gary et al., 2007; Speight, 2014; Hsu and Robinson, 2017; Speight, 2017).

The manufacture of chemicals from petroleum is based on the ready response of the various compound types to basic chemical reactions, such as oxidation, halogenation, nitration, dehydrogenation addition, polymerization, and alkylation. The low molecular weight paraffins and olefins, as found in natural gas and refinery gases, and the simple aromatic hydrocarbon derivatives have so far been of the most interest because it is individual species that can be readily be isolated and dealt with. A wide range of compounds is possible, many are being manufactured and we are now progressing to the stage in which a sizable group of products is being prepared from the heavier fractions of petroleum. For example, the various reactions of asphaltene constituents (Chapter 2) (Speight, 1994, 2014) indicate that these materials may be regarded as containing chemical functions and are therefore different and are able to participate in numerous chemical or physical conversions to, perhaps, more useful materials. The overall effect of these modifications is the production of materials that either affords good-grade aromatic cokes comparatively easily or the formation of products bearing functional groups that may be employed as a nonfuel material.

For example, the sulfonated and sulfomethylated materials and their derivatives have satisfactorily undergone tests as drilling mud thinners, and the results are comparable to those obtained with commercial mud thinners.

Here, there is the potential slow-release soil conditioners that only release the nitrogen or phosphorus after considerable weathering or bacteriological action. One may proceed a step further and suggest that the carbonaceous residue remaining after release of the hetero-elements may be a benefit to humus-depleted soils, such as the gray-wooded and solonetzic soils. It is also feasible that coating a conventional quick-release inorganic fertilizer with a water-soluble or water-dispersible

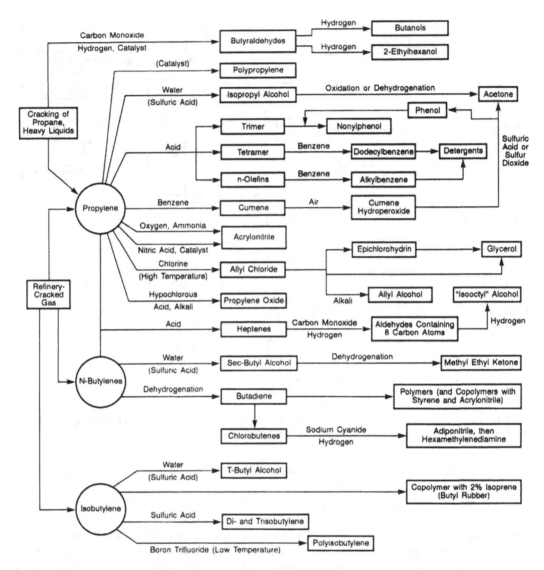

FIGURE 1.7 Chemicals from propylene and butylene. (Speight, J.G. 2007. *The Chemistry and Technology of Petroleum.* 4th Edition. CRC Press, Boca Raton, FL. Figure 27.6, p. 789.)

derivative will provide a slower-release fertilizer and an organic humus-like residue. In fact, variations on this theme are multiple.

Nevertheless, the main objective in producing chemicals from petroleum is the formation of a variety of well-defined chemical compounds that are the basis of the petrochemical industry. It must be remembered, however, that ease of separation of a particular compound from petroleum does not guarantee its use as a petrochemical building block. Other parameters, particularly the economics of the reaction sequences, including the costs of the reactant equipment, must be taken into consideration.

Petrochemicals are made, or recovered from, the entire range of petroleum fractions, but the bulk of petrochemical products are formed from the lighter (C_1–C_4) hydrocarbon gases as raw materials. These materials generally occur as natural gas, but they are also recovered from the gas streams produced during refining, especially cracking, operations. Refinery gases are also particularly valuable because they contain substantial amounts of olefins that, because of the double bonds, are much

IUPAC name	Common name	Structure	Skeletal formula
But-1-ene	1-butylene	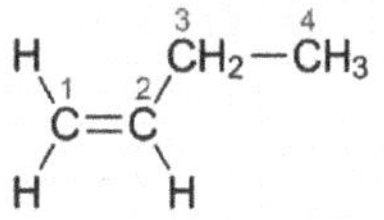	
cis-But-2-ene	cis-2-butylene	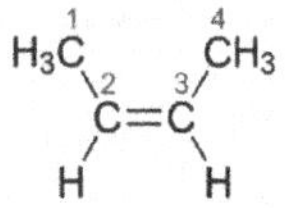	
cis-But-2-ene	trans-3-butylene	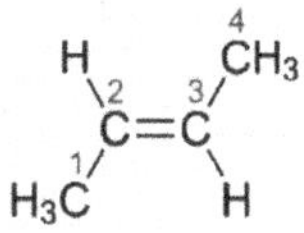	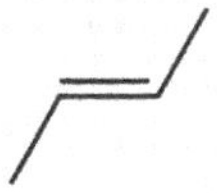
2-methylprop-1-ene	Isobutylene	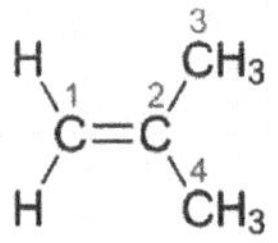	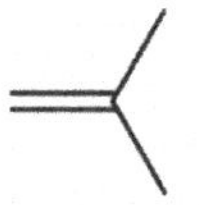

FIGURE 1.8 Representation of the various isomers of butylene (C_4H_8).

more reactive then the saturated (paraffin) hydrocarbon derivatives. Also important as raw materials are the aromatic hydrocarbon derivatives (benzene, toluene, and xylene) that are obtained in rare cases from crude oil and, more likely, from the various product streams. By means of the catalytic reforming process, nonaromatic hydrocarbon derivatives can be converted to aromatics by dehydrogenation and cyclization (Parkash, 2003; Gary et al., 2007; Speight, 2014; Hsu and Robinson, 2017; Speight, 2017).

A highly significant proportion of these basic petrochemicals is converted into plastics, synthetic rubbers, and synthetic fibers. Together these materials are known as polymers, because their molecules are high molecular weight compounds made up of repeated structural units that have combined chemically. The major products are polyethylene, polyvinyl chloride, and polystyrene, all derived from ethylene, and polypropylene, derived from monomer propylene. Major raw materials for synthetic rubbers include butadiene, ethylene, benzene, and propylene. Among synthetic fibers the polyesters, which are a combination of ethylene glycol and terephthalic acid (made from xylene), are the most widely used. They account for about one-half of all synthetic fibers. The second major synthetic fiber is nylon; it is the most important raw material being benzene. Acrylic fibers, in which the major raw material is the propylene derivative acrylonitrile, make up most of the remainder of the synthetic fibers.

1.4 PETROCHEMICALS

For the purposes of this text, there are four general types of petrochemicals: (i) aliphatic compounds, (ii) aromatic compounds, (iii) inorganic compounds, and (iv) synthesis gas (carbon monoxide and hydrogen). Synthesis gas is used to make ammonia (NH_3) and methanol (methyl alcohol, CH_3OH) as well as a variety of other chemicals (Figure 1.9). Ammonia is used primarily to form ammonium nitrate (NH_4NO_3), a source of fertilizer. Much of the methanol produced is used in making formaldehyde ($HCH=O$). The rest is used to make polyester fibers, plastics, and silicone rubber.

An aliphatic petrochemical compound is an organic compound that has an open chain of carbon atoms, be it normal (straight), e.g., n-pentane ($CH_3CH_2CH_2CH_2CH_3$) or branched, e.g., isopentane [2-methylbutane, $CH_3CH_2CH(CH_3)CH_3$]. The unsaturated compounds, olefins, include important

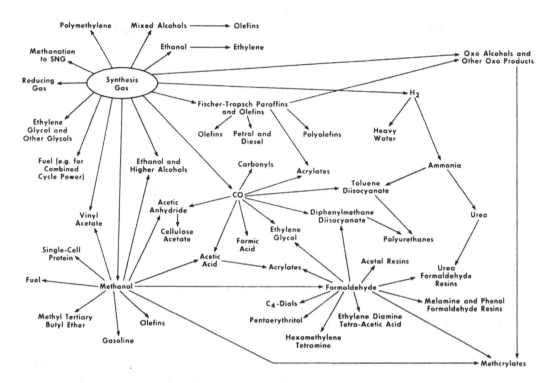

FIGURE 1.9 Production of chemicals from synthesis gas. (Speight, J.G. 2007. *The Chemistry and Technology of Petroleum.* 4th Edition. CRC Press, Boca Raton, FL. Figure 27.9, p. 802.)

starting materials such as ethylene ($CH_2=CH_2$), propylene ($CH_3 \cdot CH=CH_2$), butene-1 ($CH_3CH_2CH_2=CH_2$), isobutene (2-methylpropene [$CH_3(CH_3)C=CH_2$]), and butadiene ($CH_2=CHCH=CH_2$).

As already defined, a *petrochemical* is any chemical (as distinct from fuels and petroleum products) manufactured from petroleum (and natural gas as well as other carbonaceous sources) and used for a variety of commercial purposes (Chenier, 2002). The definition, however, has been broadened to include the whole range of aliphatic, aromatic, and naphthenic organic chemicals, as well as carbon black and such inorganic materials as sulfur and ammonia. Gasoline, kerosene, fuel oil, lubricating oil, wax, asphalt, and the like are excluded from the definition of petrochemicals, since they are not, in the true sense, chemical compounds but are in fact intimate mixtures of hydrocarbon derivatives. The classification of materials as petrochemicals is used to indicate the source of the chemical compounds, but it should be remembered that many common petrochemicals can be made from other sources, and the terminology is, therefore, a matter of source identification.

Petroleum and natural gas are made up of (predominantly) hydrocarbon constituents, which are comprised of one or more carbon atoms, to which hydrogen atoms are attached—in some cases, petroleum contains a considerable proportion of non-hydrocarbon constituents such as organic compounds containing one or more heteroatoms (such as nitrogen, oxygen, sulfur, and metals). Currently, through a variety of intermediates, petroleum and natural gas are the main sources of the raw materials because they are the least expensive, most readily available, and can be processed most easily into the primary petrochemicals. An aromatic petrochemical is also an organic chemical compound but one that contains, or is derived from, the basic benzene ring system. Furthermore, petrochemicals are often made in clusters of plants in the same area. These plants are often operated by separate companies, and this concept is known as *integrated manufacturing*. Groups of related materials are often used in adjacent manufacturing plants, to use common infrastructure and minimize transport.

1.4.1 PRIMARY PETROCHEMICALS

The primary petrochemicals are not the raw materials for the petrochemical industry. Primary raw materials are naturally occurring substances that have not been subjected to chemical changes after being recovered. Natural gas and crude oil are the basic raw materials for the manufacture of petrochemicals. Secondary raw materials, or intermediates, are obtained from natural gas and crude oils through different processing schemes. The intermediate chemicals may be low-boiling hydrocarbon compounds such as methane and ethane, or heavier hydrocarbon mixtures such as naphtha or gas oil. Both naphtha and gas oil are crude oil fractions with different boiling ranges. Coal, oil shale, and biomass are complex carbonaceous raw materials and possible future energy and chemical sources. However, they must undergo lengthy and extensive processing before they yield fuels and chemicals similar to those produced from crude oils (substitute natural gas (SNG) and synthetic crudes from coal, oil shale, and bio-oil). The term *primary petrochemical* is more specific and includes olefins (ethylene, propylene and butadiene), aromatics (benzene, toluene, and the isomers of xylene), and methanol from which petrochemical products are manufactured.

The two most common petrochemical classes are olefin derivatives (including ethylene, $CH_2=CH_2$, and propylene, $CH_3CH=CH_2$) and aromatic derivatives such as benzene (C_6H_6), toluene ($C_6H_5CH_3$), and the xylene isomers ($H_3CC_6H_4CH_3$). Olefin derivatives and aromatic derivatives are typically produced in a crude oil refinery by fluid catalytic cracking of the various crude oil distillate fractions (Speight, 1987; Parkash, 2003; Gary et al., 2007; Speight, 2014; Hsu and Robinson, 2017; Speight, 2017). Olefins are also produced by steam cracking of methane (CH_4), ethane (CH_3CH_3), and propane ($CH_3CH_2CH_3$) and aromatic derivatives are produced by steam reforming of naphtha (Speight, 1987; Parkash, 2003; Gary et al., 2007; Speight, 2014; Hsu and Robinson, 2017; Speight, 2017). Olefin derivatives and aromatic derivatives are the intermediate chemicals that lead to a substantial number (some observers would say *an innumerable number*) of products such as solvents, detergents, plastics, fibers, and elastomers.

In many instances, a specific chemical included among the petrochemicals may also be obtained from other sources, such as coal, coke, or vegetable products. For example, materials such as benzene and naphthalene can be made from either petroleum or coal, while ethyl alcohol may be of petrochemical or vegetable origin (Matar and Hatch, 2001; Meyers, 2005; Speight, 2008, 2013, 2014).

1.4.2 PRODUCTS AND END USE

Petrochemical products include such items as plastics, soaps and detergents, solvents, drugs, fertilizers, pesticides, explosives, synthetic fibers and rubbers, paints, epoxy resins, and flooring and insulating materials (Table 1.3). Petrochemicals use is also found in products as diverse as aspirin, luggage, boats, automobiles, aircraft, polyester clothes, and recording discs and tapes.

Although the petrochemical industry was showing steady growth (some observers would say "rapid growth"), the onset of World War II increased the demand for synthetic materials to replace costly and sometimes less efficient products was a catalyst for the development of petrochemicals. Before the 1940s, it was an experimental sector, starting with basic materials: (i) synthetic rubber in the 1900s; (ii) Bakelite, the first petrochemical-derived plastic in 1907; (iii) the first petrochemical solvents in the 1920s; and (iv) polystyrene in the 1930s. After this, the industry moved into a variety of areas—from household goods (kitchen appliances, textile, furniture) to medicine (heart pacemakers, transfusion bags), from leisure (such as running shoes, computers)to highly specialized fields like archaeology or crime detection.

Thus, the petrochemical industry has grown with the petroleum industry (Goldstein, 1949; Steiner, 1961; Hahn, 1970; Chenier, 2002) and is considered by some to be a mature industry. However, as is the case with the latest trends in changing crude oil types, the refining industry must also evolve to meet changing technological needs (Speight, 2011, 2014, 2017). The manufacture of chemicals or chemical intermediates from a variety of raw materials is well established (Wittcoff and Reuben,

1996; Speight, 2014)—the use of petroleum and natural gas is an excellent example of the conversion of such raw materials to more valuable products. The individual chemicals made from petroleum and natural gas are numerous and include industrial chemicals, household chemicals, fertilizers, and paints, as well as intermediates for the manufacture of products, such as synthetic rubber and plastics.

The petroleum and petrochemical industries have revolutionized modern life by providing the major basic needs of a rapidly growing, expanding, and highly technical civilization. They provide a source of products such as fertilizers, synthetic fibers, synthetic rubbers, polymers, intermediates, explosives, agrochemicals, dyes, and paints. The petrochemical industry fulfills a large number of requirements, which includes uses in the fields such as automobile manufacture, telecommunication, pesticides, fertilizers, textiles, dyes, pharmaceuticals, and explosives (Table 1.3).

1.5 PRODUCTION OF PETROCHEMICALS

For approximately 100 years, chemicals obtained as byproducts in the primary processing of coal to metallurgical coke have been the main source of a multitude of chemicals used as intermediates in the synthesis of dyes, drugs, antiseptics, and solvents. Historically, producing chemicals from coal through gasification has been used since the 1950s and, as such, dominated a large share of the chemicals industry.

Because the slate of chemical products that can be made via coal gasification, the chemical industry tends to use whatever feedstocks are most cost-effective. Therefore, interest in using coal tends to increase when oil and natural gas prices are higher and during periods of high global economic growth that may strain oil and gas production. Also, production of chemicals from coal is of much higher interest in countries like South Africa, China, India, and the United States where there are abundant coal resources. However, in recent decades, largely due to the supply of relatively cheap natural gas and crude oil, the use of coal as a source of chemicals has been superseded by the production of the chemicals from petroleum-related sources. The use of coal has also decreased because of environmental concerns without any acknowledgement that with the installation of modern process controls, coal can be a clean fuel (Speight, 2013).

Nevertheless, considering the case of natural gas and crude oil, the production of petrochemicals begins at the time the natural gas and/or the crude petroleum enters the refinery; natural gas (Katz, 1959; Kohl and Riesenfeld, 1985; Maddox et al., 1985; Newman. 1985; Kohl and Nielsen, 1997; Mokhatab et al., 2006) leading to the separation of contaminants from the hydrocarbon constituents. Petroleum refining (Parkash, 2003; Gary et al., 2007; Speight, 2014; Hsu and Robinson, 2017; Speight, 2017) begins with the distillation, or fractionation, of crude oils into separate fractions of hydrocarbon groups. The resultant products are directly related to the characteristics of the natural gas and crude oil being processed. Most of these products of distillation are further converted into more useable products by changing their physical and molecular structures through cracking, reforming, and other conversion processes. These products are subsequently subjected to various treatment and separation processes, such as extraction, hydrotreating, and sweetening, in order to produce finished products. While the simplest refineries are usually limited to atmospheric and vacuum distillation, integrated refineries incorporate fractionation, conversion, treatment, and blending with lubricant, heavy fuels, and asphalt manufacturing; they may also include petrochemical processing. It is during the refining process that other products are also produced. These products include the gaseous constituent dissolved in the crude oil that are released during the distillation processes as well as the gases produced during the various refining processes and both of these gaseous streams provide feedstocks for the petrochemical industry.

The gas (often referred to as *refinery gas* or *process gas*) varies in composition and volume, depending on the origin of the crude oil and on any additions (i.e., other crude oils blended into the refinery feedstock) to the crude oil made at the loading point. It is not uncommon to reinject light hydrocarbon derivatives such as propane and butane into the crude oil before dispatch by tanker or pipeline. This results in a higher vapor pressure of the crude, but it allows one to increase the quantity of light products obtained at the refinery. Since light ends in most petroleum markets command a premium, while in the oil field itself propane and butane may have to be reinjected or flared, the practice of *spiking* crude oil with liquefied petroleum gas is becoming fairly common. These gases are recovered by distillation (Figure 1.2). In addition to distillation, gases are also produced in the various *thermal cracking processes* (Figure 1.2) (Parkash, 2003; Gary et al., 2007; Speight, 2014; Hsu and Robinson, 2017; Speight, 2017).

Thus, in processes such as coking or visbreaking processes a variety of gases is produced. Another group of refining operations that contributes to gas production is that of the *catalytic cracking processes*. Both catalytic and thermal cracking processes result in the formation of unsaturated hydrocarbon derivatives, particularly ethylene ($CH_2=CH_2$), but also propylene (propene, $CH_3CH=CH_2$), isobutylene [isobutene, $(CH_3)_2C=CH_2$], and the *n*-butenes ($CH_3CH_2CH=CH_2$, and $CH_3CH=CHCH_3$) in addition to hydrogen (H_2), methane (CH_4) and smaller quantities of ethane (CH_3CH_3), propane ($CH_3CH_2CH_3$), and butanes [$CH_3CH_2CH_2CH_3$, $(CH_3)_3CH$]. Diolefins such as butadiene ($CH_2=CHCH=CH_2$) are also present. A further source of refinery gas is *hydrocracking*, a catalytic high-pressure pyrolysis process in the presence of fresh and recycled hydrogen. The feedstock is again heavy gas oil or residual fuel oil, and the process is mainly directed at the production of additional middle distillates and gasoline. Since hydrogen is to be recycled, the gases produced in this process again have to be separated into lighter and heavier streams; any surplus recycled gas and the liquefied petroleum gas from the hydrocracking process are both saturated (Parkash, 2003; Gary et al., 2007; Speight, 2014; Hsu and Robinson, 2017; Speight, 2017).

In a series of *reforming processes* (Parkash, 2003; Gary et al., 2007; Speight, 2014; Hsu and Robinson, 2017; Speight, 2017), commercialized under names such as *Platforming*, paraffin and naphthene (cyclic non-aromatic) hydrocarbon derivatives are converted in the presence of hydrogen and a catalyst is converted into aromatics, or isomerized to more highly branched hydrocarbon derivatives. Catalytic reforming processes thus not only result in the formation of a liquid product of higher octane number, but also produce substantial quantities of gases. The latter are not only rich in hydrogen, but also contain hydrocarbon derivatives from methane to butane isomers, with a preponderance of propane ($CH_3CH_2CH_3$), *n*-butane ($CH_3CH_2CH_2CH_3$), and isobutane [$(CH_3)_3CH$].

As might be expected, the composition of the process gas varies in accordance with reforming severity and reformer feedstock. All catalytic reforming processes require substantial recycling of a hydrogen stream. Therefore, it is normal to separate reformer gas into a propane ($CH_3CH_2CH_3$) and/or a butane stream [$CH_3CH_2CH_2CH_3$ plus $(CH_3)_3CH$], which becomes part of the refinery liquefied petroleum gas production, and a lighter gas fraction, part of which is recycled. In view of the excess of hydrogen in the gas, all products of catalytic reforming are saturated, and there are usually no olefin gases present in either gas stream.

In many refineries, naphtha in addition to other refinery gases is also used as the source of petrochemical feedstocks. In the process, naphtha crackers convert naphtha feedstock (produced by various process) (Table 1.4) into ethylene, propylene, benzene, toluene, and xylenes, as well as other byproducts in a two-step process of cracking and separating. In some cases, a combination of naphtha, gas oil, and liquefied petroleum gas may be used. The feedstock, typically naphtha, is introduced into the pyrolysis section of the naphtha where it is cracked in the presence of steam. The naphtha is converted into lower-boiling fractions, primarily ethylene and propylene. The hot gas effluent from the furnace is then quenched to inhibit further cracking and to condense higher molecular weight products. The higher molecular weight products are subsequently processed into fuel oil, light cycle oil, and pyrolysis gas byproducts. The pyrolysis gas stream can then be fed to the aromatics plants for benzene and toluene production.

In addition to recovery of gases in the distillation section of a refinery distillation, gases are also produced in the various thermal processes, thermal cracking processes, and catalytic cracking processes (Figure 1.2), and are also available in processes such as visbreaking and coking (Speight, 1987; Parkash, 2003; Gary et al., 2007; Speight, 2014; Hsu and Robinson, 2017; Speight, 2017).

Thermal cracking processes were first developed for crude oil refining, starting in 1913 and continuing the next two decades, and were focused primarily on increasing the quantity and quality of gasoline components (Parkash, 2003; Gary et al., 2007; Speight, 2014; Hsu and Robinson, 2017; Speight, 2017). As a byproduct of this process, gases were produced that included a significant proportion of lower molecular weight olefins, particularly ethylene ($CH_2=CH_2$), propylene ($CH_3CH=CH_2$), and butylenes (butenes, $CH_3CH=CH·CH_3$ and $CH_3CH_2CH=CH_2$). Catalytic cracking, introduced in 1937, is also a valuable source of propylene and butylene, but it does not account for a very significant yield of ethylene, the most important of the petrochemical building blocks (Parkash, 2003; Gary et al., 2007; Speight, 2014; Hsu and Robinson, 2017; Speight, 2017). Ethylene is polymerized to produce polyethylene or, in combination with propylene, to produce copolymers that are used extensively in food-packaging wraps, plastic household goods, or building materials. Prior to the use of petroleum and natural gas as sources of chemicals, coal was the main source of chemicals (Speight, 2013).

Once produced and separated from other product streams, the cooled gases are then compressed, treated to remove acid gases, dried over a desiccant and fractionated into separate components at low temperature through a series of refrigeration processes (Parkash, 2003; Gary et al., 2007; Speight, 2014; Hsu and Robinson, 2017; Speight, 2017). Hydrogen and methane are removed by way of a compression / expansion process after which the methane is distributed to other processes as deemed appropriate for fuel gas. Hydrogen is collected and further purified in a pressure swing unit for use in the hydrogenation (hydrotreating and hydrocracking) processes (Parkash, 2003; Gary et al., 2007; Speight, 2014; Hsu and Robinson, 2017; Speight, 2017). Polymer grade ethylene and propylene are separated in the cold section after which the ethane and propane streams are recycled back to the furnace for further cracking while the mixed butane (C_4) stream is hydrogenated prior to recycling back to the furnace for further cracking.

In many refineries, naphtha in addition to other refinery gases is also used as the source of petrochemical feedstocks. In the process, naphtha crackers convert naphtha as well as gas oil feedstocks (produced by various process) (Table 1.4) into ethylene, propylene, benzene, toluene, and xylenes as well as into other byproducts in a two-step process of cracking and separating. In some cases, a combination of naphtha, gas oil, and liquefied petroleum gas may be used. The feedstock, typically naphtha, is introduced into the pyrolysis section of the naphtha where it is cracked in the presence of steam. The naphtha is converted into lower-boiling fractions, primarily ethylene and propylene. The hot gas effluent from the furnace is then quenched to inhibit further cracking and to condense higher molecular weight products. The higher molecular weight products are subsequently processed into fuel oil, light cycle oil, and pyrolysis gas byproducts. The pyrolysis gas stream can then be fed to the aromatics plants for benzene and toluene production.

The cooled gases are then compressed, treated to remove acid gases, dried over a desiccant and fractionated into separate components at low temperature through a series of refrigeration processes (Parkash, 2003; Gary et al., 2007; Hsu and Robinson, 2017; Speight, 2017). Hydrogen and methane are removed by way of a compression / expansion process after which the methane is distributed to other process as deemed appropriate or fuel gas. Hydrogen is collected and further purified in a pressure swing unit for use in the hydrogenation process. Polymer grade ethylene and propylene are separated in the cold section after which the ethane and propane streams are recycled back to the furnace for further cracking while the mixed butane (C_4) stream is hydrogenated prior to recycling back to the furnace for further cracking.

The refinery gas (or the process gas) stream and the products of naphtha cracking are the source of a variety of *petrochemicals*. For example, thermal cracking processes (Parkash, 2003; Gary et al., 2007; Hsu and Robinson, 2017; Speight, 2017) developed for crude oil refining, starting in 1913 and continuing the next two decades, were focused primarily on increasing the quantity and quality of gasoline components. As a byproduct of this process, gases were produced that included a significant proportion of lower molecular weight olefins, particularly ethylene ($CH_2=CH_2$), propylene ($CH_3CH=CH_2$), and butylenes (butenes, $CH_3CH=CHCH_3$ and $CH_3CH_2CH=CH_2$).

Catalytic cracking (Parkash, 2003; Gary et al., 2007; Hsu and Robinson, 2017; Speight, 2017), introduced in 1937, is also a valuable source of propylene and butylene, but it does not account for a very significant yield of ethylene, the most important of the petrochemical building blocks. Ethylene is polymerized to produce polyethylene or, in combination with propylene, to produce copolymers that are used extensively in food-packaging wraps, plastic household goods, or building materials. Prior to the use of petroleum and natural gas as sources of chemicals, coal was the main source of chemicals (Speight, 2013).

The petrochemical industry has grown with the petroleum industry (Goldstein, 1949; Steiner, 1961; Hahn, 1970) and is considered by some to be a mature industry. However, as is the case with the latest trends in changing crude oil types, it must also evolve to meet changing technological needs. The manufacture of chemicals or chemical intermediates from a variety of raw materials is well established (Wittcoff and Reuben, 1996). And the use of petroleum and natural gas is an excellent example of the conversion of such raw materials to more valuable products. The individual chemicals made from petroleum and natural gas are numerous and include industrial chemicals, household chemicals, fertilizers, and paints, as well as intermediates for the manufacture of products, such as synthetic rubber and plastics.

The manufacture of chemicals from petroleum is based on the ready response of the various compound types to basic chemical reactions, such as oxidation, halogenation, nitration, dehydrogenation addition, polymerization, and alkylation. The low molecular weight paraffins and olefins, as found in natural gas and refinery gases, and the simple aromatic hydrocarbon derivatives have so far been of the most interest because it is individual species that can be readily be isolated and dealt with. A wide range of compounds is possible, many are being manufactured and we are now progressing to the stage in which a sizable group of products is being prepared from the higher molecular weight fractions of petroleum.

The various reactions of asphaltene constituents indicate that these materials may be regarded as containing chemical functions and are therefore different and are able to participate in numerous chemical or physical conversions to, perhaps, more useful materials. The overall effect of these modifications is the production of materials that either affords good-grade aromatic cokes comparatively easily or the formation of products bearing functional groups that may be employed as a nonfuel material.

For example, the sulfonated and sulfomethylated materials and their derivatives have satisfactorily undergone tests as drilling mud thinners, and the results are comparable to those obtained with commercial mud thinners (Moschopedis and Speight, 1971, 1974, 1976a, 1978). In addition, these compounds may also find use as emulsifiers for the *in situ* recovery of heavy oils. These are also indications that these materials and other similar derivatives of the asphaltene constituents, especially those containing such functions as carboxylic or hydroxyl, readily exchange cations and could well compete with synthetic zeolites. Other uses of the hydroxyl derivatives and/or the chloro-asphaltenes include high-temperature packing or heat transfer media (Moschopedis and Speight, 1976b).

Reactions incorporating nitrogen and phosphorus into the asphaltene constituents are particularly significant at a time when the effects on the environment of many materials containing these elements are receiving considerable attention. In this case, there are potential slow-release soil conditioners that only release the nitrogen or phosphorus after considerable weathering or bacteriological action. One may proceed a step further and suggest that the carbonaceous residue remaining after release of the hetero-elements may be a benefit to humus-depleted soils, such as the gray-wooded soils. It is also feasible that coating a conventional quick-release inorganic fertilizer with a water-soluble or water-dispersible derivative will provide a slower-release fertilizer and an organic humus-like residue. In fact, variations on this theme are multiple (Moschopedis and Speight, 1974, 1976a).

Petrochemicals are made, or recovered from, the entire range of petroleum fractions, but the bulk of petrochemical products are formed from the lighter (C_1–C_4) hydrocarbon gases as raw materials. These materials generally occur as natural gas, but they are also recovered from the gas streams produced during refining, especially cracking, operations. Refinery gases are also particularly valuable because they contain substantial amounts of olefins that, because of the double bonds, are much more reactive then the saturated (paraffin) hydrocarbon derivatives. Also important as raw materials are the aromatic hydrocarbon derivatives (benzene, toluene, and xylene) that are obtained in rare cases from crude oil and, more likely, from the various product streams. By means of the catalytic reforming process, nonaromatic hydrocarbon derivatives can be converted to aromatics by dehydrogenation and cyclization.

A highly significant proportion of these basic petrochemicals are converted into plastics, synthetic rubbers, and synthetic fibers. Together these materials are known as polymers, because their molecules are high molecular weight compounds made up of repeated structural units that have combined chemically. The major products are polyethylene, polyvinyl chloride, and polystyrene, all derived from ethylene, and polypropylene, derived from monomer propylene. Major raw materials for synthetic rubbers include butadiene, ethylene, benzene, and propylene. Among synthetic fibers the polyesters, which are a combination of ethylene glycol and terephthalic acid (made from xylene), are the most widely used.

Petrochemical production relies on multiphase processing of oil and associated petroleum gas. Key raw materials in the petrochemical industry include products of petroleum oil refining (primarily gases and naphtha). Petrochemical goods include ethylene, propylene, and benzene; source monomers for synthetic rubbers; and inputs for technical carbon.

The petrochemical industry has grown with the petroleum industry and is considered by some to be a mature industry. However, as is the case with the latest trends in changing crude oil types, it must also evolve to meet changing technological needs. The manufacture of chemicals or chemical intermediates from a variety of raw materials is well established. And the use of petroleum and natural gas is an excellent example of the conversion of such raw materials to more valuable products. The individual chemicals made from petroleum and natural gas are numerous and include industrial chemicals, household chemicals, fertilizers, and paints, as well as intermediates for the manufacture of products, such as synthetic rubber and plastics.

The main objective in producing chemicals from petroleum is the formation of a variety of well-defined chemical compounds that are the basis of the petrochemical industry.

1.6   THE FUTURE

The petrochemical industry is concerned with the production and trade of petrochemicals and has a direct relationship with the petroleum industry, especially the downstream sector of the industry. The petrochemical industries are specialized in the production of petrochemicals that have various industrial applications. The petrochemical industry can be considered to be a subsector of the crude oil industry since without the petroleum industry the petrochemical industry cannot exist. Thus, petroleum is the major prerequisite raw material for the production of petrochemicals either in qualities or quantities. In addition, the petrochemical industry is subject to the geopolitics of the petroleum industry, with each industry being reliant upon the other for sustained survival.

In the 1970s, as a result of various oil embargos, coal liquefaction processes seemed on the point of commercialization and would have provided new sources of coal liquids for chemical use, as well as fulfilling the principal intended function of producing alternate fuels. Because of the varying price of petroleum, this prospect is unlikely to come to fruition in the immediate future, due to the question of economic viability rather than not technical feasibility. The combination of these and other factors has contributed to sharpening the focus on the use of coal for the production of heat and power, and lessening or eclipsing its possible use as a starting point for other processes.

The growth and development of petrochemical industries depends on a number of factors and also varies from one country to another either based on technical know-how, marketability, and applicability of these petrochemicals for manufacture of petrochemical products through petrochemical processes which are made feasible by knowledge and application of petrochemistry. Moreover, petrochemistry is a branch of chemistry (chemistry being a branch of natural science concerned with the study of the composition and constitution of substances and the changes such substances undergo because of changes in the molecules that make up such substances) that deals with petroleum, natural gas, and their derivatives.

However, not all of the petrochemical or commodity chemical materials produced by the chemical industry are made in one single location, but groups of related materials are often made in adjacent manufacturing plants to induce industrial symbiosis as well as material and utility efficiency and other economies of scale (integrated manufacturing). Specialty and fine chemical companies are sometimes found in similar manufacturing locations as petrochemicals but, in most cases, they do not need the same level of large-scale infrastructure (e.g., pipelines, storage, ports and power, etc.) and therefore can be found in multi-sector business parks. This will continue as long as the refining industry continues to exist in its present form (Favennec, 2001; Speight, 2011).

The petrochemical industry continues to be impacted by the globalization and integration of the world economy. For example, world-scale petrochemical plants built during the past several years are substantially larger than those built over two decades ago. As a result, smaller, older, and less efficient units are being shut down, expanded, or, in some cases, retrofitted to produce different chemical products. In addition, crude oil prices had been on the rise during the past decade and petrochemical markets are impacted during sharp price fluctuations, creating a cloud of uncertainty in upstream and downstream investments. Also, increasing concerns over fossil fuel supply and consumption, with respect to their impact on health and the environment, have led to the passage of legislation globally that will affect chemical and energy production and processing for the foreseeable future.

The recent shift from local markets to a large global market led to an increase in the competitive pressures on petrochemical industries. Further, because of fluctuations in products' price and high price of feedstocks, economical attractiveness of petrochemical plants can be considered as a main challenge. The ever-increasing cost of energy and more stringent environmental regulations impacted the operational costs. When cheap feedstocks are not available, the best method of profitability is to apply integration and optimization in petrochemical complexes with adjacent refineries. This is valid for installed plants and plants under construction. Petrochemical-refinery integration is an important factor in reducing costs and increasing efficiencies because integration guarantees the supply of feedstock for petrochemical industries. Also, integrated schemes take the advantage of the economy of scale and an integrated complex can produce more diverse products. Petrochemical-refinery integration avoids selling crude oil, optimizes products, economizes costs, and increases benefits.

On an innovation and technological basis (Hassani et al., 2017), manufacturing processes introduced in recent years have resulted in raw material replacement, shifts in the ratio of coproduct(s) produced, and cost. This has led to a supply/demand imbalance, particularly for smaller downstream petrochemical derivatives. In addition, growing environmental concerns and higher crude oil prices have expedited the development and commercialization of renewably derived chemical products and technologies previously considered economically impractical. Among the various technological advances, the combination of vertical hydraulic fracturing ("fracking") and horizontal drilling in multistage hydraulic fracturing resulted in a considerable rise in natural gas production in the United States. This new potential has caused many countries to reexamine their natural gas reserves and pursue development of their own gas plays.

Currently, crude oil and natural gas are the main sources of the raw materials for the production of petrochemicals because they (crude oil and natural gas) are the least expensive, most readily available, and can be processed most easily into the primary petrochemicals. However, as the current century progresses and the changes in crude oil supply that might be anticipated during the next five decades (Speight, 2011), there is a continuing need to assess the potential of other sources of petrochemicals.

For example, coal could well see a revitalization of use, understanding that there is the need to adhere to the various environmental regulations that apply to the use of any fossil fuel. Coal carbonization was the earliest and most important method to produce chemicals. For many years, chemicals that have been used for the manufacture of such diverse materials as nylon, styrene, fertilizers, activated carbon, drugs, and medicine, as well as many others have been made from coal. These products will expand in the future as petroleum and natural gas resources become strained to supply petrochemical feedstocks and coal becomes a predominant chemical feedstock once more. The ways in which coal may be converted to chemicals include carbonization, hydrogenation, oxidation, solvent extraction, hydrolysis, halogenation, and gasification (followed by conversion of the synthesis gas to chemical products) (Speight, 2013, 2014). In some cases, such processing does not produce chemicals in the sense that the products are relatively pure and can be marketed as even industrial grade chemicals. Thus, although many traditional markets for coal tar chemicals have been taken over by the petrochemical industry, the position can change suddenly as oil prices fluctuate upwards. Therefore, the concept of using coal as a major source of chemicals can be very real indeed.

Compared to petroleum crude, shale oils obtained by retorting of world's oil shales in their multitude and dissimilarity, are characterized by wide boiling range and by large concentrations of hetero-elements and also by high content of oxygen-, nitrogen-, or sulfur-containing compounds. The chemical potential of oil shale as retort fuel to produce shale oil and from that liquid fuel and specialty chemicals has been used so far to a relatively small extent. While the majority of countries are discovering the real practical value of shale oil, in Estonia retorting of its national resource kukersite oil obtained for production of a variety of products is in use for 75 years already. Using stepwise cracking motor fuels have been produced and even exported before World War II. At the same time, shale oils possess molecular structures of interest to the specialty chemicals industry and also a number of nonfuel specialty products have been marketed based on functional group, broad range concentrate, or even pure compound values.

Based on large quantity of oxygen-containing compounds in heavy fraction, asphalt-blending material, road asphalt and road oils, construction mastics, anticorrosion oils, and rubber softeners are produced. Benzene and toluene for production of benzoic acid as well as solvent mixtures on pyrolysis of lighter fractions of shale oil are produced. Middle shale oil fractions having antiseptic properties are used to produce effective oil for the impregnation of wood as a major shale oil-derived specialty product. Water-soluble phenols are selectively extracted from shale oil, fractionated and crystallized for production of pure 5-methylresorcinol and other alkyl resorcinol derivatives and high-value intermediates to produce tanning agents, epoxy resins and adhesives, diphenyl ketone and phenol-formaldehyde adhesive resins, rubber modifiers, chemicals, and pesticides. Some conventional products such as coke and distillate boiler fuels are produced from shale oil as byproducts. New market opportunities for shale oil and its fractions may be found improving the oil conversion and separation techniques.

In the petrochemical industry, the organic chemicals produced in the largest volumes are methanol, ethylene, propylene, butadiene, benzene, toluene, and xylenes.

Basic chemicals and plastics are the key building blocks for manufacture of a wide variety of durable and nondurable consumer goods. The demand for chemicals and plastics is driven by global economic conditions, which are directly linked to demand for consumer goods. The petrochemical industry continues to be impacted by the globalization and integration of the world economy. In the future, manufacturing processes introduced in recent years will continue to result in the adaptation of the industry to new feedstocks which will chase shifts in the ratio of products produced. This, in turn, will lead to the potential for a supply/demand imbalance, particularly for smaller downstream petrochemical derivatives. In addition, growing environmental concerns and the variability of crude oil prices (usually upward) will expedite the development and commercialization of chemical products from sources other than crude oil and natural gas. As a result, feedstocks and technologies previously considered economically impractical will rise to meet the increasing demand.

There is, however, the ever-present political uncertainty that arise from the occurrence of natural gas and crude oil resources in countries (provider countries) other than user countries. This has serious global implications for the supply and demand of petrochemicals and raw materials. In addition, the overall expansion of the population and an increase in individual purchasing power has resulted in an increase in demand for finished goods and greater consumption of energy in China, India, and Latin America.

However, the continued development of shale gas (tight gas) resources as well as crude oil from tight formation as well as the various technological advances to recover these resources (such as the combination of vertical hydraulic fracturing and horizontal drilling) will lead to a considerable rise in natural gas production and crude oil production. This new potential will cause many countries to reexamine their natural gas reserves and crude oil reserves to pursue development of their own nationally occurring gas plays and crude oil plays.

The production of chemicals from biomass is becoming an attractive area of investment for industries in the framework of a more sustainable economy. From a technical point of view, a large fraction of industrial chemicals and materials from fossil resources can be replaced by their bio-based counterparts. Nevertheless, fossil-based chemistry is still dominant because of optimized production processes and lower costs. The best approach to maximize the valorization of biomass is the processing of biological feedstocks in integrated biorefineries, where both bio-based chemicals and energy carriers can be produced, similar to a traditional petroleum refinery. The challenge is to prove, together with the technical and economic feasibility, an environmental feasibility, in terms of lower impact over the entire production chain.

Biomass is essentially a rich mixture of chemicals and materials and, as such, has a tremendous potential as feedstock for making a wide range of chemicals and materials with applications in industries from pharmaceuticals to furniture. Various types of available biomass feedstocks, including waste, and the different pretreatment and processing technologies being developed to turn these feedstocks into platform chemicals, polymers, materials, and energy.

There are several viable biological and chemical transformation pathways from sugars to building blocks. A large number of sugars to building block transformations can be done by aerobic fermentation employing fungi, yeast, or bacteria. Chemical and enzymatic transformations are also important process options. It should be noted, however, that pathways with more challenges and barriers are less likely be considered as viable industrial processes. In addition to gasification followed by Fischer–Tropsch chemistry of the gaseous product (synthesis gas), chemical reduction, oxidation, dehydration, bond cleavage, and direct polymerization are predominated. Enzymatic biotransformations comprise the largest group of biological conversions and some biological conversions can be accomplished without the need for an intermediate building block. The 1,3-Propanediol ($HOCH_2CH_2CH_2OH$) is an example where a set of successive biological processes convert sugar directly to an end product. Each pathway has its own set of advantages and disadvantages. Biological conversions, of course, can be tailored to result in a specific molecular structure, but the operating conditions must be relatively mild. Chemical transformations can operate at high throughput but, unfortunately, less conversion specificity is achieved.

Bio-based feedstocks may present a sustainable alternative to petrochemical sources to satisfy the ever-increasing demand for chemicals. However, the conversion processes needed for these future biorefineries will likely differ from those currently used in the petrochemical industry. Biotechnology and chemo-catalysis offer routes for converting biomass into a variety of chemicals that can serve as starting-point chemicals. While a host of technologies can be leveraged for biomass upgrading, the outcome can be significant because there is the potential to upgrade the bio-derived feedstocks while minimizing the loss of carbon and the generation of byproducts.

In fact, biomass offers a source of carbon from the biosphere as an alternative to fossilized carbon laid down tens of millions of years ago. Anything that grows and is available in nonfossilized form can be classified as biomass, including arable crops, trees, bushes, animal byproducts, human and animal waste, waste food, and any other waste stream that rots quickly and which can be replenished on a rolling time frame of years or decades. One of the attractions of biomass is its versatility: under the right circumstances, it can be used to provide a sustainable supply of electricity, heat, transport fuels, or chemical feedstocks in addition to its many other uses. One of the drawbacks of biomass, especially in the face of so many potential end uses, is its limited availability, even though the precise limitation is the subject of debate. Compared with the level of attention given to biomass as a source of electricity or heat, relatively little attention has been paid to biomass as a chemical feedstock. However, in a world in which conventional feedstocks are becoming constrained and countries are endeavoring to meet targets for reducing carbon dioxide emissions, there is a question as to whether biomass is too good to burn.

Developments in homogeneous and heterogeneous catalysis have led the way to effective approaches to utilizing renewable sources; however, further advances are needed to realize technologies that are competitive with established petrochemical processes. Catalysis will play a key role, with new reactions, processes, and concepts that leverage both traditional and emerging chemo- and bio-catalytic technologies.

Thus, new knowledge and better technologies are needed in dealing with chemical transformations that involve milder oxidation conditions, selective reduction, and dehydration, better control of bond cleavage, and improvements to direct polymerization of multifunctional monomers. For biological transformations, better understanding of metabolic pathways and cell biology, lower downstream recovery costs, increased utility of mixed sugar streams, and improved molecular thermal stability are necessary. While it is possible to prepare a very large number of molecular structures from the top building blocks, there is a scarcity of information about these behaviors of the molecular products and industrial processing properties. A comprehensive database on biomolecular performance characteristics would prove extremely useful to both the public sector and private sector. Nevertheless, here is a significant market opportunity for the development of bio-based products from the four-carbon building blocks. In order to be competitive with petrochemical-derived products, there is a significant technical challenge and should be undertaken with a long-term perspective.

In summary, the petrochemical industry, which is based on crude oil and natural gas, competes with the energy providing industry for the same fossil raw material. Dwindling oil and gas reserves, concern regarding the greenhouse effect (carbon dioxide emissions) and worldwide rising energy demand raise the question of the future availability of fossil raw materials. Biotechnological, chemical and engineering solutions are needed for utilization of this second-generation bio-renewable-based supply chain. One approach consists of the concept of a biorefinery. Also gasification followed by Fischer–Tropsch chemistry is a promising pathway. In the short term and in the medium term, a feedstock mix with crude oil and natural gas dominating can most likely be expected. In the long term, due to the final limited availability of oil and gas, biomass will prevail. Prior to this change to occur, great research and development efforts must be carried out to have the necessary technology available when needed.

In summary, the petrochemical industry gives a series of value-added products to the petroleum and natural gas industry but, like any other business, suffers from issues relating to maturity. The reasons relating to the maturity of the industry are (i) expired patents, (ii) varying demand, (iii) matching demand with capacity, and (iv) intense competition. Actions to combat the aches and pains of maturity are to restructure capacity achieving mega sizes, downstream, and restructuring business practices. Strategies followed by some companies to combat maturity include exit, focus on core business, and exploit a competitive advantage.

Nevertheless, the petrochemical industry is and will remain a necessary industry for the support of modern and emerging lifestyles. In order to maintain an established petrochemical industry, strategic planning is the dominating practice to maintain the industry (replace imports, export, new products, alternate feedstocks such as the return to the chemicals-from-coal concept and the acceptance of feedstocks, such as oil shale and biomass) including developing criteria for selecting products/projects. After the oil crises of the 1970s (even though it is now four decades since these crises), it is necessary to cope with the new environment of product demand through the response to new growth markets and security of feedstock supply. Mergers, alliances, and acquisitions could well be the dominating practice to combat industry maturity and increased market demand as one of the major activities. Other strategies are the focus on core business (the production of chemicals) and, last but not least, the emergence (or, in the case of coal, the reemergence) of alternate feedstocks to ensure industry survival.

REFERENCES

Chenier, P.J. 2002. *Survey of Industrial Chemicals*. 3rd Edition. Springer, New York.

Clark, J.H., and Deswarte, F. (Editors). 2015. *Introduction to Chemicals from Biomass*. 2nd Edition. John Wiley & Sons Inc., Hoboken, NJ.

EPCA. 2016. *50 Years of Chemistry for You*. European Petrochemical Association, Brussels, Belgium. https://epca.eu/

Favennec, J.-P. (Editor). 2001. *Petroleum Refining: Refinery Operation and Management*. Editions Technip, Paris, France.

Gary, J.H., Handwerk, G.E., and Kaiser, M.J. 2007. *Petroleum Refining: Technology and Economics*. 5th Edition. CRC Press, Boca Raton, FL.

Goldstein, R.F. 1949. *The Petrochemical Industry*. E. & F. N. Spon, London, UK.

Hahn, A.V. 1970. *The Petrochemical Industry: Market and Economics*. McGraw-Hill, New York.

Hassani, H., Silva, E.S., and Al Kaabi, A.M. 2017. The Role of Innovation and Technology in Sustaining the Petroleum and Petrochemical Industry. *Technological Forecasting and Social Change*, 119 (June): 1–17.

Hsu, C.S., and Robinson, P.R. (Editors). 2017. *Handbook of Petroleum Technology*. Springer, Cham, Switzerland.

Katz, D.K. 1959. *Handbook of Natural Gas Engineering*. McGraw-Hill, New York.

Kidnay, A.J., and Parrish, W.R. 2006 *Fundamentals of Natural Gas Processing*. CRC Press, Boca Raton, FL.

Kohl, A.L., and Nielsen, R.B. 1997. *Gas Purification*. Gulf Publishing Company, Houston, TX.

Kohl, A. L., and Riesenfeld, F.C. 1985. *Gas Purification*. 4th Edition. Gulf Publishing Company, Houston, TX.

Lee, S., Speight, J.G., and Loyalka, S. 2007. *Handbook of Alternative Fuel Technologies*. CRC Press, Boca Raton, FL.

Maddox, R.N., Bhairi, A., Mains, G.J., and Shariat, A. 1985. In: *Acid and Sour Gas Treating Processes*. S.A. Newman (Editor). Gulf Publishing Company, Houston, TX. Chapter 8.

Matar, S., and Hatch, L.F. 2001. *Chemistry of Petrochemical Processes*. 2nd Edition. Butterworth-Heinemann, Woburn, MA.

Meyers, R.A. 2005. *Handbook of Petrochemicals Production Processes*. McGraw-Hill, New York.

Mokhatab, S., Poe, W.A., and Speight, J.G. 2006. *Handbook of Natural Gas Transmission and Processing*. Elsevier, Amsterdam, Netherlands.

Moschopedis, S.E., and Speight, J.G. 1971. Water-Soluble Derivatives of Athabasca Asphaltenes. *Fuel*, 50: 34.

Moschopedis, S.E., and Speight, J.G. 1974. The Chemical Modification of Bitumen and Its Non-Fuel Uses. *Preprints. Div. Fuel Chem., Am. Chem. Soc.*, 19(2): 291.

Moschopedis, S.E., and Speight, J.G. 1976a. The Chemical Modification of Bitumen Heavy Ends and Their Non-Fuel Uses. In: *Shale Oil, Tar Sands and Related Fuels Sources*. Adv. in Chem. Series No. 151, Am. Chem. Soc. T.F. Yen (Editor), p. 144.

Moschopedis, S.E., and Speight, J.G. 1976b. The Chlorinolysis of Petroleum Asphaltenes. *Chemika Chronika*, 5: 275.

Moschopedis, S.E., and Speight, J.G. 1978. Sulfoxidation of Athabasca Bitumen. *Fuel*, 857: 647.

Naderpour, N. 2008. *Petrochemical Production Processes*. SBS Publishers, Delhi, India.

Newman, S.A. 1985. *Acid and Sour Gas Treating Processes*. Gulf Publishing Company, Houston, TX.

Parkash, S. 2003. *Refining Processes Handbook*. Gulf Professional Publishing, Elsevier, Amsterdam, Netherlands.

Spitz, P.H. 1988. *Petrochemicals: The Rise of an Industry*. John Wiley & Sons Inc., Hoboken, NJ.

Speight, J.G. 1987. *Petrochemicals. Encyclopedia of Science and Technology*, Vol. 13. 6th Edition. McGraw-Hill, New York, p. 251.

Speight, J.G. 1994. Chemical and Physical Studies of Petroleum Asphaltene Constituents. In: *Asphaltene Constituents and Asphalts. I*. Developments in Petroleum Science, 40. T.F. Yen and G.V. Chilingarian (Editors). Elsevier, Amsterdam, Netherlands. Chapter 2.

Speight, J.G. 2002. *Chemical Process and Design Handbook*. McGraw-Hill, New York.

Speight, J.G. 2008. *Handbook of Synthetic Fuels Handbook: Properties, Processes, and Performance.* McGraw-Hill, New York.
Speight. J.G. 2011. *The Refinery of the Future.* Gulf Professional Publishing, Elsevier, Oxford, UK.
Speight, J.G. 2013. *The Chemistry and Technology of Coal.* 3rd Edition. CRC Press, Boca Raton, FL.
Speight, J.G. 2014. *The Chemistry and Technology of Petroleum.* 5th Edition. CRC Press, Boca Raton, FL.
Speight, J.G. 2015. *Handbook of Petroleum Product Analysis.* 2nd Edition. John Wiley & Sons Inc., Hoboken, NJ.
Speight, J.G. 2016. *Handbook of Hydraulic Fracturing.* John Wiley & Sons Inc., Hoboken, NJ.
Speight, J.G. 2017. *Handbook of Petroleum Refining.* CRC Press, Boca Raton, FL.
Steiner, H. 1961. *Introduction to Petroleum Chemicals.* Pergamon Press, New York.
Wittcoff, H.A., and Reuben, B.G. 1996. *Industrial Organic Chemicals.* John Wiley & Sons Inc., New York.

2 Feedstock Composition and Properties

2.1 INTRODUCTION

In any text related to the various aspects of petrochemical technology, it is necessary to consider the properties and behavior of the feedstocks first through the name or terminology and/or the definition of the feedstock. Because of the need for a thorough understanding of the petrochemical industry as well as crude oil and the associated feedstocks, it is essential that the definitions and the terminology of petrochemical science and technology be given prime consideration.

Terminology is the means by which various subjects are named so that reference can be made in conversations and in writings and so that the meaning is passed on. *Definitions* are the means by which scientists and engineers communicate the nature of a material to each other and to the world, either through the spoken or the written word. Thus, the definition of a material can be extremely important and have a profound influence on how the technical community and the public perceive that material. This part of the text attempts to alleviate much of the confusion that exists, but it must be remembered that the terminology of crude oil is unfortunately still open to personal choice and historical use of the various names.

While there is standard terminology that is recommended for crude oil and crude oil products (ASTM D4175, 2018), there is little in the way of standard terminology for heavy oil, extra heavy oil, and tar sand bitumen (Speight, 2013a, 2013b, 2013c, 2014a). At best, the terminology is ill-defined and subject to changes from one governing body company to another. The particularly troublesome, and more confusing, terminologies are those terms that are applied to the more viscous feedstocks, for example, the use of the terms *bitumen* (the naturally occurring carbonaceous material in tar sand deposits) and *asphalt* (refinery product produced from residua). Another example of an irrelevant terminology is the term *black oil* which, besides the color of the oil, offers nothing in the way of explanation of the properties of the oil and certainly adds nothing to any scientific and/or engineering understanding of the oil; this term (i.e., *black oil* is not used in this text).The feedstocks to be considered here are (i) natural gas, (ii) conventional crude oil, (iii) heavy oil, (iv) extra heavy oil, (v) tar sand bitumen, (vi) coal, (vii) oil shale, and (viii) biomass, including landfill gas and biogas. All of these current and potential feedstocks could continue in the petrochemical industry for the foreseeable future as the reserves of natural gas and conventional crude oil become depleted to the point of exhaustion during the next 50 years (Speight, 2011a; Speight and Islam, 2016). However, for the most part, in view of the current sources of petrochemicals, the major focus will be on petroleum with reference, where appropriate, to the other sources of petrochemicals.

2.2 NATURAL GAS

Natural gas, predominantly methane, occurs in underground reservoirs separately or in association with crude oil (Chapter 2) (Speight, 2007, 2008, 2014a). The principal types of hydrocarbon derivatives produced from natural gas are methane (CH_4) and varying amounts of higher molecular weight hydrocarbon derivatives from ethane (CH_3CH_3) to octane [$CH_3(CH_2)_6CH_3$]. Generally, the higher molecular weight liquid hydrocarbon derivatives from pentane to octane are collectively referred to as *gas condensate*.

TABLE 2.1
Composition of Associated Natural Gas from a Petroleum Well

Category	Component	Amount (% v/v)
Paraffins	Methane (CH_4)	70–98
	Ethane (C_2H_6)	1–10
	Propane (C_3H_8)	Trace-5
	Butane (C_4H_{10})	Trace-2
	Pentane (C_5H_{12})	Trace-1
	Hexane (C_6H_{14})	Trace-0.5
	Heptane and higher molecular weight (C_{7+})	Trace
Cycloparaffins	Cyclohexane (C_6H_{12})	Trace
Aromatics	Benzene (C_6H_6)+other aromatics	Trace
Non-hydrocarbons	Nitrogen (N_2)	Trace-15
	Carbon dioxide (CO_2)	Trace-1
	Hydrogen sulfide (H_2S)	Trace-1
	Helium (He)	Trace-5
	Other sulfur and nitrogen compounds	Trace
	Water (H_2O)	Trace-5

While natural gas is predominantly a mixture of combustible hydrocarbon derivatives (Table 2.1), many natural gases also contain nitrogen (N_2) as well as carbon dioxide (CO_2) and hydrogen sulfide (H_2S). Trace quantities of helium and other sulfur and nitrogen compounds may also be present. However, raw natural gas varies greatly in composition and the constituents can be several of a group of saturated hydrocarbon derivatives from methane to higher molecular weight hydrocarbon derivatives, especially natural gas that has been associated with crude oil in the reservoir, and non-hydrocarbon constituents (Table 2.1). The treatment required to prepare natural gas for distribution as an industrial or household fuel is specified in terms of the use and environmental regulations.

Briefly, natural gas contains hydrocarbon derivatives and non-hydrocarbon gases. Hydrocarbon gases are methane (CH_4), ethane (C_2H_6), propane (C_3H_8), butanes (C_4H_{10}), pentanes (C_5H_{12}), hexane (C_6H_{14}), heptane (C_7H_{16}), and sometimes trace amounts of octane (C_8H_{18}), and higher molecular weight hydrocarbon derivatives. For example:

$CH_3CH_2CH_2CH_3$
n-Butane

$(CH_3)_3CH$ or $(CH_3)_2CHCH_3$
isobutane

$CH_3CH_2CH_2CH_2CH_3$
n-Pentane

$(CH_3)_2CHCH_2CH_3$
isopentane

As illustrated above, an isoparaffin is an isomer having a methyl group branching from carbon number 2 of the main chain.

The higher-boiling hydrocarbon constituents than methane (CH_4) are often referred to as *natural gas liquids* (NGLs) and the natural gas may be referred to as *rich gas*. The constituents of natural gas liquids are hydrocarbon derivatives such as ethane (CH_3CH_3), propane ($CH_3CH_2CH_3$), butane ($CH_3CH_2CH_2CH_3$, as well as isobutane), pentane derivatives ($CH_3CH_2CH_2CH_2CH_3$, as well as isopentane) and higher molecular weight hydrocarbon derivatives which have wide use in the petrochemical industry (Chapter 6). Some aromatic derivatives [BTX—benzene (C_6H_6), toluene ($C_6H_5CH_3$), and the xylene isomers (o-, m-, and p-$CH_3C_6H_4CH_3$)] can also be present, raising safety issues due to their toxicity. The non-hydrocarbon gas portion of the natural gas contains nitrogen (N_2), carbon dioxide (CO_2), helium (He), hydrogen sulfide (H_2S), water vapor (H_2O), and other sulfur compounds (such as carbonyl sulfide (COS) and mercaptans (e.g., methyl mercaptan, CH_3SH) and trace amounts of other gases. In addition, the composition of a gas stream from a source or at a location can also vary over time which can cause difficulties in resolving the data from the application of standard test methods (Klimstra, 1978; Liss and Thrasher, 1992).

Carbon dioxide and hydrogen sulfide are commonly referred to as *acid gases* since they form corrosive compounds in the presence of water. Nitrogen, helium, and carbon dioxide are also referred to as *diluents* since none of these burn, and thus they have no heating value. Mercury can also be present either as a metal in vapor phase or as an organometallic compound in liquid fractions. Concentration levels are generally very small, but even at very small concentration levels, mercury can be detrimental due its toxicity and its corrosive properties (reaction with aluminum alloys).

The higher molecular weight constituents (i.e., the C_{5+} product) are also commonly referred to as gas condensate or natural gasoline or sometimes, on occasion, as *casinghead gas* because of the tendency of these constituents to condense at the top of the well casing. When referring to natural gas liquids in the gas stream, the term gallon per thousand cubic feet is used as a measure of high molecular weight hydrocarbon content. On the other hand, the composition of nonassociated gas (sometimes called well gas) is deficient in natural gas liquids. The gas is produced from geological formations that typically do not contain much, if any, hydrocarbon liquids. Furthermore, within the natural gas family, the composition of associated gas (a byproduct of oil production and the oil recovery process) is extremely variable, even within the gas from a petroleum reservoir (Speight, 2014a, 2018). After the production fluids are brought to the surface, they are separated at a tank battery at or near the production lease into a hydrocarbon liquid stream (crude oil or condensate), a produced water stream (brine or salty water), and a gas stream.

The gaseous mixtures considered in this volume are mixtures of various constituents that may or may not vary over narrow limits. The defining characteristics of the various gas streams in the context of this book are that the gases (i) exist in a gaseous state at room temperature; (ii) may contain hydrocarbon constituents with 1–4 carbons, i.e., methane, ethane, propane, and butane isomers; (iii) may contain diluents and inert gases; and (iv) may contain contaminants in the form of non-hydrocarbon constituents. Each constituent of the gas influences the properties.

Typically, these gases fall into the general category of fuel gases and each gas is any one of several fuels that, at standard conditions of temperature and pressure, are gaseous. Before sale of the gas to the consumer actions, it is essential to give consideration of the variability of the composition of gas streams before and after processing (Table 2.2) and the properties of the individual constituents and their effects on gas behavior, even when considering the hydrocarbon constituents only. If not, the properties of the gas may be unstable and the ability of the gas to be used for the desired purpose will be seriously affected.

2.2.1 COMPOSITION AND PROPERTIES

Natural gas is a naturally occurring gas mixture, consisting mainly of methane that is found in porous formations beneath the surface of the earth, often in association with crude oil; but while the gas from

TABLE 2.2
General Properties of Unrefined Natural Gas and Refined
Natural Gas

Property	Unrefined Gas	Refined Gas
Carbon, % w/w	73	75
Hydrogen, % w/w	27	25
Oxygen, % w/w	0.4	0
Hydrogen-to-hydrogen atomic ratio	3.5	4.0
Vapor density (air=1, 15°C)	1.5	0.6
Methane, % v/v]	80	1,000
Ethane, % v/v	5	0
Nitrogen, % v/v	15	0
Carbon dioxide, % v/v	5	0
Sulfur, ppm w/w	5	0

the various sources has a similar analysis, it is not entirely the same. In fact, variation in composition varies from field to field and may even vary within a reservoir. In addition, the variation of gas streams from different sources (Chapter 3) must also be considered when processing options are being assessed. Thus, because of the lower molecular weight constituents of these gases and their volatility, gas chromatography has been the technique of choice for fixed gas and hydrocarbon speciation and mass spectrometry is also a method of choice for compositional analysis of low molecular weight hydrocarbon derivatives (Speight, 2015; ASTM, 2018; Speight, 2018). The vapor pressure and volatility specifications will often be met automatically if the hydrocarbon composition is in order with the specification.

As with crude oil, natural gas from different wells varies widely in composition and analyses (Mokhatab et al., 2006; Speight, 2014a), and the proportion of non-hydrocarbon constituents can vary over a very wide range. The non-hydrocarbon constituents of natural gas can be classified as two types of materials: (i) diluents, such as nitrogen, carbon dioxide, and water vapors, and (ii) contaminants, such as hydrogen sulfide and/or other sulfur compounds. Thus, a particular natural gas field could require production, processing, and handling protocols different from those used for gas from another field.

Thus, there is no single composition of components which might be termed *typical* natural gas (Speight, 2007, 2014a, 2018). Methane and ethane often constitute the bulk of the combustible components; carbon dioxide (CO_2) and nitrogen (N_2) are the major noncombustible (inert) components. Thus, sour gas is natural gas that occurs mixed with higher levels of sulfur compounds (such as hydrogen sulfide (H_2S) and mercaptan derivatives, often called thiols, RSH) and which constitute a corrosive gas (Speight, 2014b). The sour gas requires additional processing for purification (Mokhatab et al., 2006; Speight, 2014a). Olefin derivatives are also present in the gas streams from various refinery processes and are not included in liquefied petroleum gas but are removed for use in petrochemical operations (Crawford et al., 1993).

The composition and properties of any gas stream depends on the characterization and properties of the hydrocarbon derivatives that make up the stream and calculation of the properties of a mixture depends on the properties of its constituents. However, calculation of the property of a mixture based on an average calculation neglects any interactions between the constituents. This makes the issue of modeling of the properties of the gas mixture a difficult one because of the frequent lack of knowledge (and omission) of any chemical or physical interactions between the gas stream constituents. Because of the lower molecular weight constituents of these gases and their volatility, gas chromatography has been the technique of choice for hydrocarbon speciation and mass spectrometry is also a method of choice for compositional analysis of low molecular weight hydrocarbon derivatives (Speight, 2015; ASTM, 2018; Speight, 2018). The vapor pressure and volatility specifications will often be met automatically if the hydrocarbon composition is in order with the specification.

Natural gas is found in petroleum reservoirs as free gas (*associated gas*) or in solution with petroleum in the reservoir (*dissolved gas*) or in reservoirs that contain only gaseous constituents and no (or little) petroleum (*unassociated gas*) (Speight, 2014a). The hydrocarbon content varies from mixtures of methane and ethane with very few other constituents (*dry* gas) to mixtures containing all the hydrocarbon derivatives from methane to pentane and even hexane (C_6H_{14}) and heptane (C_7H_{16}) (*wet* gas). In both cases, some carbon dioxide (CO_2) and inert gases, including helium (He), are present together with hydrogen sulfide (H_2S) and a small quantity of organic sulfur.

The term *petroleum gas(es)* in this context is also used to describe the gaseous phase and liquid phase mixtures comprised mainly of methane to butane (C_1–C_4 hydrocarbon derivatives) that are dissolved in the crude oil and natural gas, as well as gases produced during thermal processes in which the crude oil is converted to other products. It is necessary, however, to acknowledge that in addition to the hydrocarbon derivatives, gases such as carbon dioxide, hydrogen sulfide, and ammonia are also produced during petroleum refining and will be constituents of refinery gas that must be removed. Olefin derivatives are also present in the gas streams of various processes and are not included in liquefied petroleum gas but are removed for use in petrochemical operations (Crawford et al., 1993).

Nonassociated natural gas, which is found in reservoirs in which there is no, or at best only minimal amounts of, crude oil (Chapter 1). Nonassociated gas is usually richer in methane but is markedly leaner in terms of the higher molecular weight hydrocarbon derivatives and condensate. Conversely, there is also *associated* natural gas (*dissolved* natural gas) that occurs either as free gas or as gas in solution in the petroleum. Gas that occurs as a solution with the crude petroleum is *dissolved gas*, whereas the gas that exists in contact with the crude petroleum (*gas cap*) is *associated gas* (Chapter 1). Associated gas is usually leaner in methane than the nonassociated gas but is richer in the higher molecular weight constituents. Thus, the most preferred type of natural gas is the nonassociated gas. Such gas can be produced at high pressure, whereas associated, or dissolved, gas must be separated from petroleum at lower separator pressures, which usually involves increased expenditure for compression. Thus, it is not surprising that such gas (under conditions that are not economically favorable) is often flared or vented.

Natural gas is a naturally occurring mixture of low-boiling hydrocarbon derivatives accompanied by some non-hydrocarbon compounds. Nonassociated natural gas is found in reservoirs containing no oil (dry wells). Associated gas, on the other hand, is present in contact with and/or dissolved in crude oil and is coproduced with it. The principal component of most natural gases is methane. Higher molecular weight paraffin hydrocarbon derivatives (C_2–C_7, even to C_{10} in some cases) are usually present in smaller amounts with the natural gas mixture, and their ratios vary considerably from one gas field to another. Nonassociated gas normally contains a higher methane ratio than associated gas, while the latter contains a higher ratio of higher molecular weight hydrocarbon derivatives (Table 2.1).

Crude oil-related gases (including associated natural gas) and refinery gases (process gases) as well as product gases produced from petroleum upgrading are a category of saturated and unsaturated gaseous hydrocarbon derivatives, predominantly in the C_1–C_6 carbon number range. Some gases may also contain inorganic compounds, such as hydrogen, nitrogen, hydrogen sulfide, carbon monoxide, and carbon dioxide. As such, petroleum and refinery gases (unless produced as a salable product that must meet specifications prior to sale) are often unknown or variable composition and toxic (API, 2009; ASTM, 2018). The site-restricted petroleum and refinery gases (i.e., those not produced for sale) often serve as fuels consumed onsite, as intermediates for purification and recovery of various gaseous products, or as feedstocks for isomerization and alkylation processes within a facility. Thus, natural gas is a combustible mixture of hydrocarbon gases that, in addition to methane, also includes ethane, propane, butane, and pentane. The composition of natural gas can vary widely before it is refined (Tables 2.1 and 2.2) (Mokhatab et al., 2006; Speight, 2014a). In its purest form, such as the natural gas that is delivered to the consumer is almost pure methane.

The principal constituent of most natural gases is methane with minor amounts of heavier hydrocarbon derivatives and certain non-hydrocarbon gases such as nitrogen, carbon dioxide, hydrogen sulfide, and helium (Mokhatab et al., 2006; Speight, 2014a; ASTM, 2018; Speight, 2018). Methane can be produced in the laboratory by heating sodium acetate with sodium hydroxide and by the reaction of aluminum carbide (Al_4C_3) with water.

$$Al_4C_3 + 12H_2O \rightarrow 4Al(OH)_3 \downarrow + 3CH_4$$

$$CH_3CO_2Na + NaOH \rightarrow CH_4 + Na_2CO_3$$

The members of the hydrocarbon gases are predominantly alkane derivatives (C_nH_{2n+2}, where n is the number of carbon atoms). When inorganic constituents are present in natural gas, they consist of asphyxiant gases such as hydrogen.

Unlike other categories of crude oil products (such as naphtha, kerosene, and the higher-boiling products) (Speight, 2014a; ASTM, 2018; Speight, 2017), the constituents of the various gas streams can be evaluated, and the results of the constituent evaluation can then be used to estimate the behavior of the gas (ASTM, 2018; Speight, 2018). The constituents used to evaluate the behavior of the gas are: (i) the C_1–C_4 hydrocarbon derivatives, (ii) the C_5–C_6 hydrocarbon derivatives, although in natural gas the C_1–C_4 constituents predominate, and (iii) the asphyxiant gases, i.e., carbon dioxide, nitrogen, and hydrogen. In general, most gas streams used in this text are composed of predominantly the methane (C_1) to butane (C_4) hydrocarbon derivatives, which have extremely low melting points and boiling points.

Each of these gases has a high vapor pressures and low octanol-water partition coefficients—the octanol-water partition coefficient (K_{ow}) is a valuable parameter that represents a measure of the tendency of a chemical to move from the aqueous phase to the organic (octanol) phase. Thus:

$$K_{ow} = C_{op}C_w$$

C_{op} and C_w are the concentrations of the chemical in gm/L of the chemical in the octanol-rich phase and in the water-rich phase, respectively. In the determination of the partition coefficient at 25°C (77°F), the water-rich phase is essentially pure water (99.99 mol % water) while the octanol-rich phase is a mixture of octanol and water (79.3 mol % octanol). While not always required in the production of petrochemicals, such a property may be of some value during application of the synthetic method and as a means of product purification.

The aqueous solubility of the various constituents of gas streams varies, but the solubility of most of the hydrocarbon derivatives typically falls within a range of 22 mg/L to several hundred parts per million. There are also a few gas streams that may contain heptane derivatives and octane derivative, although such streams would necessarily be at elevated temperature and/or reduced pressure to maintain the heptane derivatives and the octane derivatives in the gaseous state. Hydrocarbon compounds containing pentane, hexane, heptane, and octane derivatives occur predominantly in low-boiling crude oil naphtha and also occur in gas condensate and natural gas.

By way of recall, in addition to methane, natural gas contains other constituents that are variously referred to as (i) natural gas liquids, (ii) natural gas condensate, and (iii) natural gasoline (Chapter 1). Also, by way of a refresher definition, natural gas liquids are hydrocarbon derivatives that occur as gases at ambient conditions (atmospheric pressure and temperature) but as liquids under higher pressure and which can also be liquefied by cooling. The specific pressure and temperature at which the gases liquefy vary by the type of gas liquids and may be described as low-boiling (*light*) or high-boiling (*heavy*) according to the number of carbon atoms and hydrogen atoms in the molecule.

In terms of the chemical reactions of natural gas, the most common reaction is combustion process which is represented as chemical reaction between methane and oxygen which results in the production of carbon dioxide (CO_2), water (H_2O) plus the exothermic liberation of energy (heat). Thus:

$$CH_4(g) + 2O_2(g) \rightarrow CO_2(g) + 2H_2O(l)$$

Higher molecular weight hydrocarbon (alkane) constituents will also participate in the combustion reaction. In an unlimited supply of oxygen and assuming that there may be traces of hydrocarbon derivatives up to octane in a natural gas stream, the combustion reactions are:

$$C_3H_8 + 5O_2 \rightarrow 3CO_2 + 4H_2O$$

$$2C_4H_{10}(g) + 13O_2(g) \rightarrow 8CO_2(g) + 10H_2O(g)$$

$$C_5H_{12}(g) + 8O_2(g) \rightarrow 5CO_2(g) + 6H_2O(g)$$

$$2C_6H_{14}(l) + 19O_2(g) \rightarrow 12CO_2(g) + 14H_2O(g)$$

$$C_7H_{16}(l) + 11O_2(g) \rightarrow 7CO_2(g) + 8H_2O(g)$$

$$2C_8H_{18}(l) + 25O_2(g) \rightarrow 16CO_2(g) + 18H_2O(g)$$

The balanced chemical equation for the complete combustion of a general hydrocarbon fuel C_xH_y is:

$$C_xH_y + (x + y/4)O_2 \rightarrow xCO_2 + x/2H_2O$$

To the purist, chemical equations do not involve fractions and to balance this final equation, the fractional numbers should be converted to whole numbers.

In an inadequate supply of air, carbon monoxide and water vapor are formed, using methane as the example:

$$2CH_4 + 3O_2 \rightarrow 2CO + 4H_2O$$

In this context of combustion, natural gas is the cleanest of all the fossil fuels. Coal and crude oil are composed of much more complex molecules, with a higher carbon ratio and as well as constituents containing nitrogen and sulfur contents. Thus, when combusted, coal and oil release higher levels of harmful emissions, including a higher ratio of carbon emissions, nitrogen oxides (NOx), and sulfur dioxide (SO_2), which under the conditions of the atmosphere, can be converted to sulfur trioxide (SO_3). Upon further reaction with the water in the atmosphere, the oxides of nitrogen and the oxides of sulfur are converted to acids and, thus, the overall result is the production of acid rain (Chapter 10):

$$SO_2 + H_2O \rightarrow H_2SO_3$$

$$2SO_2 + O_2 \rightarrow 2SO_3$$

$$SO_3 + H_2O \rightarrow H_2SO_4$$

$$2NO + H_2O \rightarrow 2HNO_2$$

$$2NO + O_2 \rightarrow 2NO_2$$

$$NO_2 + H_2O \rightarrow HNO_3$$

Substitution reactions will also occur in which the hydrocarbon derivatives in natural gas will react with, for example, chlorine to produce a range of chloro-derivatives:

$$CH_4 + Cl_2 \rightarrow CH_3Cl + HCl$$

$$CH_3Cl + Cl_2 \rightarrow CH_2Cl_2 + HCl$$

$$CH_2Cl_2 + Cl_2 \rightarrow CHCl_3 + HCl$$

$$CHCl_2 + Cl_2 \rightarrow CCl_4 + HCl$$

The reaction of chlorine with ethane may be written similar to:

$$C_2H_6 + Cl_2 \rightarrow C_2H_5Cl + HCl$$

$$C_2H_4Cl_2 + Cl_2 \rightarrow C_2H_3Cl_3 + HCl$$

The ultimate product is hexachloroethane. Both of these reactions may be used industrially. As the hydrocarbon derivatives increase in molecular size (to propane and, butane), the reaction becomes more complex.

In addition to gas streams (particularly natural gas) being used as fuel to produce heat as well as the production of hydrogen (the steam-methane reforming process) and ammonia:

$$CH_4 + H_2O \rightarrow CO + 3H_2 - steam - methane\,reforming$$

$$CO + H_2O \rightarrow CO_2 + H_2 - hydrogen\,production$$

$$3H_2 + N_2 \rightarrow 2NH_3 - Haber - Bosch\,process$$

The steam-methane reforming process is major source of hydrogen for refineries and other industries (Speight, 2016b). The general feedstock for this process is natural gas, which has a high content of methane, 85–95% (v/v). Once the feedstock gas is obtained, it is desulfurized and treated before being sent to the reformer. The feedstock must be treated first to ensure that the sulfur is not released to the environment where it can cause significant damage.

In the endothermic process, high-temperature steam (700°C–1,100°C, 1,290°F–2,010°F) is used to produce hydrogen from a methane source, such as natural gas at pressures in the order of 45–370 psi. Subsequently, in what is called the *water-gas shift reaction*, the carbon monoxide and steam are reacted using a catalyst to produce carbon dioxide and more hydrogen. In a final process step (*pressure-swing adsorption*), carbon dioxide and other impurities are removed from the gas stream, leaving essentially pure hydrogen.

$$CH_4 + H_2O \rightarrow CO + 3H_2 \quad (steam - methane\,reforming)$$

$$CO + H_2O \rightarrow CO_2 + H_2 \quad (water - gas\,shift\,reaction)$$

In most plants this reaction will occur in two different stages, a high-temperature shift reaction which is then followed by a low-temperature shift reactor. The purpose of including both reactors is to maximize the yield of hydrogen. The byproduct passes through the high-temperature shift reactor first where due to the high temperature the reaction will be able to occur rapidly. However, due to the reaction being a slightly exothermic reaction when it shifts right, it is possible to obtain a greater yield of hydrogen by passing the reformer byproducts through the low-temperature shift reactor. In both reactors, a catalyst is employed to increase the hydrogen yield.

The other low molecular weight hydrocarbon derivatives in the gas—ethane (C_2H_6), propane (C_3H_8), and the butane isomers (C_4H_{10}), either in the gas phase or liquefied, are also used for heating as well as for motor fuels and as feedstocks for chemical processing. The pentane derivatives (C_5H_{12}) are products of natural gas or crude oil fractionation or refinery operations (i.e., reforming and cracking) that are removed for use as chemical feedstocks (Table 2.3). It is only rarely that olefin derivatives occur in natural gas and they are not typically constituents of natural gas stream. However, olefin derivatives do occur in biogas produced by thermal methods (Parkash, 2003; Mokhatab et al., 2006; Gary et al., 2007; Speight, 2011a, 2011b, 2014a; Hsu and Robinson, 2017; Speight, 2017).

Because of the wide range of chemical and physical properties, a wide range of tests have been (and continue to be) developed to provide an indication of the means by which a particular gas should be processed although certain of these test methods are in more common use than others (Speight, 2015; ASTM, 2018; Speight, 2018). Initial inspection of the nature of the petroleum will provide deductions about the most logical means of refining or correlation of various properties to structural types present and hence attempted classification of the petroleum. Proper interpretation of the data resulting from the inspection of crude oil requires an understanding of their significance.

Having decided what characteristics are necessary for a gas stream, it then remains to describe the product in terms of a specification. This entails selecting suitable test methods to determine the constituents and properties of the gas stream and setting appropriate limits for any variation of the proportion of the constituents and the limits of the variation in the properties.

The hydrocarbon component distribution of liquefied petroleum gases and propene mixtures is often required for end use sale of this material. Applications such as chemical feedstocks or fuel require precise compositional data to ensure uniform quality. Trace amounts of some hydrocarbon impurities in these materials can have adverse effects on their use and processing. The component distribution data of liquefied petroleum gases and propene mixtures can be used to calculate physical properties such as relative density and vapor pressure. Precision and accuracy of compositional data are extremely important when these data are used to calculate various properties.

An issue that arises during the characterization of liquefied petroleum gas relates to the accurate determination of higher-boiling residues (i.e., higher molecular weight hydrocarbon derivatives and even oils) in the gas. Test methods using procedures similar to those employed in gas

TABLE 2.3
Possible Constituents of Natural Gas and Refinery Process Gas Streams

Gas	Molecular Weight	Boiling Point 1 atm °C (°F)	Density at 60°F (15.6°C), 1 atm	
			g/L	Relative to Air = 1
Methane	16.043	−161.5 (−258.7)	0.6786	0.5547
Ethylene	28.054	−103.7 (−154.7)	1.1949	0.9768
Ethane	30.068	−88.6 (−127.5)	1.2795	1.0460
Propylene	42.081	−47.7 (−53.9)	1.8052	1.4757
Propane	44.097	−42.1 (−43.8)	1.8917	1.5464
1,2-Butadiene	54.088	10.9 (51.6)	2.3451	1.9172
1,3-Butadiene	54.088	−4.4 (24.1)	2.3491	1.9203
1-Butene	56.108	−6.3 (20.7)	2.4442	1.9981
cis-2-Butene	56.108	3.7 (38.7)	2.4543	2.0063
trans-2-Butene	56.108	0.9 (33.6)	2.4543	2.0063
isobutene	56.104	−6.9 (19.6)	2.4442	1.9981
n-Butane	58.124	−0.5 (31.1)	2.5320	2.0698
isobutane	58.124	−11.7 (10.9)	2.5268	2.0656

chromatographic simulated distillation are becoming available. In fact, the presence of any component substantially less volatile than the main constituents of the liquefied petroleum gas will give rise to unsatisfactory performance. It is difficult to set limits to the amount and nature of the *residue* which will make a product unsatisfactory. For example, liquefied petroleum gases that contain certain anti-icing additives can give erroneous results by this test method.

Control over the residue content is of considerable importance in end use applications of liquefied petroleum gases. In liquid feed systems, residues can lead to troublesome deposits, and in vapor withdrawal systems, residues that are carried over can foul regulating equipment. Any residue that remains in the vapor-withdrawal systems will accumulate, can be corrosive, and will contaminate subsequent product. Water, particularly if alkaline, can cause failure of regulating equipment and corrosion of metals. Obviously, small amounts of oil-like material can block regulators and valves. In liquid vaporizer feed systems, the gasoline-type material could cause difficulty.

Olefin derivatives (ethylene, $CH_2=CH_2$, propylene, $CH_3CH=CH_2$, butylene derivatives, such as $CH_3CH_2CH=CH_2$, and pentylene derivatives, such as $CH_3CH_2CH_2CH=CH_2$) that occur in refinery gas (process gas) have specific characteristics and require specific testing protocols (Speight, 2015; ASTM, 2018; Speight, 2018). The amount of ethylene ($CH_2=CH_2$) in a gas stream is limited because it is necessary to restrict the number of unsaturated components to avoid the formation of deposits caused by the polymerization of the olefin constituents. In addition, ethylene (boiling point: −104°C, −155°F) is more volatile than ethane (boiling point: −88°C, −127°F) and therefore a product with a substantial proportion of ethylene will have a higher vapor pressure and volatility than one that is predominantly ethane. Butadiene is also undesirable because it may also produce polymeric products that form deposits and cause blockage of lines.

As stated above, the amount of ethylene (and other olefin derivatives) in the mixture is limited because not only it is necessary to restrict the amount of the unsaturated components (olefin derivatives) so as to avoid the formation of deposits caused by the polymerization of the olefin(s), but also to control the volatility of the sample. Ethylene (boiling point: −104°C, −155°F) is more volatile than ethane (boiling point: −88°C, −127°F) and therefore a product with a substantial proportion of ethylene will have a higher vapor pressure and volatility than one that is predominantly ethane. Butadiene is also undesirable because it may also produce polymeric products that form deposits and cause blockage of lines.

Ethylene is one of the highest volume chemicals produced in the world, with global production exceeding 100 million metric tons annually. Ethylene is primarily used in the manufacture of polyethylene, ethylene oxide, and ethylene dichloride, as well as many other lower volume products. Most of these production processes use various catalysts to improve product quality and process yield. Impurities in ethylene can damage the catalysts, resulting in significant replacement costs, reduced product quality, process downtime, and decreased yield.

Ethylene is typically manufactured through the use of steam cracking. In this process, gaseous or light liquid hydrocarbon derivatives are combined with steam and heated to 750°C–950°C (1,380°F–1,740°F) in a pyrolysis furnace. Numerous free radical reactions are initiated, and larger hydrocarbon derivatives are converted (cracked) into smaller hydrocarbon derivatives. In addition, the high temperatures used in steam cracking promote the formation of unsaturated or olefin compounds, such as ethylene. Ethylene feedstocks must be tested to ensure that only high-purity ethylene is delivered for subsequent chemical processing. Samples of high-purity ethylene typically contain only two minor impurities, methane and ethane, which can be detected in low ppm v/v concentrations.

However, steam cracking can also produce higher molecular weight hydrocarbon derivatives, especially when propane, butane, or light liquid hydrocarbon derivatives are used as starting materials and the feedstock is a heavy oil where the coking tendency of the heavy feedstock is high (Parkash, 2003; Mokhatab et al., 2006; Gary et al., 2007; Van Geem et al., 2008; Speight, 2011a, 2014a; Hsu and Robinson, 2017; Speight, 2017). Although fractionation is used in the final production stages to produce a high-purity ethylene product, it is still important to be able to identify and quantify any other hydrocarbon derivatives present in an ethylene sample. Achieving sufficient resolution of all of these compounds can be challenging due to their similarities in boiling point and chemical structure.

Since the composition of the various gases can vary so widely, no single set of specifications could cover all situations. The requirements are usually based on performances in burners and equipment, on minimum heat content, and on maximum sulfur content. Gas utilities in most states come under the supervision of state commissions or regulatory bodies, and the utilities must provide a gas that is acceptable to all types of consumers and that will give satisfactory performance in all kinds of consuming equipment. However, particularly relevant are the heating values of the various fuel gases and their constituents. For this reason, measurement of the properties of fuel gases is an important aspect of fuel gas technology.

However, the physical properties of unrefined natural gas are variable because the composition of natural gas is never constant. Therefore, the properties and behavior of natural gas are best understood by investigating the properties and behavior of the constituents. Thus, if the natural gas has been processed (i.e., any constituents such as carbon dioxide and hydrogen sulfide), have been removed and the only constituents remaining are hydrocarbon derivatives, then the properties and behavior of natural gas becomes a study of the properties and behavior of the relevant constituents.

The composition of natural gas varies depending on the field, the formation, or the character of the reservoir from which the gas is extracted and that are an artifact of its formation (Speight, 2014a). Also, the properties of other gas streams (Chapter 3) vary with the source from which the gas was produced and the process by which the gas was produced. The different hydrocarbon derivatives that form the gas streams can be separated using their different physical properties as weight, boiling point, or vapor pressure (Chapters 6 and 7). Depending on its content of higher molecular weight hydrocarbon components, natural gas can be considered as rich (five or six gallons or more of recoverable hydrocarbon components per cubic feet) or lean (less than one gallon of recoverable hydrocarbon components per cubic feet). In terms of chemical behavior, hydrocarbon derivatives are simple organic chemicals that contain only carbon and hydrogen. Thus, in this section the properties and behavior of hydrocarbon derivatives up to and include n-octane (C_8H_{18}) are presented. On the other hand, when natural gas is refined, and any remaining hydrocarbon derivatives are removed, the gas that is sold to the consumer the sole component (other than an odorizer) is methane (CH_4) and the properties are constant.

There are two major technical aspects to which gas quality relates: (i) the pipeline specification in which stringent specifications for water content and hydrocarbon dew point are stated along with limits for contaminants such as sulfur—the objective is to ensure pipeline material integrity for reliable gas transportation purpose and (ii) the interchangeability specification which may include analytical data such as calorific value and relative density that are specified to ensure satisfactory performance of end use equipment.

Gas interchangeability is a subset of the gas quality specification ensuring that gas supplied to domestic users will combust safely and efficiently. The Wobbe number is a common, but not universal, measure of interchangeability and is used to compare the rate of combustion energy output of different composition fuel gases in combustion equipment. For two fuels with identical Wobbe Indices, the energy output will be the same for given pressure and valve settings.

Finally, in terms of properties (and any test methods that are applied to natural gas) (Speight, 2015; ASTM, 2018; Speight, 2018), it is necessary to recognize the other constituents of a natural gas stream that is produced from a reservoir as well as any gas streams (Chapter 3) that may be blended into the natural gas stream. Briefly, blending is the process of mixing gases for a specific purpose where the composition of the resulting mixture is specified and controlled. Thus, natural gas liquids are products other than methane from natural gas: ethane, butane, isobutane, and propane.

Test methods for gaseous fuels have been developed over many years, extending back into the 1930s. Bulk physical property tests, such as density and heating value, as well as some compositional tests, such as the Orsat analysis and the mercuric nitrate method for the determination of unsaturation, were widely used. More recently, mass spectrometry has become a popular method of choice for compositional analysis of low molecular weight and has replaced several older methods. Also, gas chromatography is another method of choice for hydrocarbon identification in gases (Speight, 2015; ASTM, 2018; Speight, 2018).

The various gas streams (Chapter 3) are generally amenable to analytical techniques and there has been the tendency, and it remains, for the determination of both major constituents and trace constituents than is the case with the heavier hydrocarbon derivatives. The complexity of the mixtures that are evident as the boiling point of petroleum fractions and petroleum products increases make identification of many of the individual constituents difficult, if not impossible. In addition, methods have been developed for the determination of physical characteristics such as calorific value, specific gravity, and enthalpy from the analyses of mixed hydrocarbon gases, but the accuracy does suffer when compared to the data produced by methods for the direct determination of these properties.

The different methods for gas analysis include absorption, distillation, combustion, mass spectroscopy, infrared spectroscopy, and gas chromatography. Absorption methods involve absorbing individual constituents one at a time in suitable solvents and recording of contraction in volume measured. Distillation methods depend on the separation of constituents by fractional distillation and measurement of the volumes distilled. In combustion methods, certain combustible elements are caused to burn carbon dioxide and water, and the volume changes are used to calculate composition. Infrared spectroscopy is a useful application and for the most accurate analyses, mass spectroscopy and gas chromatography are the preferred methods.

However, the choice of a particular test to determine any property remains as the decision of the analyst that, then, depends upon the nature of the gas under study. For example, judgment by the analyst is necessary whether or not a test that is applied to a gas stream is suitable for that gas stream insofar as inference from the non-hydrocarbon constituents will be minimal.

The following section presents a brief illustration of the properties of natural gas hydrocarbon derivatives from methane up to and including n-octane (C_8H_{18}). This will allow the reader to understand the folly of stating the properties of natural gas as average properties rather than allowing for the composition of the gas mixture and recognition of the properties of the individual constituents.

2.2.2 Natural Gas Liquids

Natural gas liquids (lease condensate, natural gasoline) are components of natural gas that are liquid at surface in gas or oil field facilities or in gas processing plants. The composition of the natural gas liquids is dependent upon the type of natural gas and the composition of the natural gas.

Natural gas liquids are the non-methane constituents such as ethane, propane, butane, and pentanes and higher molecular weight hydrocarbon constituents which can be separated as liquids during gas processing (Chapter 7). While natural gas liquids are gaseous at underground pressure, the molecules condense at atmospheric pressure and turn into liquids. The composition of natural gas can vary by geographic region, the geological age of the deposit, the depth of the gas, and many other factors. Natural gas that contains a lot of natural gas liquids and condensates is referred to as *wet gas*, while gas that is primarily methane, with little to no liquids in it when extracted, is referred to as *dry gas*.

The higher molecular weight constituents of natural gas (i.e., the C_{5+} product) are commonly referred to as gas condensate or natural gasoline. Rich gas will have a high heating value and a high hydrocarbon dew point. When referring to natural gas liquids in the gas stream, the term gallon per thousand cubic feet is used as a measure of high molecular weight hydrocarbon content. On the other hand, the composition of nonassociated gas (sometimes called well gas) is deficient in natural gas liquids. The gas is produced from geological formations that typically do not contain much, if any, hydrocarbon liquids.

Generally, the hydrocarbon derivatives having a higher molecular weight than methane as well as any acid gases (carbon dioxide and hydrogen sulfide) are removed from natural gas prior to use of the gas as a fuel. However, since the composition of natural gas is never constant, there are standard test methods by which the composition and properties of natural gas can be determined and, thus, prepared for use. It is not the intent to cover the standard test methods in any detail in this text since descriptions of the test methods are available elsewhere (Speight, 2015; ASTM, 2018; Speight, 2018).

Natural gas liquids can be classified according to their vapor pressures as low (condensate), intermediate (natural gasoline), and high (liquefied petroleum gas) vapor pressure. Natural gas liquids include propane, butane, pentane, hexane, and heptane, but not methane and not always ethane, since these hydrocarbon derivatives need refrigeration to be liquefied.

A more general definition of natural gas liquids includes the non-methane hydrocarbon derivatives from natural gas that are separated from the gas as liquids through the process of absorption, condensation, adsorption, or other methods in gas processing or cycling plants. Generally, under this definition, such liquids consist of ethane, propane, butane, and higher molecular weight hydrocarbon derivatives. For further use, the hydrocarbon derivatives are fractionated using a system which, after de-ethanization of the natural gas liquids, produces propane, butanes, and naphtha (C_{5+}) (Mokhatab et al., 2006; Speight, 2007).

2.2.3 GAS CONDENSATE

Also, by way of a further reminder, natural gas condensate (also called *condensate*, or *gas condensate*, or *natural gasoline*) is a low-density mixture of hydrocarbon liquids that are present as gaseous components in the raw natural gas produced from many natural gas fields ((Mokhatab et al., 2006; Speight, 2007, 2014a). Some gas constituents within the raw (unprocessed) natural gas will condense to a liquid state if the temperature is reduced to below the hydrocarbon dew point temperature at a set pressure. There are many condensate sources, and each has its own unique gas condensate composition.

Natural gas condensate (condensate, gas condensate, natural gasoline) is a low-density mixture of hydrocarbon liquids that are present as gaseous components in the raw natural gas produced from many natural gas fields. Gas condensate condenses out of the raw natural gas if the temperature is reduced to below the hydrocarbon dew point temperature of the raw gas. The composition of the gas condensate liquids is dependent upon the type of natural gas and the composition of the natural gas. Similarities exist between the composition of natural gas liquids and gas condensate—to the point that the two names are often (sometimes erroneously) used interchangeably. On a strictly comparative basis, the constituents of gas condensate represent the higher-boiling constituents of natural gas liquids.

The fraction known as pentanes plus is a mixture of pentane isomers and higher molecular weight constituents (C_{5+}) that is a liquid at ambient temperature and pressure, and consists mostly of pentanes and higher molecular weight (higher carbon number) hydrocarbon derivatives. Pentanes plus includes, but is not limited to, normal pentane, isopentane, hexanes-plus (natural gasoline), and condensate.

To separate the condensate from a natural gas feedstock from a gas well or a group of wells, the stream is cooled to lower the gas temperature to below the hydrocarbon dew point at the feedstock pressure and that condenses a good part of the gas condensate hydrocarbon derivatives (Mokhatab et al., 2006; Speight, 2007, 2014a). The feedstock mixture of gas, liquid condensate, and water is then routed to a high pressure separator vessel where the water and the raw natural gas are separated and removed. The raw natural gas from the high pressure separator is sent to the main gas compressor.

The gas condensate from the high pressure separator flows through a throttling control valve to a low-pressure separator. The reduction in pressure across the control valve causes the condensate to undergo a partial vaporization referred to as a flash vaporization. The raw natural gas from the low-pressure separator is sent to a booster compressor which raises the gas pressure and sends it through a cooler and on to the main gas compressor. The main gas compressor raises the pressure of the gases from the high- and low-pressure separators to whatever pressure is required for the pipeline transportation of the gas to the raw natural gas processing plant. The main gas compressor discharge pressure will depend upon the distance to the raw natural gas processing plant and it may require that a multistage compressor be used.

At the raw natural gas processing plant, the gas will be dehydrated and acid gases and other impurities will be removed from the gas. Then the ethane, propane, butanes, and pentanes plus higher molecular weight hydrocarbon derivatives (referred to as C_{5+}) will also be removed and recovered as byproducts.

The water removed from both the high- and low-pressure separators will probably need to be processed to remove hydrogen sulfide before the water can be disposed or reused in some fashion.

2.2.4 GAS HYDRATES

The concept of natural gas production from *methane hydrate* (also called *gas hydrate, methane clathrate, natural has hydrate, methane ice, hydromethane, methane ice, fire ice*) is relatively new but does offer the potential to recover hitherto unknown reserves of methane that can be expected to extend the availability of natural gas (Giavarini et al., 2003; Giavarini and Maccioni, 2004; Giavarini et al., 2005; Makogon et al., 2007; Makogon, 2010; Wang and Economides, 2012; Yang and Qin, 2012). In terms of gas availability from this resource, 1 L of solid methane hydrate can contain up to 168 L of methane gas.

Natural gas hydrates are an unconventional source of energy and occur abundantly in nature, both in arctic regions and in marine sediments (Bishnoi and Clarke, 2006). The formation of gas hydrate occurs when water and natural gas are present at low temperature and high pressure. Such conditions often exist in oil and gas wells and pipelines. Gas hydrates offer a source of energy as well as a source of hydrocarbon derivatives for the future.

Gas hydrates are an ice-like material which is made up of methane molecules contained in a cage of water molecules and held together by hydrogen bonds. This material occurs in large underground deposits found beneath the ocean floor on continental margins and in places north of the Arctic Circle such as Siberia. It is estimated that gas hydrate deposits contain twice as much carbon as all other fossil fuels on earth. This source, if proven feasible for recovery, could be a future energy as well as chemical source for petrochemicals.

Due to its physical nature (a solid material only under high pressure and low temperature), it cannot be processed by conventional methods used for natural gas and crude oils. One approach is by dissociating this cluster into methane and water by injecting a warmer fluid such as sea water. Another approach is by drilling into the deposit. This reduces the pressure and frees methane from water. However, the environmental effects of such drilling must still be evaluated.

The methane in gas hydrates is predominantly generated by bacterial degradation of organic matter in low-oxygen environments. Organic matter in the uppermost few centimeters of sediments is first attacked by aerobic bacteria, generating carbon dioxide, which escapes from the sediments into the water column. In this region of aerobic bacterial activity, sulfate derivatives ($-SO_4$) are reduced to sulfide derivatives ($-S$). If the sedimentation rate is low (<1 cm per 1,000 years), the organic carbon content is low (<1% w/w) and oxygen is abundant, and the aerobic bacteria use up all the organic matter in the sediments. However, when sedimentation rate is high, and the organic carbon content of the sediment is high, the pore waters in the sediments are anoxic at depths of less than one foot or so and methane is produced by anaerobic bacteria.

The two major conditions that promote hydrate formation are thus: (i) high gas pressure and low gas temperature and (ii) the gas at or below its water dew point with free water present (Sloan, 1998b; Collett et al., 2009). The hydrates are believed to form by migration of gas from depth along geological faults, followed by precipitation, or crystallization, on contact of the rising gas stream with cold sea water. At high pressures methane hydrates remain stable at temperatures up to 18°C (64°F) and the typical methane hydrate contains one molecule of methane for every six molecules of water that forms the ice cage. However, the methane (hydrocarbon)-water ratio is dependent on the number of methane molecules that fit into the cage structure of the water lattice.

Chemically, gas hydrates are non-stoichiometric compounds formed by a lattice of hydrogen-bonded molecules (host), which engage low molecular weight gases or volatile liquids (guest) with specific properties that differentiate them from ice (Bishnoi and Clarke, 2006). No actual chemical bond exists between guest and host molecules. Hydrate formation is favored by low temperature and high pressure (Makogon, 1997; Sloan, 1998a; Lorenson and Collett, 2000; Carrol, 2003; Seo et al., 2009). Most methane hydrate deposits also contain small amounts of other hydrocarbon hydrates; these include ethane hydrate and propane hydrate. In fact, gas hydrates of current interest are composed of water and the following molecules: methane, ethane, propane, isobutane, normal butane, nitrogen, carbon dioxide, and hydrogen sulfide. However, other non-polar components such as argon (Ar) and ethyl cyclohexane ($C_6H_{11} \cdot C_2H_5$) can also form hydrates. Typically, gas hydrates form at temperatures in the order of 0°C (32°F) and elevated pressures (Sloan, 1998a).

The composition of natural gas hydrates is determined by the composition of the gas and water, and the pressure and temperature which existed at the time of their formation. Over geologic time, there will be changes in the thermodynamic conditions and the vertical and lateral migration of gas and water; therefore, the composition of hydrate can change both due to the absorption of free gas and the recrystallization of already-formed hydrate.

In the hydrate structure, methane is trapped within a cage-like crystal structure composed of water molecules in a structure that resembles packed snow or ice (Lorenson and Collett, 2000). The hydrate usually consists of methane with small amounts higher molecular weight components. However, in a number of cases the hydrate contains a significant volume of higher molecular weight hydrocarbon derivatives (Table 2.4) (Taylor, 2002). The presence of higher molecular weight hydrocarbon, other than methane, in the hydrates may be an indicator of the presence of petroleum reservoirs in the formation below the gas hydrate deposit.

TABLE 2.4
Composition of Gas Produced from Various Gas Hydrates

Gas Hydrate Deposit	Gas Composition mol%							
	CH_4	C_2H_6	C_3H_8	iC_4H_{10}	nC_4H_{10}	C_{5+}	CO_2	N_2
Haakon Mosby Mud volcano	99.5	0.1	0.1	0.1	0.1	0.1		
Nankai Trough, Japan	99.3						0.63	
Bush Hill White	72.1	11.5	13.1	2.4	1	0		
Bush Hill Yellow	73.5	11.5	11.6	2	1	0.3	0.1	
Green Canyon White	66.5	8.9	15.8	7.2	1.4	0.2		
Green Canyon Yellow	69.5	8.6	15.2	5.4	1.2	0		
Bush Hill	29.7	15.3	36.6	9.7	4	4.8		
Messoyakha, Russia	98.7	0.03					0.5	0.77
Mallik, Canada	99.7	0.03	0.27					
Nankai Trough-1, Japan	94.3	2.6	0.57	0.09	0.8		0.24	1.4
Blake Ridge, United States	99.9	0.02						0.08

Source: Taylor (2002).

Under the appropriate pressure, gas hydrates can exist at temperatures significantly above the freezing point of water, but the stability of the hydrate derivatives depends on pressure and gas composition and is also sensitive to temperature changes (Stern et al., 2000; Stoll and Bryan, 1979; Collett, 2001; Belosludov et al., 2007; Collett, 2010). For example, methane plus water at 600 psia forms hydrate at 5°C (41°F), while at the same pressure, methane with 1% v/v propane forms a gas hydrate at 9.4°C (49°F). Hydrate stability can also be influence by other factors, such as salinity.

Methane hydrates are restricted to the shallow lithosphere (i.e., at depths less than 6,000 ft below the surface). The necessary conditions for the formation of hydrates are found only either in polar continental sedimentary rocks where surface temperatures are less than 0°C (32°F) or in oceanic sediment at water depths greater than 1,000 ft where the bottom water temperature is in the order of 2°C (35°F).

Caution is advised when drawing generalities about the formation and the stability of gas hydrates. Methane hydrates are also formed during natural gas production operations, when liquid water is condensed in the presence of methane at high pressure. Higher molecular weight hydrocarbon derivatives such as ethane and propane can also form hydrates, although larger molecules (butane hydrocarbon derivatives and pentane hydrocarbon derivatives) cannot fit into the water cage structure and, therefore, tend to destabilize the formation of hydrates (Belosludov et al., 2007). However, for this text, the emphasis is focused on methane hydrates.

2.2.5 OTHER TYPES OF GASES

The composition and properties of any gas stream depends on the characterization and properties of the hydrocarbon derivatives that make up the stream and calculation of the properties of a mixture depends on the properties of its constituents. However, calculation of the property of a mixture based on an average calculation neglects any interactions between the constituents. This makes the issue of modeling properties a difficult one because of the frequent omission of any chemical or physical interactions between the gas stream constituents.

The defining characteristics of the various gas streams in the context of this book are that the gases (i) exist in a gaseous state at room temperature, (ii) may contain hydrocarbon constituents with 1–4 carbons, i.e., methane, ethane, propane, and butane isomers, (iii) may contain diluents and inert gases, and (iv) may contain contaminants in the form of non-hydrocarbon constituents. Each constituent of the gas influences the properties. Thus:

Hydrocarbons	Provide the calorific value of natural gas when it is burned
Diluents/inert gases	Typical gases are carbon dioxide, nitrogen, helium, and argon
Contaminants	Present in low concentrations; will affect processing operations

Many hydrocarbon gases do contain C_5 and C_6 hydrocarbon derivatives and apart from gas streams produced as processed byproducts in a refinery, the C_{5+} constituents are typically found at lower concentrations (% v/v) in gases than the C_1–C_4 constituents. There are also a few category members that may contain C_7 and even C_8 hydrocarbon derivatives, although such streams would necessarily be at elevated temperature and/or pressure to maintain the heptane (C_7H_{16}) and octane (C_8H_{18}) constituents in the gaseous state. Hydrocarbon compounds such as pentane, (C_5H_{12}), hexane (C_6H_{14}), heptane (C_7H_{16}), and octane (C_8H_{18}) derivatives are typically found predominantly in naphtha derived from crude oil.

Typically, natural gas produced from shale reservoirs and other tight reservoirs has been classified under the general title *unconventional gas*. The production process requires stimulation by horizontal drilling coupled with hydraulic fracturing because of the pack of permeability in the gas-bearing formation. The boundary between conventional gas and unconventional gas resources is not well defined, because they result from a continuum of geologic conditions. Coal seam gas, more frequently called coalbed methane (CBM), is frequently referred to as unconventional gas. Tight shale gas and gas hydrates are also placed into the category of *unconventional gas.*

In addition to gas hydrate derivatives, there are several types of unconventional gas resources that arise from different sources and/or are currently produced by methods other than those used for conventional gas production and require processing before sale to the consumer. In this section, it would be remiss not to mention prominent gases produced from biomass and waste materials, *viz*, biogas and landfill gas. Both types of gas contain methane and carbon dioxide as well as various other constituents and are often amenable to gas processing methods that are applied to natural gas (Chapter 4).

The other types of gases are listed alphabetically rather than on the basis of current importance and are: (i) biogas, (ii) coalbed methane, (iii) coal gas, (iv) flue gas, (iv) gas in geopressurized zones, (v) gas in tight formations, (vii) landfill gas, (viii) refinery gas, and (ix) shale gas (Mokhatab et al., 2006; Speight, 2011a, 2013b, 2014a).

2.2.5.1 Biogas

Biogas (often called *biogenic gas* and sometimes incorrectly known as *swamp gas*) typically refers to a biofuel gas produced by (i) anaerobic digestion by anaerobic organisms, which digest material inside a closed system or (ii) fermentation of biodegradable organic matter including manure, sewage sludge, municipal solid waste, biodegradable waste or any other biodegradable feedstock, under anaerobic conditions (Speight, 2011). Examples of biomass are: (i) wood and wood processing wastes; (ii) agricultural crops and waste materials; (iii) food, yard, and wood waste in garbage; and (iv) animal manure and human sewage which are all potential sources of biogas (biogenic gas). The process of biogas production (typically an anaerobic process) is a multistep biological process where the originally complex and big-sized organic solid wastes are progressively transformed into simpler and smaller-sized organic compounds by different bacteria strains to have a final energetically worthwhile gaseous product and a semisolid material (digestate) that is rich in nutrients and thus suitable for its utilization in farming.

Biogas production (typically an anaerobic process) is a multistep process in which originally complex organic (liquid or solid) wastes are progressively transformed into low molecular weight products by different bacteria strains (Esposito et al., 2012). Biogas can also be produced by pyrolysis of biomass (freshly harvested or as a biomass waste). Thus, the name biogas refers to a large variety of gases resulting from specific treatment processes, starting from various organic wastes, such as livestock manure, food waste, and sewage that are all potential sources of biogenic gas, or biogas, which is usually considered a form of renewable energy and is often categorized according to the source (Table 2.5). In spite of the potential differences in composition, biogas can be processed (upgraded) to the standards required for natural gas, although the choice of the relevant processing sequence depends upon the composition of the gas (Chapters 7 and 8).

During the combustion of biomass, various kinds of impurities are generated and some of them occur in the flue gas and most of the contaminants in the flue gas are related to the composition of the biomass. If the combustion is incomplete (i.e., carried out in a deficiency of oxygen), then soot, unburned matter, toxic dioxin derivatives may also occur in the flue gas.

1,2-Dioxin   1,4-Dioxin

In addition, metals (Cui et al., 2013), such as lead (Pb), also occur in the ash and may even evaporate during combustion and react, condense, and/or sublime during cooling in the boiler. While upstream of the gas cleaning installation, normally at a temperature <200°C (<390°F), all metals will occur as solid particles, except mercury which evaporates during combustion and reacts in the

TABLE 2.5
Examples of Biogas Composition

Constituents	Source-1[a]	Source-2[a]	Source-3[a]
Methane, CH_4, % v/v	50–60	60–75	60–75
Carbon dioxide, CO_2, % v/v	38–34	33–19	33–19
Nitrogen, N_2, % v/v	5–0	1–0	1–0
Oxygen, O_2, % v/v	1–0	<0,5	<0,5
Water, H_2O, % v/v	6 (@ 40°C)	6 (@ 40°C)	6 (@ 40°C)
Hydrogen sulfide, H_2S, mg/m³	100–900	1,000–4,000	3,000–10,000
Ammonia, NH_3, mg/m³	—	—	50–100

[a] Source-1: waste from domestic (household) sources; Source-2: sludge from a wastewater treatment plant; Source-3: agricultural waste.

boiler but remains mainly in its gaseous form. The impurities in the biogas are harmful if they are emitted to the atmosphere and gas cleaning units must be installed to eliminate or at least reduce this problem. The extent of the gas cleaning depends on federal, regional, and local regulations, but regional and local authorities, organizations, and individuals have often an opinion on an actual plant due to its size and location.

More generally, contaminants aside, in terms of composition, biogas is primarily a mixture of methane (CH_4) and inert carbonic gas (CO_2) but variations in the composition of the source material lead to variations in the composition of the gas (Table 2.1) (Speight, 2011). Water (H_2O), hydrogen sulfide (H_2S), and particulates are removed if present at high levels or if the gas is to be completely cleaned. Carbon dioxide is less frequently removed, but it must also be separated to achieve pipeline quality gas. If the gas is to be used without extensive cleaning, it is sometimes cofired with natural gas to improve combustion. Biogas cleaned up to pipeline quality is called renewable natural gas.

Finally, while natural gas is classified as a fossil fuel (Speight, 2014a) and biomethane is defined as a nonfossil fuel (Speight, 2011), it is further characterized or described as a green energy source. Noteworthy at this point is that methane, whatever the source (fossil fuel or nonfossil fuel) and when released into the atmosphere, is approximately 20 times more potent as a greenhouse gas than carbon dioxide. Organic matter from which biomethane is produced, would release the carbon dioxide into the atmosphere if simply left to decompose naturally, while other gases that are produced during the decomposition process, for example, nitrogen oxide(s) would make an additional contribution to the greenhouse effect.

2.2.5.2 Coalbed Methane

Just as natural gas is often located in the same reservoir as the crude oil, a gas (predominantly methane) can also be found trapped within coal seams where it is often referred to as coalbed methane (or *coal bed methane*, CBM, sometimes referred to as *coalmine methane* (CMM)). The gas occurs in the pores and cracks in the coal seam and is held there by underground water pressure. To extract the gas, a well is drilled into the coal seam and the water is pumped out (*dewatering*) which allows the gas to be released from the coal and brought to the surface.

However, the occurrence of methane in coal seams is not a new discovery and methane (also called *firedamp*) was known to coal miners for at least 150 years (or more) before it was *rediscovered* and developed as coalbed methane (Speight, 2013b). To the purist, coalmine methane is the fraction of coalbed methane that is released during the mining operation (referred to in the older literature as *firedamp* by miners because of its explosive nature). In practice, the terms coalbed methane and coalmine methane may usually refer to different sources of gas—both forms of gas, whatever the name, are equally dangerous to the miners.

Coalbed methane is relatively pure compared to conventional natural gas, containing only very small proportions of higher molecular weight hydrocarbon derivatives such as ethane and butane and other gases (such as hydrogen sulfide and carbon dioxide). Because coal is a solid, very high carbon-content mineral, there are usually no liquid hydrocarbon derivatives contained in the produced gas. The coal bed (coal seam) must first be dewatered to allow the trapped gas to flow through the formation to produce the gas. Consequently, coalbed methane usually has a lower heating value, and elevated levels of carbon dioxide, oxygen, and water that must be treated to an acceptable level, given the potential to be corrosive.

Typically, with some exceptions, coalbed gas is typically in excess of 90% v/v methane and, as subject to gas composition data, may be suitable for introduction into a commercial pipeline with little or no treatment (Mokhatab et al., 2006; Speight, 2007, 2013a). Methane within coalbeds is not structurally trapped by overlying geologic strata, as in the geologic environments typical of conventional gas deposits (Speight, 2013a, 2014a). Only a small amount (in the order of 5–10% v/v) of the coalbed methane is present as free gas within the joints and cleats of coalbeds. Most of the coalbed methane is contained within the coal itself (adsorbed to the sides of the small pores in the coal.

2.2.5.3 Coal Gas

Typically, coal gas is any gaseous product that is produced by carbonization of coal—occasionally, the term coal gas is also applied to any gas produced by the gasification of coal (Speight, 2013b). Coal carbonization is used for processing of coal to produce coke using metallurgical grade coal (Speight, 2013d). The process involves heating coal in the absence of air to produce coke, and is a multistep complex process in which a variety of solid, liquids, and gaseous products are produced which contain many valuable products. The various products from coal carbonization in addition to coke are (i) coke oven gas, (ii) coal tar, (iii) low-boiling oil, also called light oil, and (iv) aqueous solution of ammonia and ammonium salts. With the development of the steel industry there was a continuous development in coke oven plants during the later half of the 19th century to improve the process conditions and recovery of chemicals, and this continued during the 20th century to adapt to environmental pollution control strategies and energy consumption measures.

The carbonization process can be carried out at various temperatures (Table 2.6) (Speight, 2013d), although low or high temperature is preferred. Low-temperature carbonization is used to produce liquid fuels while high-temperature carbonization is used to produce gaseous products (Speight, 2013d). Low-temperature carbonization (approximately 450°C–750°C; 840°F–1380°F) is used to produce liquid fuels (with smaller amounts of gaseous products) while the high-temperature carbonization process (approximately 900°C, 1650°F) is used to produce gaseous products. The gaseous products from high-temperature carbonization process are less while liquid products are large and the production of tar is relatively low because of the cracking of the secondary (liquid products and tar) products (Speight, 2013d). Gases of high calorific value are obtained by low-temperature or medium-temperature carbonization of coal. The gases obtained by the carbonization of any given coal change in a progressive manner with increasing temperature (Table 2.6). The composition of coal gas also changes during the course of carbonization at a given temperature and secondary reactions of the volatile products are important in determining gas composition (Speight, 2013d; ASTM, 2018).

Low heat-content gas (low-Btu gas) is produced during the gasification of when the oxygen is not separated from the air and, thus, the gas product invariably has a low heat content (ca. 150–300 Btu/ft^3). Low heat-content gas is also the usual product of *in situ* gasification of coal which is used essentially as a technique for obtaining energy from coal without the necessity of mining the coal. The process is a technique for utilization of coal which cannot be mined by other techniques.

The nitrogen content of low heat-content gas ranges from somewhat less than 33% v/v to slightly more than 50% v/v and cannot be removed by any reasonable means; the presence of nitrogen at these levels renders the product gas to be low heat content. The nitrogen also strongly limits the applicability of the gas to chemical synthesis. Two other noncombustible components (water (H_2O),

TABLE 2.6
Effect of Carbonization Temperature on the Composition of Coal Gas

	Composition @ Temperature of carbonization					
	500	600	700	800	900	1,000, °C
Component	930	1,110	1,290	1,470	1,650	1,830, °F[a]
Carbon dioxide	5.7	5.0	4.4	4.0	3.2	2.5
Unsaturated hydrocarbons	3.2	4.0	5.2	5.1	4.8	4.5
Carbon monoxide	5.8	6.4	7.5	8.5	9.5	11.0
Hydrogen	20.0	29.0	40.0	47.0	50.0	51.0
Methane	49.5	47.0	36.0	31.0	29.5	29.0
Ethane	14.0	5.3	4.5	3.0	1.0	0.5
Relative yield per ton	1.0	1.64	2.83	3.82	4.46	5.01

[a] Rounded to the nearest 5°F.

and carbon dioxide (CO_2)) lower the heating value of the gas further. Water can be removed by condensation and carbon dioxide by relatively straightforward chemical means.

Medium heat-content gas (medium-Btu gas) has a heating value in the range 300–550 Btu/ft^3 and the composition is much like that of low heat-content gas, except that there is virtually no nitrogen. The primary combustible gases in medium heat-content gas are hydrogen and carbon monoxide. Medium heat-content gas is considerably more versatile than low heat-content gas; like low heat-content gas, medium heat-content gas may be used directly as a fuel to raise steam or used through a combined power cycle to drive a gas turbine, with the hot exhaust gases employed to raise steam.

Medium heat-content gas is especially amenable to the production of (i) methane, (ii) higher molecular weight hydrocarbon derivatives by the Fischer–Tropsch synthesis, (iii) methanol, and (iv) a variety of synthetic chemicals (Chadeesingh, 2011; Speight, 2013a). The reactions used to produce medium heat-content gas are the same as those employed for low heat-content gas synthesis, the major difference being the application of a nitrogen barrier (such as the use of pure oxygen) to keep diluent nitrogen out of the system.

In medium heat-content gas, the hydrogen–carbon monoxide ratio varies from 2:3 to ca. 3:1 and the increased heating value correlates with higher methane and hydrogen contents as well as with lower carbon dioxide contents. In fact, the nature of the gasification process used to produce the medium heat-content gas has an effect on the ease of subsequent processing. For example, the carbon dioxide-acceptor product is available for use in methane production because it has (i) the desired hydrogen–carbon dioxide ratio just exceeding 3:1, (ii) an initially high methane content, and (iii) relatively low carbon dioxide content and low water content.

High heat-content gas (high-Btu gas) is almost pure methane and often referred to as *synthetic natural gas* or *substitute natural gas* (SNG). However, to qualify as substitute natural gas, a product must contain at least 95% v/v methane; the energy content in the order of 980–1,080 Btu/ft^3. The commonly accepted approach to the synthesis of high heat-content gas is the catalytic reaction of hydrogen and carbon monoxide.

$$3H_2 + CO \rightarrow CH_4 + H_2O$$

The water produced by the reaction is removed by condensation and recirculated as very pure water through the gasification system. The hydrogen is usually present in slight excess to ensure that the toxic carbon monoxide is reacted.

The carbon monoxide–hydrogen reaction is not the most efficient way to produce methane because of the exothermicity of the reaction. Also, the methanation catalyst is subject to poisoning by sulfur compounds and the decomposition of metals can destroy the catalyst. Hydrogasification may be employed to minimize the need for methanation.

$$C_{coal} + 2H_2 \rightarrow CH_4$$

The product of this reaction is not pure methane and additional methanation is required after hydrogen sulfide and other impurities are removed.

2.2.5.4 Geopressurized Gas

The term *geopressure* refers to a reservoir fluid (including gas) pressure that significantly exceeds hydrostatic pressure (which is in the order of 0.4–0.5 psi per foot of depth) and may even approach overburden pressure (in the order of 1.0 psi per foot of depth). Thus, geopressurized zones are natural underground formations that are under unusually high pressure for their depth. The geopressurized zones are formed by layers of clay that are deposited and compacted very quickly on top of more porous, absorbent material such as sand or silt. Water and natural gas that are present in this clay are squeezed out by the rapid compression of the clay and enter the more porous sand or silt deposits. Geopressured reservoirs frequently are associated with substantial faulting and complex stratigraphy, which can make correlation, structural interpretation, and volumetric mapping subject to considerable uncertainty.

Geopressurized zones are typically located at great depths, usually 10,000–25,000 ft below the surface of the earth. The combination of all these factors makes the extraction of natural gas in geopressurized zones quite complicated. However, of all of the unconventional sources of natural gas, geopressurized zones are estimated to hold the greatest amount of gas.

The amount of natural gas in these geopressurized zones is uncertain although unproven estimates indicate that 5,000–49,000 trillion ft^3 (5,000–49,000$\times 10^{12}$ ft^3) of natural gas may exist in these areas. Like gas hydrates, the gas in the geopressurized zones offers an opportunity for future supplies of natural gas. However, the combination of the above factors makes the extraction of natural gas or crude oil located in geopressurized zones quite complicated (Speight, 2017).

2.2.5.5 Landfill Gas

Landfill gas, which is often included under the umbrella definition of biogas, is also produced from the decay of organic wastes (such as municipal solid waste that contains organic materials), but these wastes may not be biomass-type materials (Lohila et al., 2007; Staley and Barlaz, 2009; Speight, 2011c). Landfill sites offer another underutilized source of biogas. When municipal waste is buried in a landfill, bacteria break down the organic material contained in garbage such as newspapers, cardboard, and food waste, producing gases such as carbon dioxide and methane. Rather than allowing these gases to go into the atmosphere, where they contribute to global warming, landfill gas facilities can capture them, separate the methane, and combust it to generate electricity, heat, or both.

Landfill gas is produced by wet organic waste decomposing under anaerobic conditions in a biogas. In fact, landfill gas is a product of three processes: (i) evaporation of volatile organic compounds such as low-boiling solvents, (ii) chemical reactions between waste components, and (iii) microbial action, especially methanogenesis. The first two processes depend strongly on the nature of the waste—the most dominant process in most landfills is the third process whereby anaerobic bacteria decompose organic waste to produce biogas, which consists of methane and carbon dioxide together with traces of other compounds. Despite the heterogeneity of waste, the evolution of gases follows well-defined kinetic pattern in which the formation of methane and carbon dioxide commences approximately 6 months after depositing the landfill material. The evolution of landfill gases reaches a maximum at approximately 20 years, then declines over the course of several decades.

As should be expected, the amount of methane that is produced varies significantly based on composition of the waste (Staley and Barlaz, 2009). The efficiency of gas collection at landfills directly impacts the amount of energy that can be recovered—closed landfills (those no longer accepting waste) collect gas more efficiently than open landfills (those that are still accepting waste). The gas is a complex mix of different gases created by the action of microorganisms within a landfill. Typically, landfill gas is composed of 45–60% v/v methane, 40–60% v/v carbon dioxide, 0–1.0% v/v hydrogen sulfide, 0–0.2% v/v hydrogen (H_2), trace amounts of nitrogen (N_2), low molecular weight hydrocarbon derivatives (dry volume basis), and water vapor (saturated). The specific gravity of landfill gas is approximately 1.02–1.06. Trace amounts of other volatile organic compounds comprise the remainder (typically, 1%–2% v/v or less) and these trace gases include a large array of species, such as low molecular weight hydrocarbon derivatives. Other minor components include hydrogen sulfide, nitrogen oxides, sulfur dioxide, non-methane volatile organic compounds, polycyclic aromatic hydrocarbon derivatives, polychlorinated dibenzodioxin derivatives, and polychlorinated dibenzofuran derivatives (Brosseau, 1994; Rasi et al., 2007). All the aforementioned agents are harmful to human health at high doses.

Landfill gas collection is typically accomplished through the installation of wells installed vertically and/or horizontally in the waste mass. Design heuristics for vertical wells call for about one well per acre of landfill surface, whereas horizontal wells are normally spaced about 50–200 ft apart on center. Efficient gas collection can be accomplished at both open and closed landfills, but closed landfills have systems that are more efficient, owing to greater deployment of collection infrastructure since active filling is not occurring. On average, closed landfills have gas collection systems that capture approximately 84% v/v of produced gas, compared to approximately 67% v/v for open landfills.

Landfill gas can also be extracted through horizontal trenches instead of vertical wells. Both systems are effective at collecting. Landfill gas is extracted and piped to a main collection header, where it is sent to be treated or flared. The main collection header can be connected to the leachate collection system to collect condensate forming in the pipes. A blower is needed to pull the gas from the collection wells to the collection header and further downstream.

Landfill gas cannot be distributed through utility natural gas pipelines unless it is cleaned up to less than 3% carbon dioxide and a few parts per million of hydrogen sulfide, because carbon dioxide and hydrogen sulfide corrode the pipelines (Speight, 2014b). Thus, landfill gas must be treated to remove impurities, condensate, and particulates; hence, the need for analysis to determine the composition of the gas. However, the treatment system depends on the end use: (i) minimal treatment is needed for the direct use of gas in boiler, furnaces, or kilns and (ii) using the gas in electricity generation typically requires more in-depth treatment.

Treatment systems are divided into primary and secondary treatment processing. Primary processing systems remove moisture and particulates. Gas cooling and compression are common in primary processing. Secondary treatment systems employ multiple cleanup processes, physical and chemical, depending on the specifications of the end use. Two constituents that may need to be removed are siloxane derivatives and sulfur-containing compounds, which are damaging to equipment and significantly increase maintenance cost. Adsorption and absorption are the most common technologies used in secondary treatment processing. Also, landfill gas can be converted to high-Btu gas by reducing the amount of carbon dioxide, nitrogen, and oxygen in the gas.

The high-Btu gas can be piped into existing natural gas pipelines or in the form of compressed natural gas or liquid natural gas. Compressed natural gas and liquid natural gas can be used onsite to power hauling trucks or equipment or sold commercially. Three commonly used methods to extract the carbon dioxide from the gas are membrane separation, molecular sieve, and amine scrubbing (Chapters 7 and 8). Oxygen and nitrogen are controlled by the design and operation of the landfill since the primary cause for oxygen or nitrogen in the gas is intrusion from outside into the landfill because of a difference in pressure.

Landfill gas condensate is a liquid that is produced in landfill gas collection systems and is removed as the gas is withdrawn from landfills. Production of condensate may be through natural or artificial cooling of the gas or through physical processes such as volume expansion. The condensate is composed principally of water and organic compounds. Often the organic compounds are not soluble in water and the condensate separates into a watery (aqueous) phase and a floating organic (hydrocarbon) phase which may constitute up to 5% v/v of the liquid.

2.2.5.6 Refinery Gas

In the context of the production of petrochemicals, the most important gas streams are those produced during crude oil refining which are usually referred to as refinery gas or, on some occasions, petroleum gas. However, this latter term is not to be confused with liquefied petroleum gas. The terms *refinery gas* or *petroleum gas* are often used to identify liquefied petroleum gas or even gas that emanates as light ends (gases and volatile liquids) from the atmospheric distillation unit or from any one of several other refinery processes. For the purpose of this text, refinery gas not only describes liquefied petroleum gas but also natural gas and refinery gas (Mokhatab et al., 2006; Gary et al., 2007; Speight, 2014a; Hsu and Robinson, 2017; Speight, 2017). In this chapter, each gas is, in turn, and referenced by its name rather than the generic term *petroleum gas.* However, the composition of each gas varies and recognition of this is essential before the relevant testing protocols are selected and applied. Thus, refinery gas (fuel gas) is the non-condensable gas that is obtained during distillation of crude oil or treatment (cracking, thermal decomposition) of petroleum (Table 2.7) (Speight, 2014a).

Refinery gas is produced in considerable quantities during the different refining processes and is used as fuel for the refinery itself and as an important feedstock for the production of petrochemicals. Chemically, refinery gas consists mainly of hydrogen (H_2), methane (CH_4), ethane (C_2H_6), propane (C_3H_8), butane (C_4H_{10}), and olefin derivatives ($RCH=CHR^1$, where R and R^1 can be hydrogen or a methyl group) and may also include off-gases from petrochemical processes. Olefin derivatives such as ethylene ($CH_2=CH_2$, boiling point: $-104°C$, $-155°F$), propene (propylene, $CH_3CH=CH_2$, boiling point: $-47°C$, $-53°F$), butene (butene-1, $CH_3CH_2CH=CH_2$, boiling point: $-5°C$, $23°F$), isobutylene (($CH_3)_2C=CH_2$, boiling point: $-6°C$, $21°F$), *cis-* and *trans*-butene-2 ($CH_3CH=CHCH_3$, boiling point: ca. $1°C$, $30°F$), and butadiene ($CH_2=CHCH=CH_2$, boiling point: $-4°C$, $24°F$) as well as higher-boiling olefin derivatives are produced by various refining processes. As might be anticipated, the composition of the off-gas is variable depending on the type of crude oil, the cracking severity, and type of catalyst used for cracking (Table 2.8).

Still gas is broad terminology for low-boiling hydrocarbon mixtures and is the lowest-boiling fraction isolated from a distillation (*still*) unit in the refinery (Speight, 2014a, 2017). If the distillation unit is separating light hydrocarbon fractions, the still gas will be almost entirely methane

---

TABLE 2.7
Origin of Petroleum-Related Gases

Gas	Origin
Natural gas	Occurs naturally with or without crude oil
	A varying mixture of low-boiling hydrocarbon constituents
	Predominantly C_1 through C_4 hydrocarbon derivatives; some C_5–C_8 hydrocarbon derivatives
Gas condensate (natural gasoline)	C_5–C_8 hydrocarbon derivatives isolated from natural gas streams
Refinery gas (process gas)	A combination of gases produced by distillation
	Products from the thermal and catalytic cracking of crude oil or crude oil fraction (such as gas oil)
	Consists of C_2–C_4 hydrocarbons including olefin (>C=C<) gases
	Boiling range in the order of $-51°C$ to $-1°C$ ($-60°F$–$30°F$)
Tail gas	A combination of hydrocarbon derivatives generated from cracking processes
	Predominantly C_1–C_4 hydrocarbon derivatives

TABLE 2.8
General Composition of Refinery Gas

Component	% v/v
Hydrogen	10–50
Carbon monoxide	0.1–1
Nitrogen	2–10
Methane	30–55
Ethylene	5–18
Ethane	15–20
Propylene	1–6
Propane	1–6
Butadiene	0–0.15
Butylene	0.1–0.5
Iso- and n-butane	0.5–1
C_{5+}	0.2–2

with only traces of ethane (CH_3CH_3). If the distillation unit is handling higher-boiling fractions, the still gas might also contain propane ($CH_3CH_2CH_3$), butane ($CH_3CH_2CH_2CH_3$), and their respective isomers. *Fuel gas* and still gas are terms that are often used interchangeably but the term *fuel gas* is intended to denote the product's destination-to be used as a fuel for boilers, furnaces, or heaters.

A group of refining operations that contributes to gas production are the thermal cracking and catalytic cracking processes. The thermal cracking processes (such as the coking processes) produce a variety of gases, some of which may contain olefin derivatives (>C=C<). In the visbreaking process, fuel oil is passed through externally fired tubes and undergoes liquid phase cracking reactions, which result in the formation of lower-boiling fuel oil components. Substantial quantities of both gas and carbon are also formed in coking (both fluid coking and delayed coking) in addition to the middle distillate and naphtha. When coking a residual fuel oil or heavy gas oil, the feedstock is preheated and contacted with hot carbon (coke) which causes extensive cracking of the feedstock constituents of higher molecular weight to produce lower molecular weight products ranging from methane, via liquefied petroleum gas(es) and naphtha, to gas oil and heating oil. Products from coking processes tend to be unsaturated and olefin-type components predominating in the tail gases from coking processes.

In various catalytic cracking processes, higher boiling gas oil fractions are converted into lower boiling products by contacting the feedstock with the hot catalyst. Thus, both catalytic and thermal cracking processes, the latter being now largely used to produce chemical raw materials, result in the formation of unsaturated hydrocarbon derivatives, particularly ethylene ($CH_2=CH_2$), but also propylene (propene, $CH_3CH=CH_2$), isobutylene [isobutene, $(CH_3)_2C=CH_2$], and the n-butenes ($CH_3CH_2CH=CH_2$ and $CH_3CH=CHCH_3$) in addition to hydrogen (H_2), methane (CH_4) and smaller quantities of ethane (CH_3CH_3), propane ($CH_3CH_2CH_3$), and butane isomers [$CH_3CH_2CH_2CH_3$, $(CH_3)_3CH$]. Diolefin derivatives such as butadiene ($CH_2=CHCH=CH_2$) are also present.

In a series of reforming processes, distillation fractions which include paraffin derivatives and naphthene derivatives (cyclic nonaromatic), are treated in the presence of hydrogen and a catalyst to produce lower molecular weight products or are isomerized to more highly branched hydrocarbon derivatives. Also, the catalytic reforming process not only results in the formation of a liquid product of higher octane number, but also produces substantial quantities of gaseous products. The composition of these gases varies in accordance with process severity and the properties of the feedstock. The gaseous products are not only rich in hydrogen, but also contain hydrocarbon derivatives from methane to butane derivatives, with a preponderance of propane ($CH_3CH_2CH_3$), n-butane ($CH_3CH_2CH_2CH_3$), and isobutane [$(CH_3)_3CH$]. Since all catalytic reforming processes require substantial recycling of a hydrogen stream, it is normal to separate reformer gas into a propane ($CH_3CH_2CH_3$) and/or a butane [$CH_3CH_2CH_2CH_3/(CH_3)_3CH$] stream, which becomes part of the refinery liquefied petroleum gas production, and a lower-boiling gaseous fraction, part of which is recycled.

A further source of refinery gas is produced by the hydrocracking process which is a high pressure pyrolysis process carried out in the presence of fresh and recycled hydrogen. The feedstock is again heavy gas oil or residual fuel oil, and the process is mainly directed at the production of additional middle distillates and gasoline. Since hydrogen is to be recycled, the gases produced in this process again must be separated into lighter and heavier streams; any surplus recycle gas and the liquefied petroleum gas from the hydrocracking process are both saturated.

Both hydrocracker and catalytic reformer tail gases are commonly used in catalytic desulfurization processes (Speight, 2014a, 2017). In the latter, feedstocks ranging from light to vacuum gas oils (VGOs) are passed at pressures in the order of 500–1,000 psi with hydrogen over a hydrofining catalyst. This results mainly in the conversion of organic sulfur compounds to hydrogen sulfide:

$$[S]_{feedstock} + H_2 \rightarrow H_2S + \text{hydrocarbon derivatives}$$

The process also has the potential to produce lower-boiling hydrocarbon derivatives by hydrocracking.

Olefin derivatives are not typical constituents of natural gas but do occur in refinery gases, which can be complex mixtures of hydrocarbon gases and non-hydrocarbon gas (Speight, 2014a, 2017). Some gases may also contain inorganic compounds, such as hydrogen, nitrogen, hydrogen sulfide, carbon monoxide, and carbon dioxide. Many low molecular weight olefin derivatives (such as ethylene and propylene) and diolefin derivatives (such as butadiene) which are produced in the refinery are isolated for petrochemical use (Speight, 2014a). The individual products are: (i) ethylene, (ii) propylene, and (iii) butadiene.

Ethylene (C_2H_4) is a normally gaseous olefinic compound having a boiling point of approximately −104°C (−155°F) which may be handled as a liquid at very high pressures and low temperatures. Ethylene is made normally by cracking an ethane or naphtha feedstock in a high-temperature furnace and subsequent isolation from other components by distillation. The major uses of ethylene are in the production of ethylene oxide, ethylene dichloride, and the polyethylene polymers. Other uses include the coloring of fruit, rubber products, ethyl alcohol, and medicine (anesthetic).

Propylene concentrates are mixtures of propylene and other hydrocarbon derivatives, principally propane and trace quantities of ethylene, butylenes, and butanes. Propylene concentrates may vary in propylene content from 70% mol up to over 95% mol and may be handled as a liquid at normal temperatures and moderate pressures. Propylene concentrates are isolated from the furnace products mentioned in the preceding paragraph on ethylene. Higher purity propylene streams are further purified by distillation and extractive techniques. Propylene concentrates are used in the production of propylene oxide, isopropyl alcohol, polypropylene, and the synthesis of isoprene. As is the case for ethylene, moisture in propylene is critical.

Butylene concentrates are mixtures of butene-1, *cis*- and *trans*-butene-2, and, sometimes, isobutene (2-methyl propylene) (C_4H_8).

Butene-1

cis-Butene-2

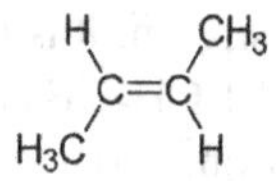

trans-Butene-2

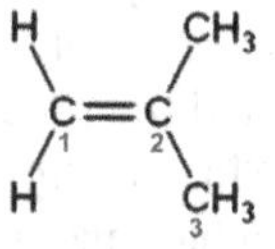

isobutene (2-methylpropene, 2-methyl propylene)

These products are stored as liquids at ambient temperatures and moderate pressures. Various impurities such as butane, butadiene, and the C_5 hydrocarbon derivatives are generally found in butylene concentrates. The majority of the butylene concentrates are used as a feedstock for either: (i) an alkylation plant, where isobutane and butylenes are reacted in the presence of either sulfuric acid or hydrofluoric acid to form a mixture of C_7–C_9 paraffins used in gasoline or (ii) butylene dehydrogenation reactors for butadiene production.

Butadiene (C_4H_6, $CH_2=CHCH=CH_2$) is a gaseous hydrocarbon at ambient temperature and pressure having a boiling point of −4.38°C (24.1°F) which may be handled as a liquid at moderate pressure. Ambient temperatures are generally used for long-term storage due to the easy formation of butadiene dimer (4-vinyl cyclohexene-l).

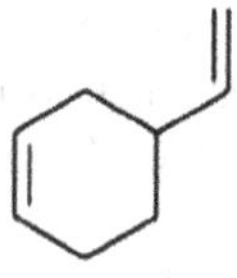

4-vinyl cyclohexene-l

Butadiene is produced by two major methods: the catalytic dehydrogenation of butane or butylenes (suing butylene-1 as the example) or both, and as a byproduct from the production of ethylene.

$$CH_3CH_2CH_2CH_3 \rightarrow CH_2=CHCH=CH_2 + 2H_2$$

$$CH_3CH_2CH=CH_2 \rightarrow CH_2=CHCH=CH_2 + H_2$$

$$2CH_3CH_3 \rightarrow CH_2=CHCH=CH_2 + 3H_2$$

In each case, the butadiene must be isolated from other components by extractive distillation techniques and subsequent purification to polymerization-grade specifications by fractional distillation. The largest end use of butadiene is as a monomer for production of GR-S synthetic rubber. Butadiene is also chlorinated to produce 2-chloro butadiene (chloroprene) ($CH_2=CHCCl=CH_2$) that is a feedstock used to produce neoprene (a polychloroprene rubber).

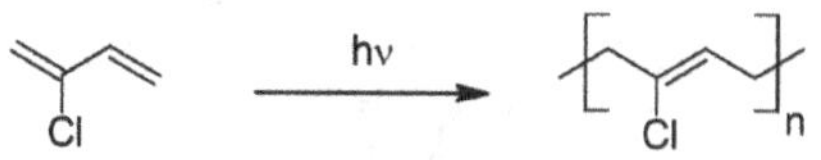

Chloroprene   neoprene

The major quality criteria for butadiene are the various impurities that may affect the polymerization reactions for which butadiene is used. The gas chromatographic examination of butadiene (ASTM, 2018) can be employed to determine the gross purity as well as C_3, C_4, and C, impurities.

Most of these hydrocarbon derivatives are innocuous to polymerization reactions, but, some, such as butadiene-1,2 and pentadiene-1,4, are capable of polymer cross-linking.

$$CH_2=C=CHCH_3 \qquad CH_2=C=CHCH_2CH_3$$

1,2-butadiene   1,2-pentadiene

2.2.5.7 Synthesis Gas

Synthesis gas (also known as *syngas*) is a mixture of carbon monoxide (CO) and hydrogen (H_2) that is used as a fuel gas but is produced from a wide range of carbonaceous feedstocks and is used to produce a wide range of chemicals. The production of synthesis gas, i.e., mixtures of carbon monoxide and hydrogen has been known for several centuries and can be produced by gasification of carbonaceous fuels. However, it is only with the commercialization of the Fischer–Tropsch reaction that the importance of synthesis gas has been realized.

Synthesis gas can be produced from any one of several carbonaceous feedstocks (such as a crude oil residuum, heavy oil, tar sand bitumen, and biomass) by gasification (partially oxidizing) of the feedstock (Speight, 2011, 2013a, 2014a, 2014b):

$$[2CH]_{feedstock} + O_2 \rightarrow 2CO + H_2$$

The initial partial oxidation step consists of the reaction of the feedstock with a quantity of oxygen insufficient to burn it completely, making a mixture consisting of carbon monoxide, carbon dioxide, hydrogen, and steam. Success in partially oxidizing heavy feedstocks such as heavy crude oil, extra heavy crude oil, and tar sand bitumen feedstocks depends mainly on the properties of the feedstock and the burner design. The ratio of hydrogen to carbon monoxide in the product gas is a function of reaction temperature and stoichiometry and can be adjusted, if desired, by varying the ratio of the steam to the feedstock.

Synthesis gas can be produced from heavy oil by partially oxidizing the oil:

$$[2CH]_{petroleum} + O_2 \rightarrow 2CO + H_2$$

Reactor temperatures vary from 1,095°C to 1,490°C (2,000°F–2,700°F), while pressures can vary from approximately atmospheric pressure to approximately 2,000 psi (13,790 kPa). The process has the capability of producing high-purity hydrogen although the extent of the purification procedure depends upon the use to which the hydrogen is to be put. For example, carbon dioxide can be removed by scrubbing with various alkaline reagents, while carbon monoxide can be removed by washing with liquid nitrogen or, if nitrogen is undesirable in the product, the carbon monoxide should be removed by washing with copper-amine solutions.

The synthesis gas generation process is a non-catalytic process for producing synthesis gas (principally hydrogen and carbon monoxide) for the ultimate production of high-purity hydrogen from gaseous or liquid hydrocarbon derivatives. In this process, a controlled mixture of preheated feedstock and oxygen is fed to the top of the generator where carbon monoxide and hydrogen emerge as the products. Soot, produced in this part of the operation, is removed in a water scrubber from the product gas stream and is then extracted from the resulting carbon water slurry with naphtha and transferred to a fuel oil fraction. The oil-soot mixture is burned in a boiler or recycled to the generator to extinction to eliminate carbon production (soot formation) as part of the process. The soot-free synthesis gas is then charged to a shift converter where the carbon monoxide reacts with steam to form additional hydrogen and carbon dioxide at the stoichiometric rate of 1 mole of hydrogen for every mole of carbon monoxide charged to the converter.

This particular partial oxidation technique has also been applied to a whole range of liquid feedstocks for hydrogen production. There is now serious consideration being given to hydrogen production by the partial oxidation of solid feedstocks such as petroleum coke (from both delayed and fluid-bed reactors), lignite, and coal, as well as petroleum residua. Although these reactions may be represented very simply using equations of this type, the reactions can be complex and result in carbon deposition on parts of the equipment thereby requiring careful inspection of the reactor.

2.2.5.8 Tight Gas

Gas from tight formations, also called tight gas and shale gas, is found in low-permeability reservoir rocks (such as shale) which prohibit natural movement of the gas to a well. The term *tight formation* refers to a formation consisting of extraordinarily impermeable, hard rock (Speight, 2013a). Tight formations are relatively low permeability, non-shale, sedimentary formations that can contain oil and gas. A *tight reservoir* (*tight sands*) is a low-permeability sandstone reservoir that produce primarily dry natural gas. A tight gas reservoir is one that cannot be produced at economic flow rates or recover economic volumes of gas unless the well is stimulated by a large hydraulic fracture treatment and/or produced using horizontal wellbores. This definition also applies to coalbed methane and tight carbonate reservoirs—shale gas reservoirs are also included by some observers (but not in this text).

By way of explanation and comparison, in a conventional sandstone reservoir the pores are interconnected so that natural gas and crude oil can flow easily through the reservoir and to the production well. Conventional gas typically is found in reservoirs with permeability greater than 1 milliDarcy (mD) and can be extracted via traditional techniques (Figure 2.1). However, in tight sandstone formations, the pores are smaller and are poorly connected (if at all) by very narrow capillaries which results in low permeability and immobility of the natural gas. Such sediments typically have an effective permeability of less than 1 mD (<1 mD). In contrast, unconventional gas is found in reservoirs with relatively low permeability (less than 1 mD) (Figure 2.1) and hence cannot be extracted via conventional methods.

Shale is a sedimentary rock characterized by low permeability, mainly compositing of mud, silts, and clay minerals but, however, this composition varies with burial depth and tectonic stresses. Shale reservoirs have a permeability which is substantially lower than the permeability of other tight reservoirs. Natural gas and low-viscosity crude oil (also known as tight oil but sometimes erroneously referred to as shale oil—by way of definition, shale oil is the liquid product produced by the decomposition of the kerogen component of oil shale) are confined in the pore spaces of these impermeable shale formations. On the other hand, oil shale is a kerogen-rich petroleum source rock that was not buried under the correct maturation conditions to experience the temperatures required to generate oil and gas (Speight, 2014a).

The natural gas that is associated with shale formations and such gas is commonly referred to as *shale gas*—to define the origin of the gas rather than the character and properties (Speight, 2013b). Thus, shale gas is natural gas produced from shale formations that typically function as both the reservoir and source rocks for the natural gas (Speight, 2013b). The gas in a shale formation is present as a free gas in the pore spaces or is adsorbed by clay minerals and organic matter. Chemically, shale gas is typically a dry gas composed primarily of methane (60–95% v/v), but some formations do produce wet gas—in the United States the Antrim and New Albany plays have typically produced water and gas.

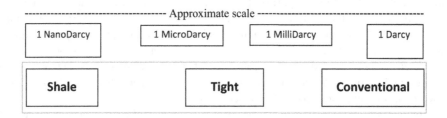

FIGURE 2.1 Representation of the differences in permeability of shale reservoirs, tight reservoirs, and conventional reservoirs.

2.3 PETROLEUM

Petroleum (also called crude oil) is a naturally occurring, unrefined liquid) which also may occur in gaseous and/or sold form) composed of hydrocarbon derivatives and other organic materials containing the so-called heteroatoms nitrogen, oxygen, sulfur, and metals such as iron, copper, nickel, and vanadium. Petroleum is found in the microscopic pores of sedimentary rocks such as sandstone and limestone (Niu and Hu, 1999; Parkash, 2003; Gary et al., 2007; Speight, 2014a; Hsu and Robinson, 2017; Speight, 2017). Not all the pores in a rock contain crude oil and some pores will be filled with water or brine that is saturated with minerals.

Petroleum can be refined to produce usable products such as gasoline, diesel fuel, fuel oils, lubricating oil, wax, and various forms of petrochemicals (Niu and Hu, 1999; Parkash, 2003; Gary et al., 2007; Speight, 2014a; Hsu and Robinson, 2017; Speight, 2017). However, crude oil is a nonrenewable resource which cannot be replaced naturally at the rate that it is consumed; it is, therefore, a limited resource but a current lifetime in the order of 50 years (Speight, 2011a, 2011b, 2011c; Speight and Islam, 2016).

2.3.1 COMPOSITION AND PROPERTIES

The molecular boundaries of crude oil cover a wide range of boiling points and carbon numbers of hydrocarbon compounds and other compounds containing nitrogen, oxygen, and sulfur, as well as metal-containing (porphyrin) constituents (Speight, 2012). However, the actual boundaries of such a *crude oil map* can only be arbitrarily defined in terms of boiling point and carbon number. In fact, crude oil is so diverse that materials from different sources exhibit different boundary limits, and for this reason alone it is not surprising that crude oil has been difficult to *map* in a precise manner.

In a very general sense, crude oil is composed of the flowing chemical types: (i) hydrocarbon compounds, compounds composed of carbon and hydrogen only, (ii) non-hydrocarbon compounds, and (iii) organometallic compounds and inorganic salts (metallic compounds). Hydrocarbon compounds are the principal constituents of most conventional crude oils and all hydrocarbon classes are present in the crude mixture, except olefin derivatives and alkynes.

Alkanes (paraffins) are saturated hydrocarbon derivatives having the general formula C_nH_{2n+2}. The simplest alkane, methane (CH_4), is the principal constituent of natural gas. Methane, ethane (CH_3CH_3), propane ($CH_3CH_2CH_2$), and butane ($CH_3CH_2CH_2CH_3$, as well as other isobutane) are gaseous hydrocarbon derivatives at ambient temperatures and atmospheric pressure. They are usually found associated with crude oils in a dissolved state. Normal alkanes (*n*-alkanes, *n*-paraffins) are straight-chain hydrocarbon derivatives having no branches. Branched alkanes are saturated hydrocarbon derivatives with an alkyl substituent or a side branch from the main chain. A branched alkane with the same number of carbons and hydrogens as an *n*-alkane is called an isomer. For example, butane (C_4H_{10}) has two isomers, *n*-butane and 2-methyl propane (isobutane). As the molecular weight of the hydrocarbon increases, the number of isomers also increases. Pentane (C_5H_{12}) has three isomers; hexane (C_6H_{14}) has five isomers. Example of the hexane isomers are: 2,2-dimethylbutane and 2,3-dimethylbutane:

2,2-Dimethylbutane   2,3-Dimethylbutane

Crude oils contain many short, medium, and long-chain normal and branched paraffins. A naphtha fraction (obtained as a low-boiling liquid stream from crude fractionation) with a narrow boiling range may contain a limited but still large number of isomers.

Saturated cyclic hydrocarbon derivatives (cycloparaffins), also known as naphthenes, are also part of the hydrocarbon constituents of crude oils. The ratio, however, depends on the type of crude oil. The lower molecular weight members of naphthenes are cyclopentane, cyclohexane, and their monosubstituted compounds. They are normally present in the light and the heavy naphtha fractions. Cyclohexane derivatives, substituted cyclopentane derivatives, and substituted cyclohexane derivatives are important precursors for aromatic hydrocarbon derivatives. The higher-boiling petroleum fractions such as kerosene and gas oil may contain two or more cyclohexane rings fused through two vicinal carbon atoms (Speight, 2014a).

Lower molecular weight aromatic compounds are present in small amounts in crude oils and light petroleum fractions. The simplest mononuclear aromatic compound is benzene (C_6H_6). Toluene ($C_6H_5CH_3$) and xylene isomers ($H_3CC_6H_4CH_3$) are also mononuclear aromatic compounds found in variable amounts in crude oils. Benzene, toluene, and the xylene isomers (BTX) are important petrochemical intermediates as well as valuable gasoline components. Separating the BTX aromatic derivatives from crude oil distillates is not feasible because they are present in low concentrations. Enriching a naphtha fraction with these aromatic derivatives is possible through a catalytic reforming process (Parkash, 2003; Gary et al., 2007; Speight, 2014a; Hsu and Robinson, 2017; Speight, 2017).

Binuclear aromatic hydrocarbon derivatives occur in the higher-boiling fractions than naphtha. Trinuclear and polynuclear aromatic hydrocarbon derivatives, in combination with heterocyclic compounds, are major constituents of heavy crude oil and the distillation residua of crude oil. The asphaltene fraction is a complex mixture of aromatic and heterocyclic compounds (Speight, 1994, 2014a). Various types of non-hydrocarbon compounds occur in crude oils and refinery streams. The most important are the organic sulfur, nitrogen, and oxygen compounds. Traces of metallic compounds are also found in all crudes.

Sulfur in crude oil is mainly present in the form of organosulfur compounds. Hydrogen sulfide is the only important inorganic sulfur compound found in crude oil but the presence of this gas is harmful because of its corrosive nature. Organosulfur compounds may generally be classified as acidic and nonacidic. Acidic sulfur compounds are the thiol derivatives (mercaptan derivatives). Thiophene derivatives, sulfide derivatives, and disulfide derivatives are examples of nonacidic sulfur compounds that occur in crude oil. Most sulfur compounds can be removed from petroleum streams through hydrotreatment processes, where hydrogen sulfide is produced and the corresponding hydrocarbon released. Hydrogen sulfide is then absorbed in a suitable absorbent and recovered as sulfur.

Organic nitrogen compounds occur in crude oils either in a simple heterocyclic form as in pyridine (C_5H_5N) derivatives and pyrrole (C_4H_5N) derivatives, or in a complex structure as in porphyrin. The nitrogen content in most crudes is very low and does not exceed 0.1% w/w. In some heavy crudes however, the nitrogen content may reach up to 0.9% w/w. Nitrogen compounds are more thermally stable than sulfur compounds and accordingly are concentrated in higher-boiling fractions and distillation residua. Low-boiling streams may contain trace amounts of nitrogen compounds, which should be removed because they poison many processing catalysts. During hydrotreatment of petroleum fractions, nitrogen compounds are hydrodenitrogenated to the corresponding hydrocarbon and ammonia. For example, using pyridine as the example, the products are n-pentane and ammonia:

$$C_5H_5N + 5H_2 \rightarrow CH_3CH_2CH_2CH_2CH_3 + NH_3$$

Nitrogen compounds in crude oil may generally be classified into basic and nonbasic categories. Basic nitrogen compounds are mainly those having a pyridine ring, and the nonbasic compounds have a pyrrole structure. Both pyridine and pyrrole are stable compounds due to their aromatic nature.

Porphyrin derivatives are nonbasic nitrogen compounds. The porphyrin ring system is composed of four pyrrole rings joined by methine (=CH-) groups and the entire ring system has aromatic character. Many metal ions can replace the pyrrole hydrogens and form chelates. The chelate is planar around the metal ion and resonance results in four equivalent bonds from the nitrogen atoms to the metal.

Iron porphyrin

Almost all crude oils and tar sand bitumen contain detectable amounts of vanadyl and nickel porphyrins. Separation of nitrogen compounds is difficult, and the compounds are susceptible to alteration and loss during handling. However, the basic low molecular weight compounds may be extracted with dilute mineral acids.

Oxygen compounds in crude oils are more complex than the sulfur types. However, their presence in petroleum streams is not poisonous to processing catalysts. Many of the oxygen compounds found in crude oils are weakly acidic. They are carboxylic acids, cresylic acid, phenol, and naphthenic acid. Naphthenic acids are mainly cyclopentane and cyclohexane derivatives having a carboxyalkyl side chain. Naphthenic acids in the naphtha fraction have a special commercial importance and can be extracted by using dilute caustic solutions. The total acid content of most crudes is generally low, but may reach as much as 3% w/w. Nonacidic oxygen compounds such as esters, ketones, and amides are less abundant than acidic compounds. They are of no commercial value.

Many metals occur in crude oils. Some of the more abundant are sodium, calcium, magnesium, aluminum, iron, vanadium, and nickel. They are present either as inorganic salts, such as sodium and magnesium chlorides, or in the form of organometallic compounds, such as those of nickel and vanadium (as in porphyrin derivatives). Calcium and magnesium can form salts or soaps with carboxylic acids. These compounds act as emulsifiers, and their presence is undesirable.

Although metals in crudes are found in trace amounts, their presence is harmful and should be removed. When crude oil is processed, sodium and magnesium chlorides produce hydrochloric acid, which is very corrosive. Desalting crude oils is a necessary step to reduce these salts. Vanadium and nickel are poisons to many catalysts and should be reduced to very low levels. Most of the vanadium and nickel compounds are concentrated in the high-boiling residua. Solvent extraction processes are used to reduce the concentration of heavy metals in petroleum residues.

Before passing on to heavy oil as a feedstock for the production of petrochemicals, there are three types of conventional crude oil that need to be addressed: (i) opportunity crude oil, (ii) high acid crude oil, and (iii) foamy oil.

2.3.1.1 Opportunity Crude Oil

Opportunity crude oils are either new crude oils with unknown or poorly understood properties relating to processing issues or are existing crude oils with well-known properties and processing concerns (Ohmes, 2014; Speight, 2014a, 2014b; Yeung, 2014). Opportunity crude oils are often, but not always, heavy crude oils but in either case are more difficult to process due to high levels of solids (and other contaminants) produced with the oil, high levels of acidity, and high viscosity. These crude oils may also be incompatible with other oils in the refinery feedstock blend and cause excessive equipment fouling when processed either in a blend or separately (Speight, 2015). There is also the need for a refinery to be configured to accommodate *opportunity crude oils* and/or *high acid crude oils* which, for many purposes are often included with heavy feedstocks.

2.3.1.2 High Acid Crude Oil

High acid crude oils are crude oils that contain considerable proportions of naphthenic acids which, as commonly used in the crude oil industry, refers collectively to all of the organic acids present in the crude oil (Shalaby, 2005; Ghoshal and Sainik, 2013; Speight, 2014b).

By the original definition, a naphthenic acid is a monobasic carboxyl group attached to a saturated cycloaliphatic structure. However, it has been a convention accepted in the oil industry that all organic acids in crude oil are called naphthenic acids. Naphthenic acids in crude oils are now known to be mixtures of low to high molecular weight acids and the naphthenic acid fraction also contains other acidic species. Naphthenic acids, which are not *user friendly* in terms of refining (Kane and Cayard, 2002; Ghoshal and Sainik, 2013; Speight, 2014c), can be either (or both) water-soluble to oil-soluble depending on their molecular weight, process temperatures, salinity of waters, and fluid pressures.

2.3.1.3   Foamy Oil

Foamy oil is oil-continuous foam that contains dispersed gas bubbles produced at the well head from heavy oil reservoirs under solution gas drive. The nature of the gas dispersions in oil distinguishes foamy oil behavior from conventional heavy oil. The gas that comes out of solution in the reservoir coalesces neither into large gas bubbles nor into a continuous flowing gas phase. Instead, it remains as small bubbles entrained in the crude oil, keeping the effective oil viscosity low while providing expansive energy that helps drive the oil toward the production. Foamy oil accounts for unusually high production in heavy oil reservoirs under solution gas drive (Sun et al., 2013).

Foamy oil behavior is a unique phenomenon associated with production of heavy oils. It is believed that this mechanism contributes significantly to the abnormally high production rate of heavy oils observed in the Orinoco Belt. During production of heavy oil from solution gas drive reservoirs, the oil is pushed into the production wells by energy supplied by the dissolved gas. As fluid is withdrawn from the production wells, the pressure in the reservoir declines and the gas that was dissolved in the oil at high pressure starts to come out of solution (hence, foamy oil). As pressure declines further with continued removal of fluids from the production wells, more gas is released from solution and the gas already released expands in volume. The expanding gas, which at this point is in the form of isolated bubbles, pushes the oil out of the pores and provides energy for the flow of oil into the production well. This process is very efficient until the isolated gas bubbles link up and the gas itself starts flowing into the production well. Once the gas flow starts, the oil has to compete with the gas for available flow energy. Thus, in some heavy oil reservoirs, due to the properties of the oil and the sand and also due to the production methods, the released gas forms foam with the oil and remains subdivided in the form of dispersed bubbles much longer.

2.3.1.4   Tight Oil

Tight oil is a low-viscosity crude oil (sometimes erroneously referred to as shale oil—by way of definition, shale oil is the liquid product produced by the decomposition of the kerogen component of oil shale) that is confined in the pore spaces of these impermeable shale formations. On the other hand, oil shale is a kerogen-rich petroleum source rock that was not buried under the correct maturation conditions to experience the temperatures required to generate oil and gas (Speight, 2014a). Economic production from tight oil formations requires the same hydraulic fracturing and often uses the same horizontal well technology used in the production of tight gas.

Tight formations (such as shale formations) are heterogeneous and vary widely over relatively short distances. Tight oil reservoirs subjected to fracking can be divided into four different groups: (i) Type I has little matrix porosity and permeability—leading to fractures dominating both storage capacity and fluid flow pathways, (ii) Type II has low matrix porosity and permeability, but here the matrix provides storage capacity while fractures provide fluid-flow paths, (iii) Type III are microporous reservoirs with high matrix porosity but low matrix permeability, thus giving induced fractures dominance in fluid-flow paths, and (iv) Type IV is macroporous reservoirs with high matrix porosity and permeability, thus the matrix provides both storage capacity and flow paths while fractures only enhance permeability.

Even in a single horizontal drill hole, the amount recovered may vary, as may recovery within a field or even between adjacent wells. This makes evaluation of plays and decisions regarding the profitability of wells on a particular lease difficult. Production of oil from tight formations requires a gas cap representing at least 15%–20% natural gas in the reservoir pore space to drive the oil toward the borehole; tight reservoirs which contain only oil cannot be economically produced but such reserves may be limited (Wachtmeister et al., 2017).

2.3.2 Other Petroleum-Derived Feedstocks

In the current context, the term *other petroleum-derived feedstocks* refers to the bulk *petroleum products*, in contrast to *petrochemicals*, which are the bulk fractions that are derived from petroleum and have commercial value as a bulk product (Parkash, 2003; Gary et al., 2007; Speight, 2014a; Hsu and Robinson, 2017; Speight, 2017). The fractions described below represent those that can be (could be) used for the production of petrochemicals remembering that the naphtha fraction, if it is high-boiling naphthas might contain some kerosene constituent, and the gas oil fraction (again, depending upon the boiling range) might contain some kerosene and fuel oil constituents.

It must also be recognized that these named fractions as produced in a refinery will have similar boiling ranges but the boiling range is often refinery specific. For example, there may (will) be minor variations (typically within 5°C–10°C, 9°F–18°F) in the boiling range of naphtha from one refinery as compared to the boiling range of naphtha from a different refinery. Thus, before using any of these fractions (including naphtha) for petrochemical production, there should be an awareness of the composition of the liquid by application of a relevant suite of analytical test methods (Speight, 2015). With this caveat in mind, the following are the description of the crude oil fractions that can be used for the production of petrochemicals.

2.3.2.1 Naphtha

Naphtha (often referred to as *naft* in the older literature) is actually a generic term applied to refined, partly refined, or an unrefined petroleum fraction. In the strictest sense of the term, not less than 10% v/v of the material should distill below 175°C (345°F) and not less than 95% v/v of the material should distill below 240°C (465°F) under standardized distillation conditions (ASTM D86, 2018, ASTM D7213, 2018). Generally (but this can be refinery dependent), naphtha is an unrefined petroleum that distills below 240°C (465°F) and is (after the gases constituents) the most volatile fraction of the petroleum. In fact, in some specifications, not less than 10% of material should distill below approximately 75°C (167°F). It is typically used as a precursor to gasoline or to a variety of solvents. Naphtha resembles gasoline in terms of boiling range and carbon number, being a precursor to gasoline.

2.3.2.1.1 Composition

Naphtha contains varying amounts of paraffins, olefin derivatives, naphthene constituents, and aromatic derivatives and olefin derivatives in different proportions in addition to potential isomers of paraffin that exist in naphtha boiling range. As a result, naphtha is divided predominantly into two main types: (i) aliphatic naphtha and (ii) aromatic (naphtha). The two types differ in two ways: first, in the kind of hydrocarbon derivatives making up the solvent, and second, in the methods used for their manufacture. Aliphatic solvents are composed of paraffinic hydrocarbon derivatives and cycloparaffins (naphthenes), and may be obtained directly from crude petroleum by distillation. The second type of naphtha contains aromatic derivatives, usually alkyl-substituted benzene, and is very rarely, if at all, obtained from petroleum as straight-run materials.

In general, naphtha may be prepared by any one of several methods, which include (i) fractionation of straight-run, cracked, and reforming distillates, or even fractionation of crude petroleum, (ii) solvent extraction, (iii) hydrogenation of cracked distillates, (iv) polymerization of unsaturated compounds such as olefin derivatives, and (v) alkylation processes. In fact, the naphtha may be a combination of product streams from more than one of these processes.

The more common method of naphtha preparation is distillation. Depending on the design of the distillation unit, either one or two naphtha steams may be produced: (i) a single naphtha with an end point of approximately 205°C (400°F) and similar to straight-run gasoline, or (ii) this same fraction divided into a light naphtha and a heavy naphtha. The end point of the light naphtha is varied to suit the subsequent subdivision of the naphtha into narrower boiling fractions and may be of the order of 120°C (250°F).

Sulfur compounds are most commonly removed or converted to a harmless form by chemical treatment with lye, doctor solution, copper chloride, or similar treating agents. Hydrorefining processes (Parkash, 2003; Gary et al., 2007; Speight, 2011, 2014a; Hsu and Robinson, 2017; Speight, 2017) are also often used in place of chemical treatment. Solvent naphtha is solvents selected for low sulfur content, and the usual treatment processes, if required, remove only sulfur compounds. Naphtha with a small aromatic content has a slight odor, but the aromatic constituents increase the solvent power of the naphtha and there is no need to remove aromatic derivatives unless an odor-free solvent is specified.

2.3.2.1.2 *Properties and Uses*

Naphtha is required to have a low level of odor to meet the specifications for use, which is related to the chemical composition—generally paraffin hydrocarbon derivatives possess the mildest odor, and the aromatic hydrocarbon derivatives have a much stronger odor. Naphtha containing higher proportions of aromatic constituents may be pale yellow—usually, naphtha is colorless (water white) and can be tested for the level of contaminants (ASTM D156, 2018). Naphtha is used as automotive fuel, engine fuel, and jet-B (naphtha type). Broadly, naphtha is classified as *light naphtha* and *heavy naphtha*. Light naphtha is used as rubber solvent, lacquer diluent, while heavy naphtha finds its application as varnish solvent, dyer's naphtha, and cleaner's naphtha. More specifically, naphtha is valuable as for solvents because of good dissolving power. The wide range of naphtha available, from the ordinary paraffin straight-run to the highly aromatic types, and the varying degree of volatility possible offer products suitable for many uses.

2.3.2.2 Kerosene

Kerosene (*kerosine*), also called paraffin or paraffin oil, is a flammable pale-yellow or colorless oily liquid with a characteristic odor. It is obtained from petroleum and used for burning in lamps and domestic heaters or furnaces, as a fuel or fuel component for jet engines, and as a solvent for greases and insecticides. Kerosene is intermediate in volatility between naphtha and gas oil. It is medium oil distilling between 150°C and 300°C (300°F–570°F). Kerosene has a flash point about 25°C (77°F) and is suitable for use as an illuminant when burned in a wide lamp. The term *kerosene* is also too often incorrectly applied to various fuel oils, but a fuel oil is actually any liquid or liquid petroleum product that produces heat when burned in a suitable container or that produces power when burned in an engine.

Kerosene was the major refinery product before the onset of the *automobile age*, but now kerosene can be termed one of several secondary petroleum products after the primary refinery product—gasoline. Kerosene originated as a straight-run petroleum fraction that boiled approximately between 205°C and 260°C (400°F and 500°F). Some crude oils, for example, those from the Pennsylvania oil fields, contain kerosene fractions of very high quality, but other crude oils, such as those having an asphalt base, must be thoroughly refined to remove aromatic derivatives and sulfur compounds before a satisfactory kerosene fraction can be obtained.

2.3.2.2.1 *Composition*

Chemically, kerosene is a mixture of hydrocarbon derivatives; the chemical composition depends on its source, but it usually consists of about ten different hydrocarbon derivatives, each containing from 10 to 16 carbon atoms per molecule; the constituents include *n*-dodecane (n-$C_{12}H_{26}$), alkyl benzenes, and naphthalene and its derivatives. Kerosene is less volatile than gasoline; it boils between 140°C (285°F) and 320°C (610°F).

Kerosene, because of its use as a burning oil, must be free of aromatic and unsaturated hydrocarbon derivatives, as well as free of the more obnoxious sulfur compounds. The desirable constituents of kerosene are saturated hydrocarbon derivatives, and it is for this reason that kerosene is manufactured as a straight-run fraction, not by a cracking process. The criteria might apply when a kerosene fraction (or a higher-boiling fraction such as gas oil) is used as a starting material for the production of petrochemical intermediates and for the direct production of petrochemical products.

2.3.2.2.2 Properties and Uses

Kerosene is by nature a fraction distilled from petroleum that has been used as a fuel oil from the beginning of the petroleum-refining industry. As such, low proportions of aromatic and unsaturated hydrocarbon derivatives are desirable to maintain the lowest possible level of smoke during burning. Although some aromatic derivatives may occur within the boiling range assigned to kerosene, excessive amounts can be removed by extraction; that kerosene is not usually prepared from cracked products almost certainly excludes the presence of unsaturated hydrocarbon derivatives.

The essential properties of kerosene are flash point, fire point, distillation range, burning, sulfur content, color, and cloud point. In the case of the flash point (ASTM D56, 2018), the minimum flash temperature is generally placed above the prevailing ambient temperature; the fire point (ASTM D92, 2018) determines the fire hazard associated with its handling and use. The boiling range (ASTM D86, 2018) is of less importance for kerosene than for gasoline, but it can be taken as an indication of the viscosity of the product, for which there is no requirement for kerosene. The ability of kerosene to burn steadily and cleanly over an extended period (ASTM D187, 2018) is an important property and gives some indication of the purity or composition of the product.

The significance of the total sulfur content of a fuel oil varies greatly with the type of oil and the use to which it is put. Sulfur content is of great importance when the oil to be burned produces sulfur oxides that contaminate the surroundings. The color of kerosene is of little significance, but a product darker than usual may have resulted from contamination or aging, and in fact a color darker than specified (ASTM D156, 2018) may be considered by some users as unsatisfactory. Finally, the cloud point of kerosene (ASTM D2500, 2018) gives an indication of the temperature at which the wick may become coated with wax particles, thus lowering the burning qualities of the oil.

2.3.2.3 Fuel Oil

Fuel oil is classified in several ways but was formally divided into two main types: *distillate fuel oil* and *residual fuel oil*, each of which was a blend of two or more refinery streams (Parkash, 2003; Gary et al., 2007; Speight, 2011, 2014a; Hsu and Robinson, 2017; Speight, 2017). Distillate fuel oil is vaporized and condensed during a distillation process and thus have a definite boiling range and do not contain high-boiling constituents. A fuel oil that contains any amount of the residue from crude distillation of thermal cracking is a residual fuel oil. The terms *distillate fuel oil* and *residual fuel oil* are losing their significance, since fuel oil is now made for specific uses and may be either distillates or residuals or mixtures of the two. The terms *domestic fuel oil*, *diesel fuel oil*, and *heavy fuel oil* are more indicative of the uses of fuel oils.

2.3.2.3.1 Composition

All of the fuel oil classes described here are refined from crude petroleum and may be categorized as either a distillate fuel or a residual fuel depending on the method of production. Fuel oil no. 1 and fuel oil no. 2 are distillate fuels which consist of distilled process streams. Residual fuel oil, such as fuel oil no. 4, is composed of the residuum remaining after distillation or cracking, or blends of such residues with distillates. Diesel fuel is approximately similar to fuel oil used for heating (fuel oils no. 1, fuel oil no. 2, and fuel oil no. 4).

All fuel oils consist of complex mixtures of aliphatic and aromatic hydrocarbon derivatives, the relative amounts depending on the source and grade of the fuel oil. The aliphatic alkanes (paraffins) and cycloalkane constituents (naphthene constituents) are hydrogen saturated and compose as much as 90% v/v of the fuel oil. Aromatic constituents (e.g., benzene) and olefin constituents compose up to 20 and 1% v/v, respectively, of the fuel oils. Fuel oil no. 1 (straight-run kerosene) is a distillate that consists primarily of hydrocarbon derivatives in the C_9–C_{16} range while fuel oil no. 2 is a higher-boiling, usually blended, distillate with hydrocarbon derivatives in the C_{11}–C_{20} range.

Residual fuel oil (and/or heavy fuel oil) is typically more complex in composition and impurities than distillate fuel oil. Therefore, a specific composition cannot always be determined—the sulfur content in residual fuel oil has been reported to vary up to 5% w/w. Residual fuel oils are complex mixtures of high molecular weight compounds having a typical boiling range from 350°C to 650°C (660°F–1200°F). They consist of aromatic, aliphatic, and naphthenic hydrocarbon derivatives, typically having carbon numbers from C_{20} to C_{50}, together with asphaltene constituents and smaller amounts of heterocyclic compounds containing sulfur, nitrogen, and oxygen.

Residual fuel oil also contains organometallic compounds from their presence in the original crude oil—the most important of which are nickel and vanadium. The metals (especially vanadium) are of particularly major significance for fuels burned in both diesel engines and boilers because when combined with sodium (perhaps from brine contamination from the reservoir or remaining after the refinery dewatering/desalting process) and other metallic compounds in critical proportions can lead to the formation of high melting point ash which is corrosive to engine parts. Other elements that occur in heavy fuel oils include iron, potassium, aluminum, and silicon—the latter two metals are mainly derived from refinery catalyst fines.

The manufacture of fuel oils at one time largely involved using what was left after removing desired products from crude petroleum. Now fuel oil manufacture is a complex matter of selecting and blending various petroleum fractions to meet definite specifications, and the production of a homogeneous, stable fuel oil requires experience backed by laboratory control.

Heavy fuel oil comprises all residual fuel oils and the constituents range from distillable constituents to residual (non-distillable) constituents that must be heated to 260°C (500°F) or more before they can be used. The kinematic viscosity is above 10 centistokes at 80°C (176°F). The flash point is always above 50°C (122°F) and the density is always higher than 0.900. In general, heavy fuel oil usually contains cracked residua, reduced crude, or cracking coil heavy product which is mixed (cut back) to a specified viscosity with cracked gas oils and fractionator bottoms. For some industrial purposes in which flames or flue gases contact the product (ceramics, glass, heat treating, and open hearth furnaces) fuel oils must be blended to contain minimum sulfur contents, and hence low-sulfur residues are preferable for these fuels.

2.3.2.3.2   Properties and Uses

No. 1 fuel oil is a petroleum distillate that is one of the most widely used of the fuel oil types. It is used in atomizing burners that spray fuel into a combustion chamber where the tiny droplets burn while in suspension. It is also used as a carrier for pesticides, as a weed killer, as a mold release agent in the ceramic and pottery industry, and in the cleaning industry. It is found in asphalt coatings, enamels, paints, thinners, and varnishes. No. 1 fuel oil is a light petroleum distillate (straight-run kerosene) consisting primarily of hydrocarbon derivatives in the range C_9–C_{16}. Fuel oil #1 is very similar in composition to diesel fuel; the primary difference is in the additives.

No. 2 fuel oil is a petroleum distillate that may be referred to as domestic or industrial fuel oil. The domestic fuel oil is usually lower boiling and a straight-run product. It is used primarily for home heating. Industrial distillate is a cracked product or a blend of both. It is used in smelting furnaces, ceramic kilns, and packaged boilers. No. 2 fuel oil is characterized by hydrocarbon chain lengths in the C_{11}–C_{20} range. The composition consists of aliphatic hydrocarbon derivatives (straight-chain alkane derivatives and cycloalkane derivatives) (64% v/v), unsaturated hydrocarbon derivatives (olefin derivatives) (1–2% v/v), and aromatic hydrocarbon derivatives (including alkyl benzenes and 2-ring, 3-ring aromatic derivatives) (35% v/v), but contains only low amounts of the polycyclic aromatic hydrocarbon derivatives (<5% v/v).

No. 6 fuel oil (also called *Bunker C oil* or *residual fuel oil*) is the residuum from crude oil after naphtha-gasoline, no. 1 fuel oil, and no. 2 fuel oil have been removed. No. 6 fuel oil can be blended directly to heavy fuel oil or made into asphalt. Residual fuel oil is more complex in composition and impurities than distillate fuels. Limited data are available on the composition of no. 6 fuel oil. Polycyclic aromatic hydrocarbon derivatives (including the alkylated derivatives) and metal-containing constituents are components of no. 6 fuel oil.

2.3.2.4 Gas Oil

Atmospheric gas oil is a fraction of crude oil recovered through distillation and is the highest-boiling fraction that can be distilled without coking short of a vacuum being pulled to lower the boiling temperature. It is often used as a feedstock for a catalytic cracking process to produce more of the valuable lighter fractions, including gases and naphtha.

The atmospheric gas oil fraction (boiling range: 215°C–345°C, 420°F–650°F) is usually taken to be a cut of straight-run distillate that boils at temperatures above those of the middle distillate range but below those of atmospheric residuals, a boiling range that is in the order of 345°C to approximately 535°C (650°C to approximately 1,000°F). Thus, there may be an overlap with the kerosene fraction. The vacuum gas oil fraction has a boiling range in the order of 345°C–535°C (650°F–1,000°F).

2.3.2.4.1 Composition

Vacuum gas oil is one of those mystery products talked about by refiners but barely understood by those of us who are not engineers. However, it is an important intermediate feedstock that can increase the output of valuable diesel and gasoline from refineries. Lighter shale crudes such as Eagle Ford can produce Vacuum gas oil material direct from primary distillation. Today we shed some light on this semifinished refinery product.

Vacuum distillation recovers gas oil from the residual oil. In layman terms, vacuum distillation involves heating the residual oil in a vacuum so that the boiling point temperature is reduced. This allows distillation at temperatures that are not possible in atmospheric distillation since otherwise coke from the heavy residual oil tends to solidify. Vacuum distillation breaks out light and heavy gas oil fractions leaving vacuum residuum that can be further processed by a coker unit or sold as fuel oil. The light and heavy gas oils output from the vacuum distillation column are known generically as vacuum gas oil or VGO. There are many different names used in the United States and worldwide for vacuum gas oil but the basic division is between light vacuum gas oil and heavy vacuum gas oil.

2.3.2.4.2 Properties and Uses

In a typical complex refinery such as are common in the United States, vacuum gas oil is further processed in one of two types of catalytic cracking units. These units use a combination of catalysts (a substance that accelerates chemical reactions) heat and pressure to crack vacuum gas oil into lower-boiling products.

2.3.2.5 Residua

A *resid* (*residuum, pl. residua*) is the black viscous residue obtained from petroleum after nondestructive distillation has removed all the volatile materials. The temperature of the distillation is usually maintained below 350°C (660°F) since the rate of thermal decomposition of petroleum constituents is minimal below this temperature but the rate of thermal decomposition of petroleum constituents is substantial above 350°C (660°F). A residuum may be liquid at room temperature (generally atmospheric residua) or almost solid (generally vacuum residua) depending upon the nature of the crude oil (Chapter 17). When a residuum is obtained from a crude oil and thermal decomposition has commenced, it is more usual to refer to this product as *pitch*. The differences between a parent petroleum and the residua are due to the relative amounts of various constituents present, which are removed or remain by virtue of their relative volatility.

2.3.2.5.1 Composition

The chemical composition of a residuum from an asphaltic crude oil is complex. Physical methods of fractionation usually indicate high proportions of asphaltenes and resins, even in amounts up to

50% w/w (or higher) of the residuum. In addition, the presence of ash-forming metallic constituents, including such organometallic compounds as those of vanadium and nickel, is also a distinguishing feature of residua and the heavier oils. Furthermore, the deeper the *cut* into the crude oil, the greater is the concentration of sulfur and metals in the residuum and the greater the deterioration in physical properties (Chapter 17).

2.3.2.6   Used Lubricating Oil

Used lubricating oil can be a raw material for converting to liquid fuels such as naphtha, kerosene, and light gas oil by using sulfated zirconia as a catalyst (Speight and Exall, 2014). Used lubricating oil—often referred to as *waste oil* without further qualification—is any lubricating oil, whether refined from crude or synthetic components, which has been contaminated by physical or chemical impurities as a result of use. Lubricating oil loses its effectiveness during operation due to the presence of certain types of contaminants. These contaminants can be divided into: (i) extraneous contaminants and (ii) products of oil deterioration.

2.3.2.6.1   Composition

Used mineral-based crankcase oil is the brown-to-black, oily liquid removed from the engine of a motor vehicle when the oil is changed. It is similar to a heavy fraction of virgin mineral oil except it contains additional chemicals from its use as an engine lubricant. The chemicals in the oil include hydrocarbon derivatives, which are distilled from crude oil to form a base oil stock, and various additives that improve the oil's performance. Used oil also contains chemicals formed when the oil is exposed to high temperatures and pressures inside an engine. It also contains some metals from engine parts and small amounts of gasoline, antifreeze, and chemicals that come from gasoline when it burns inside the engine. The chemicals found in used mineral-based crankcase oil vary, depending on the brand and type of oil, whether gasoline or diesel fuel was used, the mechanical condition of the engine that the oil came from, and the amount of use between oil changes. Used oil is not naturally found in the environment.

2.3.2.6.2   Uses

The inherent high energy content of many used lubricating oil streams may encourage the direct use of these streams as fuels, without any pretreatment and processing, and without any quality control or product specification. Such direct uses do not always constitute good practice, unless it can be demonstrated that combustion of the used oil can be undertaken in an environmentally sound manner. The use of used oil as fuel is possible because any contaminants do not present problems on combustion, or it can be burned in an environmentally sound manner without modification of the equipment in which it is being burned. However, using used oils as fuel needs to be subjected to treatments involving some form of settlement to remove sludge and suspended matter. Simple treatment of this type can substantially improve the quality of the material by removing sludge and suspended matter, carbon and to varying degrees heavy metals (Speight and Exall, 2014).

However, the constituents of used lubricating oil are the same type of constituents (with the exception of the aromatic constituents produced during service) of the gas oil fraction (Speight, 2014a; Speight and Exall, 2014). Thus, reusing used lubricating oil is preferred to disposal and might give great environmental advantages. Utilization and reusing (in this case, repurposing) the used lubricating oil as feedstock to a catalytic cracking unit to produce the starting material—hydrocarbon gases and naphtha—is preferred to disposal and also offers environmental benefits.

2.4   HEAVY OIL, EXTRA HEAVY OIL, AND TAR SAND BITUMEN

In the context of this book, heavy oil, extra heavy oil, and tar sand bitumen typically has relatively low proportions of volatile compounds with low molecular weights and quite high proportions of high molecular weight compounds of lower volatility. The high molecular weight fraction of a heavy oil is comprised of complex assortment of different molecular and chemical types—a complex mixture of compounds and not necessarily just paraffin derivatives or asphaltene constituents—with high melting points and high pour points that greatly contribute to the poor fluid properties of the heavy oil thereby contributing to low mobility (compared to conventional crude oil). The same is true for extra heavy oil and tar sand bitumen (Speight, 2013b, 2013c, 2014a).

2.4.1 Heavy Oil

The name *heavy oil* can often be misleading as it has also been used in reference to (i) fuel oil that contains residuum left over from distillation, i.e., heavy fuel oil or residual fuel oil, (ii) coal tar creosote, or (iii) viscous crude oil. Thus, for the purposes of this text the term is used to mean *viscous crude oil*. Extra heavy oil has been included in many of these categories. On the other hand, tar sand bitumen (often called simply *bitumen*) is often confused with manufactured asphalt (confusingly referred to as bitumen in many countries). To add further to this confusion, in some countries, tar sand bitumen is referred to as natural as asphalt!

Heavy oil is a viscous type of crude oil that contains higher level of sulfur than conventional crude oil and occurs in similar locations to crude oil (IEA, 2005; Ancheyta and Speight, 2007; Speight, 2016b). The nature of heavy oil is a problem for recovery operations and for refining—the viscosity of the oil may be too high thereby leading to difficulties in recovery and/or difficulties in refining the oil (Speight, 2016a, 2017). However, tar sand bitumen, in terms of properties and behavior, is far apart from conventional crude oil (Tables 2.9 and 2.10, Figure 2.2). Success with this material and with extra heavy oil depends as much on understanding the fluid (or non-fluid) properties of the material and the behavior of the fluids in the deposit in which they occur as it does on knowing the geology of the deposit (Speight, 2013a, 2013b, 2013c, 2014a). The reason is that the chemical and physical differences between heavy oil, extra heavy oil, and tar sand bitumen oil ultimately affect the viscosity and other relevant properties which, in turn, influence the individual aspects of recovery and refining operations.

Heavy oil has a much higher viscosity (and lower API gravity) than conventional crude oil and recovery of heavy oil usually requires thermal stimulation of the reservoir. The generic term *heavy oil* is often applied to a crude oil that has less than 20°API and usually, but not always, has sulfur content higher than 2% w/w (Ancheyta and Speight, 2007). Furthermore, in contrast to conventional crude oils, heavy oils are darker in color and may even be black.

2.4.2 Extra Heavy Oil

The term *heavy oil* has also been arbitrarily (incorrectly) used to describe both the heavy oils that require thermal stimulation of recovery from the reservoir and the bitumen in bituminous sand (tar sand) formations from which the heavy bituminous material is recovered by a mining operation.

Extra heavy oil is a nondescriptive term (related to viscosity) of little scientific meaning that is usually applied to tar sand bitumen-like material. The general difference is that extra heavy oil, which may have properties similar to tar sand bitumen in the laboratory but, unlike immobile tar sand bitumen in the deposit, has some degree of mobility in the reservoir or deposit (Tables 2.11 and 2.12) (Delbianco and Montanari, 2009; Speight, 2014a). An example is the extra heavy oil of the Zaca-Sisquoc extra heavy oil (sometimes referred to as the Zaca-Sisquoc bitumen) that has an API gravity in the order of 4.0°–6.0°. The reservoir has average depth of 3,500 ft, average thickness of 1,700 ft, average temperature of 51°C–71°C (125°F–160°F), and sulfur in the range of 6.8–8% w/w (Isaacs, 1992; Villarroel and Hernández, 2013). The deposit temperature is certainly equal to or above the pour point of the oil (Isaacs, 1992). This renders the oil capable of being pumped as a liquid from the deposit because of the high deposit temperature (which is higher than the pour point of the oil). The same rationale applied to the extra heavy oil found in the Orinoco deposits.

TABLE 2.9

Simplified Differentiation between Conventional Crude Oil, Tight Oil, Heavy Oil, Extra Heavy Oil, and Tar Sand Bitumen

Conventional crude oil
  Mobile in the reservoir
  API gravity: >25°
  High-permeability reservoir
  Primary recovery
  Secondary recovery
Tight oil
  Similar properties to the properties of conventional crude oil
  API gravity: >25°
  Immobile in the reservoir
  Low-permeability reservoir
  Horizontal drilling into reservoir
  Fracturing (typically multi-fracturing) to release fluids/gases
Medium crude oil
  Similar properties to the properties of conventional crude oil
  API gravity: 20°–25°
  High-permeability reservoir
  Primary recovery
  Secondary recovery
Heavy crude oil
  More viscous than conventional crude oil
  API gravity: 10°–20°
  Mobile in the reservoir
  High-permeability reservoir
  Secondary recovery
  Tertiary recovery (enhanced oil recovery—EOR; e.g., steam stimulation)
Extra heavy oil
  Fluid and/or mobile in the reservoir
  Similar properties to the properties of tar sand bitumen
  API gravity: <10°
  High-permeability reservoir
  Secondary recovery
  Tertiary recovery (enhanced oil recovery—EOR; such as steam stimulation)
Tar sand bitumen
  Immobile (solid to near-solid) in the deposit
  API gravity: <10°
  High-permeability reservoir
  Mining (often preceded by explosive fracturing)
  Steam-assisted gravity draining (SAGD)
  Solvent methods (VAPEX)
  Extreme heating methods
  Innovative methods[a]

[a] Innovative methods exclude tertiary recovery methods and methods, such as SAGD, VAPEX, but does include variants or hybrids thereof.

TABLE 2.10
Comparison of the Properties of Conventional Crude Oil with Athabasca Bitumen[a]

Property	Athabasca Bitumen	Conventional Crude Oil
Specific gravity	1.03	0.85–0.90
Viscosity, cp		
38C/100°F	7,50,000	<200
100C/212°F	11,300	
Pour point, °F	>50	ca. −20
Elemental analysis (% w/w):		
Carbon	83	86
Hydrogen	10.6	13.5
Nitrogen	0.5	0.2
Oxygen	0.9	<0.5
Sulfur	4.9	<2.0
Ash	0.8	0
Nickel (ppm)	250	<10.0
Vanadium (ppm)	100	<10.0
Fractional composition (% w/w)		
Asphaltenes (pentane)	17	<10.0
Resins	34	<20.0
Aromatics	34	>30.0
Saturates	15	>30.0
Carbon residue (% w/w)		
Conradson	14	<10.0

[a] Extra heavy oil (e.g., Zuata extra heavy oil) has a similar analysis to tar sand bitumen (Table 2.11) but some mobility in the deposit because of the relatively high temperature of the deposit.

Thus, extra heavy oil is a material that occurs in the solid or near-solid state and generally has mobility under reservoir conditions. While this type of oil resembles tar sand bitumen and does not flow easily, extra heavy oil is generally recognized as having mobility in the reservoir compared to tar sand bitumen, which is typically incapable of mobility (free flow) under reservoir conditions. It is likely that the mobility of extra heavy oil is due to a high reservoir temperature (that is higher than the pour point of the extra heavy oil) or due to other factors is variable and subject to localized conditions in the reservoir.

2.4.3 TAR SAND BITUMEN

The expression *tar sand* is commonly used in the crude oil industry to describe sandstone reservoirs that are impregnated with a heavy, viscous black crude oil that cannot be retrieved through a well by conventional production techniques (FE-76-4, above). However, the term *tar sand* is actually a misnomer; more correctly, the name *tar* is usually applied to the heavy product remaining after the destructive distillation of coal or other organic matter (Speight, 2013d). Current recovery operations of bitumen in tar sand formations are predominantly focused on a mining technique.

The term *bitumen* (also, on occasion, referred to as *native asphalt* and *extra heavy oil*) includes a wide variety of reddish brown to black materials of semisolid, viscous to brittle character that can exist in nature with no mineral impurity or with mineral matter contents that exceed 50% w/w and are often structurally dissimilar (Kam'Yanov et al., 1995; Kettler, 1995; Ratov, 1996; Parnell et al., 1996; Niu and Hu, 1999; Meyer et al., 2007; Speight, 2014a). Bitumen is found in deposits where the

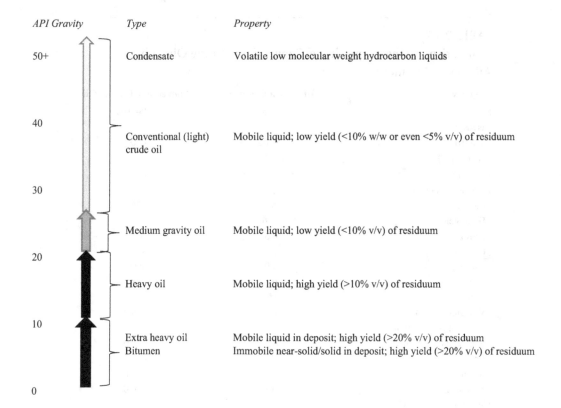

FIGURE 2.2   General description of various feedstocks by API gravity.

TABLE 2.11
Comparison of Selected Properties of Athabasca Tar Sand Bitumen (Alberta, Canada) and Zuata Extra Heavy Oil (Orinoco, Venezuela)

		Athabasca Bitumen	Zuata Extra Heavy Oil
Whole oil	API gravity	8	8
	Sulfur, % w/w	4.8	4.2
650F+	% v/v	85	86
	Sulfur, % w/w	5.4	4.6
	Ni + V, ppm	420	600
	CCR, % w/w[a]	14	15

[a] Conradson carbon residue.

---

permeability is low and passage of fluids through the deposit can only be achieved by prior application of fracturing techniques.

Tar sand bitumen is a high-boiling material with little, if any, material boiling below 350°C (660°F) and the boiling range approximately the same as the boiling range of an atmospheric residuum. *Tar sands* have been defined in the United States (FE-76-4) as:

...the several rock types that contain an extremely viscous hydrocarbon which is not recoverable in its natural state by conventional oil well production methods including currently used enhanced recovery techniques. The hydrocarbon-bearing rocks are variously known as bitumen-rocks oil, impregnated rocks, oil sands, and rock asphalt.

TABLE 2.12

Simplified Use of Pour Point and Reservoir/Deposit Temperature to Differentiate between Heavy Oil, Extra Heavy Oil, and Tar Sand Bitumen

Oil	Location	Temperature	Effect on Oil
Heavy oil	Reservoir or deposit	Higher than oil pour point	Fluid and/or mobile Mobile
	Surface/ambient	Higher than oil pour point	Fluid and/or mobile Mobile
Extra heavy oil	Reservoir or deposit	Higher than oil pour point	Fluid and/or mobile Mobile
	Surface/ambient	Lower than oil pour point	Solid to near-solid Fluidity much reduced Immobile
Tar sand bitumen	Reservoir or deposit	Lower than oil pour point	Solid to near-solid Not fluid Immobile
	Surface/ambient	Lower than oil pour point	Solid to near-solid Not fluid Immobile

The term *natural state* cannot be defined out of context and in the context of FEA Ruling 1976-4 and the term is defined in terms of the composition of the heavy oil or bitumen extra heavy oil adds a further dimension to this definition as it can be ascribed to the properties of the oil in the reservoir *vis-à-vis* the properties of the oil under ambient conditions. The final determinant of whether a reservoir is a tar sand deposit is the character of the viscous phase (bitumen) and the method that is required for recovery. From this definition and by inference crude oil and heavy oil are recoverable by well production methods and currently used enhanced recovery techniques. Fore convenience, it is assumed that before depletion of the reservoir energy, conventional crude oil is produced by primary and secondary techniques while heavy oil requires tertiary, enhanced oil recovery (EOR) techniques and recovery of tar sand bitumen requires more advanced methods (Speight, 2014a). While this is an oversimplification, it may be used as a general guide for the recovery of the different materials.

There has been the suggestion that tar sand bitumen differs from heavy oil by using an arbitrary ill-conceived limit of 10,000 centipoises as the upper limit for heavy oil and the lower limit of tar sand bitumen. Such a system based on one physical property (viscosity) is fraught with errors! For example, this requires that a tar sand bitumen could have a viscosity in the order of 10,050 centipoises and an oil with a viscosity of 9,950 centipoise is heavy oil. Both numbers fall within the limits of experimental difference of the method used to determine viscosity. The limits are the usual laboratory experimental difference be (±3%) or more likely the limits of accuracy of the method (±5% to ±10%) there is the question of accuracy when tax credits for recovery of heavy oil, extra heavy oil, and tar sand bitumen are awarded. In fact, the, inaccuracies (i.e., the limits of *experimental difference*) of the method of measuring viscosity also increase the potential for misclassification using this (or any) single property for classification purposes.

It is incorrect to refer to native bituminous materials as *tar* or *pitch*. Although the word tar is descriptive of the black, heavy bituminous material, it is best to avoid its use with respect to natural materials and to restrict the meaning to the volatile or near-volatile products produced in the destructive distillation of such organic substances as coal (Speight, 2013d). In the simplest sense, pitch is the distillation residue of the various types of tar. Thus, alternative names, such as *bituminous sand* or *oil sand*, are gradually finding usage, with the former name (bituminous sands) more technically correct. The term *oil sand* is also used in the same way as the term *tar sand*, and these terms are used interchangeably throughout this text.

Bituminous rock and *bituminous sand* are those formations in which the bituminous material is found as a filling in veins and fissures in fractured rocks or impregnating relatively shallow sand, sandstone, and limestone strata. These terms are, in fact, correct geological description of *tar sand*. The deposits contain as much as 20% w/w bituminous material, and if the organic material in the rock matrix is bitumen, it is usual (although chemically incorrect) to refer to the deposit as *rock asphalt* to distinguish it from bitumen that is relatively mineral free. A standard test (ASTM D4, 2018) is available for determining the bitumen content of various mixtures with inorganic materials, although the use of word *bitumen* as applied in this test might be questioned and it might be more appropriate to use the term *organic residues* to include *tar* and *pitch*. If the material is of the asphaltite-type or asphaltoid-type, the corresponding terms should be used: rock asphaltite or rock asphaltoid.

Since the most significant property of tar sand bitumen is its *immobility* under the conditions of temperatures and pressure in the deposit, the interrelated properties of API gravity (ASTM D287, 2018) and viscosity (ASTM D445, 2018) may present an *indication* (but only an indication) of the mobility of oil or immobility of bitumen but these properties only offer subjective descriptions of the oil in the reservoir. The most pertinent and objective representation of this oil or bitumen mobility is the *pour point* (ASTM D97, 2018) which can be compared directly to the reservoir/deposit temperature (Speight, 2014a, 2017).

The *pour point* is the lowest temperature at which oil will move, pour, or flow when it is chilled without disturbance under definite conditions (ASTM D97, 2018). In fact, the pour point of an oil when used in conjunction with the reservoir temperature give a better indication of the condition of the oil in the reservoir that the viscosity. Thus, the pour point and reservoir temperature present a more accurate assessment of the condition of the oil in the reservoir, being an indicator of the mobility of the oil in the reservoir. Indeed, when used in conjunction with reservoir temperature, the pour point gives an indication of the liquidity of the heavy oil, extra heavy oil, or bitumen and, therefore, the ability of the heavy oil extra heavy oil to flow under reservoir conditions. In summary, the pour point is an important consideration because, for efficient production, additional energy must be supplied to the reservoir by a thermal process to increase the reservoir temperature beyond the pour point.

For example, Athabasca bitumen with a pour point in the range 50°C–100°C (122°F–212°F) and a deposit temperature of 4°C–10°C (39°F–50°F) is a solid or near solid in the deposit and will exhibit little or no mobility under deposit conditions. Pour points of 35°C–60°C (95°F–140°F) have been recorded for the bitumen in Utah with formation temperatures in the order of 10°C (50°F). This indicates that the bitumen is solid within the deposit and therefore immobile. The injection of steam to raise and maintain the reservoir temperature above the pour point of the bitumen and to enhance bitumen mobility is difficult, in some cases almost impossible. Conversely, when the reservoir temperature exceeds the pour point the oil is fluid in the reservoir and therefore mobile. The injection of steam to raise and maintain the reservoir temperature above the pour point of the bitumen and to enhance bitumen mobility is possible and oil recovery can be achieved.

REFERENCES

Ancheyta, J., and Speight, J.G. 2007. *Hydroprocessing of Heavy Oils and Residua*. CRC Press, Boca Raton, FL.
API. 2009. Refinery Gases Category Analysis and Hazard Characterization. Submitted to the EPA by the American Petroleum Institute, Petroleum HPV Testing Group. HPV Consortium Registration # 1100997 United States Environmental Protection Agency, Washington, DC. June 10.
ASTM. 2018. *Annual Book of Standards*. ASTM International, West Conshohocken, PA.
ASTM D4. 2018. *Standard Test Method for Bitumen Content*. Annual Book of Standards. ASTM International, West Conshohocken, PA.
ASTM D56. 2018. *Standard Test Method for Flash Point by Tag Closed Cup Tester*. Annual Book of Standards. ASTM International, West Conshohocken, PA.

ASTM D92. 2018. *Standard Test Method for Flash and Fire Points by Cleveland Open Cup Tester.* Annual Book of Standards. ASTM International, West Conshohocken, PA.

ASTM D86. 2018. *17 Standard Test Method for Distillation of Petroleum Products and Liquid Fuels at Atmospheric Pressure.* Annual Book of Standards. ASTM International, West Conshohocken, PA.

ASTM D97. 2018. *Standard Test Method for Pour Point of Petroleum Products.* Annual Book of Standards. ASTM International, West Conshohocken, PA.

ASTM D156. 2018. *Standard Test Method for Saybolt Color of Petroleum Products (Saybolt Chromometer Method).* Annual Book of Standards. ASTM International, West Conshohocken, PA.

ASTM D187. 2018. *Standard Specification for Kerosene.* Annual Book of Standards. ASTM International, West Conshohocken, PA.

ASTM D287. 2018. *Standard Test Method for API Gravity of Crude Petroleum and Petroleum Products (Hydrometer Method).* Annual Book of Standards. ASTM International, West Conshohocken, PA.

ASTM D445. 2018. *Standard Test Method for Kinematic Viscosity of Transparent and Opaque Liquids (and Calculation of Dynamic Viscosity).* Annual Book of Standards. ASTM International, West Conshohocken, PA.

ASTM D2500. 2018. *Standard Test Method for Cloud Point of Petroleum Products and Liquid Fuels.* Annual Book of Standards. ASTM International, West Conshohocken, PA.

ASTM D7213. 2018. *Standard Test Method for Boiling Range Distribution of Petroleum Distillates in the Boiling Range from 100°C to 615°C by Gas Chromatography.* Annual Book of Standards. ASTM International, West Conshohocken, PA.

ASTM D4175. 2018. *Standard Terminology Relating to Petroleum, Petroleum Products, and Lubricants.* Annual Book of Standards. ASTM International, West Conshohocken, PA.

Belosludov, V.R., Subbotin, O.S., Krupskii, D.S., Belosludov, R.V., Kawazoe, Y., and Kudoh, J. 2007. Physical and Chemical Properties of Gas Hydrates: Theoretical Aspects of Energy Storage Application. *Materials Transactions,* 48(4): 704–710.

Bishnoi, P.R., and Clarke, M.A. 2006. Natural Gas Hydrates. In: *Encyclopedia of Chemical Processing.* CRC Press, Philadelphia, PA.

Brosseau, J. 1994. Trace Gas Compound Emissions from Municipal Landfill Sanitary Sites. *Atmospheric-Environment,* 28 (2): 285–293.

Carrol, J.J. 2003. *Natural Gas Hydrates.* Gulf Professional Publishing, Burlington, VT.

Chadeesingh, R. 2011. Chapter 5: The Fischer-Tropsch Process. In: *The Biofuels Handbook.* J.G. Speight (Editor). The Royal Society of Chemistry, London, UK. Part 3, pp. 476–517.

Collett, T.S., and Ladd, J.W. 2000. Detection of Gas Hydrate with Downhole Logs and Assessment of Gas Hydrate Concentrations (Saturations) and Gas Volumes on the Blake Ridge with Electrical Resistivity Log Data. *Proceedings of the Ocean Drilling Program, Scientific Results,* 164: 179–191.

Collett, T.S. 2001. Natural-Gas Hydrates; Resource of the Twenty-First Century? *Journal of the American Association of Petroleum Geologists,* 74: 85–108.

Collett, T.S., Johnson, A.H., Knapp, C.C., and Boswell, R. 2009, Natural Gas Hydrates: A Review. In: *Natural Gas Hydrates—Energy Resource Potential and Associated Geologic Hazards.* T.S. Collett, A.H. Johnson, C.C. Knapp, and R. Boswell (Editors). AAPG Memoir No. 89. American Association of Petroleum Geologists, Tulsa, OK, pp. 146–219.

Collett, T.S. 2010. Physical Properties of Gas Hydrates: A Review. *Journal of Thermodynamics,* 2010, Article ID 271291; doi:10.1155/2010/271291. www.hindawi.com/journals/jther/2010/271291/, accessed November 1, 2017.

Collett, T.S., Bahk, J.J., Baker, R., Boswell, R., Divins, D., Frye, M., Goldberg, D., Husebø, J., Koh, C., Malone, M., Morell, M., Myers, G., Shipp, C., and Torres, M. 2015. Methane Hydrates in Nature—Current Knowledge and Challenges. *Journal of Chemical Engineering and Data,* 60(2): 319–329.

Crawford, D.B., Durr, C.A., Finneran, J.A., and Turner, W. 1993. Chemicals from Natural Gas. In: *Chemical Processing Handbook.* J.J. McKetta (Editor). Marcel Dekker Inc., New York. p. 2.

Cui, H., Turn, S.Q., Keffer, V., Evans, D., and Foley, M. 2013. Study on the Fate of Metal Elements from Biomass in a Bench-Scale Fluidized Bed Gasifier. *Fuel,* 108: 1–12.

Delbianco, A., and Montanari, R. 2009. *Encyclopedia of Hydrocarbons, Volume III/New Developments: Energy, Transport, Sustainability.* Eni S.p.A., Rome, Italy.

Esposito, G., Frunzo, L., Liotta, F., Panico, A., and Pirozzi, F. 2012. Bio-Methane Potential Tests to Measure the Biogas Production from the Digestion and Co-Digestion of Complex Organic Substrates. *The Open Environmental Engineering Journal,* 5: 1–8.

Gary, J.G., Handwerk, G.E., and Kaiser, M.J. 2007. *Petroleum Refining: Technology and Economics.* 5th Edition. CRC Press, Boca Raton, FL.

Ghoshal, S., and Sainik, V. 2013. Monitor and Minimize Corrosion in High-TAN Crude Processing. *Hydrocarbon Processing*, 92(3): 35–38.

Giavarini, C., Maccioni, F., and Santarelli, M.L. 2003. Formation Kinetics of Propane Hydrate. *Industrial & Engineering Chemistry Research*, 42: 1517–1521.

Giavarini, C., and Maccioni, F. 2004. Self-Preservation at Low Pressure of Methane Hydrates with Various Gas Contents. *Industrial & Engineering Chemistry Research*, 43: 6616–6621.

Giavarini, C., Maccioni, F., and Santarelli, M.L. 2005. Characterization of Gas Hydrates by Modulated Differential Scanning Calorimetry. *Petroleum Science and Technology*, 23: 327–335.

Hsu, C.S., and Robinson, P.R. (Editors). 2017. *Handbook of Petroleum Technology*. Springer, Cham, Switzerland.

IEA. 2005. *Resources to Reserves: Oil & Gas Technologies for the Energy Markets of the Future*. International Energy Agency, Paris, France.

Isaacs, C.M. 1992. Preliminary Petroleum Geology Background and Well Data for Oil Samples in the Cooperative Monterey Organic Geochemistry Study, Santa Maria and Santa Barbara-Ventura Basins, CA. Open-File Report No. USGS 92-539-F. United States Geological Survey, Reston, VA.

Kam'Yanov, V.F., Braun, A.Ye., Gorbunova, L.V., and Shabotkin, L.G. 1995. Natural Bitumens of Mortuk. *Petroleum Chemistry*, 35(5): 377–389. Also: *Neftekhimiya*, 35(5): 397–409.

Kane, R.D., and Cayard, M.S. 2002. *A Comprehensive Study on Naphthenic Acid Corrosion*. NACE International, Houston, TX.

Kettler, R.M. 1995. Incipient Bitumen Generation in Miocene Sedimentary Rocks from the Japan Sea. *Organic Geochemistry*, 23(7): 699–708.

Klimstra, J. 1978. *Interchangeability of Gaseous Fuels—The Importance of the Wobbe-Index*. Report No. SAE 861578. Society of Automotive Engineers, SAE International, Warrendale, PA.

Liss, W.E., and Thrasher, W.R. 1992. *Variability of Natural Gas Composition in Select Major Metropolitan Areas of the U.S.* Report No. GRI-92/0123. Gas Research Institute, Chicago, IL.

Lohila, A., Laurila, T., Tuovinen, J.-P., Aurela, M., Hatakka, J., Thum, T., Pihlatie, M., Rinne, J., and Vesala, T. 2007. Micrometeorological Measurements of Methane and Carbon Dioxide Fluxes at a Municipal Landfill. *Environmental Science & Technology*, 41(8): 2717–2722.

Lorenson, T.D., and Collett, T.S. 2000. Gas Content and Composition of Gas Hydrate from Sediments of the Southeastern North American Continental Margin. *Proceedings of the Ocean Drilling Program, Scientific Results*. C.K. Paull, R. Matsumoto, P.J. Wallace, and W.P. Dillon. (Editors), 164: 37–46.

Makogon, Y.F. 1997. *Hydrates of Hydrocarbons*. PennWell Books, Tulsa, OK.

Makogon, Y.F., Holditch, S.A., and Makogon, T.Y. 2007. Natural Gas Hydrates—A Potential Energy Source for the 21st Century. *Journal of Petroleum Science and Engineering*, 56(1–3): 14–31.

Makogon, Y.F. 2010. Natural Gas Hydrates—A Promising Source of Energy. *Journal of Natural Gas Science and Engineering*, 2(1): 49–59.

Meyer, R.F., Attanasi, E.D., and Freeman, P.A. 2007. Heavy Oil and Natural Bitumen Resources in Geological Basins of the World. USGS Open File Report No. 2007-1084. United States Geological Survey, Reston, VA.

Mokhatab, S., Poe, W.A., and Speight, J.G. 2006. *Handbook of Natural Gas Transmission and Processing*. Elsevier, Amsterdam, Netherlands.

Niu, J., and Hu, J. 1999. Formation and Distribution of Heavy Oil and Tar Sands in China. *Marine and Petroleum Geology*, 16: 85–95.

Ohmes, R. 2014. Characterizing and Tracking Contaminants in Opportunity Crudes. *Digital Refining*. http://www.digitalrefining.com/article/1000893,Characterising_and_tracking_contaminants_in_opportunity_crudes_.html#.VJhFjV4AA, accessed November 1, 2014.

Parkash, S. 2003. *Refining Processes Handbook*. Gulf Professional Publishing, Elsevier, Amsterdam, Netherlands.

Parnell, J., Monson, B., and Geng, A. 1996. Maturity and Petrography of Bitumens in the Carboniferous of Ireland. *International Journal of Coal Geology*, 29: 23–38.

Rasi, S., Veijanen, A., and Rintala, J. 2007. Trace Compounds of Biogas from Different Biogas Production Plants. *Energy*, 32: 1375–1380.

Ratov, A.N. 1996. Physicochemical Nature of Structure Formation in High-Viscosity Crude Oils and Natural Bitumens and Their Rheological Differences. *Petroleum Chemistry*, 36: 191–206. Also: *Neftekhimiya*, 1996, 36(3): 195208.

Seo, Y., Kang, S.P., and Jang, W. 2009. Structure and Composition Analysis of Natural Gas Hydrates: 13C NMR Spectroscopic and Gas Uptake Measurements of Mixed Gas Hydrates. *Journal of Physical Chemistry*, 113(35): 9641–9649.

Shalaby, H.M. 2005. Refining of Kuwait's Heavy Crude Oil: Materials Challenges. *Proceedings. Workshop on Corrosion and Protection of Metals. Arab School for Science and Technology*. December 3–7, Kuwait.

Sloan, E.D. Jr. 1998a. Gas Hydrates: Review of Physical/Chemical Properties. *Energy & Fuels*, 12(2): 191–196.

Sloan, E.D., Jr. 1998b. *Clathrate Hydrates of Natural Gases.* 2nd Edition. Marcel Dekker Inc., New York.

Sloan, E.D. Jr. 2006. *Clathrate Hydrates of Natural Gases.* 3rd Edition. Marcel Dekker Inc., New York.

Speight, J.G. 1994. Chemical and Physical Studies of Petroleum Asphaltene constituents. In: *Asphaltene constituents and Asphalts. I.* Developments in Petroleum Science, 40. T.F. Yen and G.V. Chilingarian (Editors). Elsevier, Amsterdam, Netherlands. Chapter 2.

Speight, J.G. 2007. Natural Gas: A Basic Handbook. GPC Books, Gulf Publishing Company, Houston, TX.

Speight, J.G. 2008. *Synthetic Fuels Handbook: Properties, Processes, and Performance.* McGraw-Hill, New York.

Speight, J.G. 2011a. The Refinery of the Future. Gulf Professional Publishing, Elsevier, Oxford, UK.

Speight, J.G. 2011b. *An Introduction to Petroleum Technology, Economics, and Politics.* Scrivener Publishing, Salem, MA.

Speight, J.G. (Editor). 2011c. The Biofuels Handbook. Royal Society of Chemistry, London, UK.

Speight, J.G. 2012. *Crude Oil Assay Database.* Knovel, Elsevier, New York. 2012. Online version available at: http://www.knovel.com/web/portal/browse/display?_EXT_KNOVEL_DISPLAY_bookid=5485&VerticalID=0, accessed June 4, 2018.

Speight, J.G. 2013a. *Heavy Oil Production Processes.* Gulf Professional Publishing, Elsevier, Oxford, UK.

Speight, J.G. 2013b. *Heavy and Extra Heavy Oil Upgrading Technologies.* Gulf Professional Publishing, Elsevier, Oxford, UK.

Speight, J.G. 2013c. *Oil Sand Production Processes.* Gulf Professional Publishing, Elsevier, Oxford, UK.

Speight, J. G. 2013d. *The Chemistry and Technology of Coal.* 3rd Edition. CRC Press, Boca Raton, FL.

Speight, J.G. 2014a. *The Chemistry and Technology of Petroleum.* 5th Edition. CRC Press, Boca Raton, FL.

Speight, J.G. 2014b. *High Acid Crudes.* Gulf Professional Publishing, Elsevier, Oxford, UK.

Speight, J.G. 2014c. *Oil and Gas Corrosion Prevention.* Gulf Professional Publishing, Elsevier, Oxford, UK.

Speight, J.G., and Exall, D.I. 2014. *Refining Used Lubricating Oils.* CRC Press, Boca Raton, FL.

Speight, J.G. 2015. *Handbook of Petroleum Product Analysis.* 2nd Edition. John Wiley & Sons Inc., Hoboken, NJ.

Speight, J.G. 2016a. *Introduction to Enhanced Recovery Methods for Heavy Oil and Tar Sands.* 2nd Edition. Gulf Professional Publishing, Elsevier, Oxford, UK.

Speight, J.G. 2016b. Chapter 1: Hydrogen in Refineries. In: *Hydrogen Science and Engineering: Materials, Processes, Systems, and Technology.* D. Stolten and B. Emonts (Editors). Wiley-VCH Verlag GmbH & Co., Weinheim, Germany, pp. 3–18.

Speight, J.G., and Islam, M.R. 2016. *Peak Energy—Myth or Reality.* Scrivener Publishing, Beverly, MA.

Speight, J.G. 2017. *Handbook of Petroleum Refining.* CRC Press, Boca Raton, FL.

Speight, J.G. 2018. *Handbook of Natural Gas Analysis.* John Wiley & Sons Inc., Hoboken, NJ.

Staley, B., and Barlaz, M.A. 2009. Composition of Municipal Solid Waste in the United States and Implications for Carbon Sequestration and Methane Yield. *Journal of Environmental Engineering,* 135(10): 901–909.

Stern, L., Kirby, S., Durham, W., Circone, S., and Waite, W.F. 2000. Laboratory Synthesis of Pure Methane Hydrate Suitable for Measurement of Physical Properties and Decomposition Behavior. *Proceedings of Natural Gas Hydrate in Oceanic and Permafrost Environments.* M.D. Max (Editor). Kluwer Academic Publishers, Dordrecht, Netherlands, pp. 323–348.

Stoll, R.G., and Bryan, G.M. 1979. Physical Properties of Sediments Containing Gas Hydrates. *Journal of Geophysical Research,* 84(B4): 1629–1634.

Sun, X., Zhang, Y., Li, X., Cui, G., and Gu, J. 2013. A Case Study on Foamy Oil Characteristics of the Orinoco Belt, Venezuela. *Advances in Petroleum Exploration and Development,* 5(1): 37–41.

Taylor, C. 2002. Formation Studies of Methane Hydrates with Surfactants. *Proceedings of 2nd International Workshop On Methane Hydrates.* Washington, DC.

Van Geem, K.M., Reyniers, M.F., and Marin, G.B. 2008. Challenges of Modeling Steam Cracking of Heavy Feedstocks. *Oil & Gas Science and Technology, Rev. IFP,* 63(1): 79–94.

Villarroel, T., and Hernández, R. 2013. Technological Developments for Enhancing Extra Heavy Oil Productivity in Fields of the Faja Petrolifera del Orinoco (FPO), Venezuela. *Proceedings. AAPG Annual Convention and Exhibition,* Pittsburgh, PA. May 19–22. American Association of Petroleum Geologists, Tulsa, OK.

Wachtmeister, H., Linnea Lund, L., Aleklett, K., and Höök, M.M. 2017. Production Decline Curves of Tight Oil Wells in Eagle Ford Shale. *Natural Resources Research,* 26(3): 365–377.

Wang, X., and Economides, M.J. 2012. Natural Gas Hydrates as an Energy Source—Revisited 2012. *Proceedings of SPE International Petroleum Technology Conference 2012.* 1: 176–186. Society of Petroleum Engineers, Richardson, TX.

Yang, X., and Qin, M. 2012. Natural Gas Hydrate as Potential Energy Resources in the Future. *Advanced Materials Research,* 462: 221–224.

Yeung, T.W. 2014. Evaluating Opportunity Crude Processing. *Digital Refining.* www.digitalrefining.com/article/1000644, accessed October 25, 2014.

3 Other Feedstocks—Coal, Oil Shale, and Biomass

3.1 INTRODUCTION

Since the oil crises of the 1970s, the idea of deriving essential chemical feedstocks from renewable resources (renewable feedstocks) in a sustainable manner has been frequently suggested as an alternative to producing chemicals from petroleum-based feedstocks imported, under agreement, from unstable political regions with the accompanying geopolitics that go with such agreements. In addition to the geopolitics, the common petrochemical feedstocks that are derived from natural gas and crude oil are, in spite of discoveries of natural gas and crude oil, in tight formations (Chapter 2) and are depleting, such as petroleum and natural gas. The petrochemical industry uses petroleum and natural gas as feedstocks to make intermediates, which are later converted to final products that people use, such as plastics, paints, pharmaceuticals, and many others.

In spite of the apparent plentiful supply of oil in tight formation (such as the Bakken formation and the Eagle Ford formation in the United States), which may be limited in terms of total producible reserves (Wachtmeister et al., 2017) the oil industry is planning for the future since some of the most prolific basins have begun to experience reduced production rates and are reaching or already into maturity. At the same time, the demand for oil continues to grow every year, because of increased demands by the rapidly growing economies of China and India. This decline in the availability of conventional crude oil combined with this rise in demand for oil and oil-based products has put more pressure on the search for alternate energy sources (Speight, 2008, 2011a, 2011b, 2011c).

Several authors have correctly stated that petroleum is and will continue to be a major motivating force to the industrial society. Natural gas and natural gas liquids are important and their role will continue in the near future in the industrial economy. Some estimates suggest that relatively cheap hydrocarbon-based feedstocks will be available well into the next century, although predicting the availability of such feedstock beyond the next 50 years is risky (Speight, 2011a, 2011b; Speight and Islam, 2016). In fact, the reality is that the supply of crude oil, the basic feedstock for refineries and for the petrochemicals industry, is finite and its dominant position will become unsustainable as supply/demand issues erode its economic advantage over other alternative feedstocks. This situation will be mitigated to some extent by the exploitation of more technically challenging fossil resources and the introduction of new technologies for fuels and chemicals production from natural gas and coal (Speight, 2008, 2014)

More specifically, as crude oil prices continue to fluctuate (typically in an upward direction), C-1 chemistry, based on coal gasification and converting coal-derived synthesis gas to chemicals and other alternatives (such as biomass-derived chemical and biomass-derived synthesis gas) will become important. In fact, over the past two decades a series of technological advances has occurred that promise, in concert, to significantly improve the economic competitiveness of bio-based processes (Speight, 2008). Evaluation of this window of opportunity focuses on the inherent attributes of biological processes, application of new technology to overcome past limitations, and integration with nonbiological process steps.

In order to satisfy the demand for feedstocks for petrochemicals, it will be necessary to develop the reservoirs (of heavy oil) and deposits (of extra heavy oil and tar sand bitumen) that are predominantly located in the Western hemisphere (Chapter 2). These resources are more difficult and costly to extract, so they have barely been touched in the past. However, through these resources, the world could soon have access to oil sources almost equivalent to those of the Middle East. In fact, with the variability and uncertainty of crude oil supply due to a variety of geopolitical issues (Speight, 2011b), investments in the more challenging reservoirs tend to be on a variable acceleration-deceleration slope.

Nevertheless, the importance of heavy oil, extra heavy oil, and tar sand bitumen will continue to emerge as the demand for crude oil products remains high. As this occurs, it is worth moving ahead with heavy oil, extra heavy oil, and tar sand bitumen resources on the basis of obtaining a measure (as yet undefined and country-dependent) of oil independence. These will lead eventually—hopefully sooner rather than later—to the adoption of coal, oil shale (produced from kerogen in shale formations), and renewable feedstocks as the source materials for the production of petrochemicals. The term *renewable feedstocks* includes a huge number of materials such as agricultural crops rich in starch, lignocellulosic materials (biomass), or biomass material recovered from a variety of processing wastes.

The general term *biomass* refers to any material derived from living organisms, usually plants. In contrast to depleting feedstocks such as natural gas and crude oil, the production of bio-based chemicals which can replace the petroleum-derived chemicals will prove to be a reliable supply of resources for the future. Existing chemical technology is continually being developed to provide chemicals and end products from biomass (Bozell, 1999; Besson et al., 2014; Straathof, 2014; Khoo et al., 2015).

For bioprocesses—the conversion of biomass into useful products such as fuels and petrochemicals—one opportunity that exists is the production of butanol from bioprocessing which could be a major commodity chemical that has application as a feedstock for such products as butyl butyrate. Also, chemicals such as xylose, xylitol, furfural, tetrahydrofuran, glucose, gluconic acid, sorbitol, mannitol, levulinic acid, and succinic acid are materials that could be prepared from inexpensive cellulose and hemicellulose-derived sugars, available from clean biomass fractionation. As further examples, (i) anthraquinone, a well-known pulping catalyst and chemical intermediate, can be prepared from lignin while butadiene and a variety of pentane derivatives can be prepared using fast pyrolysis followed by catalytic upgrading on zeolite-type catalysts; (ii) acetic acid can be produced. From synthesis gas, a route that appears to be an interesting match given the unique composition of synthesis gas available from biomass, and makes possible a balanced process (through intermediate methanol or ethanol) without the costs of reforming the synthesis gas; and (iii) peracetic acid is an oxidant that is a good non-chlorine-containing pulp bleaching agent which would permit market penetration of this chemical into the pulp and paper industry.

However, a basic understanding of reactions for selectively converting biomass and biomass-derived materials into chemicals is needed. A fundamental understanding of new catalytic processes for selectively manipulating and modifying carbohydrates, lignin, and other biomass fractions will greatly improve the ability to bring biomass-derived products to market. The behavior of oxygenated molecules on zeolite and other shape-selective catalysts could lead to a better design of processes for chemicals from biomass. Developing special catalysts for biomass processing is not a high priority for the chemicals industry, but is essential for this new field if it is to compete with petroleum resources and cost-effectively produce fuels and chemicals.

Consequently, there is a renewed interest in the utilization of plant-based matter as a raw material feedstock for the chemicals industry. Plants accumulate carbon from the atmosphere via photosynthesis and the widespread utilization of these materials as basic inputs into the generation of power, fuels and chemicals is a viable route to reduce greenhouse gas emissions. As a result, the petroleum and petrochemical industries are coming under increasing pressure not only to compete effectively with global competitors utilizing more advantaged hydrocarbon feedstocks, but also to ensure that its processes and products comply with increasingly stringent environmental legislation.

Reducing dependence of any country on imported crude oil is of critical importance for long-term security and continued economic growth. Supplementing petroleum consumption with renewable biomass resources is a first step toward this goal. The realignment of the chemical industry from one of the petrochemical refining to a biorefinery concept, given time, feasibly has become a national goal of many oil-importing countries. However, clearly defined goals are necessary for construction of a biorefinery and increasing the use of biomass-derived feedstocks in industrial chemical production is important to keep the goal in perspective (Clark and Deswarte, 2008). In this context, the increased use of biofuels should be viewed as *one* of a range of possible measures for achieving self-sufficiency in energy, rather than a panacea (Crocker and Crofcheck, 2006).

Thus, in any text about the production of chemicals (petrochemicals), it would be a remiss to omit other sources of chemicals such as coal and biomass.

3.2   COAL

Coal, which is currently considered the bad boy of fossil fuels due to environmental issues some of which are real and some of which are emotional, may become more important both as an energy source and as the source of organic chemical feedstock in the 21st century.

The chemicals-from-coal industry was born in the late 18th century at the time of the Industrial Revolution when power and chemicals from coal were everyday occurrences. Thus, the coal chemicals industry refers to the conversion of coal into gas, liquid, solid fuels, and chemicals after chemical processing with coal as raw material. In the early days of the chemicals-from-coal industry, the term *chemicals* covered primarily ammonia, hydrocarbon gases, low-boiling aromatic derivatives (benzene, toluene, and xylene (BTX)), difficult-to-define tar acids, difficult-to-define tar bases, tar, pitch, and coke. In the United States, these chemicals were derived from coal almost exclusively through high-temperature byproduct carbonization. In England and Europe, these and other chemicals have been obtained to some extent through various low-temperature carbonization processes and by coal hydrogenation in England and Germany (Speight, 2013a).

Thus, the processes for the production of chemicals from coal were predominantly coking, gasification, liquefaction of coal, as well as coal tar processing, and carbide acetylene chemical engineering. The significant time frame for the production of chemicals from coal was the period from 1920 to 1940 after which World War II brought imperative demands for toluene, ammonia, and other chemicals that could not be met by the coke plants. Petroleum and natural gas were used as raw materials, and since that time they have dominated the chemical industry. However, as natural gas and crude oil resources of the world decrease (they are, of course, nonrenewable resources), the chemicals-from-coal industry may once again realize broad prospects for development. This must go along with the realization that emissions from coal plants can be reduced significantly by the installation of emissions reduction processes that have now been placed into operation in the coal-generated power plants.

In the production technology of coal processing and utilization, coking process technology is one of the earliest applications, and it is still an important part of the chemical industry. Coal gasification occupies an important position in the coal chemical industry and is used in the production of various types of gas fuel. It is a clean energy and is conducive to the improvement of living standards and environmental protections. Synthetic gas produced by coal gasification is the raw material of many products such as synthetic liquid fuel, and raw materials for the production of chemicals. The direct coal liquefaction (high-pressure coal hydrogenation liquefaction for production of naphtha and kerosene) and indirect coal liquefaction (through gasification of coal for synthesis of gasoline and diesel) can produce synthetic petroleum and chemical products (Owen, 1981; Speight, 2013a).

In fact, coal has several positive attributes when considered as a feedstock for aromatic chemicals, specialty chemicals, and carbon-based materials. Substantial progress in advanced polymer materials, incorporating aromatic and polyaromatic units in their main chains, has created new opportunities for developing value-added or specialty organic chemicals from coal and tars from coal carbonization for coke making. The decline of the coal tar industry has diminished the traditional sources of these chemicals. A new coal chemistry for chemicals and materials from coal may involve direct and indirect coal conversion strategies as well as the coproduction approach.

In addition, the needs for environmental-protection applications have also expanded market demand for carbon materials and carbon-based adsorbents.

3.2.1 COAL FEEDSTOCKS

By way of introduction, coal is a natural combustible rock composed of an organic heterogeneous substance contaminated with variable amounts of inorganic compounds. Coal is classified into different ranks according to the degree of chemical change that occurred during the decomposition of plant remains in the prehistoric period. In general, coals with a high heating value and high fixed carbon content are considered to have been subjected to more severe changes than those with lower heating values and fixed carbon contents. For example, peat, which is considered a young coal, has a low fixed carbon content and a low heating value. Important coal ranks are anthracite (which has been subjected to the most chemical change and is mostly carbon), bituminous coal, subbituminous coal, and lignite.

The birth of coal chemical industry first appeared in the late 18th century, and in the 19th century the complete system of coal chemical industry was set up. After entering 20th century, raw materials of organic chemicals were changed into coal from the former agricultural and forestry products, and then coal chemical industry became an important part of chemical industry. After World War II, the petrochemical industry saw rapid development, which weakened the position taken up by coal chemical industry by changing raw materials from coal to petroleum and natural gas. The organic matters and chemical structures of coal with condensed rings as their core units connected by bridged bonds can transform coal into various fuels and chemical products through hot working and catalytic processing.

Coal carbonization is the earliest and most important method. Coal carbonization is mainly used to produce cokes for metallurgy and some secondary products like coal gas, benzene, methylbenzene, etc. Coal gasification takes up an important position in chemical industry. City gas and varieties of fuel gases can be produced by coal gasification. The common role of low-temperature carbonization, direct coal liquefaction and indirect coal liquefaction is to produce liquid fuels.

Thus, for many years, chemicals that have been used for the manufacture of such diverse materials as nylon, styrene, fertilizers, activated carbon, drugs, and medicine, as well as many others have been made from coal (Gibbs, 1961; Mills, 1977; Pitt and Millward, 1979). These products will expand in the future as petroleum and natural gas resources become strained to supply petrochemical feedstocks and coal becomes a predominant chemical feedstock once more. Although many traditional markets for coal tar chemicals have been taken over by the petrochemical industry, the position can change suddenly as oil prices fluctuate upwards. Therefore, the concept of using coal as a major source of chemicals can be very real indeed.

A complete description of the processes to produce all the possible chemical products is beyond the scope of this text. In fact, the production of chemicals from coal has been reported in numerous texts; therefore, it is not the purpose of this text to repeat these earlier works. It is, however, the goal of this chapter to present indications of the extent to which chemicals can be produced from coal as well as indications of the variety of chemical types that arise from coal (e.g., see Lowry, 1945; Speight, 2013a).

On the basis of the thermal chemistry of coal (Speight, 2013a), many primary products of coal reactions are high molecular weight species, often aromatic in nature, that bear some relation to the carbon skeletal of coal. The secondary products (i.e., products formed by decomposition of the primary products) of the thermal decomposition of coal are lower molecular weight species but are less related to the carbon species in the original coal as the secondary reaction conditions become more severe (higher temperatures and/or longer reaction times).

In very general terms, it is these primary and secondary decomposition reactions of coal which are the means to produce chemical from coal. There is some leeway in terms of choice of the reaction conditions, and there is also the option of the complete decomposition of coal (i.e., gasification) and the production of chemicals from the synthesis gas (a mixture of carbon monoxide, CO, and hydrogen, H_2) produced by the gasification process (Chapter 5) (Speight, 2013a).

3.2.2 Properties and Composition

Coal is a combustible dark brown to black organic sedimentary rock that occurs in *coalbeds* or *coal seams*. Coal is composed primarily of carbon with variable amounts of hydrogen, nitrogen, oxygen, and sulfur and may also contain mineral matter and gases as part of the coal matrix. Coal begins as layers of plant matter that has accumulated at the bottom of a body of water after which, through anaerobic metamorphic processes, changes the chemical and physical properties of the plant remains occurred to create a solid material.

Coal is the most abundant fossil fuel in the United States, having been used for several centuries and occurs in several regions. Knowledge of the size, distribution, and quality of the coal resources is important for governmental planning; industrial planning and growth; the solution of current and future problems related to air, water, and land degradation; and for meeting the short- and long-term energy needs of the country. Knowledge of resources is also important in planning for the exportation and importation of fuel.

The types of coal, in increasing order of alteration, are lignite (brown coal), subbituminous, bituminous, and anthracite. It is believed that coal starts off as a material that is closely akin to peat, which is metamorphosed (due to thermal and pressure effects) to lignite. With the further passing of time, lignite increases in maturity to subbituminous coal. As this process of burial and alteration continues, more chemical and physical changes occur and the coal is classified as bituminous. At this point the coal is dark and hard. Anthracite is the last of the classifications, and this terminology is used when the coal has reached ultimate maturation.

The degree of alteration (or metamorphism) that occurs as a coal matures from peat to anthracite is referred to as the rank of the coal, which is the classification of a particular coal relative to other coals, according to the degree of metamorphism, or progressive alteration, in the natural series from lignite to anthracite (ASTM D388). This method of ranking coals used in the United States and Canada was developed by the American Society for Testing and Materials (ASTM; now ASTM International) and are: *(i) heating value, (ii) volatile matter, (iii) moisture, (iv) ash production by combustion which is reflective of the mineral matter content, and (v) fixed carbon* (Speight, 2005, 2013a).

Low-rank coal (such as lignite) has lower energy content because they have low carbon content. They are lighter (earthier) and have higher moisture levels. As time, heat, and burial pressure all increase, the rank does as well. High-rank coals, including bituminous and anthracite coals, contain more carbon than lower-rank coals which results in a much higher energy content. They have a more vitreous (shiny) appearance and lower moisture content then lower-rank coals.

There are many compositional differences between the coals mined from the different coal deposits worldwide. The different types of coal are most usually classified by *rank* which depends upon the degree of transformation from the original source (i.e., decayed plants) and is therefore a measure of a coal's age. As the process of progressive transformation took place, the heating value and the fixed carbon content of the coal increased and the amount of volatile matter in the coal decreased.

3.2.3 Conversion

The thermal properties of coal are important in determining the applicability of coal to a variety of conversion processes. For example, the heat content (also called the heating value or calorific value) (Chapter 8) is often considered to be the most important thermal property. However, there are other thermal properties that are of importance insofar as they are required for the design of equipment that is to be employed for the utilization (conversion, thermal treatment) of coal in processes such as combustion, carbonization, gasification, and liquefaction (Speight, 2013).

The *thermal decomposition* (which includes *pyrolysis processes* and *carbonization* processes) often may be used interchangeably. However, it is more usual to apply the term *pyrolysis* (a thermochemical decomposition of coal or organic material at elevated temperatures in the absence of oxygen which typically occurs under pressure and at operating temperatures above 430°C (800°F)) to a process which involves widespread thermal decomposition of coal (with the ensuing production of a char—carbonized residue).

The term *carbonization* is more correctly applied to the process for the production of char or coke when the coal is heated at temperatures in excess of 500°C (930°F). The ancillary terms volatilization and distillation are also used from time to time but more correctly refer to the formation and removal of volatile products (gases and liquids) during the thermal decomposition process. Thus, carbonization is the destructive distillation of coal in the absence of air accompanied by the production of carbon (coke) as well as the production of liquid and gaseous products. Next to combustion, carbonization represents one of the most popular, and oldest, uses of coal. The thermal decomposition of coal on a commercial scale is often more commonly referred to as carbonization and is more usually achieved by the use of temperatures up to 1,500°C (2,730°F). The degradation of the coal is quite severe at these temperatures and produces (in addition to the desired coke) substantial amounts of gaseous products.

Coal liquefaction is a process used to convert coal, a solid fuel, into a substitute for liquid fuels such as diesel and gasoline. Coal liquefaction has historically been used in countries without a secure supply of petroleum, such as Germany (during World War II) and South Africa (since the early 1970s). The technology used in coal liquefaction is quite old, and was first implemented during the 19th century to provide gas for indoor lighting. Coal liquefaction may be used in the future to produce oil for transportation and heating, in case crude oil supplies are ever disrupted.

The production of liquid fuels from coal is not new and has received considerable attention. In fact, the concept is often cited as a viable option for alleviating projected shortages of liquid fuels as well as offering some measure of energy independence for those countries with vast resources of coal who are also net importers of crude oil.

The gasification of coal or a derivative (i.e., char produced from coal) is, essentially, the conversion of coal (by any one of a variety of processes) to produce combustible gases (Speight, 2013). With the rapid increase in the use of coal from the 15th century onwards, it is not surprising that the concept of using coal to produce a flammable gas, especially the use of the water and hot coal, became commonplace. In fact, the production of gas from coal has been a vastly expanding area of coal technology, leading to numerous research and development programs. As a result, the characteristics of rank, mineral matter, particle size, and reaction conditions are all recognized as having a bearing on the outcome of the process; not only in terms of gas yields but also on gas composition and properties. In fact, the products of coal gasification are varied insofar as the gas composition varies with the system employed. Furthermore, it is emphasized that the gas product must be first freed from any pollutants such as particulate matter and sulfur compounds before further use, particularly when the intended use is a water-gas shift or methanation as might be necessary in the coal-to-gas-to-chemicals industry.

In terms of coal use through conversion processes, serious efforts have been made to reduce the environmental footprint left by such processes by the initiation of the Clean Coal Technology Demonstration Program that has laid the foundation for effective technologies now in use that have helped significantly lower emissions of sulfur dioxide (SO_2), nitrogen oxides (NOx), and airborne particulates. The term clean coal technology refers to a new generation of advanced coal utilization technologies that are environmentally cleaner and in many cases more efficient and less costly than the older, and more *conventional* coal-using processes (Speight, 2013). Clean coal technologies offer the potential for a more clean use of coal which will have a direct effect on the goal of the reduction of emissions and process waste into the environment thereby making a positive contribution to the resolution of issues relating to acid rain and global climate change.

3.2.4 Coal Tar Chemicals

The coal carbonization industry was established initially as a means of producing coke (Chapter 16) but a secondary industry emerged (in fact, became necessary) to deal with the secondary or byproducts (namely, gas, ammonia liquor, crude benzole, and tar) produced during carbonization (Table 3.1) (Speight, 2013a).

Coal tar is a byproduct of the carbonization of coal to produce coke and/or natural gas. Physically, coal tar is black or dark brown-colored liquid or a high-viscosity semisolid which is one of the byproducts formed when coal is carbonized (Speight, 2013a). Coal tar usually takes the form of a viscous liquid or semisolid with a naphthalene-like odor. Chemically, coal tar is a complex combination of polycyclic aromatic hydrocarbon (PAH, often represented as PNA as well) derivatives, phenol derivatives, heterocyclic oxygen, sulfur, and nitrogen compound derivatives. Because of its flammable composition coal tar is often used for fire boilers, in order to create heat. Before any heavy oil flows easily they must be heated.

Coal tar, coal tar pitch, and coal tar creosote are very similar mixtures obtained from the distillation of coal tars. The physical and chemical properties of each are similar, although limited data are available for coal tar, and coal tar pitch. By comparison, coal tar creosote is a distillation product of coal tar. They have an oily liquid consistency and range in color from yellowish-dark green to brown. The coal tar creosotes consist of aromatic hydrocarbon derivatives, anthracene, naphthalene, and phenanthrene derivatives. Typically, polycyclic aromatic hydrocarbon derivatives—two-ring naphthalene derivatives and higher condensed ring derivatives—constitute the majority of the creosote mixture. Unlike the coal tar and coal tar creosote, coal tar pitch is the nonvolatile residue produced during the distillation of coal tar. The pitch is a shiny, dark brown to black residue that contains polycyclic aromatic hydrocarbon derivatives and their methyl and polymethyl derivatives, as well as heteronuclear compounds.

As an aside, the nomenclature of the coal tar industry, like that of the petroleum industry (Speight, 2014, 2017), needs refinement and clarification. Almost any black, undefined, semisolid-to-liquid material is popularly, and often incorrectly, described as tar or pitch whether it be a manufactured product or a naturally occurring substance (Chapter 16). However, to be correct and to avoid any ambiguity, use of these terms should be applied with caution. The term "tar" is usually applied to the volatile and nonvolatile soluble products that are produced during the carbonization, or destructive distillation (thermal decomposition with the simultaneous removal of distillate), of various organic materials. By way of further definition, distillation of the tar yields an oil (volatile organic products often referred to as benzole) and a nonvolatile pitch. In addition, the origin of the tar or pitch should be made clear by the use of an appropriate descriptor, i.e., coal tar, wood tar, coal tar pitch, and the like.

Thus, the eventual primary products of the carbonization process (Chapter 16) are coke, coal tar, and crude benzole (which should not be mistaken for benzene although benzene can be isolated from benzole), ammonia liquor, and gas. The benzole fraction contains a variety of compounds, both aromatic and aliphatic in nature and can be conveniently regarded as an analog of petroleum naphtha (Speight, 2014).

TABLE 3.1
Bulk Products (% w/w) from Coal Carbonization

Product % w/w	Low-temperature Carbonization	High-temperature Carbonization
Gas	5	20
Liquids	15	2
Tar	10	3
Coke	70	75

The yield of byproduct tar from a coke oven is, on average, 8.5–9.5 U.S. gallons (32–36 L) per ton of coal carbonized but the yield from a continuous vertical retort is approximately 15.5–19.0 U.S. gallons (60–75 L) per ton of coal carbonized. In low-temperature retorts, the yield of tar varies over the range 19.0–36.0 U.S. gallons (75–135 L) per ton of coal.

Crude coal tar sometimes referred to as crude coke oven tar or simply coal tar is a byproduct collected during the carbonization of coal to make coke. Coke is used as a fuel and as a reducing agent in smelting iron ore in a blast furnace to manufacture steel and in foundry operations. Crude coal tar is a raw material that is further distilled to produce various carbon products, refined tars, and oils used as essential components in the production of aluminum, rubber, concrete, plasticizers, coatings, and specialty chemicals. Crude coal tars have been processed in the United States since Koppers Company completed the first byproduct coke ovens around 1912.

During the distillation of crude coal tar, low-density oil and medium-density oil are removed from the crude coal tar to produce various refined coal tar products. These low-density and medium-density oils represent 20–50% w/w of the crude coal tar depending upon the refined product that is desired. Coal tar contains hundreds of chemical compounds that will have varying amounts of polycyclic aromatic hydrocarbon derivatives depending upon the source.

Refined tar-based coatings have a great advantage over asphalt in that it has better chemical resistance than asphalt coatings. Refined tar-based coatings hold up better under exposures of petroleum oils and inorganic acids. Another outstanding quality of refined tar-based coatings is their extremely low permeability to moisture and there high dielectric resistance, both of which contribute to the corrosion resistance. (Munger, 1984).

Coal tar is a complex mixture and the components range from low-boiling, low molecular weight species, such as benzene, to high molecular weight polynuclear aromatic compounds. Similar classes of chemical compounds occur in the tars, usually with little regard to the method of manufacture, but there are marked variations in the proportions present in the tars due to the type of coal, the type of carbonizing equipment, and the method of recovery (Chapter 16).

Coke oven tar contains relatively low proportions (ca. 3%) of tar acids (phenols), vertical retort tars may contain up to 30% phenolic compounds. Moreover, the phenols in coke oven tars mainly comprise phenol, methyl and poly-methyl phenols (e.g. cresols and xylenols) and naphthols; those in vertical retort tar are mainly xylenols and higher-boiling phenols.

Coke oven tars contain only minor quantities of non-aromatic hydrocarbon derivatives, whilst the vertical retort tars may have up to 6% of paraffinic compounds. Low-temperature tars are more paraffinic and phenolic (as might be expected from relative lack of secondary reactions) than are the continuous vertical retort tars. Coke oven tars are comparatively rich in naphthalene and anthracene and distillation is often the means by which various chemicals can be recovered from these particular products. On the other hand, another objective of primary distillation is to obtain a pitch or refined-tar residue of the desired softening point. If the main outlet for the pitch is as a briquetting (Chapter 17) or electrode binder, primary distillation is aimed at achieving a medium-soft pitch as product or for the production of road asphalt.

In terms of composition, the compounds positively identified as pitch components consist predominantly of condensed polynuclear aromatic hydrocarbon derivatives or heterocyclic compounds containing three to six rings. Some methyl and hydroxyl substituent groups have also been observed, and it is reasonable to assume that vertical retort pitches contain paraffinic constituents in addition (McNeil, 1966). Pitches are often characterized by solvent analysis and many specifications quote limits for the amounts insoluble in certain solvents (Hoiberg, 1966).

Primary distillation of crude tar produces pitch (nonvolatile residue) and several distillate fractions, the amounts and boiling ranges of which are influenced by the nature of the crude tar (which depends upon the coal feedstock) and the processing conditions. For example, in the case of the tar from continuous vertical retorts, the objective is to concentrate the tar acids (phenol, cresols, and xylenols) into carbolic oil fractions. On the other hand, the objective with coke oven tar is to concentrate the naphthalene and anthracene components into naphthalene oil and anthracene oil, respectively.

The products of tar distillation can be divided into refined products, made by the further processing of the fractions, and bulk products which are pitch, creosote, and their blends. Coal tar low-boiling oil, or crude benzole, is similar in chemical composition to the crude benzole recovered from the carbonization gases at gas works and in coke oven plants. The main components are benzene, toluene, and xylene(s) with minor quantities of aromatic hydrocarbon derivatives, paraffins, naphthenes (cyclic aliphatic compounds), phenols, as well as sulfur and nitrogen compounds.

The first step in refining benzole is steam distillation is employed to remove compounds boiling below benzene. Low-boiling naphtha and high-boiling naphtha are the mixtures obtained when the 150°C–200°C (300°F–390°F) fraction, after removal of tar acids and tar bases, is fractionated. These naphtha fractions are used as solvents. To obtain pure products, the benzole can be distilled to yield a fraction containing benzene, toluene, and xylene(s). Benzene is used in the manufacture of numerous products including nylon, gammexane, polystyrene, phenol, nitrobenzene, and aniline. On the other hand, toluene is a starting material in the preparation of saccharin, trinitrotoluene, and polyurethane foams. The xylenes present in the low-boiling oil are not always separated into the individual pure isomers since xylene mixtures can be marketed as specialty solvents. Higher-boiling fractions of the distillate from the tar contain pyridine bases, naphtha, and coumarone resins. Other tar bases occur in the higher-boiling range and these are mainly quinoline, isoquinoline, and quinaldine.

Pyridine has long been used as a solvent, in the production of rubber chemicals, textile water-repellant agents and in the synthesis of drugs. The derivatives 2-benzylpyridine and 2- aminopyridine, are used in the preparation of antihistamines. Another market for pyridine is in the manufacture of the nonpersistent herbicides diquat and paraquat.

Alpha-picoline (C_6H_7N, 2-picoline; 2-methylpryridine) is used for the production of 2-vinylpyridine which, when copolymerized with 1,4-butadiene ($CH_2=CHCH=CH_2$) and styrene ($C_6H_5CH=CH_2$), produces a used as a latex adhesive which is used in the manufacture of automobile tires.

2-methylpyridine

Other uses are in the preparation of 2-β-methoxyethyl pyridine (known as Promintic, an anthelmintic for cattle) and in the synthesis of a 2-picoline quaternary compound (Amprolium) which is used against coccidiosis in young poultry. Beta-picoline (3-picoline; 3-methylpryridine) can be oxidized to nicotinic acid, which with the amide form (nicotinamide), belongs to the vitamin B complex; both products are widely used to fortify human and animal diets. γ-Picoline (4-picoline, 4-methylpyridine) is an intermediate in the manufacture of isonicotinic acid hydrazide (Isoniazide) which is a tuberculostatic drug. The 2,6-Lutidine (2,6-dimethylpyridine) can be converted to dipicolinic acid, which is used as a stabilizer for hydrogen peroxide and peracetic acid.

2,6-Lutidine

The tar-acid free and tar-base free coke oven naphtha can be fractionated to give a narrow boiling fraction (170°C–185°C; 340°F–365°F) containing coumarone and indene. This is treated with strong sulfuric acid to remove unsaturated components and is then washed and redistilled. The concentrate is heated with a catalyst (such as a boron fluoride/phenol complex) to polymerize the indene and part of the coumarone. Unreacted oil is distilled off and the resins obtained vary from pale amber to dark brown in color. They are used in the production of flooring tiles and in paints and polishes.

Naphthalene and several tar acids are the important products extracted from volatile oils from coal tar. It is necessary to first extract the phenolic compounds from the oils and then to process the phenol-depleted oils for naphthalene recovery. Tar acids are produced by extraction of the oils with aqueous caustic soda at a temperature sufficient to prevent naphthalene from crystallizing. The phenols react with the sodium hydroxide to give the corresponding sodium salts as an aqueous extract known variously as crude sodium phenate, sodium phenolate, sodium carbolate, or sodium cresylate. The extract is separated from the phenol-free oils which are then taken for naphthalene recovery.

Phenol (C_6H_5OH) is a key industrial chemical; however, the output of phenol from coal tar is exceeded by that of synthetic phenol. Phenol is used for the production of phenol-formaldehyde resins, while other important uses in the plastic field include the production of polyamides such as nylon, of epoxy resins and polycarbonates based on bisphenol A and of oil-soluble resins from p-t-butyl and p-octyl phenols. Phenol is used in the manufacture of pentachlorophenol which is used as a fungicide and in timber preservation. Aspirin and many other pharmaceuticals, certain detergents, and tanning agents are all derived from phenol, and another important use is in the manufacture of 2,4-dichlorophenoxyacetic acid (2,4-D) which is a selective weed killer.

2,4-Dichlorophenoxyacetic acid

Orthocresol has been used predominantly for the manufacture of the selective weed killers: 4-chloro-2-methyl-phenoxyacetic acid (MCPA) and the corresponding propionic acid (MCPP) and the butyric acid (MCPB) as well as 2,4-dinitro-o-cresol (DNOC), a general herbicide/ insecticide.

4-chloro-2-methyl-phenoxyacetic acid (MCPA)

2,4-dinitro-o-cresol (DNOC)

Paracresol (p-$HOC_6H_4CH_3$) has been used widely for the manufacture of BHT (2,6-ditertiarybutyl-4-hydroxytoluene), an antioxidant.

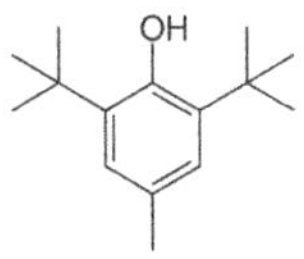

2,6-ditertiarybutyl-4-hydroxytoluene

Metacresol and paracresol mixtures are used in the production of phenoplasts, tritolyl phosphate plasticizers and petroleum additives.

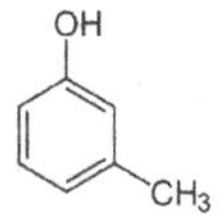

m-cresol

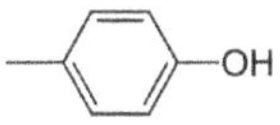

p-cresol

Other outlets for cresylic acids are as agents for froth flotation, metal degreasing, as solvents for wire-coating resins, antioxidants, cutting oils, nonionic detergents, and disinfectants.

Naphthalene is probably the most abundant component in high-temperature coal tars. The primary fractionation of the crude tar concentrates the naphthalene into oils which, in the case of coke oven tar, contain the majority (75–90% w/w) of the total naphthalene. After separation, naphthalene can be oxidized to produce phthalic anhydride which is used in the manufacture of alkyd and glyptal resins and plasticizers for polyvinyl chloride and other plastics.

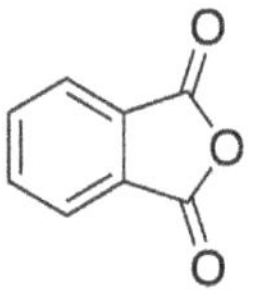

Phthalic anhydride

The main chemical extracted on the commercial scale from the higher-boiling oils (b.p. 250°C, 480°F) is crude anthracene. The majority of the crude anthracene is used in the manufacture of dyes after purification and oxidation to anthraquinone.

Coal tar creosote is the residual distillate oils obtained when the valuable components, such as naphthalene, anthracene, tar acids, and tar bases have been removed from the corresponding fractions (Figure 3.1). It is a brownish-black/yellowish-dark green oily liquid with a characteristic sharp odor, obtained by the fractional distillation of crude coal tars. The approximate distillation range is 200°C–400°C (390°F–750°F). The chemical composition of creosotes is influenced by the origin of the coal and also by the nature of the distilling process; as a result, the creosote components are rarely consistent in their type and concentration.

Major uses for creosotes have been as a timber preservative, as fluxing oils for pitch and bitumen, and in the manufacture of lampblack and carbon black. However, the use of creosote as a timber preservative has recently come under close scrutiny, as have many other ill-defined products of coal processing. Issues related to the seepage of such complex chemical mixtures into the surrounding environment have brought an awareness of the potential environmental and

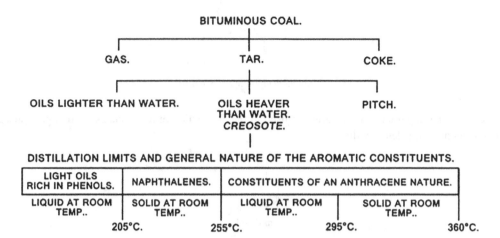

FIGURE 3.1   Representation of the production and composition of coal tar creosote.

health hazards related to the use of such chemicals. Stringent testing is now required before such chemicals can be used.

As a corollary to this section, where the emphasis has been on the production of bulk chemicals from coal, a tendency-to-be-forgotten item must also be included. That is the mineral ash from coal processes. Coal minerals are a very important part of the coal matrix and offer the potential for the recovery of valuable inorganic materials (Speight, 2013a). However, there is another aspect of the mineral content of coal that must be addressed, and that relates to the use of the ash as materials for roadbed stabilization, landfill cover, cementing (due to the content of pozzolanic materials), and wall construction to mention only a few.

3.3   OIL SHALE

Oil shale represents a large and mostly untapped hydrocarbon resource. Like tar sand (*oil sand* in Canada) and coal, oil shale is considered unconventional because oil cannot be produced directly from the resource by sinking a well and pumping. Oil has to be produced by thermal decomposition of the organic matter (kerogen) in the shale. The organic material contained in the shale is called *kerogen*, a solid material intimately bound within the mineral matrix. However, oil shale does not contain any oil—this must be produced by a process in which the kerogen is thermally decomposed (cracked) to produce the liquid product (shale oil) (Scouten, 1990; Lee, 1991; Lee et al., 2007; Speight, 2008). Compared to crude oil, shale oil obtained by retorting of oil shale is characterized by wide boiling range and by large concentrations of heteroelements and also by high content of oxygen-, nitrogen-, or sulfur-containing compounds.

3.3.1   SHALE OIL PRODUCTION

Shale oil is produced from oil shale by the thermal decomposition of the kerogen component of oil shale. Oil shale must be heated to temperatures between 400°C and 500°C (750°F–930°F). This heating process is necessary to convert the embedded sediments to kerogen oil and combustible gases. Generally, with solid fossil fuels, the yield of the volatile products depends mainly on the hydrogen content in the convertible solid fuel. Thus, compared with coal, oil shale kerogen contains more hydrogen and can produce relatively more oil and gas when thermally decomposed (Speight, 2008, 2012, 2013). From the standpoint of shale oil as a substitute for petroleum products, the composition is of great importance.

The thermal processing of oil shale to oil has quite a long history and various facilities and technologies have been used including mining of the shale followed by thermal processing as well as *in situ* decomposition of the shale (Speight, 2008, 2012). In principle, there are two ways of accomplishing the thermal decomposition of the kerogen in the shale: (i) low-temperature processing—semicoking or retorting—by heating the oil shale up to about 500°C (930°F), and (ii) high-temperature processing—coking—heating up to 1,000°C–1,200°C (1,830°F–2,190°F). A high yield deposit of oil shale will yield 25 gallons of oil per ton of oil shale.

In the mining-thermal processing option (*ex situ* production), oil shale is mined, crushed, and then subjected to thermal processing at the surface in an oil shale retort. Both pyrolysis and combustion have been used to treat oil shale in a surface retort. In the second option (*in situ* production), the shale is left in place and the retorting (e.g., heating) of the shale occurs in the ground. Generally, surface processing consists of three major steps: (i) oil shale mining and ore preparation, (ii) pyrolysis of oil shale to produce kerogen oil, and (iii) processing kerogen oil to produce refinery feedstock and high-value chemicals. For deeper, thicker deposits, not as amenable to surface or deep-mining methods, shale oil can be produced by *in situ* technology. *In situ* processes minimize, or in the case of true *in situ*, eliminate the need for mining and surface pyrolysis, by heating the resource in its natural depositional setting.

Depending on the depth and other characteristics of the target oil shale deposits, either surface mining or underground mining methods may be used. Each method, in turn, can be further categorized according to the method of heating. Another way in which the various retorting processes differ is the manner by which heat is provided to the shale by hot gas—(i) by a solid heat carrier or (ii) by conduction through a heated wall. After mining, the oil shale is transported to a facility for retorting after which the oil must be upgraded by further processing before it can be sent to a refinery, and the spent shale must be disposed, often by putting it back into the mine. Eventually, the mined land is reclaimed. Both mining and processing of oil shale involve a variety of *environmental impacts*, such as global warming and greenhouse gas emissions, disturbance of mined land, disposal of spent shale, use of water resources, and impacts on air and water quality.

3.3.2 Shale Oil Properties

Shale oil is a synthetic crude oil produced by retorting oil shale and is the pyrolysis product of the organic matter (kerogen) contained in oil shale. The raw shale oil produced from retorting oil shale can vary in properties and composition (Scouten, 1990; Lee, 1991; Lee et al., 2007; Speight, 2008). Compared with petroleum, shale oil is high in nitrogen and oxygen compounds and a higher specific gravity—on the order of 0.9–1.0 owing to the presence of high-boiling nitrogen-, sulfur-, and oxygen-containing compounds. Shale oil also has a relatively high pour point and small quantities of arsenic and iron are present.

The chemical potential of oil shale as retort fuel to produce shale oil and from that liquid fuel and specialty chemicals has been used so far to a relatively small extent. Using stepwise cracking, various liquid fuels have been produced and even exported before World War II. At the same time, shale oils possess molecular structures of interest to the specialty chemicals industry and also a number of nonfuel specialty products have been marketed based on functional group, broad range concentrate, or even pure compound values.

Shale oil (produced from kerogen-containing shale rock) is a complex mixture of hydrocarbon derivatives, and it is characterized using bulk properties of the oil. Shale oil usually contains large quantities of olefin derivatives and aromatic hydrocarbon derivatives as well as significant quantities of heteroatom compounds (nitrogen-containing compounds, oxygen-containing compounds, and sulfur-containing compounds). A typical shale oil composition includes: nitrogen 1.5–2% w/w, oxygen 0.5–1% w/w, and sulfur 0.15–1% w/w as well as mineral particles and metal-containing compounds (Scouten, 1990; Lee, 1991; Lee, 1991; Lee et al., 2007; Speight, 2008). Generally, the oil is less fluid than crude oil, which is reflected in the pour point that is in the order of 24°C–27°C

(75°F–81°F), while conventional crude oil has a pour point in the order of −60°C to +30°C (−76°F to +86°F), which affects the ability of shale oil to be transported using unheated pipelines. Shale oil also contains polycyclic aromatic hydrocarbon derivatives.

Based on large quantities of oxygen-containing compounds in the high-boiling fraction, asphalt-blending material, road asphalt, construction mastics, anticorrosion oils, rubber softeners, benzene, and toluene for production of benzoic acid as well as solvent mixtures on pyrolysis of lower-boiling fractions of shale oil are produced. Higher-boiling (mid-distillate) shale oil fractions having anti-septic properties are used to produce effective oil for the impregnation of wood as a major shale oil-derived specialty product. Water-soluble phenols are selectively extracted from shale oil, fractionated and crystallized for production of pure 5-methylresorcinol and other alkyl resorcinol derivatives and high-value intermediates to produce tanning agents, epoxy resins and adhesives, diphenyl ketone and phenol-formaldehyde adhesive resins, rubber modifiers, chemicals, and pesticides. Some conventional products such as coke and various distillate fuels are produced from shale oil as byproducts.

However, the presence of the polar constituents (containing nitrogen and oxygen functions, sulfur compounds are also issues worthy of consideration) can cause shale oil to be incompatible with conventional petroleum feedstocks and petroleum products (Speight, 2014). As a result, particular care must be taken to ensure that all the functions that cause such incompatibility are removed from the shale oil before it is blended with a conventional petroleum liquid.

3.3.2.1 Hydrocarbon Products

The fundamental structure of the organic matter in oil shale gives rise to significant quantities of waxes consisting of long normal alkanes and the alkanes are distributed throughout the raw shale oil. However, the composition of shale oil depends on the shale from which it was obtained as well as on the retorting method by which it was produced (Scouten, 1990; Lee, 1991; Lee et al., 2007; Speight, 2008). As compared with petroleum crude, shale oil is high-boiling, viscous, and is high in nitrogen and oxygen compounds.

Retorting processes, which use flash pyrolysis, produce more fragments containing high molecular weight and multi-ring aromatic structures. Processes that use slower heating conditions, with greater reaction times at low temperature 300°C–400°C (570°F–750°F), tend to produce higher concentrations of n-alkanes. Naphthene-aromatic compounds of intermediate boiling range (such as 200°C–400°C, 390°F–750°F) also tend to be formed with the slower heating processes.

Saturated hydrocarbon derivatives in the shale oil include n-alkane derivatives, iso-alkane derivatives, and cycloalkane derivatives, and the alkene derivatives consist of n-alkene derivatives, iso-alkene derivatives, and cycloalkene derivatives, while the main components of aromatic derivatives are monocyclic, bicyclic, and tricyclic aromatic derivatives and their alkyl-substituted homologues. There is a variation of the distribution of saturated hydrocarbon derivatives, alkene derivatives, and aromatic derivatives in the different boiling ranges of the shale oil product. Saturated hydrocarbon derivatives in the shale oil increase and the aromatic derivatives increase slightly with a rise in boiling range, while alkene derivatives decrease with a rise in boiling range.

A typical Green River shale oil contains 40% w/w hydrocarbon derivatives and 60% w/w heteroatomic organic compounds, which contain nitrogen, sulfur, and oxygen. The nitrogen occurs in ring compounds with nitrogen in the ring, e.g., pyridines, pyridines, pyrroles as well as in nitriles, and typically comprises 60% w/w of the heteroatomic organic components. Another 10% w/w of these components contains organically bound sulfur compounds, which exists in thiophenes as well as sulfides and disulfides. The remaining 30% w/w consists of oxygen-containing compounds, which occur as phenols and carboxylic acids.

Shale oil not only contains a large variety of hydrocarbon compounds (Table 3.2), but also has high nitrogen content compared to a nitrogen content of 0.2–0.3% w/w for a typical petroleum (Scouten, 1990; Lee, 1991; Lee et al., 2007; Speight, 2008). In addition, shale oil also has a high olefin and diolefin content—constituents, which are not present in petroleum and which require

TABLE 3.2

Major Compound Types in Shale Oil

Saturate	Heteroatom systems
paraffin	benzothiophene
cycloparaffin	dibenzothiophene
Olefin	phenol
Aromatic	carbazole
Benzene	pyridine
indan	quinoline
tetralin	nitrile
naphthalene	ketone
biphenyl	pyrrole
phenanthrene	
chrysene	

attention during processing due to their tendency to polymerize and form gums and sediments (fuel line deposits). It is the presence of these olefin derivatives and diolefin derivatives, in conjunction with high nitrogen content, which gives shale oil the characteristic difficulty in refining. Crude shale oil also contains appreciable amounts of arsenic, iron, and nickel that interfere with refining.

Other characteristic properties of shale oils are: (i) high levels of aromatic compounds, deleterious to kerosene and diesel fractions, (ii) low hydrogen-to-carbon ratio, (iii) low sulfur levels, compared with most crudes available in the world (though for some shale oils from the retorting of marine oil shale, high sulfur compounds are present), (iv) suspended solids (finely divided rock) typically due to entrainment of the rock in the oil vapor during retorting, and (v) low-to-moderate levels of metals. Thus, because of the characteristics of shale oil, further processes are needed to improve the properties of shale oil products. The basic unit operations in the oil refining are distillation, coking, hydrotreating, hydrocracking, catalytic cracking, and reforming. The process selected will largely depend on the availability of equipment and the individual economics of the particular refinery.

Although the content of asphaltene constituents and/or resin constituents in shale oil is low— shale oil being a distillate product—asphaltene constituents in shale oil may be unique since, in shale oil, it is high heteroatomic content that causes precipitation as an asphaltene fraction rather than high molecular weight—for example, the hydroxy-pyridine derivatives are insoluble in low molecular weight alkane solvents. The polarity of the nitrogen polycyclic aromatic constituents may also explain the specific properties of emulsification of water and metal complexes.

3.3.2.2 Nitrogen-Containing Compounds

Nitrogen compounds in shale oil render technological difficulties in the downstream processing of shale oil, in particular, poisoning of the refining catalysts. Such nitrogen compounds are all originated from the oil shale and the amount and types depend heavily on the geochemistry of oil shale deposits. Since direct analysis and determination of molecular forms of nitrogen-containing compounds in oil shale rock is very difficult, the analysis of shale oil that is extracted by retorting processes provides valuable information regarding the organo-nitrogen species in the oil shale.

The nitrogen content in the shale oil is relatively higher than in natural crude oil (Scouten, 1990; Lee, 1991; Lee et al., 2007; Speight, 2008). The nitrogen-containing compounds identified in shale oils can be classified as basic, weakly basic, and nonbasic. The basic nitrogen compounds in shale oils are pyridine, quinoline, acridine, amine, and their alkyl-substituted derivatives, the weakly basic ones are pyrrole, indole, carbazole and their derivatives, and the nitrile and amide homologues are the nonbasic constituents.

Most of these compounds are useful chemicals (Scouten, 1990; Lee, 1991; Lee et al., 2007; Speight, 2008), although some of them are believed to affect the stability of shale oil. Generally, basic nitrogen accounts for about one-half of the total nitrogen, and is evenly distributed in the different boiling point fractions. Nitrogen compounds occur throughout the boiling ranges of the shale oil, but have a decided tendency to exist in high-boiling point fractions. Pyrrole-type nitrogen increases with a rise in the boiling point of the shale oil fractions. Porphyrins may occur in the high-boiling point fraction of the shale oil.

Of the nitrogen-containing compounds in the low-boiling (<350°C <660°F) shale oil fraction, the majority contain one nitrogen atom. Benzoquinoline derivatives, principally acridine and alkyl-substituted homologues, could not be present significantly in the lower-boiling shale oil fractions because the boiling point of benzoquinoline and its alkyl-substituted homologues is higher than 350°C (660°F).

Organic nitrogen-containing compounds in the shale oil poison the catalysts in different catalytic processes. They also contribute to stability problems during storage of shale oil products since they induce polymerization processes, which cause an increase in the viscosity and give rise to the odor and color of the shale oil product. The high nitrogen content of shale oil could contribute to the surface and colloidal nature of shale oil, which forms emulsions with water.

3.3.2.3 Oxygen-Containing Compounds

The oxygen content of shale oil is much higher than in natural petroleum. Low molecular weight oxygen compounds in shale oil are mainly phenolic constituents—carboxylic acids and nonacidic oxygen compounds such as ketones are also present. Low molecular phenolic compounds are the main acidic oxygen-containing compounds in the low-boiling fraction of the shale oil and are usually derivatives of phenol, such as cresol and poly-methylated phenol derivatives.

The oxygen content of petroleum is typically in the order of 0.1–1.0% w/w, whereas the oxygen contents in shale oils are much higher and vary with different shale oil (Scouten, 1990; Lee, 1991; Lee et al., 2007). In addition, the oxygen content varies in different boiling point fractions of the shale oil. In general, it increases as the boiling point increases, and most of the oxygen atoms are concentrated in the high-boiling point fraction.

Other oxygen-containing constituents of shale oil include small amounts of carboxylic acids and nonacidic oxygen-containing compounds with a carbonyl functional group such as ketones, aldehydes, esters, and amides are also present in the <350°C (<660°F) fraction of shale oil. Ketones in the shale oil mainly exist as 2- and 3-alkanones. Other oxygen-containing compounds in the low-boiling (<350°C, <660°F) fraction include alcohols, naphthol, and ether constituents.

3.3.2.4 Sulfur-Containing Compounds

Sulfur compounds in the shale oils include thiols, sulfides, thiophenes, and other miscellaneous sulfur compounds. Elemental sulfur is found in some crude shale oil but is absent in others.

Generally, the sulfur content of oil-shale distillates is comparable in weight percentage to crude oil (Scouten, 1990; Lee, 1991; Lee et al., 2007; Speight, 2008). Refiners will be able to meet the current 500 ppm requirement by increasing the existing capacity of their hydrotreatment units and adding new units. However, refineries may face difficulty in treating diesel to below 500 ppm. The remaining sulfur is bound in multi-ring thiophene-type compounds that prove difficult to hydrotreat because the molecular ring structure attaches the sulfur on two sides and, if alkyl groups are present, provides steric protection for the sulfur atom. Although these compounds occur throughout the range of petroleum distillates, they are more concentrated in the residuum.

3.4 BIOMASS

Increasing attention has been (and is being) given to the possibility of utilizing photosynthetically active plants as natural solar energy-capturing devices with the subsequent conversion of available plant energy into useful fuels or chemical feedstocks (Table 3.3) (Metzger, 2006; Biddy et al., 2016; Wu et al., 2016). Acquisition of biological raw materials for energy capture follows

TABLE 3.3
Examples of Chemicals Produced from Bio-Sources

Chemical	Comment
1,3-Butadiene	The building block for the production of polybutadiene and styrene-rubber and butadiene rubber, currently produced from petroleum as a byproduct of ethylene manufacturing; can be produced through multiple biomass conversion strategies for the production of a direct renewable butadiene replacement.
1,4-Butanediol	A building block for the production of polymers, solvents, and specialty chemicals; bio-derived butanediol is being produced on a commercial scale utilizing commodity sugars; can be produced by the conversion of succinic acid to 1,4-butanediol.
Ethyl lactate	A biodegradable solvent produced by the esterification of ethanol and lactic acid; primary use for ethyl lactate is as an industrial product—the properties and performance meet or exceed those of traditional solvents such as toluene, methyl ethyl ketone, and *N*-methyl-pyrrolidone in many applications; the starting materials used to make ethyl lactate, lactic acid, and ethanol, have a high potential to be made from lignocellulosic sugars.
Fatty alcohols	Also called detergent alcohols, are linear alcohols of 12 or more carbons, used primarily to produce anionic and nonionic surfactants for household cleaners, personal care; derivatized by ethoxylation, sulfation, or sulfonation before use; can be produced from tallow, vegetable oils, or petroleum; also have the potential to be produced from renewable sources by autotrophic and heterotrophic algae, or by the microbial fermentation of carbohydrates.
Furfural	A heterocyclic aldehyde, produced by the dehydration of xylose, a monosaccharide often found in large quantities in the hemicellulose fraction of lignocellulosic biomass; any material containing a large amount of pentose (five-carbon) sugars, such as arabinose and xylose, can serve as a raw material for furfural production; converted to furfuryl alcohol which is used for the production of foundry resins; the anticorrosion properties of furfuryl alcohol is useful in the manufacture of furan fiber-reinforced plastics for piping; a broad spectrum of industrial applications, such as the production of plastics, pharmaceuticals, agrochemical products, and nonpetroleum-derived chemicals; not produced from fossil feedstocks; may be for conversion to jet and diesel fuel blend stocks.
Glycerin	A polyhydric alcohol and is a main component of triglycerides found in animal fats and vegetable oil. The word "glycerin" generally applies to commercial products containing mostly glycerol; the word "glycerol" most often refers specifically to the chemical compound 1,2,3-propanetriol and to the anhydrous content in a glycerin product or in a formulation; glycerin is the main byproduct of biodiesel production. It is also generated in the oleochemical industry during soap production and is produced synthetically from propyl; biodiesel and soap production accounts for most current glycerin production; therefore, the overall supply of glycerin is driven primarily by the demand for these products; a feedstock for conversion to more valuable products, such as epichlorohydrin and succinic acid; emerging uses include animal feed and marine fuel.
Isoprene	The building block for polyisoprene rubber, styrene co-polymers, and butyl rubber; produced by aerobic bioconversion of carbohydrates.
Lactic acid	An alpha-hydroxy acid with dual functional groups; most frequently occurring carboxylic acid in nature; produced by microbial fermentation of carbohydrates; used for applications in food, pharmaceuticals, personal care products, industrial uses, and polymers; polylactic acid has gained popularity for use in food packaging, disposable tableware, shrink wrap, and 3-D printers.
1,3-Propanediol	A linear aliphatic diol, which makes it a useful chemical building block; can be used for a variety of applications including polymers, personal care products, solvents, and lubricants; also used as a component in poly trimethylene terephthalate polymers which are used in textiles and fibers.

(Continued)

TABLE 3.3 (*Continued*)
Examples of Chemicals Produced from Bio-Sources

Chemical	Comment
Propylene glycol	Also known as 1,2-propanediol, propane-I,2-diol, and mono-propylene glycol; a viscous, colorless, odorless liquid that is nonvolatile at room temperature and is completely soluble in water; used in the production of consumer products such as antiperspirants, suntan lotions, eye drops, food flavorings, and bulking agent in oral, and topical drugs; industrial grade propylene glycol is used in the production of unsaturated polyester resins for end use markets such as residential and commercial construction, marine vessels, passenger vehicles, and consumer appliances; also used as an engine coolant and antifreeze in place of ethylene glycol, and in the airline industry as an airplane and runway de-icing agent; serves as a solvent, enzyme stabilizer, clarifying agent, and diluent; can be produced by hydrogenolysis of glycerin over mixed-metal catalysts, or hydrocracking of sorbitol.
Succinic acid	A dicarboxylic acid that can be produced from biomass and used as a precursor for the synthesis of high-value products derived from renewable resources, including commodity chemicals, polymers, surfactants, and solvents.
p-Xylene	Used to produce both terephthalic acid and dimethyl terephthalate which are raw materials for the production of polyethylene terephthalate bottles; can be produced via the traditional biochemical fermentation process followed by upgrading, thermochemical pyrolysis routes, and hybrid thermochemical/biochemical strategies of catalytic upgrading of sugars.

three main approaches: (i) purposeful cultivation of so-called clergy crops, (ii) harvesting natural vegetation, and (iii) collection of agricultural wastes. Thus, in the context of this book, *biomass* refers to (i) energy crops grown specifically to be used as fuel, such as fast-growing trees or switch grass; (ii) agricultural residues and byproducts, such as straw, sugarcane fiber, and rice hulls; and (iii) residues from forestry, construction, and other wood-processing industries (Detroy, 1981; Vasudevan et al., 2005; Wright et al., 2006; Speight, 2008). It is the term used to describe any material of recent biological origin, including plant materials such as trees, grasses, agricultural crops, and even animal manure that can be converted to a variety of feedstocks for the production of petrochemical products through primary and/or secondary conversion methods (Table 3.4).

Biomass is a renewable energy source, unlike the fossil fuel resources (natural gas, crude oil, and coal) but, like the fossil fuels, biomass is a form of stored solar energy (Speight, 2008). The energy of the sun is captured through the process of photosynthesis in growing plants. One advantage of biofuel in comparison to most other fuel types is that it is biodegradable, and thus relatively harmless to the environment if spilled.

TABLE 3.4
Methods for the Conversion of Biomass to Petrochemical Feedstocks

Feedstock	Conversion Type	Primary Method	Product	Secondary Method
Biomass	Biological conversion	Fermentation	Methane	
			Sugar	
			Protein	
	Thermochemical conversion	Pyrolysis	Gas	
			Oil	Gasification
			Char	Gasification
		Hydrocarbonization	Gas	Gasification
			Oil coke	Gasification

In order to produce fuels and chemicals, several currently available processes rely on entirely breaking down complex molecules before building up the desired compounds, such as the case with syngas production, to form alkanes and alcohols. While biomass can also be converted into syngas, an alternative and complimentary approach strategically converts biomass into chemical building blocks that retain features (e.g., electrophilic or nucleophilic character) that can be exploited in further manipulations. Such platform chemicals can be generated through either chemical routes or biological processes.

A major issue in the use of biomass is one of *feedstock diversity*. Biomass-based feedstock materials used in producing chemicals can be obtained from a large variety of sources. If considered individually, the number of potential renewable feedstocks can be overwhelming, but they tend to fall into three simple categories: (i) waste materials such as food processing wastes, (ii) dedicated feedstock crops which includes and short rotation woody crops or herbaceous energy crops such as perennials or forage crops and (iii) conventional food crops such as corn and wheat. In addition, these raw materials are composed of several similar chemical constituents, i.e., carbohydrates, proteins, lipids, lignin, and minerals.

Thermal or chemical processing of these materials is typically accomplished by novel separation and conversion methodology leading to chemicals similar to those from conventional petrochemical starting materials. Bioprocesses focus on microbiological conversion of fermentable sugars that are derived from these materials by thermal, chemical, or enzymatic means to commodity and specialty chemicals (Detroy, 1981). Thus, in choosing a feedstock for a given product, it is important not to be diverted by semantic differences that arise due to its current usage.

Biomass components, which are generally present in minor amounts, include triglycerides, sterols, alkaloids, resins, terpenes, terpenoids, and waxes. This includes everything from *primary sources* of crops and residues harvested/collected directly from the land to *secondary sources* such as sawmill residuals to *tertiary sources* of postconsumer residuals that often end up in landfills. A *fourth source*, although not usually categorized as such, includes the gases that result from anaerobic digestion of animal manure or organic waste in landfills (Wright et al., 2006; Speight, 2008).

Most present day production and use of biomass for energy is carried out in a very unsustainable manner with a great many negative environmental consequences. If biomass is to supply a greater proportion of the world's energy needs in the future, the challenge will be to produce biomass and to convert and use it without harming the natural environment. Technologies and processes exist today which, if used properly, make biomass-based fuels less harmful to the environment than fossil fuels. Applying these technologies and processes on a site-specific basis in order to minimize negative environmental impacts is a prerequisite for sustainable use of biomass energy in the future. These technologies have the ability to be coordinated in a biorefinery.

A biorefinery (Speight, 2011c) is the means by which biomass can be converted to other products—in the current context the other products are *biofuels* which have the potential to replace certain petroleum-derived fuels. In theory, a biorefinery can use all kinds of *biomass,* including wood and dedicated agricultural crops, plant- and animal-derived waste, municipal waste, and aquatic biomass (algae, seaweeds). A biorefinery produces a spectrum of marketable products and energy including intermediate and final products: food, feed, materials, chemicals, fuels, power, and/or heat. However, the differences in the various biomass feedstocks may dictate that a biorefinery be constructed and operated on the basis of the chemical composition of the feedstock and the mean by which the feedstock is to be processed.

3.4.1 Biomass Feedstocks

More generally, biomass feedstocks are recognized or classified by the specific plant content of the feedstock or the manner in which the feedstocks is produced.

For example, *primary biomass feedstocks* are thus primary biomass that is harvested or collected from the field or forest where it is grown. Examples of primary biomass feedstocks currently being used for bioenergy include grains and oilseed crops used for transportation fuel production, plus some crop residues (such as orchard trimmings and nut hulls) and some residues from logging and forest operations that are currently used for heat and power production.

Secondary biomass feedstocks differ from *primary biomass feedstocks* in that the secondary feedstocks are a byproduct of processing of the primary feedstocks. By *processing,* it is meant that there is substantial physical or chemical breakdown of the primary biomass and production of byproducts; *processors* may be factories or animals. Field processes such as harvesting, bundling, chipping, or pressing do not cause a biomass resource that was produced by photosynthesis (e.g., tree tops and limbs) to be classified as secondary biomass. Specific examples of secondary biomass includes sawdust from sawmills, black liquor (which is a byproduct of paper making), and cheese whey (which is a byproduct of cheese-making processes). Manures from concentrated animal feeding operations are collectable secondary biomass resources. Vegetable oils used for biodiesel that are derived directly from the processing of oilseeds for various uses are also a secondary biomass resource.

Tertiary biomass feedstock includes postconsumer residues and wastes, such as fats, greases, oils, construction and demolition wood debris, other waste wood from the urban environments, as well as packaging wastes, municipal solid wastes, and landfill gases. A category *other wood waste from the urban environment* includes trimmings from urban trees, which technically fits the definition of primary biomass. However, because this material is normally handled as a waste stream along with other postconsumer wastes from urban environments (and included in those statistics), it makes the most sense to consider it to be part of the tertiary biomass stream.

Tertiary biomass often includes *fats and greases*, which are byproducts of the reduction of animal biomass into component parts, since most fats and greases, and some oils, are not available for bioenergy use until after they become a postconsumer waste stream. Vegetable oils derived from processing of plant components and used directly for bioenergy (e.g., soybean oil used in biodiesel) would be a secondary biomass resource, though amounts being used for bioenergy are most likely to be tracked together with fats, greases, and waste oils.

One aspect of designing a refinery for any feedstocks is the composition of the feedstocks. For example, a heavy oil refinery would differ somewhat from a conventional refinery and a refinery for tar sand bitumen would be significantly different to both (Speight, 2008, 2014, 2017). Furthermore, the composition of biomass is variable (Speight, 2008) which is reflected in the range of heat value (heat content, calorific value) of biomass, which is somewhat lesser than for coal and much lower than the heat value for petroleum, generally falling in the range 6,000–8,500 Btu/lb (Speight, 2008). Moisture content is probably the most important determinant of heating value. Air-dried biomass typically has about 15%–20% moisture, whereas the moisture content for oven-dried biomass is around 0%. Moisture content is also an important characteristic of coals, varying in the range of 2%–30%. However, the bulk density (and hence energy density) of most biomass feedstocks is generally low, even after densification, about 10% and 40% of the bulk density of most fossil fuels.

The production of fuels and chemicals from renewable plant-based feedstocks utilizing state-of-the-art conversion technologies presents an opportunity to maintain competitive advantage and contribute to the attainment of national environmental targets. Bioprocessing routes have a number of compelling advantages over conventional petrochemicals production; however, it is only in the last decade that rapid progress in biotechnology has facilitated the commercialization of a number of plant-based chemical processes.

Plants offer a unique and diverse feedstock for chemicals and the production of *biofuels* from biomass requires some knowledge of the chemistry of biomass, the chemistry of the individual constituents of biomass, and the chemical means by which the biomass can be converted to fuel. It is widely recognized that further significant production of plant-based chemicals will only be economically viable in highly integrated and efficient production complexes producing a diverse range of chemical products. This biorefinery concept is analogous to conventional oil refineries and petrochemical complexes that have evolved over many years to maximize process synergies, energy integration, and feedstock utilization to drive down production costs.

In addition, the specific components of plants such as carbohydrates, vegetable oils, plant fiber, and complex organic molecules known as primary and secondary metabolites can be utilized to produce a range of valuable monomers, chemical intermediates, pharmaceuticals, and materials.

3.4.1.1 Carbohydrates

Plants capture solar energy as fixed carbon during which carbon dioxide is converted to water and sugars $(CH_2O)_x$:

$$CO_2 + H_2O \rightarrow (CH_2O)_x + O_2.$$

The sugars produced are stored in three types of polymeric macromolecules: (i) starch, (ii) cellulose, and (iii) hemicellulose.

In general sugar polymers such as cellulose and starch can be readily broken down to their constituent monomers by hydrolysis, preparatory to conversion to ethanol or other chemicals (Vasudevan et al., 2005; Speight, 2008). In contrast, lignin is an unknown complex structure containing aromatic groups that is totally hypothetical and is less readily degraded than starch or cellulose. Although lignocellulose is one of the cheapest and most abundant forms of biomass, it is difficult to convert this relatively unreactive material into sugars. Among other factors, the walls of lignocellulose are composed of lignin, which must be broken down in order to render the cellulose and hemicellulose accessible to acid hydrolysis. For this reason, many efforts focused on ethanol production from biomass are based almost entirely on the fermentation of sugars derived from the starch in corn grain.

Carbohydrates (starch, cellulose, sugars): starch readily obtained from wheat and potato, while cellulose is obtained from wood pulp. The structures of these polysaccharides can be readily manipulated to produce a range of biodegradable polymers with properties similar to those of conventional plastics such as polystyrene foams and polyethylene film. In addition, these polysaccharides can be hydrolyzed, catalytically or enzymatically to produce sugars, a valuable fermentation feedstock for the production of ethanol, citric acid, lactic acid, and dibasic acids such as succinic acid.

3.4.1.2 Vegetable Oils

Vegetable oil is obtained from seed oil plants such as palm, sunflower, and soya. The predominant source of vegetable oils in many countries is rapeseed oil. Vegetable oils are a major feedstock for the oleo-chemicals industry (surfactants, dispersants, and personal care products) and are now successfully entering new markets such as diesel fuel, lubricants, polyurethane monomers, functional polymer additives, and solvents.

In many cases, it has been advocated that vegetable oil, and similar feedstocks, be used as feedstocks for a catalytic cracking unit. The properties of the product(s) can be controlled by controlling the process variables including the cracking temperature as well as the type of catalyst used. The production of biodiesel by direct esterification of fatty acids with short chain alcohols occurs in one step only whereby acidic catalysts can be used to speed up the reaction (Demirbaş, 2006).

3.4.1.3 Plant Fibers

Lignocellulosic fibers extracted from plants such as hemp and flax can replace cotton and polyester fibers in textile materials and glass fibers in insulation products. Lignin is a complex chemical that is most commonly derived from wood and is an integral part of the cell wall of plants. The chemical structure of lignin is unknown and, at best, can only be represented by hypothetical formulas.

Lignin (Latin: *lignum—wood*) is one of most abundant organic chemicals on earth after cellulose and chitin. By way of clarification, chitin $[(C_8H_{13}O_5N)_n]$ is a long-chain polymeric polysaccharide of β-glucose that forms a hard, semitransparent material found throughout the natural world. Chitin is the main component of the cell walls of fungi and is also a major component of the exoskeletons of arthropods, such as the crustaceans (e.g., crab, lobster, and shrimp), and insects (e.g., ants, beetles, and butterflies), and the beaks of cephalopods (e.g., squids and octopuses).

Lignin makes up about one-quarter to one-third of the dry mass of wood and is generally considered to be a large, cross-linked hydrophobic, aromatic macromolecules with a molecular mass that is estimated to be in excess of 10,000. Lignin fills the spaces in the cell wall between cellulose, hemicellulose, and pectin components and is covalently linked (bonded) to hemicellulose. Lignin also forms covalent bonds with polysaccharides which enables cross-linking to different plant polysaccharides. Lignin confers mechanical strength to the cell wall (stabilizing the mature cell wall) and therefore the entire plant.

3.4.2 BIOREFINING

A petroleum refinery is a series of integrated unit processes by which petroleum can be converted to a slate of useful (salable) products. A petroleum refinery, as currently configured is unsuitable for processing raw, or even partially processed, biomass. A typical refinery might be suitable for processing products such as gases, liquid, or solids products from biomass processing. These products from biomass might be acceptable as a single feedstock to a specific unit or, more likely, as a feedstock to be blended with refinery streams to be coprocessed in various refinery units.

Thus, a biorefinery might, in the early stages of development, be a series of unit processes which convert biomass to a primary product that requires further processing to become the final salable product. The analogy is in the processing of bitumen from tar sand which is first processed to a synthetic crude oil (primary processing) and then sent to a refinery for conversion to salable fuel products (Speight, 2008, 2014, 2017).

Analogous, in many cases, to the thermal decomposition of crude oil constituents, in the flash pyrolysis (high-temperature cracking and short residence time), the products are ethylene, benzene, toluene, and the xylene isomers as well as carbon monoxide and carbon dioxide. The type of biomass (for example, wood) used influences the product distribution (Steinberg et al., 1992). Theoretically, the flash pyrolysis process can use a wide range of biomass sources. The process has much in common with the naphtha cracking process.

At this point in the context of flash pyrolysis, it is worthy of note that plastic waste (while not a biomass material) can also be treated by flash pyrolysis to produce starting materials for petrochemical manufacture. In the process, the mixed plastic waste is heated in an oxygen-free atmosphere. At a temperature of several hundred degrees, the constituents of the waste decompose to yield a mixture of gaseous, liquid, and solids. The composition of the product depends on temperature and pressure—the higher the temperature, the more gaseous products are formed. An important fraction of this gaseous product is ethylene if plastics are used as feedstock.

Biorefining in which biomass is converted into a variety of chemical products is not new if activities such as production of vegetable oils, beer, and wine requiring pretreatment are considered. Many of these activities are known to have been in practice for millennia. Biomass can be converted into commercial fuels, suitable to substitute for fossil fuels. These can be used for transportation, heating, electricity generation or anything else fossil fuels are used for. The conversion is accomplished through the use of several distinct processes which include both biochemical conversion and thermal conversion to produce gaseous, liquid, and solid fuels which have high energy contents, are easily transportable, and are therefore suitable for use as commercial fuels.

Biorefining offers a method to accessing the integrated production of chemicals, materials, and fuels. Although the concept of a biorefinery concept is analogous to that of an oil refinery, the differences in the various biomass feedstocks require a divergence in the methods used to convert the feedstocks to fuels and chemicals (Speight, 2014, 2017). Thus, a biorefinery, like a petroleum refinery, may need to be a facility that integrates biomass conversion processes and equipment to produce fuels, power, and chemicals from biomass. In a manner similar to the petroleum refinery, a biorefinery would integrate a variety of conversion processes to produce multiple product streams such as motor fuels and other chemicals from biomass such as the inclusion of gasification processes and fermentation processes to name only two possible processes options.

In short, a biorefinery should combine the essential technologies to transform biological raw materials into a range of industrially useful intermediates. However, the type of biorefinery would have to be differentiated by the character of the feedstock. For example, the *crop biorefinery* would use raw materials such as cereals or maize and the *lignocellulose biorefinery* would use raw material with high cellulose content, such as straw, wood, and paper waste.

As a petroleum refinery uses petroleum as the major input and processes it into many different products, a biorefinery with feedstocks such as lignocellulosic biomass as the major input and would processes it into many different products. Currently, wet-mill corn processing and pulp and paper mills can be categorized into biorefineries since they produce multiple products from biomass. Research is currently being conducted to foster new industries to convert biomass into a wide range of products, including ones that would otherwise be made from petrochemicals. The idea is for biorefineries to produce both high-volume liquid fuels and high-value chemicals or products in order to address national energy needs while enhancing operation economics.

However, the different compositional nature of the biomass feedstock, compared to crude oil, will require the application of a wider variety of processing tools in the biorefinery. Processing the individual components will utilize conventional thermochemical operations and state-of-the-art bioprocessing techniques. Although a number of new bioprocesses have been commercialized, it is clear that economic and technical barriers still exist before the full potential of this area can be realized. The biorefinery concept could significantly reduce production costs of plant-based chemicals and facilitate their substitution into existing markets. This concept is analogous to that of a modern oil refinery in that the biorefinery is a highly integrated complex that will efficiently separate biomass raw materials into individual components and convert these into marketable products such as energy, fuels, and chemicals. By analogy with crude oil, every element of the plant feedstock will be utilized including the low-value lignin components.

A key requirement for the biorefinery is the ability of the refinery to develop process technology that can economically access and convert the five- and six-membered ring sugars present in the cellulose and hemicellulose fractions of the lignocellulosic feedstock. Although engineering technology exists to effectively separate the sugar-containing fractions from the lignocellulose, the enzyme technology to economically convert the five ring sugars to useful products requires further development.

Plants are very effective chemical mini-factories or refineries insofar as they produce chemicals by specific pathways. The chemicals they produce are usually essential manufactures (called metabolites) including sugars and amino acids that are essential for the growth of the plant, as well as more complex compounds. Unlike petroleum-derived in petrochemicals where most chemicals are built from the bottom-up, bio-feedstocks already have some valuable products to skim off the top before being broken down and used to build new molecules.

As a feedstock, biomass can be converted by thermal or biological routes to a wide range of useful forms of energy including process heat, steam, electricity, as well as liquid fuels, chemicals, and synthesis gas. As a raw material, biomass is a nearly universal feedstock due to its versatility, domestic availability, and renewable character. At the same time, it also has its limitations. For example, the energy density of biomass is low compared to that of coal, liquid petroleum, or petroleum-derived fuels. The heat content of biomass, on a dry basis (7,000–9,000 Btu/lb) is at best comparable with that of a low-rank coal or lignite, and substantially (50%–100%) lower than that of anthracite, most bituminous coals, and petroleum. Most biomass, as received, has a high burden of physically adsorbed moisture, up to 50% by weight. Thus, without substantial drying, the energy content of a biomass feed per unit mass is even less. These inherent characteristics and limitations of biomass feedstocks have focused the development of efficient methods of chemically transforming and upgrading biomass feedstocks in a refinery.

The sugar-base involves breakdown of biomass into raw component sugars using chemical and biological means. The raw fuels may then be upgraded to produce fuels and chemicals that are interchangeable with existing commodities such as transportation fuels, oils, and hydrogen.

Although a number of new bioprocesses have been commercialized, it is clear that economic and technical barriers still exist before the full potential of this area can be realized. One concept gaining considerable momentum is the biorefinery which could significantly reduce production costs of plant-based chemicals and facilitate their substitution into existing markets. This concept is analogous to that of a modern oil refinery in that the biorefinery is a highly integrated complex that will efficiently separate biomass raw materials into individual components and convert these into marketable products such as energy, fuels, and chemicals.

By analogy with crude oil; every element of the plant feedstock will be utilized including the low-value lignin components. However, the different compositional nature of the biomass feedstock, compared to crude oil, will require the application of a wider variety of processing tools in the biorefinery. Processing of the individual components will utilize conventional thermochemical operations and state-of-the-art bioprocessing techniques. The production of biofuels in the biorefinery complex will service existing high-volume markets, providing economy-of-scale benefits and large volumes of byproduct streams at minimal cost for upgrading to valuable chemicals. A pertinent example of this is the production of glycerol (glycerin) as a byproduct in biodiesel plants.

Glycerol has high functionality and is a potential platform chemical for conversion into a range of higher-value chemicals. The high-volume product streams in a biorefinery need not necessarily be a fuel but could also be a large-volume chemical intermediate such as ethylene or lactic acid. In addition to a variety of methods techniques can be employed to obtain different product portfolios of bulk chemicals, fuels, and materials. Biotechnology-based conversion processes can be used to ferment the biomass carbohydrate content into sugars that can then be further processed. As one example, the fermentation path to lactic acid shows promise as a route to biodegradable plastics. An alternative is to employ thermochemical conversion processes which use pyrolysis or gasification of biomass to produce a hydrogen-rich synthesis gas which can be used in a wide range of chemical processes.

A key requirement for delivery of the biorefinery is the ability of the refinery to develop and use process technology that can economically access and convert the five- and six-membered ring sugars present in the cellulose and hemicellulose fractions of the lignocellulosic feedstock. Although engineering technology exists to effectively separate the sugar-containing fractions from the lignocellulose, the enzyme technology to economically convert the five ring sugars to useful products requires further development.

The construction of both large biofuel and renewable chemical production facilities coupled with the pace at which bioscience is being both developed and applied demonstrates that the utilization of nonfood crops will become more significant in the near term. The biorefinery concept provides a means to significantly reduce production costs such that a substantial substitution of petrochemicals by renewable chemicals becomes possible. However, significant technical challenges remain before the biorefinery concept can be realized.

If the biorefinery is truly analogous to an oil refinery in which crude oil is separated into a series of products, such as gasoline, heating oil, jet fuel, and petrochemicals, the biorefinery can take advantage of the differences in biomass components and intermediates and maximize the value derived from the biomass feedstock. A biorefinery might, for example, produce one or several low-volume, but high-value, chemical products and a low-value, but high-volume liquid transportation fuel, while generating electricity and process heat for its own use and perhaps enough for sale of electricity. The high-value products enhance profitability, the high-volume fuel helps meet national energy needs, and the power production reduces costs and avoids greenhouse gas emissions.

The basic types of processes used to generate chemicals from biomass as might be incorporated into a biorefinery are: (i) pyrolysis, (ii) gasification, (iii) anaerobic digestion, and (iv) fermentation.

3.4.2.1 Pyrolysis

Pyrolysis is a medium temperature method which produces gas, oil, and char from crops which can then be further processed into useful fuels or feedstock (Boateng et al., 2007). Pyrolysis is the direct thermochemical conversion processes which include pyrolysis, liquefaction, and solvolysis (Kavalov and Peteves, 2005).

Wood and many other similar types of biomass which contain lignin and cellulose (such as agricultural wastes, cotton gin waste, wood wastes, and peanut hulls) can be converted through a thermochemical process, such as pyrolysis, into solid, liquid, or gaseous fuels. Pyrolysis, used to produce charcoal since the dawn of civilization, is still the most common thermochemical conversion of biomass to commercial fuel.

During pyrolysis, biomass is heated in the absence of air and breaks down into a complex mixture of liquids, gases, and a residual char. If wood is used as the feedstock, the residual char is what is commonly known as charcoal. With more modern technologies, pyrolysis can be carried out under a variety of conditions to capture all the components, and to maximize the output of the desired product be it char, liquid, or gas. Pyrolysis is often considered to be the gasification of biomass in the absence of oxygen. However, the chemistry of each process may differ significantly. In general, biomass does not gasify as easily as coal, and it produces other hydrocarbon compounds in the gas mixture exiting the gasifier; this is especially true when no oxygen is used. As a result, typically an extra step must be taken to reform these hydrocarbon derivatives with a catalyst to yield a clean syngas mixture of hydrogen, carbon monoxide, and carbon dioxide.

Fast pyrolysis is a thermal decomposition process that occurs at moderate temperatures with a high heat transfer rate to the biomass particles and a short hot vapor residence time in the reaction zone. Several reactor configurations have been shown to assure this condition and to achieve yields of liquid product as high as 75% based on the starting dry biomass weight. They include bubbling fluid beds, circulating and transported beds, cyclonic reactors, and ablative reactors.

Fast pyrolysis of biomass produces a liquid product, pyrolysis oil or bio-oil that can be readily stored and transported. Pyrolysis oil is a renewable liquid fuel and can also be used for production of chemicals. Fast pyrolysis has now achieved a commercial success for production of chemicals and is being actively developed for producing liquid fuels. Pyrolysis oil has been successfully tested in engines, turbines, and boilers, and been upgraded to high-quality hydrocarbon fuels. In the 1990s, several fast pyrolysis technologies reached near-commercial status and the yields and properties of the generated liquid product, bio-oil, depend on the feedstock, the process type and conditions, and the product collection efficiency.

Direct hydrothermal liquefaction involves converting biomass to an oily liquid by contacting the biomass with water at elevated temperatures (300°C–350°C, 570°F–660°F) with sufficient pressure to maintain the water primarily in the liquid phase for residence times up to 30 min. Alkali may be added to promote organic conversion. The primary product is an organic liquid with reduced oxygen content (about 10%) and the primary byproduct is water containing soluble organic compounds.

The importance of the provisions for the supply of feedstocks as crops and other biomass are often underestimated since it is assumed that the supplies are inexhaustible. While this may be true over the long term, short-term supply of feedstocks can be as much as risk as any venture.

3.4.2.2 Gasification

Alternatively, biomass can be converted into fuels and chemicals indirectly (by gasification to syngas followed by catalytic conversion to liquid fuels) (Molino et al., 2016). Biomass gasification is a mature technology pathway that uses a controlled process involving heat, steam, and oxygen to convert biomass to hydrogen and other products, without combustion and represents an efficient process for the production of chemicals and hydrogen.

Gasification is a process that converts organic carbonaceous feedstocks into carbon monoxide, carbon dioxide, and hydrogen by reacting the feedstock at high temperatures (>700°C, 1,290°F), without combustion, with a controlled amount of oxygen and/or steam. The resulting gas mixture (*synthesis gas*, *syngas*, or *producer gas*) is itself a fuel. The power derived from carbonaceous feedstocks and gasification followed by the combustion of the product gas(es) is considered to be a source of renewable energy if the gaseous products are from a source (e.g., biomass) other than a fossil fuel. The carbon monoxide can then be reacted with water (steam) to form carbon dioxide and more hydrogen via a water-gas shift reaction. Adsorber or special membranes can separate the hydrogen from this gas stream. The simplified reaction is:

$$C_6H_{12}O_6 + O_2 + H_2O \rightarrow CO + CO_2 + H_2 + other\,species$$

$$CO + H_2O \rightarrow CO_2 + H_2\,(water\text{-}gas\,shift\,reaction)$$

This reaction scheme uses glucose as a surrogate for cellulose but it must be recognized that biomass has highly variable composition and complexity with cellulose as one major component.

Coal has, for many decades, been the primary feedstock for gasification units—coal can also be gasified *in situ* (in the underground seam) (Speight, 2013a), but that is not the subject of this text and is not discussed further. However, with the concern on the issue of environmental pollutants and the potential shortage of coal in some areas there is a move to feedstocks other than coal for gasification processes. Gasification permits the utilization of various feedstocks (coal, biomass, petroleum resids, and other carbonaceous wastes) to their fullest potential.

The advantage of the gasification process when a carbonaceous feedstock (a feedstock containing carbon) or hydrocarbonaceous feedstock (a feedstock containing carbon and hydrogen) is employed is that the product of focus—synthesis gas—is potentially more useful as an energy source and results in an overall cleaner process. The production of synthesis gas is a more efficient production of an energy source than, say, the direct combustion of the original feedstock because synthesis gas can be converted via the Fischer–Tropsch process into a range of synthesis liquid fuels suitable for using gasoline engines or diesel engines (Chapter 10) (Chadeesingh, 2011).

Biomass includes a wide range of materials that produce a variety of products which are dependent upon the feedstock (Balat, 2011; Demirbaş, 2011; Ramroop Singh, 2011; Speight, 2011a). For example, typical biomass wastes include wood material (bark, chips, scraps, and saw dust), pulp and paper industry residues, agricultural residues, organic municipal material, sewage, manure, and food processing byproducts. Agricultural residues such as straws, nut shells, fruit shells, fruit seeds, plant stalks and stover, green leaves, and molasses are potential renewable energy resources. Many developing countries have a wide variety of agricultural residues in ample quantities. Large quantities of agricultural plant residues are produced annually worldwide and are vastly underutilized. Agricultural residues, when used as fuel, through direct combustion, only a small percentage of their potential energy is available, due to inefficient burners used. Current disposal methods for these agricultural residues have caused widespread environmental concerns. For example, disposal of rice and wheat straw by open-field burning causes air pollution. In addition, the widely varying heat content of the different types of biomass varies widely and must be taken into consideration when designing any conversion process (Jenkins and Ebeling, 1985).

Raw materials that can be used to produce biomass fuels are widely available and arise from a large number of different sources and in numerous forms. Biomass can also be used to produce electricity—either blended with traditional feedstocks, such as coal or by itself. However, each of the biomass materials can be used to produce fuel but not all forms are suitable for all the different types of energy conversion technologies such as biomass gasification (Rajvanshi, 1986; Brigwater, 2003; Dasappa et al., 2004; Speight, 2011a; Basu, 2013). The main basic sources of biomass material are: (i) wood, including bark, logs, sawdust, wood chips, wood pellets and briquettes, (ii) high yield energy crops, such as wheat, that are grown specifically for energy applications, (iii) agricultural crop and animal residues, like straw or slurry, (iv) food waste, both domestic and commercial, and (v) industrial waste, such as wood pulp or paper pulp. For processing, a simple form of biomass such as untreated and unfinished wood may be cut into a number of physical forms, including pellets and wood chips, for use in biomass boilers and stoves.

Thermal conversion processes use heat as the dominant mechanism to convert biomass into another chemical form. The basic alternatives of combustion, torrefaction, pyrolysis, and gasification are separated principally by the extent to which the chemical reactions involved are allowed to proceed (mainly controlled by the availability of oxygen and conversion temperature) (Speight, 2011a).

Many forms of biomass contain a high percentage of moisture (along with carbohydrates and sugars) and mineral constituents—both of which can influence the viability of a gasification process (Chapter 3)—the presence of high levels of moisture in the biomass reduces the temperature inside the gasifier, which then reduces the efficiency of the gasifier. Therefore, many biomass gasification technologies require that the biomass be dried to reduce the moisture content prior to feeding into the gasifier. In addition, biomass can come in a range of sizes. In many biomass gasification systems, the biomass must be processed to a uniform size or shape to feed into the gasifier at a consistent rate and to ensure that as much of the biomass is gasified as possible.

Biomass, such as wood pellets, yard and crop wastes, and the so-called *energy crops* such as switch grass and waste from pulp and paper mills can be used to produce ethanol and synthetic diesel fuel. The biomass is first gasified to produce the synthetic gas (synthesis gas), and then converted via catalytic processes to these downstream products. Furthermore, most biomass gasification systems use air instead of oxygen for the gasification reactions (which is typically used in large-scale industrial and power gasification plants). Gasifiers that use oxygen require an air separation unit to provide the gaseous/liquid oxygen; this is usually not cost-effective at the smaller scales used in biomass gasification plants. Air-blown gasifiers use the oxygen in the air for the gasification reactions.

In general, biomass gasification plants are much smaller than the typical coal or petroleum coke gasification plants used in the power, chemical, fertilizer, and refining industries—the sustainability of the fuel supply is often brought into question. As such, a biomass gasification plant is less expensive to construct and has a smaller environmental footprint. For example, while a large industrial gasification plant may take up to 150 acres of land and process 2,500–15,000 tons per day of feedstock (such as coal or petroleum coke), the smaller biomass plants typically process 25–200 tons of feedstock per day and take up less than 10 acres.

Biomass gasification has been the focus of research in recent years to estimate efficiency and performance of the gasification process using various types of biomass such as sugarcane residue (Gabra et al., 2001), rice hulls (Boateng et al., 1992), pine sawdust (Lv et al., 2004), almond shells (Rapagnà and Latif, 1997; Rapagnà et al., 2000), wheat straw (Ergudenler and Ghaly, 1993), food waste (Ko et al., 2001), and wood biomass (Pakdel and Roy, 1991; Bhattacharya et al., 1999; Chen et al., 1992; Hanaoka et al., 2005). Recently, co-gasification of various biomass and coal mixtures has attracted a great deal of interest from the scientific community. Feedstock combinations including Japanese cedar wood and coal (Kumabe et al., 2007), coal and saw dust, coal and pine chips (Pan et al., 2000), coal and silver birch wood (Collot et al., 1999), and coal and birch wood (Brage et al., 2000) have been reported in gasification practice. Co-gasification of coal and biomass has some synergy—the process not only produces a low carbon footprint on the environment, but also improves the H_2/CO ratio in the produced gas which is required for liquid fuel synthesis (Sjöström et al., 1999; Kumabe et al., 2007). In addition, the inorganic matter present in biomass catalyzes the gasification of coal. However, co-gasification processes require custom fittings and optimized processes for the coal and region-specific wood residues.

While co-gasification of coal and biomass is advantageous from a chemical viewpoint, some practical problems are present on upstream, gasification, and downstream processes. On the upstream side, the particle size of the coal and biomass is required to be uniform for optimum gasification. In addition, moisture content and pretreatment (torrefaction) are very important during upstream processing.

While upstream processing is influential from a material handling point of view, the choice of gasifier operation parameters (temperature, gasifying agent, and catalysts) dictate the product gas composition and quality. Biomass decomposition occurs at a lower temperature than coal and therefore different reactors compatible to the feedstock mixture are required (Speight, 2011c; Brar et al., 2012; Speight, 2013a, 2013b). Furthermore, feedstock and gasifier type along with operating parameters not only decide product gas composition but also dictate the amount of impurities to be handled downstream.

Downstream processes need to be modified if coal is co-gasified with biomass. Heavy metal and impurities such as sulfur and mercury present in coal can make synthesis gas difficult to use and unhealthy for the environment. Alkali present in biomass can also cause corrosion problems high temperatures in downstream pipes. An alternative option to downstream gas cleaning would be to process coal to remove mercury and sulfur prior to feeding into the gasifier.

However, first and foremost, coal and biomass require drying and size reduction before they can be fed into a gasifier. Size reduction is needed to obtain appropriate particle sizes; however, drying is required to achieve moisture content suitable for gasification operations. In addition, biomass densification may be conducted to prepare pellets and improve density and material flow in the feeder areas.

It is recommended that biomass moisture content should be less than 15% w/w prior to gasification. High moisture content reduces the temperature achieved in the gasification zone, thus resulting in incomplete gasification. Forest residues or wood has a fiber saturation point at 30%–31% moisture content (dry basis) (Brar et al., 2012). Compressive and shear strength of the wood increases with decreased moisture content below the fiber saturation point. In such a situation, water is removed from the cell wall leading to shrinkage. The long-chain molecule constituents of the cell wall move closer to each other and bind more tightly. A high level of moisture, usually injected in form of steam in the gasification zone, favors formation of a water-gas shift reaction that increases hydrogen concentration in the resulting gas.

The torrefaction process is a thermal treatment of biomass in the absence of oxygen, usually at 250°C–300°C (480°F–570°F) to drive off moisture, decompose hemicellulose completely, and partially decompose cellulose (Speight, 2011a). Torrefied biomass has reactive and unstable cellulose molecules with broken hydrogen bonds and not only retains 79%–95% of feedstock energy but also produces a more reactive feedstock with lower atomic hydrogen-carbon and oxygen-carbon ratios to those of the original biomass. Torrefaction results in higher yields of hydrogen and carbon monoxide in the gasification process.

Most small- to medium-sized biomass/waste gasifiers are air blown, operated at atmospheric pressure, and at temperatures in the range 800°C–100°C (1,470°F–2,190°F). They face very different challenges compared to large gasification plants—the use of a small-scale air separation plant should oxygen gasification be preferred. Pressurized operation, which eases gas cleaning, may not be practical.

Biomass fuel producers, coal producers, and, to a lesser extent, waste companies are enthusiastic about supplying co-gasification power plants and realize the benefits of co-gasification with alternate fuels (Speight, 2008, 2011a; Lee and Shah, 2013; Speight, 2013a, 2013b). The benefits of a co-gasification technology involving coal and biomass include the use of a reliable coal supply with gate fee waste and biomass that allows the economies of scale from a larger plant to be supplied just with waste and biomass. In addition, the technology offers a future option of hydrogen production and fuel development in refineries. In fact, oil refineries and petrochemical plants are opportunities for gasifiers when the hydrogen is particularly valuable (Speight, 2011b, 2014).

In addition, while biomass may seem to some observers to be the answer to the global climate change issue, the advantages and disadvantages must be considered carefully. For example, the advantages are (i) biomass is a theoretically inexhaustible fuel source, (ii) when direct conversion of combustion of plant mass—such as fermentation and pyrolysis—is not used to generate energy there is minimal environmental impact, (iii) alcohols and other fuels produced by biomass are efficient, viable, and relatively clean-burning, and (iv) biomass is available on a worldwide basis.

On the other hand, the disadvantages include (i) the highly variable heat content of different biomass feedstocks, (ii) the high water content that can affect the process energy balance, and (iii) there is a potential net loss of energy when a biomass plant is operated on a small scale—an account of the energy put used to grow and harvest the biomass must be included in the energy balance.

3.4.2.3   Anaerobic Digestion

Anaerobic digestion is a natural process and is the microbiological conversion of organic matter to methane in the absence of oxygen. The *biochemical conversion* of biomass is completed through alcoholic fermentation to produce liquid fuels and anaerobic digestion or fermentation, resulting in biogas (hydrogen, carbon dioxide, ammonia, and methane) usually through four steps (hydrolysis, acidogenesis, acetogenesis, and methanogenesis):

Hydrolysis:

$$\text{Carbohydrates} \rightarrow \text{sugars}$$

$$\text{Fats} \rightarrow \text{fatty acids}$$

$$\text{Proteins} \rightarrow \text{amino acids}$$

Acidogenesis:

$$\text{Sugars} \rightarrow \text{carbon acids} + \text{alcohols} + \text{hydrogen} + \text{carbon dioxide} + \text{ammonia}$$

$$\text{Fatty acids} \rightarrow \text{carbon acids} + \text{alcohols} + \text{hydrogen}, \text{carbon dioxide} + \text{ammonia}$$

$$\text{Amino acids} \rightarrow \text{carbon acids} + \text{alcohols} + \text{hydrogen}, \text{carbon dioxide} + \text{ammonia}$$

Acetogenesis:

$$\text{Carbon acids} + \text{alcohols} \rightarrow \text{acetic acid} + \text{carbon dioxide} + \text{hydrogen}$$

Methanogenesis:

$$\text{Acetic acid} \rightarrow \text{methane} + \text{carbon dioxide}$$

The decomposition is caused by natural bacterial action in various stages and occurs in a variety of natural anaerobic environments, including water sediment, water-logged soils, natural hot springs, ocean thermal vents, and the stomach of various animals (e.g., cows). The digested organic matter resulting from the anaerobic digestion process is usually called *digestate*.

Symbiotic groups of bacteria perform different functions at different stages of the digestion process. There are four basic types of microorganisms involved (i) hydrolytic bacteria breakdown complex organic wastes into sugars and amino acids; (ii) fermentative bacteria then convert those products into organic acids; (iii) acidogenic microorganisms convert the acids into hydrogen, carbon dioxide, and acetate; and (iv) methanogenic bacteria produce biogas from acetic acid, hydrogen, and carbon dioxide.

The process of anaerobic digestion occurs in a sequence of stages involving distinct types of bacteria. Hydrolytic and fermentative bacteria first breakdown the carbohydrates, proteins, and fats present in biomass feedstock into fatty acids, alcohol, carbon dioxide, hydrogen, ammonia, and sulfides. This stage is called hydrolysis (or liquefaction). Next, acetogenic (acid-forming) bacteria further digest the products of hydrolysis into acetic acid, hydrogen, and carbon dioxide. Methanogenic (methane-forming) bacteria then convert these products into biogas.

The combustion of digester gas can supply useful energy in the form of hot air, hot water, or steam. After filtering and drying, digester gas is suitable as fuel for an internal combustion engine, which, combined with a generator, can produce electricity. Future applications of digester gas may include electric power production from gas turbines or fuel cells. Digester gas can substitute for natural gas or propane in space heaters, refrigeration equipment, cooking stoves, or other equipment. Compressed digester gas can be used as an alternative transportation fuel.

There are three principal byproducts of anaerobic digestion: (i) biogas, (ii) acidogenic digestate, and (iii) methanogenic digestate.

Biogas is a gaseous mixture comprising mostly methane and carbon dioxide, and also containing a small amount of hydrogen and occasionally trace levels of hydrogen sulfide. Since the gas is not released directly into the atmosphere and the carbon dioxide comes from an organic source with a short carbon cycle, biogas does not contribute to increasing atmospheric carbon dioxide concentrations; because of this, it is considered to be an environmentally friendly energy source. The production of biogas is not a steady stream; it is highest during the middle of the reaction. In the early stages of the reaction, little gas is produced because the number of bacteria is still small. Toward the end of the reaction, only the hardest to digest materials remain, leading to a decrease in the amount of biogas produced.

The second byproduct (*acidogenic digestate*) is a stable organic material comprised largely of lignin and chitin and a variety of mineral components in a matrix of dead bacterial cells; some plastic may also be present. This resembles domestic compost and can be used as compost or to make low-grade building products such as fiberboard.

The third byproduct is a liquid (*methanogenic digestate*) that is rich in nutrients and can be an excellent fertilizer dependent on the quality of the material being digested. If the digested materials include low levels of toxic heavy metals or synthetic organic materials such as pesticides or polychlorobiphenyls, the effect of digestion is to significantly concentrate such materials in the digester liquor. In such cases, further treatment will be required in order to dispose of this liquid properly. In extreme cases, the disposal costs and the environmental risks posed by such materials can offset any environmental gains provided by the use of biogas. This is a significant risk when treating sewage from industrialized catchments.

Nearly all digestion plants have ancillary processes to treat and manage all the byproducts. The gas stream is dried and sometimes sweetened before storage and use. The sludge liquor mixture has to be separated by one of a variety of ways, the most common of which is filtration. Excess water is also sometimes treated in sequencing batch reactors for discharge into sewers or for irrigation. Digestion can be either *wet* or *dry*. Dry digestion refers to mixtures which have a solid content of 30% or greater, whereas wet digestion refers to mixtures of 15% or less.

In recent years, increasing awareness that anaerobic digesters can help control the disposal and odor of animal waste has stimulated renewed interest in the technology. New digesters now are being built because they effectively eliminate the environmental hazards of dairy farms and other animal feedlots. Anaerobic digester systems can reduce fecal coliform bacteria in manure by more than 99%, virtually eliminating a major source of water pollution. Separation of the solids during the digester process removes about 25% of the nutrients from manure, and the solids can be sold out of the drainage basin where nutrient loading may be a problem. In addition, the digester's ability to produce and capture methane from the manure reduces the amount of methane that otherwise would enter the atmosphere. Scientists have targeted methane gas in the atmosphere as a contributor to global climate change.

Controlled anaerobic digestion requires an airtight chamber, called a digester. To promote bacterial activity, the digester must maintain a temperature of at least 68°F. Using higher temperatures, up to 150°F, shortens processing time and reduces the required volume of the tank by 25%–40%. However, there are more species of anaerobic bacteria that thrive in the temperature range of a standard design (mesophilic bacteria) than there are species that thrive at higher temperatures (thermophilic bacteria). High-temperature digesters also are more prone to upset because of temperature fluctuations and their successful operation requires close monitoring and diligent maintenance.

The biogas produced in a digester (*digester gas*) is actually a mixture of gases, with methane and carbon dioxide making up more than 90% of the total. Biogas typically contains smaller amounts of hydrogen sulfide, nitrogen, hydrogen, methyl mercaptans, and oxygen.

Methane is a combustible gas. The energy content of digester gas depends on the amount of methane it contains. Methane content varies from about 55% to 80%. Typical digester gas, with a methane concentration of 65%, contains about 600 Btu of energy per cubic foot. There are three basic digester designs and all of them can trap methane and reduce fecal coliform bacteria, but they differ in cost, climate suitability, and the concentration of manure solids they can digest: (i) a covered lagoon digester, (ii) a complete mix digester, (iii) a plug-flow digester.

A *covered lagoon digester*, as the name suggests, consists of a manure storage lagoon with a cover. The cover traps gas produced during decomposition of the manure. This type of digester is the least expensive of the three. Covering a manure storage lagoon is a simple form of digester technology suitable for liquid manure with less than 3% solids. For this type of digester, an impermeable floating cover of industrial fabric covers all or part of the lagoon. A concrete footing along the edge of the lagoon holds the cover in place with an airtight seal. Methane produced in the lagoon collects under the cover. A suction pipe extracts the gas for use. Covered lagoon digesters require large lagoon volumes and a warm climate. Covered lagoons have low capital cost, but these systems are not suitable for locations in cooler climates or locations where a high water table exists.

A *complete mix digester* converts organic waste to biogas in a heated tank above or below ground. A mechanical or gas mixer keeps the solids in suspension. Complete mix digesters are expensive to construct and cost more than plug-flow digesters to operate and maintain. Complete mix digesters are suitable for larger manure volumes having solids concentration of 3%–10%. The reactor is a circular steel or poured concrete container. During the digestion process, the manure slurry is continuously mixed to keep the solids in suspension. Biogas accumulates at the top of the digester. The biogas can be used as fuel for an engine-generator to produce electricity or as boiler fuel to produce steam. Using waste heat from the engine or boiler to warm the slurry in the digester reduces retention time to less than 20 days.

A *plug-flow digester* is suitable for ruminant animal manure that has a solids concentration of 11%–13%. A typical design for a plug-flow system includes a manure collection system, a mixing pit, and the digester itself. In the mixing pit, the addition of water adjusts the proportion of solids in the manure slurry to the optimal consistency. The digester is a long, rectangular container, usually built below-grade, with an airtight, expandable cover.

New material added to the tank at one end pushes older material to the opposite end. Coarse solids in ruminant manure form a viscous material as they are digested, limiting solids separation in the digester tank. As a result, the material flows through the tank in a *plug*. Average retention time (the time a manure plug remains in the digester) is 20–30 days. Anaerobic digestion of the manure slurry releases biogas as the material flows through the digester. A flexible, impermeable cover on the digester traps the gas. Pipes beneath the cover carry the biogas from the digester to an engine-generator set.

A plug-flow digester requires minimal maintenance. Waste heat from the engine-generator can be used to heat the digester. Inside the digester, suspended heating pipes allow hot water to circulate. The hot water heats the digester to keep the slurry at 25°C–40°C (77°F–104°F), a temperature range suitable for methane-producing bacteria. The hot water can come from recovered waste heat from an engine-generator fueled with digester gas or from burning digester gas directly in a boiler.

Anaerobic digestion of biomass has been practiced for almost a century, and is very popular in many developing countries such as China and India. The organic fraction of almost any form of biomass, including sewage sludge, animal wastes, and industrial effluents, can be broken down through anaerobic digestion into methane and carbon dioxide. This biogas is a reasonably clean burning fuel that can be captured and put to many different end uses such as cooking, heating, or electrical generation.

3.4.2.4 Fermentation

A number of processes allow biomass to be transformed into gaseous fuels such as methane or hydrogen (Sørensen et al., 2006). One pathway uses algae and bacteria that have been genetically modified to produce hydrogen directly instead of the conventional biological energy carriers. A second pathway uses plant material such as agricultural residues in a fermentation process leading to biogas from which the desired fuels can be isolated. This technology is established and in widespread use for waste treatment, but often with the energy produced only for on-site use, which often implies less than maximum energy yields. Finally, high-temperature gasification supplies a crude gas, which may be transformed into hydrogen by a second reaction step. In addition to biogas, there is also the possibility of using the solid byproduct as a biofuel.

Traditional fermentation plants producing biogas are in routine use, ranging from farms to large municipal plants. As feedstock they use manure, agricultural residues, urban sewage and waste from households, and the output gas is typically 64% methane. The biomass conversion process is accomplished by a large number of different agents, from the microbes decomposing and hydrolyzing plant material, over the acidophilic bacteria dissolving the biomass in aquatic solution, and to the strictly anaerobic methane bacteria responsible for the gas formation. Operating a biogas plant for a period of some months usually makes the bacterial composition stabilize in a way suitable for obtaining high conversion efficiency (typically above 60%, the theoretical limit being near to 100%), and it is found important not to vary the feedstock compositions abruptly, if optimal operation is to be maintained. Operating temperatures for the bacterial processes are only slightly above ambient temperatures, e.g., in the mesophilic region around 30°C.

The production of ethanol from corn is a mature technology that holds much potential (Nichols et al., 2006). Substantial cost reductions may be possible, however, if cellulose-based feedstocks are used instead of corn. The feed for all ethanol fermentations is sugar—traditionally a hexose (a six-carbon or "C_6" sugar) such as those present naturally in sugar cane, sugar beet, and molasses. Sugar for fermentation can also be recovered from starch, which is actually a polymer of hexose sugars (*polysaccharide*).

Biomass, in the form of wood and agricultural residues such as wheat straw, is viewed as a low cost alternative feed to sugar and starch. It is also potentially available in far greater quantities than sugar and starch feeds. As such it receives significant attention as a feed material for ethanol production. Like starch, wood and agricultural residues contain polysaccharides. However, unlike starch, while the cellulose fraction of biomass is principally a polymer of easily fermented C_6 sugars, the hemicellulose fraction is principally a polymer of C_5 sugars, with quite different characteristics for recovery, and fermentation of the cellulose and hemicellulose in biomass are bound together in a complex framework of crystalline organic material known as lignin.

There are several different methods of hydrolysis: (i) concentrated sulfuric acid, (ii) dilute sulfuric acid, (iii) nitric acid, and (iv) acid pretreatment followed by enzymatic hydrolysis.

The greatest potential for ethanol production from biomass, however, lies in enzymatic hydrolysis of cellulose. The enzyme cellulase, now used in the textile industry to stone wash denim and in detergents, simply replaces the sulfuric acid in the hydrolysis step. The cellulase can be used at lower temperatures, 30°C–50°C, which reduces the degradation of the sugar. In addition, process improvements now allow simultaneous saccharification and fermentation (SSF). In the saccharification and fermentation process, cellulase and fermenting yeast are combined, so that as sugars are produced, the fermentative organisms convert them to ethanol in the same step.

Once the hydrolysis of the cellulose is achieved, the resulting sugars must be fermented to produce ethanol. In addition to glucose, hydrolysis produces other six-carbon sugars from cellulose and five-carbon sugars from hemicellulose that are not readily fermented to ethanol by naturally occurring organisms. They can be converted to ethanol by genetically engineered yeasts that are currently available, but the ethanol yields are not sufficient to make the process economically attractive. It also remains to be seen whether the yeasts can be made hardly enough for production of ethanol on a commercial scale.

The fermentation processes to produce propanol and butanol from cellulose are fairly tricky to execute, and the Clostridium acetobutylicum currently used to perform these conversions produces an extremely unpleasant smell, and this must be taken into consideration when designing and locating a fermentation plant. This organism also dies when the butanol content of whatever it is fermenting rises to 7%. For comparison, yeast dies when the ethanol content of its feedstock hits 14%. Specialized strains can tolerate even greater ethanol concentrations—so-called turbo yeast can withstand up to 16% ethanol. However, if ordinary Saccharomyces yeast can be modified to improve its ethanol resistance, scientists may yet one day produce a strain of the Weizmann organism with a butanol resistance higher than the natural boundary of 7%. This would be useful because butanol has a higher energy density than ethanol, and because waste fiber left over from sugar crops used to make ethanol could be made into butanol, raising the alcohol yield of fuel crops without there being a need for more crops to be planted.

Wet milling and dry milling are the means by which grain and straw fractions are processed into a variety of end products. The processes encompass fermentation and distilling of grains (wheat, rye, or maize). Wet milling starts with water-soaking the grain adding sulfur dioxide to soften the kernels and loosen the hulls, after which it is ground. It uses well-known technologies and allows separation of starch, cellulose, oil, and proteins. Dry milling grinds whole grains (including germ and bran). After grinding, the flour is mixed with water to be treated with liquefying enzymes and, further, cooking the mash to breakdown the starch. This hydrolysis step can be eliminated by simultaneously adding saccharifying enzymes and fermenting yeast to the fermenter (simultaneous saccharification and fermentation).

After fermentation, the mash (called *beer*) is sent through a multicolumn distillation system, followed by concentration, purification, and dehydration of the alcohol. The residue mash (stillage) is separated into a solid (wet grains) and liquid (syrup) phase that can be combined and dried to produce *distiller's dried grains with soluble constituents*, to be used as cattle feed. Its nutritional characteristics and high vegetable fiber content make *distiller's dried grains with soluble constituents* unsuitable for other animal species.

3.4.3 CHEMICALS FROM BIOMASS

The production of biofuels to replace oil and natural gas is in active development, focusing on the use of cheap organic matter (usually cellulose, agricultural and sewage waste) in the efficient production of liquid and gas biofuels which yield high net energy gain. The carbon in biofuels was recently extracted from atmospheric carbon dioxide by growing plants, so burning it does not result in a net increase of carbon dioxide in the earth's atmosphere. As a result, biofuels are seen by many as a way to reduce the amount of carbon dioxide released into the atmosphere by using them to replace nonrenewable sources of energy.

3.4.3.1 Gaseous Products

Most biomass materials are easier to gasify than coal because they are more reactive with higher ignition stability. This characteristic also makes them easier to process thermochemically into higher-value fuels such as methanol or hydrogen. Ash content is typically lower than for most coals, and sulfur content is much lower than for many fossil fuels. Unlike coal ash, which may contain toxic metals and other trace contaminants, biomass ash may be used as a soil amendment to help replenish nutrients removed by harvest. A few biomass feedstocks stand out for their peculiar properties, such as high silicon or alkali metal contents—these may require special precautions for harvesting, processing, and combustion equipment. Note also that mineral content can vary as a function of soil type and the timing of feedstock harvest. In contrast to their fairly uniform physical properties, biomass fuels are rather heterogeneous with respect to their chemical elemental composition.

Biogas contains methane and can be recovered in industrial anaerobic digesters and mechanical biological treatment systems. Landfill gas is a less clean form of biogas which is produced in landfills through naturally occurring anaerobic digestion. Unfortunately, methane is a potent greenhouse gas and should not be allowed to escape into the atmosphere.

When biomass is heated with no oxygen or only about one-third the oxygen needed for efficient combustion (amount of oxygen and other conditions determine if biomass gasifies or pyrolyzes), it gasifies to a mixture of carbon monoxide and hydrogen (synthesis gas, syngas). Combustion is a function of the mixture of oxygen with the hydrocarbon fuel. Gaseous fuels mix with oxygen more easily than liquid fuels, which in turn mix more easily than solid fuels. Syngas, therefore, inherently burns more efficiently and cleanly than the solid biomass from which it was made.

Producing gas from biomass consists of the following main reactions, which occur inside a biomass gasifier (i) drying—biomass fuels usually contain 10–35% w/w moisture and when biomass is heated to 100°C (212°F), the moisture is converted into steam, (ii) pyrolysis—after drying, as heating continues, the biomass undergoes pyrolysis which involves thermal decomposition of the biomass without supplying any oxygen and, a result, the biomass is decomposed or separated into gases, liquids, and solids, (iii) oxidation in which air is introduced into the gasifier after the decomposition process and during oxidation, which takes place at temperatures in the order of 700°C–1,400°C (1,290°F–2,550°F), charcoal, or the solid carbonized fuel, reacts with the oxygen in the air to produce carbon dioxide and heat, and (iv) reduction that occurs at higher temperatures and under reducing conditions, that is when not enough oxygen is available, the following reactions take place forming carbon dioxide, hydrogen, and methane:

$$C + CO_2 \rightarrow 2CO$$

$$C + H_2O \rightarrow CO + H_2$$

$$CO + H_2O \rightarrow CO_2 + H_2$$

$$C + 2H_2 \rightarrow CH_4$$

Biomass gasification can thus improve the efficiency of large-scale biomass power facilities such as those for forest industry residues and specialized facilities such as black liquor recovery boilers of the pulp and paper industry, both major sources of biomass power. Like natural gas, syngas can also be burned in gas turbines, a more efficient electrical generation technology than steam boilers to which solid biomass and fossil fuels are limited.

3.4.3.2 Liquid Products

Ethanol is the predominant chemical produced from crops and has been used as fuel in the many countries such as United States since at least 1908. There are three well-known methods to convert biomass into ethanol: (i) direct fermentation of sugar/starch-rich biomass, such as sugar cane, sugar beet, or maize starch to ethanol, in which microorganisms convert carbohydrates to ethanol under anaerobic conditions; (ii) hydrolysis of lignocellulosic biomass (e.g., agricultural waste, wheat, and wood), followed by fermentation to ethanol. Here, again microorganisms convert carbohydrates to ethanol under anaerobic conditions; and (iii) gasification of lignocellulosic biomass, followed by either fermentation or chemical catalysis to ethanol.

Currently, the production of ethanol by fermentation of corn-derived carbohydrates is the main technology used to produce liquid fuels from biomass resources. Furthermore, amongst different biofuels, suitable for application in transport, bioethanol and biodiesel seem to be the most feasible ones at present. The key advantage of bioethanol and biodiesel is that they can be mixed with conventional petrol and diesel, respectively, which allows using the same handling and distribution infrastructure. Another important strong point of bioethanol and biodiesel is that when they are mixed at low concentrations (≤10% bioethanol in petrol and ≤20% biodiesel in diesel), no engine modifications are necessary.

Biologically produced alcohols, most commonly ethanol and methanol, and less commonly propanol and butanol are produced by the action of microbes and enzymes through fermentation. Methanol is a colorless, odorless, and nearly tasteless alcohol and is also produced from crops and is also used as a fuel. Methanol, like ethanol, burns more completely but releases as much or more carbon dioxide than its gasoline counterpart.

Propanol and butanol are considerably less toxic and less volatile than methanol. In particular, butanol has a high flashpoint of 35°C, which is a benefit for fire safety, but may be difficult for starting engines in cold weather.

Biodiesel is a diesel-equivalent fuel derived from biological sources (such as vegetable oils) which can be used in unmodified diesel engine vehicles. It is, thus, distinguished from the straight vegetable oils or waste vegetable oils used as fuels in some diesel vehicles. In the current context, biodiesel refers to alkyl esters made from the transesterification of vegetable oils or animal fats. Biodiesel fuel is a fuel made from the oil of certain oilseed crops such as soybean, canola, palm kernel, coconut, sunflower, safflower, corn, and a hundreds of other oil-producing crops. The oil is extracted by the use of a press and then mixed in specific proportions with other agents, which causes a chemical reaction. The results of this reaction are two products, biodiesel and soap. After a final filtration, the biodiesel is ready for use. After curing, the glycerin soap that is produced as a byproduct can be used as is, or can have scented oils added before use. In general, biodiesel compares well to petroleum-based diesel (Lotero et al., 2006). Pure biodiesel fuel (100% esters of fatty acids) is called B100. When blended with diesel fuel the designation indicates the amount of B100 in the blend, e.g., B20 is 20% v/v B100 is 80% v/v diesel, and B5 used in Europe contains 5% v/v of B100 in diesel fuel (Pinto et al., 2005).

Hydrocarbon derivatives are products from various plant species belonging to different families, which convert a substantial amount of photosynthetic products into latex. The latex of such plants contains liquid hydrocarbon derivatives of high molecular weight (10,000). These hydrocarbon derivatives can be converted into high-grade transportation fuel (i.e., petroleum). Therefore, hydrocarbon-producing plants are called petroleum plants or petroplants and their crop as petrocrop. Natural gas is also one of the products obtained from hydrocarbon derivatives. Thus, petroleum plants can be an alternative source for obtaining petroleum to be used in diesel engines. Normally, some of the latex-producing plants of families Euphorbiaceae, Apocynaceae, Asclepiadaceae, Sapotaceae, Moraceae, Dipterocarpaceae, etc. are petroplants. Similarly, sunflower (family Composiae), Hardwickia pinnata (family Leguminosae) are also petroplants. Some algae also produce hydrocarbon derivatives.

However, hydrocarbon derivatives, as such, are not usually produced from crops, there being insufficient amount of the hydrocarbon derivatives present in the plant tissue to make the process economical. However, biodiesel is produced from crops thereby offering an excellent renewable fuel for diesel engines.

Bio-oil is a product that is produced by a totally different process than that used for biodiesel production. The process (fast pyrolysis, flash pyrolysis) occurs when solid fuels are heated at temperatures between 350°C and 500°C (570°F–930°F) for a very short period of time (<2 s. The bio-oils currently produced are suitable for use in boilers for electricity generation. In another process, the feedstock is fed into a fluidized bed (at 450°C–500°C) and the feedstock flashes and vaporizes. The resulting vapors pass into a cyclone where solid particles, char, are extracted. The gas from the cyclone enters a quench tower where they are quickly cooled by heat transfer using bio-oil already made in the process. The bio-oil condenses into a product receiver and any non-condensable gases are returned to the reactor to maintain process heating. The entire reaction from injection to quenching takes only two seconds.

3.4.3.3 Solid Products

Examples of solid chemicals from biomass feedstocks include wood and wood-derived charcoal and dried dung, particularly cow dung. One widespread use of such fuels is in home cooking and heating. The biofuel may be burned on an open fireplace or in a special stove. The efficiency of this process may vary widely, from 10% for a well-made fire (even less if the fire is not made carefully) up to 40% for a custom-designed charcoal stove. Inefficient use of fuel is a cause of deforestation (though this is negligible compared to deliberate destruction to clear land for agricultural use), but more importantly it means that more work has to be put into gathering fuel, thus the quality of cooking stoves has a direct influence on the viability of biofuels.

Investigation of the products produced during thermal decomposition (pyrolysis) is worthy of investigation since the potential to produce lower molecular weight feedstocks for a petrochemical plant is high.

3.5 WASTE

It would be remiss not to mention another potential feedstock for the production of chemicals—waste material that is not included under the general category of biomass (John and Singh, 2011). Non-biomass waste is a byproduct of life and civilization; it is the material that remains after a useful component has been consumed. From an economic perspective, waste is a material involved in life or technology whose value today is less than the cost of its utilization. From a regulatory viewpoint, waste is anything discarded or that can no longer be used for its original purpose.

Waste is the general term; though the other terms are used loosely as synonyms, they have more specific meanings. The term *solid waste* includes not only solid materials, but also liquid and gases. *Domestic waste* (also known as rubbish, garbage, trash, or junk) is unwanted or undesired material. Rubbish or trash are mixed household waste including paper and packaging; food waste or garbage (North America) is kitchen and table waste; and junk or scrap is metallic or industrial material.

The thermal pyrolysis of plastic wastes produces a broad distribution of hydrocarbons, from methane to waxy products. This process takes place at high temperatures. The gaseous compounds generated can be burned out to provide the process heat requirements, but the overall yield of valuable gasoline-range hydrocarbons is poor, so that the pyrolysis process as a means for feedstock recycling of the plastic waste stream is rarely practiced on an industrial scale at present (Predel and Kaminsky, 2000; Kaminsky and Zorriqueta, 2007). In contrast, thermal cracking at low temperatures is usually aimed at the production of waxy oil fractions, which may be used in industrial units for steam cracking and in fluid catalytic cracking units (Aguado et al., 2002). An alternative to improve the yield of naphtha from the pyrolysis of plastic waste is to introduce suitable catalysts. High conversion and interesting product distribution is obtained when plastics are cracked over zeolites (Hernandez et al., 2007). Moreover, the catalytic cracking of polymers has proven itself to be a very versatile process, since a variety of products can be obtained depending on parameters such as (i) the catalyst, (ii) the polymer feedstocks, (iii) the reactor type, and (iv) the process parameters, such as temperature, pressure, and residence time of the feedstock in the hot zone, as well (v) as product removal from the hot zone (Aguado and Serrano, 1999; Demirbaş, 2004; Scheirs and Kaminsky, 2006. Marcilla et al., 2008; Al-Salem et al., 2009; Sarker et al., 2012).

In addition, urban waste (domestic and industrial) has considerable promise as a feedstock for gasification because it contains relatively more lignin, which biological processes cannot convert. Such waste is abundant in most countries and can be harnessed for production of fuels and petrochemical intermediates. Knowing the potential of the waste for gasification and subsequent fuel production is essential for reducing pressure on traditional energy sources.

Also, discarded tires can be reduced in size by grinding, chipping, pelletizing, and passed through a classifier to remove the steel belting after which the chips are pyrolyzed for 1 h at a temperature of 300°C–500°C (570°F–930°F) and then heated for 2 h in a closed retort to yield gas, distillable, and char. Discarded tires can also be shredded to 25 mm and ground to 24 mesh as a feed-preparation step for occidental flash pyrolysis that involves flash pyrolysis and product collection. The pyrolytic reaction occurs without the introduction of hydrogen or using a catalyst. This yields a gaseous stream that is passed to a quench tower from which fuel oil and gas (recycled to char fluidized and pyrolysis reactor as a supplemental fuel) and carbon black (35% w/w) is produced. In the Nippon Zeon process, crushed tire chips undergo fluidized thermal cracking (fluidized bed, 400°C–600°C, 750°F–1,110°F), which yields a gaseous stream that is passed to a quench tower from which gas and distillable oil is produced. All of the end products produced could be used directly as a supplemental fuel source at the plant or sent off-site for petrochemical manufacture.

REFERENCES

Aguado, J., and Serrano, D. 1999. *Feedstock Recycling of Plastic Wastes*. of Chemistry, Cambridge, UK.

Aguado, R., Olazar, M., San Jose, M.J., Gaisan, B., and Bilbao, J. 2002. Wax Formation in the Pyrolysis of Poly ole fins in a Conical Spouted Bed Reactor. Energy & Fuels, 16(6): 1429–1437.

Al-Salem, S.M., Lettieri, P., and Baeyens, J. 2009. Recycling and Recovery Routes of Plastic Solid Waste (PSW): A Review. Waste Management, 29(10): 2625–2643.

ASTM D388. 2018. *Standard Classification of Coal by Rank*. Annual Book of Standards. ASTM International, West Conshohocken, PA.

Balat, M. 2011. Chapter 3: Fuels from Biomass—An Overview. In: *The Biofuels Handbook*. J.G. Speight (Editor). Royal Society of Chemistry, London, UK. Part 1.

Basu, P. 2013. *Biomass Gasification, Pyrolysis and Torrefaction*. 2nd Edition. Practical Design and Theory. Academic Press, Inc., New York.

Besson, M., Gallezot, P., and Pinel, C. 2014. Conversion of Biomass into Chemicals over Metal Catalysts. Chem. Rev, 114(3): 1827–1870.

Bhattacharya, S., Mizanur Rahman Siddique, A.H.M.M.R, and Pham, H.-L. 1999. A Study in Wood Gasification on Low Tar Production. Energy, 24: 285–296.

Biddy, M.J., Scarlata, C., and Kinchin, C. 2016. *Chemicals from Biomass: A Market Assessment of Bioproducts with Near-Term Potential*. Technical Report No. NREL/TP-5100-65509. National Renewable Energy, Golden, CO. Also, United States Department of Energy, Washington, DC.

Boateng, A.A., Walawender, W.P., Fan, L.T., and Chee, C.S. 1992. Fluidized-Bed Steam Gasification of Rice Hull. Bioresource Technology, 40(3): 235–239.

Boateng, A.A., Daugaard, D.E., Goldberg, N.M., and Hicks, K.B. 2007. Bench-Scale Fluidized-Bed Pyrolysis of Switchgrass for Bio-Oil Production. Industrial & Engineering Chemistry Research, 46: 1891–1897.

Bozell, J.J. 1999. Renewable Feedstocks for the Production of Chemicals. *Preprints. Div. Fuel Chem. American Chemical Society*, 44(2): 204–209.

Brage, C., Yu, Q., Chen, G., and Sjöström, K. 2000. Tar Evolution Profiles Obtained from Gasification of Biomass and Coal. Biomass and Bioenergy, 18(1): 87–91.

Brar, J.S., Singh, K., Wang, J., and Kumar, S. 2012. Cogasification of Coal and Biomass: A Review. International Journal of Forestry Research, 2012 (2012): 1–10.

Brigwater, A.V. (Editor). 2003. *Pyrolysis and Gasification of Biomass and Waste*. CPL Press, Newbury, Berkshire.

Chadeesingh, R. 2011. Chapter 5: The Fischer-Tropsch Process. In: *The Biofuels Handbook*. J.G. Speight (Editor). The Royal Society of Chemistry, London, UK. Part 3, pp. 476–517.

Chen, G., Sjöström, K. and Bjornbom, E. 1992. Pyrolysis/Gasification of Wood in a Pressurized Fluidized Bed Reactor. Ind. Eng. Chem. Research, 31(12): 2764–2768.

Clark, J., and Deswarte, F. 2008. *Introduction to Chemicals from Biomass*. John Wiley & Sons Inc., Hoboken, NJ.

Collot, A.G., Zhuo, Y., Dugwell, D.R., and Kandiyoti, R. 1999. Co-Pyrolysis and Cogasification of Coal and Biomass in Bench-Scale Fixed-Bed and Fluidized Bed Reactors. Fuel, 78: 667–679.

Crocker, M., and Crofcheck, C. 2006. Reducing national dependence on imported oil. Energeia, 17(6).

Dasappa, S., Paul, P.J., Mukunda, H.S., Rajan, N.K.S., Sridhar, G., and Sridhar, H.V. 2004. Biomass Gasification Technology—A Route to Meet Energy Needs. Current Science, 87(7): 908–916.

Demirbaş, A. 2004. Pyrolysis of Municipal Plastic Wastes for Recovery of Gasoline-Range Hydrocarbons. Journal of Analytical and Applied Pyrolysis, 72(1): 97–102.

Demirbaş, A. 2006. Current technologies for biomass conversion into chemicals and fuels. Energy Sources Part A, 28: 1181–1188.

Demirbaş, A. 2011. Chapter 1: Production of Fuels from Crops. In: *The Biofuels Handbook*. J.G. Speight (Editor). Royal Society of Chemistry, London, UK. Part 2.

Detroy, R.W. 1981. Bioconversion of Agricultural Biomass to Organic Chemicals. In: *Organic Chemicals from Biomass*. CRC Press, Boca Raton, FL.

Ergudenler, A., and Ghaly, A.E. 1993. Agglomeration of Alumina Sand in a Fluidized Bed Straw Gasifier at Elevated Temperatures. Bioresource Technology, 43(3): 259–268.

Gabra, M., Pettersson, E., Backman, R., and Kjellström, B. 2001. Evaluation of Cyclone Gasifier Performance for Gasification of Sugar Cane Residue—Part 1: Gasification of Bagasse. Biomass and Bioenergy, 21(5): 351–369.

Gibbs, F.W. 1961. *Organic Chemistry Today*. Pergamon Books Ltd., London, UK.

Hanaoka, T., Inoue, S., Uno, S., Ogi, T., and Minowa, T. 2005. Effect of Woody Biomass Components on Air-Steam Gasification. Biomass and Bioenergy, 28(1): 69–76.

Hernandez, M.R., Garcia, A.N., and Marcilla, A. 2007. Catalytic Flash Pyrolysis of HDPE in a Fluidized Bed Reactor for Recovery of Fuel-Like Hydrocarbons. Journal of Analytical and Applied Pyrolysis, 78(2): 272–281.

Hoiberg, A.J. 1966. *Bituminous Materials: Asphalts, Tars, and Pitches*, Vol. 3. Coal Tars and Pitches. Interscience Publishers Inc., New York.

Jenkins, B.M., and Ebeling, J.M. 1985. Thermochemical Properties of Biomass Fuels. California Agriculture, 39(5): 14–18.

John, E.J., and Singh, K. 2011. Chapter 1: Production and Properties of Fuels from Domestic and Industrial Waste. In: *The Biofuels Handbook*. J.G. Speight (Editor). The Royal Society of Chemistry, London, UK. Part 3, pp. 333–376.

Kaminsky, W., and Zoriquetta, I.J.N. 2007. Catalytical and Thermal Pyrolysis of Polyolefins. Journal of Analytical and Applied Pyrolysis, 79(1–2): 368–374.

Kavalov, B., and Peteves, S.D. 2005. *Status and Perspectives of Biomass-to-Liquid Fuels in the European Union*. European Commission. Directorate General Joint Research Centre (DG JRC). Institute for Energy, Petten, The Netherlands.

Khoo, H.H., Wong, L.L., Tan, J., Isoni, V., and Sharratt, P. 2015. Synthesis of 2-Methyl Tetrahydrofuran from Various Lignocellulosic Feedstocks: Sustainability Assessment via LCA. Resour. Conserv. Recy., 95: 174.

Ko, M.K., Lee, W.Y., Kim, S.B., Lee, K.W., and Chun, H.S. 2001. Gasification of Food Waste with Steam in Fluidized Bed. Korean Journal of Chemical Engineering, 18(6): 961–964.

Kumabe, K., Hanaoka, T., Fujimoto, S., Minowa, T., and Sakanishi, K. 2007. Cogasification of Woody Biomass and Coal with Air and Steam. Fuel, 86: 684–689.

Lee, S. 1991. *Oil Shale Technology*. CRC Press, Boca Raton, FL.

Lee, S., Speight, J.G., and Loyalka, S.K. 2007. *Handbook of Alternative Fuel Technologies*. CRC Press, Boca Raton, FL.

Lee, S., and Shah, Y.T. 2013. *Biofuels and Bioenergy*. CRC Press, Boca Raton, FL.

Lotero, E., Goodwin, J.G. Jr., Bruce, D.A., Suwannakarn, K., Liu, Y., and Lopez, D.E. 2006. The Catalysis of Biodiesel Synthesis. Catalysis, 19: 41–83.

Lowry, H.H. (Editor). 1945. *Chemistry of Coal Utilization*, Vol. 3. John Wiley & Sons Inc., New York.

Lv, P.M., Xiong, Z.H., Chang, J., Wu, C.Z., Chen, Y., and Zhu, J.X. 2004. An Experimental Study on Biomass Air-Steam Gasification in a Fluidized Bed. Bioresource Technology, 95(1): 95–101.

Marcilla, A., Beltran, M.I., and Navarro, R. 2008. Evolution with the Temperature of the Compounds Obtained in the Catalytic Pyrolysis of Polyethylene over HUSY. Industrial & Engineering Chemistry Research, 47(18): 6896–6903.

McNeil, D. 1966. *Coal Carbonization Products*. Pergamon Press, London, UK.

Mills, G.A. 1977. Chem. Tech. 7(7): 418.

Metzger, J.O. 2006. Production of Liquid Hydrocarbons from Biomass. Angew. Chem. Int. Ed., 45: 696–698.

Molino, A., Chianese, S., and Musmarra, D. 2016. Biomass Gasification Technology: The State of the Art Overview. Journal of Energy Chemistry, 25(1): 10–25.

Munger, C.G. 1984. *Corrosion Prevention by Protective Coating*. NACE International, Houston, TX, p. 32.

Nichols, N.N., Dien, B.S., Bothast, R.J., and Cotta, M.A. 2006. Chapter 4: The Corn Ethanol Industry. In: *Alcoholic Fuels*. S. Minteer (Editor). CRC Press, Boca Raton, FL.

Owen, J. 1981. Conversion and Uses of Liquid Fuels from Coal. Fuel, 60(9): 755–761.

Pakdel, H., and Roy, C. 1991. Hydrocarbon Content of Liquid Products and Tar from Pyrolysis and Gasification of Wood. Energy & Fuels, 5: 427–436.

Pan, Y.G., Velo, E., Roca, X., Manyà, J.J., and Puigjaner, L. 2000. Fluidized-Bed Cogasification of Residual Biomass/Poor Coal Blends for Fuel Gas Production. Fuel, 79: 1317–1326.

Pinto, A.C., Guarieiro, L.N.N., Rezende, M.J.C., Ribeiro, N.M., Torres, E.A., Lopes, W.A., Pereira, P.A.P., and De Andrade, J.B. 2005. Biodiesel: An Overview. J. Braz. Chem. Soc., 16: 1313–1330.

Pitt, G.J., and Millward, G.R. (Editors). 1979. *Coal and Modern Coal Processing: An Introduction*. Academic Press Inc., New York.

Predel, M., and Kaminsky, W. 2000. Pyrolysis of Mixed Poly-olefins in a Fluidized-Bed Reactor and on a Pyro-GC/MS to Yield Aliphatic Waxes. Polymer Degradation and Stability, 70(3): 373–385.

Rajvanshi, A.K. 1986. Biomass Gasification. In: *Alternative Energy in Agriculture*, Vol. 2. D.Y. Goswami (Editor). CRC Press, Boca Raton, FL, pp. 83–102.

Ramroop Singh, N. 2011. Biofuel. In: *The Biofuels Handbook*. J.G. Speight (Editor). Royal Society of Chemistry, London, UK. Part 1. Chapter 5.

Rapagnà, N.J., and Latif, A. 1997. Steam Gasification of Almond Shells in a Fluidized Bed Reactor: The Influence of Temperature and Particle Size on Product Yield and Distribution. Biomass and Bioenergy, 12(4): 281–288.

Rapagnà, N.J., and, Kiennemann, A., and Foscolo, P.U. 2000. Steam-Gasification of Biomass in a Fluidized-Bed of Olivine Particles. Biomass and Bioenergy, 19(3): 187–197.

Sarker, M., Rashid, M.M., Rahman, M.S., and Molla, M. 2012. A New Kind of Renewable Energy: Production of Aromatic Hydrocarbons Naphtha Chemical by Thermal Degradation of Polystyrene (PS) Waste Plastic. American Journal of Climate Change, 2012(1): 145–153.

Scheirs, J., and Kaminsky, W. 2006. *Feedstock Recycling and Pyrolysis of Waste Plastics*. John Wiley & Sons Inc., Chichester, UK.

Scouten, C. 1990. Oil Shale. In: *Fuel Science and Technology Handbook*. J.G. Speight (Editor). Marcel Dekker Inc., New York.

Shah, S., and Gupta, M.N. 2007. Lipase catalyzed preparation of biodiesel from Jatropha oil in a solvent free system. Process Biochemistry, 42: 409–414.

Sjöström, K., Chen, G., Yu, Q., Brage, C., and Rosén, C. 1999. Promoted Reactivity of Char in Cogasification of Biomass and Coal: Synergies in the Thermochemical Process. Fuel, 78: 1189–1194.

Sørensen, B.E., Njakou, S., and Blumberga, D. 2006. Gaseous Fuels Biomass. *Proceedings. World Renewable Energy Congress IX*. WREN, London.

Speight, J.G. 1990. In: *Fuel Science and Technology Handbook*. J.G. Speight (Editor). Marcel Dekker Inc., New York. Chapter 37.

Speight, J.G. 2005. *Handbook of Coal Analysis*. John Wiley & Sons Inc., Hoboken, NJ.

Speight, J.G. 2008. *Synthetic Fuels Handbook: Properties, Processes, and Performance*. McGraw-Hill, New York.

Speight, J.G. 2011a. *The Refinery of the Future*. Gulf Professional Publishing, Elsevier, Oxford, UK.

Speight, J.G. 2011b. *An Introduction to Petroleum Technology, Economics, and Politics*. Scrivener Publishing, Salem, MA.

Speight, J.G. (Editor). 2011c. *The Biofuels Handbook*. The Royal Society of Chemistry, London, UK.

Speight, J.G. 2012. *Shale Oil Production Processes*. Gulf Professional Publishing, Elsevier, Oxford, UK.

Speight, J.G. 2013a. *The Chemistry and Technology of Coal*. 3rd Edition. CRC Press, Boca Raton, FL.

Speight. J.G. 2013b. *Coal-Fired Power Generation Handbook*. Scrivener Publishing, Salem, MA.

Speight, J.G. 2014. *The Chemistry and Technology of Petroleum*. 5th Edition. CRC Press, Boca Raton, FL.

Speight, J.G., and Islam, M.R. 2016. *Peak Energy—Myth or Reality*. Scrivener Publishing, Beverly, MA.

Speight, J.G. 2017. *Handbook of Petroleum Refining*. CRC Press, Boca Raton, FL.

Steinberg, M., Fallon, P.T., and Ssundaram, M.S. 1992. The Flash Pyrolysis and Methanolysis of Biomass (Wood) for the Production of Ethylene, Benzene, and Methanol. In: *Novel Production Methods for Ethylene, Light Hydrocarbons, and Aromatics*. R.C. von Herausgeg, L.F. Albright, B.L. Crynes, and S. Nowak (Editors). Marcel Dekker Inc., New York.

Straathof, A.J.J. 2014. Transformation of Biomass into Commodity Chemicals Using Enzymes or Cells. Chem. Rev., 114(3): 1871–1908.

US DOE. 2018. *Fossil Energy Research Benefits*. Clean Coal Technology Demonstration Program. United States Department of Energy, Washington, DC. https://www.energy.gov/sites/prod/files/cct_factcard.pdf, accessed November 5, 2018.

Vasudevan, P., Sharma, S., and Kumar, A. 2005. Liquid Fuels from Biomass: An Overview. Journal of Scientific & Industrial Research, 64: 822–831.

Wachtmeister, H., Lund, L., Aleklett, K., and Mikael Höök, M. 2017. Production Decline Curves of Tight Oil Wells in Eagle Ford Shale. Natural Resources Research, 26(3): 365–377.

Wright, L., Boundy, R., Perlack, R., Davis, S., and Saulsbury, B. 2006. *Biomass Energy Data Book.* 1st Edition. Office of Planning, Budget and Analysis, Energy Efficiency and Renewable Energy, United States Department of Energy. Contract No. DE-AC05-00OR22725. Oak Ridge National Laboratory, Oak Ridge, TN.

Wu, L., Moteki, T., Gokhale, A.A., Flaherty, D.W., and Tostel, F.D. 2016. Production of Fuels and Chemicals from Biomass: Condensation Reactions and Beyond. Chem, 1: 32–58, July 7, 2016 [a] 2016 Elsevier Inc., New York. www.sciencedirect.com/science/article/pii/S2451929416300043. www.cell.com/chem/fulltext/S2451-9294(16)30004-3

4 Feedstock Preparation

4.1 INTRODUCTION

A feedstock is raw material (unprocessed material) for a processing or manufacturing and which is an asset that is critical to the production of other products. For example, natural gas and crude oil are feedstock raw materials that provide finished products in the fuel industry. The term raw material is used to denote material in an unprocessed or minimally processed state, such as raw natural gas, crude oil, coal, shale oil, or biomass. However, coal, oil shale, and biomass tar sand are complex carbonaceous raw materials and are possible future energy and chemical sources but, like all feedstocks for petrochemical production, they must undergo sometimes lengthy and extensive processing before they become suitable gases and liquids that can be used for the production of the petrochemicals. They will, however, be synthesis gas (Chapter 5) which can be used as a precursor to a range of petrochemical products (Chapter 10).

In all cases, contaminants such as nitrogen, oxygen, sulfur, and metals must be removed before the feedstock is sent to one or more conversion units. Typically, in the natural gas and refining industries (as well as in the coal oil shale and biomass industries) the feedstock is not used directly as fuels or, in the current context, for the production of chemicals. This is due to the complex nature of the feedstock and the presence of one or more of the aforementioned impurities that are corrosive or poisonous to processing catalysts. It is, therefore, essential that any feedstock for use in the production of petrochemical product should be contaminant free when it enters any one of the various reactors (Parkash, 2003; Gary et al., 2007; Speight, 2014; Hsu and Robinson, 2017; Speight, 2017).

The petrochemical industry is concerned with the production and trade of petrochemical products whether it involves the manufacture of an intermediate product or the manufacture of a final (sales) product. The industry directly interfaces with the petroleum industry, especially the downstream sector. A petroleum refinery produces olefin derivatives and aromatic derivatives by cracking processes (such as coking processes and fluid catalytic cracking processes. In addition, the stream cracking of natural gas (methane) also produces olefin derivatives. Aromatic derivatives are produced by the catalytic reforming of naphtha. The importance of olefin derivatives and aromatic derivatives is reflected in their use as the building blocks for a wide range of materials such as solvent, detergents, adhesives, plastics, fibers, and elastomers. Moreover, the importance of the purity of the feedstocks can be tested and assured by the application of standard test methods (Speight, 2015, 2018).

Typically, the primary raw feedstocks (natural gas and crude oil) have been subjected to chemical and/or physical changes (refining) after being recovered. On the other hand, the secondary raw materials, or intermediates, are obtained from natural gas and crude oil through different processing schemes. The intermediates may be low-boiling hydrocarbon derivatives such as methane (CH_4) and ethane (C_2H_6) or higher-boiling hydrocarbon derivatives (such as propane (C_3H_8) butane (C_4H_{10}) and pentane (C_5H_{12}), even mixtures such as naphtha or gas oil that are produced from crude oil as distillation fractions (Parkash, 2003; Gary et al., 2007; Speight, 2014; Hsu and Robinson, 2017; Speight, 2017).

However, the feedstocks used for petrochemical production are varied and, in the natural state as received, are not suitable for use in petrochemical production. For example, natural gas, as it is used by consumers, is much different from the natural gas that is brought from underground formations to the wellhead (Table 4.1). Although the processing of natural gas is in many respects less complicated than the processing and refining of crude oil, it is equally necessary before its use by end users to assure the quality of the feedstocks (or product) to the end users are domestic users or commercial users as is the case with the petrochemical industry.

TABLE 4.1
Constituents of Natural Gas

Name	Formula	% v/v
Methane	CH_4	>85
Ethane	C_2H_6	3–8
Propane	C_3H_8	1–5
Butane	C_4H_{10}	1–2
Pentane[+]	$C_5H_{12}{}^+$	1–5
Carbon dioxide	CO_2	1–2
Hydrogen sulfide	H_2S	1–2
Nitrogen	N_2	1–5
Helium	He	<0.5

Pentane[+]: pentane and higher molecular weight hydrocarbon derivatives, including benzene and toluene.

Gas is often referred to as *natural gas* because it is a naturally occurring hydrocarbon mixture (that does contain some non-hydrocarbon constituents which might be labeled as impurities but often find use in other areas of technology). For the most part, natural gas consists mainly of methane, which is the simplest hydrocarbon but, nevertheless, processing (purification, refining) is required before transportation to the consumer.

In the crude oil industry, naphtha is used as a feedstock for steam cracking to produce petrochemicals (ethylene, propylene) and the production of aromatic petrochemical products (benzene, toluene, and xylenes). Also, gas oil is used as a chemical feedstock for steam cracking, although generally less preferred than naphtha and natural gas liquids (NGLs, including liquefied petroleum gases (LPGs)). While the naphtha and gas oil produced in a refinery depend on the composition of feed crude petroleum utilized and in turn the crude oil regional source, crude oil extracted from oil fields in Middle Eastern countries has different properties (and amounts of contaminants) compared with crude oil extracted from oil fields in Alaska. These differences are also reflected in the quality of the naphtha and the gas oil and the impurities in these two liquids. Furthermore, the production of feedstocks by the thermal decomposition of coal, oil shale, and biomass will, in each case, produce a variety of products (gases, liquids, and solids) and the every-present contaminants that must be removed before further processing. In addition, since the majority of the petrochemical feedstocks to produce the products rely upon the use of a gaseous or low-boiling feedstock, the focus of the chapter is on gas cleaning and petroleum refining.

Thus, it is the purpose of this chapter to present the methods by which various gaseous and liquid feedstocks can be processed and prepared for petrochemical production. This requires removal of impurities that would otherwise be deleterious to petrochemical production and analytical assurance that feedstocks are, indeed, free of deleterious contaminants (Speight, 2015, 2018).

4.2 GAS STREAMS

Raw natural gas comes from three types of wells: oil wells (*associated gas*), gas wells (*nonassociated gas*), and condensate wells (*condensate gas* but also called *nonassociated gas*). Associated gas can exist separate from oil in the formation (*free gas*), or dissolved in the crude oil (*dissolved gas*). Whatever the source of the natural gas, once separated from the crude oil (if present) it commonly exists in mixtures with other hydrocarbon derivatives—principally ethane, propane, butane, and pentane isomers (*natural gas liquids*) as well as a mixture of higher molecular weight (higher-boiling) hydrocarbon derivatives that are often referred to as *natural gasoline* (NG). In addition, raw natural gas contains water vapor, hydrogen sulfide (H_2S), carbon dioxide, helium, nitrogen, and other compounds. Natural gas liquids are sold separately and have a variety of different uses such as providing feedstocks for oil refineries or petrochemical plants.

Trace quantities of sulfur compounds in hydrocarbon products can be harmful to many catalytic chemical processes in which these products are used. Maximum permissible levels of total sulfur are normally included in specifications for such hydrocarbon derivatives. It is recommended that this test method be used to provide a basis for agreement between two laboratories when the determination of sulfur in hydrocarbon gases is important. In the case of liquefied petroleum gas, total volatile sulfur is measured on an injected gas sample. One test method (ASTM D3246, 2018) describes a procedure for the determination of sulfur in the range from 1.5 to 100 mg/kg (ppm w/w) in hydrocarbon products that are gaseous at normal room temperature and pressure.

Acidic constituents such as carbon dioxide and hydrogen sulfide as well as mercaptan derivatives (also called thiols, RSH) can contribute to corrosion of refining equipment, harm catalysts, pollute the atmosphere, and prevent the use of hydrocarbon components in petrochemical manufacture (Mokhatab et al., 2006; Speight, 2007, 2014). When the amount of hydrogen sulfide is high, it may be removed from a gas stream and converted to sulfur or sulfuric acid; a recent option for hydrogen sulfide removal is the use of chemical scavengers. Some natural gases contain sufficient carbon dioxide to warrant recovery as dry ice.

Gas streams produced during petroleum and natural gas refining are not always hydrocarbon in nature and may contain contaminants, such as carbon oxides (CO_x, where x = 1 and/or 2), sulfur oxides (SO_x, where x = 2 and/or 3), as well as ammonia (NH_3), carbonyl sulfide (COS), and mercaptan derivatives (RSH). The presence of these impurities may eliminate some of the sweetening processes from use since some of these processes remove considerable amounts of acid gas but not to a sufficiently low concentration. On the other hand, there are those processes not designed to remove (or incapable of removing) large amounts of acid gases whereas they are capable of removing the acid gas impurities to very low levels when the acid gases are present only in low-to-medium concentration in the gas (Katz, 1959; Mokhatab et al., 2006; Speight, 2007, 2014).

4.2.1 Sources

The sources of the various gas streams that are used as petrochemical feedstocks are varied. However, in terms of gas cleaning (i.e., removal of the contaminants before petrochemical production), the processes are largely the same but it is a question of degree. For example, gas streams for some sources may produce gases that may contain higher amounts of carbon dioxide and/or hydrogen sulfide and therefore the processes will have to be selected accordingly (Table 4.2). The same selection criteria apply to liquid streams whether the streams originate from natural gas or from crude oil.

4.2.1.1 Gas Streams from Natural Gas

In addition to its primary importance as a fuel, natural gas is also a source of hydrocarbon derivatives for petrochemical feedstocks. Although natural gas is mostly considered as a "clean" fuel as compared to other fossil fuels, the natural gas found in reservoirs deposit is not necessarily "clean" and free of impurities. Furthermore, the natural gas processed at the wells will have different range of composition depending on type, depth, and location of the underground reservoirs of porous sedimentary deposit and the geology of the area. Most often, oil and natural gas are found together in a reservoir. When the natural gas is produced from oil wells, it is categorized as associated with (dissolved in) crude oil or nonassociated. It is apparent that two gas wells producing from the same reservoir may have different compositions. Further, the composition of the gas produced from a given reservoir may differ with time as the small hydrocarbon molecules (two to eight carbons) in addition to methane that exist in a gaseous state at underground pressures will become liquid (condense) at normal atmospheric pressure in the reservoir. Generally, they are called condensates or natural gas liquids.

TABLE 4.2
Brief Descriptions of the Major Unit Operations

Unit	Function
Gas-oil separator	Separation of the gas stream and the crude oil at the top and bottom part of the cylindrical shell, respectively, by the action of pressure at the wellhead where gravity separates the gas hydrocarbon derivatives from the heavier oil.
Condensate separator	Removal of condensate from the gas stream by mechanical separators at the wellhead. In condensate treatment section, two main operations, namely water washing and condensate stabilization are performed. Based on the quality of the associated water, the condensate may require water wash to remove salts and additives.
Dehydrator	Removal of water vapor using dehydration process so that the natural gas will be free from the formation of hydrates, corrosion problem, and dew point. In this treatment, process of absorption using ethylene glycol is used to remove water and other particles from the feed stream. As another option, adsorption process can also be used for water removal using dry-bed dehydration towers.
Acid gas removal unit	Removal contaminates in the dry gas such as CO_2, H_2S, some remaining water vapor, inert gases such as helium, and oxygen. The use of alkanolamines or Benfield solution processes is mostly common to absorb CO_2 and H_2S from the feed gas.
Nitrogen extractor	Removal of nitrogen from the stream using two common ways. In the first type, nitrogen is cryogenically separated from the gas stream by the difference in their boiling point. In the second type, separation of methane from nitrogen takes place using physical absorption process. Usually regeneration is done by reducing the pressure. If there were trace amounts of inert gases like helium then pressure swing adsorption unit can be used to extract them from the gas stream. Also called the nitrogen rejection unit.
Demethanizer	Separation of e methane from natural gas liquids using cryogenic processing or absorption techniques. The demethanization process can take place in the plant or as nitrogen extraction process. As compared to absorption method, the cryogenic method is more efficient for the lighter liquids separation, such as ethane.
Fractionator	Separation of natural gas liquids present in the gas stream by varying the volatility of the hydrocarbon derivatives present in the stream. In fractionation, the natural gas liquids after the demethanizer is subjected to rise through towers and heated to increase the temperature of the gas stream in stages, assisting the vapor and liquid phases thoroughly contacted, allowing the components to vaporize and condense easily and separate and flow into specific holding tanks.

While the major constituent of natural gas is methane, there are components such as carbon dioxide (CO_2), hydrogen sulfide (H_2S), and mercaptan derivatives (thiols; RSH), as well as trace amounts of sundry other emissions such as carbonyl sulfide (COS). The fact that methane has a foreseen and valuable end use makes it a desirable product, but in several other situations it is considered a pollutant, having been identified as a greenhouse gas.

In practice, heaters and scrubbers are usually installed at or near to the wellhead. The scrubbers serve primarily to remove sand and other large-particle impurities and the heaters ensure that the temperature of the gas does not drop too low. With natural gas that contains even low quantities of water, natural gas hydrates ($C_nH_{2n+2} \cdot xH_2O$) tend to form when temperatures drop. These hydrates are solid or semisolid compounds, resembling ice-like crystals. If the hydrates accumulate, they can impede the passage of natural gas through valves and gathering systems. To reduce the occurrence of hydrates, small natural gas-fired heating units are typically installed along the gathering pipe wherever it is likely that hydrates may form.

Natural gas hydrates are usually considered as possible nuisances in the development of oil and gas fields, caution in handling the hydrates cannot be overemphasized because of their tendency to explosively decompose. On the other hand, if handled correctly and with caution, hydrates can be used for the safe and economic storage of natural gas. In remote offshore areas, the use of hydrates for natural gas transportation is also presently considered as an economic alternative to the processes based either on liquefaction or on compression.

4.2.1.2   Natural Gas Liquids and Liquefied Petroleum Gas

Natural gas coming directly from a well contains higher molecular weight hydrocarbon derivatives, often referred to as *natural gas liquids* that, in most instances (depending upon the market demand), have a higher value as separate products and making it worthwhile to remove these constituents from the gas stream (Mokhatab et al., 2006; Speight, 2007; Abdel-Aal et al., 2016). The removal of natural gas liquids usually takes place in a relatively centralized processing plant, and uses techniques similar to those used to dehydrate natural gas. There are two basic steps to the treatment of natural gas liquids in the natural gas stream. In the first step, the liquids must be extracted from the natural gas and in the second step the natural gas liquids must be separated into the base constituents. These two processes account for approximately 90% v/v of the total production of natural gas liquids.

Natural gas liquids are the non-methane constituents such as ethane, propane, butane, and pentanes and higher molecular weight hydrocarbon constituents which can be separated as liquids during gas processing (Figures 4.1 and 4.2). The higher molecular weight constituents (i.e., the C_{5+} product) are commonly referred to as gas condensate or natural gasoline. Rich gas will have a high heating value and a high hydrocarbon dew point. When referring to natural gas liquids in the gas stream, the term gallon per thousand cubic feet is used as a measure of high molecular weight hydrocarbon content. On the other hand, the composition of nonassociated gas (sometimes called well gas) is deficient in natural gas liquids. The gas is produced from geological formations that typically do not contain much, if any, hydrocarbon liquids.

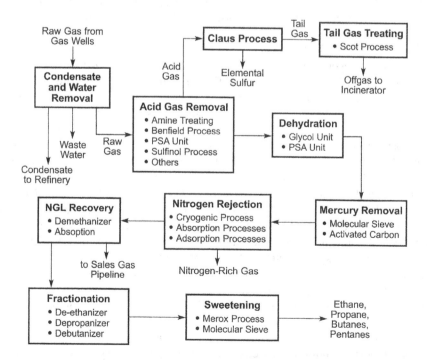

FIGURE 4.1   Schematic diagram for the flow of natural gas cleaning options.

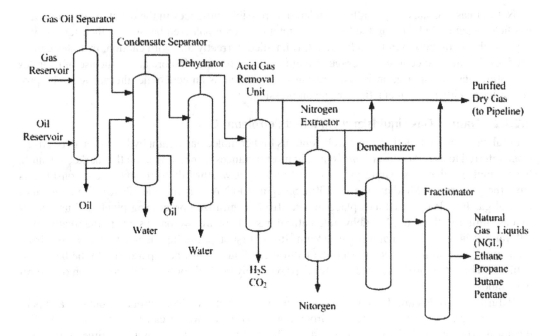

FIGURE 4.2   Representation of the integrated processing units in a gas processing plant.

Generally, the hydrocarbon derivatives having a higher molecular weight than methane as well as any acid gases (carbon dioxide and hydrogen sulfide) are removed from natural gas prior to use of the gas as a fuel. However, since the composition of natural gas is never constant, there are standard test methods by which the composition and properties of natural gas can be determined and, thus, prepared for use. It is not the intent to cover the standard test methods in any detail in this text since descriptions of the test methods are available elsewhere (Speight, 2015; Speight, 2018).

4.2.1.3 Gas Streams from Crude Oil

There are two broad categories of gas that is produced from crude oil. The first category is the associated gas that originated from crude oil formations and also from condensate wells (*condensate gas* but also called *nonassociated gas*). Associated gas can exist separate from oil in the formation (*free gas*), or dissolved in the crude oil (*dissolved gas*). The second category of the gas produced during crude oil refining and the terms *refinery gas* and *process gas* are also often used to include all the gaseous products and byproducts that emanate from a variety of refinery processes.

Organic sulfur compounds and hydrogen sulfide are common contaminants that must be removed prior to most uses. Gas with a significant amount of sulfur impurities, such as hydrogen sulfide, is termed sour gas and often referred to as acid gas. Processed natural gas that is available to end users is tasteless and odorless. However, before gas is distributed to end users, it is odorized by adding small amounts of thiols (RSH, also called mercaptans) to assist in leak detection. Processed natural gas is harmless to the human body but natural gas is a simple asphyxiant and can kill if it displaces air to the point where the oxygen content will not support life.

Once the composition of a mixture has been determined it is possible to calculate various properties such as specific gravity, vapor pressure, calorific value, and dew point. In liquefied petroleum gas where the composition is such that the hydrocarbon dew point is known to be low, a dew point method will detect the presence of traces of water.

Typically, natural gas samples are analyzed for molecular composition by gas chromatography and for stable isotopic composition by isotope ratio mass spectrometry. Carbon isotopic composition was determined for methane (CH_4), ethane (C_2H_6), propane (C_3H_8), and butane, particularly isobutane (C_4H_{10}) (ASTM D3246, 2018). Another important property of the gas streams discussed in this text is the hydrocarbon dew point. The hydrocarbon dew point is reduced to such a level that retrograde condensation, i.e., condensation resulting from pressure drop, cannot occur under the worst conditions likely to be experienced in the gas transmission system. Similarly, the water dew point is reduced to a level sufficient to preclude formation of C_1–C_4 hydrates in the system. Generally, pipeline owners prefer that the specifications for the transmission of natural gas limit the maximum concentration of water vapor allowed. Excess water vapor can cause corrosive conditions, degrading pipelines, and equipment. The water can also condense and freeze or form methane hydrates (Chapter 7) causing blockages. Water vapor content also affects the heating value of natural gas, thus influencing the quality of the gas.

In order to process-associated dissolved natural gas for further use (petrochemical or other), the gas must be separated from the oil in which it is dissolved and is most often performed using equipment installed at or near the wellhead. The actual process used to separate oil from natural gas, as well as the equipment that is used, can vary widely. Although dry pipeline quality natural gas is virtually identical across different geographic areas, raw natural gas from different regions will vary in composition (Table 4.1) (Chapter 2) and therefore separation requirements may emphasize or de-emphasize the optional separation processes. In many instances, natural gas is dissolved in oil underground primarily due to the formation pressure. When this natural gas and oil is produced, it is possible that it will separate on its own and but, in general a separator is required. The conventional type of separator is consisting of a simple closed tank, where the force of gravity serves to separate the liquids like oil from the natural gas.

In certain instances, however, specialized equipment is necessary to separate oil and natural gas. An example of this type of equipment is the low-temperature separator. This is most often used for wells producing high-pressure gas along with light crude oil or condensate. These separators use pressure differentials to cool the wet natural gas and separate the oil and condensate. Wet gas enters the separator, being cooled slightly by a heat exchanger. The gas then travels through a high-pressure liquid *knockout pot* that serves to remove any liquids into a low-temperature separator. The gas then flows into this low-temperature separator through a choke mechanism, which expands the gas as it enters the separator. This rapid expansion of the gas allows for the lowering of the temperature in the separator. After removal of the liquids, the dry gas is sent back through the heat exchanger where it is warmed by the incoming wet gas. By varying the pressure of the gas in various sections of the separator, it is possible to vary the temperature, which causes the crude oil and some water to be condensed out of the wet gas stream.

On the other hand, petroleum refining produces gas streams that contain substantial amounts of acid gases such as hydrogen sulfide and carbon dioxide. These gas streams are produced during initial distillation of the crude oil and during the various conversion processes. Of particular interest is the hydrogen sulfide (H_2S) that arises from the hydrodesulfurization (Chapter 10) and hydrocracking (Chapter 11) of feedstocks that contain organically bound sulfur:

$$[S]_{feedstock} + H_2 \rightarrow H_2S + hydrocarbon\, derivatives$$

Petroleum refining involves, with the exception of heavy crude oil, *primary distillation* (Chapter 7) that results in separation into fractions differing in carbon number, volatility, specific gravity, and other characteristics. The most volatile fraction that contains most of the gases which are generally dissolved in the crude, is referred to as *pipestill gas* or *pipestill light ends* and consists essentially of hydrocarbon gases ranging from methane to butane(s), or sometimes pentane(s).

The gas varies in composition and volume, depending on crude origin and on any additions to the crude made at the loading point. It is not uncommon to reinject light hydrocarbon derivatives such as propane and butane into the crude oil before dispatch by tanker or pipeline. This results in a higher vapor pressure of the crude, but it allows one to increase the quantity of light products obtained at the refinery. Since light ends in most petroleum markets command a premium, while in the oil field itself propane and butane may have to be reinjected or flared, the practice of *spiking* crude oil with liquefied petroleum gas is becoming fairly common.

In addition to the gases obtained by distillation of petroleum, more highly volatile products result from the subsequent processing of naphtha and middle distillate to produce gasoline. Hydrogen sulfide is produced in the desulfurization processes involving hydrogen treatment of naphtha, distillate, and residual fuel; and from the coking or similar thermal treatments of vacuum gas oils (VGOs) and heavy feedstocks (Chapter 8). The most common processing step in the production of gasoline is the catalytic reforming of hydrocarbon fractions in the heptane (C_7) to decane (C_{10}) range.

Additional gases are produced in *thermal cracking processes*, such as the coking or visbreaking processes (Chapter 8) for the processing of heavy feedstocks. In the visbreaking process, fuel oil is passed through externally fired tubes and undergoes liquid-phase cracking reactions, which result in the formation of lighter fuel oil components. Oil viscosity is thereby reduced, and some gases, mainly hydrogen, methane, and ethane, are formed. Substantial quantities of both gas and carbon are also formed in coking (both fluid coking and delayed coking) in addition to the middle distillate and naphtha. When coking a residual fuel oil or heavy gas oil, the feedstock is preheated and contacted with hot carbon (coke) which causes extensive cracking of the feedstock constituents of higher molecular weight to produce lower molecular weight products ranging from methane, via liquefied petroleum gas(es) and naphtha, to gas oil and heating oil. Products from coking processes tend to be unsaturated and olefin components predominate in the tail gases from coking processes.

Another group of refining operations that contributes to gas production is that of the *catalytic cracking processes* (Chapter 9). These consists of fluid-bed catalytic cracking and there are many process variants in which heavy feedstocks are converted into cracked gas, liquefied petroleum gas, catalytic naphtha, fuel oil, and coke by contacting the heavy hydrocarbon with the hot catalyst. Both catalytic and thermal cracking processes, the latter being now largely used for the production of chemical raw materials, result in the formation of unsaturated hydrocarbon derivatives, particularly ethylene ($CH_2=CH_2$), but also propylene (propene, $CH_3CH=CH_2$), isobutylene [isobutene, $(CH_3)_2C=CH_2$] and the *n*-butenes ($CH_3CH_2CH=CH_2$, and $CH_3CH=CHCH_3$) in addition to hydrogen (H_2), methane (CH_4) and smaller quantities of ethane (CH_3CH_3), propane ($CH_3CH_2CH_3$), and butanes [$CH_3CH_2CH_2CH_3$, $(CH_3)_3CH$]. Diolefin derivatives such as butadiene ($CH_2=CHCH=CH_2$) are also present.

A further source of refinery gas is *hydrocracking*, a catalytic high-pressure pyrolysis process in the presence of fresh and recycled hydrogen (Chapter 11). The feedstock is again heavy gas oil or residual fuel oil, and the process is mainly directed at the production of additional middle distillates and gasoline. Since hydrogen is to be recycled, the gases produced in this process again have to be separated into lighter and heavier streams; any surplus recycle gas and the liquefied petroleum gas from the hydrocracking process are both saturated.

In a series of *reforming processes*, commercialized under names such as *platforming*, paraffin and naphthene (cyclic non-aromatic) hydrocarbon derivatives are converted in the presence of hydrogen, and a catalyst is converted into aromatic derivatives, or isomerized to more highly branched hydrocarbon derivatives (Parkash, 2003; Gary et al., 2007; Speight, 2014; Hsu and Robinson, 2017; Speight, 2017). Catalytic reforming processes thus not only result in the formation of a liquid product of higher octane number, but also produce substantial quantities of gases. The latter are rich in hydrogen, but also contain hydrocarbon derivatives from methane to butanes, with a preponderance of propane ($CH_3CH_2CH_3$), *n*-butane ($CH_3CH_2CH_2CH_3$), and isobutane [$(CH_3)_3CH$].

The composition of the process gas varies in accordance with reforming severity and reformer feedstock. All catalytic reforming processes require substantial recycling of a hydrogen stream. Therefore, it is normal to separate reformer gas into a propane ($CH_3CH_2CH_3$) and/or a butane stream [$CH_3CH_2CH_2CH_3$ plus $(CH_3)_3CH$], which becomes part of the refinery liquefied petroleum gas production, and a lighter gas fraction, part of which is recycled. In view of the excess of hydrogen in the gas, all products of catalytic reforming are saturated, and there are usually no olefin gases present in either gas stream.

Gases from hydrocracking units and from catalytic reformer gas units are commonly used in catalytic desulfurization processes (Parkash, 2003; Gary et al., 2007; Speight, 2014; Hsu and Robinson, 2017; Speight, 2017). In the latter, feedstocks ranging from light to vacuum gas oils are passed at pressures of 500–1,000 psi with hydrogen over a hydrofining catalyst. This results mainly in the conversion of organic sulfur compounds to hydrogen sulfide:

$$[S]_{feedstock} + H_2 \rightarrow H_2S + hydrocarbons$$

This process also produces some light hydrocarbon derivatives by hydrocracking. Thus, refinery gas streams, while ostensibly being hydrocarbon in nature, may contain large amounts of acid gases such as hydrogen sulfide and carbon dioxide. Most commercial plants employ hydrogenation to convert organic sulfur compounds into hydrogen sulfide. Hydrogenation is effected by means of recycled hydrogen-containing gases or external hydrogen over a nickel molybdate or cobalt molybdate catalyst.

The presence of impurities in gas streams may eliminate some of the sweetening processes, since some processes remove large amounts of acid gas but not to a sufficiently low concentration. On the other hand, there are those processes not designed to remove (or incapable of removing) large amounts of acid gases whereas they are capable of removing the acid gas impurities to very low levels when the acid gases are present only in low-to-medium concentration in the gas.

Finally, another acid gas, hydrogen chloride (HCl), although not usually considered to be a major emission, is produced from mineral matter and the brine that often accompany petroleum during production and is gaining increasing recognition as a contributor to acid rain. However, hydrogen chloride may exert severe local effects because it does not need to participate in any further chemical reaction to become an acid. Under atmospheric conditions that favor a buildup of stack emissions in the areas where hydrogen chloride is produced, the amount of hydrochloric acid in rain water could be quite high.

In summary, refinery process gas, in addition to hydrocarbon derivatives, may contain other contaminants, such as carbon oxides (CO_x, where x = 1 and/or 2), sulfur oxides (SO_x, where x = 2 and/or 3), as well as ammonia (NH_3), mercaptan derivatives (R-SH), and carbonyl sulfide (COS). From an environmental viewpoint, petroleum processing can result in a variety of gaseous emissions. It is a question of degree insofar as the composition of the gaseous emissions may vary from process to process but the constituents are, in the majority of cases, the same.

4.2.2 Gas Processing

Treated natural gas consists mainly of methane; the properties of both gases (natural gas and methane) are nearly similar. However, natural gas is not pure methane, and its properties are modified by the presence of impurities, such as nitrogen, carbon dioxide, and small amounts of unrecovered higher-boiling (nongaseous at STP) hydrocarbon derivatives. An important property of natural gas is its heating value—relatively high amounts of nitrogen and/or carbon dioxide reduce the heating value of the gas. Pure methane has a heating value of 1,009 Btu/ft^3. This value is reduced to approximately 900 Btu/ft^3 if the gas contains approximately 10% v/v nitrogen and carbon dioxide—the heating value of either nitrogen or carbon dioxide is zero. On the other hand, the heating value of natural gas could exceed methane's due to the presence of higher molecular weight hydrocarbon derivatives, which have higher heating values. For example, the heating value of ethane is 1,800 Btu/ft^3 and the heating value of a product gas is a function of the constituents present in the mixture. In the natural gas trade, a heating value of 1 million Btu is approximately equivalent to 1,000 ft^3 of natural gas.

For petrochemical use, methane must be purified and sent to the petrochemical production site in a condition of acceptability so as not to interfere with (deactivate or destroy) the activity of any processes catalyst(s). To reach the condition of acceptability, the methane must be the end product of treated by a series of processes that have successfully removed the contaminants (Chapter 2) that are present in the raw (untreated) gas. Thus, gas processing (also called *gas cleaning* or *gas refining*) consists of separating all the various hydrocarbon derivatives and fluids from the pure natural gas (Kidnay and Parrish, 2006; Mokhatab et al., 2006; Speight, 2007, 2014). While often assumed to be hydrocarbon derivatives in nature, there are also components of the gaseous products that must be removed prior to release of the gases to the atmosphere or prior to use of the gas in another part of the refinery, i.e., as a fuel gas or as a process feedstock.

The processes that have been developed to accomplish gas purification vary from a simple once-through wash operation to complex multistep recycling systems. In many cases, the process complexities arise because of the need for recovery of the materials used to remove the contaminants or even recovery of the contaminants in the original, or altered, form (Katz, 1959; Kohl and Riesenfeld, 1985; Newman, 1985; Speight, 2007; Mokhatab et al., 2006; Speight, 2014). In addition to the corrosion of equipment by acid gases (Speight, 2014) the escape into the atmosphere of sulfur-containing gases can eventually lead to the formation of the constituents of acid rain, i.e., the oxides of sulfur (sulfur dioxide (SO_2) and sulfur trioxide (SO_3)). Similarly, the nitrogen-containing gases can also lead to nitrous and nitric acids (through the formation of the oxides NO_x, where $x = 1$ or 2), which are the other major contributors to acid rain. The release of carbon dioxide and hydrocarbon derivatives as constituents of refinery effluents can also influence the behavior and integrity of the ozone layer.

Gas processing involves the use of several different types of processes to remove contaminants from gas streams, but there is always overlap between the various processing concepts. In addition, the terminology used for gas processing can often be confusing and/or misleading because of the overlap (Curry, 1981; Maddox, 1982). Gas processing is necessary to ensure that the natural gas prepared for transportation (usually by pipeline) and for sales must be as clean and pure as the specifications dictate. Thus, natural gas, as it is used by consumers, is much different from the natural gas that is brought from underground formations up to the wellhead. Moreover, although natural gas produced at the wellhead is composed primarily of methane, it is by no means is pure.

The processes that have been developed to accomplish gas purification vary from a simple single-stage once-through washing-type operation to complex multistep recycling systems (Speight, 2007; Mokhatab et al., 2006; Speight, 2014). In many cases, the process complexities arise because of the need for recovery of the materials used to remove the contaminants or even recovery of the contaminants in the original, or altered, form (Katz, 1959; Kohl and Riesenfeld, 1985; Newman. 1985; Kohl and Nielsen, 1997; Mokhatab et al., 2006). In addition, the precise area of application, of a given process is difficult to, define and several factors must be considered: before process selection: (i) the types of contaminants in the gas, (ii) the concentrations of contaminants in the gas, (iii) the degree of contaminant removal desired, (iv) the selectivity of acid gas removal required, (v) the temperature of the gas to be processed, (vi) the pressure of the gas to be processed, (vii) the volume of the gas to be processed, (viii) the composition of the gas to be processed, (ix) the ratio of carbon dioxide to hydrogen sulfide ratio in the gas feedstock, and (x) the desirability of sulfur recovery due to environmental issues or economic issues.

4.2.2.1 Acid Gas Removal

In addition to water and natural gas liquids removal, one of the most important parts of gas processing involves the removal of hydrogen sulfide and carbon dioxide, which are generally referred to as contaminants. Natural gas from some wells contains significant amounts of hydrogen sulfide and carbon dioxide and is usually referred to as *sour gas*. Sour gas is undesirable because the sulfur compounds it contains can be extremely harmful, even lethal, to breathe and the gas can also be extremely corrosive. The process for removing hydrogen sulfide from sour gas is commonly referred to as *sweetening* the gas.

There are four general processes used for emission control (often referred to in another, more specific context as flue gas desulfurization: (i) physical adsorption in which a solid adsorbent is used, (ii) physical absorption in which a selective absorption solvent is used, (iii) chemical absorption is which a selective absorption solvent is used, (iv) and catalytic oxidation thermal oxidation (Soud and Takeshita, 1994; Speight, 2007; Mokhatab et al., 2006; Speight, 2014).

4.2.2.1.1 Adsorption

Adsorption is a physical-chemical phenomenon in which the gas is concentrated on the surface of a solid or liquid to remove impurities. It must be emphasized that *absorption* differs from *adsorption* in that absorption is not a physical-chemical surface phenomenon but a process in which the absorbed gas is ultimately distributed throughout the absorbent (liquid). The process depends only on physical solubility and may include chemical reactions in the liquid phase (*chemisorption*). Common absorbing media used are water, aqueous amine solutions, caustic, sodium carbonate, and nonvolatile hydrocarbon oils, depending on the type of gas to be absorbed.

On the other hand, adsorption is usually a gas–solid interaction in which an adsorbent such as activated carbon (the *adsorbent* or *adsorbing medium*) which can be regenerated upon *desorption* (Mokhatab et al., 2006; Speight, 2007, 2014). The quantity of material adsorbed is proportional to the surface area of the solid and, consequently, adsorbents are usually granular solids with a large surface area per unit mass. Subsequently, the captured (adsorbed) gas can be desorbed with hot air or steam either for recovery or for thermal destruction. Adsorber units are widely used to increase a low-gas concentration prior to incineration unless the gas concentration is very high in the inlet air stream and the process is also used to reduce problem odors (or obnoxious odors) from gases. There are several limitations to the use of adsorption systems, but it is generally the case that the major limitation is the requirement for minimization of particulate matter and/or condensation of liquids (e.g., water vapor) that could mask the adsorption surface and drastically reduce its efficiency.

In these processes, a solid with a high-surface area is used. Molecular sieves (zeolites) are widely used and are capable of adsorbing large amounts of gases. In practice, more than one adsorption bed is used for continuous operation. One bed is in use while the other is being regenerated. Regeneration is accomplished by passing hot dry fuel gas through the bed. Molecular sieves are competitive only when the quantities of hydrogen sulfide and carbon disulfide are low. Molecular sieves are also capable of adsorbing water in addition to the acid gases.

Noteworthy commercial processes used are the Selexol, the Sulfinol, and the Rectisol processes. In these processes, no chemical reaction occurs between the acid gas and the solvent. The solvent, or absorbent, is a liquid that selectively absorbs the acid gases and leaves out the hydrocarbons. In the Selexol process, for example, the solvent is dimethyl ether of polyethylene glycol. Raw natural gas passes countercurrently to the descending solvent. When the solvent becomes saturated with the acid gases, the pressure is reduced, and hydrogen sulfide and carbon dioxide are desorbed. The solvent is then recycled to the absorption tower.

4.2.2.1.2 Absorption

Absorption is achieved by dissolution (a physical phenomenon) or by reaction (a chemical phenomenon) (Barbouteau and Dalaud, 1972; Mokhatab et al., 2006; Speight, 2007, 2014). In addition to economic issues or constraints, the solvents used for gas processing should have: (i) a high capacity for acid gas, (ii) a low tendency to dissolve hydrogen, (iii) a low tendency to dissolve low molecular weight hydrocarbon derivatives, (iv) low vapor pressure at operating temperatures to minimize solvent losses, (v) low viscosity, (vi) low thermal stability, (vii) absence of reactivity toward gas components, (viii) low tendency for fouling, (ix) a low tendency for corrosion, and (x) economically acceptable (Mokhatab et al., 2006; Speight, 2007, 2014).

The processes using ethanolamine and potassium phosphate are now widely used. The ethanolamine process, known as the *Girbotol* process, removes acid gases (hydrogen sulfide and carbon dioxide) from liquid hydrocarbon derivatives as well as from natural and from refinery gases. The Girbotol process uses an aqueous solution of ethanolamine ($H_2NCH_2CH_2OH$) that reacts with hydrogen sulfide at low temperatures and releases hydrogen sulfide at high temperatures. The etha-nolamine solution fills a tower called an absorber through which the sour gas is bubbled. Purified gas leaves the top of the tower, and the ethanolamine solution leaves the bottom of the tower with the absorbed acid gases. The ethanolamine solution enters a reactivator tower where heat drives the acid gases from the solution. Ethanolamine solution, restored to its original condition, leaves the bottom of the reactivator tower to go to the top of the absorber tower, and acid gases are released from the top of the reactivator.

The process using potassium phosphate is known as phosphate desulfurization, and it is used in the same way as the Girbotol process to remove acid gases from liquid hydrocarbon derivatives as well as from gas streams. The treatment solution is a water solution of potassium phosphate (K_3PO_4), which is circulated through an absorber tower and a reactivator tower in much the same way as the ethanolamine is circulated in the Girbotol process; the solution is regenerated thermally.

4.2.2.1.3 Chemisorption

Chemisorption (chemical absorption processes are characterized by a high capability of absorbing large amounts of acid gases. They use a solution of a relatively weak base, such as monoethanol-amine (MEA). The acid gas forms a weak bond with the base which can be regenerated easily. Mono- and diethanolamine (DEA) derivatives are frequently used for this purpose. The amine con-centration normally ranges between 15% and 30%. Natural gas is passed through the amine solution where sulfides, carbonates, and bicarbonates are formed. Diethanolamine is a favored absorbent due to its lower corrosion rate, smaller amine loss potential, fewer utility requirements, and minimal reclaiming needs. Diethanolamine also reacts reversibly with 75% of carbonyl sulfides (COS), while the mono- reacts irreversibly with 95% of the carbonyl sulfide and forms a degradation product that must be disposed in an environmentally acceptable manner.

Treatment of gas to remove the acid gas constituents (hydrogen sulfide and carbon dioxide) is most often accomplished by contact of the natural gas with an alkaline solution. The most commonly used treating solutions are aqueous solutions of the ethanolamine or alkali carbonates, although a considerable number of other treating agents have been developed in recent years (Mokhatab et al., 2006; Speight, 2007, 2014). Most of these newer treating agents rely upon physical absorption and chemical reaction. When only carbon dioxide is to be removed in large quantities or when only

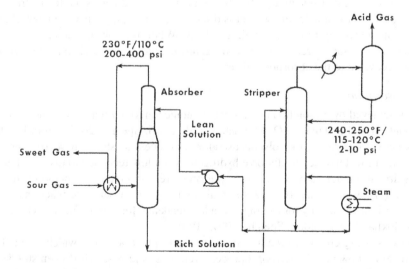

FIGURE 4.3   The amine (olamine) process. (Speight, J.G. 2014. *The Chemistry and Technology of Petroleum.* 5th Edition. CRC Press: Boca Raton, FL. Figure 4.6, p. 708.)

partial removal is necessary, a hot carbonate solution or one of the physical solvents is the most economical selection.

The primary process (Figure 4.3) for sweetening sour natural gas uses an amine (*olamine*) solution to remove the hydrogen sulfide (the *amine process*). The sour gas is run through a tower, which contains the olamine solution. There are two principle amine solutions used, monoethanolamine and diethanolamine. Either of these compounds, in liquid form, will absorb sulfur compounds from natural gas as it passes through. The effluent gas is virtually free of sulfur compounds, and thus loses its sour gas status. Like the process for the extraction of natural gas liquids and glycol dehydration, the amine solution used can be regenerated for reuse. Although most sour gas sweetening involves the amine absorption process, it is also possible to use solid desiccants like iron sponge to remove hydrogen sulfide and carbon dioxide (Mokhatab et al., 2006; Speight, 2007; Abdel-Aal et al., 2016).

Diglycolamine (DGA) is another amine solvent used in the Econamine process (Figures 4.1 and 4.2). Absorption of acid gases occurs in an absorber containing an aqueous solution of diglycolamine, and the heated rich solution (saturated with acid gases) is pumped to the regenerator. Diglycolamine solutions are characterized by low freezing points, which make them suitable for use in cold climates.

The most well-known hydrogen sulfide removal process is based on the reaction of hydrogen sulfide with iron oxide (often also called the iron sponge process or the dry box method) in which the gas is passed through a bed of wood chips impregnated with iron oxide.

The iron oxide process (which was implemented during the 19th century and also referred to as the iron sponge process) is the oldest and still the most widely used batch process for sweetening natural gas and natural gas liquids (Mokhatab et al., 2006; Speight, 2006, 2014). In the process (Figure 4.4) the sour gas is passed down through the bed. In the case where continuous regeneration is to be utilized a small concentration of air is added to the sour gas before it is processed. This air serves to continuously regenerate the iron oxide, which has reacted with hydrogen sulfide, which serves to extend the onstream life of a given tower but probably serves to decrease the total amount of sulfur that a given weight of bed will remove.

The process is usually best applied to gases containing low to medium concentrations (300 ppm) of hydrogen sulfide or mercaptan derivatives. This process tends to be highly selective and does not normally remove significant quantities of carbon dioxide. As a result, the hydrogen sulfide stream from the process is usually high purity. The use of iron oxide process for sweetening sour gas is

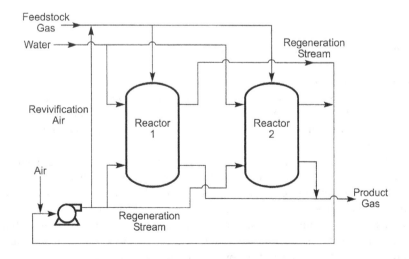

FIGURE 4.4 Iron oxide process.

based on adsorption of the acid gases on the surface of the solid sweetening agent followed by chemical reaction of ferric oxide (Fe_2O_3) with hydrogen sulfide:

$$2Fe_2O_3 + 6H_2S \rightarrow 2Fe_2S_3 + 6H_2O$$

The reaction requires the presence of slightly alkaline water and a temperature below 43°C (110°F) and bed alkalinity (pH + 8 to 10) should be checked regularly, usually on a daily basis. The pH level is to be maintained through the injection of caustic soda with the water. If the gas does not contain sufficient water vapor, water may need to be injected into the inlet gas stream.

The ferric sulfide produced by the reaction of hydrogen sulfide with ferric oxide can be oxidized with air to produce sulfur and regenerate the ferric oxide:

$$2Fe_2S_3 + 3O_2 \rightarrow 2Fe_2O_3 + 6S$$

$$2S + 2O_2 \rightarrow 2SO_2$$

The regeneration step is exothermic and air must be introduced slowly so the heat of reaction can be dissipated. If air is introduced quickly the heat of reaction may ignite the bed. Some of the elemental sulfur produced in the regeneration step remains in the bed. After several cycles this sulfur will form a cake over the ferric oxide, decreasing the reactivity of the bed. Typically, after ten cycles the bed must be removed and a new bed should be introduced into the vessel.

The iron oxide process is one of several metal oxide-based processes that scavenge hydrogen sulfide and organic sulfur compounds (mercaptan derivatives) from gas streams through reactions with the solid-based chemical adsorbent (Kohl and Riesenfeld, 1985). They are typically non-regenerable, although some are partially regenerable, losing activity upon each regeneration cycle. Most of the processes are governed by the reaction of a metal oxide with hydrogen sulfide to form the metal sulfide. For regeneration, the metal oxide is reacted with oxygen to produce elemental sulfur and the regenerated metal oxide. In addition, to iron oxide, the primary metal oxide used for dry sorption processes is zinc oxide.

In the zinc oxide process, the zinc oxide media particles are extruded cylinders 3–4 mm in diameter and 4–8 mm in length (Kohl and Nielsen, 1997; Mokhatab et al., 2006; Speight, 2007; Abdel-Aal et al., 2016) and react readily with the hydrogen sulfide:

$$ZnO + H_2S \rightarrow ZnS + H_2O$$

At increased temperatures (205°C–370°C, 400°F–700°F) zinc oxide has a rapid reaction rate, therefore, providing a short mass transfer zone, resulting in a short length of unused bed and improved efficiency.

Removal of larger amounts of hydrogen sulfide from gas streams requires a continuous process, such as the *Ferrox* process or the Stretford process. The *Ferrox process* is based on the same chemistry as the iron oxide process except that it is fluid and continuous. The *Stretford* process employs a solution containing vanadium salts and anthraquinone disulfonic acid (Mokhatab et al., 2006; Abdel-Aal et al., 2016).

Most hydrogen sulfide removal processes return the hydrogen sulfide unchanged, but if the quantity involved does not justify installation of a sulfur recovery plant (usually a Claus plant), it is necessary to select a process that directly produces elemental sulfur. In the *Beavon-Stretford* process, a hydrotreating reactor converts sulfur dioxide in the off-gas to hydrogen sulfide which is contacted with Stretford solution (a mixture of vanadium salt, anthraquinone disulfonic acid, sodium carbonate, and sodium hydroxide) in a liquid-gas absorber. The hydrogen sulfide reacts stepwise with sodium carbonate and anthraquinone disulfonic acid to produce elemental sulfur, with vanadium serving as a catalyst. The solution proceeds to a tank where oxygen is added to regenerate the reactants. One or more froth or slurry tanks are used to skim the product sulfur from the solution, which is recirculated to the absorber.

4.2.2.1.4   Other Processes

There is a series of alternate processes that involve (i) the use of chemical reactions to remove contaminants from gas streams or (ii) the use of specialized equipment to physically remove contaminants from gas streams.

As example of the first category, i.e., the use of chemical reactions to remove contaminants from gas streams, strong basic solutions are effective solvents for acid gases. However, these solutions are not normally used for treating large volumes of natural gas because the acid gases form stable salts, which are not easily regenerated. For example, carbon dioxide and hydrogen sulfide react with aqueous sodium hydroxide to yield sodium carbonate and sodium sulfide, respectively.

$$CO_2 + 2NaOH \rightarrow Na_2CO_3 + H_2O$$

$$H_2S + 2NaOH \rightarrow Na_2S + 2H_2O$$

However, a strong caustic solution is used to remove mercaptans from gas and liquid streams. In the Merox Process, for example, a caustic solvent containing a catalyst such as cobalt, which is capable of converting mercaptans (RSH) to caustic insoluble disulfides (RSSR), is used for streams rich in mercaptans after removal of H_2S. Air is used to oxidize the mercaptans to disulfides. The caustic solution is then recycled for regeneration. The Merox process is mainly used for treatment of refinery gas streams

As one of the major contaminates in natural gas feeds, carbon dioxide must optimally be removed as it reduces the energy content of the gas and affect the selling price of the natural gas. Moreover, it becomes acidic and corrosive in the presence of water that has a potential to damage the pipeline and the equipment system. In addition, when the issue of transportation of the natural gas to a very far distance is a concern, the use of pipelines will be too expensive so that liquefied natural gas (LNG), gas to liquid and chemicals are considered to be an alternative option. In a liquefied natural gas processing plant, while cooling the natural gas to a very low temperature, the CO_2 can be frozen and block pipeline systems and cause transportation drawback. Hence, the presence of CO_2 in natural gas remains one of the challenging gas separation problems in process engineering for CO_2/CH_4 systems. Therefore, the removal of CO_2 from the natural gas through the purification processes is vital for an improvement in the quality of the product (Mokhatab et al, 2006; Speight, 2007, 2014).

Amine washing of gas emissions involves chemical reaction of the amine with any acid gases with the liberation of an appreciable amount of heat and it is necessary to compensate for the absorption of heat. Amine derivatives such as ethanolamine (monoethanolamine), diethanolamine, triethanolamine, methyl diethanolamine (MDEA), diisopropanolamine (DIPA), and diglycolamine have been used in commercial applications (Katz, 1959; Kohl and Riesenfeld, 1985; Maddox et al., 1985; Polasek and Bullin, 1985; Jou et al., 1985; Pitsinigos and Lygeros, 1989; Kohl and Nielsen, 1997; Mokhatab et al., 2006; Speight, 2007, 2014; Abdel-Aal et al., 2016). The chemistry of the amine process (also called the *olamine process*) can be represented by simple equations for low partial pressures of the acid gases:

$$2RNH_2 + H_2S \rightarrow (RNH_3)_2 S$$

$$2RHN_2 + CO_2 + H_2O \rightarrow (RNH_3)_2 CO_3$$

At high acid gas partial pressure, the reactions will lead to the formation of other products:

$$(RNH_3)_2 S + H_2S \rightarrow 2RNH_3HS$$

$$(RNH_3)_2 CO_3 + H_2O \rightarrow 2RNH_3HCO_3$$

The reaction is extremely rapid and the absorption of hydrogen sulfide is limited only by mass transfer—this is not the case for carbon dioxide—and the reaction is also more complex than these equations would indicate and can lead to a series on unwanted side reactions and byproducts (Mokhatab et al., 2006; Speight, 2007). Regeneration of the amine (olamine) solution leads to near complete desorption of carbon dioxide and hydrogen sulfide. A comparison between monoethanolamine, diethanolamine, and diisopropanolamine shows that monoethanolamine is the cheapest of the three olamines but exhibits the highest heat of reaction and corrosion. On the other hand, diisopropanolamine is the most expensive of the three olamines but exhibits the lowest heat of reaction with a lower propensity for corrosion.

Carbonate washing is a mild alkali process (typically the alkali is potassium carbonate, K_2CO_3) for gas processing for the removal of acid gases (such as carbon dioxide and hydrogen sulfide) from gas streams and uses the principle that the rate of absorption of carbon dioxide by potassium carbonate increases with temperature (Mokhatab et al., 2006; Speight, 2007, 2014). It has been demonstrated that the process works best near the temperature of reversibility of the reactions:

$$K_2CO_3 + CO_2 + H_2O \rightarrow 2KHCO_3$$

$$K_2CO_3 + H_2S \rightarrow KHS + KHCO_3$$

The Fluor process uses propylene carbonate to remove carbon dioxide, hydrogen sulfide, carbonyl sulfide, water, and higher-boiling hydrocarbon derivatives (C_2^+) from natural gas (Abdel-Aal et al., 2016).

Water washing, in terms of the outcome, is almost analogous to (but often less effective than) washing with potassium carbonate (Kohl and Riesenfeld, 1985; Kohl and Nielsen, 1997), and it is also possible to carry out the desorption step by pressure reduction. The absorption is purely physical and there is also a relatively high absorption of hydrocarbon derivatives, which are liberated at the same time as the acid gases.

In *chemical conversion processes*, contaminants in gas emissions are converted to compounds that are not objectionable or that can be removed from the stream with greater ease than the original constituents. For example, a number of processes have been developed that remove hydrogen sulfide and sulfur dioxide from gas streams by absorption in an alkaline solution.

Catalytic oxidation is a chemical conversion process that is used predominantly for destruction of volatile organic compounds and carbon monoxide. These systems operate in a temperature regime in the order of 205°C–595°C (400°F–1,100°F) in the presence of a catalyst—in the absence of the catalyst, the system would require a higher operating temperature. The catalysts used are typically a combination of noble metals deposited on a ceramic base in a variety of configurations (e.g., honeycomb-shaped) to enhance good surface contact. Catalytic systems are usually classified on the basis of bed types such as *fixed bed* (or *packed bed*) and *fluid bed* (*fluidized bed*). These systems generally have very high destruction efficiencies for most volatile organic compounds, resulting in the formation of carbon dioxide, water, and varying amounts of hydrogen chloride (from halogenated hydrocarbon derivatives). The presence in emissions of chemicals such as heavy metals, phosphorus, sulfur, chlorine, and most halogens in the incoming air stream act as poison to the system and can foul up the catalyst. Thermal oxidation systems, without the use of catalysts, also involve chemical conversion (more correctly, chemical destruction) and operate at temperatures in excess of 815°C (1,500°F), or 220°C–610°C (395°F–1,100°F) higher than catalytic systems.

Other processes include the *Alkazid process* for removal of hydrogen sulfide and carbon dioxide using concentrated aqueous solutions of amino acids. The hot potassium carbonate process decreases the acid content of natural and refinery gas from as much as 50% to as low as 0.5% and operates in a unit similar to that used for amine treating. The *Giammarco-Vetrocoke* process is used for hydrogen sulfide and/or carbon dioxide removal. In the hydrogen sulfide removal section, the reagent consists of sodium carbonate (Na_2CO_3) or potassium carbonate (K_2CO_3) or a mixture of the carbonates which contains a mixture of arsenite derivatives and arsenate derivatives; the carbon dioxide removal section utilizes hot aqueous alkali carbonate solution activated by arsenic trioxide (As_2O_3) or selenous acid (H_2SeO_3) or tellurous acid (H_2TeO_3). A word of caution might be added about the last three chemicals which are toxic and can involve stringent environmental-related disposal protocols.

Molecular sieves are highly selective for the removal of hydrogen sulfide (as well as other sulfur compounds) from gas streams and over continuously high absorption efficiency. They are also an effective means of water removal and thus offer a process for the simultaneous dehydration and desulfurization of gas. Gas that has excessively high water content may require upstream dehydration (Mokhatab et al., 2006; Speight, 2007, 2014; Abdel-Aal et al., 2016). The *molecular sieve process* is similar to the iron oxide process. Regeneration of the bed is achieved by passing heated clean gas over the bed. As the temperature of the bed increases, it releases the adsorbed hydrogen sulfide into the regeneration gas stream. The sour effluent regeneration gas is sent to a flare stack, and up to 2% v/v of the gas seated can be lost in the regeneration process. A portion of the natural gas may also be lost by the adsorption of hydrocarbon components by the sieve (Mokhatab et al., 2006; Speight, 2007, 2014).

In this process, unsaturated hydrocarbon components, such as olefin derivatives and aromatic derivatives, tend to be strongly adsorbed by the molecular sieve. Molecular sieves are susceptible to poisoning by such chemicals as glycols and require thorough gas cleaning methods before the adsorption step. Alternatively, the sieve can be offered some degree of protection by the use of *guard beds* in which a less expensive catalyst is placed in the gas stream before contact of the gas with the sieve, thereby protecting the catalyst from poisoning. This concept is analogous to the use of guard beds or attrition catalysts in the petroleum industry. Other processes worthy of note include: (i) the Selexol process, (ii) the Sulfinol process, (iii) the LOCAT process, and (iv) the Sulferox process (Mokhatab et al., 2006; Abdel-Aal et al., 2016).

The Selexol process uses a mixture of the dimethyl ether of propylene glycol as a solvent. It is nontoxic and its boiling point is not high enough for amine formulation. The selectivity of the solvent for hydrogen sulfide (H_2S) is much higher than that for carbon dioxide (CO_2), so it can be used to selectively remove these different acid gases, minimizing carbon dioxide content in the hydrogen sulfide stream sent to the sulfur recovery unit (SRU) and enabling regeneration of solvent for carbon dioxide recovery by economical flashing. In the process, a stream of natural gas is injected in the bottom of the absorption tower operated at 1,000 psi. The rich solvent is flashed in a flash drum (flash reactor) at 200 psi where methane is flashed and recycled back to the absorber and joins the sweet (low-sulfur or no-sulfur) gas stream. The solvent is then flashed at atmospheric pressure and acid gases are flashed off. The solvent is then stripped by steam to completely regenerate the solvent, which is recycled back to the absorber. Any hydrocarbon derivatives are condensed and any remaining acid gases are flashed from the condenser drum. This process is used when there is a high acid gas partial pressure and no heavy hydrocarbon derivatives. Diisopropanolamine can be added to this solvent to remove carbon dioxide to a level suitable for pipeline transportation.

The Sulfinol process uses a solvent that is a composite solvent, consisting of a mixture of diisopropanolamine (30–45% v/v) or MDEA, sulfolane (tetrahydrothiophene dioxide) (40–60% v/v), and water (5–15% v/v). The acid gas loading of the Sulfinol solvent is higher and the energy required for its regeneration is lower than those of purely chemical solvents. At the same time, it has the advantage over purely physical solvents that severe product specifications can be met more easily and co-absorption of hydrocarbon derivatives is relatively low. Aromatic compounds, higher molecular weight hydrocarbon derivatives, and carbon dioxide are soluble to a lesser extent. The process is typically used when the hydrogen sulfide–carbon dioxide ratio is greater than 1:1 or where carbon dioxide removal is not required to the same extent as hydrogen sulfide removal. The process uses a conventional solvent absorption and regeneration cycle in which the sour gas components are removed from the feed gas by countercurrent contact with a lean solvent stream under pressure. The absorbed impurities are then removed from the rich solvent by stripping with steam in a heated regenerator column. The hot lean solvent is then cooled for reuse in the absorber. Part of the cooling may be by heat exchange with the rich solvent for partial recovery of heat energy. The solvent reclaimer is used in a small ancillary facility for recovering solvent components from higher-boiling products of alkanolamine degradation or from other high-boiling or solid impurities.

The LOCAT process uses an extremely dilute solution of iron chelates. A small portion of the chelating agent is depleted in some side reactions and is lost with precipitated sulfur. In this process, sour gas is contacted with the chelating reagent in the absorber and H_2S reacts with the dissolved iron to form elemental sulfur.

$$H_2S + 2Fe^{3+} \rightarrow S + 2Fe^{2+} + 2H^+$$

The sulfur is removed from the regenerator to centrifugation and melting. Application of heat is not required because of the exothermic reaction. The reduced iron ion is regenerated in the regenerator by air blowing:

$$4Fe^{2+} + O_2 + 2H_2O \rightarrow 4Fe^{3+} + 4OH$$

In the Sulferox process, chelating iron compounds are the heart of the process. Sulferox is a redox technology, as is the LOCAT; however, in this case, a concentrated iron solution is used to oxidize hydrogen sulfide to elemental sulfur. Chelating agents are used to increase the solubility of iron in the operating solution. As a result of high iron concentrations in the solution the rate of liquid circulation can be kept low and, consequently, the equipment is small. As in the LOCAT process, there are two basic reactions; the first takes place in the absorber and the second takes place in the regenerator, as in reaction. The key to the Sulferox technology is the ligand used in the process which allows the process to use high total iron concentrations (>1% w/w) in the process, the acid gas enters the contactor, where hydrogen sulfide is oxidized to produce elemental sulfur. The treated gas and the Sulferox solution flow to the separator, where sweet gas exits at the top and the solution is sent to the regenerator where ferrous iron (Fe^{2+}) is oxidized by air to ferric iron (Fe^{3+}) and the solution is regenerated and sent back to the contactor. Sulfur settles in the regenerator and is taken from the bottom to filtration, where sulfur cake is produced. At the top of the regenerator, spent air is released. A makeup Sulferox solution is added to replace the degradation of the ligands. Control of this degradation rate and purging of the degradation products ensures smooth operation of the process.

As an example of the second category, i.e., the use of specialized equipment to physically remove contaminants from gas streams, the removal of *particulate matter* (dust control) from gas streams is an absolute necessity if the stream is to be purified for use as a feedstock for petrochemical production. Historically, particulate matter control has been one of the primary concerns of industries, since the emission of particulate matter is readily observed through the deposition of fly ash and soot as well as in impairment of visibility (Mody and Jakhete, 1988). Different degrees of control can be achieved by use of various types of equipment but selection of the process equipment, which depends upon proper characterization of the particulate matter emitted by a specific process, the appropriate piece of equipment can be selected, sized, installed, and performance tested. The general classes of control devices for particulate matter are categorized as: (i) cyclone collectors, (ii) fabric filters, and (iii) wet scrubbers.

Cyclone collectors are the most common of the inertial collector class and are effective in removing coarser fractions of particulate matter and operate by contacting the particles in the gas stream with a liquid. In principle, the particles are incorporated in a liquid bath or in liquid particles which are much larger and therefore more easily collected. In the process, the particle-laden gas stream enters an upper cylindrical section tangentially and proceeds downward through a conical section. Particles migrate by centrifugal force generated by providing a path for the carrier gas to be subjected to a vortex-like spin. The particles are forced to the wall and are removed through a seal at the apex of the inverted cone. A reverse-direction vortex moves upward through the cyclone and discharges through a top-center opening. Cyclones are often used as primary collectors because of their relatively low efficiency (50%–90% is usual).

Fabric filters are typically designed with non-disposable filter bags. As the gaseous (dust-containing) emissions flow through the filter media (typically cotton, polypropylene, fiberglass, or Teflon), particulate matter is collected on the bag surface as a dust cake. Fabric filters operate with collection efficiencies up to 99.9%, although other advantages are evident but there are several issues that arise during use of such equipment.

Wet scrubbers are devices in which a countercurrent spray liquid is used to remove particles from an air stream. Device configurations include plate scrubbers, packed bed scrubbers, orifice scrubbers, venturi scrubbers, and spray towers, individually or in various combinations. Wet scrubbers can achieve high collection efficiencies at the expense of prohibitive pressure drops. The *foam scrubber* is a modification of a wet scrubber in which the particle-laden gas is passed through a foam generator, where the gas and particles are enclosed by small bubbles of foam.

Other methods include use of high-energy input *venturi scrubbers* or electrostatic scrubbers, where particles or water droplets are charged and flux force/condensation scrubbers, where a hot humid gas is contacted with cooled liquid or where steam is injected into saturated gas. In the latter scrubber the movement of water vapor toward the cold water surface carries the particles with it (*diffusiophoresis*), while the condensation of water vapor on the particles causes the particle size to increase, thus facilitating collection of fine particles.

Electrostatic precipitators operate on the principle of imparting an electric charge to particles in the incoming air stream, which are then collected on an oppositely charged plate across a high-voltage field. Particles of high resistivity create the most difficulty in collection. Conditioning agents such as sulfur trioxide (SO_3) have been used to lower resistivity. Important parameters include design of electrodes, spacing of collection plates, minimization of air channeling, and collection-electrode rapping techniques (used to dislodge particles). Techniques under study include the use of high-voltage pulse energy to enhance particle charging, electron-beam ionization, and wide plate spacing. Electrical precipitators are capable of efficiencies >99% under optimum conditions, but performance is still difficult to predict in new situations.

4.2.2.2   Recovery of Condensable Hydrocarbon Derivatives

Hydrocarbon derivatives that are of higher molecular weight than methane that are present in natural gases are valuable raw materials and important fuels. They can be recovered by lean oil extraction. The first step in this scheme is to cool the treated gas by exchange with liquid propane. The cooled gas is then washed with a cold hydrocarbon liquid, which dissolves most of the condensable hydrocarbons. The uncondensed gas is dry natural gas and is composed mainly of methane with small amounts of ethane and heavier hydrocarbon derivatives. The condensed hydrocarbon derivatives or natural gas liquids are stripped from the rich solvent, which is recycled. Dry natural gas may then be used either as a fuel or as a chemical feedstock. Another way to recover natural gas liquids is by using cryogenic cooling (cooling to very low temperatures in the order of −100°C to −115°C (−150°F to −175°F), which are achieved primarily through lowering the temperatures to below the dew point.

To prevent hydrate formation, natural gas may be treated with glycols, which dissolve water efficiently. Ethylene glycol (EG), diethylene glycol (DEG), and triethylene glycol (TEG) are typical solvents for water removal. Triethylene glycol is preferable in vapor phase processes because of its low vapor pressure, which results in less glycol loss. The triethylene glycol absorber unit typically contains 6–12 bubble-cap trays to accomplish the water absorption. However, more contact stages may be required to reach dew points below −40°F. Calculations to determine the number of trays or feet of packing, the required glycol concentration, or the glycol circulation rate require vapor-liquid equilibrium data. Predicting the interaction between triethylene glycol and water vapor in natural gas over a broad range allows the designs for ultra-low dew point applications to be made.

One alternative to using bubble-cap trays is adiabatic expansion of the inlet gas. The inlet gas is first treated to remove water and acid gases, and then cooled via heat exchange and refrigeration. Further cooling of the gas is accomplished through turbo expanders, and the gas is sent to a demethanizer to separate methane from the higher-boiling hydrocarbon derivatives (often referred to as natural gas liquids, NGLs). Improved recovery of the higher-boiling hydrocarbon derivatives could be achieved through better control strategies and use of online gas chromatographic analysis.

Membrane separation process are very versatile and are designed to process a wide range of feedstocks and offer a simple solution for removal and recovery of higher-boiling hydrocarbon derivatives (natural gas liquids) from natural gas (Abdel-Aal et al., 2016). The separation process is based on high-flux membranes that selectively permeates higher-boiling hydrocarbon derivatives (compared to methane) and are recovered as a liquid after recompression and condensation. The residue stream from the membrane is partially depleted of higher-boiling hydrocarbon derivatives, and is then sent to sales gas stream. Gas permeation membranes are usually made with vitreous polymers that exhibit good selectivity but, to be effective, the membrane must be very permeable with respect to the separation process.

4.2.2.2.1 Extraction

There are two principle techniques for removing natural gas liquids from the natural gas stream: the absorption method and the cryogenic expander process. In the process, a turboexpander is used to produce the necessary refrigeration and very low temperatures and high recovery of light components, such as ethane and propane, can be attained. The natural gas is first dehydrated using a molecular sieve followed by cooling of the dry stream (Figure 4.5). The separated liquid containing most of the heavy fractions is then demethanized, and the cold gases are expanded through a turbine that produces the desired cooling for the process. The expander outlet is a two-phase stream that

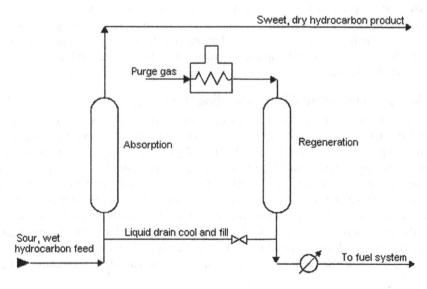

FIGURE 4.5 Drying using a molecular sieve. (Speight, J.G. 2014. *The Chemistry and Technology of Petroleum.* 5th Edition. CRC Press: Boca Raton, FL. Figure 4.4, p. 706.)

is fed to the top of the demethanizer column. This serves as a separator in which: (i) the liquid is used as the column reflux and the separator vapors combined with vapors stripped in the demethanizer are exchanged with the feed gas and (ii) the heated gas, which is partially recompressed by the expander compressor, is further recompressed to the desired distribution pressure in a separate compressor.

The extraction of natural gas liquids from the natural gas stream produces both cleaner, purer natural gas, as well as the valuable hydrocarbon derivatives that are the natural gas liquids themselves. This process allows for the recovery of approximately 90–95% v/v of the ethane originally in the gas stream. In addition, the expansion turbine is able to convert some of the energy released when the natural gas stream is expanded into recompressing the gaseous methane effluent, thus saving energy costs associated with extracting ethane.

4.2.2.2.2 Absorption

The absorption method of high molecular weight recovery of hydrocarbon derivatives is very similar to using absorption for dehydration (Speight, 2007; Mokhatab et al., 2006; Speight, 2014). The main difference is that, in the absorption of natural gas liquids, absorbing oil is used as opposed to glycol. This absorbing oil has an affinity for natural gas liquids in much the same manner as glycol has an affinity for water. Before the oil has picked up any natural gas liquids, it is termed *lean absorption oil*.

The *oil absorption process* involves the countercurrent contact of the lean (or stripped) oil with the incoming wet gas with the temperature and pressure conditions programmed to maximize the dissolution of the liquefiable components in the oil. The *rich* absorption oil (sometimes referred to as *fat* oil), containing natural gas liquids, exits the absorption tower through the bottom. It is now a mixture of absorption oil, propane, butanes, pentanes, and other higher-boiling hydrocarbon derivatives. The rich oil is fed into lean oil stills, where the mixture is heated to a temperature above the boiling point of the natural gas liquids but below that of the oil. This process allows for the recovery of around 75% v/v of the butane isomers and 85–90% v/v of the pentane isomers and higher-boiling constituents from the natural gas stream.

The basic absorption process is subject to modifications that improve process effectiveness and even to target the extraction of specific natural gas liquids. In the refrigerated oil absorption method, where the lean oil is cooled through refrigeration, propane recovery can be in the order of 90%+ v/v and approximately 40% v/v of the ethane can be extracted from the natural gas stream. Extraction of the other, higher-boiling natural gas hydrocarbon derivatives is typically near-quantitative using this process.

The AET process for recovery of liquefied petroleum gas utilizes non-cryogenic absorption to recover ethane, propane, and higher-boiling constituents from natural gas streams. The absorbed gases in the rich solvent from the bottom of the absorber column are fractionated in the solvent regenerator column which separates gases (as an overhead fraction) and lean solvent (as a bottoms fraction). After heat recuperation, the lean solvent is pre-saturated with absorber overhead gases. The chilled solvent flows in the top of the absorber column. The separated gases are sent to storage.

4.2.2.2.3 Fractionation

Fractionation processes are very similar to those processes classed as *liquid removal* processes but often appear to be more specific in terms of the objectives; hence, the need to place the fractionation processes into a separate category. The fractionation processes are those processes that are used (i) to remove the more significant product stream first or (ii) to remove any unwanted light ends from the higher-boiling liquid products.

In the general practice of natural gas processing, the first unit is a de-ethanizer followed by a depropanizer then by a debutanizer and, finally, a butane fractionator. Thus, each column can operate at a successively lower pressure, thereby allowing the different gas streams to flow from column to column by virtue of the pressure gradient, without necessarily the use of pumps.

The purification of hydrocarbon gases by any of these processes is an important part of refinery operations, especially in regard to the production of liquefied petroleum gas. This is actually a mixture of propane and butane, which is an important domestic fuel, as well as an intermediate material in the manufacture of petrochemicals (Parkash, 2003; Gary et al., 2007; Speight, 2014; Hsu and Robinson, 2017; Speight, 2017). The presence of ethane in liquefied petroleum gas must be avoided because of the inability of this lighter hydrocarbon to liquefy under pressure at ambient temperatures and its tendency to register abnormally high pressures in the liquefied petroleum gas containers. On the other hand, the presence of pentane in liquefied petroleum gas must also be avoided, since this particular hydrocarbon (a liquid at ambient temperatures and pressures) may separate into a liquid state in the gas lines.

Typically, natural gas liquids are fractionated to produce three separate streams: (i) an ethane-rich stream, which is used for producing ethylene, (ii) liquefied petroleum gas, which is a propane-butane mixture that is mainly used as a fuel or a chemical feedstock and is evolving into an important feedstock for olefin production, and (iii) natural gasoline is mainly constituted of C_{5+} hydrocarbon derivatives that is added to gasoline to raise its vapor pressure which is also evolving into an important feedstock for olefin production. Natural gas liquids may contain significant amounts of cyclohexane, a precursor for nylon. Recovery of cyclohexane from natural gas liquids by conventional distillation is difficult and not economical because heptane (C_7H_{16}) isomers are also present, which boil at temperatures nearly identical to that of cyclohexane—an extractive distillation process is preferred for cyclohexane recovery.

Thus, after separation of the natural gas liquids from the natural gas stream, they must be separated (fractionated) into the individual constituents prior to sales. The process of fractionation (which is based on the different boiling points of the hydrocarbon derivatives that constitute the natural gas liquids) occurs in stages with each stage involving separation of the hydrocarbon derivatives as individual products. The process commences with the removal of the lower-boiling hydrocarbon derivatives from the feedstock. The particular fractionators are used in the following order: (i) the de-ethanizer, which is used to separate the ethane from the stream of natural gas liquids; (ii) the depropanizer, which is used to separate the propane from the de-ethanized stream; (iii) the debutanizer, which is used to separate the butane isomers, leaving the pentane isomers and higher-boiling hydrocarbon derivatives in the stream; and (iv) the butane splitter or de-isobutanizer, which is used to separate *n*-butane and isobutane.

After the recovery of natural gas liquids, sulfur-free dry natural gas (methane) may be liquefied for transportation through cryogenic tankers. Further treatment may be required to reduce the water vapor below 10 ppm and carbon dioxide and hydrogen sulfide to less than 100 and 50 ppm, respectively. Two methods are generally used to liquefy natural gas: the expander cycle and mechanical refrigeration. In the expander cycle, part of the gas is expanded from a high transmission pressure to a lower pressure. This lowers the temperature of the gas. Through heat exchange, the cold gas cools the incoming gas, which in a similar way cools more incoming gas until the liquefaction temperature of methane is reached.

In mechanical refrigeration, a multicomponent refrigerant consisting of nitrogen, methane, ethane, and propane is used through a cascade cycle. When these liquids evaporate, the heat required is obtained from natural gas, which loses energy/temperature till it is liquefied. The refrigerant gases are recompressed and recycled.

4.2.2.2.4　Enrichment

The natural gas product fed into a petrochemical production system must meet specific quality measures in order for the pipeline grid to operate properly. Consequently, natural gas produced at the wellhead, which in most cases contains contaminants and natural gas liquids, must be processed, i.e., cleaned, before it can be safely delivered to the high-pressure, long-distance pipelines that transport the product to the consuming public. Natural gas that is not within certain specific gravities, pressures, Btu content range, or water content levels will cause operational problems, pipeline deterioration, or can even cause pipeline rupture. Thus, the purpose of *enrichment* is to produce natural gas for sale and enriched tank oil. The tank oil contains more light hydrocarbon liquids than natural petroleum, and the residue gas is drier (leaner, i.e., has lesser amounts of the higher molecular weight hydrocarbon derivatives). Therefore, the process concept is essentially the separation of hydrocarbon liquids from the methane to produce a lean, dry gas.

The natural gas received and transported must (especially in the United States and many other countries) meet the quality standards specified by pipeline. These quality standards vary from pipeline to pipeline and are usually a function of (i) the design of the pipeline system, (ii) the design of any downstream interconnecting pipelines, and (iii) the requirements of the customer. In general, these standards specify that the natural gas should (i) be within a specific Btu content range, typically 1,035 Btu ft^3 ± 50 Btu ft^3, (ii) be delivered at a specified hydrocarbon dew point temperature level to prevent any vaporized gas liquid in the mix from condensing at pipeline pressure, (iii)) contain no more than trace amounts of elements such as hydrogen sulfide, carbon dioxide, nitrogen, water vapor, and oxygen, (iv) be free of particulate solids and liquid water that could be detrimental to the pipeline or its ancillary operating equipment. Gas processing equipment, whether in the field or at processing/treatment plants, assures that these specifications can be met.

In most cases, processing facilities extract contaminants and higher-boiling hydrocarbon derivatives from the gas stream but, in some cases, the gas processors blend some higher-boiling hydrocarbon derivatives into the gas stream in order to bring it within acceptable Btu levels. For instance, in some areas if the produced gas (including coalbed methane (CBM)) does not meet (is below) the Btu requirements of the pipeline operator, in which case a blend of higher Btu-content natural gas or a propane-air mixture is injected to enrich the heat content (Btu value) prior for delivery to the pipeline. In other instances, such as at liquefied natural gas import facilities where the heat content of the re-gasified gas may be too high for pipeline receipt, vaporized nitrogen may be injected into the natural gas stream to lower its Btu content.

Briefly, and because it is sometimes combined with petroleum-based natural gas for processing purposes, coalbed methane is the generic term given to methane gas held in coal and released or produced when the water pressure within the buried coal is reduced by pumping from either vertical or inclined to horizontal surface holes. Thermogenic coalbed methane is predominantly formed during the coalification process whereby organic matter is slowly transformed into coal by increasing temperature and pressure as the organic matter is buried deeper and deeper by additional deposits of organic and inorganic matter over long periods of geological time. On the other hand, late-stage biogenic coalbed methane is formed by relatively recent bacterial processes (involving naturally occurring bacteria associated with meteoric water recharge at outcrop or sub-crop) can dominate the generation of coalbed methane. The amount of methane stored in coal is closely related to the rank and depth of the coal, the higher the coal rank and the deeper the coal seam is presently buried (causing pressure on coal) the greater its capacity to produce and retain methane gas. Gas derived from coal is generally pure and requires little or no processing because it is solely methane and not mixed with heavier hydrocarbon derivatives, such as ethane, which is often present in conventional natural gas.

The number of steps and the type of techniques used in the process of creating pipeline-quality natural gas most often depends upon the source and makeup of the wellhead production stream. Among the several stages of gas processing are: (i) gas-oil separation, (ii) water removal, (iii) liquids removal, (iv) nitrogen removal, (v) acid gas removal, and (vi) fractionation.

In many instances, pressure relief at the wellhead will cause a natural separation of gas from oil (using a conventional closed tank, where gravity separates the gas hydrocarbon derivatives from the heavier oil). In some cases, however, a multistage gas-oil separation process is needed to separate the gas stream from the crude oil. These gas-oil separators are commonly closed cylindrical shells, horizontally mounted with inlets at one end, an outlet at the top for removal of gas, and an outlet at the bottom for removal of oil. Separation is accomplished by alternately heating and cooling (by compression) the flow stream through multiple steps. However, the number of steps and the type of techniques used in the process of creating pipeline-quality natural gas most often depends upon the source and makeup of the gas stream. In some cases, several of the steps may be integrated into one unit or operation, performed in a different order or at alternative locations (lease/plant), or not required at all.

4.2.2.3 Water Removal

Water is a common impurity in gas streams, and removal of water is necessary to prevent condensation of the water and the formation of ice or the formation of gas hydrates. Gas hydrates are solid white compounds formed from a physical-chemical reaction between hydrocarbon derivatives and water under the high pressures and low temperatures used to transport natural gas via pipeline. Hydrates reduce pipeline efficiency.

Water in the liquid phase causes corrosion or erosion problems in pipelines and equipment, particularly when carbon dioxide and hydrogen sulfide are present in the gas. The simplest method of water removal (refrigeration or cryogenic separation) is to cool the gas to a temperature at least equal to or (preferentially) below the dew point (Figure 4.6) (Mokhatab et al., 2006; Speight, 2007, 2014).

In addition to separating petroleum and some condensate from the wet gas stream, it is necessary to remove most of the associated water. Most of the liquid, free water associated with extracted natural gas is removed by simple separation methods at or near the wellhead. However, the removal of the water vapor that exists in solution in natural gas requires a more complex treatment. This treatment consists of dehydrating the natural gas, which usually involves one of two processes: either absorption or adsorption.

Moisture may be removed from hydrocarbon gases at the same time as hydrogen sulfide is removed. Moisture removal is necessary to prevent harm to anhydrous catalysts and to prevent the formation of hydrocarbon hydrates (e.g., $C_3H_8 \cdot 18H_2O$) at low temperatures. A widely used dehydration and desulfurization process is the glycolamine process, in which the treatment solution is a mixture of ethanolamine and a large amount of glycol. The mixture is circulated through an absorber and a reactivator in the same way as ethanolamine is circulated in the Girbotol process. The glycol absorbs moisture from the hydrocarbon gas passing up the absorber; the ethanolamine absorbs hydrogen sulfide and carbon dioxide. The treated gas leaves the top of the absorber; the spent ethanolamine-glycol mixture enters the reactivator tower, where heat drives off the absorbed acid gases and water.

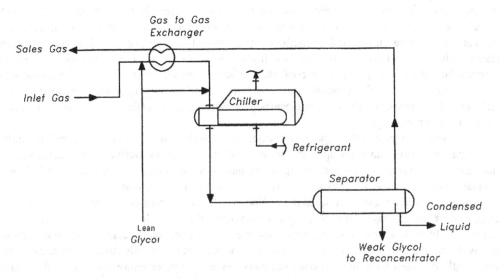

FIGURE 4.6 The glycol refrigeration process. (Speight, J.G. 2014. *The Chemistry and Technology of Petroleum*. 5th Edition. CRC Press: Boca Raton, FL. Figure 4.2, p. 702.)

Absorption occurs when the water vapor is taken out by a dehydrating agent. Adsorption occurs when the water vapor is condensed and collected on the surface. In a majority of cases, cooling alone is insufficient and, for the most part, impractical for use in field operations. Other more convenient water removal options use (i) *hygroscopic* liquids (e.g., diethylene glycol or triethylene glycol) and (ii) solid adsorbents or desiccants (e.g., alumina, silica gel, and molecular sieves). Ethylene glycol can be directly injected into the gas stream in refrigeration plants.

4.2.2.3.1 Absorption

An example of absorption dehydration is known as *glycol dehydration*—the principal agent in this process is diethylene glycol which has a chemical affinity for water (Mokhatab et al., 2006; Speight, 2007; Abdel-Aal et al., 2016). Glycol dehydration involves using a solution of a glycol such as diethylene glycol or triethylene glycol, which is brought into contact with the wet gas stream in a *contactor*. In practice, absorption systems recover 90%–99% by volume of methane that would otherwise be flared into the atmosphere.

In the process, a liquid desiccant dehydrator serves to absorb water vapor from the gas stream. The glycol solution absorbs water from the wet gas and, once absorbed, the glycol particles become heavier and sink to the bottom of the contactor where they are removed. The dry natural gas is then transported out of the dehydrator. The glycol solution, bearing all of the water stripped from the natural gas, is recycled through a specialized boiler designed to vaporize only the water out of the solution. The boiling point differential between water (100°C, 212°F) and glycol (204°C, 400°F) makes it relatively easy to remove water from the glycol solution.

As well as absorbing water from the wet gas stream, the glycol solution occasionally carries with it small amounts of methane and other compounds found in the wet gas. In order to decrease the amount of methane and other compounds that would otherwise be lost, flash tank separator-condensers are employed to remove these compounds before the glycol solution reaches the boiler. The flash tank separator (Chapter 6) consists of a device that reduces the pressure of the glycol solution stream, allowing the methane and other hydrocarbon derivatives to vaporize (*flash*). The glycol solution then travels to the boiler, which may also be fitted with air- or water-cooled condensers, which serve to capture any remaining organic compounds that may remain in the glycol solution. The regeneration (stripping) of the glycol is limited by temperature: diethylene glycol and triethylene glycol decompose at or even before their respective boiling points. Such techniques as stripping of hot triethylene glycol with dry gas (e.g., heavy hydrocarbon vapors, the *Drizo process*) or vacuum distillation are recommended.

Another absorption process, the Rectisol process, is a physical acid gas removal process using an organic solvent (typically methanol) at subzero temperatures, and characteristic of physical acid gas removal processes, it can purify synthesis gas down to 0.1 ppm total sulfur, including hydrogen sulfide (H_2S) and carbonyl sulfide (COS), and carbon dioxide (CO_2) in the ppm range (Mokhatab et al., 2006; Abdel-Aal et al., 2016). The process uses methanol as a wash solvent and the wash unit operates under favorable at temperatures below 0°C (32°F). To lower the temperature of the feed gas temperatures, it is cooled against the cold product streams, before entering the absorber tower. At the absorber tower, carbon dioxide and hydrogen sulfide (with carbonyl sulfide) are removed. By use of an intermediate flash, co-absorbed products such as hydrogen and carbon monoxide are recovered, thus increasing the product recovery rate. To reduce the required energy demand for the carbon dioxide compressor, the carbon dioxide product is recovered in two different pressure steps (medium pressure and lower pressure). The carbon dioxide product is essentially sulfur-free (H_2S-free, COS-free) and water free. The carbon dioxide products can be used for enhanced oil recovery (EOR) and/or sequestration or as pure carbon dioxide for other processes.

4.2.2.3.2 Solid Adsorbents

Adsorption is a physical-chemical phenomenon in which the gas is concentrated on the surface of a solid or liquid to remove impurities. It must be emphasized that *adsorption* differs from *absorption* in that absorption is not a physical-chemical surface phenomenon but a process in which the absorbed gas is ultimately distributed throughout the absorbent (liquid). Dehydration using a solid adsorbent or solid desiccant is the primary form of dehydrating natural gas using adsorption, and usually consists of two or more adsorption towers, which are filled with a solid desiccant (Mokhatab et al., 2006; Speight, 2007; Abdel-Aal et al., 2016). Typical desiccants include activated alumina or a granular silica gel material. Wet natural gas is passed through these towers, from top to bottom. As the wet gas passes around the particles of desiccant material, water is retained on the surface of these desiccant particles. Passing through the entire desiccant bed, almost all the water is adsorbed onto the desiccant material, leaving the dry gas to exit the bottom of the tower. There are several solid desiccants which possess the physical characteristic to adsorb water from natural gas. These desiccants are generally used in dehydration systems consisting of two or more towers and associated regeneration equipment.

Molecular sieves—a class of aluminosilicates which produce the lowest water dew points and which can be used to simultaneously sweeten, dry gases and liquids (Mokhatab et al., 2006; Speight, 2007; Abdel-Aal et al., 2016)—are commonly used in dehydrators ahead of plants designed to recover ethane and other natural gas liquids. These plants operate at very cold temperatures and require very dry feed gas to prevent formation of hydrates. Dehydration to −100°C (−148°F) dew point is possible with molecular sieves. Water dew points less than −100°C (−148°F) can be accomplished with special design and definitive operating parameters (Mokhatab et al., 2006).

Molecular sieves are commonly used to selectively adsorb water and sulfur compounds from light hydrocarbon streams such as liquefied petroleum gas, propane, butane, pentane, light olefin derivatives, and alkylation feed. Sulfur compounds that can be removed are hydrogen sulfide, mercaptan derivatives, sulfide derivatives, and disulfide derivatives. In the process, the sulfur-containing feedstock is passed through a bed of sieves at ambient temperature. The operating pressure must be high enough to keep the feed in the liquid phase. The operation is cyclic in that the adsorption step is stopped at a predetermined time before sulfur breakthrough occurs. Sulfur and water are removed from the sieves by purging with fuel gas at 205°C–315°C (400°F–600°F).

Solid-adsorbent dehydrators are typically more effective than liquid absorption dehydrators (e.g., glycol dehydrators) and are usually installed as a type of straddle system along natural gas pipelines. These types of dehydration systems are best suited for large volumes of gas under very high pressure, and are thus usually located on a pipeline downstream of a compressor station. Two or more towers are required due to the fact that after a certain period of use, the desiccant in a particular tower becomes saturated with water. To regenerate and recycle the desiccant, a high-temperature heater is used to heat gas to a very high temperature and passage of the heated gas stream through a saturated desiccant bed vaporizes the water in the desiccant tower, leaving it dry and allowing for further natural gas dehydration.

Although two-bed adsorbent treaters have become more common (while one bed is removing water from the gas, the other undergoes alternate heating and cooling), on occasion, a three-bed system is used: one bed adsorbs, one is being heated, and one is being cooled. An additional advantage of the three-bed system is the facile conversion of a two-bed system so that the third bed can be maintained or replaced, thereby ensuring continuity of the operations and reducing the risk of a costly plant shutdown.

Silica gel (SiO_2) and alumina (Al_2O_3) have good capacities for water adsorption (up to 8% by weight). Bauxite (crude alumina, Al_2O_3) adsorbs up to 6% by weight water, and molecular sieves adsorb up to 15% by weight water. Silica is usually selected for dehydration of sour gas because of its high tolerance to hydrogen sulfide and to protect molecular sieve beds from plugging by sulfur. Alumina *guard beds* serve as protectors by the act of attrition and may be referred to as an *attrition reactor* containing an *attrition catalyst* (Chapter 6) (Speight, 2014) may be placed ahead of the molecular sieves to remove the sulfur compounds. Downflow reactors are commonly used for adsorption processes, with an upward flow regeneration of the adsorbent and cooling using gas flow in the same direction as adsorption flow.

Solid desiccant units generally cost more to buy and operate than glycol units. Therefore, their use is typically limited to applications such as gases having a high hydrogen sulfide content, very low water dew point requirements, simultaneous control of water, and hydrocarbon dew points. In processes where cryogenic temperatures are encountered, solid desiccant dehydration is usually preferred over conventional methanol injection to prevent hydrate and ice formation (Kidnay and Parrish, 2006).

4.2.2.4 Nitrogen Removal

Nitrogen may often occur in sufficient quantities in natural gas and, consequently, lower the heating value of the gas. Thus, several plants for *nitrogen removal* from natural gas have been built, but it must be recognized that nitrogen removal requires liquefaction and fractionation of the entire gas stream, which may affect process economics. In some cases, the nitrogen-containing natural gas is blended with a gas having a higher heating value and sold at a reduced price depending upon the thermal value (Btu/ft^3).

For high flow-rate gas streams, a cryogenic process is typical and involves the use of the different volatility of methane (b.p. −161.6°C/−258.9°F) and nitrogen (b.p. −195.7°C/−320.3°F) to achieve separation. In the process, a system of compression and distillation columns drastically reduces the temperature of the gas mixture to a point where methane is liquefied and the nitrogen is not. On the other hand, for smaller volumes of gas, a system utilizing pressure swing adsorption (PSA) is a more typical method of separation. In pressure swing adsorption method, methane and nitrogen can be separated by using an adsorbent with an aperture size very close to the molecular diameter of the larger species (the methane) which allows nitrogen to diffuse through the adsorbent. This results in a purified natural gas stream that is suitable for pipeline specifications. The adsorbent can then be regenerated, leaving a highly pure nitrogen stream. The pressure swing adsorption method is a flexible method for nitrogen rejection, being applied to both small and large flow rates.

4.2.2.5 The Claus Process

The Claus process is not so much a gas cleaning process but a process for the disposal of hydrogen sulfide, a toxic gas that originates in natural gas as well as during crude oil processing such as in the coking, catalytic cracking, hydrotreating, and hydrocracking processes. Burning hydrogen sulfide as a fuel gas component or as a flare gas component is precluded by safety and environmental considerations since one of the combustion products is the highly toxic sulfur dioxide (SO$_2$), which is also toxic. As described above, hydrogen sulfide is typically removed from the refinery light ends gas streams through an olamine process after which application of heat regenerates the olamine and forms an acid gas stream. Following from this, the acid gas stream is treated to convert the hydrogen sulfide elemental sulfur and water. The conversion process utilized in most modern refineries is the Claus process, or a variant thereof.

The Claus process (Figure 4.7) involves combustion of approximately one-third of the hydrogen sulfide to sulfur dioxide and then reaction of the sulfur dioxide with the remaining hydrogen sulfide in the presence of a fixed bed of activated alumina, cobalt molybdenum catalyst resulting in the formation of elemental sulfur:

$$2H_2S + 3O_2 \rightarrow 2SO_2 + 2H_2O$$

$$2H_2S + SO_2 \rightarrow 3S + 2H_2O$$

Different process flow configurations are in use to achieve the correct hydrogen sulfide/sulfur dioxide ratio in the conversion reactors.

In a split-flow configuration, one-third split of the acid gas stream is completely combusted and the combustion products are then combined with the non-combusted acid gas upstream of the conversion reactors. In a once-through configuration, the acid gas stream is partially combusted by only providing sufficient oxygen in the combustion chamber to combust one-third of the acid gas.

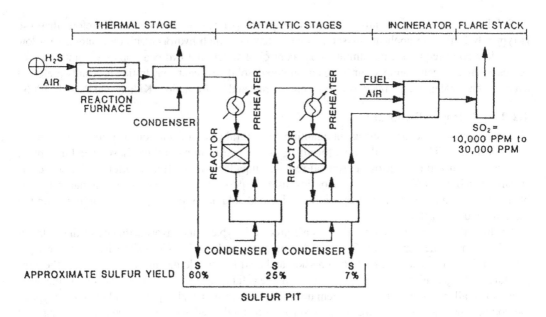

FIGURE 4.7   Claus process. (Speight, J.G. 2014. *The Chemistry and Technology of Petroleum*. 5th Edition. CRC Press: Boca Raton, FL. Figure 4.8, p. 712.)

Two or three conversion reactors may be required depending on the level of hydrogen sulfide conversion required. Each additional stage provides incrementally less conversion than the previous stage. Overall, conversion of 96–97% v/v of the hydrogen sulfide to elemental sulfur is achievable in a Claus process. If this is insufficient to meet air quality regulations, a Claus process tail gas treater is utilized to remove essentially the entire remaining hydrogen sulfide in the tail gas from the Claus unit. The tail gas treater may employ employs a proprietary solution to absorb the hydrogen sulfide followed by conversion to elemental sulfur (Table 4.3).

The SCOT (Shell Claus Off-gas Treating) unit is a most common type of tail gas unit and uses a hydrotreating reactor followed by amine scrubbing to recover and recycle sulfur, in the form of hydrogen, to the Claus unit. In the process, tail gas (containing hydrogen sulfide and sulfur dioxide) is contacted with hydrogen and reduced in a hydrotreating reactor to form hydrogen sulfide

TABLE 4.3
Examples of Tail Gas Treating Processes

Unit	Function
Caustic scrubbing	An incinerator converts trace sulfur compounds in the off-gas to sulfur dioxide that is contacted with caustic which is sent to the wastewater treatment system.
Polyethylene glycol	Off-gas from the Claus unit is contacted with this solution to generate an elemental sulfur product; unlike the Beavon Stretford process, no hydrogenation reactor is used to convert sulfur dioxide to hydrogen sulfide.
Selectox	A hydrogenation reactor converts sulfur dioxide in the off-gas to hydrogen sulfide; a solid catalyst in a fixed bed reactor converts the hydrogen sulfide to elemental sulfur which is recovered for sales.
Sulfite/Bisulfite	Following Claus reactors, an incinerator converts trace sulfur compounds to sulfur dioxide which is then contacted with sulfite solution in an absorber, where the sulfur dioxide reacts with the sulfite to produce a bisulfite solution; the gas is then emitted to the stack; the bisulfite is regenerated and liberated sulfur dioxide is sent to the Claus units for recovery.

TABLE 4.4

Sulfur Removal/Recovery Processes

Sodium hydrosulfide	Fuel gas containing hydrogen sulfide is contacted with sodium hydroxide in an absorption column. The resulting liquid is the product of sodium hydrosulfide (NaHS).
Iron chelate	Fuel gas containing hydrogen sulfide is contacted with iron chelate catalyst dissolved in solution; hydrogen sulfide is converted to elemental sulfur, which is recovered.
Stretford	Similar to iron chelate, except Stretford solution is used instead of iron chelate solution.
Ammonium thiosulfate	In this process, hydrogen sulfide is contacted with air to form sulfur dioxide, which is contacted with ammonia in a series of absorption column to produce ammonium thiosulfate for offsite sale.
Hyperion	Fuel gas is contacted over a solid catalyst to form elemental sulfur; the sulfur is collected and sold. The catalyst is comprised of iron and naphthoquinone sulfonic acid.
Sulfatreat	The Sulfatreat material is a black granular solid powder; the hydrogen sulfide forms a chemical bond with the solid; when the bed reaches capacity, the Sulfatreat solids are removed and replaced with fresh material. The sulfur is not recovered.
Hysulf	Hydrogen sulfide is contacted with a liquid quinone in an organic solvent such as n-methyl-2-pyrolidone (NMP), forming sulfur; the sulfur is removed and the quinone reacted to its original state, producing hydrogen gas.

and water. The catalyst is typically cobalt/molybdenum on alumina. The gas is then cooled in a water contractor. The hydrogen sulfide-containing gas enters an amine absorber which is typically in a system segregated from the other refinery amine systems. The purpose of segregation is two-fold: (i) the tail gas treater frequently uses a different amine than the rest of the plant and (ii) the tail gas is frequently cleaner than the refinery fuel gas (in regard to contaminants) and segregation of the systems reduces maintenance requirements for the SCOT unit. Amines chosen for use in the tail gas system tend to be more selective for hydrogen sulfide and are not affected by the high levels of carbon dioxide in the off-gas.

The hydrotreating reactor converts sulfur dioxide in the off-gas to hydrogen sulfide that is then contacted with a Stretford solution (a mixture of a vanadium salt, anthraquinone disulfonic acid, sodium carbonate, and sodium hydroxide) in a liquid-gas absorber (Abdel-Aal et al., 2016). The hydrogen sulfide reacts stepwise with sodium carbonate and the anthraquinone sulfonic acid to produce elemental sulfur, with vanadium serving as a catalyst. The solution proceeds to a tank where oxygen is added to regenerate the reactants. One or more froth or slurry tanks are used to skim the product sulfur from the solution, which is recirculated to the absorber. Other tail gas treating processes include: (i) caustic scrubbing, (ii) polyethylene glycol treatment, (iii) Selectox process, and (iv) a sulfite/bisulfite tail gas treating (Mokhatab et al., 2006; Speight, 2007, 2014).

A sulfur removal process (Table 4.4) must be very precise, since natural gas contains only a small quantity of sulfur-containing compounds that must be reduced several orders of magnitude. Most consumers of natural gas require less than 4 ppm in the gas—a characteristic feature of natural gas that contains hydrogen sulfide is the presence of carbon dioxide (generally in the range of 1–4% v/v). In cases where the natural gas does not contain hydrogen sulfide, there may also be a relative lack of carbon dioxide.

4.3 PETROLEUM STREAMS

In a very general sense, crude oil refining can be traced back over 5,000 years to the times when asphalt materials and oils were isolated from areas where natural seepage occurred and the resulting bitumen was send for construction purposes (Abraham, 1945; Forbes, 1958a, 1958b, 1959; Hoiberg, 1960; Forbes, 1964; Speight, 1978). Any treatment of the asphalt (such as hardening in the air prior to use) or of the oil (such as allowing for more volatile components to escape prior to use in lamps) may be considered to be refining under the general definition of refining. However, crude oil refining as it is now practiced is a very recent science and many innovations evolved during the 20th century.

Briefly, crude oil refining is the separation of crude oil into fractions and the subsequent treating of these fractions to yield marketable products (Figure 4.8) (Parkash, 2003; Gary et al., 2007; Speight, 2008, 2011, 2014, 2015; Hsu and Robinson, 2017; Speight, 2017). In fact, a refinery is essentially a group of manufacturing plants, which vary in number with the variety of products produced. However, in addition to the simplified schematic of a refinery, the refinery (for the present purposes) can actually be considered as two refineries—(i) a section for low-viscosity feedstocks and (ii) a section for high-viscosity feedstocks (Parkash, 2003; Gary et al., 2007; Speight, 2014; Hsu and Robinson, 2017; Speight, 2017).

In this way, the processes can be selected and products manufactured to give a balanced operation in which the refinery feedstock oil is converted into a variety of products in amounts that are in accord with the demand for each. For example, the manufacture of products from the lower-boiling portion of crude oil automatically produces a certain amount of higher-boiling components using distillation and various thermal processes (Parkash, 2003; Gary et al., 2007; Speight, 2014; Hsu and Robinson, 2017; Speight, 2017). If the latter cannot be sold as, say, heavy fuel oil, these products will accumulate until refinery storage facilities are full. To prevent the occurrence of such a situation, the refinery must be flexible and be able to change operations as needed. This usually means more processes are required for refining the heavier feedstocks: (i) thermal processes to change an excess of heavy fuel oil into more gasoline with coke as the residual product or (ii) a vacuum distillation process to separate the viscous feedstock into lubricating oil stocks and asphalt.

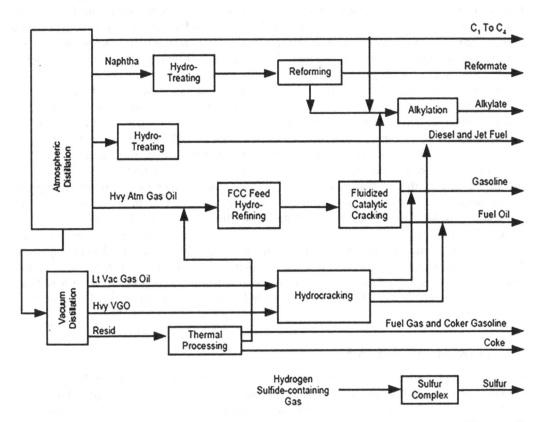

FIGURE 4.8   Schematic diagram of a conversion refinery showing the relative placement of the various processing units.

To convert crude oil into desired products in an economically feasible and environmentally acceptable manner. Refinery process for crude oil are generally divided into three categories: (i) separation processes, of which distillation is the prime example; (ii) conversion processes, of which coking and catalytic cracking are prime example; and (iii) finishing processes, of which hydrotreating to remove sulfur is a prime example.

4.3.1 REFINERY CONFIGURATION

The simplest refinery configuration is the *topping refinery*, which is designed to prepare feedstocks for petrochemical manufacture or for production of industrial fuels (Table 4.5). The topping refinery consists of tankage, a distillation unit, recovery facilities for gases and light hydrocarbon derivatives, and the necessary utility systems (steam, power, and water treatment plants). Topping refineries produce large quantities of unfinished oils and are highly dependent on local markets, but the addition of hydrotreating and reforming units to this basic configuration results in a more flexible *hydroskimming refinery*, which can also produce desulfurized distillate fuels and high-octane gasoline. These refineries may produce up to half of their output as residual fuel oil, and they face increasing market loss as the demand for low-sulfur (even no-sulfur) and high-sulfur fuel oil increases.

The most versatile refinery configuration is the *conversion refinery* which incorporates all the basic units found in both the topping and hydroskimming refineries, but it also features gas oil conversion plants such as catalytic cracking and hydrocracking units, olefin conversion plants such as alkylation or polymerization units, and, frequently, coking units for sharply reducing or eliminating the production of residual fuels (Parkash, 2003; Gary et al., 2007; Speight, 2014; Hsu and Robinson, 2017; Speight, 2017). The predominant steps in a deep conversion are (i) catalytic cracking and (ii) hydrocracking—modern conversion refineries may produce two-thirds of their output as unleaded gasoline, with the balance distributed between liquefied petroleum gas, jet fuel, diesel fuel, and a small quantity of coke. Many such refineries also incorporate solvent extraction processes for

TABLE 4.5
Examples of Refinery Types

Refinery Type	Processes	Type	Complexity	Comparative[a]
Topping	Distillation	Skimming	Low	1
Hydroskimming	Distillation	Hydroskimming	Moderate	3
	Reforming			
	Hydrotreating			
Conversion	Distillation	Cracking	High	6
	Fluid catalytic cracking			
	Hydrocracking			
	Reforming			
	Alkylation			
	Hydrotreating			
Deep conversion	Distillation	Coking	Very high	10
	Coking			
	Fluid catalytic cracking			
	Hydrocracking			
	Reforming			
	Alkylation			
	Hydrotreating			

[a] Indicates complexity on an arbitrary numerical scale of 1–10 with "1" being the least complex and "10" being the most complex.

manufacturing lubricants and petrochemical units with which to recover propylene, benzene, toluene, and xylenes for further processing into polymers.

Since a refinery is a group of integrated manufacturing plants (Parkash, 2003; Gary et al., 2007; Speight, 2014; Hsu and Robinson, 2017; Speight, 2017) that are selected to give a balanced production of salable products in amounts that are in accord with the demand for each, it is necessary to prevent the accumulation of nonsalable products, the refinery must be flexible and be able to change operations as needed. The complexity of petroleum is emphasized insofar as the actual amount of the products vary significantly from one crude oil to another (Parkash, 2003; Gary et al., 2007; Speight, 2014; Hsu and Robinson, 2017; Speight, 2017). In addition, the configuration of refineries may vary from refinery to refinery. Some refineries may be more oriented toward the production of gasoline (large reforming and/or catalytic cracking) whereas the configuration of other refineries may be more oriented toward the production of middle distillates such as jet fuel, and gas oil that can also lead to the production of petrochemical intermediates.

In addition, the predominant processes that are used to produce starting materials for the production of petrochemicals are the cracking process by which the molecular size of the crude oil constituents is reduced (cracked) to the required molecular dimensions of the petrochemical starting materials.

The term *cracking* applies to the decomposition of petroleum constituents, which is induced by elevated temperatures (>350°C, >660°F) whereby the higher molecular weight constituents of petroleum are converted to lower molecular weight products. Cracking reactions involve carbon-carbon bond rupture and are thermodynamically favored at high temperature.

4.3.2 CRACKING PROCESSES

4.3.2.1 Thermal Cracking Processes

With the dramatic increases in the number of gasoline-powered vehicles, distillation processes (Chapters 4 and 7) were not able to completely fill the increased demand for gasoline. In 1913, the *thermal cracking process* was developed and is the phenomenon by which higher-boiling (higher molecular weight) constituents in petroleum are converted into lower-boiling (lower molecular weight) products application of elevated temperatures (usually in the order of >350°C, >660°F).

Thermal cracking is the oldest and in principle the simplest refinery conversion process. The temperature and pressure depends on the type of feedstock and the product requirements as well as the residence time. Thermal cracking processes allow the production of lower molecular weight products such as the constituents of liquefied petroleum gas and naphtha/gasoline constituents from higher molecular weight fraction such as gas oils and residua. The simplest thermal cracking process—the visbreaking process (Chapters 4 and 8)—is used to upgrade fractions such as distillation residua and other heavy feedstocks (Chapters 4 and 7) to produce fuel oil that meets specifications or feedstocks for other refinery processes.

Thus, cracking is a phenomenon by which higher-boiling constituents (higher molecular weight constituents) in petroleum are converted into lower-boiling (lower molecular weight) products. However, certain products may interact with one another to yield products having higher molecular weights than the constituents of the original feedstock. Some of the products are expelled from the system as, say, gases, gasoline-range materials, kerosene-range materials, and the various intermediates that produce other products such as coke. Materials that have boiling ranges higher than gasoline and kerosene may (depending upon the refining options) be referred to as *recycle* stock, which is recycled in the cracking equipment until conversion is complete.

In thermal cracking processes, some of the lower molecular weight products are expelled from the system as gases, gasoline-range materials, kerosene-range materials, and the various intermediates that produce other products such as coke. Materials that have boiling ranges higher than gasoline and kerosene may (depending upon the refining options) be referred to as *recycle* stock, which is recycled in the cracking equipment until conversion is complete.

Thermal cracking is a *free radical* chain reaction. A free radical (in which an atom or group of atoms possessing an unpaired electron) is very reactive (often difficult to control) and it is the mode of reaction of free radicals that determines the product distribution during thermal cracking (i.e., non-catalytic thermal decomposition). In addition, a significant feature of hydrocarbon free radicals is the resistance to isomerization during the existence of the radical. For example, thermal cracking does not produce any degree of branching in the products (by migration of an alkyl group) other than that already present in the feedstock. Nevertheless, the classical chemistry of free radical formation and behavior involves the following chemical reactions—it can only be presumed that the formation of free radicals during thermal (non-catalytic) cracking follows similar paths:

1. *Initiation reaction*, where a single molecule breaks apart into two free radicals. Only a small fraction of the feedstock constituents may actually undergo initiation, which involves breaking the bond between two carbon atoms, rather than the thermodynamically stronger bond between a carbon atom and a hydrogen atom.

$$CH_3CH_3 \rightarrow 2CH_3\bullet$$

2. *Hydrogen abstraction reaction* in which the free radical abstracts a hydrogen atom from another molecule:

$$CH_3\bullet + CH_3CH_3 \rightarrow CH_4 + CH_3CH_2\bullet$$

3. *Radical decomposition reaction* in which a free radical decomposes into an alkene:

$$CH_3CH_2\bullet \rightarrow CH_2=CH_2 + H\bullet$$

4. *Radical addition reaction* in which a radical reacts with an alkene to form a single, larger free radical:

$$CH_3CH_2\bullet + CH_2=CH_2 \rightarrow CH_3CH_2CH_2CH_2\bullet$$

5. *Termination reaction* in which two free radicals react with each other to produce the products—two common forms of termination reactions are *recombination reactions* (in which two radicals combine to form one molecule) and *disproportionation reactions* (in which one free radical transfers a hydrogen atom to the other to produce an alkene and an alkane):

$$CH_3\bullet + CH_3CH_2\bullet \rightarrow CH_3CH_2CH_3$$

$$CH_3CH_2\bullet + CH_3CH_2\bullet \rightarrow CH_2=CH_2 + CH_3CH_3$$

The smaller free radicals, hydrogen, methyl, and ethyl are more stable than the larger radicals. They will tend to capture a hydrogen atom from another hydrocarbon, thereby forming a saturated hydrocarbon and a new radical. In addition, many thermal cracking processes and many different chemical reactions occur simultaneously. Thus, an accurate explanation of the mechanism of the thermal cracking reactions is difficult. The primary reactions are the decomposition of higher molecular weight species into lower molecular weight products.

As the molecular weight of the hydrocarbon feedstock increases the reactions become much more complex lading to a wider variety of products. For example, using a more complex hydrocarbon (dodecane, $C_{12}H_{26}$) as the example, two general types of reaction occur during cracking:

1. The decomposition of high molecular weight constituents into lower molecular weight constituents (*primary reactions*):

$$CH_3(CH_2)_{10}CH_3 \rightarrow CH_3(CH_2)_8CH_3 + CH_2=CH_2$$

$$CH_3(CH_2)_{10}CH_3 \rightarrow CH_3(CH_2)_7CH_3 + CH_2=CHCH_3$$

$$CH_3(CH_2)_{10}CH_3 \rightarrow CH_3(CH_2)_6CH_3 + CH_2=CHCH_2CH_3$$

$$CH_3(CH_2)_{10}CH_3 \rightarrow CH_3(CH_2)_5CH_3 + CH_2=CH(CH_2)_2CH_3$$

$$CH_3(CH_2)_{10}CH_3 \rightarrow CH_3(CH_2)_4CH_3 + CH_2=CH(CH_2)_3CH_3$$

$$CH_3(CH_2)_{10}CH_3 \rightarrow CH_3(CH_2)_3CH_3 + CH_2=CH(CH_2)_4CH_3$$

$$CH_3(CH_2)_{10}CH_3 \rightarrow CH_3(CH_2)_2CH_3 + CH_2=CH(CH_2)_5CH_3$$

$$CH_3(CH_2)_{10}CH_3 \rightarrow CH_3CH_2CH_3 + CH_2=CH(CH_2)_6CH_3$$

$$CH_3(CH_2)_{10}CH_3 \rightarrow CH_3CH_3 + CH_2=CH(CH_2)_7CH_3$$

$$CH_3(CH_2)_{10}CH_3 \rightarrow CH_4 + CH_2=CH(CH_2)_8CH_3$$

2. Reactions by which some of the primary products interact to form higher molecular weight materials (secondary reactions):

$$CH_2=CH_2 + CH_2=CH_2 \rightarrow CH_3CH_2CH=CH_2$$

$$RCH=CH_2 + R^1CH=CH_2 \rightarrow cracked\,residuum + coke + other\,products$$

Thus, from the chemistry of the thermal decomposing of pure compounds (and assuming little interference from other molecular species in the reaction mixture), it is difficult but not impossible to predict the product types that arise from the thermal cracking of various feedstocks. However, during thermal cracking, all the reactions illustrated above can and do occur simultaneously and to some extent are uncontrollable. However, one of the significant features of hydrocarbon-free radicals is their resistance to isomerization, for example, migration of an alkyl group and, as a result, thermal cracking does not produce any degree of branching in the products other than that already present in the feedstock.

Data obtained from the thermal decomposition of pure compounds indicate certain decomposition characteristics that permit predictions to be made of the product types that arise from the thermal cracking of various feedstocks. For example, normal paraffin derivatives are believed to form, initially, higher molecular weight material, which subsequently decomposes as the reaction progresses. Other paraffinic materials and (terminal) olefin derivatives are produced. An increase in pressure inhibits the formation of low molecular weight gaseous products and therefore promotes the formation of higher molecular weight materials.

Furthermore, for saturated hydrocarbon derivatives, the connecting link between gas-phase pyrolysis and liquid-phase thermal degradation is the concentration of alkyl radicals. In the gas phase, alkyl radicals are present in low concentration and undergo unimolecular radical decomposition reactions to form α-olefin derivatives and smaller alkyl radicals. In the liquid phase, alkyl radicals are in much higher concentration and prefer hydrogen abstraction reactions to radical decomposition reactions. It is this preference for hydrogen abstraction reactions that gives liquid-phase thermal degradation a broad product distribution.

Branched paraffin derivatives react somewhat differently to the normal paraffin derivatives during cracking processes and produce substantially higher yields of olefin derivatives having one fewer carbon atoms than the parent hydrocarbon. Cycloparaffin derivatives (naphthenes) react differently to their noncyclic counterparts and are somewhat more stable. For example, cyclohexane produces hydrogen, ethylene, butadiene, and benzene: alkyl-substituted cycloparaffin derivatives decompose by means of scission of the alkyl chain to produce an olefin and a methyl or ethyl cyclohexane. The aromatic ring is considered fairly stable at moderate cracking temperatures (350°C–500°C, 660°F–930°F). Alkylated aromatic derivatives, like the alkylated naphthenes, are more prone to dealkylation than to ring destruction. However, ring destruction of the benzene derivatives occurs above 500°C (930°F), but condensed aromatic derivatives may undergo ring destruction at somewhat lower temperatures (450°C, 840°F).

Generally, the relative ease of cracking of the various types of hydrocarbon derivatives *of the same molecular weight* is given in the following descending order: (i) paraffin derivatives, (ii) olefin derivatives, (iii) naphthene derivatives, and (iv) aromatic derivatives. To remove any potential confusion, paraffin derivatives are the least stable and aromatic derivatives are the most stable. Also, within any type of hydrocarbon, the higher molecular weight hydrocarbon derivatives tend to crack easier than the lighter ones. Paraffin derivatives are by far the easiest hydrocarbon derivatives to crack with the rupture most likely to occur between the first and second carbon bonds in the lighter paraffin derivatives. However, as the molecular weight of the paraffin molecule increases rupture tend to occur nearer the middle of the molecule. The main secondary reactions that occur in thermal cracking are polymerization and condensation.

Two extremes of the thermal cracking in terms of product range are represented by high-temperature processes: (i) steam cracking or (ii) pyrolysis. Steam cracking is a process in which feedstock is decomposed into lower molecular weight (often unsaturated) products saturated hydrocarbon derivatives. Steam cracking is the key process in the petrochemical industry, producing ethylene ($CH_2=CH_2$), propylene ($CH_3CH=CH_2$), butylene [$CH_3CH_2CH=CH_2$ and/or $CH_3CH=CHCH_3$ and/or $(CH_3)_2C=CH_2$], benzene (C_6H_6), toluene ($C_6H_5CH_3$), ethylbenzene ($C_6H_5CH_2CH_3$), and the xylene isomers (1,2-$CH_3C_6H_4CH_3$, 1,3-$CH_3C_6H_4CH_3$, and 1,4-$CH_3C_6H_4CH_3$). These intermediates are converted into a variety of polymers (plastics), solvents, resins, fibers, detergents, ammonia, and other synthetic organic compounds.

In the process, a gaseous or liquid hydrocarbon feed such as ethane or naphtha is diluted with steam and briefly heated in a furnace (at approximately 850°C, 1,560°F) in the absence of oxygen at a short residence time (often in the order of milliseconds). After the cracking temperature has been reached, the products are rapidly quenched in a heat exchanger. The products produced in the reaction depend on the composition of the feedstock, the feedstock/steam ratio, the cracking temperature, and the residence time. Pyrolysis processes require temperatures in the order of 750°C–900°C (1,380°F–1,650°F) to produce high yields of low molecular weight products, such as ethylene, for petrochemical use. Delayed coking, which uses temperature in the order of 500°C (930°F) is used to produced distillates from nonvolatile residua as well as coke for fuel and other uses—such as the production of electrodes for the steel and aluminum industries.

4.3.2.2 Catalytic Cracking Processes

Catalytic cracking is the thermal decomposition of petroleum constituents in the presence of a catalyst. Thermal cracking has been superseded by catalytic cracking as the process for gasoline manufacture. Indeed, gasoline produced by catalytic cracking is richer in branched paraffin derivatives, cycloparaffin derivatives, and aromatic derivatives, which all serve to increase the quality of the gasoline. Catalytic cracking also results in production of the maximum amount of butene derivatives and butane derivatives (C_4H_8 and C_4H_{10}) rather than production of ethylene and ethane (C_2H_4 and C_2H_6).

Catalytic cracking processes evolved in the 1930s from research on petroleum and coal liquids. The petroleum work came to fruition with the invention of acid cracking. The work to produce liquid fuels from coal, most notably in Germany, resulted in metal sulfide hydrogenation catalysts. In the 1930, a catalytic cracking catalyst for petroleum that used solid acids as catalysts was developed using acid-treated clay minerals. Clay minerals are a family of crystalline aluminosilicate solids, and the acid treatment develops acidic sites by removing aluminum from the structure. The acid sites also catalyze the formation of coke, and Houdry developed a moving bed process that continuously removed the cooked beads from the reactor for regeneration by oxidation with air.

Although thermal cracking is a free radical (neutral) process, catalytic cracking is an ionic process involving carbonium ions, which are hydrocarbon ions having a positive charge on a carbon atom. The formation of carbonium ions during catalytic cracking can occur by: (i) addition of a proton from an acid catalyst to an olefin and/or (ii) abstraction of a hydride ion (H^-) from a hydrocarbon by the acid catalyst or by another carbonium ion. However, carbonium ions are not formed by cleavage of a carbon-carbon bond.

In essence, the use of a catalyst permits alternate routes for cracking reactions, usually by lowering the free energy of activation for the reaction. The acid catalysts first used in catalytic cracking were amorphous solids composed of approximately 87% silica (SiO_2) and 13% alumina (Al_2O_3) and were designated as low-alumina catalysts. However, this type of catalyst is now being replaced by crystalline aluminosilicates (zeolites) or molecular sieves.

The first catalysts used for catalytic cracking were acid-treated clay minerals, formed into beads. In fact, clay minerals are still employed as catalyst in some cracking processes (Speight, 2014). Clay minerals are a family of crystalline aluminosilicate solids, and the acid treatment develops acidic sites by removing aluminum from the structure. The acid sites also catalyze the formation of coke, and the development of a moving bed process that continuously removed the cooked beads from the reactor reduced the yield of coke; clay regeneration was achieved by oxidation with air.

Clays are natural compounds of silica and alumina, containing major amounts of the oxides of sodium, potassium, magnesium, calcium, and other alkali and alkaline earth metals. Iron and other transition metals are often found in natural clays, substituted for the aluminum cations. Oxides of virtually every metal are found as impurity deposits in clay minerals.

Clay minerals are layered crystalline materials. They contain large amounts of water within and between the layers. Heating the clays above 100°C can drive out some or all of this water; at higher temperatures, the clay structures themselves can undergo complex solid state reactions. Such behavior makes the chemistry of clays a fascinating field of study in its own right. Typical clays include kaolinite, montmorillonite, and illite. They are found in most natural soils and in large, relatively pure deposits, from which they are mined for applications ranging from adsorbents to paper making.

Once the carbonium ions are formed, the modes of interaction constitute an important means by which product formation occurs during catalytic cracking. For example, isomerization either by hydride ion shift or by methyl group shift, both of which occur readily. The trend is for stabilization of the carbonium ion by *movement* of the charged carbon atom toward the center of the molecule, which accounts for the isomerization of α-olefin derivatives to internal olefin derivatives when carbonium ions are produced. Cyclization can occur by internal addition of a carbonium ion to a double bond which, by continuation of the sequence, can result in aromatization of the cyclic carbonium ion.

Like the paraffin derivatives, naphthenes do not appear to isomerize before cracking. However, the naphthenic hydrocarbon derivatives (from C_9 upward) produce considerable amounts of aromatic hydrocarbon derivatives during catalytic cracking. Reaction schemes similar to that outlined here provide possible routes for the conversion of naphthenes to aromatic derivatives. Alkylated benzenes undergo nearly quantitative dealkylation to benzene without apparent ring degradation below 500°C (930°F). However, polymethly benzene derivatives undergo disproportionation and isomerization with very little benzene formation.

Coke formation is considered, with just cause to a malignant side reaction of normal carbenium ions. However, while chain reactions dominate events occurring on the surface, and produce the majority of products, certain less desirable bimolecular events have a finite chance of involving the same carbenium ions in a bimolecular interaction with one another. Of these reactions, most will produce a paraffin and leave carbene/carboid-type species on the surface. This carbene/carboid-type species can produce other products but the most damaging product will be one which remains on the catalyst surface and cannot be desorbed and results in the formation of coke, or remains in a non-coke form but effectively blocks the active sites of the catalyst.

A general reaction sequence for coke formation from paraffin derivatives involves oligomerization, cyclization, and dehydrogenation of small molecules at active sites within zeolite pores:

$$Alkanes \rightarrow alkenes$$

$$Alkenes \rightarrow oligomers$$

$$Oligomers \rightarrow naphthenes$$

$$Naphthenes \rightarrow aromatics$$

$$Aromatics \rightarrow coke$$

Whether or not these are the true steps to coke formation can only be surmised. The problem with this reaction sequence is that it ignores sequential reactions in favor of consecutive reactions. And it must be accepted that the chemistry leading up to coke formation is a complex process, consisting of many sequential and parallel reactions.

There is a complex and little understood relationship between coke content, catalyst activity, and the chemical nature of the coke. For instance, the atomic hydrogen/carbon ratio of coke depends on how the coke was formed; its exact value will vary from system to system. And it seems that catalyst decay is not related in any simple way to the hydrogen-to-carbon atomic ratio of the coke, or to the total coke content of the catalyst, or any simple measure of coke properties. Moreover, despite many and varied attempts, there is currently no consensus as to the detailed chemistry of coke formation. There is, however, much evidence and good reason to believe that catalytic coke is formed from carbenium ions which undergo addition, dehydrogenation and cyclization, and elimination side reactions in addition to the mainline chain propagation processes.

4.3.3 Dehydrogenation Processes

Dehydrogenation processes involve the use of chemical reactions by means of which less saturated and more reactive compounds can be produced (Parkash, 2003; Gary et al., 2007; Speight, 2014; Hsu and Robinson, 2017; Speight, 2017). There are many important conversion processes in which hydrogen is directly or indirectly removed. In the current context, the largest-scale dehydrogenations are those of hydrocarbon derivatives such as the conversion of paraffin derivatives to olefin derivatives, olefin derivatives to diolefin derivatives:

$$-CH_2CH_2CH_2CH_3 \rightarrow -CH_2CH_2CH=CH_2$$

$$-CH_2CH_2CH=CH_2 \rightarrow -CH=CHCH=CH_2$$

Another example is the conversion of cycloparaffin derivatives to aromatic derivatives—the simplest example of which is the conversion of cyclohexane to benzene:

$$C_6H_{12} \rightarrow C_6H_6 + 3H_2$$

Dehydrogenation reactions of less specific character occur frequently in the refining and petrochemical industries, where many of the processes have names of their own. Some in which dehydrogenation plays a large part are pyrolysis, cracking, gasification by partial combustion, carbonization, and reforming.

The common primary reactions of pyrolysis are dehydrogenation and carbon bond scission. The extent of one or the other varies with the starting material and operating conditions, but because of its practical importance, methods have been found to increase the extent of dehydrogenation and, in some cases, to render it almost the only reaction.

Dehydrogenation is essentially the removal of hydrogen from the parent molecule. For example, at 550°C (1,025°F) n-butane loses hydrogen to produce butene-1 and butene-2 The development of selective catalysts, such as chromic oxide (chromia, Cr_2O_3) on alumina (Al_2O_3) has rendered the dehydrogenation of paraffin derivatives to olefin derivatives particularly effective, and the formation of higher molecular weight material is minimized. The extent of dehydrogenation (vis-à-vis carbon-carbon bond scission) during the thermal cracking of petroleum varies with the starting material and operating conditions, but because of its practical importance, methods have been found to increase the extent of dehydrogenation and, in some cases, to render it almost is the only reaction.

Naphthenes are somewhat more difficult to dehydrogenate, and cyclopentane derivatives form only aromatic derivatives if a preliminary step to form the cyclohexane structure can occur. Alkyl derivatives of cyclohexane usually dehydrogenate at 480°C–500°C (895°F–930°F), and polycyclic naphthenes are also quite easy to dehydrogenate thermally. In the presence of catalysts, cyclohexane and its derivatives are readily converted into aromatic derivatives; reactions of this type are prevalent in catalytic cracking and reforming. Benzene and toluene are prepared by the catalytic dehydrogenation of cyclohexane and methyl cyclohexane, respectively.

Polycyclic naphthenes can also be converted to the corresponding aromatic derivatives by heating at 450°C (840°F) in the presence of a chromia-alumina (Cr_2O_3-Al_2O_3) catalyst. Alkyl aromatic derivatives also dehydrogenate to various products. For example, styrene is prepared by the catalytic dehydrogenation of ethylbenzene. Other alkylbenzenes can be dehydrogenated similarly; isopropylbenzene yields α-methyl styrene.

In general, dehydrogenation reactions are difficult reactions which require high temperatures for favorable equilibria as well as for adequate reaction velocities. Dehydrogenation reactions—using reforming reactions as the example—are endothermic and, hence, have high heat requirements and active catalysts are usually necessary. Furthermore, since permissible hydrogen partial pressures are inadequate to prevent coke deposition, periodic regenerations are often necessary. Because of these problems with pure dehydrogenations, many efforts have been made to use oxidative dehydrogenations in which oxygen or another oxidizing agent combines with the hydrogen removed. This expedient has been successful with some reactions where it has served to overcome thermodynamic limitations and coke-formation problems.

The endothermic heat of pure dehydrogenation may be supplied through the walls of tubes (2–6 in. id), by preheating the feeds, adding hot diluents, reheaters between stages, or heat stored in periodically regenerated fixed or fluidized solid catalyst beds. Usually, fairly large temperature gradients will have to be tolerated, either from wall to center of tube, from inlet to outlet of bed, or from start to finish of a processing cycle between regenerations. The ideal profile of a constant temperature (or even a rising temperature) is seldom achieved in practice. In oxidative dehydrogenation reactions the complementary problem of temperature rise because of exothermic nature of the reaction is encountered. Other characteristic problems met in dehydrogenations are the needs for rapid heating and quenching to prevent side reactions, the need for low pressure drops through catalyst beds, and the selection of reactor materials that can withstand the operating conditions.

Selection of operating conditions for a straight dehydrogenation reaction often requires a compromise. The temperature must be high enough for a favorable equilibrium and for a good reaction rate, but not as high as to cause excessive cracking or catalyst deactivation. The rate of the dehydrogenation reaction diminishes as conversion increases, not only because equilibrium is approached more closely, but also because in many cases reaction products act as inhibitors. The ideal temperature profile in a reactor would probably show an increase with distance, but practically attainable profiles normally are either flat or show a decline. Large adiabatic beds in which the decline is steep are often used. The reactor pressure should be as low as possible without excessive recycle costs or equipment size. Usually the pressure is close to near atmospheric pressure but reduced pressures have been used in the Houdry butane dehydrogenation process. In any case, the catalyst bed must be designed for a low pressure drop.

Rapid preheating of the feed is desirable to minimize cracking. Usually this is done by mixing preheated feedstock with superheated diluent just as the two streams enter the reactor. Rapid cooling or quenching at the exit of the reactor is usually necessary to prevent condensation reactions of the olefinic products. Materials of construction must be resistant to attack by hydrogen, capable of prolonged operation at high temperature, and not be unduly active for conversion of hydrocarbon derivatives to carbon. Alloy steels containing chromium are usually favored although steel alloys containing nickel are also used but these latter alloys can cause problems arising from carbon formation. If steam is not present, traces of sulfur compounds may be needed to avoid carbonization. Both steam and sulfur compounds act to keep metal walls in a passive condition.

In fact, fluid catalytic cracking has been the second major supplier of propylene after steam cracking, and has proven high flexibility in feedstock and product slate. Crude oil cracking in a fluid catalytic cracking process may appear as an ideal candidate to fulfill petrochemical producer's needs. Fluid catalytic cracking units usually run on vacuum distillation products namely vacuum gas oil and vacuum residue (VR). Also, atmospheric residue (AR) can be used as a feedstock for the fluid catalytic cracking unit. In some small refineries, it was shown that the fluid catalytic cracking unit could substitute the main distillation unit, separating and converting the heavy part of the crude oil all in once. Problems associated with heavy material or metals in crude oil are readily addressed by residuum fluid catalytic cracking technology (which treats, precisely, the heaviest part of the crude). Lighter fractions of the crude, especially the paraffinic naphtha, will crack to a lower extent under traditional fluid catalytic cracking conditions. This problem has also been studied by most of the refiners with the aim of increasing propylene (and ethylene) yield in the fluid catalytic cracking unit. All the technologies developed to enhance olefin yield from fluid catalytic cracking are of high interest for converting crude to petrochemicals (Parkash, 2003; Gary et al., 2007; Speight, 2008, 2011, 2014, 2015; Hsu and Robinson, 2017; Speight, 2017). Such a technology may probably be based on a conversion unit that can handle the high-boiling constituents of the crude oil, converting it partially to light olefin derivatives and reducing the amount of high-boiling products to minimum. A modified fluid catalytic cracking process would be an ideal candidate and other units—such as a steam cracking unit—may also be added to complement the fluid catalytic cracking unit to produce low molecular weight olefin derivatives from the lower-boiling fractions from the fluid catalytic unit.

4.3.4 DEHYDROCYCLIZATION PROCESSES

Catalytic aromatization involving the loss of 1 mol of hydrogen followed by ring formation and further loss of hydrogen has been demonstrated for a variety of paraffin derivatives (typically *n*-hexane and *n*-heptane) (Parkash, 2003; Gary et al., 2007; Speight, 2014; Hsu and Robinson, 2017; Speight, 2017). Thus, *n*-hexane can be converted to benzene, heptane is converted to toluene, and octane is converted to ethyl benzene and o-xylene. Conversion takes place at low pressures, even atmospheric, and at temperatures above 300°C (570°F), although 450°C–550°C (840°F–1,020°F) is the preferred temperature range.

The catalysts are metals (or their oxides) of the titanium, vanadium, and tungsten groups and are generally supported on alumina; the mechanism is believed to be dehydrogenation of the paraffin to an olefin, which in turn is cyclized and dehydrogenated to the aromatic hydrocarbon. In support of this, olefin derivatives can be converted to aromatic derivatives much more easily than the corresponding paraffin derivatives.

4.4 STREAMS FROM COAL, OIL SHALE, AND BIOMASS

4.4.1 COAL

4.4.1.1 Coal Gas

Gases produced from coal invariably contain constituents that are damaging to the climate or environment—these will be constituents such as carbon dioxide (CO_2), nitrogen oxides (NOx), sulfur oxides (SOx), dust and particles; and toxins such as dioxin and mercury. The processes that have been developed for gas cleaning vary from a simple once-through wash operation to complex multistep systems with options for recycle of the gases (Speight, 2007; Mokhatab et al., 2006; Speight, 2014). In some cases, process complexities arise because of the need for recovery of the materials used to remove the contaminants or even recovery of the contaminants in the original form or in the altered form.

The purpose of preliminary cleaning of gases which arise from coal utilization is the removal of materials such as mechanically carried solid particles (either process products and/or dust) as well as liquid vapors (i.e., water, tars, and aromatics such as benzenes and/or naphthalene derivatives); in some instances, preliminary cleaning might also include the removal of ammonia gas. For example, cleaning of town gas is the means by which the crude tar-carrying gases from retorts or coke ovens are (first), in a preliminary step, freed from tarry matter, condensable aromatics (such as naphthalene) and (second) purified by removal of materials such as hydrogen sulfide, other sulfur compounds, and any other unwanted components that will adversely affect the use of the gas. In more general terms, gas cleaning is divided into (i) removal of particulate impurities and (ii) removal of gaseous impurities. For the purposes of this section, the latter operation includes the removal of hydrogen sulfide, carbon dioxide, sulfur dioxide, and the like.

There is also need for subdivision of these two categories as dictated by needs and process capabilities: (i) coarse cleaning whereby substantial amounts of unwanted impurities are removed in the simplest, most convenient manner; (ii) fine cleaning for the removal of residual impurities to a degree sufficient for the majority of normal chemical plant operations, such as catalysis or preparation of normal commercial products, or cleaning to a degree sufficient to discharge an effluent gas to atmosphere through a chimney; (iii) ultra-fine cleaning where the extra step is justified by the nature of the subsequent operations or the need to produce a particularly pure product.

Since coal is a complex, heterogeneous material, there is a wide variety of constituents that are not required in a final product and must be removed during processing. Coal composition and characteristics vary significantly; there are varying amounts of sulfur, nitrogen, and trace metal species which must be disposed in the correct manner (Speight, 2008). Thus, whether the process be gasification to produce a pipeline quality gas or a series of similar steps for gas cleaning before use as a petrochemical feedstocks, the stages required during this processing are numerous and can account for a major portion of a gas cleaning facility.

Generally, the majority of the sulfur that occurs naturally in the coal is driven into the product gas. Thermodynamically, the majority of the sulfur should exist as hydrogen sulfide, with smaller amounts of carbonyl sulfide (COS) and carbon disulfide (CS_2). However, data from some operations show higher than expected (from thermodynamic considerations) concentrations of carbonyl sulfide and carbon disulfide.

The existence of mercaptan derivatives, thiophene derivatives, and other organic sulfur compounds in (gasifier) product gas will probably be a function of the degree of severity of the process, contacting schemes, and heat-up rate. Those processes that tend to produce tarry products and oil products may also tend to drive off high molecular weight organic sulfur compounds into the raw product gas.

4.4.1.2 Coal Liquids

The conversion of coal to liquids and the upgrading of the coal liquids have been considered as future alternatives of petroleum to produce synthetic liquid fuels. Since the late 1970s, coal liquefaction processes have been developed into integrated two-stage processes, in which coal is hydro-liquefied in the first stage and the coal liquids are upgraded in the second stage. Upgrading of the coal liquids is an important aspect of this approach and may determine whether such liquefaction can be economically feasible.

Liquid products from coal are generally different from those produced by petroleum refining, particularly as they can contain substantial amounts of phenols. Therefore, there will always be some question about the place of coal liquids in refining operations. In more generic terms, the liquid products from coal may be classified as neutral oils (essentially pure hydrocarbons), tar acids (phenols), and tar bases (basic nitrogen compounds) that a refinery must accommodate to produce the necessary hydrocarbon derivatives for the production of petrochemicals.

The neutral oils making up 80–85% v/v of hydrogenated coal distillate are approximately half aromatic compounds, including polycyclic aromatic hydrocarbons. Typical components of the neutral oils are benzene, naphthalene, and phenanthrene. Hydroaromatic compounds (cycloparaffins, also called *naphthenes*) are another important component of neutral oils. Hydroaromatic compounds are formed at high hydrogen pressures in the presence of a catalyst, but in the presence of another species capable of accepting hydrogen (such as unreacted coal), hydroaromatic species lose hydrogen to form the thermodynamically more stable aromatic compounds and are important intermediates in the transfer of hydrogen to unreacted coal during liquid-phase coal hydrogenation and solvent refining of coal.

The next most abundant component of neutral oil consists of liquid olefins. The olefins are reactive and tend to undergo polymerization, oxidation, and other reactions causing changes in the properties of the product with time. On the other hand, olefins are excellent raw materials for the manufacture of synthetic polymers (Chapter 11) and other chemicals and, thus, can be valuable chemical byproducts in coal liquids.

In addition to neutral oil, coal liquids contain tar acids (consisting of phenolic compounds) which may constitute from 5 to 15% w/w of many coal liquids. They constitute one of the major differences between coal liquids and natural petroleum, which has a much lower content of oxygen-containing compounds and, although tar acids are valuable chemical raw materials, they are troublesome to catalysts in refining processes. Tar bases containing basic nitrogen make up 2–4% w/w of coal hydrogenation liquids. Tar bases are made of a variety of compounds, such as pyridine, quinoline, aniline, and higher molecular weight analogs.

Because of these products, coal liquids have remained largely unacceptable as refinery feedstocks because of their high concentrations of aromatic compounds and high heteroatom and metals content (Speight, 2008, 2014). Successful upgrading process will have to achieve significant reductions in the content of the aromatic components.

4.4.2 OIL SHALE

4.4.2.1 Oil Shale Gas

Oil shale gas is produced by retorting (pyrolysis) of oil shale (2012). In the pyrolysis process, oil shale is heated until the kerogen in the shale decomposes. There is no exact formula for oil shale gas—the composition of the gas depends of retorted oil shale and exploited technology. Typical components of oil shale gas are usually methane, hydrogen, carbon monoxide, carbon dioxide, and nitrogen as well as hydrocarbon derivatives such as ethylene. The gas may also contain hydrogen sulfide and other impurities.

The initial composition of the crude shale oil produced in the retorting step is the primary influence in the design of the subsequent upgrading operation. In particular, nitrogen compounds, sulfur compounds, and other non-hydrocarbon constituents dictate the cleaning processes that are selected (Mokhatab et al., 2006; Speight, 2018).

This being the case, the gas can be subjected to cleaning in a natural gas processing plant but only after detailed analysis of the gas. The analysis would assist in the determination of not only the constituents of the gas, but also the relative amounts of each constituents and the necessary adjustment that would have to be made to accept the gas for cleaning in a conventional natural gas processing plant.

4.4.2.2 Shale Oil

Shale oil is a synthetic crude oil produced by retorting oil shale and is the pyrolysis product of the organic matter (kerogen) contained in oil shale. The raw shale oil produced from retorting oil shale can vary in properties and composition and, as the oil exits the retort, it is by no means a pure distillate and usually contains emulsified water and suspended solids (Speight, 2008, 2012). Therefore, the first step in upgrading is usually dewatering and desalting. Furthermore, if not removed, the arsenic and iron in shale oil would poison and foul the supported catalysts used in hydrotreating. Because these materials are soluble, they cannot be removed by filtration. Several methods have been used specifically to remove arsenic and iron. Other methods involve hydrotreating; these also lower sulfur, olefin, and diolefin contents and thereby make the upgraded product less prone to gum formation. After these steps the shale oil may be suitable for admittance to typical refinery processing.

Compared with petroleum, shale oil is high in nitrogen and oxygen compounds and a higher specific gravity—in the order of 0.9–1.0 owing to the presence of high-boiling nitrogen-, sulfur-, and oxygen-containing compounds. Shale oil also has a relatively high pour point and small quantities of arsenic and iron are also present.

The chemical potential of oil shales is as a retort fuel to produce shale oil and from that liquid fuel and specialty chemicals have been used so far to a relatively small extent. Using stepwise cracking, various liquid fuels have been produced and even exported before World War II. At the same time, shale oils possess molecular structures of interest to the specialty chemicals industry and also a number of nonfuel specialty products have been marketed based on functional group, broad range concentrate, or even pure compound values.

Shale oil is a complex mixture of hydrocarbon derivatives, and it is characterized using bulk properties of the oil. Shale oil usually contains large quantities of olefin derivatives and aromatic hydrocarbon derivatives as well as significant quantities of heteroatom compounds (nitrogen-containing compounds, oxygen-containing compounds, and sulfur-containing compounds). A typical shale oil composition includes: nitrogen 1.5–2% w/w, oxygen 0.5–1% w/w, and sulfur 0.15–1% w/w as well as mineral particles and metal-containing compounds (Speight, 2008). Generally, the oil is less fluid than crude oil, and becoming which is reflected in the pour point that is in the order of 24°C–27°C (75°F–81°F), while conventional crude oil has a pour point in the order of −60°C to +30°C (−76°F to +86°F) that affects the ability of shale oil to be transported using unheated pipelines. Shale oil also contains polycyclic aromatic hydrocarbon derivatives.

The initial composition of the crude shale oil produced in the retorting step is the primary influence in the design of the subsequent upgrading operation. In particular, nitrogen compounds, sulfur compounds, and organometallic compounds dictate the upgrading process that is selected. Crude shale oil typically contains nitrogen compounds (throughout the total boiling range of shale oil) in concentrations that are 10–20 times the amounts found in typical crude oils (Speight, 2012). Removal of the nitrogen-bearing compounds is an essential requirement of the upgrading effort, since nitrogen is poisonous to most catalysts used in subsequent refining steps and creates unacceptable amounts of NOx pollutants when nitrogen-containing fuels are burned.

As with shale oil gas, the shale oil can be subjected to refining in a petroleum refinery but only after detailed analysis of the oil. The analysis would assist in the determination of not only the constituents of the oil but also the relative amounts of each constituents and the necessary adjustment that would have to be made to accept the oil by the refinery.

Thus, upgrading activities are dictated by factors such as the initial composition of the oil shale, the compositions of retorting products, the composition and quality of desired petroleum feedstocks or petroleum end products of market quality, and the decision to develop other byproducts such as sulfur and ammonia into salable products.

4.4.3 BIOMASS

The utilization of biomass to produce valuable products by thermal processes is an important aspect of biomass technology (Speight, 2008). Biomass pyrolysis gives usually rise to three phases: (i) gases, (ii) condensable liquids, and (iii) char/coke. However, there are various types of related kinetic pathways ranging from very simple paths to more complex paths and all usually include several elementary processes occurring in series or in competition. As anticipated, the kinetic paths are different for cellulose, lignin, and hemicelluloses (biomass main basic components) and also for usual biomasses according to their origin, composition, and inorganic contents.

The main biomass constituents—hemicellulose, cellulose, and lignin—can be selectively devolatilized into value-added chemicals. This thermal breakdown is guided by the order of thermochemical stability of the biomass constituents that ranges from hemicellulose (as the least stable constituent) to the more stable—lignin exhibits an intermediate thermal degradation behavior. Thus, wood constituents are decomposed in the order of hemicellulose–cellulose–lignin, with a restricted decomposition of the lignin at relatively low temperatures. With prolonged heating, condensation of the lignin takes place, whereby thermally large stable macromolecules develop. Whereas both hemicellulose and cellulose exhibit a relatively high devolatilization rate over a relatively narrow temperature range, thermal degradation of lignin is a slow-rate process that commences at a lower temperature when compared to cellulose.

Since the thermal stabilities of the main biomass constituents partially overlap and the thermal treatment is not specific, a careful selection of temperatures, heating rates, and gas and solid residence times is required to make a discrete degasification possible when applying a stepwise increase in temperature. Depending on these process conditions and parameters such as composition of the biomass and the presence of catalytically active materials, the product mixture is expected to contain degradation products from hemicellulose, cellulose, or lignin.

As stated elsewhere, a major issue in the use of biomass is one of *feedstock diversity*. Biomass-based feedstock materials used in producing chemicals can be obtained from a large variety of sources. If considered individually, the number of potential renewable feedstocks can be overwhelming, but they tend to fall into three simple categories: (i) waste materials such as food processing wastes, (ii) dedicated feedstock crops which includes, and) short rotation woody crops or herbaceous energy crops such as perennials or forage crops), and (iii) conventional food crops such as corn and wheat. In addition, these raw materials are composed of several similar chemical constituents, i.e., carbohydrates, proteins, lipids, lignin, and minerals.

4.4.3.1 Biogas

Most biomass materials are easier to convert to gas than coal because they are more reactive with higher ignition stability. This characteristic also makes them easier to process thermochemically into higher-value fuels such as methanol or hydrogen. Ash content is typically lower than in most coals and the sulfur content of the biomass is much lower than in many fossil fuels. The mineral content of biomass can vary as a function of soil type and the timing of feedstock harvest. Biogas contains methane and can be recovered in industrial anaerobic digesters and mechanical biological treatment systems. Landfill gas is a less clean form of biogas which is produced in landfills through naturally occurring anaerobic digestion.

This being the case, the gas can only be subjected to cleaning in a natural gas processing plant after detailed analysis of the gas. The analysis would assist in the determination of not only the constituents of the gas but also the relative amounts of each constituents and the necessary adjustment that would have to be made to accept the gas for cleaning in a conventional natural gas processing plant.

4.4.3.2 Bio-liquids

Hydrocarbon derivatives are products of various plant species belonging to different families which convert a substantial amount of photosynthetic products into latex. The latex of such plants contains liquid hydrocarbon derivatives of high molecular weight. These hydrocarbon derivatives can be converted into high-grade transportation fuel.

Bio-oil is a product that is produced by pyrolysis (flash pyrolysis) that occurs when solid fuels are heated at temperatures between 350°C and 500°C (570°F–930°F) for a very short period of time (<2 s). The bio-oils currently produced are suitable for use in boilers for electricity generation. In another process, the feedstock is fed into a fluidized bed (at 450°C–500°C, 840°F–930°F) and the feedstock flashes and vaporizes. The resulting vapors pass into a cyclone where solid particles, char, are extracted. The gas from the cyclone enters a quench tower where they are quickly cooled by heat transfer using bio-oil already made in the process. The bio-oil condenses into a product receiver and any non-condensable gases are returned to the reactor to maintain process heating.

Thus, petroleum refineries can be an alternative source for obtaining petroleum to be used in diesel engines. However, hydrocarbon derivatives, as such, are not usually produced from crops, there being insufficient amount of the hydrocarbon derivatives present in the plant tissue to make the process economical.

REFERENCES

Abdel-Aal, H.K., Aggour, M.A., and Fahim, M.A. 2016. *Petroleum and Gas Field Processing*. CRC Press, Boca Raton, FL.

Abraham, H. 1945. *Asphalts and Allied Substances*. Van Nostrand Scientific Publishers, New York.

ASTM D3246. 2018. *Standard Test Method for Sulfur in Petroleum Gas by Oxidative Microcoulometry*. ASTM International, West Conshohocken, PA.

Barbouteau, L., and Dalaud, R. 1972. Chapter 7. In: *Gas Purification Processes for Air Pollution Control*. G. Nonhebel (Editor). Butterworth and Co., London, UK.

Curry, R.N. 1981. *Fundamentals of Natural Gas Conditioning*. PennWell Publishing Co., Tulsa, OK.

Forbes, R. J. 1958a. *A History of Technology*. Oxford University Press, Oxford, UK.

Forbes, R. J. 1958b. *Studies in Early Petroleum Chemistry*. E. J. Brill, Leiden, The Netherlands.

Forbes, R.J. 1959. *More Studies in Early Petroleum Chemistry*. E.J. Brill, Leiden, The Netherlands.

Forbes, R. J. 1964. *Studies in Ancient Technology*. E. J. Brill, Leiden, The Netherlands.

Gary, J.G., Handwerk, G.E., and Kaiser, M.J. 2007. *Petroleum Refining: Technology and Economics*. 5th Edition. CRC Press, Boca Raton, FL.

Hoiberg A.J. 1960. *Bituminous Materials: Asphalts, Tars and Pitches, I & II*. Interscience, New York.

Hsu, C.S., and Robinson, P.R. (Editors). 2017. *Handbook of Petroleum Technology*. Springer, Cham, Switzerland.

Jou, F.Y., Otto, F.D., and Mather, A.E. 1985. Chapter 10. In *Acid and Sour Gas Treating Processes*. S.A. Newman (Editor). Gulf Publishing Company, Houston, TX.

Katz, D.K. 1959. *Handbook of Natural Gas Engineering*. McGraw-Hill Book Company, New York.

Kidnay, A.J., and Parrish, W.R. 2006. *Fundamentals of Natural Gas Processing*. CRC Press, Boca Raton, FL.

Kohl, A.L., and Riesenfeld, F.C. 1985. *Gas Purification*. 4th Edition. Gulf Publishing Company, Houston, TX.

Kohl, A.L., and Nielsen, R.B. 1997. *Gas Purification*. Gulf Publishing Company, Houston, TX.

Maddox, R.N. 1982. *Gas Conditioning and Processing. Volume 4. Gas and Liquid Sweetening*. Campbell Publishing Co., Norman, OK.

Maddox, R.N., Bhairi, A., Mains, G.J., and Shariat, A. 1985. Chapter 8. In: *Acid and Sour Gas Treating Processes*. S.A. Newman (Editor). Gulf Publishing Company, Houston, TX.

Mody, V., and Jakhete, R. 1988. *Dust Control Handbook*. Noyes Data Corp., Park Ridge, NJ.

Mokhatab, S., Poe, W.A., and Speight, J.G. 2006. *Handbook of Natural Gas Transmission and Processing*. Elsevier, Amsterdam, Netherlands.

Newman, S.A. 1985. *Acid and Sour Gas Treating Processes*. Gulf Publishing, Houston, TX.

Parkash, S. 2003. *Refining Processes Handbook*. Gulf Professional Publishing, Elsevier, Amsterdam, Netherlands.

Pitsinigos, V.D., and Lygeros, A.I. 1989. Predicting H_2S-MEA Equilibria. *Hydrocarbon Processing* 58(4): 43–44.

Polasek, J., and Bullin, J. 1985. Chapter 7. In: *Acid and Sour Gas Treating Processes*. S.A. Newman (Editor). Gulf Publishing Company, Houston, TX.

Soud, H., and Takeshita, M. 1994. *FGD Handbook*. No. IEACR/65. International Energy Agency Coal Research, London, UK.

Speight, J.G. 1978. *Personal Observations at Archeological Digs at The Cities of Babylon, Calah, Nineveh, and Ur*. College of Science, University of Mosul, Iraq.

Speight, J.G. 2007. *Natural Gas: A Basic Handbook*. GPC Books, Gulf Publishing Company, Houston, TX.

Speight, J.G. 2008. *Synthetic Fuels Handbook: Properties, Processes, and Performance*. McGraw-Hill, New York.

Speight, J.G. (Editor), 2011. *The Biofuels Handbook*. Royal Society of Chemistry, London, UK.

Speight, J.G. 2012. *Shale Oil Production Processes*. Gulf Professional Publishing, Elsevier, Oxford, UK.

Speight, J.G. 2014. *The Chemistry and Technology of Petroleum*. 4th Edition. CRC Press, Boca Raton, FL.

Speight, J.G. 2015. *Handbook of Petroleum Product Analysis*. 2nd Edition. John Wiley & Sons Inc., Hoboken, NJ.

Speight, J.G. 2017. *Handbook of Petroleum Refining*. CRC Press, Boca Raton, FL.

Speight, J.G. 2018. *Handbook of Natural Gas Analysis*. John Wiley & Sons Inc., Hoboken, NJ.

5 Feedstock Preparation by Gasification

5.1 INTRODUCTION

The influx of viscous feedstocks such as heavy oil, extra heavy oil, and tar sand bitumen into refineries creates, and will continue to create challenges but, at the same time, it also creates opportunities by improving the ability of refineries to handle viscous feedstocks thereby enhancing refinery flexibility to meet the increasingly stringent product specifications for refined fuels. (Speight, 2013a, 2014a, 2014b). Upgrading viscous feedstocks is an increasingly prevalent means of extracting the maximum amount of liquid fuels from each barrel of crude oil that enters the refinery. Although solvent deasphalting processes (Chapter 14) and coking processes (Chapter 10) are used in refineries to upgrade viscous feedstocks to intermediate products (which are then processed further) to produce transportation fuels, the integration of viscous feedstock processing units and gasification presents some unique synergies that will enhance the performance of the future refinery by preparing otherwise difficult to refine feedstocks to feedstocks that are suitable for the production of petrochemicals (Figures 5.1 and 5.2) (Wallace et al., 1998; Furimsky, 1999; Penrose et al., 1999; Gray and Tomlinson, 2000; Abadie and Chamorro, 2009; Wolff and Vliegenthart, 2011; Speight, 2011b, 2014a).

Gasification offers more scope for recovering products from waste than incineration. When waste is burnt in a modern incinerator the only practical product is energy, whereas the gases, oils, and solid char from pyrolysis and gasification cannot only be used as a fuel but also purified and used as a feedstock for petrochemicals and other applications. Many processes also produce a stable granulate instead of an ash that can be more easily and safely utilized. In addition, some processes are targeted at producing specific recyclables such as metal alloys and carbon black. From waste gasification, in particular, it is feasible to produce hydrogen that many see as an increasingly valuable resource.

Gasification can be used in conjunction with gas engines (and potentially gas turbines) to obtain higher conversion efficiency than conventional fossil fuel energy generation. By displacing fossil fuels, waste pyrolysis and gasification can help meet renewable energy targets, address concerns

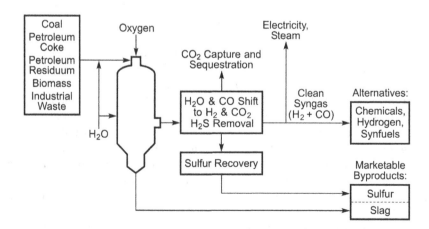

FIGURE 5.1 The gasification process can accommodate a variety of carbonaceous feedstocks.

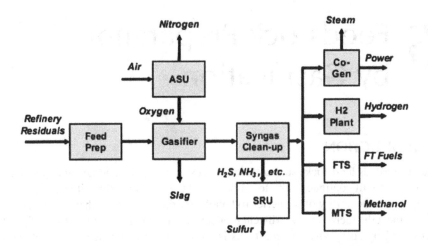

FIGURE 5.2 Gasification as might be employed on-site in a refinery. (*Source*: National Energy Technology Laboratory, United States Department of Energy, Washington, DC. www.netl.doe.gov/technologies/coalpower/gasification/gasifipedia/7-advantages/7-3-4_refinery.html)

about global warming and contribute to achieving Kyoto Protocol commitments. Conventional incineration, used in conjunction with steam-cycle boilers and turbine generators, achieves lower efficiency. Many of the processes fit well into a modern integrated approach to waste management. They can be designed to handle the waste residues and are fully compatible with an active program of composting for the waste fraction that is subject to decay and putrefaction.

A wide range of materials can be handled by gasification technologies and specific processes have been optimized to handle particular feedstock (e.g., tire pyrolysis and sewage sludge gasification), while others have been designed to process mixed wastes. For example, recovering energy from agricultural and forestry residues, household and commercial waste, materials recycling (auto-shredder residue, electrical and electronic scrap, tires, mixed plastic waste, and packaging residues) are feasible processes.

Briefly, gasification is a process in which combustible materials are partially oxidized or partially combusted. The product of gasification is a combustible synthesis gas (also referred to as syngas). Because gasification involves the partial, rather than complete, oxidization of the feed, gasification processes operate in an oxygen-lean environment. The process has grown from a predominately coal conversion process used for making *town gas* for industrial lighting to an advanced process for the production of multiproduct, carbon-based fuels from a variety of feedstocks such as crude oil viscous feedstocks, biomass, or other carbonaceous feedstocks (Figure 5.1) (Kumar et al., 2009; Speight, 2013a, 2014a, 2014b; Luque and Speight, 2015).

Gasification is an appealing process for the utilization of relatively inexpensive feedstocks that might otherwise be declared as waste and sent to a landfill (where the production of methane—a so-called greenhouse gas—will be produced) or combusted which may not (depending upon the feedstock) be energy efficient. Overall, use of a gasification technology (Speight, 2013a, 2014b) with the necessary gas cleanup options can have a smaller environmental footprint and lesser effect on the environment than landfill operations or combustion of the waste. In fact, there are strong indications that gasification is a technically viable option for the waste conversion, including residual waste from separate collection of municipal solid waste (MSW). The process can meet existing emission limits and can have a significant effect on the reduction of landfill disposal using known gasification technologies (Arena, 2012; Speight, 2014b; Luque and Speight, 2015) or thermal plasma (Fabry et al., 2013).

In the gasification process, organic (carbonaceous) feedstocks into carbon monoxide, carbon dioxide, and hydrogen by reacting the feedstock at high temperatures (>700°C, 1,290°F), without combustion, with a controlled amount of oxygen and/or steam (Marano, 2003; Lee et al., 2007; Higman and Van der Burgt, 2008; Speight, 2008; Sutikno and Turini, 2012; Speight, 2013a, 2014b). Unconventional carbonaceous feedstocks include solids, liquids, and gases such as heavy oil, extra heavy oil, tar sand bitumen, residua, and biomass (Speight, 2014b). The gasification is not a single-step process, but involves multiple subprocesses and reactions. The generated synthesis gas has wide range of applications ranging from power generation to chemicals production. The power derived from the gasification of carbonaceous feedstocks followed by the combustion of the product gas(es) is considered to be a source of renewable energy of derived gaseous products (Table 5.1) that are generated from a carbonaceous source (e.g., biomass) other than a fossil fuel (Speight, 2008).

Indeed, the increasing interest in gasification technology reflects a convergence of changes in providing energy to the marketplace: (i) the maturity of gasification technology, and (ii) the extremely low emissions from integrated gasification combined cycle (IGCC) plants, especially air emissions, and (iii) the potential for control of greenhouse gases (Speight, 2014b). Another advantage of gasification is the use of synthesis gas is potentially more efficient as compared to direct combustion of the original fuel because it can be (i) combusted at higher temperatures, (ii) used in fuel cells, (iii) used to produce methanol and hydrogen, (iv) converted via the Fischer–Tropsch (FT) process into a range of synthesis liquid fuels suitable for use of gasoline engines or diesel engines (Chadeesingh, 2011; Luque and Speight, 2015).

Coal has been the primary feedstock for gasification units for many decades. However, there is a move to feedstocks other than coal for gasification processes with the concern on the issue of environmental pollutants and the potential shortage for coal in some area (except at the United States) (Speight, 2014b). Nevertheless, coal still prevails as a gasification feedstock and will remain so for at least several decades into the future, if not well into the next century (Speight, 2011b, 2013a, Luque and Speight, 2015). The gasification process can also utilize carbonaceous feedstocks which would otherwise have been disposed (e.g., biodegradable waste).

Coal gasification plants are cleaner with respect to standard pulverized coal combustion facilities, producing fewer sulfur and nitrogen byproducts, which contribute to smog and acid rain. For this reason, gasification appeals as a way to utilize relatively inexpensive and expansive coal reserves, while reducing the environmental impact. Indeed, the increasing mounting interest in coal gasification technology reflects a convergence of two changes in the electricity generation marketplace: (i) the maturity of gasification technology and (ii) the extremely low emissions from IGCC plants, especially air emissions, and the potential for lower cost control of greenhouse gases than other coal-based systems. Fluctuations in the costs associated with natural gas-based power, which is viewed as a major competitor to coal-based power, can also play a role. Furthermore, gasification

TABLE 5.1
Gasification Products

Product	Properties
Low Btu gas	150–300 Btu/scf
	Approximately 50% v/v nitrogen
	Smaller amounts of carbon monoxide and hydrogen
	Some carbon dioxide
	Trace amounts of methane
Medium Btu gas	300–550 Btu/scf
	Predominantly carbon monoxide and hydrogen
	Small amounts of methane
	Some carbon dioxide
High Btu gas	980–1080 Btu/scf
	Predominantly methane—typically >85% v/v

permits the utilization of various feedstocks (coal, biomass, crude oil resids, and other carbonaceous wastes) to their fullest potential (Speight, 2013a, 2014b; Orhan et al., 2014). Thus, power developers would be well advised to consider gasification as a means of converting coal to gas.

Liquid fuels, including gasoline, diesel, naphtha and jet fuel, are usually processed from crude oil in the refinery (Speight, 2014a). However, with fluctuating availability and varying prices of crude oil, liquid fuels from coal (coal-to-liquids (CTL)) and liquid fuels from biomass (biomass-to-liquids (BTL)) are always under consideration as alternative routes used for liquid fuel production. Both coal and biomass are converted to synthesis gas which is subsequently converted into a mixture of liquid products by FT processes (Chadeesingh, 2011; Speight, 2013a; Adhikari et al., 2015). The liquid fuel obtained after FT synthesis is eventually upgraded using known crude oil refinery technologies to produce gasoline, naphtha, diesel fuel, and jet fuel (Chadeesingh, 2011; Speight, 2014a).

5.2 GASIFICATION CHEMISTRY

It is important to distinguish gasification from pyrolysis. The main difference between pyrolysis and gasification is the absence of a gasifying agent in the case of pyrolysis. Pyrolysis is a thermal degradation of organic compounds, at a range of temperatures in the order of 300°C–900°C (570°F–1,650°F), under oxygen-deficient circumstances to produce various forms of products such as gases (often referred to as biogas), a liquid product (referred to as called bio-oil), and a solid product (referred to as biochar) whereas gasification is a thermal cracking of solid carbonaceous material into a combustible gas mixture, mainly composed of hydrogen, carbon monoxide (CO), carbon dioxide (CO_2), and methane (CH_4), and other gases with some byproducts (solid char or slag, oils, and water). The produced gaseous product has chemical composition and properties that are largely affected by the operational conditions throughout pyrolysis and gasification (such as reactor temperature, residence time, pressure) that are affected by the feedstock type and composition as well as the reactor geometry. Thus, pyrolysis and gasification are complex chemical mechanisms, which incorporate several operational and environmental challenges of carbon-based feedstock.

In the current context, gasification involves the thermal decomposition of the feedstock and the reaction of the feedstock carbon and other pyrolysis products with oxygen, water, and fuel gases such as methane and is represented by a sequence of simple chemical reactions (Table 5.2). However, the gasification process is often considered to involve two distinct chemical stages: (i) devolatilization

---

TABLE 5.2

Reactions that Occur During Gasification of a Carbonaceous Feedstock

$2C + O_2 \rightarrow 2CO$

$C + O2 \rightarrow CO_2$

$C + CO_2 \rightarrow 2CO$

$CO + H_2O \rightarrow CO_2 + H_2 \,(\text{shift reaction})$

$C + H_2O \rightarrow CO + H_2 \,(\text{water gas reaction})$

$C + 2H_2 \rightarrow CH_4$

$2H_2 + O_2 \rightarrow 2H_2O$

$CO + 2H2 \rightarrow CH_3OH$

$CO + 3H2 \rightarrow CH_4 + H_2O \,(\text{methanation reaction})$

$CO_2 + 4H_2 \rightarrow CH_4 + 2H_2O$

$C + 2H_2O \rightarrow 2H_2 + CO_2$

$2C + H_2 \rightarrow C_2H_2$

$CH_4 + 2H_2O \rightarrow CO_2 + 4H_2$

---

of the feedstock to produce volatile matter and char; (ii) followed by char gasification, which is complex and specific to the conditions of the reaction—both processes contribute to the complex kinetics of the gasification process (Sundaresan and Amundson, 1978).

Gasification of a carbonaceous material in an atmosphere of carbon dioxide can be divided into two stages: (i) pyrolysis and (ii) gasification of the pyrolytic char. In the first stage, pyrolysis (removal of moisture content and devolatilization) occurs at comparatively lower temperature. In the second stage, gasification of the pyrolytic char is achieved by reaction with oxygen/carbon dioxide mixtures at high temperature. In nitrogen and carbon dioxide environments from room temperature to 1,000°C (1,830°F), the mass loss rate of pyrolysis in nitrogen may be significant differently (sometime lower, depending on the feedstock) to mass loss rate in carbon dioxide, which may be due (in part) to the difference in properties of the bulk gases.

5.2.1 General Aspects

Generally, the gasification of carbonaceous feedstocks (such as heavy oil, extra heavy oil, tar sand bitumen, crude oil residua, biomass, and waste) includes a series of reaction steps that convert the feedstock into *synthesis gas* (carbon monoxide (CO) plus hydrogen (H_2)) and other gaseous products. This conversion is generally accomplished by introducing a gasifying agent (air, oxygen, and/or steam) into a reactor vessel containing the feedstock where the temperature, pressure, and flow pattern (moving bed, fluidized, or entrained bed) are controlled.

The gaseous products—other than carbon monoxide and hydrogen—and the proportions of these product gases (such as carbon dioxide (CO_2) methane (CH_4) water vapor (H_2O) hydrogen sulfide (H_2S), and sulfur dioxide (SO_2)) depends on (i) the type of feedstock, (ii) the chemical composition of the feedstock, (iii) the gasifying agent or gasifying medium, as well as, (iv) the thermodynamics and chemistry of the gasification reactions as controlled by the process-operating parameters (Singh et al., 1980; Pepiot et al., 2010; Shabbar and Janajreh, 2013; Speight, 2013a, 2013b, 2014b). In addition, the kinetic rates and extents of conversion for the several chemical reactions that are a part of the gasification process are variable and are typically functions of (i) temperature, (ii) pressure, (iii) reactor configuration, (iv) gas composition of the product gases, and (v) whether or not these gases influence the outcome of the reaction (Johnson, 1979; Speight, 2013a, 2013b, 2014b).

In a gasifier, the feedstock is exposed to high temperatures generated from the partial oxidation of the carbon. As the particle is heated, any residual moisture (assuming that the feedstock has been pre-fired) is driven off and further heating of the particle begins to drive off the volatile gases. Discharge of the volatile products will generate a wide spectrum of hydrocarbon derivatives ranging from carbon monoxide and methane to long-chain hydrocarbon derivatives comprising distillable tar and non-distillable pitch. The complexity of the products will also affect the progress and rate of the reaction when each product is produced by a different chemical process at a different rate. At a temperature above 500°C (930°F) the conversion of the feedstock to char and ash is completed. In most of the early gasification processes, this was the desired byproduct but for gas generation the char provides the necessary energy to effect further heating and—typically, the char is contacted with air or oxygen and steam to generate the product gases. Furthermore, with an increase in heating rate, feedstock particles are heated more rapidly and are burned in a higher temperature region, but the increase in heating rate has almost no substantial effect on the mechanism (Irfan, 2009).

Most notable effects in the physical chemistry of the gasification process are those effects due to the chemical character of the feedstock as well as the physical composition of the feedstock (Speight, 2011a, 2013a, 2014a, 2014b). In more general terms of the character of the feedstock, gasification technologies generally require some initial processing of the feedstock with the type and degree of pretreatment, a function of the process and/or the type of feedstock. Another factor, often presented as very general *rule of thumb*, is that optimum gas yields and gas quality are obtained at operating temperatures of approximately 595°C–650°C (1,100°F–1200°F). A gaseous product with a higher heat content (BTU/ft^3) can be obtained at lower system temperatures but the overall yield of gas (determined as the *fuel-to-gas ratio*) is reduced by the unburned char fraction.

With some feedstocks, the higher the amounts of volatile material produced in the early stages of the process the higher the heat content of the product gas. In some cases, the highest gas quality may be produced at the lowest temperatures but when the temperature is too low, char oxidation reaction is suppressed and the overall heat content of the product gas is diminished. All such events serve to complicate the reaction rate and make derivative of a global kinetic relationship applicable to all types of feedstock subject to serious question and doubt.

Depending on the type of feedstock being processed and the analysis of the gas product desired, pressure also plays a role in product definition. In fact, some (or all) of the following processing steps will be required: (i) pretreatment of the feedstock; (ii) primary gasification of the feedstock; (iii) secondary gasification of the carbonaceous residue from the primary gasifier; (iv) removal of carbon dioxide, hydrogen sulfide, and other acid gases; (v) shift conversion for adjustment of the carbon monoxide/hydrogen mole ratio to the desired ratio; and (vi) catalytic methanation of the carbon monoxide/hydrogen mixture to form methane. If high heat content (high-Btu) gas is desired, all of these processing steps are required since gasifiers do not typically yield methane in the significant concentration.

5.2.2 PRETREATMENT

While feedstock pretreatment for introduction into the gasifier is often considered to be a physical process in which the feedstock is prepared for gasifier—typically as pellets or finely ground feedstock—there are chemical aspects that must also be considered.

Some feedstocks, especially certain types of coal, display caking, or agglomerating, characteristics when heated (Speight, 2013a) and these coal types are usually not amenable to treatment by gasification processes employing fluidized bed or moving-bed reactors; in fact, caked coal is difficult to handle in fixed bed reactors. The pretreatment involves a mild oxidation treatment that destroys the caking characteristics of coals and usually consists of low-temperature heating of the coal in the presence of air or oxygen.

While this may seemingly be applicable to coal gasification only, this form of coal pretreatment is particularly important when a non-coal feedstock is co-gasified with coal. Co-gasification of other feedstocks, such as coal and especially biomass, with crude oil coke offers a bridge between the depletion of crude oil stocks when coal is used as well as a supplementary feedstock based on renewable energy sources (biomass). These options can contribute to reduce the crude oil dependency and carbon dioxide emissions since biomass is known to be neutral in terms of carbon dioxide emissions. The high reactivity of biomass and the accompanying high production of volatile products suggest that some synergetic effects might occur in simultaneous thermochemical treatment of petcoke and biomass, depending on the gasification conditions such as (i) feedstock type and origin, (ii) reactor type, and (iii) process parameters (Penrose et al., 1999; Gray and Tomlinson, 2000; McLendon et al., 2004; Lapuerta et al., 2008; Fermoso et al., 2009; Shen et al., 2012; Khosravi and Khadse, 2013; Speight, 2013a, 2014a, 2014b; Luque and Speight, 2015).

For example, carbonaceous fuels are gasified in reactors, a variety of gasifiers such as the fixed or moving bed, fluidized bed, entrained flow, and molten bath gasifiers have been developed that have differing feedstock requirements (Table 5.3) (Shen et al., 2012; Speight, 2014b). If the flow patterns are considered, the fixed bed and fluidized bed gasifiers intrinsically pertain to a countercurrent reactor in which the fuels are usually sent into the reactor from the top of the gasifier, whereas the oxidant is blown into the reactor from the bottom. With regard to the entrained flow reactor, it is necessary to pulverize the feedstock (such as coal and petcoke). On the other hand, when the feedstock is sent into an entrained flow gasifier, the fuels can be in either form of dry feed or slurry feed. In general, dry-feed gasifiers have the advantage over slurry-feed gasifiers in that the former can be operated with lower oxygen consumption. Moreover, dry-feed gasifiers have an additional degree of freedom that makes it possible to optimize synthesis gas production (Shen et al., 2012).

TABLE 5.3

Characteristics of the Different Types of Gasifiers

Gasifier Type	Fuel Properties
Fixed/moving bed	Particle size: 1–10 cm
	Mechanically stable fuel particles (unblocked passage of gas through the bed)
	Pellets or briquettes preferred
	Updraft configuration more tolerant to biomass moisture content (up to 40–50% w/w)
	Drying occurs as biomass moves down the gasifier
Fluidized bed	Ash melting temperature of fuel: higher limit for operating temperature.
	Fuel particle size relatively small to ensure good contact with bed material; typically <40 mm
	Good fuel flexibility due to high thermal inertia of the bed.
Entrained bed	Fuel particle size: <50 μm.
	Pulverized for high fuel conversion in short residence times
	Low moisture content
	Ash melting behavior can influence for reactor/process design.

5.2.3 REACTIONS

Gasification involves the thermal decomposition of feedstock and the reaction of the feedstock carbon and other pyrolysis products with oxygen, water, and fuel gases such as methane. The presence of oxygen, hydrogen, water vapor, carbon oxides, and other compounds in the reaction atmosphere during pyrolysis may either support or inhibit numerous reactions with carbonaceous feedstocks and with the products evolved. The distribution of weight and chemical composition of the products are also influenced by the prevailing conditions (i.e., temperature, heating rate, pressure, and residence time) and, last but by no means least, the nature of the feedstock (Speight, 2014a, 2014b).

If air is used for combustion, the product gas will have a heat content of ca. 150–300 Btu/ft³ (depending on process design characteristics) and will contain undesirable constituents such as carbon dioxide, hydrogen sulfide, and nitrogen. The use of pure oxygen, although expensive, results in a product gas having a heat content in the order of 300–400 Btu/ft³ with carbon dioxide and hydrogen sulfide as byproducts (both of which can be removed from low or medium heat content, low-Btu or medium-Btu) gas by any of several available processes (Speight, 2013a, 2014a).

If a high heat content (high-Btu) gas (900–1,000 Btu/ft³) is required, efforts must be made to increase the methane content of the gas. The reactions which generate methane are all exothermic and have negative values, but the reaction rates are relatively slow and catalysts may therefore, be necessary to accelerate the reactions to acceptable commercial rates. Indeed, the overall reactivity of the feedstock and char may be subject to catalytic effects. It is also possible that the mineral constituents of the feedstock (such as the mineral matter in coal and biomass) may modify the reactivity by a direct catalytic effect (Davidson, 1983; Baker and Rodriguez, 1990; Mims, 1991; Martinez-Alonso and Tascon, 1991).

In the process, the feedstock undergoes three processes in its conversation to synthesis gas—the first two processes, pyrolysis and combustion, occur very rapidly—all of which are highly dependent upon the properties of the biomass (Figure 5.3). In pyrolysis, char is produced as the feedstock heats up and volatiles are released. In the combustion process, the volatile products and some of the char reacts with oxygen to produce various products (primarily carbon dioxide and carbon monoxide) and the heat required for subsequent gasification reactions. Finally, in the gasification process, the feedstock char reacts with steam to produce hydrogen (H_2) and carbon monoxide (CO).

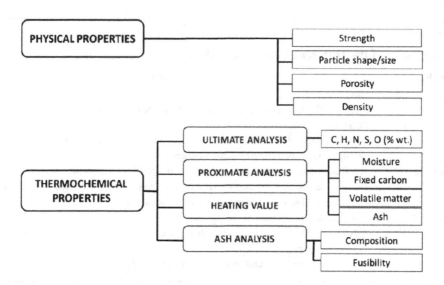

FIGURE 5.3 Biomass properties that influence the gasification process.

Combustion:

$$2C_{feedstock} + O_2 \rightarrow 2CO + H_2O$$

Gasification:

$$C_{feedstock} + H_2O \rightarrow H_2 + CO$$

$$CO + H_2O \rightarrow H_2 + CO_2$$

The resulting synthesis gas is approximately 63% v/v carbon monoxide, 34% v/v hydrogen, and 3% v/v carbon dioxide. At the gasifier temperature, the ash and other feedstock mineral matter liquefies and exits at the bottom of the gasifier as slag, a sand-like inert material that can be sold as a coproduct to other industries (e.g., road building). The synthesis gas exits the gasifier at pressure and high temperature and must be cooled prior to the synthesis gas cleaning stage.

Although processes that use the high temperature to raise high-pressure steam are more efficient for electricity production, full-quench cooling, by which the synthesis gas is cooled by the direct injection of water, is more appropriate for hydrogen production. Full-quench cooling provides the necessary steam to facilitate the water-gas shift reaction, in which carbon monoxide is converted to hydrogen and carbon dioxide in the presence of a catalyst:

Water-Gas Shift Reaction:

$$CO + H_2O \rightarrow CO_2 + H_2$$

This reaction maximizes the hydrogen content of the synthesis gas, which consists primarily of hydrogen and carbon dioxide at this stage. The synthesis gas is then scrubbed of particulate matter and sulfur is removed via physical absorption (Speight, 2013a, 2014a). The carbon dioxide is captured by physical absorption or a membrane and either vented or sequestered.

Thus, in the initial stages of gasification, the rising temperature of the feedstock initiates devolatilization and the breaking of weaker chemical bonds to yield volatile tar, volatile oil, phenol derivatives, and hydrocarbon gases. These products generally react further in the gaseous phase to form hydrogen, carbon monoxide, and carbon dioxide. The char (fixed carbon) that remains after devolatilization reacts with oxygen, steam, carbon dioxide, and hydrogen. Overall, the chemistry of gasification is complex, but can be conveniently (and simply) represented by the following reactions:

$$C + O_2 \rightarrow CO_2 \qquad \Delta H_r = -393.4 \, \text{MJ/kmol} \tag{5.1}$$

$$C + \tfrac{1}{2}O_2 \rightarrow CO \quad \Delta H_r = -111.4 \, \text{MJ/kmol} \tag{5.2}$$

$$C + H_2O \rightarrow H_2 + CO \quad \Delta H_r = 130.5 \, \text{MJ/kmol} \tag{5.3}$$

$$C + CO_2 \leftrightarrow 2CO \quad \Delta H_r = 170.7 \, \text{MJ/kmol} \tag{5.4}$$

$$CO + H_2O \leftrightarrow H_2 + CO_2 \quad \Delta H_r = -40.2 \, \text{MJ/kmol} \tag{5.5}$$

$$C + 2H_2 \rightarrow CH_4 \quad \Delta H_r = -74.7 \, \text{MJ/kmol} \tag{5.6}$$

The designation C represents carbon in the original feedstock as well as carbon in the char formed by devolatilization of the feedstock. Reactions (5.1) and (5.2) are exothermic oxidation reactions and provide most of the energy required by the endothermic gasification reactions (5.3) and (5.4). The oxidation reactions occur very rapidly, completely consuming all the oxygen present in the gasifier, so that most of the gasifier operates under reducing conditions. Reaction (5.5) is the water-gas shift reaction, where water (steam) is converted to hydrogen—this reaction is used to alter the hydrogen/carbon monoxide ratio when synthesis gas is the desired product, such as for use in FT processes. Reaction (5.6) is favored by high pressure and low temperature and is, thus, mainly important in lower temperature gasification systems. Methane formation is an exothermic reaction that does not consume oxygen and, therefore, increases the efficiency of the gasification process and the final heat content of the product gas. Overall, approximately 70% of the heating value of the product gas is associated with the carbon monoxide and hydrogen but this varies depending on the gasifier type and the process parameters (Speight, 2011a; Chadeesingh, 2011; Speight, 2013a).

In essence, the direction of the gasification process is subject to the constraints of thermodynamic equilibrium and variable reaction kinetics. The combustion reactions (reaction of the feedstock or char with oxygen) essentially go to completion. The thermodynamic equilibrium of the rest of the gasification reactions are relatively well defined and collectively have a major influence on thermal efficiency of the process as well as on the gas composition. Thus, thermodynamic data are useful for estimating key design parameters for a gasification process, such as: (i) calculating of the relative amounts of oxygen and/or steam required per unit of feedstock, (ii) estimating the composition of the produced synthesis gas, and (iii) optimizing process efficiency at various operating conditions.

Other deductions concerning gasification process design and operations can also be derived from the thermodynamic understanding of its reactions. Examples include: (i) production of synthesis gas with low methane content at high temperature, which requires an amount of steam in excess of the stoichiometric requirement; (ii) gasification at high temperature, which increases oxygen consumption and decreases the overall process efficiency; (iii) production of synthesis gas with a high methane content, which requires operation at low temperature (approximately 700°C, 1,290°F), but the methanation reaction kinetics will be poor without the presence of a catalyst.

Relative to the thermodynamic understanding of the gasification process, the kinetic behavior is much more complex. In fact, very little reliable global kinetic information on gasification reactions exists, partly because it is highly dependent on (i) the chemical nature of the feed, which varies significantly with respect to composition, mineral impurities; (ii) feedstock reactivity; and (iii) process parameters, such as temperature, pressure, and residence time. In addition, physical characteristics of the feedstock (or char) also play a role in phenomena such boundary-layer diffusion, pore diffusion, and ash layer diffusion which also influence the kinetic outcome. Furthermore, certain impurities, in fact, are known to have catalytic activity on some of the gasification reactions which can have further influence on the kinetic imprint of the gasification reactions.

5.2.3.1 Primary Gasification

Primary gasification involves thermal decomposition of the raw feedstock via various chemical processes and many schemes involve pressures ranging from atmospheric to 1,000 psi. Air or oxygen may be admitted to support combustion to provide the necessary heat. The product is usually a low heat content (low-Btu) gas ranging from a carbon monoxide/hydrogen mixture to mixtures containing varying amounts of carbon monoxide, carbon dioxide, hydrogen, water, methane, hydrogen sulfide, nitrogen, and typical tar-like products of thermal decomposition of carbonaceous feedstocks are complex mixtures and include hydrocarbon oils and phenolic products (Dutcher et al., 1983; Speight, 2011a, 2013a, 2014b).

Devolatilization of the feedstock occurs rapidly as the temperature rises above 300°C (570°F). During this period, the chemical structure is altered, producing solid char, tar products, condensable liquids, and low molecular weight gases. Furthermore, the products of the devolatilization stage in an inert gas atmosphere are very different from those in an atmosphere containing hydrogen at elevated pressure. In an atmosphere of hydrogen at elevated pressure, additional yields of methane or other low molecular weight gaseous hydrocarbon derivatives can result during the initial gasification stage from reactions such as: (i) direct hydrogenation of feedstock or semi-char because of any reactive intermediates formed and (ii) the hydrogenation of other gaseous hydrocarbon derivatives, oils, tars, and carbon oxides. Again, the kinetic picture for such reactions is complex due to the varying composition of the volatile products which, in turn, are related to the chemical character of the feedstock and the process parameters, including the reactor type.

A solid char product may also be produced, and may represent the bulk of the weight of the original feedstock, which determines (to a large extent) the yield of char and the composition of the gaseous product.

5.2.3.2 Secondary Gasification

Secondary gasification usually involves the gasification of char from the primary gasifier, which is typically achieved by reaction of the hot char with water vapor to produce carbon monoxide and hydrogen:

$$C_{char} + H_2O \rightarrow CO + H_2$$

The reaction requires heat input (endothermic) for the reaction to proceed in its forward direction. Usually, an excess amount of steam is also needed to promote the reaction. However, excess steam used in this reaction has an adverse effect on the thermal efficiency of the process. Therefore, this reaction is typically combined with other gasification reactions in practical applications. The hydrogen-carbon monoxide ratio of the product synthesis gas depends on the synthesis chemistry as well as process engineering.

The mechanism of this reaction section is based on the reaction between carbon and gaseous reactants, not for reactions between feedstock and gaseous reactants. Hence, the equations may oversimply the actual chemistry of the steam gasification reaction. Even though carbon is the dominant atomic species present in feedstock, feedstock is more reactive than pure carbon. The presence of various reactive organic functional groups and the availability of catalytic activity via naturally occurring mineral ingredients can enhance the relative reactivity of the feedstock—for example, anthracite, which has the highest carbon content among all ranks of coal (Speight, 2013a) is most difficult to gasify or liquefy.

After the rate of devolatilization has passed a maximum of another reaction becomes important— in this reaction which the semi-char is converted to char (sometimes erroneously referred to as *stable char*) primarily through the evolution of hydrogen. Thus, the gasification process occurs as the char reacts with gases such as carbon dioxide and steam to produce carbon monoxide and hydrogen. The resulting gas (producer gas or synthesis gas) may be more efficiently converted to electricity than is typically possible by direct combustion of the char. Also, corrosive elements in the ash may be refined out by the gasification process, allowing high-temperature combustion of the gas from otherwise problematic feedstocks (Speight, 2011a, 2013a, 2014b).

Oxidation and gasification reactions consume the char and the oxidation and the gasification kinetic rates follow Arrhenius-type dependence on temperature. On the other hand, the kinetic parameters are feedstock-specific and there is no true global relationship to describe the kinetics of char gasification—the characteristics of the char are also feedstock-specific. The complexity of the reactions makes the reaction initiation and the subsequent rates subject to many factors, any one of which can influence the kinetic aspects of the reaction.

Although the initial gasification stage (devolatilization) is completed in seconds or even less at elevated temperature, the subsequent gasification of the char produced at the initial gasification stage is much slower, requiring minutes or hours to obtain significant conversion under practical conditions and reactor designs for commercial gasifiers are largely dependent on the reactivity of the char and also on the gasification medium (Johnson, 1979; Sha, 2005). Thus, the distribution and chemical composition of the products are also influenced by the prevailing conditions (i.e., temperature, heating rate, pressure, residence time, etc.) and, last but not least, the nature of the feedstock. Also, the presence of oxygen, hydrogen, water vapor, carbon oxides, and other compounds in the reaction atmosphere during pyrolysis may either support or inhibit numerous reactions with feedstock and with the products evolved.

The reactivity of char produced in the pyrolysis step depends on nature of the feedstock and increases with oxygen content of the feedstock but decreases with carbon content. In general, char produced from a low-carbon feedstock is more reactive than char produced from a high-carbon feedstock. The reactivity of char from a low-carbon feedstock may be influenced by catalytic effect of mineral matter in char. In addition, as the carbon content of the feedstock increases, the reactive functional groups present in the feedstock decrease and the char becomes more aromatic and cross-linked in nature (Speight, 2013a). Therefore, char obtained from high-carbon feedstock contains a lesser number of functional groups and higher proportion of aromatic and cross-linked structures, which reduce reactivity. The reactivity of char also depends upon thermal treatment it receives during formation from the parent feedstock—the gasification rate of char decreases as the char preparation temperature increases due to the decrease in active surface areas of char. Therefore, a change of char preparation temperature may change the chemical nature of char, which in turn may change the gasification rate.

Typically, char has a higher surface area compared to the surface area of the parent feedstock, even when the feedstock has been pelletized, and the surface area changes as the char undergoes gasification—the surface area increases with carbon conversion, reaches maximum and then decreases. These changes in turn affect gasification rates—in general, reactivity increases with the increase in surface area. The initial increase in surface area appears to be caused by cleanup and widening of pores in the char. The decrease in surface area at high-carbon conversion may be due to coalescence of pores, which ultimately leads to collapse of the pore structure within the char.

Heat transfer and mass transfer processes in fixed or moving bed gasifiers are affected by complex solids flow and chemical reactions. Coarsely crushed feedstock settles while undergoing heating, drying, devolatilization, gasification, and combustion. Also, the feedstock particles change in diameter, shape, and porosity—nonideal behavior may result from certain types of chemical structures in the feedstock, gas bubbles, and channel and a variable void fraction may also change heat and mass transfer characteristics.

An important factor is the importance of the pyrolysis temperature as a major factor in the thermal history, and consequently in the thermodynamics of the feedstock char. However, the thermal history of a char should also depend on the rate of temperature rise to the pyrolysis temperature and on the length of time the char is kept at the pyrolysis temperature (soak time), which might be expected to reduce the residual entropy of the char by employing a longer soak time.

Alkali metal salts are known to catalyze the steam gasification reaction of carbonaceous materials, including coal. The process is based on the concept that alkali metal salts (such as potassium carbonate, sodium carbonate, potassium sulfide, sodium sulfide, and the like) will catalyze the steam gasification of feedstocks. The order of catalytic activity of alkali metals on the gasification reaction is:

$$Cesium(Cs) > rubidium(Rb) > potassium(K) > sodium(Na) > lithium(Li)$$

Catalyst amounts in the order of 10–20% w/w potassium carbonate will lower bituminous coal gasifier temperatures from 925°C (1,695°F) to 700°C (1,090°F) and that the catalyst can be introduced to the gasifier impregnated on coal or char. In addition, tests with potassium carbonate showed that this material also acts as a catalyst for the methanation reaction. In addition, the use of catalysts can reduce the amount of tar formed in the process. In the case of catalytic steam gasification of coal, carbon deposition reaction may affect catalyst life by fouling the catalyst-active sites. This carbon deposition reaction is more likely to take place whenever the steam concentration is low.

Ruthenium-containing catalysts are used primarily in the production of ammonia. It has been shown that ruthenium catalysts provide 5–10 times higher reactivity rates than other catalysts. However, ruthenium quickly becomes inactive due to its necessary supporting material, such as activated carbon, which is used to achieve effective reactivity. However, during the process, the carbon is consumed, thereby reducing the effect of the ruthenium catalyst.

Catalysts can also be used to favor or suppress the formation of certain components in the gaseous product by changing the chemistry of the reaction, the rate of reaction, and the thermodynamic balance of the reaction. For example, in the production of synthesis gas (mixtures of hydrogen and carbon monoxide), methane is also produced in small amounts. Catalytic gasification can be used to either promote methane formation or suppress it.

5.2.3.3 Water-Gas Shift Reaction

The water-gas shift reaction (shift conversion) is necessary because the gaseous product from a gasifier generally contains large amounts of carbon monoxide and hydrogen, plus lesser amounts of other gases. Carbon monoxide and hydrogen (if they are present in the mole ratio of 1:3) can be reacted in the presence of a catalyst to produce methane. However, some adjustment to the ideal ratio (1:3) is usually required and, to accomplish this, all or part of the steam is treated according to the waste-gas shift (shift conversion) reaction. This involves reacting carbon monoxide with steam to produce a carbon dioxide and hydrogen whereby the desired 1:3 mole ratio of carbon monoxide to hydrogen may be obtained:

$$CO(g) + H_2O(g) \rightarrow CO_2(g) + H_2(g)$$

Even though the water-gas shift reaction is not classified as one of the principal gasification reactions, it cannot be omitted in the analysis of chemical reaction systems that involve synthesis gas. Among all reactions involving synthesis gas, this reaction equilibrium is least sensitive to the temperature variation—the equilibrium constant is least strongly dependent on the temperature. Therefore, the reaction equilibrium can be reversed in a variety of practical process conditions over a wide range of temperature.

The water-gas shift reaction in its forward direction is mildly exothermic and although all the participating chemical species are in gaseous form, the reaction is believed to be heterogeneous insofar as the chemistry occurs at the surface of the feedstock and the reaction is actually catalyzed by carbon surfaces. In addition, the reaction can also take place homogeneously as well as heterogeneously and a generalized understanding of the water-gas shift reaction is difficult to achieve. Even the published kinetic rate information is not immediately useful or applicable to a practical reactor situation.

Synthesis gas from a gasifier contains a variety of gaseous species other than carbon monoxide and hydrogen. Typically, they include carbon dioxide, methane, and water (steam). Depending on the objective of the ensuing process, the composition of synthesis gas may need to be preferentially readjusted. If the objective of the gasification process is to obtain a high yield of methane, it would be preferred to have the molar ratio of hydrogen to carbon monoxide at 3:1:

$$CO(g) + 3H_2(g) \rightarrow CH_4(g) + H_2O(g)$$

On the other hand, if the objective of generating synthesis gas is the synthesis of methanol via a vapor phase low-pressure process, the stoichiometrically consistent ratio between hydrogen and carbon monoxide would be 2:1. In such cases, the stoichiometrically consistent synthesis gas mixture is often referred to as *balanced gas*, whereas a synthesis gas composition that is substantially deviated from the principal reaction's stoichiometry is called *unbalanced gas*. If the objective of synthesis gas production is to obtain a high yield of hydrogen, it would be advantageous to increase the ratio of hydrogen to carbon monoxide by further converting carbon monoxide (and water) into hydrogen (and carbon dioxide) via the water-gas shift reaction.

The water-gas shift reaction is one of the major reactions in the steam gasification process, where both water and carbon monoxide are present in ample amounts. Although the four chemical species involved in the water-gas shift reaction are gaseous compounds at the reaction stage of most gas processing, the water-gas shift reaction, in the case of steam gasification of feedstock, predominantly takes place on the solid surface of feedstock (heterogeneous reaction). If the product synthesis gas from a gasifier needs to be reconditioned by the water-gas shift reaction, this reaction can be catalyzed by a variety of metallic catalysts.

Choice of specific kinds of catalysts has always depended on the desired outcome, the prevailing temperature conditions, composition of gas mixture, and process economics. Typical catalysts used for the reaction include catalysts containing iron, copper, zinc, nickel, chromium, and molybdenum.

5.2.3.4 Carbon Dioxide Gasification

The reaction of carbonaceous feedstocks with carbon dioxide produces carbon monoxide (*Boudouard reaction*) and (like the steam gasification reaction) is also an endothermic reaction:

$$C(s) + CO_2(g) \rightarrow 2CO(g)$$

The reverse reaction results in carbon deposition (carbon fouling) on many surfaces including the catalysts and results in catalyst deactivation.

This gasification reaction is thermodynamically favored at high temperatures (>680°C, >1,255°F), which is also quite similar to the steam gasification. If carried out alone, the reaction requires high temperature (for fast reaction) and high pressure (for higher reactant concentrations) for significant conversion but as a separate reaction a variety of factors come into play: (i) low conversion, (ii) slow kinetic rate, and (iii) low thermal efficiency.

Also, the rate of the carbon dioxide gasification of a feedstock is different to the rate of the carbon dioxide gasification of carbon. Generally, the carbon-carbon dioxide reaction follows a reaction order based on the partial pressure of the carbon dioxide that is approximately 1.0 (or lower) whereas the feedstock-carbon dioxide reaction follows a reaction order based on the partial pressure of the carbon dioxide that is 1.0 (or higher). The observed higher reaction order for the feedstock reaction is also based on the relative reactivity of the feedstock in the gasification system.

5.2.3.5 Hydrogasification

Not all high heat content (high-Btu) gasification technologies depend entirely on catalytic methanation and, in fact, a number of gasification processes use hydrogasification, that is, the direct addition of hydrogen to feedstock under pressure to form methane.

$$C_{char} + 2H_2 \rightarrow CH_4$$

The hydrogen-rich gas for hydrogasification can be manufactured from steam and char from the hydrogasifier. Appreciable quantities of methane are formed directly in the primary gasifier and the heat released by methane formation is at a sufficiently high temperature to be used in the steam-carbon reaction to produce hydrogen so that less oxygen is used to produce heat for the steam-carbon reaction. Hence, less heat is lost in the low-temperature methanation step, thereby leading to higher overall process efficiency.

Hydrogasification is the gasification of feedstock in the presence of an atmosphere of hydrogen under pressure. Thus, not all high heat content (high-Btu) gasification technologies depend entirely on catalytic methanation and, in fact, a number of gasification processes use hydrogasification, that is, the direct addition of hydrogen to feedstock under pressure to form methane:

$$[C]_{feedstock} + H_2 \rightarrow CH_4$$

The hydrogen-rich gas for hydrogasification can be manufactured from steam by using the char that leaves the hydrogasifier. Appreciable quantities of methane are formed directly in the primary gasifier and the heat released by methane formation is at a sufficiently high temperature to be used in the steam-carbon reaction to produce hydrogen so that less oxygen is used to produce heat for the steam-carbon reaction. Hence, less heat is lost in the low-temperature methanation step, thereby leading to higher overall process efficiency.

The hydrogasification reaction is exothermic and is thermodynamically favored at low temperatures ($<670°C$, $<1,240°F$), unlike the endothermic both steam gasification and carbon dioxide gasification reactions. However, at low temperatures, the reaction rate is inevitably too slow. Therefore, a high temperature is always required for kinetic reasons, which in turn requires high pressure of hydrogen, which is also preferred for equilibrium considerations. This reaction can be catalyzed by salts such as potassium carbonate (K_2CO_3), nickel chloride ($NiCl_2$), iron chloride ($FeCl_2$), and iron sulfate ($FeSO_4$). However, use of a catalyst in feedstock gasification suffers from difficulty in recovering and reusing the catalyst and the potential for the spent catalyst becoming an environmental issue.

In a hydrogen atmosphere at elevated pressure, additional yields of methane or other low molecular weight hydrocarbon derivatives can result during the initial feedstock gasification stage from direct hydrogenation of feedstock or semi-char because of active intermediate formed in the feedstock structure after pyrolysis. The direct hydrogenation can also increase the amount of feedstock carbon that is gasified as well as the hydrogenation of gaseous hydrocarbon derivatives, oil, and tar.

The kinetics of the rapid-rate reaction between gaseous hydrogen and the active intermediate depends on hydrogen partial pressure (P_{H2}). Greatly increased gaseous hydrocarbon derivatives produced during the initial feedstock gasification stage are extremely important in processes to convert feedstock into methane (synthetic natural gas, substitute natural gas (SNG)).

5.2.3.6 Methanation

Several exothermic reactions may occur simultaneously within a methanation unit. A variety of metals have been used as catalysts for the methanation reaction; the most common, and to some extent the most effective methanation catalysts, appear to be nickel and ruthenium, with nickel being the most widely used (Cusumano et al., 1978):

Ruthenium (Ru) > nickel (Ni) > cobalt (Co) > iron (Fe) > molybdenum (Mo).

Nearly all the commercially available catalysts used for this process are, however, very susceptible to sulfur poisoning and efforts must be taken to remove all hydrogen sulfide (H_2S) before the catalytic reaction starts. It is necessary to reduce the sulfur concentration in the feed gas to less than 0.5 ppm v/v in order to maintain adequate catalyst activity for a long period of time.

The synthesis gas must be desulfurized before the methanation step since sulfur compounds will rapidly deactivate (poison) the catalysts. A processing issue may arise when the concentration of carbon monoxide is excessive in the stream to be methanated since large amounts of heat must be removed from the system to prevent high temperatures and deactivation of the catalyst by sintering as well as the deposition of carbon. To eliminate this problem, temperatures should be maintained below 400°C (750°F).

The methanation reaction is used to increase the methane content of the product gas, as needed for the production of high Btu gas.

$$4H_2 + CO_2 \rightarrow CH_4 + 2H_2O$$

$$4H_2 + CO_2 \rightarrow CH_4 + 2H_2O$$

$$2CO \rightarrow C + CO_2$$

$$CO + H_2O \rightarrow CO_2 + H_2$$

Among these, the most dominant chemical reaction leading to methane is the first one. Therefore, if methanation is carried out over a catalyst with a synthesis gas mixture of hydrogen and carbon monoxide, the desired hydrogen-carbon monoxide ratio of the feed synthesis gas is around 3:1. The large amount of water (vapor) produced is removed by condensation and recirculated as process water or steam. During this process, most of the exothermic heat due to the methanation reaction is also recovered through a variety of energy integration processes.

Whereas all the reactions listed above are quite strongly exothermic except the forward water-gas shift reaction, which is mildly exothermic, the heat release depends largely on the amount of carbon monoxide present in the feed synthesis gas. For each 1% v/v carbon monoxide in the feed synthesis gas, an adiabatic reaction will experience a 60°C (108°F) temperature rise, which may be termed as *adiabatic temperature rise*.

5.3 GASIFICATION PROCESSES

Gasification is an established tried-and-true method that can be used to convert crude oil coke (petcoke), heavy oil, extra heavy oil, tar sand bitumen, and other refinery viscous feedstocks streams (such as vacuum residua, visbreaker tar, and deasphalter pitch) into power, steam, and hydrogen for use in the production of cleaner transportation fuels. The main requirement for a gasification feedstock is that it contains both hydrogen and carbon.

A number of factors have increased the interest in gasification applications in crude oil refinery operations: (i) coking capacity has increased with the shift to heavier, more sour crude oils being supplied to the refiners; (ii) hazardous waste disposal has become a major issue for refiners in many countries; (iii) there is strong emphasis on the reduction of emissions of criteria pollutants and greenhouse gases; (iv) requirements to produce ultra-low-sulfur fuels are increasing the hydrogen needs of the refineries; and (v) the requirements to produce low-sulfur fuels and other regulations could lead to refiners falling short of demand for lower-boiling products such as gasoline and jet and diesel fuel. The typical gasification system incorporated into the refinery consists of several process plants including (i) a feedstock preparation area, (ii) the type of gasifier, (iii) a gas cleaning section, (iv) a sulfur recovery unit (SRU), and (v) downstream process options that are dependent on the nature of the products.

The gasification process can provide high-purity hydrogen for a variety of uses within the refinery. Hydrogen is used in the refinery to remove sulfur, nitrogen, and other impurities from intermediate to finished product streams and in hydrocracking operations for the conversion of high-boiling distillates and oils into low-boiling products such as naphtha, kerosene, and diesel (Parkash, 2003; Gary et al., 2007; Speight, 2014a; Hsu and Robinson, 2017; Speight, 2017). Furthermore, electric power and high-pressure steam can be generated by the gasification of crude oil coke and viscous feedstocks to drive mostly small and intermittent loads such as compressors, blowers, and pumps. Steam can also be used for process heating, steam tracing, partial pressure reduction in fractionation systems, and stripping low-boiling components to stabilize process streams. Also, the gasification system and refinery operations can share common process equipment. This usually includes an amine stripper or sulfur plant, waste water treatment, and cooling water systems (Mokhatab et al., 2006; Speight, 2007, 2014a).

5.3.1 Gasifiers

A gasifier differs from a combustor in that the amount of air or oxygen available inside the gasifier is carefully controlled so that only a relatively small portion of the fuel burns completely. The *partial oxidation* process provides the heat and rather than combustion, most of the carbon-containing feedstock is chemically broken apart by the heat and pressure applied in the gasifier resulting in the chemical reactions that produce synthesis gas. However, the composition of the synthesis gas will vary because of dependence upon the conditions in the gasifier and the type of feedstock. Minerals in the fuel (i.e., the rocks, dirt, and other impurities which do not gasify) separate and leave the bottom of the gasifier either as an inert glass-like slag or other marketable solid products.

Four types of gasifier are currently available for commercial use: (i) the countercurrent fixed bed, (ii) cocurrent fixed bed, (iii) the fluidized bed, and (iv) the entrained flow (Speight, 2008, 2013a).

In a fixed bed process, the coal is supported by a grate and combustion gases (steam, air, oxygen, etc.) pass through the supported coal whereupon the hot produced gases exit from the top of the reactor. Heat is supplied internally or from an outside source, but caking coals cannot be used in an unmodified fixed bed reactor. Due to the liquid-like behavior, the fluidized beds are very well mixed, which effectively eliminates the concentration and temperature gradients inside the reactor. The process is also fairly simple and reliable to operate as the bed acts as a large thermal reservoir that resists rapid changes in temperature and operation conditions. The disadvantages of the process include the need for recirculation of the entrained solids carried out from the reactor with the fluid, and the nonuniform residence time of the solids that can cause poor conversion levels. The abrasion of the particles can also contribute to serious erosion of pipes and vessels inside the reactor (Kunii and Levenspiel, 2013).

The *countercurrent fixed bed* (*up draft*) gasifier consists of a fixed bed of carbonaceous fuel (e.g., coal or biomass) through which the *gasification agent* (steam, oxygen, and/or air) flows in countercurrent configuration. The ash is either removed dry or as a slag. The nature of the gasifier means that the fuel must have high mechanical strength and must be noncaking so that it will form a permeable bed, although recent developments have reduced these restrictions to some extent. The throughput for this type of gasifier is relatively low. Thermal efficiency is high as the gas exit temperatures are relatively low and, as a result, tar and methane production is significant at typical operation temperatures, so product gas must be extensively cleaned before use or recycled to the reactor.

The cocurrent fixed bed (down draft) gasifier is similar to the countercurrent type, but the gasification agent gas flows in cocurrent configuration with the fuel (downwards, hence the name down draft gasifier). Heat needs to be added to the upper part of the bed, either by combusting small amounts of the fuel or from external heat sources. The produced gas leaves the gasifier at a high temperature, and most of this heat is often transferred to the gasification agent added in the top of the bed. Since all tars must pass through a hot bed of char in this configuration, tar levels are much lower than the countercurrent type.

In the *fluidized bed* gasifier, the fuel is fluidized in oxygen (or air) and steam. The temperatures are relatively low in dry ash gasifiers, so the fuel must be highly reactive; low-grade coals are particularly suitable. The fluidized bed system uses finely sized coal particles and the bed exhibits liquid-like characteristics when a gas flows upward through the bed. Gas flowing through the coal produces turbulent lifting and separation of particles and the result is an expanded bed having greater coal surface area to promote the chemical reaction, but such systems have a limited ability to handle caking coals. The agglomerating gasifiers have slightly higher temperatures, and are suitable for higher-rank coals. Fuel throughput is higher than the fixed bed, but not as high as for the entrained flow gasifier. The conversion efficiency is typically low, so recycle or subsequent combustion of solids is necessary to increase conversion. Fluidized bed gasifiers are most useful for fuels that form highly corrosive ash that would damage the walls of slagging gasifiers. The ash is removed dry or as high molecular weight agglomerated materials—a disadvantage of biomass feedstocks is that they generally contain high levels of corrosive ash.

In the *entrained flow* gasifier, a dry pulverized solid, an atomized liquid fuel or a fuel slurry is gasified with oxygen (much less frequent: air) in cocurrent flow. The high temperatures and pressures also mean that a higher throughput can be achieved but thermal efficiency is somewhat lower as the gas must be cooled before it can be sent to a gas processing facility. All entrained flow gasifiers remove the major part of the ash as a slag as the operating temperature is well above the ash fusion temperature; the entrained system is suitable for both caking and noncaking coals.

Entrained flow reactors use atomized liquid, slurry or dry pulverized solid as a feedstock. Once pumped inside the gasifier, the feedstock is gasified with oxygen in a cocurrent flow. The temperatures are usually very high in comparison to fluidized beds ranging from 1,300°C to 1,500°C (2,370°F–2,730°F). High-temperature cracks the feedstock into lower-boiling products.

In *IGCC* systems, the synthesis gas is cleaned of its hydrogen sulfide, ammonia, and particulate matter and is burned as fuel in a combustion turbine (much like natural gas is burned in a turbine). The combustion turbine drives an electric generator. Hot air from the combustion turbine can be channeled back to the gasifier or the air separation unit (ASU), while exhaust heat from the combustion turbine is recovered and used to boil water, creating steam for a steam turbine generator. The use of these two types of turbines—a combustion turbine and a steam turbine—in combination, known as a *combined cycle*, is one reason why gasification-based power systems can achieve unprecedented power generation efficiencies.

Gasification also offers more scope for recovering products from waste than incineration (Speight, 2014b). When waste is burnt in an incinerator the only practical product is energy, whereas the gases, oils, and solid char from pyrolysis and gasification can not only be used as a fuel but also purified and used as a feedstock for petrochemicals and other applications. Many processes also produce a stable granulate instead of an ash which can be more easily and safely utilized. In addition, some processes are targeted at producing specific recyclables such as metal alloys and carbon black. From waste gasification, in particular, it is feasible to produce hydrogen, which many see as an increasingly valuable resource.

IGCC is used to raise power from viscous feedstocks. The value of these refinery residuals, including crude oil coke, will need to be considered as part of an overall upgrading project. Historically, many delayed coking projects have been evaluated and sanctioned on the basis of assigning zero value to crude oil coke having high-sulfur and high metal content.

While there are many alternate uses for the synthesis gas produced by gasification, and a combination of products/utilities can be produced in addition to power. A major benefit of the IGCC concept is that power can be produced with the lowest sulfur oxide (Sox) and nitrogen oxide (NOx) emissions of any liquid/solid feed power generation technology.

5.3.2 FT Synthesis

The synthesis reaction is dependent on a catalyst, mostly an iron or cobalt catalyst where the reaction takes place. There is either a low-temperature Fischer–Tropsch process (LTFT process) or high-temperature Fischer–Tropsch process (HTFT process), with temperatures ranging between 200°C and 240°C (390°F and 465°F) for LTFT and between 300°C and 350°C (570°F and 660°F) for HTFT. The HTFT uses an iron catalyst and the LTFT either an iron or a cobalt catalyst. The different catalysts include also nickel-based and ruthenium-based catalysts, which also have enough activity for commercial use in the process.

The reactors are the *multi-tubular fixed bed*, the *slurry* or the *fluidized bed* (with either fixed or circulating bed) reactor. The fixed bed reactor consists of thousands of small tubes with the catalyst as surface-active agent in the tubes. Water surrounds the tubes and regulates the temperature by settling the pressure of evaporation. The slurry reactor is widely used and consists of fluid and solid elements, where the catalyst has no particularly position, but flows around as small pieces of catalyst together with the reaction components. The slurry and fixed bed reactor are used in LTFT. The fluidized bed reactors are diverse, but characterized by the fluid behavior of the catalyst.

The HTFT technology uses a fluidized catalyst at 300°C–330°C. Originally circulating fluidized bed units were used (Synthol reactors). Since 1989 a commercial-scale classical fluidized bed unit has been implemented and improved upon. The LTFT technology has originally been used in tubular fixed bed reactors at 200°C–230°C. This produces a more paraffinic and waxy product spectrum than the *high-temperature* technology. A new type of reactor (the Sasol slurry phase distillate reactor has been developed and is in commercial operation. This reactor uses a slurry phase system rather than a tubular fixed bed configuration and is currently the favored technology for the commercial production of synfuels.

Under most circumstances, the production of synthesis gas by reforming natural gas will be more economical than from coal gasification, but site-specific factors need to be considered. In fact, any technological advance in this field (such as better energy integration or the oxygen transfer ceramic membrane reformer concept) will speed up the rate at which the synfuels technology will become common practice.

There are large coal reserves which may increasingly be used as a fuel source during oil depletion. Since there are large coal reserves in the world, this technology could be used as an interim transportation fuel if conventional oil were to become more expensive. Furthermore, combination of biomass gasification and FT synthesis is a very promising route to produce transportation fuels from renewable or *green* resources.

Although the focus of this section has been on the production of hydrocarbon derivatives from synthesis gas, it is worthy of note that clean synthesis gas can also be used (i) as chemical *building blocks* to produce a broad range of chemicals using processes well established in the chemical and petrochemical industry; (ii) as a fuel producer for highly efficient fuel cells (which run off the hydrogen made in a gasifier) or perhaps in the future, hydrogen turbines and fuel cell-turbine hybrid systems; and (iii) as a source of hydrogen that can be separated from the gas stream and used as a fuel or as a feedstock for refineries (which use the hydrogen to upgrade crude oil products).

The aim of underground (*or in situ*) gasification of coal is the conversion into combustible gases by combustion of a coal seam in the presence of air, oxygen, or oxygen and steam. Thus, seams that were considered to be inaccessible, unworkable, or uneconomical to mine could be put to use. In addition, strip mining and the accompanying environmental impacts, the problems of spoil banks, acid mine drainage, and the problems associated with use of high-ash coal are minimized or even eliminated.

The principles of underground gasification are very similar to those involved in the aboveground gasification of coal. The concept involves the drilling and subsequent linking of two boreholes so that gas will pass between the two. Combustion is then initiated at the bottom of one borehole (injection well) and is maintained by the continuous injection of air. In the initial reaction zone (combustion zone), carbon dioxide is generated by the reaction of oxygen (air) with the coal:

$$[C]_{coal} + O_2 \rightarrow CO_2$$

The carbon dioxide reacts with coal (partially devolatilized) further along the seam (reduction zone) to produce carbon monoxide:

$$[C]_{coal} + CO_2 \rightarrow 2CO$$

In addition, at the high temperatures that can frequently occur, moisture injected with oxygen or even moisture inherent in the seam may also react with the coal to produce carbon monoxide and hydrogen:

$$[C]_{coal} + H_2O \rightarrow CO + H_2$$

The gas product varies in character and composition but usually falls into the low heat (low Btu) category ranging from 125 to 175 Btu/ft^3.

5.3.3 FEEDSTOCKS

For many decades, coal has been the primary feedstock for gasification units but recent concerns about the use of fossil fuels and the resulting environmental pollutants, irrespective of the various gas cleaning processes and gasification plant environmental cleanup efforts, there is a move to feedstocks other than coal for gasification processes (Speight, 2013a, 2014b). But more pertinent to the present text, the gasification process can also use carbonaceous feedstocks which would otherwise have been discarded and unused, such as waste biomass and other similar biodegradable wastes. Various feedstocks such as biomass, crude oil resids, and other carbonaceous wastes can be used to their fullest potential. In fact, the refining industry has seen fit to use viscous feedstock gasification as a source of hydrogen for the past several decades (Speight, 2014a).

Gasification processes can accept a variety of feedstocks but the reactor must be selected on the basis of feedstock properties and behavior in the process. The advantage of the gasification process when a carbonaceous feedstock (a feedstock containing carbon) or hydrocarbonaceous feedstock (a feedstock containing carbon and hydrogen) is employed is that the product of focus—synthesis gas—is potentially more useful as an energy source and results in an overall cleaner process. The production of synthesis gas is a more efficient production of an energy source than, say, the direct combustion of the original feedstock because synthesis gas can be (i) combusted at higher temperatures, (ii) used in fuel cells, (iii) used to produce methanol, (iv) used as a source of hydrogen, and (v) particularly because the synthesis gas can be converted via the FT process into a range of synthesis liquid fuels suitable for use gasoline engines, for diesel engines, or for wax production.

5.3.3.1 Heavy Feedstocks

Gasification is the only technology that makes possible a zero residue target for refineries, contrary to all conversion technologies (including thermal cracking, catalytic cracking, cooking, deasphalting, hydroprocessing, etc.) which can only reduce the bottom volume, with the complication that the residue qualities generally get worse with the degree of conversion (Speight, 2014a).

The flexibility of gasification permits to handle any type of refinery residue, including crude oil coke, tank bottoms, and refinery sludge and make available a range of value-added products including electricity, steam, hydrogen, and various chemicals based on synthesis gas chemistry: methanol, ammonia, methyl tert-butyl ether (MTBE), tert-amyl methyl ether (TAME), acetic acid, and formaldehyde (Speight, 2008, 2013a). The environmental performance of gasification is unmatched. No other technology processing low-value refinery residues can come close to the emission levels achievable with gasification (Speight, 2013a, 2013b, 2014a, 2014b).

Gasification is also a method for converting crude oil coke and other refinery nonvolatile waste streams (often referred to as refinery residuals and include but not limited to atmospheric residuum, vacuum residuum, visbreaker tar, and deasphalter pitch) into power, steam, and hydrogen for use in the production of cleaner transportation fuels. The main requirement for a gasification feedstock is that the feedstock it contains both hydrogen and carbon and several suitable feedstocks are produced on-site as part of typical refinery processing (Speight, 2011b). The typical gasification system incorporated into a refinery consists of several process units including feed preparation, the gasifier, an ASU, synthesis gas cleanup, SRU, and downstream process options depending on target products.

The benefits of the addition of a gasification system in a refinery to process crude oil coke or other residuals include: (i) production of power, steam, oxygen, and nitrogen for refinery use or sale; (ii) source of synthesis gas for hydrogen to be used in refinery operations as well as for the production of lower-boiling refinery products through the FT synthesis; (iii) increased efficiency of power generation, improved air emissions, and reduced waste stream versus combustion of crude oil coke or residues or incineration; (iv) no off-site transportation or storage for crude oil coke or residuals; and (v) the potential to dispose of waste streams including hazardous materials (Marano, 2003).

Gasification can provide high-purity hydrogen for a variety of uses within the refinery (Speight, 2014a). Hydrogen is used in refineries to remove sulfur, nitrogen, and other impurities from intermediate to finished product streams and in hydrocracking operations for the conversion of high-boiling distillates into lower-boiling products, naphtha, kerosene, and low-boiling gas oil. Hydrocracking and severe hydrotreating require hydrogen which is at least 99% v/v, while less severe hydrotreating can work with gas streams containing 90% v/v pure hydrogen.

Electric power and high-pressure steam can be generated via gasification of crude oil coke and residuals to drive mostly small and intermittent loads such as compressors, blowers, and pumps. Steam can also be used for process heating, steam tracing, partial pressure reduction in fractionation systems, and stripping low-boiling components to stabilize process streams.

Carbon soot is produced during gasification, which ends up in the quench water. The soot is transferred to the feedstock by contacting, in sequence, the quench water blowdown with naphtha, and then the naphtha-soot slurry with a fraction of the feed. The soot mixed with the feed is finally recycled into the gasifier, thus achieving 100% conversion of carbon to gas.

5.3.3.2 Solvent Deasphalter Bottoms

The deasphalting unit (deasphalter) is a unit in a petroleum refinery for bitumen upgrader that separates an asphalt-like product from petroleum, heavy oil, or bitumen. The deasphalter unit is usually placed after the vacuum distillation tower where, by the use of a low-boiling liquid hydrocarbon solvent (such as propane or butane under pressure), the insoluble asphalt-like product (*deasphalter bottoms*) is separated from the feedstock—the other output from the deasphalter is deasphalted oil (DAO).

The solvent deasphalting process has been employed for more than six decades to separate high molecular weight fractions of crude oil boiling beyond the range of economical commercial distillation. The earliest commercial applications of solvent deasphalting used liquid propane as the solvent to extract high-quality lubricating oil bright stock from vacuum residue. The process has been extended to the preparation of catalytic cracking feeds, hydrocracking feeds, hydrodesulfurization feedstocks, and asphalts. The latter product (asphalt, also called *deasphalter bottoms*) is used for (i) road asphalt manufacture, (ii) refinery fuel, or (iii) gasification feedstock for hydrogen production.

In fact, the combination of ROSE solvent deasphalting and gasification has been commercially proven at the ERG Petroli refinery (Bernetti et al., 2000). The combination is very synergistic and offers a number of advantages including a low-cost feedstock to the gasifier, thus enhancing the refinery economics, and converts low-value feedstock to high-value products such as power, steam, hydrogen, and chemical feedstock. The process also improves the economics of the refinery by eliminating/reducing the production of low-value fuel oil and maximizing the production of transportation fuel.

5.3.3.3 Asphalt, Tar, and Pitch

The terms asphalt, tar, and pitch are nondescript terms that are often applied in a refinery to any viscous, black, difficult-to-identify product. The terms are often applied the insoluble product from a deasphalting unit (also called deasphalter bottoms). The terms will be covered in this subsection because of the application of the nomenclature to the products of other processes.

Asphalt does not occur naturally but is manufactured from crude oil and is a black or brown material that has a consistency varying from a viscous liquid to a glassy solid (Speight, 2014a). To a point, asphalt can resemble bitumen (isolated form tar sand formation), hence the tendency to refer to bitumen (incorrectly) as *native asphalt*. It is recommended that there be differentiation between asphalt (manufactured) and bitumen (naturally occurring) other than by use of the qualifying terms *crude oil* and *native* since the origins of the materials may be reflected in the resulting physicochemical properties of the two types of materials. It is also necessary to distinguish between the asphalts which originate from crude oil by refining and the product in which the source of the asphalt is a material other than crude oil, e.g., *Wurtzilite asphalt* (Speight, 2014a). In the absence of a qualifying word, it should be assumed that the word *asphalt* (with or without qualifiers such as *cutback*, *solvent*, and *blown*, which indicate the process used to produce the asphalt) refers to the product manufactured from crude oil.

When the asphalt is produced simply by distillation of an asphaltic crude oil, the product can be referred to as *residual asphalt* or *straight-run asphalt*. For example, if the asphalt is prepared by *solvent* extraction of viscous feedstock or by lower-boiling hydrocarbon (propane) precipitation, or if *blown* or otherwise treated, the term should be modified accordingly to qualify the product (e.g., *solvent asphalt, propane asphalt, blown asphalt*).

Asphalt softens when heated and is elastic under certain conditions and has many uses. For example, the mechanical properties of asphalt are of particular significance when it is used as a binder or adhesive. The principal application of asphalt is in road surfacing that may be done in a variety of ways. Other important applications of asphalt include canal and reservoir linings, dam facings, and sea works. The asphalt so used may be a thin, sprayed membrane, covered with earth for protection against weathering and mechanical damage, or thicker surfaces, often including riprap (crushed rock). Asphalt is also used for roofs, coatings, floor tiles, soundproofing, waterproofing, and other building-construction elements and in a number of industrial products, such as batteries. For certain applications an asphaltic emulsion is prepared, in which fine globules of asphalt are suspended in water.

Tar is a product of the destructive distillation of many bituminous or other organic materials and is a brown to black, oily, viscous liquid to semisolid material. However, *tar* is most commonly produced from *bituminous coal* and is generally understood to refer to the product from coal, although it is advisable to specify *coal tar* if there is the possibility of ambiguity. The most important factor in determining the yield and character of the coal tar is the carbonizing temperature. Three general temperature ranges are recognized, and the products have acquired the designations: *low-temperature tar* (approximately 450°C–700°C; 540°F–1,290°F), *mid-temperature tar* (approximately 700°C–900°C; 1,290°F–1,650°F), and *high-temperature tar* (approximately 900°C–1,200°C; 1,650°F–2,190°F). Tar released during the early stages of the decomposition of the organic material is called *primary tar* since it represents a product that has been recovered without the secondary alteration that results from prolonged residence of the vapor in the heated zone.

Treatment of the distillate (boiling up to 250°C, 480°F) of the tar with caustic soda causes separation of a fraction known as *tar acids*; acid treatment of the distillate produces a variety of organic nitrogen compounds known as *tar bases*. The residue left following removal of the high-boiling distillate, is *pitch*, a black, hard, and highly ductile material that is the dark brown-to-black non-distillable residue.

Coal tar pitch is a soft to hard and brittle substance containing chiefly aromatic resinous compounds along with aromatic and other hydrocarbon derivatives. Pitch is used chiefly as road tar, in waterproofing roofs and other structures, and to make electrodes. *Wood tar pitch* is a bright, lustrous substance containing resin acids; it is used chiefly in the manufacture of plastics and insulating materials and in caulking seams. *Pitch* derived from fats, fatty acids, or fatty oils by distillation are usually soft substances containing polymers and decomposition products; they are used chiefly in varnishes and paints and in floor coverings.

Any of the above derivatives can be used as a gasification feedstock. The properties of asphalt change markedly during the aging process (oxidation in service) to the point where the asphalt fails to perform the task for which it was designed. In some case, the asphalt is recovered and reprocessed for additional use or it may be sent to a gasifier.

5.3.3.4 Petroleum Coke

Coke is the solid carbonaceous material produced from crude oil during thermal processing. More particularly, coke is the residue left by the destructive distillation (i.e., thermal cracking such as the delayed coking process) of crude oil residua. The coke formed in catalytic cracking operations is usually non-recoverable because of the materials deposited on the catalyst during the process and such coke is often employed as fuel for the process (Gray and Tomlinson, 2000; Speight, 2014a). It is often characterized as a solid material with a honeycomb-type of appearance having high-carbon content (95%+ w/w) with some hydrogen and, depending on the process, as well as sulfur and nitrogen. The color varies from gray to black, and the material is insoluble in organic solvents.

Typically, the composition of crude oil coke varies with the source of the crude oil, but in general, large amounts of high molecular weight complex hydrocarbon derivatives (rich in carbon but correspondingly poor in hydrogen) make up a high proportion. The solubility of crude oil *coke* in carbon disulfide has been reported to be as high as 50%–80%, but this is in fact a misnomer and is due to soluble product adsorbed on the coke—by definition coke is the insoluble, honeycomb material that is the end product of thermal processes. However, coke is not always a product with little use—three physical structures of coke can be produced by delayed coking: (i) shot coke, (ii) sponge coke, or (iii) needle coke, which finds different uses within the industry.

Shot coke is an abnormal type of coke resembling small balls. Due to mechanisms not well understood the coke from some coker feedstocks forms into small, tight, nonattached clusters that look like pellets, marbles, or ball bearings. It usually is a very hard coke, i.e., low Hardgrove grindability index (Speight, 2013a). Such coke is less desirable to the end users because of difficulties in handling and grinding. It is believed that feedstocks high in asphaltene constituents and low API gravity favor shot coke formation. Blending aromatic materials with the feedstock and/or increasing the recycle ratio reduces the yield of shot coke. Fluidization in the coke drums may cause formation of shot coke. Occasionally, the smaller *shot coke* may agglomerate into ostrich egg-sized pieces. Such coke may be more suitable as a gasification feedstock.

Sponge coke is the common type of coke produced by delayed coking units. It is in a form that resembles a sponge and has been called honeycombed. Sponge coke, mostly used for anode-grade carbon, is dull and black, having porous, amorphous structure. *Needle coke* (*acicular coke*) is a special quality coke produced from aromatic feed stocks is silver-gray, having crystalline broken needle structure, and is believed to be chemically produced through cross-linking of condensed aromatic hydrocarbon derivatives during coking reactions. It has a crystalline structure with more unidirectional pores and is used in the production of electrodes for the steel and aluminum industries and is particularly valuable because the electrodes must be replaced regularly.

Crude oil coke is employed for a number of purposes, but its chief use is (depending upon the degree of purity—i.e., contains a low amount of contaminants) for the manufacture of carbon electrodes for aluminum refining, which requires a high-purity carbon low in ash and sulfur free; the volatile matter must be removed by calcining. In addition to its use as a metallurgical reducing agent, crude oil coke is employed in the manufacture of carbon brushes, silicon carbide abrasives, and structural carbon (e.g., pipes and Raschig rings), as well as calcium carbide manufacture from which acetylene is produced:

$$Coke \rightarrow CaC_2$$

$$CaC_2 + H_2O \rightarrow HC\equiv CH$$

The flexibility of the gasification technology permits the refinery to handle any kind of refinery residue, including crude oil coke, tank bottoms, and refinery sludge and makes available a range of value-added products, electricity, steam, hydrogen, and various chemicals based on synthesis gas chemistry: methanol, ammonia, MTBE, TAME, acetic acid, and formaldehyde (Speight, 2008, 2013a). With respect to gasification, no other technology processing low-value refinery residues can come close to the emission levels achievable with gasification (Speight, 2014a) and is projected to be a major part of the refinery of the future (Speight, 2011b).

Gasification is also a method for converting crude oil coke and other refinery nonvolatile waste streams (often referred to as refinery residuals and include but not limited to atmospheric residuum, vacuum residuum, visbreaker tar, and deasphalter pitch) into power, steam, and hydrogen for use in the production of cleaner transportation fuels. And as for the gasification of coal and biomass (Speight, 2013a; Luque and Speight, 2015), the main requirement for a feedstock to a gasification unit is that the feedstock contains both hydrogen and carbon, of which a variety of feedstocks are available from the throughput of a typical refinery (Table 5.4).

The typical gasification system incorporated into the refinery consists of several process plants including (i) feed preparation, (ii) the gasifier, (iii) an ASU, (iv) synthesis gas cleanup, (v) SRU, and (vi) downstream process options such as Fischer-Tropsch synthesis (FTS) and methanol synthesis, depending on the desired product slate. The benefits to a refinery for adding a gasification system for crude oil coke or other residuals are: (i) production of power, steam, oxygen, and nitrogen for refinery use or sale; (ii) source of synthesis gas for hydrogen to be used in refinery operations and for the production of lower-boiling refinery products through FTS; (iii) increased efficiency of power generation, improved air emissions, and reduced waste stream versus combustion of crude oil coke or viscous feedstock or incineration; (iv) no off-site transportation or storage for crude oil coke or viscous feedstock; and (v) the potential to dispose waste streams including hazardous materials.

Gasification of coke can provide high-purity hydrogen for a variety of uses within the refinery such as (i) sulfur removal, (ii) nitrogen removal, as well as removal of other impurities from intermediate to finished product streams and in hydrocracking operations for the conversion of high-boiling distillates into lower-boiling products, such as naphtha, kerosene, and low-boiling gas oil (Speight, 2014a). Hydrocracking and severe hydrotreating require hydrogen which is at least 99% v/v pure, while less severe hydrotreating can require gas stream containing hydrogen in the order of 90% v/v purity.

Electric power and high-pressure steam can be generated by the gasification of crude oil coke and viscous feedstocks to drive mostly small and intermittent loads such as compressors, blowers, and pumps. Steam can also be used for process heating, steam tracing, partial pressure reduction in fractionation systems, and stripping low-boiling components to stabilize process streams.

During gasification some soot (typically 99%+ carbon) is produced, which ends up in the quench water. The soot is transferred to the feedstock by contacting, in sequence, the quench water blowdown with naphtha, and then the naphtha-soot slurry with a fraction of the feed. The soot mixed with the feed is recycled to the gasifier, thus achieving 100% conversion of carbon to gas.

TABLE 5.4
Types of Feedstocks Produced On-Site that Are Available for Gasification

Ultimate Analysis	Vacuum Resid	Visbreaker Bottoms	Asphalt	Petroleum Coke
Carbon, % w/w	84.9	86.1	85.1	88.6
Hydrogen % w/w	10.4	10.4	9.1	2.8
Nitrogen% w/w	0.5	0.6	0.7	1.1
Sulfur, % w/w	42	2.4	5.1	7.3
Oxygen, % w/w		0.5		0.0
Ash, % w/w	0.0		0.1	02

5.3.3.5 Coal

Coal is a fossil fuel formed in swamp ecosystems where plant remains were saved from oxidation and biodegradation by water and mud (Chapter 3) (Speight, 2013a). Coal is a combustible organic sedimentary rock (composed primarily of carbon, hydrogen, and oxygen as well as other minor elements including sulfur) formed from ancient vegetation and consolidated between other rock strata to form coal seams. The harder forms can be regarded as organic metamorphic rock (e.g., anthracite coal) because of a higher degree of maturation.

Coal is the largest single source of fuel for the generation of electricity worldwide (Speight, 2013b), as well as the largest source of carbon dioxide emissions, which have been implicated as the primary cause of global climate change, although the debate still rages as to the actual cause (or causes) of climate change. Coal is found as successive layers, or seams, sandwiched between strata of sandstone and shale and extracted from the ground by coal mining—either underground coal seams (underground mining) or by open-pit mining (surface mining).

Coal remains in adequate supply and at current rates of recovery and consumption, the world global coal reserves have been variously estimated to have a reserves/production ratio of at least 155 years. However, as with all estimates of resource longevity, coal longevity is subject to the assumed rate of consumption remaining at the current rate of consumption and, moreover, to technological developments that dictate the rate at which the coal can be mined. But most importantly, coal is a fossil fuel and an *unclean* energy source that will only add to global warming. In fact, the next time electricity is advertised as a clean energy source just consider the means by which the majority of electricity is produced—almost 50% of the electricity generated in the United States derives from coal (EIA, 2007; Speight, 2013a).

Coal occurs in different forms or *types* (Speight, 2013a). Variations in the nature of the source material and local or regional the variations in the coalification processes cause the vegetal matter to evolve differently. Various classification systems thus exist to define the different types of coal. Using the American Society for Testing and Materials (ASTM; now ASTM International) system of classification (ASTM D388, 2015), the coal precursors are transformed over time (as geological processes increase their effect over time) into:

i. Lignite: also referred to as brown coal, is the lowest rank of coal and used almost exclusively as fuel for steam-electric power generation. Jet is a compact form of lignite that is sometimes polished and has been used as an ornamental stone since the Iron Age.
ii. Subbituminous coal: the properties range from those of lignite to those of bituminous coal and is used primarily as fuel for steam-electric power generation.
iii. Bituminous coal—a dense coal, usually black, sometimes dark brown, often with well-defined bands of brittle and dull material, used primarily as fuel in steam-electric power generation, with substantial quantities also used for heat and power applications in manufacturing and to make coke.
iv. Anthracite—the highest rank; a harder, glossy, black coal used primarily for residential and commercial space heating.

Chemically, coal is a hydrogen-deficient hydrocarbon with an atomic hydrogen-to-carbon ratio near 0.8, as compared to crude oil hydrocarbon derivatives, which have an atomic hydrogen-to-carbon ratio approximately equal to 2, and methane (CH_4) that has an atomic carbon-to-hydrogen ratio equal to 4. For this reason, any process used to convert coal to alternative fuels must add hydrogen or redistribute the hydrogen in the original coal to generate hydrogen-rich products and coke (Speight, 2013a).

The chemical composition of the coal is defined in terms of its proximate and ultimate (elemental) analyses (Speight, 2013a). The parameters of proximate analysis are moisture, volatile matter, ash, and fixed carbon. Elemental analysis (ultimate analysis) encompasses the quantitative determination of carbon, hydrogen, nitrogen, sulfur, and oxygen within the coal. Additionally, specific physical and mechanical properties of coal and particular carbonization properties are also determined.

Carbon monoxide and hydrogen are produced by the gasification of coal in which a mixture of gases is produced. In addition to carbon monoxide and hydrogen, methane and other hydrocarbon derivatives are also produced depending on conditions. Gasification may be accomplished either *in situ* or in processing plants. *In situ* gasification is accomplished by controlled, incomplete burning of a coalbed underground while adding air and steam. The gases are withdrawn and may be burned to produce heat, generate electricity, or are utilized as synthesis gas in indirect liquefaction as well as for the production of chemicals.

Producing diesel and other fuels from coal can be performed through the conversion of coal to synthesis gas, a combination of carbon monoxide, hydrogen, carbon dioxide, and methane. Synthesis gas is subsequently reacted through FTS processes to produce hydrocarbon derivatives that can be refined into liquid fuels. By increasing the quantity of high-quality fuels from coal (while reducing costs), research into this process could help mitigating the dependence on ever-increasingly expensive and depleting stocks of crude oil.

While coal is an abundant natural resource, its combustion or gasification produces both toxic pollutants and greenhouse gases. By developing adsorbents to capture the pollutants (mercury, sulfur, arsenic, and other harmful gases), scientists are striving not only to reduce the quantity of emitted gases but also to maximize the thermal efficiency of the cleanup.

Gasification thus offers one of the cleanest and versatile ways to convert the energy contained in coal into electricity, hydrogen, and other sources of power. Turning coal into synthesis gas is not a new concept, in fact the basic technology dates back to pre-World War II. In fact, a gasification unit can process virtually all the viscous feedstock and wastes that are produced in refineries leading to enhanced yields of high-value products (and hence their competitiveness in the market) by deeper upgrading of their crude oil.

5.3.3.6 Biomass

Biomass can be considered as any renewable feedstock that is in principle be *carbon neutral* (while the plant is growing, it uses the sun's energy to absorb the same amount of carbon from the atmosphere as it releases into the atmosphere) (Speight, 2008, 2011a).

Raw materials that can be used to produce biomass derived fuels are widely available; they come from a large number of different sources and in numerous forms (Rajvanshi, 1986). The main basic sources of biomass include: (i) wood, including bark, logs, sawdust, wood chips, wood pellets and briquettes; (ii) high yield energy crops, such as wheat, grown specifically for energy applications; (iii) agricultural crops and residues (e.g., straw); and (iv) industrial waste, such as wood pulp or paper pulp. For processing, a simple form of biomass such as untreated and unfinished wood may be converted into a number of physical forms, including pellets and wood chips, for use in biomass boilers and stoves.

Biomass includes a wide range of materials that produce a variety of products which are dependent upon the feedstock (Balat, 2011; Demirbaş, 2011; Ramroop Singh, 2011; Speight, 2011a). In addition, the heat content of the different types of biomass widely varies and has to be taken into consideration when designing any conversion process (Jenkins and Ebeling, 1985).

Thermal conversion processes use heat as the dominant mechanism to convert biomass into another chemical form. The basic alternatives of combustion, torrefaction, pyrolysis, and gasification are separated principally by the extent to which the chemical reactions involved are allowed to proceed (mainly controlled by the availability of oxygen and conversion temperature) (Speight, 2011a).

Energy created by burning biomass (fuel wood), also known as dendrothermal energy, is particularly suited for countries where fuel wood grows more rapidly, e.g., tropical countries. There is a number of other less common, more experimental or proprietary thermal processes that may offer benefits including hydrothermal upgrading and hydroprocessing. Some have been developed to be compatible with high moisture content biomass (e.g., aqueous slurries) and allow them to be converted into more convenient forms.

Some of the applications of thermal conversion are combined heat and power (CHP) and cofiring. In a typical dedicated biomass power plant, efficiencies range from 7% to 27%. In contrast, biomass cofiring with coal, typically occurs at efficiencies close to those of coal combustors (30%–40%) (Baxter, 2005; Liu et al., 2011).

Many forms of biomass contain a high percentage of moisture (along with carbohydrates and sugars) and mineral constituents—both of which can influence the economics and viability of a gasification process. The presence of high levels of moisture in biomass reduces the temperature inside the gasifier, which then reduces the efficiency of the gasifier. Many biomass gasification technologies, therefore, require dried biomass to reduce the moisture content prior to feeding into the gasifier. In addition, biomass can come in a range of sizes. In many biomass gasification systems, biomass must be processed to a uniform size or shape to be fed into the gasifier at a consistent rate as well as to maximize gasification efficiency.

Biomass such as wood pellets, yard, and crop waste and "energy crops" including switch grass and waste from pulp and paper mills can also be employed to produce bioethanol and synthetic diesel. Biomass is first gasified to produce synthesis gas and then subsequently converted via catalytic processes to the aforementioned downstream products. Biomass can also be used to produce electricity—either blended with traditional feedstocks, such as coal or by itself (Shen et al., 2012; Khosravi and Khadse, 2013; Speight, 2014b).

Most biomass gasification systems use air instead of oxygen for gasification reactions (which is typically used in large-scale industrial and power gasification plants). Gasifiers that use oxygen require an ASU to provide the gaseous/liquid oxygen; this is usually not cost-effective at the smaller scales used in biomass gasification plants. Air-blown gasifiers utilize oxygen from air for gasification processes.

In general, biomass gasification plants are comparatively smaller to those of typical coal or crude oil coke plants used in the power, chemical, fertilizer, and refining industries. As such, they are less expensive to build and have a smaller environmental footprint. While a large industrial gasification plant may take up 150 acres of land and process 2,500–15,000 tons/day of feedstock (e.g., coal or crude oil coke), smaller biomass plants typically process 25–200 tons of feedstock per day and take up less than 10 acres.

Finally, while biomass may seem to some observers to be the answer to the global climate change issue, advantages and disadvantages of biomass as feedstock must be considered carefully (Table 5.5). Also, while taking the issues of global climate change into account, it must not be

---

TABLE 5.5

The Advantages and Disadvantages of Using Biomass as a Feedstock for Energy Production and Chemicals Production

Advantages

Theoretically inexhaustible fuel source.

Minimal environmental impact when processes such as fermentation and pyrolysis are used.

Alcohols and other fuels produced by biomass are efficient, viable, and relatively clean-burning.

Biomass is available on a worldwide basis.

Disadvantages

Could contribute to global climate change and particulate pollution when direct combustion is employed.

Production of biomass and the technological conversion to alcohols or other fuels can be expensive.

Life cycle assessments should be considered to address energy input and output.

Possibly a net loss of energy when operated on a small scale—energy is required to grow the biomass.

ignored that the earth is in an interglacial period where warming will take place. The extent of this warming is not known—no one was around to measure the temperature change in the last interglacial period—and by the same token the contribution of anthropological sources to global climate change cannot be measured accurately.

5.3.3.7 Solid Waste

Waste may be MSW which had minimal presorting, or refuse-derived fuel (RDF) with significant pretreatment, usually mechanical screening and shredding. Other more specific waste sources (excluding hazardous waste) and possibly including crude oil coke may provide niche opportunities for co-utilization (Bridgwater, 2003; Arena, 2012; Speight, 2013a, 2014b).

The traditional waste-to-energy plant, based on mass-burn combustion on an inclined grate, has a low public acceptability despite the very low emissions achieved over the last decade with modern flue gas cleanup equipment. This has led to difficulty in obtaining planning permissions to construct needed new waste-to-energy plants. After much debate, various governments have allowed options for advanced waste conversion technologies (gasification, pyrolysis, and anaerobic digestion), but will only give credit to the proportion of electricity generated from nonfossil waste.

Use of waste materials as co-gasification feedstocks may attract significant disposal credits (Ricketts et al., 2002). Cleaner biomass materials are renewable fuels and may attract premium prices for the electricity generated. Availability of sufficient fuel locally for an economic plant size is often a major issue, as is the reliability of the fuel supply. Use of more predictably available coal alongside these fuels overcomes some of these difficulties and risks. Coal could be regarded as the base feedstock that keeps the plant running when the fuels producing the better revenue streams are not available in sufficient quantities.

Coal characteristics are very different to younger hydrocarbon fuels such as biomass and waste. Hydrogen-to-carbon ratios are higher for younger fuels, as is the oxygen content. This means that reactivity is very different under gasification conditions. Gas cleaning issues can also be very different, being sulfur a major concern for coal gasification and chlorine compounds and tars more important for waste and biomass gasification. There are no current proposals for adjacent gasifiers and gas cleaning systems, one handling biomass or waste and one coal, alongside each other and feeding the same power production equipment. However, there are some advantages to such a design as compared with mixing fuels in the same gasifier and gas cleaning systems.

Electricity production or combined electricity and heat production remain the most likely area for the application of gasification or co-gasification. The lowest investment cost per unit of electricity generated is the use of the gas in an existing large power station. This has been done in several large utility boilers, often with the gas fired alongside the main fuel. This option allows a comparatively small thermal output of gas to be used with the same efficiency as the main fuel in the boiler as a large, efficient steam turbine can be used. It is anticipated that addition of gas from a biomass or wood gasifier into the natural gas feed to a gas turbine to be technically possible, but there will be concerns as to the balance of commercial risks to a large power plants and the benefits of using the gas from the gasifier.

Furthermore, the disposal of municipal and industrial waste has become an important problem because the traditional means of disposal, landfill, are much less environmentally acceptable than previously. Much stricter regulation of these disposal methods will make the economics of waste processing for resource recovery much more favorable. One method of processing waste streams is to convert the energy value of the combustible waste into a fuel. One type of fuel attainable from waste is a low heating value gas, usually 100–150 Btu/scf, which can be used to generate process steam or to generate electricity. Coprocessing such waste with coal is also an option (Speight, 2008, 2013a, 2014b).

Co-gasification technology varies, being usually site-specific and high feedstock dependent. At the largest scale, the plant may include the well proven fixed bed and entrained flow gasification processes. At smaller scales, emphasis is placed on technologies which appear closest to commercial operation. Pyrolysis and other advanced thermal conversion processes are included where power generation is practical using the on-site feedstock produced. However, the needs to be addressed are (i) core fuel handling and gasification/pyrolysis technologies, (ii) fuel gas cleanup, and (iii) conversion of fuel gas to electric power (Ricketts et al., 2002).

Waste may be MSW that had minimal presorting or RDF with significant pretreatment, usually mechanical screening and shredding. Other more specific waste sources (excluding hazardous waste) and possibly including crude oil coke, may provide niche opportunities for co-utilization.

Co-utilization of waste and biomass with coal may provide economies of scale that help achieve the above identified policy objectives at an affordable cost. In some countries, governments propose co-gasification processes as being *well suited for community-sized developments* suggesting that waste should be dealt with in smaller plants serving towns and cities, rather than moved to large, central plants (satisfying the so-called *proximity principal*).

In fact, neither biomass nor wastes are currently produced, or naturally gathered at sites in sufficient quantities to fuel a modern large and efficient power plant. Disruption, transport issues, fuel use, and public opinion all act against gathering hundreds of megawatts (MWe) at a single location. Biomass or waste-fired power plants are therefore inherently limited in size and hence in efficiency (labor costs per unit electricity produced) and in other economies of scale. The production rates of municipal refuse follow reasonably predictable patterns over time periods of a few years. Recent experience with the very limited current *biomass for energy* harvesting has shown unpredictable variations in harvesting capability with long periods of zero production over large areas during wet weather.

The situation is very different for coal. This is generally mined or imported and thus large quantities are available from a single source or a number of closely located sources, and supply has been reliable and predictable. However, the economics of new coal-fired power plants of any technology or size have not encouraged any new coal-fired power plant in the gas generation market.

The potential unreliability of biomass, longer-term changes in refuse and the size limitation of a power plant using only waste and/or biomass can be overcome combining biomass, refuse, and coal. It also allows benefit from a premium electricity price for electricity from biomass and the gate fee associated with waste. If the power plant is gasification-based, rather than direct combustion, further benefits may be available. These include a premium price for the electricity from waste, the range of technologies available from the gas to electricity part of the process, gas cleaning prior to the main combustion stage instead of after combustion and public image, which is generally better for gasification as compared to combustion. These considerations lead to current studies of co-gasification of wastes/biomass with coal (Speight, 2008).

For large-scale power generation (>50 MWe), the gasification field is dominated by plant based on the pressurized, oxygen-blown, entrained flow or fixed bed gasification of fossil fuels. Entrained gasifier operational experience to date has largely been with well-controlled fuel feedstocks with short-term trial work at low co-gasification ratios and with easily handled fuels.

Analyses of the composition of MSW indicate that plastics do make up measureable amounts (5%–10% or more) of solid waste streams. Many of these plastics are worth recovering as energy. In fact, many plastics, particularly the polyolefin derivatives, have high calorific values and simple chemical constitutions of primarily carbon and hydrogen. As a result, waste plastics are ideal candidates for the gasification process. Because of the myriad of sizes and shapes of plastic products size reduction is necessary to create a feed material of a size less than 2 in. in diameter. Some forms of waste plastics such as thin films may require a simple agglomeration step to produce a particle of higher bulk density to facilitate ease of feeding. A plastic, such as high-density polyethylene, processed through a gasifier is converted to carbon monoxide and hydrogen and these materials in turn may be used to form other chemicals including ethylene from which the polyethylene is produced—*closed the loop recycling.*

5.3.3.8 Black Liquor

Black liquor is the spent liquor from the Kraft process in which pulpwood is converted into paper pulp by removing lignin and hemicellulose constituents as well as other extractable materials from wood to free the cellulose fibers. The equivalent spent cooking liquor in the sulfite process is usually called *brown liquor*, but the terms *red liquor*, *thick liquor*, and *sulfite liquor* are also used. Approximately seven units of black liquor are produced in the manufacture of one unit of pulp (Biermann, 1993).

Black liquor comprises an aqueous solution of lignin residues, hemicellulose, and the inorganic chemical used in the process and 15% w/w solids of which 10% w/w are inorganic and 5% w/w are organic. Typically, the organic constituents in black liquor are 40–45% w/w soaps, 35–45% w/w lignin, and 10–15% w/w other (miscellaneous) organic materials.

The organic constituents in the black liquor are made up of water-/alkali-soluble degradation components from the wood. Lignin is partially degraded to shorter fragments with sulfur contents in the order of 1–2% w/w and sodium content at approximately 6% w/w of the dry solids. Cellulose (and hemicellulose) is degraded to aliphatic carboxylic acid soaps and hemicellulose fragments. The extractable constituents yield *tall oil soap* and crude turpentine. The tall oil soap may contain up to 20% w/w sodium. Lignin components currently serve for hydrolytic or pyrolytic conversion or combustion. Alternatively, hemicellulose constituents may be used in fermentation processes.

Gasification of black liquor has the potential to achieve higher overall energy efficiency as compared to those of conventional recovery boilers, while generating an energy-rich synthesis gas. The synthesis gas can then be burned in a gas turbine combined cycle system (*BLGCC—black liquor gasification combined cycle*—and similar to *IGCC*) to produce electricity or converted (through catalytic processes) into chemicals or fuels (e.g., methanol, dimethyl ether, FT hydrocarbon derivatives and diesel fuel).

5.4 GASIFICATION IN A REFINERY

Gasification in the refinery is a known method for converting petroleum coke (petcoke) and other refinery waste streams and residuals (vacuum residual, visbreaker tar, and deasphalter pitch) into power, steam, and hydrogen for use in the production of cleaner transportation fuels (Table 5.4). The main requirement for a gasification feedstock is that it contains both hydrogen and carbon.

The gasification of refinery feedstocks and other carbonaceous feedstocks has been used for many years to convert organic solids and liquids into useful gaseous, liquid, and cleaner solid fuels (Speight, 2011a; Brar et al., 2012). In the current context (Figures 5.1 and 5.2) there are a large number of different feedstock types for use in a refinery-based gasifier, each with different characteristics, including size, shape, bulk density, moisture content, energy content, chemical composition, ash fusion characteristics, and homogeneity of all these properties (Speight, 2013a, 2014a, 2014b). Coal and crude oil coke are used as primary feedstocks for many large gasification plants worldwide. Additionally, a variety of biomass and waste-derived feedstocks can be gasified, with wood pellets and chips, waste wood, plastics, MSW, RDF, agricultural and industrial wastes, sewage sludge, switch grass, discarded seed corn, corn stover, and other crop residues all being used. Moreover, gasification is (i) a well-established technology, (ii) has broad flexibility of feedstocks and operation, and (iii) is the most environmentally friendly route for handling these feedstocks for power production.

Typically, like all gasification processes, the process is carried out at high temperature (>1,000°C, >1,830°F) producing synthesis gas (*syngas*), some carbon black, and ash as major products; the amount of ash depends upon the amount of mineral matter in the feedstock. IGCC is an alternative process for residua conversion and is a known and used technology within the refining industry for (1) hydrogen production, (2) fuel gas production, and (3) power generation which, when coupled with efficient gas cleaning methods has minimum effect on the environment (low SOx and NOx) (Wolff and Vliegenthart, 2011; Speight, 2013a, 2013b, 2014b).

The gasification of coal, biomass, crude oil, or any carbonaceous residues is generally aimed to feedstock conversion to gaseous products. In fact, depending on the previously described type of gasifier (e.g., air-blown, enriched oxygen-blown) and the operating conditions, gasification can be used to produce a fuel gas that is suitable for several applications. Thus, gasification offers one of the most versatile methods (with a reduced environmental impact with respect to combustion) to convert carbonaceous feedstocks into electricity, hydrogen, and other valuable energy products.

The ability of the gasification process to handle non-coal unconventional feedstocks such as heavy crude oil, extra heavy crude oil, tar sand bitumen, or any refinery residual stream enhances the economic potential of most refineries and oil fields. Upgrading heavy crude oil—either in the oil field at the source or residua in the refinery—is (and will continue to be) an increasingly prevalent means of extracting maximum value from each barrel of oil produced (Speight, 2011a, 2014). Upgrading can convert marginal heavy crude oil into light, higher-value crude, and can convert heavy, sour refinery bottoms into valuable transportation fuels. On the other hand, most upgrading techniques leave behind an even heavier residue and the costs deposition of such a byproduct may approach the value of the production of liquid fuels and other salable products. In short, the gasification of residua, petroleum coke, or other heavy feedstocks to generate synthesis gas produces a clean fuel for firing in a gas turbine.

Gasification for electric power generation enables the use of a common technology in modern gas-fired power plants (*combined cycle*) to recover more of the energy released by burning the fuel. The use of these two types of turbines in the combined cycle system involves (i) a combustion turbine and (ii) a steam turbine. The increased efficiency of the combined cycle for electrical power generation results in a 50% v/v decrease in carbon dioxide emissions compared to conventional coal plants. Gasification units could be modified to further reduce their climate change impact because a large part of the carbon dioxide generated can be separated from the other product gas *before* combustion (e.g., carbon dioxide can be separated/sequestered from gaseous byproducts by using adsorbents (e.g., metal-organic frameworks, MOFs) to prevent its release to the atmosphere). Gasification has also been considered for many years as an alternative to combustion of solid or liquid fuels. Gaseous mixtures are simpler to clean as compared to solid or high-viscosity liquid fuels. Cleaned gases can be used in internal combustion-based power plants that would suffer from severe fouling or corrosion if solid or low-quality liquid fuels were burned inside them.

In fact, the hot synthesis gas produced by gasification of carbonaceous feedstocks can then be processed to remove sulfur compounds, mercury, and particulate matter prior to its use as fuel in a combustion turbine generator to produce electricity. The heat in the exhaust gases from the combustion turbine is recovered to generate additional steam. This steam, along with the steam produced by the gasification process, drives a steam turbine generator to produce additional electricity. In the past decade, the primary application of gasification to power production has become more common due to the demand for high efficiency and low environmental impact.

As anticipated, the quality of the gas generated in a system is influenced by feedstock characteristics, gasifier configuration as well as the amount of air, oxygen, or steam introduced into the system. The output and quality of the gas produced is determined by the equilibrium established when the heat of oxidation (combustion) balances the heat of vaporization and volatilization plus the sensible heat (temperature rise) of the exhaust gases. The quality of the outlet gas (BTU/ft^3) is determined by the amount of volatile gases (such as hydrogen, carbon monoxide, water, carbon dioxide, and methane) in the gas stream. With some feedstocks, the higher the amounts of volatile produced in the early stages of the process, the higher the heat content of the product gas. In some cases, the highest gas quality may be produced at lower temperatures. However, char oxidation reaction is suppressed when the temperature is too low, and the overall heat content of the product gas is diminished.

Gasification agents are normally air, oxygen-enriched air, or oxygen. Steam is sometimes added for temperature control, heating value enhancement, or allowing the use of external heat (*allothermal gasification*). The major chemical reactions break and oxidize hydrocarbon derivatives to give a product gas containing carbon monoxide, carbon dioxide, hydrogen, and water. Other important components include hydrogen sulfide, various compounds of sulfur and carbon, ammonia, low-boiling hydrocarbon derivatives, and high-boiling tars.

Depending on the employed gasifier technology and operating conditions, significant quantities of water, carbon dioxide, and methane can be present in the product gas, as well as a number of minor and trace components. Under reducing conditions in the gasifier, most of the feedstock sulfur converts to hydrogen sulfide (H_2S), but 3%–10% converts to carbonyl sulfide (COS). Organically bound nitrogen in the coal feedstock is generally converted to gaseous nitrogen (N_2), but some ammonia (NH_3) and a small amount of hydrogen cyanide (HCN) are also formed. Any chlorine in the coal is converted to hydrogen chloride (HCl), with some chlorine present in the particulate matter (fly ash). Trace elements, such as mercury and arsenic, are released during gasification and partition among the different phases (e.g., fly ash, bottom ash, slag, and product gas).

5.4.1 Gasification of Heavy Feedstocks

The gasification process can be used to convert viscous feedstocks such as heavy oil, extra heavy oil, tar sand bitumen, vacuum residua, and deasphalter bottoms into synthesis gas which is primarily hydrogen and carbon monoxide (Wallace et al., 1998; Speight, 2014a, 2017). The heat generated by the gasification reaction is recovered as the product gas is cooled. For example, when the quench version of Texaco gasification is employed, the steam generated is of medium and low pressure. Note that the low-level heat used for deasphalting integration is the last stage of cooling the synthesis gas.

In addition, integration of solvent deasphalting/gasification facility is an alternative for upgrading viscous oils economically (Wallace et al., 1998). An integrated solvent deasphalting/gasification unit can increase the throughput or the crude flexibility of the refinery without creating a new, highly undesirable viscous oil stream. Typically, the addition of a solvent deasphalting unit to process vacuum tower bottoms increases a refinery's production of diesel oil. The DAO is converted to diesel using hydrotreating and catalytic cracking (Chapter 11). Unfortunately, the deasphalter bottoms often need to be blended with product diesel oil to produce a viable outlet for these bottoms. A gasification process is capable of converting these deasphalter bottoms to synthesis gas which can then be converted to hydrogen for use in hydrotreating and hydrocracking processes. The synthesis gas may also be used by in cogeneration facilities to provide low-cost power and steam to the refinery. If the refinery is part of a petrochemical complex, the synthesis gas can be used as a chemical feedstock. The heat generated by the gasification reaction is recovered as the product gas is cooled.

5.4.2 Gasification of Heavy Feedstocks with Coal

For many decades, coal has been the primary feedstock for gasification units—coal can also be gasified *in situ* (in the underground seam) (Speight, 2013a; Luque and Speight, 2015), but that is not the subject of this text and is not discussed further. However, with the concern on the issue of environmental pollutants and the potential shortage of coal in some areas there is a move to feedstocks other than coal for gasification processes. Gasification permits the utilization of various feedstocks (coal, biomass, crude oil resids, and other carbonaceous wastes) to their fullest potential. Thus, power developers would be well advised to consider gasification as a means of converting coal to gas.

Coal is a combustible organic sedimentary rock (composed primarily of carbon, hydrogen, and oxygen) formed from ancient vegetation and consolidated between other rock strata to form coal seams. The harder forms, such as anthracite coal, can be regarded as organic metamorphic rocks because of a higher degree of maturation (Speight, 2013a). Coal is the largest single source of fuel for the generation of electricity worldwide (EIA, 2007; Speight, 2013b), as well as the largest source of carbon dioxide emissions, which have been implicated (rightly or wrongly) as the primary cause of global climate change (Speight, 2013b; Speight and Islam, 2016). Many of the proponents of global climate change forget (or refuse to acknowledge) that the earth is in an interglacial period when warming and climate change can be expected—this was reflected in the commencement of the melting of the glaciers approximately 11,000 years ago. Thus, considering the geological sequence of events, the contribution of carbon dioxide from anthropogenic sources is not known with any degree of accuracy.

Coal occurs in different forms or *types* (Speight, 2013a). Variations in the nature of the source material and local or regional the variations in the coalification processes cause the vegetal matter to evolve differently. Thus, various classification systems exist to define the different types of coal. Thus, as geological processes increase their effect over time, the coal precursors are transformed over time into: (i) lignite—also referred to as brown coal and is the lowest rank of coal that is used almost exclusively as fuel for steam-electric power generation; *jet* is a compact form of lignite that is sometimes polished and has been used as an ornamental stone since the Iron Age; (ii) subbituminous coal, which exhibits properties ranging from those of lignite to those of bituminous coal, are used primarily as fuel for steam-electric power generation; (iii) bituminous coal, which is a dense coal, usually black, sometimes dark brown, often with well-defined bands of bright and dull material, is used primarily as fuel in steam-electric power generation, with substantial quantities also used for heat and power applications in manufacturing and to make coke; and (iv) anthracite, which is the highest rank coal and is a hard, glossy, black coal used primarily for residential and commercial space heating. Chemically, coal is a hydrogen-deficient hydrocarbon with an atomic hydrogen-to-carbon ratio near 0.8, as compared to crude oil hydrocarbon derivatives, which have an atomic hydrogen-to-carbon ratio approximately equal to 2, and methane (CH_4) that has an atomic carbon-to-hydrogen ratio equal to 4. For this reason, any process used to convert coal to alternative fuels must add hydrogen or redistribute the hydrogen in the original coal to produce hydrogen-rich products and coke (Speight, 2013a).

Gas turbine improvements lead to a number of power plants where fuels (usually coal) are gasified with a viscous feedstock and the gas is cleaned and used in a combined cycle gas turbine power plant. Such power plants generally have higher capital cost, higher operating cost, and lower availability than conventional combustion and steam cycle power plant on the same fuel. Efficiencies of the most sophisticated plants have been broadly similar to the best conventional steam plants with losses in gasification and gas cleaning being balanced by the high efficiency of combined cycle power plants. Environmental aspects resulting from the gas cleaning before the main combustion stage have often been excellent, even in plants with exceptionally high levels of contaminants in the feedstock fuels.

5.4.3 Gasification of Heavy Feedstocks with Biomass

Gasification is an established technology (Hotchkiss, 2003; Speight, 2013a). Comparatively, biomass gasification has been the focus of research in recent years to estimate efficiency and performance of the gasification process using various types of biomass such as sugarcane residue (Gabra et al., 2001), rice hulls (Boateng et al., 1992), pine sawdust (Lv et al., 2004), almond shells (Rapagnà and Latif, 1997; Rapagnà et al., 2000), wheat straw (Ergudenler and Ghaly, 1993), food waste (Ko et al., 2001), and wood biomass (Pakdel and Roy, 1991; Bhattacharya et al., 1999; Chen et al., 1992; Hanaoka et al., 2005). Recently, co-gasification of various biomass and coal mixtures has attracted a great deal of interest from the scientific community. Feedstock combinations including Japanese cedar wood and coal (Kumabe et al., 2007), coal and saw dust, coal and pine chips (Pan et al., 2000), coal and silver birch wood (Collot et al., 1999), and coal and birch wood (Brage et al., 2000) have been reported in gasification practices. Co-gasification of coal and biomass has some synergy—the process not only produces a low carbon footprint on the environment, but also improves the H_2/CO ratio in the produced gas which is required for liquid fuel synthesis (Sjöström et al., 1999; Kumabe et al., 2007). In addition, the inorganic matter present in biomass catalyzes the gasification of coal. However, co-gasification processes require custom fittings and optimized processes for the coal and region-specific wood residues.

While co-gasification of coal and biomass is advantageous from a chemical viewpoint, some practical problems are present on upstream, gasification, and downstream processes. On the upstream side, the particle size of the coal and biomass is required to be uniform for optimum gasification. In addition, moisture content and pretreatment (torrefaction) are very important during upstream processing.

While upstream processing is influential from a material handling point of view, the choice of gasifier operation parameters (temperature, gasifying agent, and catalysts) dictate the product gas composition and quality. Biomass decomposition occurs at a lower temperature than coal and therefore different reactors compatible to the feedstock mixture are required (Brar et al., 2012). Furthermore, feedstock and gasifier type along with operating parameters not only decide product gas composition but also dictate the amount of impurities to be handled downstream. Downstream processes need to be modified if coal is co-gasified with biomass. Heavy metals and other impurities such as sulfur-containing compounds and mercury present in coal can make synthesis gas difficult to use and unhealthy for the environment. Alkali present in biomass can also cause corrosion problems high temperatures in downstream pipes. An alternative option to downstream gas cleaning would be to process coal to remove mercury and sulfur prior to feeding into the gasifier.

However, first and foremost, coal and biomass require drying and size reduction before they can be fed into a gasifier. Size reduction is needed to obtain appropriate particle sizes; however, drying is required to achieve moisture content suitable for gasification operations. In addition, biomass densification may be conducted to prepare pellets and improve density and material flow in the feeder areas. It is recommended that biomass moisture content should be less than 15% w/w prior to gasification. High moisture content reduces the temperature achieved in the gasification zone, thus resulting in incomplete gasification. Forest residues or wood has a fiber saturation point at 30%–31% moisture content (dry basis) (Brar et al., 2012). Compressive and shear strength of the wood increases with decreased moisture content below the fiber saturation point. In such a situation, water is removed from the cell wall leading to shrinkage. The long-chain molecule constituents of the cell wall move closer to each other and bind more tightly. A high level of moisture, usually injected in form of steam in the gasification zone, favors formation of a water-gas shift reaction that increases hydrogen concentration in the resulting gas.

The torrefaction process is a thermal treatment of biomass in the absence of oxygen, usually at 250°C–300°C to drive off moisture, decompose hemicellulose completely, and partially decompose cellulose (Speight, 2011a). Torrefied biomass has reactive and unstable cellulose molecules with broken hydrogen bonds and not only retains 79%–95% of feedstock energy but also produces a more reactive feedstock with lower atomic hydrogen-carbon and oxygen-carbon ratios to those of the original biomass. Torrefaction results in higher yields of hydrogen and carbon monoxide in the gasification process.

Finally, the presence of mineral matter in the coal-biomass feedstock is not appropriate for fluidized bed gasification. Low melting point of ash present in woody biomass leads to agglomeration that causes defluidization of the ash and sintering, deposition as well as corrosion of the gasifier construction metal bed. Biomass containing alkali oxides and salts are likely to produce clinkering/slagging problems from ash formation (McKendry, 2002). Thus, it is imperative to be aware of the melting of biomass ash, its chemistry within the gasification bed (no bed, silica/sand, or calcium bed), and the fate of alkali metals when using fluidized bed gasifiers.

Most small- to medium-sized biomass/waste gasifiers are air blown operating at atmospheric pressure and at temperatures in the range of 800°C–100°C (1,470°F–2,190°F). They face very different challenges compared to large gasification plants—the use of small-scale air separation plant should oxygen gasification be preferred. Pressurized operation, which eases gas cleaning, may not be practical.

Biomass fuel producers, coal producers and, to a lesser extent, waste companies are enthusiastic about supplying co-gasification power plants and realize the benefits of co-gasification with alternate fuels (Lee, 2007; Speight, 2008, 2011a; Lee and Shah, 2013; Speight, 2013a, 2013b). The benefits of a co-gasification technology involving coal and biomass include the use of a reliable coal supply with gate-fee waste and biomass that allows the economies of scale from a larger plant to be supplied just with waste and biomass. In addition, the technology offers a future option of hydrogen production and fuel development in refineries. In fact, oil refineries and petrochemical plants are opportunities for gasifiers when the hydrogen is particularly valuable (Speight, 2011b, 2014a).

5.4.4 GASIFICATION OF HEAVY FEEDSTOCKS WITH WASTE

Waste may be MSW that had minimal presorting or RDF with significant pretreatment, usually mechanical screening and shredding. Other more specific waste sources (excluding hazardous waste) and possibly including crude oil coke may provide niche opportunities for co-utilization (John and Singh, 2011).

For large-scale power generation (>50 MWe), the gasification field is dominated by plant based on the pressurized, oxygen-blown, entrained flow or fixed bed gasification of fossil fuels.

The use of fuel cells with gasifiers is frequently discussed but the current cost of fuel cells is such that their use for mainstream electricity generation is uneconomic. In summary, coal may be co-gasified with waste or biomass for environmental, technical, or commercial reasons. It allows larger, more efficient plants than those sized for grown biomass or arising waste within a reasonable transport distance; specific operating costs are likely to be lower and fuel supply security is assured.

5.5 GAS PRODUCTION AND OTHER PRODUCTS

The gasification of a carbonaceous feedstock (i.e., char produced from the feedstock) is the conversion of the feedstock (by any one of a variety of processes) to produce gaseous products that are combustible as well as a wide range of chemical products from synthesis gas (Figure 5.4).

With the rapid increase in the use of coal from the 15th century onwards, it is not surprising that the concept of using coal to produce a flammable gas, especially the use of the water and hot coal became common place (van Heek and Muhlen, 1991). As a result, the characteristics of rank, mineral matter, particle size, and reaction conditions are all recognized as having a bearing on the outcome of the process not only in terms of gas yields but also on gas properties (van Heek and Muhlen, 1991). The products from the gasification of the process may be of low, medium, or high heat content (high-Btu) as dictated by the process as well as by the ultimate use for the gas

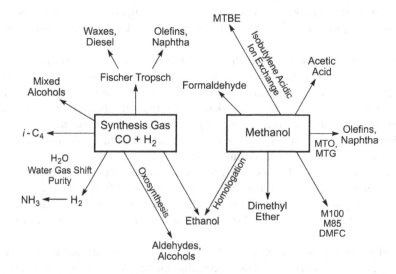

FIGURE 5.4 Potential products from heavy feedstock gasification.

(Baker and Rodriguez, 1990; Probstein and Hicks, 1990; Lahaye and Ehrburger, 1991; Matsukata et al., 1992; Speight, 2013a).

The ability of a refinery to efficiently accommodate heavy crude oils or heavy bottom streams (such as deasphalter bottoms and visbreaker bottoms) enhances the economic potential of the refinery and the development of heavy oil and tar sand resources. A refinery with the flexibility to meet the increasing product specifications for fuels through the ability to upgrade heavy feedstocks is an increasingly attractive means of extracting maximum value from each barrel of oil produced. Upgrading can convert marginal heavy crude oil into light, higher value crude, and can convert heavy, sour refinery bottoms into valuable transportation fuels. On the downside, most upgrading processes also produce an even heavier residue whose disposition costs may approach the value of the upgrade itself.

For example, solvent deasphalting and residue coking are used in heavy crude-based refineries to upgrade heavy bottom streams to intermediate products that may be processed to produce transportation fuels. The technology may also be used in the oil field to enhance the value of heavy crude oil before the feedstock reaches the refinery and a beneficial use is often difficult to find for the byproducts from these processes.

5.5.1 Gaseous Products

The products of gasification are varied insofar as the gas composition varies with the system employed (Speight, 2013a). It is emphasized that the gas product must be first freed from any pollutants such as particulate matter and sulfur compounds before further use, particularly when the intended use is a water-gas shift or methanation (Cusumano et al., 1978; Probstein and Hicks, 1990).

5.5.1.1 Synthesis Gas

Synthesis gas is comparable in its combustion efficiency to natural gas (Speight, 2008; Chadeesingh, 2011), which reduces the emissions of sulfur, nitrogen oxides, and mercury, resulting in a much cleaner fuel (Nordstrand et al., 2008; Sondreal et al., 2004; Yang et al., 2007; Wang et al., 2008). The resulting hydrogen gas can be used for electricity generation or as a transport fuel. The gasification process also facilitates capture of carbon dioxide emissions from the combustion effluent (see discussion of carbon capture and storage below).

Although synthesis gas can be used as a standalone fuel, the energy density of synthesis gas is approximately half that of natural gas and is therefore mostly suited for the production of transportation fuels and other chemical products. Synthesis gas is mainly used as an intermediary building block for the final production (synthesis) of various fuels such as SNG, methanol, and synthetic crude oil fuel (dimethyl ether—synthesized gasoline and diesel fuel) (Chadeesingh, 2011; Speight, 2013a). At this point, and in order to dismiss any confusion that may arise, synthesis gas, as generated from biomass, is not the same as biogas. Biogas is a clean and renewable form of energy generated from biomass that could very well substitute for conventional sources of energy. The gas is generally composed of methane (55%–65%), carbon dioxide (35%–45%), nitrogen (0%–3%), hydrogen (0%–1%), and hydrogen sulfide (0%–1%).

The use of synthesis gas offers the opportunity to furnish a broad range of environmentally clean fuels and chemicals and there has been steady growth in the traditional uses of synthesis gas. Almost all hydrogen gas is manufactured from synthesis gas and there has been an increase in the demand for this basic chemical. In fact, the major use of synthesis gas is in the manufacture of hydrogen for a growing number of purposes, especially in crude oil refineries (Speight, 2014a). Methanol not only remains the second largest consumer of synthesis gas but has shown remarkable growth as part of the methyl ethers used as octane enhancers in automotive fuels.

The FTS remains the third largest consumer of synthesis gas, mostly for transportation fuels but also as a growing feedstock source for the manufacture of chemicals, including polymers.

The hydroformylation of olefin derivatives (the Oxo reaction), a completely chemical use of synthesis gas, is the fourth largest use of carbon monoxide and hydrogen mixtures. A direct application of synthesis gas as fuel (and eventually also for chemicals) that promises to increase is its use for IGCC units for the generation of electricity (and also chemicals), crude oil coke, or viscous feedstocks (Holt, 2001). Finally, synthesis gas is the principal source of carbon monoxide, which is used in an expanding list of carbonylation reactions, which are of major industrial interest.

5.5.1.2 Low Btu Gas

During the production of gas by oxidation with air, the oxygen is not separated from the air and, as a result, the gas product invariably has a low Btu content (low heat content, 150–300 Btu/ft^3). Several important chemical reactions and a host of side reactions, are involved in the manufacture of low heat content gas under the high-temperature conditions employed (Chadeesingh, 2011; Speight, 2013a). Low heat content gas contains several components, four of which are always major components present at levels of at least several percent; a fifth component, methane, is marginally a major component.

The nitrogen content of low heat content gas ranges from somewhat less than 33% v/v to slightly more than 50% v/v and cannot be removed by any reasonable means; the presence of nitrogen at these levels makes the product gas *low heat content* by definition. The nitrogen also strongly limits the applicability of the gas to chemical synthesis. Two other noncombustible components (water, H_2O, and carbon dioxide, CO) further lower the heating value of the gas; water can be removed by condensation and carbon dioxide by relatively straightforward chemical means.

The two major combustible components are hydrogen and carbon monoxide; the H_2/CO ratio varies from approximately 2:3 to 3:2. Methane may also make an appreciable contribution to the heat content of the gas. Of the minor components hydrogen sulfide is the most significant and the amount produced is, in fact, proportional to the sulfur content of the feedstock. Any hydrogen sulfide present must be removed by one, or more, of several procedures (Mokhatab et al., 2006; Speight, 2007, 2014a).

Low heat content gas is of interest to industry as a fuel gas or even, on occasion, as a raw amaterial from which ammonia, methanol, and other compounds may be synthesized.

5.5.1.3 Medium Btu Gas

Medium Btu gas (medium heat content gas) has a heating value in the range 300–550 Btu/ft^3 and the composition is much like that of low heat content gas, except that there is virtually no nitrogen. The primary combustible gases in medium heat content gas are hydrogen and carbon monoxide. Medium heat content gas is considerably more versatile than low heat content gas; like low heat content gas, medium heat content gas may be used directly as a fuel to raise steam, or used through a combined power cycle to drive a gas turbine, with the hot exhaust gases employed to raise steam, but medium heat content gas is especially amenable to synthesize methane (by methanation), higher hydrocarbon derivatives (by FTS), methanol, and a variety of synthetic chemicals.

The reactions used to produce medium heat content gas are the same as those employed for low heat content gas synthesis, the major difference being the application of a nitrogen barrier (such as the use of pure oxygen) to keep diluent nitrogen out of the system.

In medium heat content gas, the H_2/CO ratio varies from 2:3 to 3:1 and the increased heating value correlates with higher methane and hydrogen contents as well as with lower carbon dioxide contents. Furthermore, the very nature of the gasification process used to produce the medium heat content gas has a marked effect upon the ease of subsequent processing. For example, the CO_2–acceptor product is quite amenable to use for methane production because it has (i) the desired H_2/CO ratio just exceeding 3:1, (ii) an initially high methane content, and (iii) relatively low water and carbon dioxide contents. Other gases may require appreciable shift reaction and removal of large quantities of water and carbon dioxide prior to methanation.

5.5.1.4 High Btu Gas

High Btu) gas (high heat content gas) is essentially pure methane and often referred to as SNG (Speight, 1990, 2013a). However, to qualify as substitute natural gas, a product must contain at least 95% methane, giving an energy content (heat content) of synthetic natural gas in the order of 980–1,080 Btu/ft^3.

The commonly accepted approach to the synthesis of high heat content gas is the catalytic reaction of hydrogen and carbon monoxide:

$$3H_2 + CO \rightarrow CH_4 + H_2O$$

To avoid catalyst poisoning, the feed gases for this reaction must be quite pure and, therefore, impurities in the product are rare. The large quantities of water produced are removed by condensation and recirculated as very pure water through the gasification system. The hydrogen is usually present in slight excess to ensure that the toxic carbon monoxide is reacted; this small quantity of hydrogen will lower the heat content to a small degree.

The carbon monoxide/hydrogen reaction is somewhat inefficient as a means of producing methane because the reaction liberates large quantities of heat. In addition, the methanation catalyst is troublesome and prone to poisoning by sulfur compounds and the decomposition of metals can destroy the catalyst. Hydrogasification may be thus employed to minimize the need for methanation:

$$[C]_{feedstock} + 2H2 \rightarrow CH_4$$

The product of hydrogasification is far from pure methane and additional methanation is required after hydrogen sulfide and other impurities are removed.

5.5.2 LIQUID PRODUCTS

The production of liquid fuels from a carbonaceous feedstock via gasification is often referred to as the *indirect liquefaction* of the feedstock (Speight, 2013a, 2014a). In these processes, the feedstock is not converted directly into liquid products but involves a two-stage conversion operation in which the feedstock is first converted (by reaction with steam and oxygen) to produce a gaseous mixture that is composed primarily of carbon monoxide and hydrogen (synthesis gas). The gas stream is subsequently purified (to remove sulfur, nitrogen, and any particulate matter) after which it is catalytically converted to a mixture of liquid hydrocarbon products.

The synthesis of hydrocarbon derivatives from carbon monoxide and hydrogen (synthesis gas) (the FTS) is a procedure for the indirect liquefaction of various carbonaceous feedstocks (Speight, 2011a, 2011b). This process is the only liquefaction scheme currently in use on a relatively large commercial scale (for the production of liquid fuels from coal using the FT process.

Thus, the feedstock is converted to gaseous products at temperatures in excess of 800°C (1,470°F), and at moderate pressures, to produce synthesis gas:

$$[C]_{feedstock} + H_2O \rightarrow CO + H_2$$

In practice, the FT reaction is carried out at temperatures of 200°C–350°C (390°F–660°F) and at pressures of 75–4,000 psi. The hydrogen/carbon monoxide ratio is typically in the order of 2/2:1 or 2/5:1, since up to three volumes of hydrogen may be required to achieve the next stage of the liquid production, the synthesis gas must then be converted by means of the water-gas shift reaction to the desired level of hydrogen:

$$CO + H_2O \rightarrow CO_2 + H_2$$

After this, the gaseous mix is purified and converted to a wide variety of hydrocarbon derivatives:

$$nCO + (2n+1)H_2 \rightarrow C_nH_{2n+2} + nH_2O$$

These reactions result primarily in low- and medium-boiling aliphatic compounds suitable for gasoline and diesel fuel.

5.5.3 SOLID PRODUCTS

The solid product (solid waste) of a gasification process is typically ash which is the oxides of metals containing constituents of the feedstock. The amount and type of solid waste produced is very much feedstock dependent. The waste is a significant environmental issue due to the large quantities produced; chiefly fly ash if coal is the feedstock or a co-feedstock, and the potential for leaching of toxic substances (such as heavy metals such as lead and arsenic) into the soil and groundwater at disposal sites.

At the high temperature of the gasifier, most of the mineral matter of the feedstock is transformed and melted into slag, an inert glass-like material and, under such conditions, nonvolatile metals and mineral compounds are bound together in molten form until the slag is cooled in a water bath at the bottom of the gasifier, or by natural heat loss at the bottom of an entrained bed gasifier. Slag production is a function of mineral matter content of the feedstock—coal produces much more slag per unit weight than crude oil coke. Furthermore, as long as the operating temperature is above the fusion temperature of the ash, slag will be produced. The physical structure of the slag is sensitive to changes in operating temperature and pressure of the gasifier and a quick physical examination of the appearance of the slag can often be an indication of the efficiency of the conversion of feedstock carbon to gaseous product in the process.

Slag is comprised of black, glassy, silica-based materials and is also known as *frit*, which is a high density, vitreous, and abrasive material low in carbon and formed in various shapes from jagged and irregular pieces to rod and needle-like forms. Depending upon the gasifier process parameters and the feedstock properties, there may also be residual carbon char. Vitreous slag is much preferable to ash, because of its habit of encapsulating toxic constituents (such as heavy metals) into a stable, non-leachable material. Leachability data obtained from different gasifiers unequivocally shows that gasifier slag is highly non-leachable, and can be classified as nonhazardous. Because of its particular properties and nonhazardous, nontoxic nature, slag is relatively easily marketed as a byproduct for multiple advantageous uses, which may negate the need for its long-term disposal.

The physical and chemical properties of gasification slag are related to (i) the composition of the feedstock, (ii) the method of recovering the molten ash from the gasifier, and (iii) the proportion of devolatilized carbon particles (char) discharged with the slag. The rapid water-quench method of cooling the molten slag inhibits recrystallization, and results in the formation of a granular, amorphous material. Some of the differences in the properties of the slag may be attributed to the specific design and operating conditions prevailing in the gasifiers.

Char is the finer component of the gasifier solid residuals, composed of unreacted carbon with various amounts of siliceous ash. Char can be recycled back into the gasifier to increase carbon usage and has been used as a supplemental fuel source for use in a combustor. The irregularly shaped particles have a well-defined pore structure and have excellent potential as an adsorbent and precursor to activated carbon. In terms of recycling char to the gasifier, a property that is important to fluidization is the effective particle density. If the char has a large internal void space, the density will be much less than that of the feedstock (especially coal) or char from slow carbonization of a carbonaceous feedstock.

5.6 THE FUTURE

The future depends very much on the effect of gasification processes on the surrounding environment. It is these environmental effects and issues that will direct the success of gasification.

In fact, there is the distinct possibility that within the foreseeable future the gasification process will increase in popularity in crude oil refineries—some refineries may even be known as gasification refineries (Speight, 2011b). A gasification refinery would have, as the center piece, gasification technology as is the case of the Sasol refinery in South Africa (Couvaras, 1997). The refinery would produce synthesis gas (from the carbonaceous feedstock) from which liquid fuels would be manufactured using the FTS technology.

In fact, gasification to produce synthesis gas can proceed from any carbonaceous material, including biomass. Inorganic components of the feedstock, such as metals and minerals, are trapped in an inert and environmentally safe form as char, which may be used as a fertilizer. Biomass gasification is, therefore, one of the most technically and economically convincing energy possibilities for a potentially carbon neutral economy.

The manufacture of gas mixtures of carbon monoxide and hydrogen has been an important part of chemical technology for about a century. Originally, such mixtures were obtained by the reaction of steam with incandescent coke and were known as *water gas*. Eventually, steam reforming processes, in which steam is reacted with natural gas (methane) or crude oil naphtha over a nickel catalyst, found wide application for the production of synthesis gas.

A modified version of steam reforming known as autothermal reforming, which is a combination of partial oxidation near the reactor inlet with conventional steam reforming further along the reactor, improves the overall reactor efficiency and increases the flexibility of the process. Partial oxidation processes using oxygen instead of steam also found wide application for synthesis gas manufacture, with the special feature that they could utilize low-value feedstocks such as viscous crude oil residues. In recent years, catalytic partial oxidation employing very short reaction times (milliseconds) at high temperatures (850°C–1,000°C) is providing still another approach to synthesis gas manufacture (Hickman and Schmidt, 1993).

In a gasifier, the carbonaceous material undergoes several different processes: (i) pyrolysis of carbonaceous fuels, (ii) combustion, and (iii) gasification of the remaining char. The process is very dependent on the properties of the carbonaceous material and determines the structure and composition of the char, which will then undergo gasification reactions.

As crude oil supplies decrease, the desirability of producing gas from other carbonaceous feedstocks will increase, especially in those areas where natural gas is in short supply. It is also anticipated that costs of natural gas will increase, thereby allowing the gasification process to compete as an economically viable process.

The conversion of the gaseous products of gasification processes to synthesis gas, a mixture of hydrogen (H_2) and carbon monoxide (CO), in a ratio appropriate to the application, needs additional steps, after purification. The product gases—carbon monoxide, carbon dioxide, hydrogen, methane, and nitrogen—can be used as fuels or as raw materials for chemical or fertilizer manufacture.

Gasification by means other than the conventional methods has also received some attention and has provided rationale for future processes (Rabovitser et al., 2010). In the process, a carbonaceous material and at least one oxygen carrier are introduced into a nonthermal plasma reactor at a temperature in the range of approximately 300°C to 700°C (570°F–1,290°F) and a pressure in a range from atmospheric pressure to approximate 1,030 psi and a nonthermal plasma discharge is generated within the nonthermal plasma reactor. The carbonaceous feedstock and the oxygen carrier are exposed to the nonthermal plasma discharge, resulting in the formation of a product gas which comprises substantial amounts of hydrocarbon derivatives, such as methane, hydrogen, and/or carbon monoxide.

Furthermore, gasification and conversion of carbonaceous solid fuels to synthesis gas for application of power, liquid fuels, and chemicals is practiced worldwide. Crude oil coke, coal, biomass, and refinery waste are major feedstocks for an on-site refinery gasification unit. The concept of blending of a variety of carbonaceous feedstocks (such as coal, biomass, or refinery waste) with a viscous feedstock of the coke from the thermal processing of the viscous feedstock is advantageous in order to obtain the highest value of products as compared to gasification of crude oil coke alone.

Furthermore, based on gasifier type, co-gasification of carbonaceous feedstocks can be an advantageous and efficient process. In addition, the variety of upgrading and delivery options that are available for application to synthesis gas enable the establishment of an integrated energy supply system whereby synthesis gases can be upgraded, integrated, and delivered to a distributed network of energy conversion facilities, including power, CHP, and combined cooling, heating and power (sometime referred to as *tri-generation*) as well as used as fuels for transportation applications.

As a final note, the production of chemicals from biomass is based on thermochemical conversion routes which are, in turn, based on biomass gasification (Roddy and Manson-Whitton, 2012). The products are (i) a gas, which is the desired product and (ii) a solid ash residue whose composition depends on the type of biomass. Continuous gasification processes for various feedstocks have been under development since the early 1930s. Ideally, the gas produced would be a mixture of hydrogen and carbon monoxide, but, in practice, it also contains methane, carbon dioxide, and a range of contaminants.

A variety of gasification technologies is available across a range of sizes, from small updraft and downdraft gasifiers through a range of fluidized bed gasifiers at an intermediate scale and on to larger entrained flow and plasma gasifiers (Bridgwater, 2003; Roddy and Manson-Whitton, 2012). In an updraft gasifier, the oxidant is blown up through the fixed gasifier bed with the syngas exiting at the top whereas in a downdraft gasifier, the oxidant is blown down through the reactor with the synthesis gas exiting at the bottom.

Gasification processes tend to operate either above the ash melting temperature (typically 1,200°C, 2,190°F) or below the ash melting temperature (typically >1,000°C, >1,830°F). In the higher temperature processes, there is little methane or tar formation. The question of which gasification technology is the most appropriate depends on whether the priority is to (i) produce a very pure synthesis gas, (ii) accommodate a wide range of feedstock types, (iii) avoid preprocessing of biomass, or (iv) operate at a large scale, and produce chemical products from a variety of feedstocks.

In summary, a refinery that is equipped with a gasifier is a suitable refinery for the complete conversion of heavy feedstocks and (including petroleum coke) to valuable products (including petrochemicals). In fact, integration between bottoms processing units and gasification, presents some unique synergies including the production of feedstocks for a petrochemical complex.

REFERENCES

Abadie, L.M., and Chamorro, J.M. 2009. The Economics of Gasification: A Market-Based Approach. *Energies*, 2: 662–694.

Adhikari, U., Eikeland, M.S., and Halvorsen, B.M. 2015. Gasification of Biomass for Production of Syngas for Biofuel. *Proceedings of 56th SIMS*. October 7–9. Linköping, Sweden. pp. 255–260. www.ep.liu.se/ecp/119/025/ecp15119025.pdf

Arena, U. 2012. Process and Technological Aspects of Municipal Solid Waste Gasification. A Review. *Waste Management*, 32: 625–639.

ASTM D388. 2015. *Standard Classification of Coal by Rank*. Annual Book of Standards. ASTM International, West Conshohocken, PA.

Baker, R.T.K., and Rodriguez, N.M. 1990. *Fuel Science and Technology Handbook*. Marcel Dekker Inc., New York. Chapter 22.

Balat, M. 2011. Chapter 3: Fuels from Biomass—An Overview. In: *The Biofuels Handbook*. J.G. Speight (Editor). Royal Society of Chemistry, London, UK. Part 1.

Bhattacharya, S., Siddique, A.H.M.M.R., and Pham, H.-L. 1999. A Study in Wood Gasification on Low Tar Production. *Energy*, 24: 285–296.

Baxter, L. 2005. Biomass-Coal Co-Combustion: Opportunity for Affordable Renewable Energy. *Fuel*, 84(10): 1295–1302.

Bernetti, A., De Franchis, M., Moretta, J.C., and Shah, P.M. 2000. Solvent Deasphalting and Gasification: A Synergy. *Petroleum Technology Quarterly*, Q4: 1–7.

Biermann, C.J. 1993. *Essentials of Pulping and Papermaking*. Academic Press Inc., New York.

Boateng, A.A., Walawender, W.P., Fan, L.T., and Chee, C.S. 1992. Fluidized-Bed Steam Gasification of Rice Hull. *Bioresource Technology*, 40(3): 235–239.

Brage, C., Yu, Q., Chen, G., and Sjöström, K. 2000. Tar Evolution Profiles Obtained from Gasification of Biomass and Coal. *Biomass and Bioenergy*, 18(1): 87–91.

Brar, J.S., Singh, K., Wang, J., and Kumar, S. 2012. Cogasification of Coal and Biomass: A Review. *International Journal of Forestry Research*, 2012: 1–10.

Bridgwater, A.V. 2003. Renewable Fuels and Chemicals by Thermal Processing of Biomass. *Chemical Engineering Journal*, 91: 87–102.

Chadeesingh, R. 2011. Chapter 5: The Fischer-Tropsch Process. In: *The Biofuels Handbook*. J.G. Speight (Editor). The Royal Society of Chemistry, London, UK. Part 3, pp. 476–517.

Chen, G., Sjöström, K. and Bjornbom, E. 1992. Pyrolysis/Gasification of Wood in a Pressurized Fluidized Bed Reactor. *Industrial and Engineering Chemistry Research*, 31(12): 2764–2768.

Collot, A.G., Zhuo, Y., Dugwell, D.R., and Kandiyoti, R. 1999. Co-Pyrolysis and Cogasification of Coal and Biomass in Bench-Scale Fixed-Bed and Fluidized Bed Reactors. *Fuel*, 78: 667–679.

Couvaras, G. 1997. Sasol's Slurry Phase Distillate Process and Future Applications. *Proceedings of Monetizing Stranded Gas Reserves Conference*. December 1997, Houston, TX.

Cusumano, J.A., Dalla Betta, R.A., and Levy, R.B. 1978. *Catalysis in Coal Conversion*. Academic Press Inc., New York.

Davidson, R.M. 1983. *Mineral Effects in Coal Conversion*. Report No. ICTIS/TR22. International Energy Agency, London, UK.

Demirbaş, A. 2011. Chapter 1: Production of Fuels from Crops. In: *The Biofuels Handbook*. J.G. Speight (Editor). Royal Society of Chemistry, London, UK. Part 2.

Dutcher, J.S., Royer, R.E., Mitchell, C.E., and Dahl, A.R. 1983. In: *Advanced Techniques in Synthetic Fuels Analysis*. C.W. Wright, W.C. Weimer, and W.D. Felic (Editors). Technical Information Center, United States Department of Energy, Washington, DC. p. 12.

EIA. 2007. *Net Generation by Energy Source by Type of Producer*. Energy Information Administration, United States Department of Energy, Washington, DC. www.eia.doe.gov/cneaf/electricity/epm/table1_1.html

Ergudenler, A., and Ghaly, A.E. 1993. Agglomeration of Alumina Sand in a Fluidized Bed Straw Gasifier at Elevated Temperatures. *Bioresource Technology*, 43(3): 259–268.

Fabry, F., Rehmet, C., Rohani, V-J., and Fulcheri, L. 2013. Waste Gasification by Thermal Plasma: A Review. *Waste and Biomass Valorization*, 4(3): 421–439.

Fermoso, J., Plaza, M.G., Arias, B., Pevida, C., Rubiera, F., and Pis, J.J. 2009. Co-Gasification of Coal with Biomass and Petcoke in a High-Pressure Gasifier for Syngas Production. *Proceedings of 1st Spanish National Conference on Advances in Materials Recycling and Eco-Energy*. November 12–13, Madrid, Spain.

Furimsky, E. 1999. Gasification in a Petroleum Refinery of the 21st Century. *Oil & Gas Science and Technology, Revue Institut Français du Pétrole*, 54(5): 597–618.

Gabra, M., Pettersson, E., Backman, R., and Kjellström, B. 2001. Evaluation of Cyclone Gasifier Performance for Gasification of Sugar Cane Residue—Part 1: Gasification of Bagasse. *Biomass and Bioenergy*, 21(5): 351–369.

Gary, J.G., Handwerk, G.E., and Kaiser, M.J. 2007. *Crude oil Refining: Technology and Economics*. 5th Edition. CRC Press, Boca Raton, FL.

Gray, D., and Tomlinson, G. 2000. Opportunities For Petroleum Coke Gasification Under Tighter Sulfur Limits For Transportation Fuels. *Proceedings of 2000 Gasification Technologies Conference*. October 8–11, San Francisco, CA.

Hanaoka, T., Inoue, S., Uno, S., Ogi, T., and Minowa, T. 2005. Effect of Woody Biomass Components on Air-Steam Gasification. *Biomass and Bioenergy*, 28(1): 69–76.

Hickman, D.A., and Schmidt, L.D. 1993. Syngas Formation by Direct Catalytic Oxidation of Methane. *Science*, 259: 343–346.

Higman, C., and Van der Burgt, M. 2008. *Gasification*. 2nd Edition. Gulf Professional Publishing, Elsevier, Amsterdam, The Netherlands.

Holt, N.A.H. 2001. *Integrated Gasification Combined Cycle Power Plants*. Encyclopedia of Physical Science and Technology. 3rd Edition. Academic Press Inc., New York.

Hotchkiss, R. 2003. Coal Gasification Technologies. *Proceedings of Institute of Mechanical Engineers Part A*, 217(1): 27–33.

Hsu, C.S., and Robinson, P.R. (Editors). 2017. *Handbook of Petroleum Technology*. Springer, Cham, Switzerland.

Irfan, M.F. 2009. Research Report: Pulverized Coal Pyrolysis & Gasification in $N_2/O_2/CO_2$ Mixtures by Thermo-Gravimetric Analysis. Novel Carbon Resource Sciences Newsletter, Kyushu University, Fukuoka, Japan. 2: 27–33.

Jenkins, B.M., and Ebeling, J.M. 1985. Thermochemical Properties of Biomass Fuels. *California Agriculture* (May-June), pp. 14–18.

John, E., and Singh, K. 2011. Chapter 2: Properties of Fuels from Domestic and Industrial Waste. In: *The Biofuels Handbook*. J.G. Speight (Editor). The Royal Society of Chemistry, London, UK. Part 3, pp. 377–407.

Johnson J.L. 1979. *Kinetics of Coal Gasification*. John Wiley and Sons Inc., Hoboken, NJ.

Khosravi, M., and Khadse, A., 2013. Gasification of Petcoke and Coal/Biomass Blend: A Review. *International Journal of Emerging Technology and Advanced Engineering*, 3(12): 167–173.

Ko, M.K., Lee, W.Y., Kim, S.B., Lee, K.W., and Chun, H.S. 2001. Gasification of Food Waste with Steam in Fluidized Bed. *Korean Journal of Chemical Engineering*, 18(6): 961–964.

Kumabe, K., Hanaoka, T., Fujimoto, S., Minowa, T., and Sakanishi, K. 2007. Cogasification of Woody Biomass and Coal with Air and Steam. *Fuel*, 86: 684–689.

Kumar, A., Jones, D.D., and Hanna, M.A. 2009. Thermochemical Biomass Gasification: A Review of the Current Status of the Technology. *Energies*, 2: 556–581.

Kunii, D., and Levenspiel, O. 2013. *Fluidization Engineering*. 2nd Edition. Butterworth-Heinemann, Elsevier, Amsterdam, The Netherlands.

Lahaye, J., and Ehrburger, P. (Editors). 1991. *Fundamental Issues in Control of Carbon Gasification Reactivity*. Kluwer Academic Publishers, Dordrecht, The Netherlands.

Lapuerta M., Hernández, J.J., Pazo A., and López, J. 2008. Gasification and co-gasification of biomass wastes: Effect of the biomass origin and the gasifier operating conditions. *Fuel Processing Technology*, 89(9): 828–837.

Lee, S. 2007. Gasification of Coal. In: *Handbook of Alternative Fuel Technologies*. S. Lee, J.G. Speight, and S. Loyalka (Editors). CRC Press, Boca Raton, FL.

Lee, S., Speight, J.G., and Loyalka, S. 2007. *Handbook of Alternative Fuel Technologies*. CRC Press, Boca Raton, FL.

Lee, S., and Shah, Y.T. 2013. *Biofuels and Bioenergy*. CRC Press, Boca Raton, FL.

Liu, G., Larson, E.D., Williams, R.H., Kreutz, T.G., Guo, X. 2011. Making Fischer-Tropsch Fuels and Electricity from Coal and Biomass: Performance and Cost Analysis. *Energy & Fuels*, 25: 415–437.

Luque, R., and Speight, J.G. (Editors). 2015. *Gasification for Synthetic Fuel Production: Fundamentals, Processes, and Applications*. Woodhead Publishing, Elsevier, Cambridge, UK.

Lv, P.M., Xiong, Z.H., Chang, J., Wu, C.Z., Chen, Y., and Zhu, J.X. 2004. An Experimental Study on Biomass Air-Steam Gasification in a Fluidized Bed. *Bioresource Technology*, 95(1): 95–101.

Marano, J.J. 2003. Refinery Technology Profiles: Gasification and Supporting Technologies. Report prepared for the United States Department of Energy, National Energy Technology Laboratory. United States Energy Information Administration, Washington, DC. June.

Martinez-Alonso, A., and Tascon, J.M.D. 1991. In: *Fundamental Issues in Control of Carbon Gasification Reactivity*. Lahaye, J., and Ehrburger, P. (Editors). Kluwer Academic Publishers, Dordrecht, The Netherlands.

Matsukata, M., Kikuchi, E., and Morita, Y. 1992. A New Classification of Alkali and Alkaline Earth Catalysts for Gasification of Carbon. *Fuel*, 71: 819–823.

McKendry, P. 2002. Energy Production from Biomass Part 3: Gasification Technologies. *Bioresource Technology*, 83(1): 55–63.

McLendon T.R., Lui A.P., Pineault R.L., Beer S.K., and Richardson S.W. 2004. High-Pressure Co-Gasification of Coal and Biomass in a Fluidized Bed. *Biomass and Bioenergy*, 26(4): 377–388.

Mims, C.A. 1991. In: Fundamental Issues in Control of Carbon Gasification Reactivity. J. Lahaye and P. Ehrburger (Editors). Kluwer Academic Publishers, Dordrecht, The Netherlands, p. 383.

Mokhatab, S., Poe, W.A., and Speight, J.G. 2006. *Handbook of Natural Gas Transmission and Processing*. Elsevier, Amsterdam, The Netherlands.

Nordstrand D., Duong D.N.B., Miller, B.G. 2008. Chapter 9: Post-combustion Emissions Control. In: *Combustion Engineering Issues for Solid Fuel Systems*. B.G. Miller and D. Tillman (Editors). Elsevier, London, UK.

Orhan, Y., İs, G., Alper, E., McApline, K., Daly, S., Sycz, M., and Elkamel, A. 2014. Gasification of Oil Refinery Waste for Power and Hydrogen Production. *Proceedings of 2014 International Conference on Industrial Engineering and Operations Management*. January 7–9, Bali, Indonesia.

Pakdel, H., and Roy, C. 1991. Hydrocarbon Content of Liquid Products and Tar from Pyrolysis and Gasification of Wood. *Energy & Fuels*, 5: 427–436.

Pan, Y.G., Velo, E., Roca, X., Manyà, J.J., and Puigjaner, L. 2000. Fluidized-Bed Cogasification of Residual Biomass/Poor Coal Blends for Fuel Gas Production. *Fuel*, 79: 1317–1326.

Parkash, S. 2003. *Refining Processes Handbook*. Gulf Professional Publishing, Elsevier, Amsterdam, The Netherlands.

Penrose, C.F., Wallace, P.S., Kasbaum, J.L., Anderson, M.K., and Preston, W.E. 1999. *Enhancing Refinery Profitability by Gasification, Hydroprocessing and Power Generation. Proceedings of Gasification Technologies Conference*. October, San Francisco, CA. www.globalsyngas.org/uploads/eventLibrary/GTC99270.pdf

Pepiot, P., Dibble, C.J., and Foust, C.G. 2010. Computational Fluid Dynamics Modeling of Biomass Gasification and Pyrolysis. In: *Computational Modeling in Lignocellulosic Biofuel Production*. M.R. Nimlos and M.F. Crowley (Editors). ACS Symposium Series. American Chemical Society, Washington, DC.

Probstein, R.F., and Hicks, R.E. 1990. *Synthetic Fuels*. pH Press, Cambridge, MA. Chapter 4.

Rabovitser, I.K., Nester, S., and Bryan, B. 2010. Plasma Assisted Conversion of Carbonaceous Materials into A Gas. United States Patent 7736400. June 25.

Rajvanshi, A.K. 1986. Biomass Gasification. In: *Alternative Energy in Agriculture*, Vol. 2. D.Y. Goswami (Editor). CRC Press, Boca Raton, FL. pp. 83–102.

Ramroop Singh, N. 2011. Chapter 5: Biofuel. In: *The Biofuels Handbook*. J.G. Speight (Editor). Royal Society of Chemistry, London, UK. Part 1.

Rapagnà, N.J., and Latif, A. 1997. Steam Gasification of Almond Shells in a Fluidized Bed Reactor: The Influence of Temperature and Particle Size on Product Yield and Distribution. *Biomass and Bioenergy*, 12(4): 281–288.

Rapagnà, N.J., and, Kiennemann, A., and Foscolo, P.U. 2000. Steam-Gasification of Biomass in a Fluidized-Bed of Olivine Particles. *Biomass and Bioenergy*, 19(3): 187–197.

Ricketts, B., Hotchkiss, R., Livingston, W., and Hall, M. 2002. Technology Status Review of Waste/Biomass Co-Gasification with Coal. *Proceedings of Institution of Chemical Engineers Fifth European Gasification Conference*. April 8–10, Noordwijk, Netherlands.

Roddy, D.J., and Manson-Whitton, C. 2012. Biomass Gasification and Pyrolysis. In: *Comprehensive Renewable Energy*. Vol. 5. Biomass and Biofuels. D.J. Roddy (Editor). Elsevier, Amsterdam, The Netherlands.

Sha, X. 2005. Coal Gasification. In: *Coal, Oil Shale, Natural Bitumen, Heavy Oil and Peat*. Encyclopedia of Life Support Systems (EOLSS), Developed under the Auspices of the UNESCO, EOLSS Publishers, Oxford, UK, www.eolss.net.

Shabbar, S., and Janajreh, I. 2013. Thermodynamic Equilibrium Analysis of Coal Gasification Using Gibbs Energy Minimization Method. *Energy Conversion and Management*, 65: 755–763.

Shen, C-H., Chen, W-H., Hsu, H-W., Sheu, J-Y., and Hsieh, T-H. 2012. Co-Gasification Performance of Coal and Petroleum Coke Blends in A Pilot-Scale Pressurized Entrained-Flow Gasifier. *International Journal of Energy Research*, 36: 499–508.

Singh, S.P., Weil, S.A., and Babu, S.P. 1980. Thermodynamic Analysis of Coal Gasification Processes. *Energy*, 5(8–9): 905–914.

Sjöström, K., Chen, G., Yu, Q., Brage, C., and Rosén, C. 1999. Promoted Reactivity of Char in Cogasification of Biomass and Coal: Synergies in the Thermochemical Process. *Fuel*, 78: 1189–1194.

Sondreal, E. A., Benson, S. A., Pavlish, J. H., and Ralston, N.V.C. 2004. An Overview of Air Quality III: Mercury, Trace Elements, and Particulate Matter. *Fuel Processing Technology*, 85: 425–440.

Speight, J.G. 1990. Chapter 33. In: *Fuel Science and Technology Handbook*. J.G. Speight (Editor). Marcel Dekker Inc., New York.

Speight, J.G. 2007. *Natural Gas: A Basic Handbook*. GPC Books, Gulf Publishing Company, Houston, TX.

Speight, J.G. 2008. *Synthetic Fuels Handbook: Properties, Processes, and Performance*. McGraw-Hill, New York.

Speight, J.G. (Editor). 2011a. *Biofuels Handbook*. Royal Society of Chemistry, London, UK.

Speight, J.G. 2011b. *The Refinery of the Future*. Gulf Professional Publishing, Elsevier, Oxford, UK.

Speight, J.G. 2013a. *The Chemistry and Technology of Coal*. 3rd Edition. CRC Press, Boca Raton, FL.

Speight, J.G. 2013b. *Coal-Fired Power Generation Handbook*. Scrivener Publishing, Salem, MA.

Speight, J.G. 2014a. *The Chemistry and Technology of Petroleum*. 5th Edition. CRC Press, Boca Raton, FL.

Speight, J.G. 2014b. *Gasification of Unconventional Feedstocks*. Gulf Professional Publishing, Elsevier, Oxford, UK.

Speight, J.G., and Islam, M.R. 2016. *Peak Energy—Myth or Reality*. Scrivener Publishing, Beverly, MA.

Speight, J.G. 2017. *Handbook of Petroleum Refining*. CRC Press, Boca Raton, FL.

Sundaresan, S., and Amundson, N.R. 1978. Studies in Char Gasification—I: A lumped Model. *Chemical Engineering Science*, 34: 345–354.

Sutikno, T., and Turini, K. 2012. Gasifying Coke to Produce Hydrogen in Refineries. *Petroleum Technology Quarterly*, Q3: 105.

Van Heek, K.H., Muhlen, H-J. 1991. In: *Fundamental Issues in Control of Carbon Gasification Reactivity.* J. Lahaye and P. Ehrburger (Editors). Kluwer Academic Publishers Inc., Dordrecht, The Netherlands. p. 1.

Wallace, P.S., Anderson, M.K., Rodarte, A.I., and Preston, W.E. 1998. Heavy Oil Upgrading by the Separation and Gasification of Asphaltenes. *Proceedings of Presented at the Gasification Technologies Conference.* San Francisco, CA. October. www.globalsyngas.org/uploads/eventLibrary/gtc9817p.pdf

Wang, Y., Duan Y., Yang L., Jiang Y., Wu C., Wang Q., and Yang X. 2008. Comparison of Mercury Removal Characteristic between Fabric Filter and Electrostatic Precipitators of Coal-Fired Power Plants. *Journal of Fuel Chemistry and Technology*, 36(1): 23–29.

Wolff, J., and Vliegenthart, E. 2011. Gasification of Heavy Ends. *Petroleum Technology Quarterly*, Q2: 1–5.

Yang, H., Xua, Z., Fan, M., Bland, A.E., and Judkins, R.R. 2007. Adsorbents for Capturing Mercury in Coal-Fired Boiler Flue Gas. *Journal of Hazardous Materials*, 146: 1–11.

6 Chemicals from Paraffin Hydrocarbons

6.1 INTRODUCTION

Natural gas and crude oil are primary feedstocks, and continue to be, the main sources of secondary feedstocks for the production of petrochemicals. For example, methane from natural gas as well as other low-boiling (low molecular weight) hydrocarbon derivatives is recovered for use as feedstocks for the production of olefin derivatives and diolefin derivatives. In addition, the gaseous constituents from crude oil (associated natural gas) as well as refinery gases from different crude oil processing schemes—such as cracking and reforming processes (Parkash, 2003; Gary et al., 2007; Speight, 2014a; Hsu and Robinson, 2017; Speight, 2017)—are important sources for olefin derivatives. Paraffin hydrocarbon derivatives (i.e., hydrocarbon derivatives with the general formula C_nH_{2n+2}) that are used for producing petrochemical products range from methane (CH_4) to the higher molecular weight hydrocarbon derivatives that exist in various distillate fractions such as naphtha, kerosene, and gas oil as well as the nonvolatile residua (resids, residues) (Table 6.1). The proportion of pure hydrocarbon derivatives in residua is typically at a low level and most of the constituents of residua also contain heteroatoms (nitrogen, oxygen, sulfur, and metals) in various molecular locations (Speight, 2014a).

Chemically, paraffin derivatives are relatively inactive compared to olefin derivatives, diolefin derivatives, and aromatic derivatives. However, these compounds (paraffin derivatives) are the precursors for olefin derivatives through a variety of cracking processes; the C_6–C_9 paraffin derivatives and cycloparaffin derivatives are especially important for the production of aromatic products

---

TABLE 6.1

Names of the Simple Saturated (Paraffin and Cycloparaffin) and Unsaturated (Olefin and Acetylene) Hydrocarbon Derivatives

Number of Carbon Atoms	Alkane (Single Bond) (C_nH_{2n+2})	Alkene (Double Bond) (C_nH_{2n})	Alkyne (Triple Bond) (C_nH_{2n-2})	Cycloalkane (C_nH_{2n})	Diene (C_nH_{2n-2})
1	Methane	N/A	N/A	N/A	N/A
2	Ethane	Ethylene (ethene)	Acetylene (ethyne)	N/A	N/A
3	Propane	Propylene (propene)	Propyne (methylacetylene)	Cyclopropane	Propadiene
4	Butane	Butylene (butene)	Butyne	Cyclobutane	Butadiene
5	Pentane	Pentene	Pentyne	Cyclopentane	Pentadiene
6	Hexane	Hexene	Hexyne	Cyclohexane	Hexadiene
7	Heptane	Heptene	Heptyne	Cycloheptane	Heptadiene
8	Octane	Octene	Octyne	Cyclooctane	Octadiene
9	Nonane	Nonene	Nonyne	Cyclononane	Nonadiene
10	Decane	Decene	Decyne	Cyclodecane	Decadiene
11	Undecane	Undecene	Undecyne	Cycloundecane	Undecadiene
12	Dodecane	Dodecene	Dodecyne	Cyclododecane	Dodecadiene

N/A: not applicable to this formula.

---

through reforming (Parkash, 2003; Gary et al., 2007; Speight, 2014a; Hsu and Robinson, 2017; Speight, 2017).

Briefly, and by way of introduction, the term *steam cracking unit* refers to all processes inside the battery limits of a steam cracker and typically consist of three sections: (i) the pyrolysis section, (ii) the primary fractionation/compression section, and (iii) the product recovery/separation.

The pyrolysis section is the heart of a steam cracking unit. The feedstock first enters the convection section of a pyrolysis furnace in which the feedstock (such as naphtha) is vaporized with slightly superheated steam and is passed into long (12–25 m), narrow tubes (25–125 mm in diameter) which are fabricated from iron, chromium, and nickel alloys. The steam is added to reduce partial pressure of products and to prevent unwanted side reactions. Pyrolysis takes place mainly in the radiant section of the furnace (in the absence of a catalyst) where tubes are externally heated from 750°C to 900°C (up to 1,100°C, 1,380°F–1,650°F, and up to 2,010°F) where the feedstock is cracked (thermally decomposed) to lower molecular weight products. After leaving the furnace, the hot gas mixture is quenched in a series of heat exchangers to approximately about 350°C (650°F). This avoids degradation by secondary reactions, but the heat exchangers are prone to fouling (i.e., coke formation on the walls). Further cooling of the gas mixture is achieved using a liquid quench oil.

Primary fractionation applies to liquid feedstocks such as naphtha and gas oil only and not for gaseous feedstocks such as ethane. In the primary fractionator, gas mixtures are first cooled to around 150°C (300°F) and most benzene, toluene, and xylenes (BTX) and fuel oils are condensed. In the quench water tower, water from dilution steam is recovered for recycling. The steam is passed through four or five stages of gas compression with temperatures at approximately 15°C–100°C (59°F–212°F), then cooling and finally cleanup to remove acid gases, carbon dioxide, and water. A common issue with compression is fouling (buildup of solid byproducts) in the cracked gas compressors and aftercoolers. Wash oil and water are used to reduce fouling.

The product recovery/separation section is essentially a separation process through distillation, refrigeration, and extraction. The equipment includes chilling trains and fractionation towers, which include refrigeration, demethanizer, de-ethanizer, depropanizer, and finally a debutanizer. Demethanization requires very low temperatures (in the order of −114°C (−173°F). Ethylene and ethane separation often requires large distillation columns with 120–180 trays and high reflux ratios. Undesired acetylene is removed through catalytic hydrogenation or extractive distillation. Similarly, propane and propylene are reboiled with quench water at approximately 80°C (176°F) and separated into the depropanizer. Ethylene and propylene refrigeration systems can be operated at low temperatures within −10°C to −150°C (14 and −238°F) for cooling and high pressure (220–450 psi) for compression. Ethane and propane are either recycled as feedstocks, or burned (or exported) as fuels. A considerable amount of ethylene is condensed and recovered through turbo-expanders. Methane and hydrogen are separated at cryogenic temperatures by turbo-compressors. Generally, ethane steam cracking also requires three sections that are similar to those in the case of naphtha steam cracking process. However, ethane steam cracking requires slightly higher temperature in the furnace, a higher capacity of the de-ethanizer but less infrastructure facilities.

An additional issue is the potential for coking to occur. The reactions leading to the formation of coke results in the deposition of the coke (carbon) on the furnace coils and therefore reduces the effectiveness of the furnace. Although the presence of steam reduces some coking, regular decoking still is required in various parts of the pyrolysis section. Before decoking, feedstocks are removed from the furnace after which high-pressure steam and air are fed to the furnace as it is heated up to 880°C–900°C or even up to 1,100°C (up to 1,100°C, 1,615°F–1,650°F and up to 2,010°F). Coke on the inner surfaces of the wall and tubes is either washed away with high-pressure water or removed mechanically. A typical steam cracker has six to eight furnaces to accommodate onstream operation and off-stream maintenance. Similar decoking, though far less frequent, is also required for the heat exchangers that are associated with furnaces; heat exchanger foiling is a common occurrence in high-temperature operations (Parkash, 2003; Gary et al., 2007; Speight, 2014a, 2014b; Hsu and Robinson, 2017; Speight, 2017).

This chapter presents to the reader a selection of the chemical and physical properties of the methane (CH_4) and butane (C_4H_{10}) paraffin derivatives. The chemical and physical properties of the higher molecular weight paraffin derivatives that are typically present as mixtures in the various distillates fractions of crude oil are also presented.

6.2 METHANE

Methane (CH_4) is the simplest alkane, and the principal component of natural gas (usually 70–90% v/v) (Katz, 1959; Kohl and Riesenfeld, 1985; Maddox et al., 1985; Newman. 1985; Kohl and Nielsen, 1997; Mokhatab et al., 2006; Speight, 2014a). Methane (CH_4) commonly (often incorrectly) known as natural gas, is colorless and naturally odorless, and burns efficiently without many byproducts. It is also known as marsh gas or methyl hydride and is easily ignited. The vapor is lighter than air (Table 6.2). Under prolonged exposure to fire or intense heat the containers may rupture in a violent explosion. Methane is used in making other chemicals and as a constituent of the fuel, natural gas.

In addition, there is a large, but unknown, amount of methane in gas hydrates (methane clathrates) in the ocean floors and significant amounts of methane are produced anaerobically by methanogenesis. Other sources include mud volcanoes, such as those that occur regularly in Trinidad, which are connected with deep geological faults, landfills, and livestock (primarily ruminants) from enteric fermentation.

Natural gas can be used as a source of hydrocarbon derivatives (e.g., ethane and propane) that are higher molecular weight than methane and that are important chemical intermediates.

The preparation of chemicals and chemical intermediates from methane (natural gas) should not be restricted to those described here, but should be regarded as some of the building blocks of the petrochemical industry.

In addition to methane being the major constituent of natural gas, the products of the gasification of carbonaceous feedstocks also contain methane. The commonly accepted approach to the synthesis of methane from the carbonaceous feedstock is the catalytic reaction of hydrogen and carbon monoxide:

$$CO + 3H_2 \rightarrow CH_4 + H_2O$$

TABLE 6.2
Properties of Methane

Chemical formula	CH_4
Molar mass	16.04 g/mol
Appearance	Colorless gas
Odor	Odorless
Density	0.656 g/L (gas, 25°C, 1 atm); 0.716 g/L (gas, 0°C, 1 atm); 0.42262 g/cm³ (liquid, −162°C)
Liquid density	0.4226
Vapor density (air = 1)	0.55
Melting point	−182.5°C; −296.4°F; 90.7°K
Boiling point	−161.49°C; −258.68°F; 111.66°K
Solubility in water	22.7 mg/L
Solubility	Soluble in ethanol, diethyl ether, benzene, toluene, methanol, acetone
Flash point	−188°C (−306.4°F; 85.1°K)
Autoignition temperature	537°C (999°F; 810 K)
Explosive limits	4.4–17% v/v in air

A variety of metals have been used as catalysts for the methanation reaction; the most common, and to some extent the most effective methanation catalysts, appear to be nickel and ruthenium, with nickel being the most widely used. The synthesis gas must be desulfurized before the methanation step since sulfur compounds will rapidly deactivate (poison) the catalysts. Also, the composition of the products of gasification processes are varied insofar as the gas composition varies with the feedstock and the system employed (Speight, 2014c). It is emphasized that the gas product must be first freed from any pollutants such as particulate matter and sulfur compounds before further use, particularly when the intended use is a water-gas shift or methanation (Mokhatab et al., 2006; Speight, 2007, 2013, 2014a, 2014a).

The production of methane from the carbonaceous feedstock does not depend entirely on catalytic methanation and, in fact, a number of gasification processes use hydrogasification, that is, the direct addition of hydrogen to the carbonaceous feedstock under pressure to form methane:

$$C_{coal} + 2H_2 \rightarrow CH_4$$

The hydrogen-rich gas for the hydrogasification process can be manufactured from steam by using the char that leaves the hydrogasifier. Appreciable quantities of methane are formed directly in the primary gasifier and the heat released by methane formation is at a sufficiently high temperature to be used in the steam-carbon reaction to produce hydrogen so that less oxygen is used to produce heat for the steam-carbon reaction. Hence, less heat is lost in the low-temperature methanation step, thereby leading to higher overall process efficiency.

Methane is a major raw material for many chemical processes and the potential number of chemicals that can be produced from methane is almost limitless. Indeed, methane can be converted to a wide variety of chemicals, in addition to serving as a source of synthesis gas. This leads to a wide variety of chemicals which involve chemistry (i.e., the chemistry of methane and other one-carbon compounds). In this, aspect, the use of coal and other carbonaceous feedstocks for the production of chemicals is similar to the chemistry employed in the synthesis of chemicals from the gasification products of the carbonaceous feedstock.

In the chemical industry, methane is the feedstock of choice for the production of hydrogen, methanol, acetic acid, and acetic anhydride. To produce any of these chemicals, methane is first made to react with steam in the presence of a nickel catalyst at high temperatures (700°C–1,100°C, 1,290°F–2,010°F):

$$CH_4 + H_2O \rightarrow CO + 3H_2$$

The synthesis gas is then reacted in various ways to produce a wide variety of products.

In addition, acetylene is prepared by passing methane through an electric arc. When methane is made to react with chlorine (gas), various chloromethane derivatives are produced: chloromethane (CH_3Cl), dichloromethane (CH_2Cl_2), chloroform ($CHCl_3$), and carbon tetrachloride (CCl_4). However, the use of these chemicals is declining—acetylene may be replaced by less costly substitutes, and the chloromethane derivatives are used less often because of health and environmental concerns.

It must be recognized that there are many other options for the formation of chemical intermediates and chemicals from methane by indirect routes, i.e., other compounds are prepared from the methane that are then used as further sources of petrochemical products.

In summary, methane can be an important source of petrochemical intermediates and solvents.

6.2.1 Physical Properties

At room temperature and standard pressure (STP) methane is a colorless and odorless gas. Methane has a boiling point of −161°C (−257.8°F) at a pressure of one atmosphere (14.7 psi). Methane is lighter than air, having a specific gravity of 0.554 (Table 6.2) and burns readily in air, forming carbon dioxide and water vapor; the flame is pale, slightly luminous, and very hot. The boiling point of methane is −162°C (−259.6°F) and the melting point is −182.5°C (−296.5°F). Methane in general is very stable, but mixtures of methane and air (with the methane content between 5% and 14% v/v) are explosive. Explosions of such mixtures have been a frequent occurrence in coal mines and have been the cause of many mine disasters.

Methane is lighter than air, having a specific gravity of 0.554 (Table 6.2) and burns readily in air, forming carbon dioxide and water vapor; the flame is pale, slightly luminous, and very hot. The boiling point of methane is −162°C (−259.6°F) and the melting point is −182.5°C (−296.5°F).

Methane is a relatively potent greenhouse gas and compared with carbon dioxide, it has a high global warming potential of 72 (calculated over a period of 20 years) or 25 (for a time period of 100 years). Methane in the atmosphere is eventually oxidized, producing carbon dioxide and water. As a result, methane in the atmosphere has a half-life of 7 years.

6.2.2 CHEMICAL PROPERTIES

Structurally, methane is a tetrahedral molecule with four equivalent carbon-hydrogen (C-H) bonds. The primary chemical reactions of methane are combustion, steam reforming to synthesis gas (syngas, mixtures of carbon monoxide, and hydrogen), and halogenation. Typically, the chemical reactions of methane are difficult to control. Although there is great interest in converting methane into useful or more easily liquefied compounds, the only practical processes are relatively unselective.

There are two types of routes through which methane gas can be converted into olefin derivatives: (i) indirect routes via syngas or ethane and (ii) direct routes (directly from methane to low-boiling olefin derivatives. Another indirect route is methane to olefin derivatives via Fischer–Tropsch liquids and the subsequent conversion to high-value chemicals by means of steam cracking. The direct route from methane to olefin derivatives is a modified Fischer–Tropsch reaction, but this route is technically difficult because of low selectivity to low-boiling olefin derivatives and the high yield of high molecular weight hydrocarbon derivatives (Wang et al., 2003).

Typically, in the chemical industry, methane is converted by the steam reforming process to synthesis gas (also called syngas) which is a mixture of carbon monoxide and hydrogen which is an important building block for many chemicals using the Fischer–Tropsch process (Chapter 10) (Speight, 2013, 2014a). The process that employs a nickel-based catalyst and requires high temperature that is in the order of 700°C–1,100°C (1,290°F–2,010°F):

$$CH_4 + H_2O \rightarrow CO + 3H_2 \text{ — steam reforming process}$$

Similarly, in the Haber–Bosch synthesis of ammonia from air, natural gas (methane) is reduced to a mixture of carbon dioxide, water, and ammonia:

$$CH_4 + H_2O \rightarrow CO + 3H_2 \text{ — steam-methane reforming}$$

$$CO + H_2O \rightarrow CO_2 + H_2 \text{ — hydrogen production}$$

$$3H_2 + N_2 \rightarrow 2NH_3 \text{ — Haber–Bosch process}$$

Other commercially –viable processes that use methane as a chemical feedstock include the catalytic oxidation to methanol, which is based on the oxidative coupling of methane, and the direct reaction of methane with sulfur trioxide to produce methane sulfonic acid:

$$2CH_4 + O_2 \rightarrow 2CH_3OH$$

$$CH_4 + SO_3 \rightarrow CH_3SO_3H$$

The combustion of methane is an exothermic reaction that can be represented by a simple equation that reflects the thermal oxidation of methane to carbon dioxide and water:

$$CH_4 + 2O_2 \rightarrow CO_2 + 2H_2O$$

The reaction, however, is a multiple step reaction and can be generally represented by the following equations in which the species M^* signifies an energetic third body, from which energy is transferred during a molecular collision. Thus:

$$CH_4 + M^* \rightarrow CH_3 + H + M$$

$$CH_4 + O_2 \rightarrow CH_3 + HO_2$$

$$CH_4 + HO_2 \rightarrow CH_3 + 2OH$$

$$CH_4 + OH \rightarrow CH_3 + H_2O$$

$$O_2 + H \rightarrow O + OH$$

$$CH_4 + O \rightarrow CH_3 + OH$$

$$CH_3 + O_2 \rightarrow CH_2O + OH$$

$$CH_2O + O \rightarrow CHO + OH$$

$$CH_2O + OH \rightarrow CHO + H_2O$$

$$CH_2O + H \rightarrow CHO + H_2$$

$$CHO + O \rightarrow CO + OH$$

$$CHO + OH \rightarrow CO + H_2O$$

$$CHO + H \rightarrow CO + H_2$$

$$H_2 + O \rightarrow H + OH$$

$$H_2 + OH \rightarrow H + H_2O$$

$$CO + OH \rightarrow CO_2 + H$$

$$H + OH + M \rightarrow H_2O + M^*$$

$$H + H + M \rightarrow H_2 + M^*$$

$$H + O_2 + M \rightarrow HO_2 + M^*$$

As illustrated, formaldehyde (HCHO) is an early intermediate product and oxidation of formaldehyde gives the formyl radical (HCO), which then produce carbon monoxide (CO). Any resulting hydrogen oxidizes to water or other intermediates. Finally, the carbon monoxide is oxidized to carbon dioxide. The overall reaction rate is a function of the concentration of the various entities during the combustion process. The higher the temperature, the greater the concentration of radical species, and more rapid the combustion process.

In the partial oxidation reaction (Arutyunov, 2007), methane and other hydrocarbon derivatives in natural gas react with a limited amount of oxygen (typically from air) that is not enough to completely oxidize the hydrocarbon derivatives to carbon dioxide and water. With less than the stoichiometric amount of oxygen available, the reaction products contain primarily hydrogen and carbon monoxide (and nitrogen, if the reaction is carried out with air rather than pure oxygen), and a relatively small amount of carbon dioxide and other compounds. Subsequently, in a water-gas shift reaction, the carbon monoxide reacts with water to form carbon dioxide and more hydrogen.

$$2CH_4 + O_2 \rightarrow 2CO + 2H_2 \quad \text{(partial oxidation of methane)}$$

$$CO + H_2O \rightarrow CO_2 + H_2 \quad \text{(water-gas shift)}$$

The partial oxidation reaction is an exothermic process and the process is, typically, much faster than steam reforming and requires a smaller reactor vessel. This process initially produces less hydrogen per unit of the input fuel than is obtained by steam reforming of the same fuel.

6.2.3 Chemicals from Methane

Methane (CH_4) is a one-carbon paraffinic hydrocarbon that is not very reactive under normal conditions. It is a colorless gas that is insoluble in water—is the first member of the alkane series (C_nH_{2n+2}) and is the main component of natural gas. It is also a byproduct in all gas streams from processing crude oils. It is a colorless, odorless gas that is lighter than air (Table 6.2).

Only a few chemicals can be produced directly from methane under relatively severe conditions. Chlorination of methane is only possible by thermal or photochemical initiation. Methane can be partially oxidized with a limited amount of oxygen or in presence of steam to a synthesis gas mixture. Many chemicals can be produced from methane via the more reactive synthesis gas mixture. Synthesis gas (Chapter 10) is the precursor for two major chemicals, ammonia, and methanol. Both compounds are the hosts for many important petrochemical products. A few chemicals are based on the direct reaction of methane with other reagents. These are carbon disulfide, hydrogen cyanide chloromethanes, and synthesis gas mixture. Currently, a redox fuel cell based on methane is being developed.

The availability of hydrogen from catalytic reforming operations has made its application economically feasible in a number of petroleum-refining operations. Previously, the chief sources of large-scale hydrogen (used mainly for ammonia manufacture) were the cracking of methane (or natural gas) and the reaction between methane and steam. In the latter, at 900°C–1,000°C (1,650°F–1,830°F) conversion into carbon monoxide and hydrogen results:

$$CH_4 + H_2O \rightarrow CO + 3H_2$$

If this mixture is treated further with steam at 500°C over a catalyst, the carbon monoxide present is converted into carbon dioxide and more hydrogen is produced:

$$CO + H_2O \rightarrow H_2 + CO_2$$

The reduction of carbon monoxide by hydrogen is the basis of several syntheses, including the manufacture of methanol and higher alcohols (Chapter 8). Indeed, the synthesis of hydrocarbon derivatives by the Fischer–Tropsch reaction has received considerable attention:

$$nCO + 2nH_2 \rightarrow (CH_2)_n + nH_2O$$

This occurs in the temperature range 200°C–350°C (390°F–660°F), which is sufficiently high for the water-gas shift to take place in presence of the catalyst:

$$CO + H_2O \rightarrow CO_2 + H_2$$

The major products are olefin derivatives and paraffin derivatives, together with some oxygen-containing organic compounds in the product mix may be varied by changing the catalyst or the temperature, pressure, and carbon monoxide-hydrogen ratio.

The hydrocarbon derivatives formed are mainly aliphatic, and on a molar basis methane is the most abundant; the amount of higher hydrocarbon derivatives usually decreases gradually with increase in molecular weight. Isoparaffin formation is more extensive over zinc oxide (ZnO) or thoria (ThO$_2$) at 400°C–500°C (750°F–930°F) and at higher pressure. Paraffin waxes are formed over ruthenium catalysts at relatively low temperatures (170°C–200°C, 340°F–390°F), high pressures (1,500 psi), and with a carbon monoxide-hydrogen ratio. The more highly branched product made over the iron catalyst is an important factor in a choice for the manufacture of automotive fuels. On the other hand, a high-quality diesel fuel (paraffin character) can be prepared over cobalt.

Secondary reactions play an important part in determining the final structure of the product. The olefin derivatives produced are subjected to both hydrogenation and double-bond shifting toward the center of the molecule; *cis* and *trans* isomers are formed in about equal amounts. The proportions of straight-chain molecules decrease with rise in molecular weight, but even so they are still more abundant than branched-chain compounds up to approximately C$_{10}$.

The small amount of aromatic hydrocarbon derivatives found in the product covers a wide range of isomer possibilities. In the C$_6$–C$_9$ range, benzene, toluene, ethylbenzene, xylene, *n*-propyl- and iso-propylbenzene, methyl ethyl benzene derivatives and trimethyl benzene derivatives have been identified; naphthalene derivatives and anthracene derivatives are also present.

Paraffin hydrocarbon derivatives are less reactive than olefin derivatives; only a few chemicals are directly based on them. Nevertheless, paraffinic hydrocarbon derivatives are the starting materials for the production of olefin derivatives. Methane's relation with petrochemicals is primarily through synthesis gas (Chapter 5). Ethane, on the other hand, is a major feedstock for steam crackers for the production of ethylene. Few chemicals could be obtained from the direct reaction of ethane with other reagents. The higher paraffin derivatives—propane, butanes, pentanes, and higher molecular weight paraffin derivatives—also have limited direct use in the chemical industry except for the production of light olefin derivatives through steam cracking.

6.2.3.1   Carbon Disulfide

Methane reacts with sulfur (an active nonmetal element of group 6A) at high temperatures to produce carbon disulfide (CS$_2$). Activated alumina or clay is used as the catalyst at approximately 675°C (1,245°F) and 30 psi. The process starts by vaporizing pure sulfur, mixing it with methane, and passing the mixture over the alumina catalyst:

$$CH_4 + 4S \rightarrow CS_2 + 2H_2S$$

Hydrogen sulfide, a coproduct, is used to recover sulfur by the Claus reaction. A carbon disulfide yield of 85%–90% based on methane is anticipated. An alternative route for carbon disulfide is by the reaction of liquid sulfur with charcoal. However, this method is not used very much.

Carbon disulfide is primarily used to produce rayon and cellophane (regenerated cellulose) and is also used to produce carbon tetrachloride using iron powder as a catalyst at 30°C (86°F):

$$CS_2 + 3Cl_2 \rightarrow CCl_4 + S_2Cl_2$$

The sulfur chloride is an intermediate that is then reacted with carbon disulfide to produce more carbon tetrachloride and sulfur:

$$2S_2Cl_2 + CS_2 \rightarrow CCl_4 + 6S$$

Thus, the overall reaction is:

$$CS_2 + 2Cl_2 \rightarrow CCl_4 + 2S$$

Carbon disulfide is also used to produce xanthate derivatives [ROC(=S)S-Na$^+$] as an ore flotation agent and ammonium thiocyanate (NH$_4$SCH) as a corrosion inhibitor in ammonia handling systems.

6.2.3.2   Ethylene

Ethylene is considered to be one of the most important raw materials in the chemical industry. Its significance is driven by its molecular structure, i.e., carbon-carbon double bonds (H$_2$C=CH$_2$). This π-bond is responsible for its chemical reactivity. The double bond is also a place of high electron density; therefore, it is susceptible to attack by electrophiles. It is a volatile substance, colorless at room temperature, noncorrosive, nontoxic, flammable gas, slightly soluble in water, and soluble in most organic solvents. It has boiling and melting points of −104°C and −169.2°C, respectively, at a pressure of 1 atm. Ethylene is a very active chemical; as exemplified by the reaction between ethylene and water to produce ethyl alcohol. Most of the ethylene reactions are catalyzed by transition metals, which bind transiently to the ethylene using both the π and π* orbitals. Ethylene is also an active alkylating agent, which can be used for the production of important monomers, such as ethyl benzene (EB), which is dehydrogenated to styrene.

Ethylene is an important petrochemical starting material and is used extensively for production of polyethene (high-density polyethylene, HDPE, low-density polyethylene, LDPE, and linear low-density polyethylene, LLDPE) as well as a major feedstock for the manufacture of ethylene dichloride (CH$_2$ClCH$_2$Cl), vinyl chloride (CH$_2$=CHCl), and the cyclic ethylene oxide. In the modern petroleum refinery, ethylene is generated from various process units along with liquid products such as naphtha, kerosene, and gas oil. Naphtha from the refineries widely transported to petrochemical plants to produce ethylene and convert into above valuable chemical products. One of the technologies used in present industries for ethylene production is by naphtha cracking technology.

Traditionally, ethylene is produced by way of naphtha cracking and steam cracking of ethane (Salkuyeh and Adams, 2015). Ethylene can also be produced using natural gas as a feedstock either via the direct route that involves oxidative coupling of methane or via the indirect route that involves methanol to olefin (MTO) process (Ortiz-Espinoza et al., 2015, Salkuyeh and Adams, 2015; Dutta et al., 2017).

In the process, the feedstock preheated by a heat exchanger, mixed with steam, and then further heated to its incipient cracking temperature (500°C–680°C, 930°F–1,255°F, depending upon the feedstock). At this point, the heated feedstock enters a reactor (typically, a fired tubular reactor) where it is heated to cracking temperatures (750°C–785°C, 1,380°F–1,605°F). During this reaction, hydrocarbons in the feedstock are cracked to produce lower molecular weight products, including ethylene. The cracking reaction is highly endothermic and, therefore, high energy rates are needed. The cracking coils are designed to optimize the temperature and pressure profiles in order to maximize the yield of desired or value products. Short residence times in the furnace are also important as they increase the yields of primary products such as ethylene and propylene. Long residence times will favor the secondary reactions. Thus:

Feedstock/steam	Primary Reactions	Secondary Reactions
	Ethylene	C_4 products
	Propylene	C_5 products
	Acetylene	C_6 products
	Hydrogen	Aromatic derivatives
	Methane	C_7 products
	Hydrocarbons	Higher molecular weight products

Methane conversion and selectivity toward ethylene for the oxidative coupling reaction depends on the methane/oxygen ratio. A lower methane/oxygen ratio decreases the selectivity toward ethylene production but does increase methane conversion. Because of the low conversion and poor selectivity of the process, unreacted methane and byproducts produced (such as ethane and higher molecular weight hydrocarbon derivatives, as well as carbon monoxide, hydrogen, carbon dioxide, and water) need to be separated to obtain a high-purity ethylene as the final product. Water is removed partially during the multistage compression, and a molecular sieve unit to remove the remaining water.

6.2.3.3 Hydrogen Cyanide

Hydrogen cyanide (hydrocyanic acid, HCN) is a colorless liquid (boiling point: 25.6°C, 7.8.1°F) that is miscible with water, producing a weakly acidic solution. It is a highly toxic compound, but a very useful chemical intermediate with high reactivity. It is used in the synthesis of acrylonitrile and adiponitrile, which are important monomers for plastic and synthetic fiber production.

Hydrogen cyanide is produced by the Andrussaw process which involves the high-temperature reaction of ammonia, methane, and air over a platinum catalyst. The reaction is exothermic, and the released heat is used to supplement the required catalyst-bed energy:

$$2CH_4 + 2NH_3 \rightarrow 2HCN + 6H_2O$$

A platinum-rhodium alloy is used as a catalyst at 1,100°C (2,010°F). Approximately equal amounts of ammonia and methane with 75 vol% air are introduced to the preheated reactor. The catalyst has several layers of wire gauze with a special mesh size (approximately 100 mesh).

On the other hand, the Degussa process involves the reaction of ammonia with methane in absence of air using a platinum, aluminum-ruthenium alloy as a catalyst at approximately 1,200°C (2,190°F). The reaction produces hydrogen cyanide and hydrogen, and the yield is over 90%:

$$CH_4 + NH_3 \rightarrow HCN + 3H_2$$

Hydrogen cyanide may also be produced by the reaction of ammonia and methanol in presence of oxygen:

$$NH_3 + CH_3OH + O_2 \rightarrow HCN + 3H_2O$$

Hydrogen cyanide is a reactant in the production of acrylonitrile, methyl methacrylates (from acetone), adiponitrile, and sodium cyanide. It is also used to make oxamide, a long-lived fertilizer that releases nitrogen steadily over the vegetation period. Oxamide is produced by the reaction of hydrogen cyanide with water and oxygen using a copper nitrate catalyst at about 70°C (158°F) and atmospheric pressure:

6.2.3.4 Chloromethane Derivatives

The ease with which chlorine can be introduced into the molecules of all the hydrocarbon types present in petroleum has resulted in the commercial production of a number of widely used compounds.

With saturated hydrocarbon derivatives the reactions are predominantly substitution of hydrogen by chloride and are strongly exothermic, difficult to control, and inclined to become explosively violent:

$$RH + Cl_2 \rightarrow RCl + HCl$$

Moderately high temperatures are used, about 250°C–300°C (480°F–570°F) for the thermal chlorination of methane, but as the molecular weight of the paraffin increases the temperature may generally be lowered. A mixture of chlorinated derivatives is always obtained, and many variables, such as choice of catalyst, dilution of inert gases, and presence of other chlorinating agents (antimony pentachloride, sulfuryl chloride, and phosgene), have been tried in an effort to direct the path of the reaction.

Methane yields four compounds upon chlorination in the presence of heat or light:

$$CH_4 + Cl_2 \rightarrow CH_3Cl, CH_2Cl_2, CHCl_3, CCl_4.$$

These compounds, known as chloromethane or methyl chloride, dichloromethane or methylene chloride, trichloromethane or chloroform, and tetrachloromethane or carbon tetrachloride, are used as solvents or in the production of chlorinated materials.

The successive substitution of methane hydrogens with chlorine produces a mixture of four chloromethanes: (i) Methyl chloride, CH_3Cl, also known as monochloromethane, (ii) methylene dichloride, CH_2Cl_2, also known as dichloromethane, (iii) chloroform, $CHCl_3$, also known as trichloromethane, and (iv) carbon tetrachloride, CCl_4, also known as tetrachloromethane. Each of these four compounds has many industrial applications.

Methane is the most difficult alkane to chlorinate. The reaction is initiated by chlorine free radicals obtained via the application of heat (thermal) or light (hv). Thermal chlorination (more widely used industrially) occurs at approximately 350°C–370°C (660°F–700°F) and atmospheric pressure. A typical product distribution for a feedstock (methane-chlorine) ratio of 1.7 v/v is: methyl chloride 58.7% v/v, methylene dichloride 29.3% v/v, chloroform 9.7% v/v, and carbon tetrachloride 2.3% v/v.

The first step in the process involves the breaking of the chlorine-chlorine bond which forms two chlorine (Cl•) free radicals after which the chlorine free radical reacts with methane to form a methyl free radical (CH_3•) and hydrogen chloride. The methyl radical then reacts in a subsequent step with a chlorine molecule, forming methyl chloride and a chlorine radical:

$$Cl• + CH_4 \rightarrow CH_3 + HCl$$

$$CH_3• + Cl_2 \rightarrow CH_3Cl + Cl•$$

The freshly generated chlorine radical Cl atom either attacks another methane molecule and repeats the above reaction, or it reacts with a methyl chloride molecule to form a chloromethyl free radical (•CH_2Cl, in which the free electron resides on the carbon atom) and hydrogen chloride:

$$Cl• + CH_3Cl \rightarrow •CCH_2Cl$$

The chloromethyl free radical then attacks another chlorine molecule and produces dichloromethane along with a Cl atom:

$$•CH_2Cl + Cl_2 \rightarrow CH_2Cl_2 + Cl•$$

This formation of chlorine free radicals continues until all chlorine is consumed. Chloroform and carbon tetrachloride are formed in a similar way by reaction of dichloromethyl (•$CHCl_2$) and trichloromethyl (•CCl_3) free radicals with chlorine.

Product distribution among the products depends primarily on the mole ratio of the reactants in the feedstock. For example, the yield of methyl chloride (CH_3Cl) chloromethane could be increased to 80% v/v by increasing the methane-chlorine mole ratio to 10:1 at 450°C (840°F). If dichloromethane is the desired product, the methane-chlorine mole ratio is lowered and the monochloromethane recycled. Decreasing the methane-chlorine mole ratio generally increases polysubstitution and, hence, the yield of chloroform and carbon tetrachloride.

An alternative way to produce methyl chloride (monochloromethane) is the reaction of methanol with hydrogen chloride (HCl). Methyl chloride could be further chlorinated to give a mixture of chloromethanes (methylene dichloride, chloroform, and carbon tetrachloride).

The major use of methyl chloride is to produce silicon polymers. Other uses include the synthesis of tetramethyl lead as a gasoline octane booster, a methylating agent in methyl cellulose production, a solvent, and a refrigerant. Methylene dichloride has a wide variety of markets, for example, a paint remover as well as a degreasing solvent, a blowing agent for polyurethane foams, and a solvent for cellulose acetate. Chloroform is mainly used to produce chlorodifluoromethane (also known as Fluorocarbon 22) by the reaction with hydrogen fluoride:

$$CHCl_3 + 2HF \rightarrow CHCl_2F_2 + 2HCl$$

This compound is used as a refrigerant and as an aerosol propellant. It is also used to synthesize tetrafluoroethylene, which is polymerized to a heat-resistant polymer (Teflon):

$$CHCl_2F_2 \rightarrow CF_2=CF_2 + 2HCl$$

Carbon tetrachloride is used to produce chlorofluorocarbons (CFCs), such as trichlorofluoromethane (CCl_3F) and dichlorodifluoromethane (CCl_2F_2)) by the reaction with hydrogen fluoride using an antimony pentachloride ($SbCl_5$) catalyst:

$$CCl_4 + HF \rightarrow CCl_3F + 2HCl$$

$$CCl_4 + 2HF \rightarrow CCl_2F_2 + 2HCl$$

The product mixture is composed of trichlorofluoromethane (Freon-11) and dichlorodifluoromethane (Freon-12). These compounds are used as aerosols and as refrigerants. However, because of the depleting effect of chlorofluorocarbons on the ozone layer, the production of these compounds has been reduced considerably.

6.2.3.5 Synthesis Gas

Synthesis gas may be produced from a variety of feedstocks, such as from natural gas (CH_4) by the steam reforming process. The first step in the process is to ensure that the methane feedstock is free from hydrogen sulfide. The purified gas is then mixed with steam and introduced to the first reactor (primary reformer). The reactor is constructed from vertical stainless steel tubes lined in a refractory furnace. The steam to natural gas ratio varies from 4 to 5 depending on natural gas composition (natural gas may contain ethane and heavier hydrocarbon derivatives) and the pressure used.

A promoted nickel-type catalyst contained in the reactor tubes is used at temperature and pressure ranges of 700°C–800°C (1,290°F–1,470°F) and 450–750 psi, respectively. The reforming reaction is equilibrium limited. It is favored at high temperatures, low pressures, and a high steam to carbon ratio. These conditions minimize methane slip at the reformer outlet and yield an equilibrium mixture that is rich in hydrogen.

The product gas from the primary reformer is a mixture of hydrogen (H_2), carbon monoxide (CO), carbon dioxide (CO_2), unreacted methane (CH_4), and steam (H_2O). The predominant process reactions are:

$$CH_4 + H_2O \rightarrow CO + 3H_2$$

$$CH_4 + 2H_2O \rightarrow CO_2 + 4H_2$$

For the production of methanol, this mixture could be used directly with no further treatment except adjusting the hydrogen/(CO + CO$_2$) ratio to approximately 2:1. For the production of hydrogen for ammonia synthesis, however, further treatment steps are needed. First, the required amount of nitrogen for ammonia must be obtained from atmospheric air. This is achieved by partially oxidizing unreacted methane in the exit gas mixture from the first reactor in another reactor (secondary reforming).

The main reaction occurring in the secondary reformer is the partial oxidation of methane with a limited amount of air. The product is a mixture of hydrogen, carbon dioxide, carbon monoxide, plus nitrogen, which does not react under these conditions:

$$2CH_4 + O_2 \rightarrow 2CO + 2H_2$$

The reactor temperature can reach over 900°C (1,650°F) in the secondary reformer due to the exothermic reaction heat.

The second step after secondary reforming is removing carbon monoxide, which poisons the catalyst used for ammonia synthesis. This is done in three further steps: (i) shift conversion, (ii) carbon dioxide removal, and (iii) methanation of the remaining carbon monoxide and carbon dioxide.

In the *shift converter*, carbon monoxide is reacted with steam to give carbon dioxide and hydrogen. The feed to the shift converter contains large amounts of carbon monoxide which should be oxidized. An iron catalyst promoted with chromium oxide is used at a temperature range of 425°C–500°C (795°F–930°F) to enhance the oxidation.

$$CO + H_2O \rightarrow CO_2 + H_2$$

Exit gases from the shift conversion are treated to remove carbon dioxide. This may be done by absorbing carbon dioxide in a physical or chemical absorption solvent or by adsorbing it using a special type of molecular sieves. Carbon dioxide, recovered from the treatment agent as a byproduct, is mainly used with ammonia to produce urea. The product is a pure hydrogen gas containing small amounts of carbon monoxide and carbon dioxide, which are further removed by methanation.

Catalytic *methanation* is the reverse of the steam reforming reaction. Hydrogen reacts with carbon monoxide and carbon dioxide, converting them to methane. Methanation reactions are exothermic, and methane yield is favored at lower temperatures:

$$3H_2 + CO \rightarrow CH_4 + H_2O$$

$$4H_2 + CO_2 \rightarrow CH_4 + 2H_2O$$

The forward reactions are also favored at higher pressures. However, the space velocity becomes high with increased pressures, and contact time becomes shorter, decreasing the yield. The actual process conditions of pressure, temperature, and space velocity are practically a compromise of several factors. Raney nickel is the preferred catalyst. Typical methanation reactor operating conditions are 200°C–300°C (390°F–570°F) and approximately 150 psi. The product is a gas mixture of hydrogen and nitrogen having an approximate ratio of 3:1 for ammonia production.

Many chemicals are produced from synthesis gas as a consequence of the high reactivity associated with hydrogen and carbon monoxide gases, the two constituents of synthesis gas. The reactivity of this mixture was demonstrated during World War II, when it was used to produce alternative hydrocarbon fuels using Fischer–Tropsch technology (Chapter 10).

Synthesis gas is also an important building block for aldehydes from olefin derivatives. The catalytic hydroformylation reaction (Oxo reaction) is used with many olefin derivatives to produce aldehydes and alcohols of commercial importance. The two major chemicals based on synthesis gas are ammonia and methanol. Each compound is a precursor for many other chemicals. From ammonia, urea, nitric acid, hydrazine, acrylonitrile, methylamines, and many other minor chemicals are produced. Each of these chemicals is also a precursor to many other chemicals.

Methanol, the second major product from synthesis gas, is a unique compound of high chemical reactivity as well as good fuel properties. It is a building block for many reactive compounds such as formaldehyde, acetic acid, and methylamine. It also offers an alternative way to produce hydrocarbon derivatives in the gasoline range (Mobil to gasoline process, also called the MTG process).

In the Mobil to gasoline process, methanol is the feedstock for the production of gasoline, which represents a competing technology to the traditional Fischer–Tropsch process. Instead of the traditional Fischer–Tropsch technology to convert syngas to liquids to be further refined into end products, such as gasoline, the Mobil to gasoline process follows a methanol synthesis unit with a methanol to gasoline synthesis process that yields gasoline very close to the final fuel specifications, requiring minimal end processing. This process may also prove to be a competitive source for producing light olefin derivatives in the future.

Synthesis gas conversion processes are significantly more advanced in development than direct or other two-step methane conversion schemes. Indirect gas conversion processes are currently practiced for methanol synthesis and for hydrocarbon formation via the Fischer–Tropsch synthesis. Diesel-range hydrocarbon derivatives (via Fischer–Tropsch synthesis) and gasoline (via Mobil to gasoline processes) can be produced. These indirect processes have continued to evolve as continuous improvements have come about from advances in synthesis gas generation, from the design and deployment of three-phase bubble columns for synthesis gas conversion, and from the development of improved catalytic materials for the selective synthesis of paraffin derivatives, intermediate size α-olefin derivatives, and higher alcohols. Small modular gas conversion plants using catalytic partial oxidation in monolith reactors and carbon monoxide hydrogenation in bubble columns may also create future opportunities for combining hydrogen and power generation with the synthesis of commodity petrochemicals and even of liquid fuels.

Selective pathways for ethylene and propylene from synthesis gas remain unavailable, because Flory-type chain growth kinetics leads to broad carbon number distributions. The most promising approach uses acid-catalyzed chain growth reactions of methanol within shape-selective channels in pentasil zeolites and silico-alumino-phosphate microporous materials in a three-step process requiring synthesis gas generation, methanol synthesis, and methanol-to-olefin derivatives or methanol-to-gasoline (MTG) conversion. The latter must be carried out in fluid bed or moving bed reactors because of the need for frequent regeneration. Currently available technologies use silico-aluminophosphate materials or modified pentasil zeolites in order to provide optimum shape-selective environments for the synthesis of light olefin derivatives. Intermediate range α-olefin derivatives (C_5–C_{15}) can be produced with modest selectivity from CO and H_2 on promoted iron-based catalysts. The valuable mid-range paraffin derivatives and the smaller olefin derivatives formed as byproducts are much more valuable than the paraffin derivatives formed and useful only as fuels.

Higher alcohol synthesis is also restricted to broad product distributions governed by stochastic chain growth kinetics. Recently, bifunctional catalysts consisting of metal sites for hydrogenation reactions and basic sites for alcohol coupling steps have led to high selectivity to branched alcohols. Chain growth appears to be restricted to C_4 and C_5 alcohols by the chemical constraints of base-catalyzed aldol condensation reactions. High 2-methyl-1-butanol yields (200–300 g/kg-cat-h) have been recently reported on Pd-based bifunctional catalysts, but at very high pressures (>200 bar) and temperatures (400°C–450°C, 750°F–840°F). Dimethyl ether (DME) isomerization and aldol condensation of olefin derivatives with methanol using acid-based bifunctional catalysts provide alternate but unexplored pathways to overcome the C_1–C_2 conversion bottleneck during CO hydrogenation to form higher alcohols.

6.2.3.6 Urea

The major end use of ammonia is the fertilizer field for the production of urea, ammonium nitrate and ammonium phosphate, and sulfate. Anhydrous ammonia could be directly applied to the soil as a fertilizer. Urea is gaining wide acceptance as a slow-acting fertilizer.

Ammonia is the precursor for many other chemicals such as nitric acid, hydrazine, acrylonitrile, and hexamethylenediamine. Ammonia, having three hydrogen atoms per molecule, may be viewed as an energy source. It has been proposed that anhydrous liquid ammonia may be used as a clean fuel for the automotive industry. The oxidation reaction (the combustion reaction) is:

$$4NH_3 + 3O_2 \rightarrow 2N_2 + 6H_2O$$

Compared with hydrogen, anhydrous ammonia is more manageable and can be stored in iron or steel containers and could be transported commercially via pipeline, railroad tanker cars, and highway tanker trucks.

Only nitrogen and water are produced. However, many factors must be considered such as the coproduction of nitrogen oxides, the economics related to retrofitting of auto engines, etc. The following describes the important chemicals based on ammonia.

The highest fixed nitrogen-containing fertilizer 46.7% w/w, urea is a white solid that is soluble in water and alcohol. It is usually sold in the form of crystals, prills, flakes, or granules. Urea is an active compound that reacts with many reagents. It forms adducts and clathrates with many substances such as phenol and salicylic acid. By reacting with formaldehyde, it produces an important commercial polymer (urea formaldehyde resins) that is used as glue for particle board and plywood.

The technical production of urea is based on the reaction of ammonia with carbon dioxide. The reaction occurs in two steps: ammonium carbamate is formed first, followed by a decomposition step of the carbamate to urea and water. The first reaction is exothermic and the equilibrium is favored at lower temperatures and higher pressures. Higher operating pressures are also desirable for the separation absorption step that results in a higher carbamate solution concentration. A higher ammonia ratio than stoichiometric is used to compensate for the ammonia that dissolves in the melt. The reactor temperature ranges between 170°C and 220°C (340°F–395°F) at a pressure of about 3,000 psi. The second reaction represents the decomposition of the carbamate. The reaction conditions are 200°C (390°F) and 450 psi:

$$2NH_3 + CO_2 \rightarrow H_2NCOONH_4$$

$$H_2NCOONH_4 \rightarrow H_2NCONH_2 + H_2O$$

Decomposition in presence of excess ammonia limits corrosion problems and inhibits the decomposition of the carbamate to ammonia and carbon dioxide. The urea solution leaving the carbamate decomposer is expanded by heating at low pressures and ammonia recycled. The resultant solution is further concentrated to a melt, which is then prilled by passing it through special sprays in an air stream.

The major use of urea is the fertilizer field. About 10% of urea is used for the production of adhesives and plastics (urea formaldehyde and melamine formaldehyde resins). Animal feed accounts for about 5% of the urea produced. Urea possesses a unique property of forming adducts with n-paraffin derivatives. This is used in separating C_{12}–C_{14} n-paraffin derivatives from kerosene for detergent production.

6.2.3.7 Methyl Alcohol

Methyl alcohol (methanol, CH_3OH) is the first member of the aliphatic alcohol family. Methanol was originally produced by the destructive distillation of wood (wood alcohol) for charcoal production. Currently, it is mainly produced from synthesis gas.

As a chemical compound, methanol is highly polar, and hydrogen bonding is evidenced by its relatively high-boiling temperature (65°C, 149°F), high heat of vaporization, and low volatility. Due to the high oxygen content of methanol (50% w/w), it is being considered as a gasoline blending compound to reduce carbon monoxide and hydrocarbon emissions in automobile exhaust gases. It was also tested for blending with gasolines due to the high-octane number (RON = 112). During the late 1970s and early 1980s, many experiments tested the possible use of pure (straight) methanol as an alternative fuel for gasoline cars. Several problems were encountered, however, in its use as a fuel (such as starting a cold engine) due to its high vaporization heat (heat of vaporization is 3.7 times that of gasoline), its lower heating value, which is approximately half that of gasoline, and its corrosive properties. However, methanol is a potential fuel for gas turbines because it burns smoothly and has exceptionally low nitrogen oxide emission levels.

Due to the high reactivity of methanol, many chemicals could be derived from it. For example, it could be oxidized to formaldehyde, an important chemical building block, carbonylated to acetic acid, and dehydrated and polymerized to hydrocarbon derivatives in the gasoline range (methanol-to-gasoline) process. Much of the current work is centered on the use of shape-selective catalysts to convert methanol to light olefin derivatives as a possible future source of ethylene and propylene.

Methanol is produced by the catalytic reaction of carbon monoxide and hydrogen (synthesis gas). Because the ratio of CO:H_2 in synthesis gas from natural gas is approximately 1:3, and the stoichiometric ratio required for methanol synthesis is 1:2, carbon dioxide is added to reduce the surplus hydrogen. An energy-efficient alternative to adjusting the carbon monoxide/hydrogen ratio is to combine the steam reforming process with autothermal reforming (combined reforming) so that the amount of natural gas fed is that required to produce a synthesis gas with a stoichiometric ratio of approximately 1:2.05. If an autothermal reforming step is added, pure oxygen should be used. (This is a major difference between secondary reforming in case of ammonia production, where air is used to supply the needed nitrogen.)

An added advantage of combined reforming is the decrease in the emissions of nitrogen oxides (NOx). The following reactions are representative for methanol synthesis:

$$CO + 2H_2 \rightarrow CH_3OH$$

$$CO_2 + 3H_2 \rightarrow CH_3OH + H_2O$$

Old processes use a zinc-chromium oxide catalyst at a high-pressure range of approximately 4,000–6,200 psi for methanol production. A low-pressure process has been developed by ICI operating at about 700 psi using an active copper-based catalyst at 240°C (430°F). The synthesis reaction occurs over a bed of heterogeneous catalyst arranged in either sequential adiabatic beds or placed within heat transfer tubes. The reaction is limited by equilibrium, and methanol concentration at the converter's exit rarely exceeds 7%. The converter effluent is cooled to 40°C (104°F) to condense product methanol, and the unreacted gases are recycled. Crude methanol from the separator contains water and low levels of byproducts, which are removed using a two-column distillation system.

As a methylating agent, it is used with many organic acids to produce the methyl esters such as methyl acrylate, methyl methacrylate, methyl acetate, and methyl terephthalate. Methanol is also used to produce dimethyl carbonate (DMC) and methyl-t-butyl ether, an important gasoline additive. It is also used to produce synthetic gasoline using a shape-selective catalyst (methanol-to-gasoline process). Olefin derivatives from methanol may be a future route for ethylene and propylene in competition with steam cracking of hydrocarbon derivatives. The use of methanol in fuel cells is being investigated. Fuel cells are theoretically capable of converting the free energy of oxidation of a fuel into electrical work. In one type of fuel cells, the cathode is made of vanadium which catalyzes the reduction of oxygen, while the anode is iron (III) which oxidizes methane to CO_2 and iron (II) is formed in aqueous sulfuric acid (H_2SO_4).

Commercial methanol synthesis processes use indirect routes based on synthesis gas intermediates. These processes lead to much higher methanol yields (25%–30% yield per pass) and >99% CH_3OH selectivity. Synthesis gas generation requires high temperatures and large capital investments but recent process improvements that combine partial oxidation and steam reforming in a nearly thermoneutral process have led to practical thermal efficiencies (71%–72%) for the overall methanol synthesis process very close to theoretical thermal efficiency values (84.2%). Recent development in the use of adiabatic monolith reactors for partial methane oxidation to synthesis gas have provided a novel and useful approach, especially for small-scale methanol synthesis applications that cannot exploit the beneficial economies of scale of steam reforming and autothermal reforming processes. The chemical and structural integrity of noble metal catalytic coatings in these monolith reactors and the mixing and handling of explosive CH_4/O_2 mixtures remain challenging issues in the design of the short contact time reactors for the production of synthesis gas. Recent advances in the design of ceramic membranes for oxygen and hydrogen transport may become useful in decreasing or eliminating air purification costs and in driving endothermic steam reforming reactions to higher conversions or lower operating temperatures. Such ceramic membranes have advanced beyond their status as a laboratory curiosity and into developmental consortia, but their reliable practical implementation will require additional advances in the synthesis and stabilization of thin metal oxide films within novel reactor geometries as well as the development of faster proton conductors for the case of hydrogen separation schemes.

Methanol can also be formed via other indirect routes, such as via processes involving the formation of methyl bisulfate on Hg complexes followed by its conversion to methanol via hydrolysis, and by the regeneration of the sulfuric acid oxidant by SO_2 oxidation. Methyl bisulfate yields can reach 70%–80%, because of the low reactivity of methyl bisulfate relative to methanol and even methane. Turnover rates are very low in the temperature tolerated by these homogeneous catalysts and Hg organometallic complexes and sulfuric acid are very toxic and difficult to handle and regenerate. More recent studies have increased the stability of these catalysts by introducing more stable ligands and eliminated Hg-based materials with Pt-based homogeneous catalysts. This approach involves a new type of protected intermediate (methyl bisulfate) and leads to higher methanol yields than in commercial routes based on synthesis gas. It is, however, neither a direct route to methanol and involves three steps, including a very costly oxidant regeneration step and a deprotection step limited by thermodynamics.

Oxychlorination provides another indirect route to methanol and to hydrocarbon derivatives via acid-catalyzed hydration or oligomerization of methyl chloride. It involves the use and the costly regeneration of Cl_2 and it requires corrosion-resistant vessels and significant temperature cycling of process streams. CH_3Cl yields of 25%–30% have been achieved with yield losses predominately to CO_2 and CH_2Cl_2. An additional chemical hurdle in these processes is the low selectivity to monochlorinated products. Higher yields will require more selective monochlorination and lower combustion rates of these desired intermediates. These improvements appear unlikely because of the kinetic instability of CH_3Cl relative to CH_4 and of the thermodynamic stability of the sequential CO_2 and CH_2Cl_2 products.

Once methanol is produced, it can be converted to dimethyl ether (CH_3OCH_3):

$$2CH_3OH \rightarrow CH_3OCH_3 + H_2O$$

The dimethyl ether can then be converted into olefin derivatives through olefin synthesis reactions. In the process, a fluidized or fixed bed reactor is used and the temperature is maintained below 600°C (1,110°F)—compared to a temperature regime of 750°C–900°C (1,380°F–1,650°F) in steam cracking—and a pressure in the order of 15–45 psi. As in the steam cracking process, high severity (high temperature, low pressure, and short resident time) favors ethylene production over propylene production. In this process step, dehydration catalysts are used. There are basically two major catalyst families, ZSM (zeolite silicon microspores doped with metal ions such as Mn, Sb, Mg, or Ba) and silico-alumino-phosphate molecular sieve doped with metal ions such as Mn, Ni, or Co. The main differences between ZSM and silico-alumino-phosphate catalysts are pore sizes and acidity, which are the main causes for shape selectivity. ZSM catalysts have shape selectivity favoring propylene and heavy hydrocarbon derivatives over ethylene. Also, they reportedly lead to less formation of aromatic coke and carbon oxides than silico-alumino-phosphate catalysts and, moreover, silico-alumino-phosphate catalysts have a shape selectivity favoring light olefin derivatives over high-boiling hydrocarbon derivatives.

The subsequent cooling, recovery, and separation processes are quite similar to those of steam cracking. One difference is that after the recovery and separation of C_4/C_5, the olefin upgrading (sometimes referred to as olefin conversion) process converts C_4/C_5 to ethylene and propylene. The composition and yield of final products depend on catalysts, reactor configurations, and severity (such as temperature, residence time). Polymer-grade light olefin derivatives of high purity (97%–99%) are the major products.

6.2.3.8 Formaldehyde

The main industrial route for producing formaldehyde (HCHO) is the catalyzed air oxidation of methanol.

$$CH_3OH + O_2 \rightarrow HCHO + H_2O$$

A silver-gauze catalyst is still used in some older processes that operate at a relatively higher temperature (about 500°C). Some processes use an iron-molybdenum oxide catalyst and chromium oxide or cobalt oxide can be used to dope the catalyst. The oxidation reaction is exothermic and occurs at approximately 400°C–425°C (750°F–800°F) and atmospheric pressure. Excess air is used to keep the methanol air ratio below the explosion limits.

One-step homogeneous oxidation of methane at high pressures and temperatures has led to the synthesis of methanol with yields in the order of 4%. The rapid subsequent combustion reactions of methanol to form CO_2 limit conversions in practice, while the explosive nature of the required reactant mixtures creates engineering challenges in the design of mixing schemes and pressure vessels. Solid catalysts have not led to yield improvements, because surface sites activate C-H bonds in both methane and methanol. Staged O_2 introduction using multiple injectors or dense oxygen conducting membranes are also unlikely to increase yields, because the rates of activation of C-H bonds in CH_3OH and CH_4 appear to depend similarly on O_2 concentrations. Continuous extraction of methanol can prevent combustion, but requires its selective absorption, adsorption, or permeation above ambient temperatures, none of which are currently possible, as well as a low conversion per pass. The introduction of sites for methanol conversion to hydrocarbon derivatives into homogeneous methane to methanol oxidation reactors may provide a less costly product separation scheme, as well as the protection of activated CH_4 (as methanol) as aromatic molecules that are considerably less reactive than methanol.

Formaldehyde is the simplest and most reactive aldehyde. Condensation polymerization of formaldehyde with phenol, urea, or melamine produces phenol-formaldehyde, urea formaldehyde, and melamine formaldehyde resins, respectively. These are important glues used in producing particle board and plywood.

A catalyst system based on SiO_2 prepared by wet impregnation method with vanadium and molybdenum were prepared and tested for (biogas) methane partial oxidation to formaldehyde. In general, a vanadium pentoxide/silica (V_2O_5/SiO_2) is a catalyst for the partial oxidation of biogas methane to formaldehyde (Singh et al., 2010). As methane conversion increases, formaldehyde selectivity also increases up to a reaction temperature in the order of 600°C (1,110°F) and then decreases because of the decline in the selectivity for formaldehyde formation at high temperatures because of the decomposition of formaldehyde:

$$HCHO \rightarrow CO + H_2$$

The vanadium pentoxide catalyst exhibits a heterolytic oxygen exchange mechanism similar to molybdenum trioxide (MoO_3) (Singh et al., 2010).

Condensation of formaldehyde with acetaldehyde in presence of a strong alkali produces pentaerythritol, a polyhydric alcohol that is used for alkyd resin production:

$$4HCHO + CH_3CHO + NaOH \rightarrow \underset{\text{Pentaerythritol}}{C(CH_2OH)_4} + \underset{\text{sodium formate}}{HCOONa}$$

Formaldehyde reacts with ammonia and produces hexamethylene tetramine (hexamine) which is a cross-linking agent for phenolic resins:

$$6HCHO + 4NH_3 \rightarrow (CH_2)_6 N_4 + 6H_2O$$

Methyl chloride is produced by the vapor-phase reaction of methanol and hydrogen chloride:

$$CH_3OH + HCl \rightarrow CH_3Cl + H_2O$$

Many catalysts are used to effect the reaction, such as zinc chloride on pumice, cuprous chloride, and ignited alumina gel. The reaction conditions are 350°C (660°F) at nearly atmospheric pressure. Methyl chloride may also be produced directly from methane with other chloromethanes.

Zinc chloride is also a catalyst for a liquid-phase process using concentrated hydrochloric acid at 100°C–150°C (212°F–300°F). Hydrochloric acid may be generated *in situ* by reacting sodium chloride with sulfuric acid. However, methyl chloride from methanol may be further chlorinated to produce dichloromethane, chloroform, and carbon tetrachloride.

Methyl chloride is primarily an intermediate for the production of other chemicals. Other uses of methyl chloride have been mentioned with chloromethane derivatives.

The carbonylation of methanol is currently one of the major routes for acetic acid production. The basic liquid-phase process developed by BASF uses a cobalt catalyst at 250°C and a high pressure of about 70 atm. The newer process uses a rhodium complex catalyst in presence of methyl iodide (CH_3I), which acts as a promoter. The reaction occurs at 150°C (300°F) and atmospheric pressure:

$$CH_3OH + CO \rightarrow CH_3COOH$$

Acetic acid is also produced by the oxidation of acetaldehyde and the oxidation of *n*-butane. However, acetic acid from the carbonylation route has an advantage over the other commercial processes because both methanol and carbon monoxide come from synthesis gas, and the process conditions are quite mild.

The main use of acetic acid is to produce vinyl acetate (44%), followed by acetic acid esters (13%) and acetic anhydride (12%). Vinyl acetate is used for the production of adhesives, film, paper, and textiles. Acetic acid is also used to produce pharmaceuticals, dyes, and insecticides. Chloroacetic acid (from acetic acid) is a reactive intermediate used to manufacture many chemicals such as glycine and carboxymethyl cellulose.

Dimethyl carbonate [$CO(OCH_3)_2$] is a colorless liquid with a pleasant odor. It is soluble in most organic solvents but insoluble in water. The classical synthesis of dimethyl carbonate is the reaction of methanol with phosgene. Because phosgene is toxic, a non-phosgene route may be preferred. The new route reacts methanol with urea over a tin catalyst. However, the yield is low. Using electron donor solvents such as trimethylene glycol dimethyl ether and continually distilling off the product increases the yield.

$$H_2NCONH_2 + 2CH_3OH \rightarrow CH_3OC(=O)OCH_3 + 2NH_3$$

Dimethyl carbonate is used as a specialty solvent. It could be used as an oxygenate to replace methyl tertiary butyl ether (MTBE). It has almost three times the oxygen content as methyl tertiary butyl ether. It has also a high-octane rating. However, it must be evaluated in regard to economics and toxicity.

Methylamine derivatives can be synthesized by alkylating ammonia with methyl halides or with methyl alcohol. The reaction with methanol usually occurs at approximately 500°C (930°F) and 300 psi in the presence of an aluminum silicate or phosphate catalyst. The alkylation does not stop at the monomethylamine stage, because the produced amine is a better nucleophile than ammonia. The product distribution at equilibrium is: monomethylamine (43%), dimethylamine (24%), and trimethylamine (33%):

$$CH_3OH + NH_3 \rightarrow CH_3NH_2 + H_2O$$

$$CH_3OH + CH_3NH_2 \rightarrow (CH_3)_2 NH + H_2O$$

$$CH_3OH + (CH_3)_2 NH \rightarrow (CH_3)_2 NH + H_2O$$

To improve the yield of monomethylamine and dimethylamine, a shape-selective catalyst has been tried. Carbogenic sieves are microporous materials (similar to zeolites), which have catalytic as well as shape-selective properties. Combining the amorphous aluminum silicate catalyst (used for producing the amines) with carbogenic sieves gave higher yields of the more valuable monomethylamine and dimethylamine.

Dimethylamine is the most widely used of the three amines. Excess methanol and recycling monomethylamine increases the yield of dimethylamine. The main use of dimethylamine is the synthesis of dimethylformamide and dimethylacetamide, which are solvents for acrylic and polyurethane fibers. Monoethylamine is used in the synthesis of Sevin, an important insecticide. Trimethylamine has only one major use, the synthesis of choline, a high-energy additive for poultry feed.

Methanol may have a more important role as a basic building block in the future because of the multiple sources of synthesis gas. The reaction of methanol over a ZSM-5 catalyst could be considered as dehydration, oligomerization reaction. It may be simply represented as:

$$nCH_3OH \rightarrow H(CH_2)_n H + nH_2O$$

In this equation, $(CH_2)_n$ represents the hydrocarbon derivatives (paraffin derivatives + olefin derivatives + aromatic derivatives). The hydrocarbon derivatives obtained are in the gasoline range.

Converting methanol to hydrocarbon derivatives is not as simple as it looks from the previous equation. Many reaction mechanisms have been proposed, and most of them are centered on the intermediate formation of dimethyl ether followed by olefin formation. Olefin derivatives are thought to be the precursors for paraffin derivatives and aromatic derivatives. The product distribution is influenced by the catalyst properties as well as the various reaction parameters. The catalyst activity and selectivity are functions of acidity, crystalline size, silica/alumina ratio, and even the synthetic procedure.

The important property of ZSM-5 and similar zeolites is the intercrystalline catalyst sites, which allow one type of reactant molecule to diffuse, while denying diffusion to others. This property, which is based on the shape and size of the reactant molecules as well as the pore sizes of the catalyst, is called shape selectivity. Chen and Garwood document investigations regarding the various aspects of ZSM-5 shape selectivity in relation to its intercrystalline structure and pore structure.

In general, two approaches have been found that enhance selectivity toward light olefin formation. One approach is to use catalysts with smaller pore sizes such as crionite, chabazite, and zeolite T. The other approach is to modify ZSM-5 and similar catalysts by reducing the pore size of the catalyst through incorporation of various substances in the zeolite channels and/or by lowering the acidity by decreasing the alumina (Al_2O_3)/silica (SiO_3) ratio. This latter approach is used to stop the reaction at the olefin stage, thus limiting the steps up to the formation of olefin derivatives and suppressing the formation of higher hydrocarbon derivatives.

6.2.3.9 Aldehyde Derivatives

Hydroformylation of olefin derivatives (Oxo reaction) produces aldehydes with one more carbon than the reacting olefin. For example, when ethylene is used, propionaldehyde is produced. This reaction is especially important for the production of higher aldehydes that are further hydrogenated to the corresponding alcohols. The reaction is catalyzed with cobalt or rhodium complexes. Olefin derivatives with terminal double bonds are more reactive and produce aldehydes which are hydrogenated to the corresponding primary alcohols. With olefin derivatives other than ethylene, the hydroformylation reaction mainly produces a straight chain aldehyde with variable amounts of branched chain aldehydes. The reaction could be generally represented as:

$$2RCH{=}CH_2 + 2H_2 + 2CO \rightarrow RCH_2CH_2CHO + RCH(CH_3)CHO$$

The largest commercial process is the hydroformylation of propene, which yields n-butyraldehyde and iso-butyraldehyde. The n-Butyraldehyde (n-butanal) is either hydrogenated to n-butanol or transformed to 2-ethyl hexanol via aldol condensation and subsequent hydrogenation. The 2-ethyl hexanol is an important plasticizer for polyvinyl chloride (PVC). Other olefin derivatives applied in the hydroformylation process with subsequent hydrogenation are propylene trimer and tetramer for the production of decyl and tridecyl alcohols, respectively, and C_7 olefin derivatives (from copolymers of C_3 and C_4 olefin derivatives) for iso-decyl alcohol production.

Several commercial processes are currently operative. Some use rhodium catalyst complex incorporating phosphine ligands at relatively lower temperatures and pressures and produce less branched aldehydes. Older processes use a cobalt carbonyl complex at higher pressures and temperatures and produce a higher ratio of the branched aldehydes. The hydroformylation reaction using phosphine ligands occurs in an aqueous medium. A higher catalyst activity is anticipated in aqueous media than in hydrocarbon derivatives. Selectivity is also higher. Having more than one phase allows for complete separation of the catalyst and the products.

6.2.3.10 Ethylene Glycol

Ethylene glycol could be produced directly from synthesis gas using a rhodium catalyst at 230°C (445°F) at very high pressure (50,000 psi):

$$3H_2 + 2CO \rightarrow HOCH_2CH_2OH$$

Other routes have been tried starting from formaldehyde or paraformaldehyde. One process reacts formaldehyde with carbon monoxide and hydrogen (hydroformylation) at approximately 110°C (230°F) and 4,000 psi using a rhodium triphenyl phosphine catalyst with the intermediate formation of glycolaldehyde. Glycolaldehyde is then reduced to ethylene glycol:

$$2CO + 2H_2 \rightarrow HOCH_2CHO$$

$$HOCH_2CHO + H_2 \rightarrow HOCH_2CH_2OH$$

In the DuPont process (the oldest syngas process to produce ethylene glycol), formaldehyde is reacted with carbon monoxide in the presence of a strong mineral acid. The intermediate is glycolic acid, which is esterified with methanol. The ester is then hydrogenated to ethylene glycol and methanol, which is recovered.

$$HCHO + CO + H_2O \rightarrow HOCH_2CO_2H$$

$$HOCH_2CO_2H + CH_3OH \rightarrow HOCH_2CO_2CH_3 + H_2O$$

$$HOCH_2CO_2CH_3 + 2H_2 \rightarrow HOCH_2CH_2OH + CH_3OH$$

6.2.3.11   Nitration

Hydrocarbon derivatives that are usually gaseous (including normal and isopentane) react smoothly in the vapor phase with nitric acid to give a mixture of nitro-compounds, but there are side reactions, mainly of oxidation. Only mononitro-derivatives are obtained with the lower paraffin derivatives as high temperatures, and they correspond to those expected if scission of a C-C and C-H bond occurs. Ethane, for example, yields nitromethane and nitroethane:

$$CH_4 + HNO_3 \rightarrow CH_3NO_2 + H_2O$$

On the other hand, more complex chemicals yield a more complex product mix—propane yields nitromethane, nitroethane, 1-nitropropane, and 2-nitropropane.

The nitro-derivatives of the lower paraffin derivatives are colorless and noncorrosive and are used as solvents or as starting materials in a variety of syntheses. For example, treatment with inorganic acids and water yields fatty acids (RCO_2H) and hydroxylamine (NH_2OH) salts and condensation with an aldehyde (RCH=O) yields nitro-alcohols [$RCH(NO_2)OH$].

6.2.3.12   Oxidation

The oxidation of hydrocarbon derivatives and hydrocarbon mixtures has received considerable attention, but the uncontrollable nature of the reaction and the mixed character of the products have made resolution of the reaction sequences extremely difficult. Therefore, it is not surprising that, except for the preparation of mixed products having specific properties, such as fatty acids, hydrocarbon derivatives higher than pentanes are not employed for oxidation because of the difficulty of isolating individual compounds.

Methane undergoes two useful reactions at 90°C (195°F) in the presence of iron oxide (Fe_3O_4) as a catalyst:

$$CH_4 + H_2O \rightarrow CO + 3H_2$$

$$CO + H_2O \rightarrow CO_2 + H_2$$

Alternatively, partial combustion of methane can be used to provide the required heat and steam. The carbon dioxide produced then reacts with methane at 900°C (1,650°F) in the presence of a nickel catalyst:

$$CH_4 + 2O_2 \rightarrow O_2 + 2H_2O$$

$$CO_2 + CH_4 \rightarrow 2CO + 2H_2$$

$$CH_4 + H_2O \rightarrow CO + 3H_2$$

Methanol (methyl alcohol, CH_3OH) is the second major product produced from methane. Synthetic methanol has virtually completely replaced methanol obtained from the distillation of wood, its original source material. One of the older trivial names used for methanol was wood alcohol. The synthesis reaction takes place at 350°C and 300 atm in the presence of ZnO as a catalyst:

$$2CH_4 + O_2 \rightarrow 2CH_3OH$$

An example of a methane-to-methanol is the Lurgi Mega-Methanol process in which methane is first fed into a pre-reforming reactor where it is partially reformed with steam to syngas (with a hydrogen-carbon dioxide ratio in the order of 3–5). Pre-reforming reduces coking in the subsequent steps. Unreformed methane is further converted to synthesis gas in the autothermal reforming reactor with oxygen as a reforming agent at approximately 1,000°C (1,830°F). Autothermal reforming has two stages: (i) a partial oxidative non-catalytic process in which methane is partially oxidized to produce syngas and (ii) a catalytic steam reforming process in which unconverted methane is further reformed into synthesis gas. After these two stages, the synthesis syngas is converted into raw methanol (not yet dewatered) through an exothermic synthesis process with a temperature range of 200°C–280°C (390°F–535°F).

Methanol is then oxidized by oxygen from air to formaldehyde (sometimes referred to as methanal):

$$2CH_3OH + O_2 \rightarrow 2CH_2O + 2H_2O$$

Formaldehyde is used to produce synthetic resins either alone or with phenol, urea, or melamine; other uses are minor.

By analogy to the reaction with oxygen, methane reacts with sulfur in the presence of a catalyst to give the carbon disulfide used in the rayon industry:

$$CH_4 + 4S(g) \rightarrow CS_2 + 2H_2S$$

The major non-petrochemical use of methane is in the production of hydrogen for use in the Haber synthesis of ammonia. Ammonia synthesis requires nitrogen, obtained from air, and hydrogen. The most common modern source of the hydrogen consumed in ammonia production, about 95% of it, is methane.

When propane and butane are oxidized in the vapor phase, without a catalyst, at 270°C–350°C (520°F–660°F) and at 50–3,000 psi, a wide variety of products is obtained, including C_1–C_4 acids, C_2–C_7 ketones, ethylene oxide, esters, formals, acetals, and others.

Cyclohexane is oxidized commercially and is somewhat selective in its reaction with air at 150°C–250°C (300°F–480°F) in the liquid phase in the presence of a catalyst, such as cobalt acetate. Cyclohexanol derivatives are the initial products, but prolonged oxidation produces adipic acid. On the other hand, oxidation of cyclohexane and methylcyclohexane over vanadium pentoxide at 450°C–500°C (840°F–930°F) affords maleic and glutaric acids.

6.2.3.13 Carboxylic Acids

The preparation of carboxylic acids from petroleum, particularly from paraffin wax, for esterification to fats or neutralization to form soaps has been the subject of a large number of investigations. Wax oxidation with air is comparatively slow at low temperature and normal pressure, very little reaction taking place at 110°C (230°F), with a wax melting at 55°C (130°F) after 280h. At higher temperatures the oxidation proceeds more readily; maximum yields of mixed alcohol and high molecular weight acids are formed at 110°C–140°C (230°F–285°F) at 60–150 psi; higher temperatures (140°C–160°C, 285°F–320°F) result in more acid formation.

6.2.3.14 Alkylation

Alkylation chemistry contributes to the efficient utilization of C_4 olefin derivatives generated in the cracking operations. Isobutane has been added to butenes (and other low-boiling olefin derivatives) to give a mixture of highly branched octane derivatives (e.g., heptane derivatives) by alkylation. The reaction is thermodynamically favored at low temperatures (<20°C), and thus very powerful acid catalysts are employed. Typically, sulfuric acid (85%–100%), anhydrous hydrogen fluoride, or a solid sulfonic acid is employed as the catalyst in these processes. The first step in the process is the formation of a carbocation by combination of an olefin with an acid proton:

$$(CH_3)_2 C=CH_2 + H^+ \rightarrow (CH_3)_3 C^+$$

Step 2 is the addition of the carbocation to a second molecule of olefin to form a dimer carbocation. The extensive branching of the saturated hydrocarbon results in high-octane. In practice, mixed butenes are employed (isobutylene, 1-butene, and 2-butene), and the product is a mixture of isomeric octanes that has an octane number of 92–94. With the phaseout of leaded additives in our motor gasoline pools, octane improvement is a major challenge for the refining industry. Alkylation is one answer.

6.2.3.15 Thermolysis

Although there are relatively unreactive organic molecules, paraffin hydrocarbon derivatives are known to undergo thermolysis when treated under high-temperature, low-pressure vapor-phase conditions. The cracking chemistry of petroleum constituents has been extensively studied (Albright et al., 1992). Cracking is the major process for generating ethylene and the other olefin derivatives that are the reactive building blocks of the petrochemical industry (Chenier, 2002; Al-Megren and Xiao, 2016). In addition to thermal cracking, other very important processes that generate sources of hydrocarbon raw materials for the petrochemical industry include catalytic reforming, alkylation, dealkylation, isomerization, and polymerization.

Cracking reactions involve the cleavage of carbon-carbon bonds with the resulting redistribution of hydrogen to produce smaller molecules. Thus, cracking of petroleum or petroleum fractions is a process by which larger molecules are converted into smaller, lower-boiling molecules. In addition, cracking generates two molecules from one, with one of the product molecules saturated (paraffin) and the other unsaturated (olefin).

At the high temperatures of refinery crackers (usually >500°C, 950°F), there is a thermodynamic driving force for the generation of more molecules from fewer molecules; that is, cracking is favored. Unfortunately, in the cracking process certain products interact with one another to produce products of increased molecular weight over that in the original feedstock. Thus, some products are taken off from the cracker as useful light products (olefin derivatives, gasoline, and others), but other products include heavier oil and coke.

$$\text{Paraffin wax} \rightarrow \underset{\text{alcohol}}{ROH} + \underset{\text{acid}}{RCO_2H}$$

Acids from formic (HCO_2H) to that with a 10-carbon atom chain [$CH_3(CH_2)_9CO_2H$] have been identified as products of the oxidation of paraffin wax. Substantial quantities of water-insoluble acids are also produced by the oxidation of paraffin wax, but apart from determination of the average molecular weight (ca. 250), very little has been done to identify individual numbers of the product mixture.

The pyrolysis of methane to form acetylene (alkyne) derivatives, olefin derivatives, and arenes is highly endothermic and requires very high temperatures and concurrent combustion of methane as a fuel in order to provide the enthalpy of reaction in a heat exchange furnace. At these high temperatures, homogeneous pathways lead to acetylene, polynuclear aromatic derivatives, and soot as the preferred products for both kinetics and thermodynamic reasons. Thermal efficiencies are low because of the extreme temperature cycling required for process streams and because of the rapid quenching protocols used to restrict chain growth and soot formation. The low pressures required by thermodynamics and the slow homogeneous reactions lead to large reactor vessels, which must be protected against carbon deposition and metal dusting corrosion by using coatings or specialized materials of construction. Recently, several approaches have addressed these limitations and they have led to significant control of methane pyrolysis selectivity. One approach involves the synthesis of benzene and ethylene with 30%–40% yields per pass using homogeneous pyrolysis reactors and the rapid thermal quenching of reaction products. Another improvement uses shape-selective catalysts based on molybdenum and tungsten species held within shape-selective channels in pentasil zeolites (H-ZSM5), where chain growth to form polynuclear aromatic derivatives is spatially restricted and the presence of CH_4 activation sites on the surface of carbide clusters leads to CH_4 pyrolysis reactions at much lower temperatures (700°C) than for homogeneous pathways. The addition of small amounts of CO_2 during CH_4 pyrolysis on these catalysts and the selective deactivation of acid sites on external zeolite surfaces have led to marked improvements in catalyst stability and to lower selectivity to hydrocarbon derivatives larger than naphthalene. The combination of such catalysts with hydrogen removal by ceramic membranes remains a promising but challenging approach to the direct conversion of methane to larger hydrocarbon derivatives.

A two-step cyclic process involving the deposition of CHx fragments from methane on metal surfaces at 200°C–300°C (390°F–570°F) and the coupling of such fragments during a subsequent hydrogenation cycle has led to very low yields of C_{2+} hydrocarbon derivatives. Product yields are constrained by the unfavorable thermodynamics of this overall process at low temperatures and by the selectivity losses to surface carbon, which also cause rapid deactivation and loss of methane reactants. A more promising and thermodynamically feasible cyclic strategy involves the formation of relatively pure H_2 streams using CH_4-H_2O cyclic processes on solids that can generate reactive carbon and H_2 during the CH_4 cycle and remove the carbon as CO while producing additional H_2 during subsequent contact with steam.

6.2.4 OXIDATIVE COUPLING

In the oxidative coupling process, methane (CH_4) and oxygen react over a catalyst to form water and a methyl radical ($CH_3\bullet$) (often referred to as *partial oxidation*). The methyl radicals combine to form a higher molecular weight alkane, mostly ethane (C_2H_6), which dehydrogenates into ethylene ($CH_2=CH_2$). Complete oxidation (rapid formation of carbon dioxide before the radicals link up to form ethane and ethylene) is an undesired reaction. The function of the catalyst is to control the oxidation so that reactions can be kept on the desired path and catalysts used are mostly oxides of alkali, alkaline earth, and other rare earth metals. Hydrogen and steam are sometimes added in order to reduce coking on catalysts.

The compression, separation, and recovery sections of these processes are similar to those of ethane steam cracking except for the sections for water/carbon dioxide removal and methanization. Ethylene-containing gas streams are compressed and water is condensed after which the gases pass through an acid gas removal system where carbon dioxide is removed. Additional water is removed in a refrigeration unit and then completely removed along with carbon dioxide. In the methanization section, carbon monoxide, carbon dioxide, and hydrogen are converted to methane, which is recycled as a feedstock to increase the total yield. From the remaining stream, ethylene, ethane, propylene, and propane are separated through C_2 and C_3 separation, respectively.

Oxidative coupling combines at the molecular level methane dimerization via radical-like pathways with the removal of the hydrogen formed, often before recombinative desorption to form H_2, via oxidation steps. Ethane and ethylene are the predominant hydrocarbon derivatives formed. Heterogeneous catalysts improve the yields and selectivity attainable in homogeneous coupling reactions, but the active surface sites in these materials also activate C-H bonds in ethane and ethylene, leading to secondary combustion reactions, which limit attainable yields to 20%–25% and lead to significant formation of CO_2. These secondary pathways are very exothermic and they place significant heat transfer loads on the catalytic reactors. The use of cyclic strategies using lattice oxygen and moving or fluid bed reactors provides an attractive alternate option, which avoids unselective homogeneous combustion pathways and uses the heat capacity of the solids to increase the efficiency to heat removal during reaction. Protecting strategies in oxidative coupling can lead to two-step processes similar to those for methanol synthesis. In principle, synthesis gas, methyl bisulfate, or methyl chloride intermediates can be used to form ethylene, but ethylene formation from these protected intermediates is not very selective and the overall processes is not environmentally benign.

Other protecting strategies have been attempted, or at least proposed, in order to increase C_2 yields in oxidative methane coupling. The separation of ethylene from the reactor effluent using cation-exchange zeolites has been carried out and it has led to slight C_2 yield improvements. The low adsorption temperatures required and the lack of adsorption selectivity for ethane, however, limit the practical applications of these approaches. The *in situ* conversion of ethane and ethylene to aromatic derivatives using cation-exchange zeolites after oxidative coupling can lead to the simpler separation of aromatic derivatives from the product stream, but in a process for the synthesis of aromatic derivatives instead of ethylene. Finally, yields improvements can result from the separation of methane and oxygen reactants in space (via membranes) or in time (via cyclic reactors). The first approach exploits any differences in kinetic oxygen response between the coupling and the product combustion reactions. O_2 is introduced along the reactor via multiple injectors or oxygen-conducting membrane walls. Low O_2 concentrations can also be achieved using back-mixed fluidized beds operated at high oxygen conversion levels.

The staging of the oxidant along a tubular reactor has not led to significant improvements, apparently as a result of the similar kinetic dependences of CH_4 and C_2 activation steps on O_2 concentration. Detailed kinetic simulations have shown that distributed oxygen introduction is unlikely to give C_2 yields above 35%–40%. Laboratory tests have shown that poor radial dispersion of the oxygen feed can lead to high local O_2 concentrations, which can cause stable flames and structural damage at membrane walls. Detailed kinetic and process simulations have suggested that continuous ethylene removal from a recycle stream can lead to 75%–85% C_2 yields during oxidative coupling of methane. When the removal of ethylene requires temperatures significantly lower than for oxidative coupling, the required recycle ratios become impractical, because extensive thermal cycling leads to second-law inefficiencies and to large capital and operating costs associated with heat exchange and recompression. These constraints can be overcome by designing absorbers or membranes that remove ethylene selectively from dilute streams containing ethane, CH_4, CO_2, H_2O, and O_2 at typical oxidative coupling temperatures. A reaction-separation protocol using simulated chromatographic reactors has led to 65%–70% ethylene yields. The practical use of these reactors as moving beds, however, requires porous solids that separate C_2 from methane and oxygen at elevated temperatures using adsorption, capillary condensation, or diffusion differences among these components. Cation-exchanged zeolites and microporous carbons are promising as materials for the separation of ethylene from such mixtures, but optimum operating temperatures remain well below those required for oxidative coupling.

Oxidative coupling of methane can be carried out without contact between hydrocarbon derivatives and O_2 using cyclic reactors or hydrogen-conducting membranes. High C_2 yields (25%–28%) can be achieved by cycling reducible oxides between a methane activation reactor and a solids reoxidation vessel. This cyclic process can maintain constant temperatures during solids recycle and uses air (instead of pure O_2) as the oxidant, but the process requires complex handling of solids in fluidized beds. Such redox cycles can be carried out within a single vessel by internally segregating a fluid bed into a reaction zone and an oxidative regeneration zone. The appropriate design oxygen donor solids can be used to the amount of and rate of reaction of the available lattice oxygen. These cyclic oxidative coupling schemes remain of fundamental and practical interest in the activation and conversion of methane via oxidative routes.

Another type of cyclic reactor uses hydrogen absorption into a solid during methane pyrolysis and the removal of the absorbed hydrogen as water in subsequent oxidation cycles. This approach has been recently reported for dehydrogenation reactions of higher alkanes. In such cyclic schemes, pyrolysis and oxidation are separated temporally. In the mathematically analogous catalytic membrane process, reactants are separated spatially using a diffusion barrier that permits only hydrogen transport. Experimental tests using thick membrane disks of hydrogen-conducting perovskites led to very low methane conversion rates and C_2 yields. Significant improvements are possible by combining cation-exchanged zeolites for selective methane pyrolysis at low temperatures with much thinner oxide films, the successful synthesis of which has been recently reported. Hydrogen transport rates at the H_2 pressures prevalent during methane pyrolysis, however, must be improved significantly before practical applications of such schemes can be seriously considered.

6.3 ETHANE

Ethane (C_2H_6) is a two-carbon alkane that, at standard temperature and pressure, is a colorless, odorless gas.

Ethane is isolated on an industrial scale from natural gas and as a byproduct of petroleum refining. Its chief use is as petrochemical feedstock for ethylene production, usually by pyrolysis (Vincent et al., 2008):

$$CH_3CH_3 \rightarrow CH_2{=}CH_2 + H_2$$

After methane, ethane is the second-largest component of natural gas. Natural gas from different gas fields varies in ethane content from less than 1% to more than 6% v/v. Prior to the 1960s, ethane and larger molecules were typically not separated from the methane component of natural gas, but simply burnt along with the methane as a fuel. Currently, ethane is an important petrochemical feedstock, and it is separated from the other components of natural gas in most gas processing plants. Ethane can also be separated from petroleum gas, a mixture of gaseous hydrocarbon derivatives that arises as a byproduct of petroleum refining.

The main source for ethane is natural gas liquids (NGLs). Approximately 40% of the available ethane is recovered for chemical use. The only large consumer of ethane is the steam cracking process for ethylene production. Thus, ethane is most efficiently separated from methane by liquefying it at cryogenic temperatures. Various refrigeration strategies exist: the most economical process presently in wide use employs turbo-expansion, and can recover over 90% of the ethane in natural gas. In this process, chilled gas expands through a turbine; as it expands, its temperature drops to about −100°C. At this low temperature, gaseous methane can be separated from the liquefied ethane and heavier hydrocarbon derivatives by distillation. Further distillation then separates ethane from the propane and heavier hydrocarbon derivatives.

6.3.1 Physical Properties

Ethane (C_2H_6, CH_3CH_3) is a colorless, odorless, gaseous hydrocarbon belonging to the paraffin series that is structurally the simplest hydrocarbon that contains a single carbon-carbon bond (Table 6.3). Ethane is an important constituent of natural gas that also occurs dissolved in crude petroleum and as a byproduct of petroleum refining; it is also produced by the carbonization of coal.

Ethane has a boiling point of −88.5°C (−127.3°F) and melting point of −182.8°C (−297.0°F). Solid ethane exists in several modifications. On cooling under normal pressure, the first modification to appear is a plastic crystal, crystallizing in the cubic system. In this form, the positions of

TABLE 6.3

Properties of Ethane

Chemical formula	C_2H_6
Molar mass	30.07 g/mol
Appearance	Colorless gas
Odor	Odorless
Density	1.3562 mg/cm^3 (at 0°C), 0.5446 g/cm^3 (at 184 K)
Liquid density	0.446 at 0°C
Vapor density (air = 1)	1.05
Melting point	−182.8°C; −296.9°F; 90.4°K
Boiling point	−88.5°C; −127.4°F; 184.6°K
Flash point	−94.4°C, 137.9°F
Solubility in water	56.8 mg/L
Explosive limits	3–12% v/v in air

the hydrogen atoms are not fixed; the molecules may rotate freely around the long axis. Cooling this ethane below approximately 89.9°K (−183.2°C; −297.8°F) changes it to monoclinic metastable ethane. Ethane is only very sparingly soluble in water.

When ethane is combusted in excess air, it produces carbon dioxide and water with a heating value of 1,800 Btu/ft^3 (approximately double that produced from methane). As a constituent of natural gas, ethane is normally burned with methane as a fuel gas. Ethane's relation with petrochemicals is mainly through its cracking to ethylene.

6.3.2 CHEMICAL PROPERTIES

Chemically, ethane can be considered as two methyl groups joined, that is, a dimer of methyl groups. The chemistry of ethane involves chiefly free radical reactions. Ethane can react with the halogens, especially chlorine and bromine, by free radical halogenation which proceeds through the propagation of the ethyl radical:

$$C_2H_5 \bullet + Cl_2 \rightarrow C_2H_5Cl + Cl \bullet$$

$$Cl \bullet + C_2H_6 \rightarrow C_2H_5 \bullet + HCl$$

Because halogenated ethane derivatives can undergo further free radical halogenation, this process results in a mixture of several halogenated products. In the chemical industry, more selective chemical reactions are used for the production of any particular two-carbon haloalkane.

The complete combustion of ethane produces carbon dioxide and water according to the chemical equation:

$$2C_2H_6 + 7O_2 \rightarrow 4CO_2 + 6H_2O$$

Combustion may also occur without an excess of oxygen, forming a mix of amorphous carbon and carbon monoxide:

$$2C_2H_6 + O_2 \rightarrow 4C + 6H_2O$$

$$2C_2H_6 + 5O_2 \rightarrow 4CO + 6H_2O$$

$$2C_2H_6 + 4O_2 \rightarrow 2C + 2CO + 6H_2O$$

Combustion occurs by a complex series of free radical reactions. An important series of reaction in ethane combustion is the combination of an ethyl radical with oxygen, and the subsequent breakup of the resulting peroxide into ethoxy and hydroxyl radicals. Thus:

$$C_2H_5\bullet + O_2 \rightarrow C_2H_5OO\bullet$$

$$C_2H_5OO\bullet + HR \rightarrow C_2H_5OOH + R\bullet$$

$$C_2H_5OOH \rightarrow C_2H_5O\bullet + \bullet OH$$

The principal carbon-containing products of incomplete ethane combustion are single-carbon compounds such as carbon monoxide and formaldehyde (HCHO). One important route by which the carbon-carbon bond in ethane is broken, to yield these single-carbon products, is the decomposition of the ethoxy radical into a methyl radical and formaldehyde, which can in turn undergo further oxidation:

$$C_2H_5O\bullet \rightarrow CH_3\bullet + HCHO$$

Minor products in the incomplete combustion of ethane include acetaldehyde (CH_3CHO), methane (CH_4), methanol (CH_3OH), and ethanol (CH_3CH_2OH). At higher temperatures, especially in the range 600°C–900°C (1,110–1,650°F), ethylene ($CH_2=CH_2$) is a significant product:

$$C_2H_5\bullet + O_2 \rightarrow CH_2=CH_2 + \bullet OOH$$

Similar reactions (with agents other than oxygen as the hydrogen abstractor) are involved in the production of ethylene from ethane in steam cracking (Speight, 2014a).

The chief use of ethane is as a feedstock for ethylene production by steam cracking in which the ethane is diluted with steam and briefly heated to very high temperatures (typically 900°C, 1,650°F, or even higher).

$$CH_3CH_3 \rightarrow CH_2=CH_2 + H_2$$

Ethane is favored for ethylene production—a basic petrochemical feedstock—the steam cracking of ethane is selective for ethylene, while the steam cracking of higher molecular weight hydrocarbon derivatives yields a product mixture that contains less ethylene but more of the higher molecular weight olefin derivatives, such as propylene ($CH_3CH=CH_2$), butadiene ($CH=CHCH=CH_2$), and aromatic hydrocarbon derivatives.

6.3.3 Chemicals from Ethane

In addition to the chlorination of methane, other examples of the chlorination reaction include the formation of ethyl chloride by the chlorination of ethane:

$$CH_3CH_3 + Cl_2 \rightarrow CH_3CH_2Cl + HCl$$

The byproduct hydrogen chloride may be used for the hydrochlorination of ethylene to produce more ethyl chloride. Hydrochlorination of ethylene, however, is the main route for the production of ethyl chloride:

$$CH_2=CH_2 + HCl \rightarrow CH_3CH_2Cl$$

Ethyl chloride (CH_3CH_2Cl) is also prepared by the direct addition of hydrogen chloride (HCl) to ethylene ($CH_2=CH_2$) or by reacting ethyl ether ($CH_3CH_2OCH_2CH_3$) or ethyl alcohol (CH_3CH_2OH) with hydrogen chloride. The chlorination of *n*-pentane and isopentane does not take place in the liquid or vapor phase below 100°C (212°F) in the absence of light or a catalyst, but above 200°C (390°F), and it proceeds smoothly by thermal action alone. The hydrolysis of the mixed chlorides obtained yields all the isomeric amyl (C_5) alcohols except isoamyl alcohol. Reaction with acetic acid produces the corresponding amyl acetates, which find wide use as solvents. Major uses of ethyl chloride are the manufacture of tetraethyl lead and the synthesis of insecticides. It is also used as an alkylating agent and as a solvent for fats and wax.

A small portion of vinyl chloride is produced from ethane by means of the Transcat process in which a combination of chlorination, oxychlorination, and dehydrochlorination reactions occur in a molten salt reactor. The reaction occurs over a copper oxychloride catalyst at a wide temperature range of 310°C–640°C (590°F–1,185°F). During the reaction, the copper oxychloride is converted to copper(I) chloride (CuCl) and copper(II) chloride ($CuCl_2$), which are air oxidized to regenerate the catalyst.

Vinyl chloride is an important monomer for polyvinyl chloride. The main route for obtaining this monomer, however, is via ethylene. An approach to utilize ethane as an inexpensive chemical intermediate is to ammoxidize it to acetonitrile (CH_3CN). The reaction takes place in presence of a cobalt-B-zeolite.

$$2CH_3CH_3 + 2NH_3 + 3O_2 \rightarrow 4CH_3CN + 3H_2O$$

6.4 PROPANE

Propane is produced as a byproduct of two other processes: (i) natural gas processing and (ii) petroleum refining. The processing of natural gas involves removal of butane, propane, and large amounts of ethane from the raw gas to prevent condensation of these volatiles in natural gas pipelines. Additionally, crude oil refineries produce some propane as a byproduct of cracking processes (Parkash, 2003; Gary et al., 2007; Speight, 2007, 2014a; Hsu and Robinson, 2017; Speight, 2017). Propane can also be produced as a biofuel by the thermal conversion of various types of biomass (Speight, 2011).

Propane is produced from both crude oil refining and natural gas processing. Propane is not produced for its own sake, but is a byproduct of these two other processes. Natural gas plant production of propane primarily involves extracting materials such as propane and butane from natural gas to prevent these liquids from condensing and causing operational problems in natural gas pipelines. Similarly, when oil refineries make major products such as motor gasoline, diesel and heating oil, some propane is produced as a byproduct of those processes.

Propane has a wide variety of uses worldwide including small domestic heating applications to large industrial and manufacturing processes. Some of the more common uses of propane are for residential and commercial heating and cooking, motor fuel use in vehicles, irrigation pumps, and power generation, agricultural crop drying and weed control, and as a raw material in the petrochemical industry to make things such as plastics, alcohol, fibers, and cosmetics.

6.4.1 Physical Properties

Propane (C_3H_8, $CH_3CH_2CH_3$) is a three-carbon alkane that is a gas at standard temperature and pressure, but compressible to a transportable liquid. It is a gaseous paraffin hydrocarbon (C_3H_8) having a boiling point of −42.3 (−44.0°F) and a melting point of −187.7°C (−305.8°F) (Table 6.4). Propane may be handled as a liquid at ambient temperatures and moderate pressures. Commercial propane as sold on the various markets may include varying amounts of ethane, butanes, and liquefied refinery gases.

TABLE 6.4
Properties of Propane

Chemical formula	C_3H_8
Molar mass	44.10 g/mol
Appearance	Colorless gas
Odor	Odorless
Density	2.0098 kg/m³ (at 0°C, 101.3 kPa)
Liquid density	0.493 at 25°C
Vapor density (air=1)	2.05
Melting point	−187.7°C; −305.8°F; 85.5°K
Boiling point	−42.25 to −42.04°C; −44.05 to −43.67°F
Flash point	−104°C, −155°F
Solubility in water	47 mg/L (at 0°C)
Explosive limits	2.3–9.5% v/v in air

Liquid propane is a selective hydrocarbon solvent used to separate paraffinic constituents in lube oil base stocks from harmful asphaltic materials. It is also a refrigerant for liquefying natural gas and used for the recovery of condensable hydrocarbon derivatives from natural gas.

6.4.2 CHEMICAL PROPERTIES

Propane is a more reactive paraffin than ethane and methane. This is due to the presence of two secondary hydrogens that could be easily substituted (Chapter 6). Propane is obtained from natural gas liquids or from refinery gas streams. Liquefied petroleum gas (LPG) is a mixture of propane and butane and is mainly used as a fuel. The heating value of propane is 2,300 Btu/ft³. Liquefied petroleum gas is currently an important feedstock for the production of olefin derivatives for petrochemical use.

Propane is an odorless, nontoxic hydrocarbon (C_3H_8) gas at normal pressures and temperatures (ASTM D2163, 2016). When pressurized, it is a liquid with an energy density 270 times greater than its gaseous form. A gallon of liquid propane has about 25% less energy than a gallon of gasoline.

Propane is a simple asphyxiant and since, unlike methane, propane is denser than air, it may accumulate in low spaces, such as depressions in the surface of the earth, and near the floor in domestic and industrial building. If a leak in a propane fuel system occurs, the gas will tend to sink into any enclosed area and thus poses a risk of explosion and fire. The typical scenario is a leaking cylinder stored in a basement; the propane leak drifts across the floor to the pilot light on the furnace or water heater, and results in an explosion or fire. This property makes propane generally unsuitable as a fuel for boats. One hazard associated with propane storage and transport is known as a BLEVE (*b*oiling *l*iquid *e*xpanding *v*apor *e*xplosion).

Propane is stored under pressure at room temperature and, propane and its mixtures, will flash evaporate at atmospheric pressure and cool well below the freezing point of water. The cold gas, which appears white due to moisture condensing from the air, may cause frostbite.

Propane undergoes dehydrogenation to propylene by a catalytic cracking process:

$$CH_3CH_2CH_3 \rightarrow CH_3CH=CH_2 + H_2$$

Propane undergoes combustion reactions in a similar fashion to other alkanes but exhibits several different degrees of complexity (Qin et al., 2000). Put simply, in the presence of excess oxygen, propane burns to form water and carbon dioxide:

$$C_3H_8 + 5O_2 \rightarrow 3CO_2 + 4H_2O$$

When insufficient oxygen is present for complete combustion, incomplete combustion occurs, allowing carbon monoxide and/or soot (carbon) to be formed as well:

$$2C_3H_8 + 9O_2 \rightarrow 4CO_2 + 2CO + 8H_2O$$

$$C_3H_8 + 2O_2 \rightarrow 3C + 4H_2O$$

Propane combustion is much cleaner than gasoline combustion, though not as clean as natural gas (methane) combustion. The presence of carbon-carbon bonds, plus the multiple bonds of propylene ($CH_3CH=CH_2$) and butylene ($CH_3CH=CHCH_3$, $CH_3CH_2CH=CH_2$) create organic exhausts besides carbon dioxide and water vapor during typical combustion. These bonds also cause propane to burn with a visible flame.

Chemicals directly based on propane are few, although as mentioned, propane and liquefied petroleum gas are important feedstocks for the production of olefin derivatives. Propylene has always been obtained as a coproduct with ethylene from steam cracking processes.

6.4.3 Chemicals from Propane

A major use of propane recovered from natural gas is the production of light olefin derivatives by steam cracking processes. However, more chemicals can be obtained directly from propane by reaction with other reagents than from ethane. This may be attributed to the relatively higher reactivity of propane than ethane due to presence of two secondary hydrogens, which are easily substituted.

6.4.3.1 Oxidation

The non-catalytic oxidation of propane in the vapor phase is nonselective and produces a mixture of oxygenated products. Oxidation at temperatures below 400°C (750°F) produces a mixture of aldehydes (acetaldehyde and formaldehyde) and alcohols (methyl alcohol and ethyl alcohol):

$$CH_3CH_2CH_3 + [O] \rightarrow CH_3CHO + HCHO + CH_3OH + CH_3CH_2OH$$

At higher temperatures, propylene and ethylene are obtained in addition to hydrogen peroxide.

$$CH_3CH_2CH_3 + [O] \rightarrow CH_3CH=CH_2 + CH_2=CH_2 + H_2O_2$$

Due to the nonselectivity of this reaction, separation of the products is complex, and the process is not industrially attractive.

6.4.3.2 Chlorination

Chemically, methane (typical of alkanes) undergoes very few reactions. One of these reactions is halogenation, or the substitution of hydrogen with halogen to form a *halomethane*. This is a very important reaction providing alternative pathway for methane activation for the production of synthetic crude oil, fuels, and chemicals. Industrial use of this process will not only eliminate the expensive air separation plants, but as well produce far less greenhouse gases. Gas-phase thermal oxidation and catalytic oxidative methanation process are suitable for industrial application. The proposed process is based on elimination of need for air separation for oxygen production; hence, the gas-phase thermal chlorination is selected (Rozanov and Treger, 2010; Alvarez-Galvan et al., 2011; Treger et al., 2012; Rabiu and Yusuf, 2013).

Methane chlorination is a radical reaction characterized by poor selectivity (Rozanov and Treger, 2010), forming a products stream consisting of equilibrium concentration of all the chloromethane derivatives:

$$CH_4 + Cl_2 \rightarrow CH_3Cl + HCl$$

$$CH_3Cl + Cl_2 \rightarrow CH_2Cl_2 + HCl$$

$$CH_2Cl_2 + Cl_2 \rightarrow CHCl_3 + HCl$$

$$CHCl_3 + Cl_2 \rightarrow CCl_4 + HCl$$

The process conditions can be selected to maximize the proportions of di- and trichloromethanes. To produce the higher chloro-derivatives, methyl chloride is separated from the products and recycled with unreacted methane. When iron-based catalysts are employed, the polymerization of methylene chloride (CH_2Cl_2) and chloroform ($CHCl_3$) to higher molecular weight hydrocarbon derivatives (mainly olefin derivatives) can be achieved. Hence, the emphasis is to maximize the yield and recovery of these compounds for the feasibility of this process.

Chlorination of propane with chlorine at 480°C–640°C (895°F–1,185°F) yields a mixture of perchloroethylene (Perchlor) and carbon tetrachloride:

$$CH_3CH_2CH_3 + 8Cl_2 \rightarrow CCl_2{=}CCl_2 + CCl_4 + 8HCl$$

Carbon tetrachloride is usually recycled to produce more perchloroethylene:

$$2CCl_4 \rightarrow CCl_2{=}CCl_2 + 2Cl_2$$

Perchlor may also be produced from ethylene dichloride (1,2-dichloroethane) through an oxychlorination-oxyhydrochlorination process; trichloroethylene (Trichlor) is coproduced. Perchlor and Trichlor are used as metal degreasing agents and as solvents in dry cleaning. Perchlor is also used as a cleaning and drying agent for electronic equipment and as a fumigant.

Further to the chlorination of methane, a modified process for the conversion of natural gas to transportation fuels and chemicals consists of three principal steps: (i) production of chloromethane compounds, (ii) conversion of the chloromethane derivatives to hydrocarbon derivatives, and (iii) chlorine recovery. The first step involves gas-phase thermal or catalytic selective chlorination of methane to predominantly dichloromethane and trichloromethane after which the monochloromethane is separated and recycled. In the second step, the chloromethane is fed into a moving bed reactor packed with an iron-based Fischer–Tropsch catalyst and wherein it is converted to predominantly olefin hydrocarbon derivatives, Fischer–Tropsch products, and hydrogen chloride gas. The hydrogen chloride byproduct is separated from the Fischer–Tropsch products to obtain premium fuels. The process features a close chlorine loop and the Deacon reaction (a reaction for obtaining chlorine gas by passing air and hydrogen chloride over a heated catalyst (as copper chloride) is used to recover chlorine from the hydrogen chloride byproduct, so that effectively there is no net consumption of chlorine in the overall process. Finally, the plant employs a hydrolyser to regenerate the chloride catalyst (Rabiu and Yusuf, 2013). The overall reaction can be represented as:

$$nCH_4 + O_2 \rightarrow C_nH_{2n} + nH_2O$$

6.4.3.3 Dehydrogenation

Dehydrogenation of paraffin derivatives yields olefins. However, in the petrochemical industry, olefins are not a final product, but rather building blocks for manufacturing the most used chemical commodities. A reliable dehydrogenation technology allows for the design of integrated schemes for fuels and petrochemicals from natural gas, and is becoming a promising alternative feedstock for the new century due to its abundant reserves and low cost. In a refinery, the availability of a dehydrogenation technology permits innovation in the design of new process schemes, by formulating new components for gasoline and diesel fuel. Petrochemical intermediates (propylene, high-purity isobutylene, butadiene, butylene isomers), and components for the blending in the final product that will become gasoline (isooctane, alkylates) or for the diesel pool (high cetane diesel, long-chain linear oxygenates) are made conveniently available through paraffin dehydrogenation (Parkash, 2003; Gary et al., 2007; Speight, 2014a; Hsu and Robinson, 2017; Speight, 2017).

In order to dehydrogenate a saturated noncyclic hydrocarbon, process temperatures are typically in the order of 500°C–600°C (930°F–1,110°F). In general terms, various metals are used as catalysts—examples are nickel (Ni), platinum (Pt), palladium (Pd), and iron (Fe) are suitable, and also the oxides zinc (ZnO), chromium (Cr_2O_3), and iron (Fe_2O_3). Under certain conditions, a diene rather than an alkene may form from n-butane during dehydrogenation.

Again, in general terms, carbon-hydrogen bonds are broken to form a double bond. For example, in the dehydrogenation of n-butane, a mixture of isomers may form. From isobutane, isobutylene may be obtained.

$$2CH_3CH_2CH_2CH_3 \rightarrow =CH_3CH=CHCH_3 + CH_2=CHCH_2CH_3 + 2H_2$$

$$(CH_3)_2 CHCH_3 \rightarrow (CH_3)_2 C=CH_2 + H_3$$

The reactor for obtaining the two butylene isomers from n-butane is typically a tubular device. In the dehydrogenation of butane, the feedstocks is introduced into the reactor in compressed form (in normal conditions, butane is in the gaseous form which is easily converted to the liquid form at −0.5°C (31.1°F). The butane is moved to the heat exchanger—by means of a piston device, where it is heated, evaporated, and thus changes back to the gaseous form. Then, in the reactor, the butane is heated to the temperature required for the reaction (approximately 500°C (932°F)). On contacting the catalyst, the butane vapor dehydrogenates, forming a mixture of unreacted n-butane, butylenes (butene-1 and butene-2), hydrogen, and secondary products. Typically, the contact time of the butane with the catalyst is in the order of less than 3 s, otherwise a large quantity of secondary products (byproducts, including soot) may form, which affects the yield of the desired product(s). After the separation of any secondary products, butane and butylene in the mixture, butylene may be subjected to further dehydrogenation to produce 1,3-butadiene ($CH_2=CHCH=CH_2$).

With the chromia-alumina catalyst, at a temperature of 450°C–650°C (840°F–1,200°F), butadiene-1,3 forms from butane. Thus:

$$CH_3CH_2CH_2CH_3 \rightarrow CH_2=CHCH=CH_2 + 2H_2$$

In dehydrogenation, butane does not form in a cycle and does not form cyclobutene because cyclobutene has an unstable structure; at the reaction temperature, it is also capable of thermally breaking down to ethylene ($CH_2=CH_2$).

Dehydrogenation reactions find wide application in production of a variety of products such as hydrogen, olefin derivatives, polymers, and oxygenates, i.e., production of light (C_3–C_4), olefin derivatives, higher-boiling olefin derivatives (C_4–C_8) for detergents, polypropylene, styrene, aldehydes, and ketones. The demand for basic chemicals such as acrylonitrile, oxo alcohols, ethylene, and propylene oxides are rapidly growing and as a result the dehydrogenation of lower alkanes is a rapidly expanding business.

The dehydrogenation process and, thus, the extent of the equilibrium and the rate of the reaction are favored at high temperatures and at a low pressure because the volume of reaction products exceeds that of reactants. Removal of hydrogen from the products improves the equilibrium extent and reaction rate of dehydrogenation. Gas-phase dehydrogenation is favored by low partial pressures of the reactants and dehydrogenation catalysts are less sensitive than hydrogenation catalysts to poisons such as deactivation by coke formation and deposition of the coke on the catalyst.

This leads to irreversible deactivation due to phase transformation, sintering, and volatilization of the components of the catalyst at the high temperatures involved.

In addition to the examples presented above, more specifically the dehydrogenation process involves the following parameters (the reactants are in the gas phase):

Supported noble metals: Pt, Pd, Rh, Ru, Re, Pt-Re
Supported transition metals: Ni, Co, Fe, Cu, Mo
Catalyst supports: y–Al_2O_3, SiO_2, TiO_2, zeolites, kieselguhr
Raney type metal catalyst: Ni, Cu-Ni
Oxide catalysts: Cr_2O_3, Fe_2O_3, Al_2O_3–Cr_2O_3, Fe_2O_3–K_2CO_3–$C_{r2}O_3$ $Ca_3Ni(PO_4)_3$–Cr_2O_3
Reactors:
  Tubular reactor
  Multi-tray fixed bed reactor
  Moving bed reactor
  Fluidized bed reactor

Important industrial dehydrogenation process includes the following:

1. The catalytic dehydrogenation of propane is a selective reaction that produces mainly propene:

$$CH_3CH_2CH_3 \rightarrow CH_3CH=CH_2$$

The process could also be used to dehydrogenate butane, isobutane, or mixed liquefied petroleum gas feedstocks. It is a single-stage system operating at a temperature range of 540°C–680°C. Conversions in the range of 55%–65% are attainable, and selectivity may reach up to 95%.

As an example, the UOP Oleflex process that can be used for the catalytic dehydrogenation of isobutane, normal butane, or mixed butanes to make iso, normal, or mixed butylenes. Traditionally, this process has been used throughout the world to enable the production of gasoline blending components. The process employs a proprietary platinum on alumina catalyst doped with tin and alkali metals between 500°C and 700°C (930°F and 1,290°F). Dehydrogenation is highly selective, resulting in yields in excess of 90% v/v. This process was based on the Pacol process, in which normal paraffins (C_{10}–C_{14} alkane derivatives) are dehydrogenated in a vapor-phase reaction to corresponding mono-olefins over a highly selective and active catalyst for detergent manufacture.

The Phillips steam active reforming (STAR) process is used to dehydrogenate lower paraffins (propane or butanes) into their corresponding olefins (propylene or butylenes), which can be further processed to valuable downstream products. In the process, as an example, a 0.2–0.6% w/w platinum on alumina (Al_2O_3) catalyst doped with zinc and tin is used to dehydrogenate propane diluted with steam. The catalyst is importantly water-stable, allowing the steam-dilution to drive the equilibrium toward dehydrogenation.

On the other hand, the Catofin and Linde-BASF processes employ chromium-based catalysts. The Catofin dehydrogenation process uses fixed bed reactors with a catalyst and operating conditions that are selected to optimize the complex relationship among conversion, selectivity, and energy consumption. The overall selectivity of isobutane to isobutylene via the CATOFIN process is greater than 90% w/w and the selectivity of propane to propylene is greater than 86% w/w. The Linde catalyst is composed of 18 parts chromia (Cr_2O_3) and 0.25 parts zirconia (ZrO_2) on alumina (Al_2O_3) with a trace of potassium ions (K^+). The active site structure is controversial, but the catalytic cycle involves chromium (Cr^{3+} and Cr^{4+}).

2. Preparation of butadiene by dehydrogenation of *n*-butane and *n*-butenes using several different catalyst types:

$$CH_3CH_2CH_2CH_3 \rightarrow CH_3CH_2CH=CH_2 + H_2$$

Al_2O_3–Cr_2O_3 catalyst
   Fluidized bed reactor
   560°C–600°C (1,040°F–1,110°F)

$$CH_3CH_2CH=CH_2 \rightarrow CH_2=CH\text{-}CH=CH_2 + H_2$$

Fe_2O_3–K_2CO_3–Cr_2O_3 or $Ca_3Ni(PO_4)_3$–Cr_2O_3 catalyst
   600°C–660°C (1,110°F–1,220°F)

The Houdry dehydrogenation process was originally designed to produce butenes at less than atmospheric pressure, for the production of butenes and was also used for butadiene production in 1940s using chromia-alumina catalyst (Catadiene process):

$$CH_3CH_2CH_2CH_3 \rightarrow CH_2=CHCH=CH_2 + 2H_2$$

Cr_2O_3 supported on Al_2O_3 catalyst
   Adiabatic reactor, 620°C–700°C (1,150°F–1,290°F)
   Also: Cr_2O_3 supported on Al_2O_3 catalyst
   fluidized bed
   550°C–600°C (1,020°F–1,110°F)

The Catadiene dehydrogenation process is a reliable, proven route for the production of 1,3 butadiene from *n*-butane or a mix of *n*-butane and *n*-butenes. Lummus Technology has exclusive worldwide licensing rights to this technology. The catalyst is produced by Clariant, a leading company in the development of process catalysts. Also, the Catadiene is the only commercial technology available for on-purpose production of *n*-butylene isomers and butadiene from *n*-butane. Due to the lower process temperature, the process provides high conversion and selectivity for conversion of *n*-butane to *n*-butylenes and butadiene. The process employs multiple reactors operating in a cyclic manner with an automated program so that the flow of process streams is continuous. In addition, the process unit can be operated to coproduce butylenes and butadiene or to produce only butadiene.

3. Dehydrogenation of isopentane to isoprene—a two-step process:

The dehydrogenation of isopentane to isoprene can be achieved by two stages in which isopentane in the first stage is dehydrogenated to amylene derivatives which is further dehydrogenated to isoprene in the second stage of the process as shown below.

$$CH_3CH(CH_3)CH_2CH_3 \rightarrow C_5H_{10}(\text{mono} - \text{olefin derivatives}) + H_2$$

$$CH_3CH(CH_3)CH_2CH_3 \rightarrow C_5H_{10}(\text{mono} - \text{olefin derivatives}) + H_2$$

Cr_2O_3 supported on Al_2O_3 catalyst
   Fluidized bed reactor
   530°C–610°C (985°F–1,030°F)

$$C_5H_{10}(\text{mono} - \text{olefin derivatives}) \rightarrow CH_2=C(CH_3) - CH=CH_2 + H_2$$

$Ca_3Ni(PO_4)_3$–Cr_2O_3 catalyst
   550°C–650°C (1,020°F–1,200°F)

A one-step method is also available. In this method, the dehydrogenation of isopentane and isopentane/isoamylene mixtures is carried out on the same catalyst without intermediate separation of isopentane and isoamylene derivatives. An important advantage of the two-step process is the possibility of the use of highly selective catalyst at each stage and high energy consumption significantly undermines the competitiveness of the two-step method in comparison with the one-step method.

4. Dehydrogenation of ethylbenzene to styrene.

The development of commercial processes for the manufacture of styrene based on the dehydrogenation of ethylbenzene was achieved in the 1930s. The need for synthetic styrene-butadiene rubber (SBR) during World War II provided the impetus for large-scale production. After 1946, this capacity became available for the manufacture of a high-purity monomer that could be polymerized to a stable, clear, colorless, and cheap plastic (polystyrene and styrene copolymers). Peacetime uses of styrene-based plastics expanded rapidly, and polystyrene is now one of the least expensive thermoplastics on a cost-per-volume basis. Styrene itself is a liquid that can be handled easily and safely. The activity of the vinyl group makes styrene easy to polymerize and copolymerize.

The direct dehydrogenation of ethylbenzene to styrene is carried out in the vapor phase with steam over a catalyst consisting primarily of iron oxide. The reaction is endothermic, and can be accomplished either adiabatically or isothermally. Both methods are used in practice:

$$C_6H_5CH_2CH_3 \rightarrow C_6H_5CH=CH_2$$

$Fe_2O_3–Cr_2O_3$ catalyst
   Adiabatic reactor
   580°C–650°C (1,075°F–1,200°F)
   or isothermal tubular reactor
   580°C–610°C (1,075°F–1,130°F)

The major reaction is the reversible, endothermic conversion of ethylbenzene to styrene and hydrogen, that is,

$$C_6H_5CH_2CH_3 \rightarrow C_6H_5CH=CH_2 + H_2$$

This reaction proceeds thermally with low yield and catalytically with high yield. As it is a reversible gas-phase reaction producing 2 mol of product from 1 mol of starting material, low pressure favors the forward reaction. Competing thermal reactions degrade ethylbenzene to benzene, and also to carbon and styrene as well as to toluene:

$$C_6H_5CH_2CH_3 \rightarrow C_6H_6 + CH_2=CH_2$$

$$C_6H_5CH_2CH_3 \rightarrow 8C + 5H_2$$

$$C_6H_5CH_2CH_3 + H_2 \rightarrow C_6H_5CH_3 + CH_4$$

The issue with the production of carbon is that the carbon is a catalyst poison. When potassium is incorporated into the iron oxide catalyst, the catalyst becomes self-cleaning (through enhancement of the reaction of carbon with steam to give carbon dioxide, which is removed in the reactor vent gas):

$$C + 2H_2O \rightarrow CO_2 + 2H_2$$

Typical operating conditions in commercial reactors are approximately 620°C (1,150°F) and as low a pressure as practicable. The overall yield depends on the relative amounts of catalytic conversion to styrene and thermal cracking to byproducts. At equilibrium under typical conditions, the reversible reaction results in about 80% conversion of ethylbenzene. The dehydrogenation of ethylbenzene is carried out in the presence of steam because: (i) the steam lowers the partial pressure of ethylbenzene, shifting the equilibrium toward styrene and minimizing the loss to thermal cracking, (ii) the steam supplies the necessary heat of reaction, and (iii) the steam cleans the catalyst by reacting with carbon to produce carbon dioxide and hydrogen.

5. Oxidative dehydrogenation of *n*-butylene to butadiene.

Butadiene is produced as a byproduct of the steam cracking process used to produce ethylene and other olefin derivatives. When mixed with steam and briefly heated to very high temperatures (often over 900°C, 1,650°F), aliphatic hydrocarbon derivatives give up hydrogen to produce a complex mixture of unsaturated hydrocarbon derivatives, including butadiene. The quantity of butadiene produced depends on the hydrocarbon derivatives used as the feedstock. Low-boiling feedstocks feeds, such as ethane, yield primarily ethylene but higher molecular weight feedstocks favor the formation of higher molecular weight olefins, butadiene, and aromatic hydrocarbon derivatives. The butadiene is typically isolated from the other four-carbon hydrocarbon derivatives produced in steam cracking by extractive distillation using a polar aprotic such as acetonitrile, *N*-methyl-2-pyrrolidone, furfural, or dimethylformamide from which it is then recovered by distillation.

Acetonitrile, boiling point: 81°C (178°F)

N-methyl-2-pyrrolidone, boiling point: 202°C (396°F)

Furfural, boiling point: 162°C (324°F)

Dimethylformamide, boiling point: 152°C (305°F)

In the 1960s, the process to produce butadiene from normal butene derivatives by oxidative dehydrogenation using a catalyst was developed. Since that time various dehydrogenation processes have been described and developed (Passmann, 1970; Dumez and Froment, 1976; Park et al., 2016). In fact, the gradually increasing demand on 1,3-butadiene (CH_2=CHCH=CH_2) has led to further development of this alternative production routes. Thus, catalytic oxidative dehydrogenation of 1-butene to 1,3-butadiene using carbon dioxide as a mild oxidant has been systematically studied over Fe_2O_3/Al_2O_3 catalysts (Yan et al., 2015).

$$2CH_3CH_2CH=CH_2 + [O] \rightarrow 2CH_2=CHCH=CH_2 + 2H_2O$$

Fe_2O_3–ZnO–Cr_2O_3 or $Ca_3Ni(PO_4)_3$–Cr_2O_3 catalyst
350°C–450°C (660°F–840°F)

The loaded ferric oxide (Fe_2O_3) promotes oxygen mobility and modifies the surface acidity of the alumina (Al_2O_3) which leads to a higher 1-butene conversion and butadiene selectivity.

6.4.3.4 Nitration

Nitrating propane produces a complex mixture of nitro-compounds ranging from nitromethane to nitropropanes. The presence of lower nitroparaffin derivatives is attributed to carbon-carbon bond fission occurring at the temperature used. Temperatures and pressures are in the order of 390°C–440°C (735°F–825°F) and 100–125 psi, respectively.

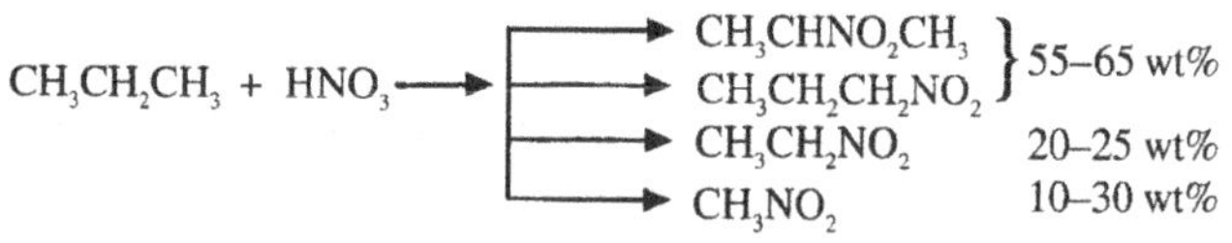

$$CH_3CH_2CH_3 + HNO_3 \longrightarrow \begin{array}{ll} CH_3CHNO_2CH_3 & \\ CH_3CH_2CH_2NO_2 & \end{array} \Big\} \; 55\text{–}65 \; wt\%$$
$$\begin{array}{ll} CH_3CH_2NO_2 & 20\text{–}25 \; wt\% \\ CH_3NO_2 & 10\text{–}30 \; wt\% \end{array}$$

Increasing the mole ratio of propane to nitric acid increases the yield of nitropropane derivatives.

Nitropropane derivatives are good solvents for vinyl and epoxy resins and are also used to manufacture rocket propellants.

Nitropropane reacts with formaldehyde producing nitro-alcohol derivatives:

$$CH_3CH_2CH_2NO_2 + HCHO \rightarrow CH_3CH_2CH(NO_2)CH_2OH$$

These difunctional compounds are versatile solvents.

6.5 BUTANE ISOMERS

Like propane, butanes are obtained from natural gas liquids and from refinery gas streams. The C4 acyclic paraffin consists of two isomers: n-butane and isobutane (2-methylpropane). The physical as well as the chemical properties of the two isomers are quite different due to structural differences. There are two isomers of butane:

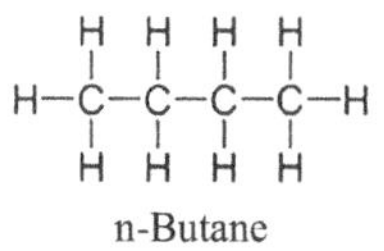

n-Butane

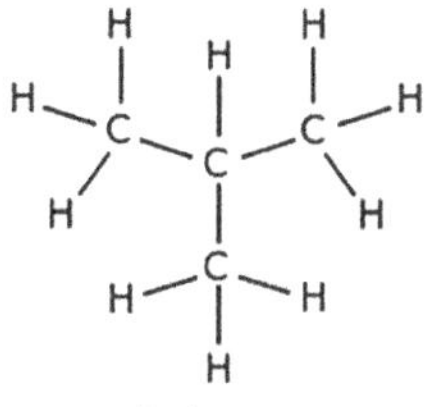

isobutane

In the IUPAC system of nomenclature, however, the name butane refers only to the *n*-butane isomer ($CH_3CH_2CH_2CH_3$). Butane derivatives are highly flammable, colorless, easily liquefied gases that quickly vaporize at room temperature.

The butane isomers present in natural gas can be separated from the large quantities of lower-boiling gaseous constituents, such as methane and ethane, by absorption in a light oil. The butane thus obtained can be stripped from the absorbent along with propane and marketed as liquefied petroleum gas that meets the required specifications or they can be separated from the propane and then from each other by fractional distillation: *n*-butane boils at −0.5° C (31.1° F) (Table 6.5); isobutane boils at −11.7° C (10.9°F) (Table 6.6).

Butane derivatives that are formed by catalytic cracking and other refinery processes can be recovered by absorption into a light oil. Commercially, *n*-butane can be added to gasoline to increase its volatility (as an aid to ignition) in cold climates. Transformed to isobutane in a refinery process known as isomerization, it can be reacted with certain other hydrocarbon derivatives such as butylene to form valuable high-octane constituents of gasoline.

Like propane, *n*-butane is mainly obtained from natural gas liquids. It is also a byproduct from different refinery operations. Currently, the major use of *n*-butane is to control the vapor pressure of product gasoline. Due to new regulations restricting the vapor pressure of gasoline, this use is expected to be substantially reduced. Surplus *n*-butane could be isomerized to isobutane, which is currently in high demand for producing isobutene. Isobutene is a precursor for methyl and ethyl tertiary butyl ethers (ETBEs), which are important octane number boosters. Another alternative

TABLE 6.5
Properties of *n*-Butane

Chemical formula	C_4H_{10}
Molar mass	58.12 g/mol
Appearance	Colorless gas
Odor	Gasoline-like or natural gas-like
Density	2.48 kg/m³ (at 15°C, 59°F)
Liquid density	0.573 at 25°C
Vapor density (air = 1)	2.1
Melting point	−140 to −134°C; −220 to −209°F
Boiling point	−1 to 1°C; 30°F–34°F
Solubility in water	61 mg/L (at 20°C, 68°F)
Explosive limits	1.9–8.5% v/v in air

TABLE 6.6
Properties of isobutane

Chemical formula	C_4H_{10}
Molar mass	58.12 g/mol
Appearance	Colorless gas
Odor	Odorless
Density	2.51 kg/m³ (at 15°C, 100 kPa)
Liquid density	0.551 at 25°C
Vapor density	2.01
Melting point	−159.42°C (−254.96°F)
Boiling point	−11.7°C (10.9°F)
Solubility in water	48.9 mg/L (at 25°C, 77°F)
Explosive limits	1.8%–8.4% in air

outlet for surplus *n*-butane is its oxidation to maleic anhydride. Almost all new maleic anhydride processes are based on butane oxidation. *n*-Butane has been the main feedstock for the production of butadiene. However, this process has been replaced by steam cracking hydrocarbon derivatives, which produce considerable amounts of byproduct butadiene.

6.5.1 PHYSICAL PROPERTIES

n-Butane ($CH_3CH_2CH_2CH_3$) is a colorless gas with a boiling point of −1°C (30°F) that, unlike the first three *alkanes*, is very soluble in water (Table 6.6). The principal raw materials for its production are petroleum and liquefied natural gas. It forms an explosive and flammable mixture with air at low concentrations. Its main uses in industry are as a raw material in the production of butadiene and acetic acid. It is also used as a domestic fuel, as a gasoline blending component, as a solvent and as a refrigerant.

The isobutane (($CH_3)_2CHCH_3$) is also a colorless gas with a boiling point of −11.7°C (10.9°F) that is also soluble in water (Table 6.6). Although the physical properties of isobutane are similar to the properties of *n*-butane, isobutane exhibits markedly different chemical behavior. Isobutane is obtained by petroleum fractionation of natural gas or by isomerization of butane. It forms an explosive and flammable mixture with air at low concentrations. Its main uses are as a raw material in organic synthesis, for the production of synthetic rubber, and in the production of branched hydrocarbon derivatives of high-octane grading.

6.5.2 CHEMICAL PROPERTIES

In the presence of excess oxygen, butane burns to form carbon dioxide and water vapor:

$$2C_4H_{10} + 13O_2 \rightarrow 8CO_2 + 10H_2O$$

On the other hand, when the supply of oxygen is limited, carbon (soot) or carbon monoxide may also be formed:

$$2C_4H_{10} + 9O_2 \rightarrow 8CO + 10H_2O$$

n-Butane is the feedstock for the DuPont catalytic process for the preparation of maleic anhydride:

Maleic anhydride

Thus:

$$2CH_3CH_2CH_2CH_3 + 7O_2 \rightarrow 2C_2H_2(CO)_2O + 8H_2O$$

Maleic anhydride is a solid compound that melts at 53°C (127°F), is soluble in water, alcohol, and acetone, but insoluble in hydrocarbon solvents. The production of maleic anhydride from *n*-butenes is a catalyzed reaction occurring at approximately 400°C–440°C and 30–45 psi. A catalyst, consisting of a mixture of oxide of molybdenum, vanadium, and phosphorous, may be used.

n-Butane, like all hydrocarbon derivatives, undergoes free radical chlorination providing both 1-chlorobutane ($CH_3CH_2CH_2CH_2Cl$) and 2-chlorobutane ($CH_3CH_2CH_2CClCH_3$), as well as more highly chlorinated derivatives. The relative rates of the chlorination are partially explained by the differing bond dissociation energy for the two types of C-H bonds. Isobutane, on the other hand, is a much more reactive compound due to the presence of a tertiary hydrogen:

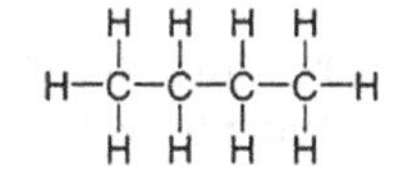

n-Butane

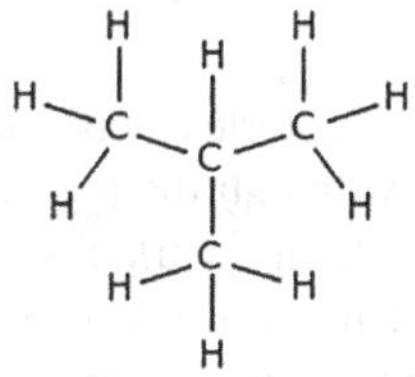

isobutane

Butane is primarily used as a fuel gas within liquefied petroleum gas. Like ethane and propane, the main chemical use of butane is as feedstock for steam cracking units for olefin production. Dehydrogenation of *n*-butane to butenes and to butadiene is an important route for the production of synthetic rubber. *n*-Butane is also a starting material for acetic acid and maleic anhydride production.

Due to its higher reactivity, isobutane is an alkylating agent of light olefin derivatives for the production of alkylates. Alkylates are a mixture of branched hydrocarbon derivatives in the gasoline range having high-octane ratings (Chapter 3). Dehydrogenation of isobutane produces isobutene, which is a reactant for the synthesis of methyl tertiary butyl ether. This compound is currently in high demand for preparing unleaded gasoline due to its high octane rating and clean burning properties. (Octane ratings of hydrocarbon derivatives are noted later in this chapter.)

The chemistry of *n*-butane is more varied than that of propane, partly because *n*-butane has four secondary hydrogen atoms available for substitution and three carbon-carbon bonds that can be cracked at high temperatures, *viz*:

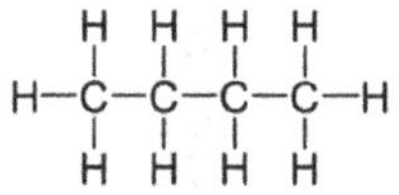

Like propane, the non-catalytic oxidation of butane yields a variety of products including organic acids, alcohols, aldehydes, ketones, and olefin derivatives. Although the non-catalytic oxidation of butane produces mainly aldehyde derivatives and alcohol derivatives, the catalyzed oxidation yields predominantly acid derivatives.

6.5.3   CHEMICALS FROM BUTANE

6.5.3.1   Oxidation

The oxidation of *n*-butane represents a good example illustrating the effect of a catalyst on the selectivity for a certain product. The noncatalytic oxidation of *n*-butane is non-selective and produces a mixture of oxygenated compounds including formaldehyde, acetic acid, acetone, and alcohols.

Typical weight % yields when n-butane is oxidized in the vapor phase at a temperature range of 360°C–450°C (680°F–840°F) and approximately 100 psi are formaldehyde 33%, acetaldehyde 31%, methanol 20%, acetone 4%, and mixed solvents 12%.

On the other hand, the catalytic oxidation of n-butane, either using cobalt or manganese acetate, produces acetic acid at 75%–80% yield. Byproducts of commercial value are obtained in variable amounts. In the Celanese process, the oxidation reaction is performed at a temperature range of 150°C–225°C (300°F–435°F) and a pressure of approximately 800 psi.

$$CH_3CH_2CH_2CH_3 + 3O_2 \rightarrow 2CH_3CO_2H + H_2O$$

The main byproducts are formic acid, ethanol, methanol, acetaldehyde, acetone, and methyl ethyl ketone (MEK). When manganese acetate is used as a catalyst, more formic acid (25%) is obtained at the expense of acetic acid.

Catalytic oxidation of n-butane at 490° (915°F) over a cerium chloride, Co-Mo oxide catalyst produces maleic anhydride:

$$2\ CH_3CH_2CH_2CH_3 + 7\ O_2 \rightarrow 2\ \underset{}{\overset{}{\text{(maleic anhydride ring structure)}}} + 8H_2O$$

Other catalyst systems such as iron–vanadium pentoxide–phosphorous pentoxide over silica alumina are used for the oxidation. In the Monsanto process, n-butane and air are fed to a multi-tube fixed bed reactor, which is cooled with molten salt. The catalyst used is a proprietary modified vanadium oxide. The exit gas stream is cooled, and crude maleic anhydride is absorbed then recovered from the solvent in the stripper.

Another process for the partial oxidation of n-butane to maleic anhydride (the DuPont) process uses a circulating fluidized bed reactor. Solids flux in the riser reactor is high and the superficial gas velocities are also high, which encounters short residence times usually in seconds. The developed catalyst for this process is based on vanadium-phosphorous oxides that provide the oxygen needed for oxidation. The selective oxidation of n-butane to maleic anhydride involves a redox mechanism where the removal of eight hydrogen atoms as water and the insertion of three oxygen atoms into the butane molecule occur. The reaction temperature is approximately 500°C (930°F). Subsequent hydrogenation of maleic anhydride produces tetrahydrofuran.

Oxidation of n-butane to maleic anhydride is becoming a major source for this important chemical. Maleic anhydride could also be produced by the catalytic oxidation of n-butene isomers and benzene. The principal use of maleic anhydride is in the synthesis of unsaturated polyester resins. These resins are used to fabricate glass fiber-reinforced materials. Other uses include fumaric acid, alkyd resins, and pesticides. Maleic acid esters are important plasticizers and lubricants. Maleic anhydride could also be a precursor for 1,4-butanediol.

Maleic anhydride

Thus:

$$2C_2H_2(CO)_2O \rightarrow HOCH_2CH_2CH_2CH_2OH$$

6.5.3.2 Production of Aromatics

Liquefied petroleum gas, a mixture of propane and butane isomers, is catalytically reacted to produce an aromatic-rich product. The first step is assumed to be the dehydrogenation of propane and butane to the corresponding olefin derivatives followed by oligomerization to C_{6-}, C_{7-}, and C_8 olefin derivatives. These compounds are then dehydrocyclized to benzene-toluene-xylene aromatic derivatives. The following reaction sequence illustrates the formation of benzene from 2-propane:

$$2CH_3CH_2CH_3CH_3CH_2CH_2CH_2CH{=}CH_2 + 2H_2$$

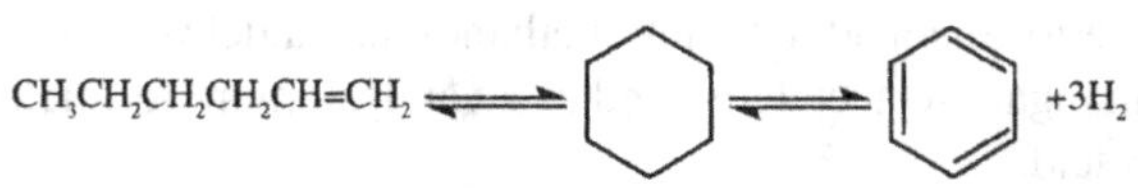

$$CH_3CH_2CH_2CH_2CH{=}CH_2 \rightleftharpoons \hexagon \rightleftharpoons \benzene \;\; +3H_2$$

Although olefin derivatives are intermediates in this reaction, the final product contains a very low olefin concentration. The overall reaction is endothermic due to the predominance of dehydrogenation and cracking. Methane and ethane are byproducts from the cracking reaction.

The process consists of a reactor section, continuous catalyst regeneration unit, and product recovery section. Stacked radial-flow reactors are used to minimize pressure drop and to facilitate catalyst recirculation to and from the continuous catalyst regeneration unit. The reactor feed consists solely of liquefied petroleum gas plus the recycle of unconverted feed components; no hydrogen is recycled. The liquid product contains about 92% w/w benzene, toluene, and xylenes, with a balance of C_{9+} aromatic derivatives and a low nonaromatic content. Therefore, the product could be used directly for the recovery of benzene by fractional distillation (without the extraction step needed in catalytic reforming).

6.5.3.3 Isomerization

Because of the increasing demand for isobutylene for the production of oxygenates as gasoline additives, a substantial amount of n-butane is isomerized to isobutane, which is further dehydrogenated to isobutene. The Butamer process has a fixed bed reactor containing a highly selective catalyst that promotes the conversion of n-butane to isobutane equilibrium mixture. Isobutane is then separated in a deisobutanizer tower. The n-butane is recycled with makeup hydrogen. The isomerization reaction occurs at a relatively low temperature:

$$CH_3CH_2CH_2CH_3 \rightarrow \underset{\text{Isobutane}}{(CH_3)_2 CHCH_3}$$

6.5.4 Chemicals from Isobutane

Isobutane is mainly used as an alkylating agent to produce different compounds (alkylates) with a high-octane number for blending with other constituents to manufacture gasoline pool. Isobutane is in high demand as an isobutene precursor for producing oxygenates such as methyl and ethyl tertiary butyl ethers. Accordingly, greater amounts of isobutane are produced from n-butane through isomerization followed by dehydrogenation to isobutene. The Catofin process is currently used to dehydrogenate isobutane to isobutene. Alternatively, isobutane could be thermally cracked to yield predominantly isobutene plus propane. Other byproducts are fuel gas and C_{5+} liquid. The steam cracking process is made of three sections: (i) a cracking furnace, (ii) a vapor recovery section, and (iii) a product fractionation section.

6.6 LIQUID PETROLEUM FRACTIONS AND RESIDUES

Liquid petroleum fractions and residua are not typically composed of hydrocarbon derivatives, but are more likely composed of hydrocarbonaceous derivatives in which a high proportion of the molecular constituents may contain derivatives of sulfur, nitrogen, oxygen, and metals. Nevertheless, it is opportune at this time in the text to present a description of these various fractions and the means by which they can be used as feedstocks to produce precursors that are suitable for the manufacture of petrochemical products.

In the modern refinery, the typical product split between fuels and chemicals for traditional medium conversion fuel refineries has been about 95% fuels to 5% chemicals. However, the modern refinery is exhibiting a trend that is to change the refining complex product slate, and crude oil refining is shifting from an emphasis on transportation fuels to higher margin chemical products. With this focus, each component of the petrochemical complex is evaluated for its potential to contribute to increased production of the desired chemical product slate. Meanwhile, environmental regulations continue to become more onerous, affecting suitability of technologies and configuration choices. As a result, the technologies and the process configurations chosen for a modern refinery/chemical complex must evolve to meet these emerging challenges. Since crude oil fraction, particularly the higher-boiling fractions, can be a significant portion of the refinery feedstock, conversion of these fractions has become an important factor in maximizing chemical production and meeting evolving environmental standards. Conversion processes convert relatively low-value fractions, such as gas oil, fuel oil, and resid, to higher-value products.

Liquid petroleum fractions are light naphtha, heavy naphtha, kerosene, and gas oil which are also sources of starting chemicals for petrochemical products (Table 6.7). The bottom product from distillation units is the residue. These mixtures are intermediates through which other reactive intermediates are obtained. Heavy naphtha is a source of aromatic derivatives via catalytic reforming and of olefin derivatives from steam cracking units. Gas oils and residues are sources of olefin derivatives through cracking and pyrolysis processes (Parkash, 2003; Gary et al., 2007; Speight, 2014a; Hsu and Robinson, 2017; Speight, 2017).

High molecular weight n-paraffin derivatives are obtained from different petroleum fractions through physical separation processes. Those in the range of C_8–C_{14} are usually recovered from the kerosene fraction. Vapor-phase adsorption using a molecular sieve is used to achieve the separation. The n-paraffin derivatives are then desorbed by the action of ammonia. Continuous operation is possible by using two adsorption sieve columns: one bed is onstream while the other bed is being desorbed.

n-Paraffin derivatives could also be separated by forming an adduct with urea. For a paraffinic hydrocarbon to form an adduct under ambient temperature and atmospheric pressure, the compound must contain a long unbranched chain of at least six carbon atoms. Ease of adduct formation and adduct stability increases with increase of chain length.

As with shorter-chain n-paraffin derivatives, the longer-chain compounds are not highly reactive. However, they may be oxidized, chlorinated, dehydrogenated, sulfonated, and fermented under special conditions. The C_5–C_{17} paraffin derivatives are used to produce olefin derivatives or monochlorinated paraffin derivatives for the production of detergents.

TABLE 6.7
Crude Oil Fractions as Sources of Petrochemicals

Petroleum Fraction	Source	Intermediate Feedstock
Naphtha	Distillation thermal and catalytic cracking	Ethylene, propylene butane, butadiene, benzene, toluene, xylenes
Kerosene	Distillation thermal and catalytic cracking	Linear n-C_{10}–C_{14} alkanes
Gas oil	Distillation thermal and catalytic cracking	Ethylene, propylene, Butylenes, butadiene
Wax	Dewaxing	C_6–C_{20} alkanes

6.6.1 NAPHTHA

Naphtha is a generic term normally used in the petroleum refining industry for the overhead liquid fraction obtained from atmospheric distillation units (Parkash, 2003; Gary et al., 2007; Speight, 2014a; Hsu and Robinson, 2017; Speight, 2017). The approximate boiling range of light straight-run naphtha (LSR) is 35°C–90°C, while it is about 80°C–200°C for heavy straight-run naphtha. Naphtha is also obtained from other refinery processing units such as catalytic cracking, hydrocracking, and coking units. The composition of naphtha, which varies appreciably, depends mainly on the crude type and whether it is obtained from atmospheric distillation or other processing units.

In terms of alternate feedstocks (Chapter 3), coal can be converted into naphtha via either direct or indirect liquefaction (Speight, 2013). Subbituminous coal (geologically young) and brown coal (lignite) are more suitable for direct coal liquefaction than bituminous coal. A well-known process is the Bergius process in which coal is first ground into fine particles and then mixed with a high-boiling, aromatic solvent (recovered later) at about 450°C to form a slurry that is rich in aromatic constituents. Through a low-severity, catalytic hydrogenation process, the slurry is refined into liquid products, including naphtha. In contrast to Fischer–Tropsch naphtha (with no aromatic constituents), the naphtha produced by this route is rich in aromatic constituents.

Coal can also be converted into naphtha via Fischer–Tropsch processes (indirect liquefaction). In the process, coal is first converted into syngas (through gasification) and the synthesis gas is then converted into Fischer–Tropsch liquids that are similar to those from natural gas-to-liquid processes. As with Fischer–Tropsch naphtha derived from methane, steam cracking of Fischer–Tropsch naphtha derived from coal can lead to a high yield of low-boiling olefin derivatives. In contrast to natural gas-to-liquid processes, Fischer–Tropsch naphtha production from coal requires extensive gas cleanup after coal gasification (e.g., removal of sulfur and other impurities such as metals with the use of solvents and absorbents).

Similar to methane and coal, biomass can also be converted into Fischer-Tropsch naphtha through Fischer–Tropsch processes. As with Fischer-Tropsch naphtha derived from methane and coal, the Fischer–Tropsch naphtha derived from biomass has a high paraffin content and leads to high yields of lower-boiling olefin derivatives if it is used in steam cracking. However, Fischer–Tropsch naphtha production from raw biomass must deal with the high water content of the biomass.

Plastic waste, especially polyolefin derivatives (e.g., the polypropylene used in plastic bags), can be converted into naphtha and other hydrocarbon derivatives (e.g., mostly high-boiling oils) through a series of liquefaction, pyrolysis, and separation processes, which involve the use of hydrogen, steam, and catalysts. The naphtha produced is similar to naphtha derived from crude oil and, if used in steam cracking, can lead to a similar yield of light olefin derivatives. Currently, the dominant methods to dispose plastic waste are landfills, incineration, or making secondary plastics. The utilization of plastic waste for the production of naphtha and petrochemicals is a potential method for plastic waste disposal.

Stoddard solvent is a chemical mixture containing hydrocarbon derivatives that range from C_7–C_{12} with the majority of hydrocarbon derivatives in the C_9–C_{11} range and, therefore, relates to a medium-to-high-boiling naphtha. The hydrocarbon derivatives composing Stoddard solvent are 30–50% v/v alkane derivatives, 30–40% v/v cycloalkane derivatives, and 10–20% v/v aromatic derivatives. Stoddard solvent is considered to be a form of naphtha but not all forms of naphtha are considered to be Stoddard solvent. Stoddard solvent is produced from straight-run distillate of paraffinic or mixed base crude oil and must meet the specifications of the American Society for Testing and Materials designation for Type I mineral spirits (Stoddard solvent) (ASTM D235, 2018).

6.6.1.1 Physical Properties

Naphtha is often divided into two main types, aliphatic and aromatic. The two types differ in two ways: first, in the kind of hydrocarbon derivatives making up the solvent, and second, in the methods used for their manufacture. Aliphatic solvents are composed of paraffinic hydrocarbon derivatives and cycloparaffin derivatives (naphthene derivatives), and may be obtained directly from crude petroleum by distillation. The second type of naphtha contains aromatic derivatives, usually alkyl-substituted benzene, and is very rarely, if at all, obtained from petroleum as straight-run materials.

In terms of the physical properties—in this case, the boiling range—there are three types of naphtha that are being used in basic petrochemicals production today (they are differentiated here because different kinds of naphtha lead to different mixes of light olefin derivatives) but which can vary in boiling range and composition depending upon the refinery processes used to produce the naphtha.

Light naphtha (*low-boiling naphtha*) is also called paraffinic naphtha contains hydrocarbon derivatives in the molecular range C_5H_{12}–C_6H_{14} and is a byproduct of a petroleum refinery. A small amount of light naphtha also comes from natural gas condensates in oil and natural gas fields. Steam cracking of light naphtha leads to a high yield of light olefin derivatives. Naphtha made from Fischer–Tropsch processes (often referred to as Fischer–Tropsch naphtha or F-T naphtha) is also a low-boiling naphtha that leads to a higher ethylene yield than regular low-boiling naphtha.

Heavy naphtha (*high-boiling naphtha*) is also called non-normal paraffinic naphtha (contains hydrocarbon derivatives in the molecular in the range of C_7H_{16}–C_9H_{20}) and is richer in aromatic derivatives than low-boiling naphtha. Since the octane number of this naphtha is low, it cannot directly be used as transportation fuel and, therefore, is often converted through a reforming step into high-octane naphtha that is a suitable blend stock for gasoline manufacture. However, it can also be used for petrochemicals production.

Full range naphtha is a mixture of light and heavy naphtha which contains hydrocarbon derivatives in the molecular range of C_5H_{12}–C_9H_{20} (in the range of C_5H_{12}–$9H_{20}$). It is the most common type of naphtha used in steam cracking.

6.6.1.2 Chemical Properties

Naphtha from atmospheric distillation is characterized by an absence of olefinic compounds. Its main constituents are straight and branched chain paraffin derivatives, cycloparaffin derivatives (naphthene derivatives), and aromatic derivatives, and the ratios of these components are mainly a function of the crude origin. Naphtha obtained from cracking units generally contain variable amounts of olefin derivatives, higher ratios of aromatic derivatives, and branched paraffin derivatives. Due to presence of unsaturated compounds, they are less stable than straight-run naphthas. On the other hand, the absence of olefin derivatives increases the stability of naphthas produced by hydrocracking units. In refining operations, however, it is customary to blend one type of naphtha with another to obtain a required product or feedstock.

Selecting the naphtha type can be an important processing procedure. For example, a paraffinic-base naphtha is a better feedstock for steam cracking units because paraffin derivatives are cracked at relatively lower temperatures than cycloparaffin derivatives. Alternately, a naphtha rich in cyclo-paraffin derivatives would be a better feedstock to catalytic reforming units because cycloparaffin derivatives are easily dehydrogenated to aromatic compounds.

The main use of naphtha in the petroleum industry is in gasoline production. Light naphtha is normally blended with reformed gasoline (from catalytic reforming units) to increase its volatility and to reduce the aromatic content of the product gasoline.

Heavy naphtha from atmospheric distillation units or hydrocracking units has a low-octane rating, and it is used as a feedstock to catalytic reforming units. Catalytic reforming is a process of upgrading low octane naphtha to a high-octane reformate by enriching it with aromatic derivatives and branched paraffin derivatives. The octane rating of gasoline fuels is a property related to the spontaneous ignition of unburned gases before the flame front and causes a high pressure. A fuel with a low-octane rating produces a strong knock, while a fuel with a high-octane rating burns smoothly without detonation. Octane rating is measured by an arbitrary scale in which isooctane (2,2,4-trimethylpentane) is given a value of 100 and n-heptane a value of zero. A fuel's octane number equals the percentage of isooctane in a blend with *n*-heptane.

The octane number is measured using a single-cylinder engine (CFR engine) with a variable compression ratio. The octane number of a fuel is a function of the different hydrocarbon constituents present. In general, aromatic derivatives and branched paraffin derivatives have higher octane ratings than straight-chain paraffin derivatives and cycloparaffin derivatives.

Naphtha is also a major feedstock to steam cracking units for the production of olefin derivatives. This route to olefin derivatives is especially important in places such as Europe, where ethane is not readily available as a feedstock because most gas reservoirs produce nonassociated gas with a low ethane content.

6.6.1.3 Chemicals from Naphtha

Low-boiling naphtha containing hydrocarbon derivatives in the C_5–C_7 range is a feedstock in Europe for producing acetic acid by oxidation. Similar to the catalytic oxidation of n-butane, the oxidation of low-boiling naphtha is performed at approximately the same temperature and pressure ranges (170°C–200°C, 340°F–390°F, 700 psi) in the presence of manganese acetate catalyst. The yield of acetic acid is approximately 40% w/w.

$$\text{Low-boiling naphtha } (C_5 - C_7 \text{ hydrocarbons}) + O_2 \rightarrow CH_3COOH + \text{by-products} + H_2O$$

The product mixture contains essentially oxygenated compounds (such as carboxylic acid derivatives, alcohol derivatives, ester derivatives, aldehyde derivatives, and ketone derivatives). As many as 13 distillation columns are used to separate the complex mixture. The number of products could be reduced by recycling most of them to extinction.

Naphtha is also a commonly used feedstock for the production of synthesis gas, which is used to synthesize methanol and ammonia (Chapter 10). Another important role for naphtha is its use as a feedstock for steam cracking units for the production of low-boiling olefin derivatives. On the other hand, high-boiling naphtha, on the other hand, is a major feedstock for catalytic reforming. The product reformate containing a high percentage of C_6–C_8 aromatic hydrocarbon derivatives is used to make gasoline. Reformates are also extracted to separate the aromatic derivatives as intermediates for petrochemicals.

In the ethylene production process, straight-run naphtha or hydrocracked naphtha used as feedstock in commercially established ethylene production industries. A cracking furnace having a separate convection section for preheating and a radiant section is used. The interior of the furnace contains burners placed along the sidewalls or at the bottom of the furnace. Temperature in the furnace continuously maintained between 950°C and 1,000°C (1,740°F–1,830°F) by the series of burners controlling; a relatively low pressure (75 psi) pressure is maintained in the tubes by the naphtha feed pumps.

The fresh naphtha feedstock is preheated by a heat exchanger that uses the cracked products stream that comes out from the furnace. The preheated naphtha is mixed with steam and passes to the convective section. Its temperature is raised to 300°C (570°F) temperature and pass to the radiation section of the furnace for further increasing to 800°C (1,470°F). This is the condition where naphtha is cracked into simple compounds. Steam is added to dilute the feedstock and to prevent the coke formation at the cracking zone. High-temperature product gas is cooled by removing the latent heat of water in steam generators and transfer line heat exchangers operate with high thermal efficiency during cooling the product gas. Products from C_2 to C_4 are formed during cracking along with some quantity of benzene, toluene, xylene isomers, ethyl benzene, hydrogen, and fuel oil. Optimum values of residence time, steam ratio as well as temperature and pressure effect the byproducts formation. An oil quenching mechanism is used to cool the furnace effluent gas and the recovered heat is used to produce low-pressure steam in the plant utility. Gas oil and fuel oil are obtained when the gas is passed to the primary fractionators.

The volatile components from the primary fractionators are cooled where the high molecular weight hydrocarbon derivatives (above C_3) are liquefied and separated through separators.

The cracked products are passed through coolers and compressors at 30°C (86°F) and 450 psi where separation of C_3, C_4, C_5, and C_6 components takes place by partial fractionation and liquefaction. This can be done in four stages: (i) cooling the whole mass of gas to 30°C (86°F) under 525 psi to liquefy C_4 and heavier constituents; (ii) the uncondensed gas is subjected to severe conditions, i.e., up to +30°C (86°F) and 300 psi whereby propane condenses leaving ethane and ethylene in gaseous form; (iii) dry gas constituting methane and hydrogen is separated from ethane/ethylene mixture this mixture is used as a refrigerant. Acidic constituents such as carbon monoxide, carbon dioxide, hydrogen sulfide, and sulfur dioxide are removed in acid gas removing unit. After the treatment the outlet gas is predominantly a mixture of methane, hydrogen, ethylene, ethane, and slight amount of acetylene (1%). Hydrogen is separated initially then the gas is liquefied. Hydrogen gas is purified and part of it is sent to hydrogenation units to convert acetylene and propadiene to ethylene and propane, respectively. Tail gas obtained from demethanizer is abundant in methane and used in fuel system. The gases are sent to acid gas removal unit, demethanizer for methane removal and hydrogen purification unit. Tail gas is removed and then traces of acetylene are converted into ethylene by hydrogenation. Then it is sent to ethylene splitter for separation of ethylene. The ethane and ethylene are liquefied and fractionated. The heavy bottoms of first stage unit are processed for C_3 and heavy ends.

6.6.2 KEROSENE

Kerosene, a distillate fraction higher-boiling than naphtha, is normally a product from distilling crude oils under atmospheric pressures (Parkash, 2003; Gary et al., 2007; Speight, 2014a; Hsu and Robinson, 2017; Speight, 2017). It may also be obtained as a product from thermal and catalytic cracking or hydrocracking units. Kerosene from cracking units is usually less stable than kerosene produced from atmospheric distillation and hydrocracking units due to presence of variable amounts of olefinic constituents.

6.6.2.1 Physical Properties

Kerosene (kerosine), also called paraffin or paraffin oil, is usually a clear colorless liquid (but often a pale yellow liquid) which does not stop flowing except at very low temperature (normally below −30°C). However, kerosene containing high olefin and nitrogen contents may develop some color (pale yellow) after being produced. It is obtained from petroleum and used for burning in lamps and domestic heaters or furnaces, as a fuel or fuel component for jet engines, and as a solvent for greases and insecticides.

Kerosene is intermediate in volatility between naphtha gas oil. It is a medium oil distilling between 150°C and 300°C (300°F–570°F). Kerosene has a flash point about 25°C (77°F) and is suitable for use as an illuminant when burned in a wide lamp. The term *kerosene* is also too often incorrectly applied to various fuel oils, but a fuel oil is actually any liquid or liquid petroleum product that produces heat when burned in a suitable container or that produces power when burned in an engine.

6.6.2.2 Chemical Properties

Chemically, kerosene is a mixture of hydrocarbon derivatives; the chemical composition depends on its source, but it usually consists of about ten different hydrocarbon derivatives, each containing 10 to 16 carbon atoms per molecule; the constituents include *n*-dodecane (n-$C_{12}H_{26}$), alkyl benzenes, and naphthalene and its derivatives. Kerosene is less volatile than gasoline; it boils between 140°C (285°F) and 320°C (610°F).

Kerosene, because of its use as a burning oil, must be free of aromatic and unsaturated hydrocarbons, as well as free of the more obnoxious sulfur compounds. The desirable constituents of kerosene are saturated hydrocarbons, and it is for this reason that kerosene is manufactured as a straight-run fraction, not by a cracking process.

Although the kerosene constituents are predominantly saturated materials, there is evidence for the presence of substituted tetrahydronaphthalene. Dicycloparaffin derivatives also occur in substantial amounts in kerosene. Other hydrocarbons with both aromatic and cycloparaffin rings in the same molecule, such as substituted indan, also occur in kerosene. The predominant structure of the dinuclear aromatics appears to be that in which the aromatic rings are condensed, such as naphthalene whereas the *isolated* two-ring compounds, such as biphenyl, are only present in traces, if at all.

The main constituents of kerosene obtained from atmospheric and hydrocracking units are paraffin derivatives, cycloparaffin derivatives, and aromatic derivatives. Kerosines with high-normal paraffin content are suitable feedstocks for extracting C C_{14} n-paraffin derivatives, which are used for producing biodegradable detergents (Chapter 6). Currently, kerosene is mainly used to produce jet fuels, after it is treated to adjust its burning quality and freezing point. Before the widespread use of electricity, kerosene was extensively used to fuel lamps, and is still used for this purpose in remote areas. It is also used as a fuel for heating purposes.

6.6.2.3   Chemicals from Kerosene

Kerosene has been an important household fuel since the mid-19th century. In developed countries, its use has greatly declined because of electrification. However, in developing countries, kerosene's use for cooking and lighting remains widespread. This review focuses on household kerosene uses, mainly in developing countries, their associated emissions, and their hazards. Kerosene is often advocated as a cleaner alternative to solid fuels, biomass, and coal, for cooking, and kerosene lamps are frequently used when electricity is unavailable.

Although present in varying quantities, depending on fuel source and quality, naphthalene, benzene, n-hexane, toluene, and the xylene isomers are among several chemicals present in kerosene. However, in the present context, chemicals can be obtained from kerosene in a manner similar to the production of chemicals from naphtha—by means of the steam cracking process.

6.6.3   Gas Oil

Gas oil is a higher-boiling petroleum fraction than kerosene (Parkash, 2003; Gary et al., 2007; Speight, 2014a; Hsu and Robinson, 2017; Speight, 2017). It can be obtained from the atmospheric distillation of crude oils (atmospheric gas oil (AGO)), from vacuum distillation of topped crudes (vacuum gas oil (VGO)), or from cracking and hydrocracking units.

The conventional process for olefin is steam cracking of C_2–C_4 low-boiling paraffin derivatives from natural gas or from refinery gas streams. However, the increasing demand for gaseous fuel and the rising price of natural gas have limited the supply of light hydrocarbon derivatives. As an answer to this increasing demand, fluid catalytic cracking (FCC) is traditionally the dominant refinery conversion process for producing high-octane gasoline. Driven by an increased demand for light olefin derivatives worldwide, fluid catalytic cracking is also an option to yield petrochemical feedstocks from heavy oils through the innovation of hardware, operating parameters, and catalyst formulation. In this respect, a number of fluid catalytic cracking technologies have been developed including: (i) deep catalytic cracking (DCC), (ii) the catalytic pyrolysis process (CPP), (iii) ultimate catalytic cracking (UCC), and (iv) high-severity fluid catalytic cracking (HSFCC) (Parkash, 2003; Gary et al., 2007; Speight, 2014a; Hsu and Robinson, 2017; Speight, 2017).

6.6.3.1   Physical Properties

Atmospheric gas oil has a relatively lower density and sulfur content than vacuum gas oil produced from the same crude. The aromatic content of gas oils varies appreciably, depending mainly on the crude type and the process to which it has been subjected. For example, the aromatic content is approximately 10% for light gas oil and may reach up to 50% for vacuum and cracked gas oil.

6.6.3.2   Chemical Properties

Atmospheric gasoil is a distillation fraction derived from an atmospheric distillation unit, and it is primarily made up of molecules with 14–20 carbon atoms. The atmospheric gas oil distillation has a boiling range between 215°C and 343°C (420°F–650°F).

The primary use of atmospheric gas oil is as a blend-stock to produce diesel fuel or heating oil. However, it must typically go through a distillate hydrotreating unit first to remove sulfur. The higher-boiling fraction of the gas oil can also be fed into a catalytic cracking unit when a refinery is trying to maximize the yield of naphtha over kerosene as well as low-boiling products for petrochemical manufacture.

6.6.3.3   Chemicals from Gas Oil

The primary uses for gas oil are the production of fuels. A secondary use is as a feedstock for steam cracking to produce petrochemicals (ethylene, propylene) and the production of aromatic petrochemical products (benzene, toluene, and xylene isomers). Gas oil is used as a chemical feedstock for steam cracking, although generally less preferred than naphtha and natural gas liquids (including liquefied petroleum gases). The gas oil output of a refinery depends on the composition of crude oil feedstock which, in turn, is dependent upon the crude oil regional source. In addition, naphthenic crude oils tend to produce relatively greater quantities of naphtha than the paraffinic crudes of the same specific gravity, which produce higher relative amounts of gas oils.

Many compound classes have been identified by GC×GC–TOFMS, such as tri-, tetra-, and pentacyclic terpane derivatives, sterane derivatives, and hopane derivatives. Several polycyclic aromatic hydrocarbons (PAHs), such as fluorene, phenanthrene, pyrene, and benzo[g,h,i]perylene; sulfur compounds, such as alkyl benzothiophene derivatives, alkyl dibenzothiophene derivatives, and alkyl benzonaphthothiophene derivatives; and alkylphenol derivatives (Avila et al., 2012). The separation of individual chemicals (or even chemical streams) from such a mixture is an indomitable task and, as a result, gas oil (like other complex petroleum products) best serves as a cracking stock to produce the starting materials for petrochemical production.

A major use of gas oil is as a fuel for diesel engines. Another important use is as a feedstock to cracking and hydrocracking units. Gases produced from these units are suitable sources for light olefin derivatives and liquefied petroleum gas which may be used as a fuel, as a feedstock to steam cracking units for olefin production, or as a feedstock for a Cyclar unit, which can be used for the production of aromatic derivatives from liquefied petroleum gas. In the UOP-BP process, benzene, toluene, and xylenes are produced by dearomatization of propane and butane. The process consists of reaction system, continuous regeneration of catalyst, and product recovery. The catalyst is a zeolite-type catalyst with a non-noble metal promoter (Gosling et al., 1999).

The Cyclar process is used to convert liquefied petroleum gas directly into a liquid aromatics product in a single operation. The process is divided into three major sections: (i) the reactor section includes a radial-flow reactor stack, combined feed exchanger, and heaters; (ii) the catalyst regenerator section includes a regenerator stack and catalyst transfer system; and (iii) the product recovery section which includes product separators, compressors, stripper, and gas recovery equipment. Fresh feedstock and recycle are combined and heat exchanged against reactor effluent after which the combined feedstock is then raised to reaction temperature in the charge heater and sent to the reactor section where four adiabatic, radial-flow reactors are arranged in one or more vertical stacks. The catalyst flows by gravity down the stack, while the charge flows radially across the annular catalyst beds. Between each reactor, the charge is reheated to reaction temperature in an inter-reactor heater. The effluent from the last reactor is split into vapor and liquid products in a separator. The liquid is sent to a stripper where low-boiling saturates are removed from the C_6 aromatic product. The vapor from the separator is compressed and sent to a gas recovery section, typically a cryogenic unit, for separation into a 95% pure hydrogen product stream, a fuel gas stream of light saturates, and a recycle stream of unconverted liquefied petroleum gas.

As expected under the process parameters, coke is deposited on the catalyst and, to combat this deactivation effect, the partially deactivated catalyst is continually withdrawn from the bottom of the reactor stack and transferred to the catalyst regenerator. The catalyst flows down through the regenerator where the accumulated carbon is burned off and the regenerated catalyst is lifted with hydrogen to the top of the reactor stack. The principal operating variables for the Cyclar process are temperature, space velocity, pressure, and feedstock composition. The temperature must be high enough to ensure nearly complete conversion of reaction intermediates in order to produce a liquid product that is essentially free of nonaromatic impurities, but low enough to minimize nonselective thermal reactions. Space velocity is optimized against conversion within this temperature range to obtain high product yields with minimum operating costs. Reaction pressure has a major impact on process performance.

The RZ-Platforming process is a fixed bed system that is well suited for use in aromatics production facilities, particularly for those producers who require large amounts of benzene. The process uses the RZ-100 catalyst to convert feedstock components (C_6 and C_7 paraffins) into aromatic derivatives. The process is primarily used for situations where higher yields of benzene and toluene are desired. The ability of the process to handle low-boiling paraffin feedstocks and its flexibility in processing straight-run naphtha fractions provide many options for improving aromatics production and supplying needed hydrogen, either in new units or in existing aromatics facilities.

6.6.4 FUEL OIL

Fuel oil is classified in several ways but generally may be divided into two main types: *distillate fuel oil* and *residual fuel oil*. Distillate fuel oil is vaporized and condensed during a distillation process and thus have a definite boiling range and do not contain high-boiling constituents. A fuel oil that contains any amount of the residue from crude distillation of thermal cracking is a residual fuel oil. The terms *distillate fuel oil* and *residual fuel oil* are losing their significance, since fuel oil is now made for specific uses and may be either distillates or residuals or mixtures of the two. The terms *domestic fuel oil*, *diesel fuel oil*, and *heavy fuel oil* are more indicative of the uses of fuel oils.

Heavy fuel oil comprises all residual fuel oils (including those obtained by blending). Heavy fuel oil constituents range from distillable constituents to residual (non-distillable) constituents that must be heated to 260°C (500°F) or more before they can be used. The kinematic viscosity is above 10 centistokes at 80°C (176°F). The flash point is always above 50°C (122°F) and the density is always higher than 0.900. In general, heavy fuel oil usually contains cracked residua, reduced crude, or cracking coil heavy product which is mixed (cut back) to a specified viscosity with cracked gas oils and fractionator bottoms. For some industrial purposes in which flames or flue gases contact the product (ceramics, glass, heat treating, and open hearth furnaces) fuel oils must be blended to contain minimum sulfur contents, and hence low-sulfur residues are preferable for these fuels. Example of fuel oil types are:

No. 1 fuel oil is a petroleum distillate that is one of the most widely used of the fuel oil types. It is used in atomizing burners that spray fuel into a combustion chamber where the tiny droplets burn while in suspension. It is also used as a carrier for pesticides, a weed killer, a mold release agent in the ceramic and pottery industry, and in the cleaning industry. It is found in asphalt coatings, enamels, paints, thinners, and varnishes. No. 1 fuel oil is a light petroleum distillate (straight-run kerosene) consisting primarily of hydrocarbons in the range C_9-C_{16}. Fuel oil #1 is very similar in composition to diesel fuel; the primary difference is in the additives.

No. 2 fuel oil is a petroleum distillate that may be referred to as domestic or industrial. The domestic fuel oil is usually lower-boiling and a straight-run product. It is used primarily for home heating. Industrial distillate is a cracked product or a blend of both. It is used in smelting furnaces, ceramic kilns, and packaged boilers. No. 2 fuel oil is characterized by hydrocarbon chain lengths in the C_{11}-C_{20} range. The composition consists of aliphatic hydrocarbon derivatives (straight chain alkanes and cycloalkanes) (64%), 1%–2% unsaturated hydrocarbon derivatives (olefin derivatives)

(1%–2%), and aromatic hydrocarbon derivatives (including alkyl benzenes and 2-ring, 3-ring aromatic derivatives) (35%) but contains only low amounts (<5%) of the polycyclic aromatic hydrocarbon derivatives.

No. 6 fuel oil (also called *Bunker C oil* or *residual fuel oil*) is the residuum from crude oil after naphtha-gasoline, no. 1 fuel oil, and no. 2 fuel oil have been removed. No. 6 fuel oil can be blended directly to heavy fuel oil or made into asphalt. Residual fuel oil is more complex in composition and impurities than distillate fuels. Limited data are available on the composition of no. 6 fuel oil. Polycyclic aromatic hydrocarbon derivatives (including the alkylated derivatives) and metal-containing constituents are components of no. 6 fuel oil.

Since the boiling ranges, sulfur contents, and other properties of even the same fraction vary from crude oil to crude oil and with the way the crude oil is processed, it is difficult to specify which fractions are blended to produce specific fuel oils. In general, however, furnace fuel oil is a blend of straight-run gas oil and cracked gas oil to produce a product boiling in the 175°C–345°C (350°F–50°F) range.

Residual fuel oil is generally known as the bottom product from atmospheric distillation units (Parkash, 2003; Gary et al., 2007; Speight, 2014a; Hsu and Robinson, 2017; Speight, 2017). Fuel oils produced from cracking units are unstable. When used as fuels, they produce smoke and deposits that may block the burner orifices. The constituents of residual fuels are more complex than those of gas oils. A major part of the polynuclear aromatic compounds, asphaltenes, and heavy metals found in crude oils is concentrated in the residue. The main use of residual fuel oil is for power generation. It is burned in direct-fired furnaces and as a process fuel in many petroleum and chemical companies. Due to the low market value of fuel oil, it is used as a feedstock to catalytic and thermal cracking units.

6.6.4.1 Physical Properties

The physical properties of fuel oil are dependent upon the grade and method of production. In general terms, fuel oil is any liquid fuel that is burned in a furnace or boiler for the generation of heat or used in an engine for the generation of power. The term *fuel oil* is also used in a stricter sense to refer only to the highest-boiling commercial fuel that can be obtained from crude oil.

6.6.4.2 Chemical Properties

As with the physical properties, the chemical properties of fuel oil are dependent upon the grade and method of production. Typically, fuels oil grades consist of higher molecular weight hydrocarbon derivatives, particularly alkane derivative, cycloalkane derivatives, and aromatic derivatives. The chain length varies with the type of fuel oil. For example:

Name	Type	Chain Length[a]
No. 1 fuel oil	Distillate	9–16
No. 2 fuel oil	Distillate	10–20
No. 6 fuel oil	Residual	20–70

[a] For illustrative purposes only.

More specifically, all fuel oils consist of complex mixtures of aliphatic and aromatic hydrocarbons. The aliphatic alkanes (paraffins) and cycloalkanes (naphthenes) are hydrogen saturated and compose approximately 80%–90% of the fuel oils. Aromatics (e.g., benzene) and olefins (e.g., styrene and indene) compose 10%–20% and 1%, respectively, of the fuel oils. Fuel oil no. 1 (straight-run kerosene) is a light distillate which consists primarily of hydrocarbons in the C_9–C_{16} range; fuel oil no. 2 is a heavier, usually blended, distillate with hydrocarbons in the C_{11}–C_{20} range. Straight-run distillates may also be used to produce fuel oil no. 1 and diesel fuel oil no. 1. Diesel fuel no. 1 and no. 2 are similar in chemical composition to fuel oil no. 1 and fuel oil no. 2, respectively, with the exception of the additives.

Diesel fuels predominantly contain a mixture of C_{10}–C_{19} hydrocarbons, which include approximately 64% aliphatic hydrocarbons, 1%–2% olefinic hydrocarbons, and 35% aromatic hydrocarbons. Jet fuels are based primarily on straight-run kerosene, as well as additives. All of the above fuel oils contain less than 5% polycyclic aromatic hydrocarbons. Fuel no. 4 (marine diesel fuel) is less volatile than diesel fuel no. 2 and may contain up to 15% residual process streams, in addition to more than 5% polycyclic aromatic hydrocarbons. Residual fuel oils are generally more complex in composition and impurities than distillate fuel oils; therefore, a specific composition cannot be determined. Sulfur content in residual fuel oils has been reported to be from 0.18% to 4.36% by weight.

6.6.4.3 Chemicals from Fuel Oil

Many compound classes have been identified in fuel oils but the separation of individual chemicals (or even chemical streams) from such a mixture is an indomitable task and, as a result, gas oil (like other complex petroleum products) best serves as a cracking stock to produce the starting materials for petrochemical production.

Residues containing high levels of heavy metals are not suitable for catalytic cracking units. These feedstocks may be subjected to a demetallization process to reduce their metal contents. For example, the metal content of vacuum residues could be substantially reduced by using a selective organic solvent such as pentane or hexane, which separates the residue into oil (with a low metal and asphaltene content) and asphalt (with high metal content). Demetallized oils could be processed by direct hydrocatalysis (Parkash, 2003; Gary et al., 2007; Speight, 2014a; Hsu and Robinson, 2017; Speight, 2017)

Another approach used to reduce the harmful effects of heavy metals in petroleum residues is metal passivation. In this process, an oil-soluble treating agent containing antimony is used that deposits on the catalyst surface in competition with contaminant metals, thus reducing the catalytic activity of these metals in promoting coke and gas formation. Metal passivation is especially important in fluid catalytic cracking processes. Additives that improve fluid catalytic cracking processes were found to increase catalyst life and improve the yield and quality of products.

Residual fuel oils (Chapter 2) with high heavy metal content can serve as feedstocks for thermal cracking units such as delayed coking. Low-metal fuel oils are suitable feedstocks to catalytic cracking units. Product gases from cracking units may be used as a source for light olefin derivatives and liquefied petroleum gas for petrochemical production. Residual fuel oils are also feedstocks for steam cracking units for the production of olefin derivatives.

6.6.5 Resids

A *resid* (*residuum, pl. residua*) is the residue obtained from petroleum after nondestructive distillation has removed all the volatile materials (Parkash, 2003; Gary et al., 2007; Speight, 2014a; Hsu and Robinson, 2017; Speight, 2017). The temperature of the distillation is usually maintained below 350°C (660°F) since the rate of thermal decomposition of petroleum constituents is minimal below this temperature but the rate of thermal decomposition of petroleum constituents is substantial above 350°C (660°F) (Parkash, 2003; Gary et al., 2007; Speight, 2014a; Hsu and Robinson, 2017; Speight, 2017). *Resids* are black, viscous materials and are obtained by distillation of a crude oil under atmospheric pressure (atmospheric residuum) or under reduced pressure (vacuum residuum). They may be liquid at room temperature (generally atmospheric residua) or almost solid (generally vacuum residua) depending upon the nature of the crude oil.

When a residuum is obtained from a crude oil and thermal decomposition has commenced, it is more usual to refer to this product as *pitch* (Speight, 2014a)) The differences between a parent petroleum and the residua are due to the relative amounts of various constituents present, which are removed or remain by virtue of their relative volatility.

6.6.5.1 Physical Properties

The chemical composition of a residuum from an asphaltic crude oil is complex. Physical methods of fractionation usually indicate high proportions of asphaltenes and resins, even in amounts up to 50% (or higher) of the residuum. In addition, the presence of ash-forming metallic constituents, including such organometallic compounds as those of vanadium and nickel, is also a distinguishing feature of residua and the heavier oils. Furthermore, the deeper the *cut* into the crude oil, the greater is the concentration of sulfur and metals in the residuum and the greater the deterioration in physical properties (Chapter 17).

6.6.6 USED LUBRICATING OIL

Used lubricating oil—often referred to as *waste oil* without further qualification—is any lubricating oil, whether refined from crude or synthetic components, which has been contaminated by physical or chemical impurities as a result of use (Speight and Exall, 2014). Lubricating oil loses its effectiveness during operation due to the presence of certain types of contaminants. These contaminants can be divided into: (i) extraneous contaminants and (ii) products of oil deterioration. Extraneous contaminants are introduced from the surrounding air and by metallic particles from the engine. Contaminants from the air are dust, dirt, and moisture—in fact, air itself may be considered as a contaminant since it can cause foaming of the oil. The contaminants from the engine are: (i) metallic particles resulting from wear of the engine, (ii) carbonaceous particles due to incomplete fuel combustion, (iii) metallic oxides present as corrosion products of metals, (iv) water from leakage of the cooling system, (v) water as a product of fuel combustion, and (vi) fuel or fuel additives or their byproducts, which might enter the crankcase of engines.

In terms of the products of oil deterioration, many products are formed during oil deterioration. Some of these important products are: (i) sludge, which is a mixture of oil, water, dust, dirt, and carbon particles that results from the incomplete combustion of the fuels. Sludge may deposit on various parts of the engine or remain in colloidal dispersion in the oil; (ii) lacquer, which is a hard or gummy substance that deposits on engine parts as a result of subjecting sludge in the oil to high temperature operation; and (iii) oil-soluble products, which result from oxidation and remain in the oil and cannot be filtered out and deposit on the engine parts. The quantity and distribution of engine deposits vary widely depending on the conditions at which the engine is operated. At low temperatures, carbonaceous deposits originate mainly from incomplete combustion products of the fuel and not from the lubricating oil. At high temperature, the increase in lacquer and sludge deposits may be caused by the lubricating oil.

6.6.7 NAPHTHENIC ACIDS

Naphthenic acids are a mixture of cycloparaffins with alkyl side chains ending with a carboxylic group (Speight, 2014d). The low molecular weight naphthenic acids (8–12 carbons) are compounds having either a cyclopentane or a cyclohexane ring with a carboxyalkyl side chain. These compounds are normally found in middle distillates such as kerosene and gas oil. Naphthenic acids constitute about 50% w/w of the total acidic compounds in crude oils. Naphthenic-based crude oils contain a higher percentage of naphthenic acids. Consequently, it is more economical to isolate these acids from naphthenic-based crude oils.

The production of naphthenic acids from middle distillates occurs by extraction with 7%–10% caustic solution.

The sodium salts, which are soluble in the lower aqueous layer, are separated from the hydrocarbon layer and treated with a mineral acid to spring out the acids. The free acids are then dried and distilled. Using strong caustic solutions for the extraction may create separation problems because naphthenic acid salts are emulsifying agents.

Free naphthenic acids are corrosive and are mainly used as their salts and esters. The sodium salts are emulsifying agents for preparing agricultural insecticides, additives for cutting oils, and emulsion breakers in the oil industry. Other metal salts of naphthenic acids have many varied uses. For example, calcium naphthenate is a lubricating oil additive, and zinc naphthenate is an antioxidant. Lead, zinc, and barium naphthenate derivatives are wetting agents used as dispersion agents for paints. Some oil-soluble metal naphthenate derivatives, such as those of zinc, cobalt, and lead, are used as driers in oil-based paints. Among the diversified uses of naphthenate derivatives is the use of aluminum naphthenate derivatives as gelling agents for gasoline flame throwers (napalm). Manganese naphthenate derivatives are well-known oxidation catalysts.

Cresylic acid is a commercial mixture of phenolic compounds including phenol, cresol derivatives, and xylenol derivatives. This mixture varies widely according to its source. Cresylic acid derivatives constitute part of the oxygen compounds found in crudes that are concentrated in the naphtha fraction obtained principally from naphthenic and asphaltic-based crudes. Phenolic compounds, which are weak acids, are extracted with relatively strong aqueous caustic solutions.

Originally, cresylic acid was obtained from caustic waste streams that resulted from treating light distillates with caustic solutions to reduce H_2S and mercaptans. Currently, most of these streams are hydrodesulfurized, and the product streams practically do not contain phenolic compounds. However, cresylic acid is still obtained to a lesser extent from petroleum fractions, especially cracked gasolines, which contain higher percentages of phenols. It is also extracted from coal liquids. Strong alkaline solutions are used to extract cresylic acid. The aqueous layer contains, in addition to sodium phenate and cresylate, a small amount of sodium naphthenate derivatives and sodium mercaptide derivatives. The reaction between cresols and sodium hydroxide gives sodium cresylate.

Mercaptans in the aqueous extract are oxidized to the disulfides, which are insoluble in water and can be separated from the cresylate solution by decantation:

$$2RSH + O_2 \rightarrow RSSR + H_2O$$

Free cresylic acid is obtained by treating the solution with a weak acid or dilute sulfuric acid. Refinery flue gases containing carbon dioxide are sometimes used to release cresylic acid. Aqueous streams with low cresylic acid concentrations are separated by adsorption by passing them through one or more beds containing a high adsorbent resin. The resin is regenerated with 1% sodium hydroxide solution. It should be noted that the extraction of cresylic acid does not create an isolation problem with naphthenic acids which are principally present in heavier fractions. Naphthenic acids, which are relatively stronger acids (lower pKa value), are extracted with less concentrated caustic solution.

Cresylic acid is mainly used as degreasing agent and as a disinfectant of a stabilized emulsion in a soap solution. Cresols are used as flotation agents and as wire enamel solvents. Tricresyl phosphate derivatives are produced from a mixture of cresols and phosphorous oxychloride. The esters are plasticizers for vinyl chloride polymers and are also used as gasoline additives for reducing carbon deposits in the combustion chamber.

6.6.8 CHEMICALS FROM LIQUID PETROLEUM FRACTIONS AND RESIDUES

In texts on the production of petrochemical products, much space is typically given to the use of the gaseous hydrocarbon derivatives as feedstocks for petrochemical processes. The higher-boiling fraction of petroleum (naphtha, kerosene, fuel oil, gas oil, and residua) is not always included. It is, therefore, appropriate at this point of the text to include such fractions as feedstocks for petrochemical production.

6.6.8.1 Oxidation

Oxidation is a process in which a chemical change because of the addition of oxygen or the interaction of oxygen with the chemical to remove hydrogen (i.e., oxidative dehydrogenation). Oxidation can occur in the presence or absence of a catalyst.

The catalytic oxidation of long-chain paraffin derivatives (C_{18}–C_{30} derivatives) over manganese salts produces a mixture of fatty acids with different chain lengths. Temperature and pressure ranges of 105°C–120°C (220°F–250°F) and 220–900 psi, respectively, are used. About 60% w/w yield of fatty acid derivatives up to the C_{14} fatty acid derivatives is obtained.

$$2RCH_2 \left(CH_2\right)_n CH_2CH_2R + 5O_2 \rightarrow R\left(CH_2\right)_n CO_2H + RCH_2CO_2H + H_2O$$

These acids are used for making soaps. The main source for fatty acids for soap manufacture, however, is the hydrolysis of fats and oils (a nonpetroleum source).

n-Paraffin derivatives can also be oxidized to alcohols by a dilute oxygen stream (3%–4%, oxygen by volume) in the presence of a mineral acid. The acid converts the alcohols to esters, which prohibit further oxidation of the alcohols to fatty acids. The obtained alcohols are also secondary. These alcohols are of commercial importance for the production of nonionic detergents (ethoxylate derivatives).

6.6.8.2 Chlorination

Chlorination is a reaction that falls under the groups of reactions known as *halogenation*. The ease of halogenation is influenced by the halogen—fluorine (fluorination) and chlorine are more electrophilic and are more aggressive halogenating agents while bromine (bromination) is a weaker halogenating agent than both fluorine and chlorine, and iodine (iodination) is the least reactive halogen. The facility of dehydrohalogenation follows the reverse trend: iodine is most easily removed from organic compounds, and organo-fluorine compounds are highly stable.

In the current context, both saturated and unsaturated compounds react directly with chlorine, the former usually requiring UV light to initiate homolysis of chlorine. Chlorination is conducted on a large-scale industrially; major processes include routes to 1,2-dichloroethane (a precursor to polyvinyl chloride), as well as various chlorinated ethane derivatives as solvents.

Chlorination of n-paraffin derivatives (C_{10}–C_{14}) in the liquid phase produces a mixture of chloroparaffin derivatives. Selectivity to monochlorination could be increased by limiting the reaction to a low conversion and by decreasing the chlorine to hydrocarbon ratio. Substitution of secondary hydrogen predominates. Thus:

$$RCH_2CH_2R' + Cl_2 \rightarrow RCHClCH_2R' + HCl$$

Monochloro-paraffin derivatives in this range may be dehydrochlorinated to the corresponding mono-olefin derivatives and used as alkylating agents for the production of biodegradable detergents. Alternatively, the Monochloro-paraffin derivatives are used directly to alkylate benzene in presence of a Lewis acid catalyst to produce alkylates for the detergent production. On the other hand, polychlorination can be carried out on the whole range of n-paraffin derivatives from C_{10} to C_{30} at a temperature range of 80°C–120°C (176°F–248°F) using a high chlorine-paraffin ratio. The product has a chlorine content of approximately 70%. Polychloro paraffin derivatives are used as cutting oil additives, plasticizers, and retardant chemicals.

6.6.8.3 Sulfonation

Sulfonation is a chemical reaction in which the sulfonic acid functional group (SO_3H) is introduced into a molecule (Michael and Weiner, 1936). For example, sulfonation with sulfur trioxide and sulfuric acid converts benzene into benzene sulfonic acid:

Linear secondary alkane sulfonates are produced by the reaction between sulfur dioxide and C_{15}–C_{17} n-paraffin derivatives:

$$RH + SO_2 + O_2 + H_2O \rightarrow RSO_3H + H_2SO_4$$

The reaction is catalyzed by ultraviolet light with a wavelength between 3,300 and 3,600 A. The sulfonate derivatives are nearly 100% biodegradable, soft and stable in hard water, and have good washing properties.

Sodium alkanesulfonates can also be produced from the free radical addition of sodium bisulfite and alpha olefin derivatives:

$$RCH{=}CH_2 + NaHSO_3 \rightarrow RCH_2CH_2SO_3Na$$

The sulfonation reaction is an important reaction in chemistry and is used in many aspects of the petrochemical industry, such as color developers, flame retardants, and pharmaceutical products.

6.6.8.4   Other Products

Cresylic acid is mainly used as degreasing agent and as a disinfectant of a stabilized emulsion in a soap solution. Cresol derivatives are used as flotation agents and as wire enamel solvents. Tricresyl phosphate derivatives are produced from a mixture of cresols and phosphorous oxychloride. The esters are plasticizers for vinyl chloride polymers and are also used as gasoline additives for reducing carbon deposits in the combustion chamber. Cresylic acid is also used in resins, disinfectants, solvents, and electrical insulation.

Naphthenic acid is removed from petroleum fractions not only to minimize corrosion, but also to recover commercially useful products. The greatest current and historical usage of naphthenic acid is in metal naphthenate derivatives. Naphthenic acids are recovered from petroleum distillates by alkaline extraction and then regenerated via an acidic neutralization process and then distilled to remove impurities. Naphthenic acids sold commercially are categorized by acid number, impurity level, and color, and used to produce metal naphthenate derivatives and other derivatives such as esters and amides.

Salts of naphthenic acids are widely used as hydrophobic sources of metal ions in diverse applications. Aluminum salts of naphthenic acids and palmitic acid [hexadecanoic acid, $CH_3(CH_2)_{14}CO_2H$] were combined during World War II to produce napalm.

REFERENCES

Albright, L.F., Crynes, B.L., and Nowak, S. 1992. *Novel Production Methods for Ethylene, Light Hydrocarbons, and Aromatic derivatives*. Marcel Dekker Inc., New York.

Al-Megren, H., and Xiao, T. 2016. *Petrochemical Catalyst Materials, Processes, and Emerging Technologies*. IGI Global, Hershey, PA.

Alvarez-Galvan, M.C., Mota, N., and Ojeda, M. 2011. Direct Methane Conversion Routes to Chemicals and Fuels. *Catalysis Today*, 171(1): 15–23.

Arutyunov, V.S. 2007. Partial Oxidation of Hydrocarbon Gases as a Base for New Technological Processes in Gas and Power Production. *Studies in Surface Science and Catalysis*, 167: 269–274.

ASTM D235. 2018. *Standard Specification for Mineral Spirits (Petroleum Spirits) (Hydrocarbon Dry Cleaning Solvent)*. Annual Book of Standards. ASTM International, West Conshohocken, PA.

ASTM D3246. 2018. *Standard Test Method for Sulfur in Petroleum Gas by Oxidative Microcoulometry*. Annual Book of Standards. ASTM International, West Conshohocken, PA.

Avila, B.M.F., Vaz, B.G., Pereira, R., Gomes, A.O., Pereira, R.C.L., Corilo, Y.E., Simas, R.C., Lopes Nascimento, H.D., Eberlin, M.N., and Azevedo, D.A. 2012. Comprehensive Chemical Composition of Gas Oil Cuts Using Two-Dimensional Gas Chromatography with Time-of-Flight Mass Spectrometry and Electrospray Ionization Coupled to Fourier Transform Ion Cyclotron Resonance Mass Spectrometry. *Energy & Fuels*, 26(8): 5069–5079.

Chenier, P.J. 2002. *Survey of Industrial Chemicals*. 3rd Edition. Springer, New York.

Dumez, F.J., and Froment, G.F. 1976. Dehydrogenation of 1-Butene into Butadiene. Kinetics, Catalyst Coking, and Reactor Design. *Industrial and Engineering Chemistry Process Design and Development*, 15(2): 291–301.

Dutta, A., Chit, C.W., Iftekhar A Karimi, I.A., and Farooq, S. 2017. Ethylene from Natural Gas Via Oxidative Coupling of Methane and Cold Energy of LNG. *Proceedings of 27th European Symposium on Computer Aided Process Engineering—ESCAPE 27*. A. Espuña, M. Graells, and L. Puigjaner (Editors). October 1–5, Barcelona, Spain. Elsevier B.V., Amsterdam, Netherlands. www.sciencedirect.com/science/article/pii/B9780444639653503111

Gary, J.G., Handwerk, G.E., and Kaiser, M.J. 2007. *Petroleum Refining: Technology and Economics*. 5th Edition. CRC Press, Boca Raton, FL.

Gosling, C.D., Wilcher, F.P., Sullivan, L., and Mountiford, R.A. 1999. Process LPG to BTX Products. *Hydrocarbon Processing*, 69, December 1991.

Hsu, C.S., and Robinson, P.R. (Editors). 2017. *Handbook of Petroleum Technology*. Springer, Cham, Switzerland.

Katz, D.K. 1959. *Handbook of Natural Gas Engineering*. McGraw-Hill, New York.

Kohl, A.L., and Nielsen, R.B., 1997. *Gas Purification*. Gulf Publishing Company, Houston, TX.

Kohl, A. L., and Riesenfeld, F.C. 1985. *Gas Purification*. 4th Edition, Gulf Publishing Company, Houston, TX.

Maddox, R.N., Bhairi, A., Mains, G.J., and Shariat, A. 1985. Chapter 8. In: *Acid and Sour Gas Treating Processes*. S.A. Newman (Editor). Gulf Publishing Company, Houston, TX.

Michael, A., and Weiner, N. 1936. The Mechanism of the Sulfonation Process. *Journal of the American Chemical Society*, 58(2): 294–299.

Mokhatab, S., Poe, W.A., and Speight, J.G. 2006. *Handbook of Natural Gas Transmission and Processing*. Elsevier, Amsterdam, The Netherlands.

Newman, S.A. 1985. *Acid and Sour Gas Treating Processes*. Gulf Publishing Company, Houston, TX.

Ortiz-Espinoza, A.P., El-Halwagi, M.M., and Jiménez-Gutiérrez, A. 2015, Analysis of Two Alternatives to Produce Ethylene from Shale Gas. *Computer Aided Chemical Engineering*, 37: 485–490.

Park, S., Lee, Y., Kim, G., and Hwang, S. 2016. Production of Butene and Butadiene by Oxidative Dehydrogenation of Butane over Carbon Nanomaterial Catalysts. *Korean Jounral of Chemical Engineering*, 33: 3417–3424.

Parkash, S. 2003. *Refining Processes Handbook*. Gulf Professional Publishing, Elsevier, Amsterdam, The Netherlands.

Passmann, W. 1970. Modern Production Methods Based on 1,3-Butadiene and 1-Butene. *Industrial and Engineering Chemistry*, 62(5): 48–51.

Qin, Z., Lissianski, V.V., Yang, H., Gardiner, W.C., Davis, S.G., and Wang, H. 2000. Combustion Chemistry of Propane: A Case Study Of Detailed Reaction Mechanism Optimization. *Proceedings of the Combustion Institute*, 28: 1663–1669.

Rabiu, A.M., and Yusuf, I.M. 2013. Industrial Feasibility of Direct Methane Conversion to Hydrocarbons over Fe-Based Fischer Tropsch Catalyst. *Journal of Power and Energy Engineering*, 1: 41–46.

Rozanov, V.N., and Treger, Y.A. 2010. Kinetics of the Gas- Phase Thermal Chlorination of Methane. *Kinetics and Catalysis*, 51(5): 635–643.

Salkuyeh, Y.K., and Adams, T.A. 2015, A Novel Polygeneration Process to Co-Produce Ethylene and Electricity from Shale Gas with Zero CO2 Emissions Via Methane Oxidative Coupling. *Energy Conservation and Management*, 92: 406–420.

Singh, D, Pratap, D., Vashishtha, M., and Mathur, A.K. 2010. Direct Catalytic Conversion of Biogas Methane to Formaldehyde. *International Journal of ChemTech Research*, 2(1): 476–482.

Speight, J.G. 2007. *Natural Gas: A Basic Handbook*. GPC Books, Gulf Publishing Company, Houston, TX.

Speight, J.G. (Editor). 2011. *The Biofuels Handbook*. Royal Society of Chemistry, London, UK.

Speight, J.G. 2013. *The Chemistry and Technology of Coal*. 3rd Edition. CRC Press, Boca Raton, FL.

Speight, J.G. 2014a. *The Chemistry and Technology of Petroleum*. 4th Edition. CRC Press, Boca Raton, FL.

Speight, J.G. 2014b. *Oil and Gas Corrosion Prevention*. Gulf Professional Publishing, Elsevier, Oxford, UK.

Speight, J.G. 2014c. *Gasification of Unconventional Feedstocks*. Gulf Professional Publishing, Elsevier, Oxford, UK.

Speight, J.G. 2014d. *High Acid Crudes*. Gulf Professional Publishing, Elsevier, Oxford, UK.

Speight, J.G., and Exall, D.I. 2014. *Refining Used Lubricating Oils*. CRC Press, Boca Raton, FL.

Speight, J.G. 2017. *Handbook of Petroleum Refining*. CRC Press, Boca Raton, FL.

Treger, Y.A., Rozanov, V.N., and Sokolova, S.V. 2012. Producing Ethylene and Propylene from Natural Gas via the Intermediate Synthesis of Methyl Chloride and Its Sub- sequent Catalytic Pyrolysis. *Catalysis in Industry*, 4(4): 231–235.

Vincent, R.S., Lindstedt, R.P., Malika, N.A., Reid, I.A.B., and Messenger, B.E. 2008. The Chemistry of Ethane Dehydrogenation over a Supported Platinum Catalyst. *Journal of Catalysis*, 260: 37–64.

Wang, C., Xu, L., and Wang, Q. 2003. Review of Directly Producing Light Olefins via Co-Hydrogenation. *Journal of Natural Gas Chemistry*, 12(1): 10–16.

Yan, W., Luo. J., Kouk, Q-Y., Zheng, J.E., Zhong, Z., Liu, Y., and Borgna, A. 2015. Improving Oxidative Dehydrogenation of 1-Butene to 1,3-Butadiene on Al_2O_3 by Fe_2O_3 Using CO_2 as a Soft Oxidant. *Applied Catalysis A; General*, 508(11): 61–67.

7 Chemicals from Olefin Hydrocarbons

7.1 INTRODUCTION

Olefin derivatives (C_nH_{2n}, such as ethylene $CH_2=CH_2$) are the basic building blocks for a host of chemical products. These unsaturated materials enter into polymers and rubbers and with other reagents and react to form a wide variety of useful compounds, including alcohols, epoxides, amines, and halides.

Olefin derivatives are present in the gaseous products of catalytic cracking processes (Parkash, 2003; Gary et al., 2007; Speight, 2014; Hsu and Robinson, 2017; Speight, 2017) that offer promising source materials. Cracking paraffin hydrocarbon derivatives and heavy oils also produces olefin derivatives. For example, cracking ethane, propane, butane, and other feedstock such as gas oil, naphtha, and residua produces ethylene. Propylene (also known as propene or methyl ethylene) is produced from thermal and catalytic cracking of naphtha and gas oils, as well as propane and butane.

The most important olefin derivatives used for the production of petrochemicals are ethylene ($CH_2=CH_2$), propylene ($CH_2CH=CH_2$), the butylene isomers ($CH_3CH_2CH=CH_2$ and $CH_3CH=CHCH_3$), and isoprene [$CH_2=C(CH_3)CH=CH_2$]. Olefin derivatives are not typical constituents of natural gas but do occur in refinery gas, which can be complex mixtures of hydrocarbon gases (Table 7.1) and non-hydrocarbon gases (Chapter 1). Many low molecular weight olefin derivatives and diolefin derivatives which are produced in the refinery are isolated for petrochemical use (Speight, 2014). The individual products are: (i) ethylene, (ii) propylene, and (iii) 1,3-butadiene ($CH_2=CHCH=CH_2$).

Butadiene can be recovered from refinery streams as butadiene, as butylene derivatives, or as butane derivatives; the latter two on appropriate heated catalysts dehydrogenate to give 1,3-butadiene:

$$CH_2=CHCH_2CH_3 \rightarrow CH_2=CHCH=CH_2 + H_2.$$

$$CH_3CH_2CH_2CH_3 \rightarrow CH_2=CHCH=CH_2$$

An alternative source of butadiene is ethanol, which on appropriate catalytic treatment also gives the compound diolefin:

$$2C_2H_5OH \rightarrow CH_2=CHCH=CH_2 + 2H_2O$$

Olefin derivatives present in the gaseous product streams from catalytic cracking processes offer promising source of these materials. Cracking paraffin hydrocarbon derivatives and heavy oils also produces olefin derivatives. For example, cracking ethane, propane, butane, and other feedstock such as gas oil, naphtha, and residua produces ethylene. Propylene is produced from thermal and catalytic cracking of naphtha and gas oils, as well as propane and butane.

As far as can be determined, the first large-scale petrochemical process was the sulfuric acid absorption of propylene ($CH_3CH=CH_2$) from refinery cracked gases to produce isopropyl alcohol [$(CH_3)_2CHOH$].

$$CH_3CH=CH_2 + H_2O \rightarrow (CH_3)_2 CHOH$$

TABLE 7.1

Possible Constituents of Natural Gas and Refinery Process Gas Streams

Gas	Molecular Weight	Boiling Point 1 atm °C (°F)	Density at 60°F (15.6°C), 1 atm	
			g/L	Relative to Air = 1
Methane	16.043	−161.5 (−258.7)	0.6786	0.5547
Ethylene	28.054	−103.7 (−154.7)	1.1949	0.9768
Ethane	30.068	−88.6 (−127.5)	1.2795	1.0460
Propylene	42.081	−47.7 (−53.9)	1.8052	1.4757
Propane	44.097	−42.1 (−43.8)	1.8917	1.5464
1,2-Butadiene	54.088	10.9 (51.6)	2.3451	1.9172
1,3-Butadiene	54.088	−4.4 (24.1)	2.3491	1.9203
1-Butene	56.108	−6.3 (20.7)	2.4442	1.9981
cis-2-Butene	56.108	3.7 (38.7)	2.4543	2.0063
trans-2-Butene	56.108	0.9 (33.6)	2.4543	2.0063
isobutene	56.104	−6.9 (19.6)	2.4442	1.9981
n-Butane	58.124	−0.5 (31.1)	2.5320	2.0698
isobutane	58.124	−11.7 (10.9)	2.5268	2.0656

The interest in thermal reactions of hydrocarbon derivatives has been high since the 1920s when alcohols were produced from the ethylene and propylene formed during petroleum cracking. The range of products formed from petroleum pyrolysis has widened over the past six decades to include the main chemical building blocks. These include ethane, ethylene, propane, propylene, butane derivatives, butadiene, and aromatic derivatives. Additionally, other commercial products from thermal reactions of petroleum include coke, carbon, and asphalt.

Ethylene manufacture is achieved using a variety of processes (Table 7.2); of which the steam cracking process is in widespread practice throughout the world. The operating facilities are similar to gas oil cracking units, operating at temperatures of 840°C (1550°F) and at low pressures (24 psi) (Parkash, 2003; Gary et al., 2007; Speight, 2014; Hsu and Robinson, 2017; Speight, 2017). Steam is added to the vaporized feed to achieve a 50–50 mixture, and furnace residence times are only 0.2–0.5 s. Ethane extracted from natural gas is the predominant feedstock for ethylene cracking units. Propylene and butylene are largely derived from catalytic cracking units and from cracking a naphtha or low-boiling gas oil fraction to produce a full range of olefin products.

TABLE 7.2

Example of Processes By Which Ethylene Is Produced

- Thermal cracking
- Fluidized bed cracking
- Catalytic pyrolysis and catalytic partial oxidation
- Membrane dehydrogenation of ethane
- Oxidative dehydrogenation of ethane by using nickel oxide-based catalyst
- Methane oxidative coupling technology
- Dehydration of ethanol
- Methanol conversion to ethylene
- Disproportionation of propylene
- Ethylene from coal by the Fischer–Tropsch process
- Ethylene reclamation from the refinery byproduct and off-gases

The majority of the propylene used in the petrochemical industry is made from propane, which is obtained from natural gas stripper plants or from refinery gases (Parkash, 2003; Gary et al., 2007; Speight, 2014; Hsu and Robinson, 2017; Speight, 2017):

$$CH_3CH_2CH_3 \rightarrow CH_3CH=CH_2 + H_2.$$

The uses of propylene include gasoline, polypropylene, isopropanol, trimers, and tetramers for detergents, propylene oxide (PO), cumene, and glycerin.

Two butylene derivatives (1-butylene or 1-butene, $CH_3CH_2CH=CH_2$, and 2-butylene or 2-butene, $CH_3CH=CHCH_3$) are industrially significant. The latter has end uses in the production of butyl rubber and polybutylene plastics. On the other hand, 1-butylene is used in the production of 1,3-butadiene ($CH_2=CHCH-CH_2$) for the synthetic rubber industry. Butylene derivatives arise primarily from refinery gases or from the cracking of other fractions of crude oil.

Butadiene can be recovered from refinery streams as butadiene, as butylene derivatives, or as butanes; the latter two on appropriate heated catalysts dehydrogenate to give 1,3-butadiene:

$$CH_2=CHCH_2CH_3 \rightarrow CH_2=CHCH=CH_2 + H_2.$$

$$CH_3CH_2CH_2CH_3 \rightarrow CH_3=CHCH=CH_2$$

An alternative source of butadiene is ethanol, which on appropriate catalytic treatment also gives the compound diolefin:

$$2C_2H_5OH \rightarrow CH_2=CHCH=CH_2 + 2H_2O$$

Olefin derivatives containing more than four carbon atoms are in little demand as petrochemicals and thus are generally used as fuel. The single exception to this is 2-methyl-1,3-butadiene or isoprene, which has a significant use in the synthetic rubber industry. It is more difficult to make than is 1,3-butadiene. Some is available in refinery streams, but more is manufactured from refinery stream 2-butylene by reaction with formaldehyde:

$$CH_3CH=CHCH_3 + HCHO \rightarrow CH_2=CH(CH_3)CH=CH_2 + H_2O.$$

7.2 CHEMICALS FROM ETHYLENE

Ethylene (ethene, C_2H_4), the first member of the olefin series ($RCH=CH_2$, where R can be hydrogen atom or an alkyl group starting with the methyl group, CH_3), is a colorless gas with a sweet odor. It is slightly soluble in water and alcohol. Ethylene is a normally gaseous olefinic compound having a boiling point of approximately $-104°C$ ($-155°F$). It may be handled as a liquid at very high pressures and low temperatures (Table 7.3).

Ethylene is a valuable starting chemical because it is the source of a vast array of commercial chemicals. This unique position of ethylene among other hydrocarbon intermediates is due to some favorable properties inherent in the ethylene molecule such as: (i) simple structure with high reactivity, (ii) relatively inexpensive compound, (iii) easily produced from any hydrocarbon source through steam cracking and in high yields, and (iv) less byproducts generated from ethylene reactions with other compounds than from other olefin derivatives.

Ethylene is a constituent of refinery gases, especially those produced from catalytic cracking units. Ethylene is made normally by cracking an ethane or naphtha feedstock in a high-temperature furnace and subsequent isolation from other components by distillation. The major uses of ethylene are in the production of ethylene oxide, ethylene dichloride, and the polyethylene polymers. Other uses include the coloring of fruit, rubber products, ethyl alcohol, and medicine (anesthetic).

TABLE 7.3
Properties of Ethylene

Chemical formula	C_2H_4
Molar mass	28.05 g/mol
Appearance	Colorless gas
Density	1.178 kg/m³ at 15°C, gas
Melting point	−169.2°C (−272.6°F; 104.0K)
Boiling point	−103.7°C (−154.7°F; 169.5K)
Solubility in water	3.5 mg/100 mL (17°C); 2.9 mg/L
Solubility in ethanol	4.22 mg/L
Solubility in diethyl ether	Good

Ethylene manufacture via the steam cracking process is in widespread practice throughout the world. The operating facilities are similar to gas oil cracking units, operating at temperatures in the order of 840°C (1550°F) and at low pressure (24 psi).

Ethylene is a highly active compound that reacts easily by addition to many chemical reagents. For example, ethylene with water forms ethyl alcohol. Addition of chlorine to ethylene produces ethylene dichloride (1,2-dichloroethane, CH_2ClCH_2Cl), which is cracked to vinyl chloride ($CH_2=CHCl$), which is an important precursor for the manufacture of plastics.

$$CH_2ClCH_2Cl \rightarrow CH_2=CHCl + HCl$$

Ethylene is also an active alkylating agent. For example, the alkylation of benzene with ethylene produces ethyl benzene (EB, $C_6H_5C_2H_5$), which is dehydrogenated to styrene. Styrene is a monomer used in the manufacture of many commercial polymers and copolymers. Ethylene can be polymerized to different grades of polyethylene derivatives [$H(CH_2CH_2)_nH$] or copolymerized with other olefin derivatives. Catalytic oxidation of ethylene produces ethylene oxide, which is hydrolyzed to ethylene glycol (Figure 7.1). Ethylene glycol (CH_2OHCH_2OH) is a monomer for the production of synthetic fibers.

Ethylene reacts by addition to many inexpensive reagents such as water, chlorine, hydrogen chloride, and oxygen to produce valuable chemicals (Figure 7.2). It can be initiated by free radicals or

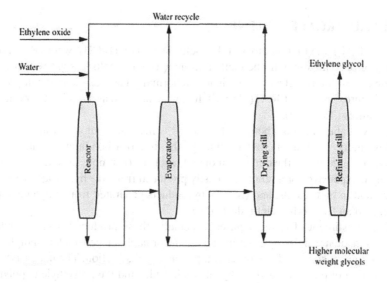

FIGURE 7.1   Manufacture of ethylene glycol.

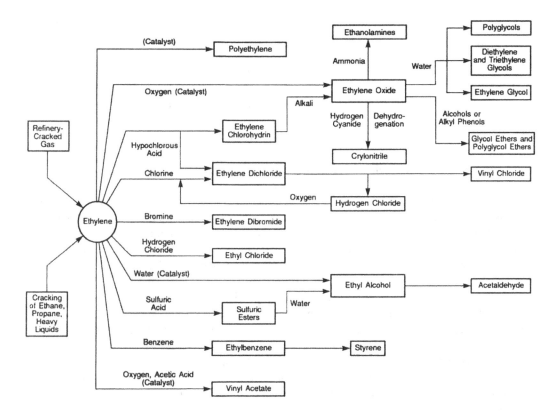

FIGURE 7.2 Chemicals from ethylene.

by coordination catalysts to produce polyethylene, the largest-volume thermoplastic polymer. It can also be copolymerized with other olefin derivatives producing polymers with improved properties. For example, when ethylene is polymerized with propylene, a thermoplastic elastomer is obtained.

7.2.1 ALCOHOLS

Most ethanol is produced from sugar (e.g., starch from maize grains or sucrose from sugarcane) fermentation using the yeast species *Saccharomyces cerevisiae*. Recent efforts have suggested that edible sugars be replaced with lignocellulosic biomass (e.g., corn stover) as the process feedstock, which may reduce the cost of bioethanol and decrease emissions of greenhouse gases simultaneously.

The earliest method for conversion of olefin derivatives into alcohols involved their absorption in sulfuric acid to form esters, followed by dilution and hydrolysis, generally with the aid of steam. In the case of ethyl alcohol, the direct catalytic hydration of ethylene can be employed. Ethylene is readily absorbed in 98%–100% sulfuric acid at 75°C–80°C (165°F–175°F), and both ethyl and diethyl sulfate are formed; hydrolysis takes place readily on dilution with water and heating.

The direct hydration of ethylene to ethyl alcohol is practiced over phosphoric acid on diatomaceous earth or promoted tungsten oxide under 100 psi pressure and at 300°C (570°F):

$$CH_2{=}CH_2 + H_2O \rightarrow C_2H_5OH$$

Ethylene of high purity is required in direct hydration than in the acid absorption process and the conversion per pass is low, but high yields are possible by recycling. Propylene and the normal butylene derivatives can also be hydrated directly.

Ethylene, produced from ethane by cracking, is oxidized in the presence of a silver catalyst to ethylene oxide:

$$2H_2C{=}CH_2 + O_2 \rightarrow C_2H_4O.$$

The vast majority of the ethylene oxide produced is hydrolyzed at 100°C to ethylene glycol:

$$C_2H_4O + H_2O \rightarrow HOCH_2CH_2OH.$$

Approximately 70% of the ethylene glycol produced is used as automotive antifreeze and a majority of the remainder is used in the synthesis of polyesters.

Of the higher olefin derivatives, one of the first alcohol syntheses practiced commercially was that of isopropyl alcohol from propylene. Sulfuric acid absorbs propylene more readily than it does ethylene, but care must be taken to avoid polymer formation by keeping the mixture relatively cool and using acid of about 85% strength at 300–400 psi pressure; dilution with inert oil may also be necessary. Acetone is readily made from isopropyl alcohol, either by catalytic oxidation or by dehydrogenation over metal (usually copper) catalysts.

1-Butanol may be produced from syngas via methanol and subsequent alcohol homologation; however, the currently favored route involves the stereo-selective rhodium-catalyzed hydroformylation of propylene to n-butyraldehyde followed by hydrogenation to 1-butanol (Scheme 1A). Alternatively, 1-butanol may be produced by microbial fermentation using organisms such as Clostridium acetobutylicum, which provides mixtures of acetone, 1-butanol, and ethanol (ABE fermentation), or other species that produce 1-butanol exclusively. The Guerbet reaction of bioethanol, which is more easily produced by fermentation and separated at higher titers, provides an alternative route for 1-butanol production.

Secondary butyl alcohol is formed on absorption of 1-butylene or 2-butylene by 78%–80% sulfuric acid, followed by dilution and hydrolysis. Secondary butyl alcohol is converted into methyl ethyl ketone (MEK) by catalytic oxidation or dehydrogenation.

There are several methods for preparing higher alcohols. One method in particular the so-called Oxo reaction and involves the direct addition of carbon monoxide (CO) and a hydrogen (H) atom across the double bond of an olefin to form an aldehyde (RCH=O), which in turn is reduced to the alcohol (RCH_2OH). Hydroformylation (the Oxo reaction) is brought about by contacting the olefin with synthesis gas (1:1 carbon monoxide-hydrogen) at 75°C–200°C (165°F–390°F) and 1500–4500 psi over a metal catalyst, usually cobalt. The active catalyst is held to be cobalt hydrocarbonyl $HCO(CO)_4$, formed by the action of the hydrogen on dicobalt octacarbonyl $[CO_2(CO)_8]$.

A wide variety of olefin derivatives enter the reaction, those containing terminal unsaturated being the most active. The hydroformylation is not specific; the hydrogen and carbon monoxide added across each side of the double bond. Thus, propylene gives a mixture of 60% n-butyraldehyde and 40% iso-butyraldehyde. Terminal ($RCH{=}CH_2$) and nonterminal ($R^1CH{=}CHR^2$, where R^1 and R^2 may be the same or different groups) olefin derivatives, such as 1-pentene ($CH_3CH_2CH_2CH{=}CH_2$) and 2-pentene ($CH_3CH_2CH{=}CHCH_3$), give essentially the same distribution of straight-chain and branched-chain C_6 aldehydes, indicating that rapid isomerization takes place. Simple branched structures add mainly at the terminal carbon; isobutylene forms 95% iso-valeraldehyde and 5% trimethyl acetaldehyde also called pivaldehyde.

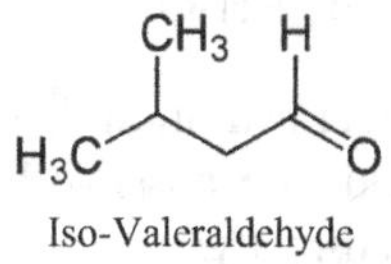

Iso-Valeraldehyde

Trimethyl aldehyde

Commercial application of the synthesis has been most successful in the manufacture of iso-octyl alcohol from a refinery C_3–C_4 copolymer, decyl alcohol from propylene trimer, and tridecyl alcohol from propylene tetramer. Important outlets for the higher alcohols lie in their sulfonation to make detergents and the formation of esters with dibasic acids for use as plasticizers and synthetic lubricants.

The hydrolysis of ethylene chlorohydrin ($HOCH_2CH_2Cl$) or the cyclic ethylene oxide produces ethylene glycol ($HOCH_2CH_2OH$). The main use for this chemical is for antifreeze mixtures in automobile radiators and for cooling aviation engines; considerable amounts are used as ethylene glycol dinitrate in low-freezing dynamite. Propylene glycol is also made by the hydrolysis of the respective chlorohydrin or oxide.

Glycerin ($CH_2OHCHOHCH_2OH$) can be derived from propylene by high-temperature chlorination to produce alkyl chloride, followed by hydrolysis to allyl alcohol and then conversion with aqueous chloride to glycerol chlorohydrin, a product that can be easily hydrolyzed to glycerol (glycerin). Glycerin has found many uses over the years; important among these are as solvent, emollient, sweetener, in cosmetics, and as a precursor to nitroglycerin and other explosives.

The hydrogenolysis of renewable triglycerides derived from vegetable oils offers a pathway to desirable C8+ alcohols. At high H_2 pressures, Zn- and Cu-based heterogeneous catalysts reduce both carboxylic groups as well as C=C bonds in unsaturated fatty acids and esters to give a range of higher alcohols. Alcohols can also be generated from fatty acids by oxidative cleavage using Ru, Os, or Pd catalysts and oxidants such as ozone (O_3), sodium periodate ($NaIO_4$), or hydrogen peroxide (H_2O_2). The resulting shorter-chain aldehydes and acids can be readily hydrogenated to form the desired alcohols.

7.2.2 ALKYLATION

Alkylation is the transfer of an alkyl group from one molecule to another. The alkyl group may be transferred as an alkyl carbocation, a free radical, a carbanion, or a carbene and any equivalent of these groups. An alkyl group is a piece of a molecule with the general formula C_nH_{2n+1}, where n is the integer depicting the number of carbons linked together. For example, a methyl group ($n=1, CH_3$) is a fragment of a methane (CH_4) molecule. Alkylating agents utilize selective alkylation by adding the desired aliphatic carbon chain to the previously chosen starting molecule. Alkyl groups can also be removed (dealkylation). Alkylating agents are often classified according to their nucleophilic or electrophilic character. In the context of refining operations, alkylation refers to, for example, the alkylation of isobutane with an olefin such as the alkylation of isobutane with propylene to produce 2,4-dimethyl pentane which is used as a blend stock to increase the octane number of gasoline:

Isobutane   propylene   2,4-dimethyl pentane

Ethylene is an active alkylating agent. It can be used to alkylate aromatic compounds using Friedel–Crafts type catalysts. Commercially, ethylene is used to alkylate benzene using a zeolite catalyst for the production of ethyl benzene, a precursor for styrene:

$$C_6H_6 \qquad CH_2=CH_2 \qquad\qquad C_6H_5CH_2CH_3$$

$$C_6H_5CH_2CH_3 \qquad\qquad C_6H_5CH=CH_2$$

Alkylation chemistry contributes to the efficient utilization of C_4 olefin derivatives generated in the cracking operations (Speight, 2007). Isobutane has been added to butylene derivatives (and other low-boiling olefin derivatives) to give a mixture of highly branched octanes (e.g., heptanes) by a process called alkylation. The reaction is thermodynamically favored at low temperatures (<20°C), and thus very powerful acid catalysts are employed. Typically, sulfuric acid (85%–100%), anhydrous hydrogen fluoride, or a solid sulfonic acid is employed as the catalyst in these processes. The first step in the process is the formation of a carbocation by combination of an olefin with an acid proton:

$$(CH_3)_2 C=CH_2 + H^+ \rightarrow (CH_3)_3 C^+$$

Step 2 is the addition of the carbocation to a second molecule of olefin to form a dimer carbocation. The extensive branching of the saturated hydrocarbon results in high octane. In practice, mixed butylenes are employed (isobutylene, 1-butylene, and 2-butylene), and the product is a mixture of isomeric octanes that has an octane number of 92–94. With the phaseout of leaded additives in our motor gasoline pools, octane improvement is a major challenge for the refining industry. Alkylation is one option. Hydroalkylation reactions can increase the carbon number of furans and phenols and are an appealing alternative to simple hydrodeoxygenation of the reactants.

In addition, a combination of hydrogen transfer and acid-catalyzed alkylation reaction produced bicyclohexane derivatives from the reaction of phenol and substituted phenol derivatives over a number of solid Brønsted acid catalysts including Amberlyst-15, sulfated zirconia, heteropolyacids, and zeolites. Importantly, H-b-zeolites gave high yields of polycyclic alkylation products even within liquid water, whereas meso- and macroporous solid acids showed little reactivity. The hydroalkylation of m-cresol over Pt- and Pd-containing zeolites (H-Y and H-MOR) gave a distribution of products containing two or more six-carbon rings (e.g., dimethyl bicyclohexane derivatives). Yields for the alkylation products and the related intermediates, including methylcyclohexanone, approached 80% after the ratio of the metal to acid sites was tuned to optimize the rates of hydrogen transfer, dehydration, and alkylation steps. Overall, such alkylation reactions of substituted furan derivatives and phenol derivatives, obtained from pyrolysis of lignin, can produce polycyclic hydrocarbon derivatives (e.g., C_{14+}) that may be ring opened and used as fuels. Recent and promising work shows that transalkylation reactions of 2,5-dimethylfuran (from glucose isomerization and dehydration) with ethylene (obtained from ethanol dehydration) followed by isomerization can produce p-xylene (1,4-$CH_3C_6H_4CH_3$) with selectivity of 75% and 90% at acid sites within H-Y and H-b-zeolites, respectively. This chemistry provides a renewable pathway from sugars to the production of building block aromatics, which are critical for the production of polyesters, among other polymers, and have higher value than precursors to fuels.

7.2.3 HALOGEN DERIVATIVES

Halogenation is a chemical reaction that involves the addition of one or more halogens (fluorine, chlorine, bromine, or iodine) to a compound. The reaction pathway and the stoichiometry of the reaction depend on the structural features and functional groups of the organic substrate, as well as on the specific halogen. Inorganic compounds such as metals also undergo halogenation.

The ease of halogenation (i.e., the reaction rate) is influenced by the halogen. Fluorine and chlorine are more electrophilic and are more aggressive halogenating agents. Bromine is a weaker halogenating agent than both fluorine and chlorine, while iodine is the least reactive of the halogens. The ease (reaction rate) of dehydrohalogenation follows the reverse trend: iodine is most easily removed from organic compounds while and organo-fluorine compounds are very stable.

Several pathways exist for the halogenation of organic compounds, including free radical halogenation, electrophilic halogenation, and halogen addition. The structure of the substrate is one factor that determines the pathway. Saturated hydrocarbons typically do not add halogens but undergo free radical halogenation, involving substitution of hydrogen atoms by halogen. The chemistry of the halogenation of alkanes is usually determined by the relative weakness of the available carbon-hydrogenation (C–H) bonds. The preference for reaction at tertiary and secondary positions results from greater stability of the corresponding free radicals and the transition state leading to the products.

Generally, at ordinary temperatures, chlorine reacts with olefin derivatives by addition. Thus, ethylene is chlorinated to 1,2-dichloroethane (dichloroethane) or to ethylene dichloride:

$$H_2C=CH_2 + Cl_2 \rightarrow H_2ClCCH_2Cl.$$

$$H_2C=CH_2 + Cl_2 \rightarrow H_2ClCCH_2Cl.$$

There are some minor uses for ethylene dichloride, but about 90% of it is cracked to vinyl chloride, the monomer of polyvinyl chloride (PVC):

$$H_2ClCCH_2Cl \rightarrow HCl + H_2C=CHCl.$$

At slightly higher temperatures, olefin derivatives and chlorine react by substitution of a hydrogen atom by a chlorine atom. Thus, in the chlorination of propylene, a rise of 50°C (90°F) changes the product from propylene dichloride [$CH_3CH(Cl)CH_2Cl$] to allyl chloride ($CH_2=CHCH_2Cl$).

7.2.4 OXYGEN DERIVATIVES

Oxidation is a process in which a chemical substance changes because of the addition of oxygen or the removal of hydrogen.

The most striking industrial olefin oxidation process involves ethylene, which is air oxidized over a silver catalyst at 225°C–325°C (435°F–615°F) to give pure ethylene oxide in yields ranging from 55% to 70%. Also, esters ($R^1CO_2R^2$, where R^1 and R^2 can be the same alkyl groups or different alkyl groups) are formed directly by the addition of acids to olefin derivatives, mercaptans by the addition of hydrogen sulfide to olefin derivatives, sulfides by the addition of mercaptans to olefin derivatives, and amines by the addition of ammonia and other amines to olefin derivatives, represented simply as:

$$RCH=CH_2 + CH_3CO_2H \rightarrow RCH_2CH_2CO_2CH_3$$
$$\text{Acetate ester}$$

$$RCH=CH_2 + H_2S \rightarrow RCH_2CH_2SH$$
$$\text{Mercaptan}$$

$$R^1CH=CH_2 + R^2SH \rightarrow R^1CH_2CH_2SR^2$$
$$\text{Sulfide}$$

$$RCH=CH_2 + NH_3 \rightarrow RCH_2CH_2NH_2$$
$$\text{Amine}$$

Analogous higher olefin oxides can be prepared from propylene, butadiene, octene, dodecene, and styrene via the chlorohydrin route or by reaction with peracetic acid. Acrolein is formed by air oxidation or propylene over a supported cuprous oxide catalyst or by condensing acetaldehyde and formaldehyde.

$$\underset{\text{Acrolein}}{\ce{H2C=CH-CHO}}$$

$$CH_3CH=CH_2 + [O] \rightarrow CH_2=CHCHO$$

$$CH_3CHO + HCHO \rightarrow CH_2=CHCHO + H_2O$$

When acrolein and air are passed over a catalyst, such as cobalt molybdate, acrylic acid is produced or if acrolein is reacted with ammonia and oxygen over molybdenum oxide, the product is acrylonitrile.

$$2CH_2=CHCHO + O_2 \rightarrow 2CH_2=CHCOOH$$

$$2CH_2=CHCHO + O_2 + 2NH_3 \rightarrow 2CH_2=CHC\equiv N + 3H_2O$$

Similarly, propylene may be converted to acrylonitrile.

$$2CH_3-CH=CH_2 + 2NH_3 + 3O_2 \rightarrow 2CH_2=CH-C\equiv N + 6H_2O$$

Acrolein and acrylonitrile are important starting materials for the synthetic materials (acrylates). Acrylonitrile is also used in plastics, which are made by copolymerization of acrylonitrile with styrene or with a styrene-butadiene mixture.

Oxidation of the higher olefin derivatives by air is difficult to control, but at temperatures between 350°C and 500°C (660°F and 930°F) maleic acid is obtained from amylene and a vanadium pentoxide catalyst; higher yields of the acid are obtained from hexene, heptene, and octene. Ethylene can be oxidized to a variety of useful chemicals. The oxidation products depend primarily on the catalyst used and the reaction conditions although ethylene oxide is considered to be the most important oxidation product of ethylene. Acetaldehyde ($CH_3CH=O$) and vinyl acetate ($CH_3CO_2CH=CH_2$) are also oxidation products obtained from ethylene under special catalytic conditions.

Ethylene oxide is a colorless gas that liquefies when cooled below 12°C (54°F) which is highly soluble in water and in organic solvents.

$$\underset{\text{Ethylene oxide}}{\overset{\displaystyle O}{(CH_2\!-\!CH_2)}}$$

Ethylene oxide (also called oxirane by the International Union of Pure and Applied Chemistry, IUPAC) is a cyclic ether and the simplest epoxide (a three-membered ring consisting of one oxygen atom and two carbon atoms). Ethylene oxide is a colorless and flammable gas with a faintly sweet odor. Because it is a strained ring conformation, ethylene oxide easily participates in several of additional reactions that result in ring-opening. Ethylene oxide is a precursor for many chemicals of great commercial importance, including ethylene glycols, ethanolamine derivatives, and alcohol ethoxylate derivatives.

Ethylene oxide was first reported in 1859 by the French chemist Charles Adolph Wurtz, who prepared it by treating 2-chloroethanol with potassium hydroxide:

$$ClCH_2CH_2OH + KOH \rightarrow (CH_2CH_2)O + KCl + H_2O$$

In the current petrochemical industry, the main route to ethylene oxide is oxygen or air oxidation of ethylene over a silver catalyst.

$$2CH_2=CH_2 + O_2 \rightarrow 2(CH_2CH_2)O$$

The formation of ethylene oxide by these routes is reaction that is highly exothermic; the excessive temperature increase reduces ethylene oxide yield and causes catalyst deterioration. Thus, a concomitant reaction is the complete oxidation of ethylene to carbon dioxide and water:

$$CH_2=CH_2 + 3O_2 \rightarrow 2CO_2 + 2H_2O$$

Excessive oxidation can be minimized by using modifiers such as organic chlorides. It seems that silver is a unique epoxidation catalyst for ethylene. All other catalysts are relatively ineffective, and the reaction to ethylene is limited among lower olefin derivatives. Propylene and butylene isomers do not form epoxides through this route.

Using oxygen as the oxidant versus air is currently favored because it is more economical. In the process, compressed oxygen, ethylene, and recycled gas are fed to a multi-tubular reactor. The temperature of the oxidation reaction is controlled by boiling water in the shell side of the reactor. Effluent gases are cooled and passed to the scrubber where ethylene oxide is absorbed as a dilute aqueous solution. Unreacted gases are recycled. Epoxidation reaction occurs at approximately 200°C–300°C (390°F–570°F) with a short residence time of 1 s. A selectivity of 70%–75% can be reached for the oxygen-based process. Selectivity is the ratio of moles of ethylene oxide produced per mole of ethylene reacted. Ethylene oxide selectivity can be improved when the reaction temperature is lowered and the conversion of ethylene is decreased (higher recycle of unreacted gases).

Ethylene oxide is a highly active intermediate. It reacts with all compounds that have a labile hydrogen such as water, alcohols, organic acids, and amines. The epoxide ring opens, and a new compound with a hydroxyethyl group is produced. The addition of a hydroxyethyl group increases the water solubility of the resulting compound. Further reaction of ethylene oxide produces polyethylene oxide derivatives with increased water solubility.

7.2.4.1 Ethylene Glycol

Ethylene glycol ($HOCH_2CH_2OH$) is colorless syrupy liquid that is readily soluble in water. The boiling and the freezing points of ethylene glycol are 197.2°C (386.9°F) and −13.2°C (8.2°F), respectively. Ethylene glycol is one of the monomers for polyesters, and the most widely used synthetic fiber polymers.

The main route for producing ethylene glycol (also known in the gas processing industry as MEG, Chapter 4) is the hydration of ethylene oxide in presence of dilute sulfuric acid.

$$CH_2=CH_2 \rightarrow HOCH_2CH_2OH$$

The hydrolysis reaction occurs at a temperature range of 50°C–100°C (112°F–212°F) and the reaction time is approximately 30 min. Diethylene glycol (also known in the gas processing industry as DEG) and triethylene glycol (also known in the gas processing industry as TEG) are coproducts with ethylene (the mono-glycol).

$$2CH_2=CH_2 \rightarrow HOCH_2CH_2OCH_2CH_2OH$$

$$2CH_2=CH_2 \rightarrow HOCH_2CH_2OCH_2CH_2OCH_2CH_2OH$$

In the process, increasing the water/ethylene oxide ratio and decreasing the contact time decreases the formation of higher glycols. A water/ethylene oxide ratio in the order of 10 is typically used to produce a yield in the order of 90% w/w yield of the mono-glycol. In addition, the diethylene glycol and the triethylene glycol derivatives have wide commercial use in the gas processing industry (Chapter 4).

The reaction occurs at approximately 80°C–130°C (176°F–266°F) using a catalyst. Many catalysts have been tried for this reaction, and there is an indication that the best catalyst types are those of the tertiary amine and quaternary ammonium functionalized resins. This route produces ethylene glycol of a high purity and avoids selectivity problems associated with the hydrolysis of ethylene oxide. The coproduct dimethyl carbonate is a liquid soluble in organic solvents. It is used as a specialty solvent, a methylating agent in organic synthesis, and a monomer for polycarbonate resins. It may also be considered as a gasoline additive due to its high oxygen content and its high-octane rating.

Ethylene glycol could also be obtained directly from ethylene by two methods, the Oxirane acetoxylation and the Teijin oxychlorination processes. In the Oxirane process, ethylene is reacted in the liquid phase with acetic acid in the presence of a tellurium oxide (TeO_2) catalyst at approximately 160°C (320°F) and 400 psi. The product is a mixture of mono-acetate and the diacetate of ethylene glycol after which the acetates are hydrolyzed to ethylene glycol and acetic acid. The hydrolysis reaction occurs at approximately 107°C–130°C (214°F–266°F) and 80 psi. Acetic acid is then recovered for further use:

$$2CH_2=CH_2 \rightarrow 3CH_3CO_2H(+TeO_2) \rightarrow CH_3CO_2CH_2CH_2OH + CH_3CO_2CH_2CH_2O_2CCH_3 + 3H_2O$$

$$CH_3CO_2CH_2CH_2OH + CH_3CO_2CH_2CH_2O_2CCH_3 \rightarrow 2HOCH_2CH_2OH + 3CH_3CO_2H$$

A higher glycol yield (approximately 94%) than from the ethylene oxide process is anticipated. However, there are certain problems inherent in the Oxirane process such as corrosion caused by acetic acid and the incomplete hydrolysis of the acetates. Also, the separation of the glycol from unhydrolyzed monoacetate is hard to accomplish.

The Teijin oxychlorination, on the other hand, is considered a modern version of the Wurtz chlorohydrin process (now obsolete) for the production of ethylene oxide. In this process, ethylene chlorohydrin is obtained by the catalytic reaction of ethylene with hydrochloric acid in presence of thallium chloride ($TlCl_3$) catalyst. Ethylene chlorohydrin is then hydrolyzed *in situ* to ethylene glycol:

$$CH_2=CH_2 + H_2O(+TlCl_3) \rightarrow HOCH_2CH_2Cl(+TlCl + HCl)$$

$$HOCH_2CH_2Cl + H_2O \rightarrow HOCH_2CH_2OH + HCl$$

Catalyst regeneration occurs by the reaction of thallium (I) chloride (TlCl) with copper(II) chloride in the presence of oxygen or air. The formed cuprous chloride (CuCl) is reoxidized by the action of oxygen in the presence of hydrogen chloride (HCl).

A new route to ethylene glycol from ethylene oxide via the intermediate formation of ethylene carbonate has recently been developed. Ethylene carbonate [$(CH_2O)_2CO$] is classified as the carbonate ester of ethylene glycol and carbonic acid (H_2CO_3). At room temperature (25°C, 77°F) ethylene carbonate is a transparent crystalline solid, practically odorless and colorless and somewhat soluble in water. In the liquid state (m.p. 34°C–37°C, 94°F–99°F), it is a colorless odorless liquid.

Ethylene carbonate is produced by the catalyzed reaction between ethylene oxide and carbon dioxide:

$$(CH_2)_2O + CO_2 \rightarrow (CH_2O)_2CO$$

Similarly, ethylene carbonate may also be formed by the reaction of carbon monoxide, ethylene oxide, and oxygen. Alternatively, it can be obtained by the reaction of phosgene and methanol.

$$2(CH_2)_2O + 2CO + O_2 \rightarrow 2(CH_2O)_2CO$$

Ethylene carbonate can also be produced from the reaction of urea and ethylene glycol using zinc oxide (ZnO) as a catalyst at a temperature of 150°C (300°F):

$$H_2NCOCNH_2 + HOCH_2CH_2OH \rightarrow (CH_2O)_2CO + 2NH_3$$

Ethylene carbonate is a reactive chemical and may be converted to dimethyl carbonate (a useful solvent that also finds use as a methylating agent) by means of a transesterification reaction with methyl alcohol:

$$C_2H_4CO_3 + 2CH_3OH \rightarrow CH_3OCO_2CH_3 + HOC_2H_4OH$$

Dimethyl carbonate may itself be similarly converted (by transesterification) to diphenyl carbonate:

$$CH_3OCO_2CH_3 + 2C_6H_5OH \rightarrow C_6H_5OCO_2C_6H_5 + 2CH_3OH$$

7.2.4.2 Ethoxylates

Ethoxylates are the products of ethoxylation reactions in which ethylene oxide is added to a substrate. The most widely used reaction relative to the petrochemical industry is alkoxylation which involves the addition of epoxides to substrates. Typically, esters $(R^1CO_2R^2)$ are formed directly by the addition of acids to olefin derivatives, mercaptans by the addition of hydrogen sulfide to olefin derivatives, sulfides by the addition of mercaptans to olefin derivatives, and amines by the addition of ammonia and other amines to olefin derivatives.

In the usual application, alcohol derivatives and phenol derivatives are converted into ester-type products with the general formula $R(OC_2H_4)_nOH$ where n ranges from 1 to as high as 10. Such compounds are called alcohol ethoxylates. Alcohol ethoxylate derivatives are often converted to related species called ethoxy sulfate derivatives. Alcohol ethoxylates and ethoxy sulfate derivatives are surfactants that are used widely in many commercial products, including cosmetic products.

The reaction between ethylene oxide and long-chain fatty alcohols or fatty acids is called ethoxylation. Ethoxylation of C_{10} to C_{14} linear alcohol derivatives and linear alkylphenol derivatives produces nonionic detergents. The reaction with alcohols could be represented as:

$$ROH + (CH_2CH_2)O \rightarrow RO[(CH_2CH_2)O]_n H$$

The solubility of the ethoxylate derivatives can be varied by adjusting the number of ethylene oxide units in the molecule. The solubility is also a function of the chain length of the alkyl group in the alcohol or in the phenol and longer-chain alkyl groups reduce water solubility. In practice, the number of ethylene oxide units and the chain length of the alkyl group are varied to either produce water-soluble or oil-soluble surface-active agents.

Linear alcohols used for the production of ethoxylates are produced by the oligomerization of ethylene using Ziegler catalysts or by the Oxo reaction using alpha olefin derivatives. Similarly, esters of fatty acids and polyethylene glycols are produced by the reaction of long-chain fatty acids and ethylene oxide.

7.2.4.3 Ethanolamines

The ethanolamine derivatives (also called *olamine* derivatives) comprise a group of amino alcohol derivatives and contain both a primary amine (–NH$_2$) function and a primary alcohol (–CH$_2$OH) function. Ethanolamine is a colorless viscous liquid with an odor that is reminiscent to the odor of ammonia. The olamine family includes ethanolamine (HOCH$_2$CH$_2$NH$_2$, 2-aminoethanol, also called monoethanolamine, MEA), diethanolamine (DEA, HOCH$_2$CH$_2$NHCH$_2$CH$_2$OH), and triethanolamine (TEA). These ethanolamine derivatives have been, and continue to be, used widely in the gas processing industry for the removal of acid gases from gas streams (Chapter 4) (Kohl and Riesenfeld, 1985; Maddox et al., 1985; Newman, 1985; Kohl and Nielsen, 1997; Kidnay and Parrish, 2006; Mokhatab et al., 2006; Speight, 2007, 2014). They are also used to manufacture detergents, metalworking fluids, and as gas sweetening. Triethanolamine is used in detergents and cosmetics applications and as a cement additive.

Ethanolamine derivatives are prepared by the reaction of aqueous ammonia and ethylene oxide and the product (monoethanolamine) reacts with a second and third equivalent of ethylene oxide to give diethanolamine and triethanolamine:

$$(CH_2CH_2)O + NH_3 \rightarrow HOCH_2CH_2NH_2$$

$$H_2NCH_2CH_2OH + (CH_2CH_2)O \rightarrow HN(CH_2CH_2OH)_2$$

$$HN(CH_2CH_2OH)_2 + (CH_2CH_2)O \rightarrow N(CH_2CH_2OH)_3$$

The reaction conditions are approximately 30°C–40°C (86°F–104°F) and atmospheric pressure (14.7 psi):

The relative ratios of the ethanolamine derivatives produced depend principally on the ethylene oxide/ammonia ratio. A low ethylene oxide-ammonia ratio increases monoethanolamine yield. Increasing this ratio increases the yield of diethanolamine and triethanolamine derivatives.

7.2.4.4 1,3-Propanediol

1,3-Propanediol, although a product related to propylene is included here because of the production of this product from an ethylene derivative, namely ethylene oxide.

1,3-Propanediol is a colorless liquid that boils at 210°C (410°F) which is soluble in water, alcohol, and ether and is used as an intermediate for polyester production. This diol can be produced via the hydroformylation of ethylene oxide which yields 3-hydroxypropionaldehyde. Hydrogenation of the product produces 1,3-propanediol.

$$(CH_2CH_2)O + HCHO \rightarrow HOCH_2CH_2CHO$$

$$HOCH_2CH_2CHO + H_2 \rightarrow HOCH_2CH_2CH_2OH$$

The catalyst is a cobalt carbonyl that is prepared *in situ* from cobaltous hydroxide, and nonyl pyridine as the promotor. Oxidation of the aldehyde produces 3-hydroxypropionic acid.

$$HOCH_2CH_2CHO + O_2 \rightarrow HOCH_2CH_2CO_2H + H_2O$$

1,3-Propanediol and 3-hydroxypropionic acid could also be produced from acrolein by hydrolysis of the acrolein followed by hydrogenation of the 3-hydroxypropionaldehyde:

$$CH_2{=}CHCHO + H_2O \rightarrow HOCH_2CH_2CHO$$

$$HOCH_2CH_2CHO + H_2 \rightarrow HOCH_2CH_2CH_2OH$$

7.2.4.5   Acetaldehyde

Acetaldehyde (CH_3CHO) is an intermediate for many chemicals such as acetic acid, n-butanol, pentaerythritol, and polyacetaldehyde. It is a colorless liquid with a pungent odor. It is a reactive compound with no direct use except for the synthesis of other compounds. For example, it is oxidized to acetic acid and acetic anhydride.

$$CH_3CHO + [O] \rightarrow CH_3CO_2H + (CH_3CO)_2 O$$

It is a reactant in the production of 2-ethylhexanol for the synthesis of plasticizers and also in the production of pentaerythritol, a polyhydric compound used in alkyd resins.

There are many ways to produce acetaldehyde. Historically, it was produced either by the silver-catalyzed oxidation or by the chromium-activated copper-catalyzed dehydrogenation of ethanol.

$$2CH_3CH_2OH + O_2 \rightarrow 2CH_3CHO + 2H_2O$$

Currently, acetaldehyde is obtained from ethylene by using a homogeneous catalyst (Wacker catalyst). The catalyst allows the reaction to occur at much lower temperatures (typically 130°C, 266°F) than those used for the oxidation or the dehydrogenation of ethanol (approximately 500°C for the oxidation and 250°C for the dehydrogenation).

Ethylene oxidation is carried out through oxidation-reduction (redox). The overall reaction is the oxidation of ethylene by oxygen as represented by:

$$2CH_2{=}CH_2 + O_2 \rightarrow 2CH_3CHO$$

The Wacker process uses an aqueous solution of palladium(II) chloride, copper(II) chloride catalyst system. In the course of the reaction, the ethylene is oxidized to acetaldehyde and the palladium (Pd^{2+}) ions are reduced to palladium metal:

$$CH_2{=}CH_2 + H_2O + PdCl_2 \rightarrow CH_3CHO + 2HCl + Pd^{\circ}$$

The formed palladium (Pd°) is then reoxidized by the action of Cu(II) ions, which are reduced to Cu(I) ions:

$$Pd^{\circ} + 2CuCl_2 \rightarrow PdCl_2 + 2CuCl$$

The reduced Cu(I) ions are reoxidized to Cu(II) ions by reaction with oxygen and HCl:

$$4CuCl + O_2 + 4HCl \rightarrow 4CuCl_2 + H_2O$$

The oxidation reaction may be carried out in a single-stage or a twostage process. In the single-stage, ethylene, oxygen, and recycled gas are fed into a vertical reactor containing the catalyst solution. Heat is controlled by boiling off some of the water. The reaction conditions are approximately 130°C (266°F) and 45 psi. In the two-stage process, the reaction occurs under relatively higher pressure (approximately 120 psi) to ensure higher ethylene conversion. The reaction temperature is approximately 130°C (266°F). The catalyst solution is then withdrawn from the reactor to a tube-oxidizer to enhance the oxidation of the catalyst at approximately 150 psi. The yield of acetaldehyde from either process is about 95%. Byproducts from this reaction include acetic acid, ethyl chloride, chloroacetaldehyde, and carbon dioxide. The Wacker reaction can also be carried out for other olefin derivatives with terminal double bonds. With propylene, for example, approximately 90% yield of acetone is obtained. 1-Butylene gave approximately 80% yield of methyl ethyl ketone.

Acetic acid is obtained from different sources. Carbonylation of methanol is currently the major route. Oxidation of butane derivatives and butylene derivatives is an important source of acetic acid. It is also produced by the catalyzed oxidation of acetaldehyde:

$$2CH_3CHO + O_2 \rightarrow 2CH_3CO_2H$$

The reaction occurs in the liquid phase at approximately 65°C (149°F) using manganese acetate [Mn(OCOCH$_3$)$_2$] as a catalyst.

Manganese acetate

Vinyl acetate (CH$_3$COOCH=CH$_2$) is a reactive colorless liquid that polymerizes easily if not stabilized. It is an important monomer for the production of polyvinyl acetate, polyvinyl alcohol, and vinyl acetate copolymers. Vinyl acetate was originally produced by the reaction of acetylene and acetic acid in the presence of mercury(II) acetate. Currently, it is produced by the catalytic oxidation of ethylene with oxygen, with acetic acid as a reactant and palladium as the catalyst:

$$2CH_2=CH_2 + 2CH_3CO_2H + O_2 \rightarrow 2CH_2=CHOCOCH_3 + H_2O$$

The process is similar to the catalytic liquid-phase oxidation of ethylene to acetaldehyde. The difference between the two processes is the presence of acetic acid. In practice, acetaldehyde is a major coproduct. The mole ratio of acetaldehyde to vinyl acetate can be varied from 0.3:1 to 2.5:1.13. The liquid-phase process is not used extensively due to corrosion problems and the formation of a fairly wide variety of byproducts.

In the vapor-phase process, oxyacylation of ethylene is carried out in a tubular reactor at approximately 116°C (240°F) and 75 psi. The palladium acetate is supported on carriers resistant to attack by acetic acid. Conversions of about 10%–15% based on ethylene are normally used to operate safely outside the explosion limits (approximately 10% v/v). A selectivity in the order of 91%–94% based on ethylene is attainable.

n-Butanol is normally produced from propylene by the Oxo reaction (sometimes known as hydroformylation). It may also be obtained from the aldol condensation of acetaldehyde in presence of a base. Hydroformylation, also known as oxo synthesis or oxo process, is an industrial process for the production of aldehyde derivatives from alkene derivatives. This chemical reaction results in the addition of a formyl group (HC=O) and a hydrogen atom to a carbon–carbon double bond.

It is an important reaction because aldehydes are easily converted into many secondary products. For example, the resulting aldehyde derivatives are hydrogenated to alcohol derivatives that are converted to detergent products. Hydroformylation is also used in the specialty chemicals industry and is especially relevant to the synthesis of fragrances and drugs.

In the process, an alkene derivative is treated with carbon monoxide and hydrogen at high pressure (in the order of 150–1,500 psi) at a temperature in the range between 40°C to 200°C (104°F–390°F). The reaction is an example of homogeneous catalysis since the transition metal catalyst is invariably soluble in the reaction medium.

By way of clarification, the IUPAC definition defines a transition metal as (quote) an element whose atom has a partially filled d subshell, or which can give rise to cations with an incomplete d subshell (end quote). Most scientists describe a transition metal as any element in the d-block of the periodic table, which includes groups 3–12 on the periodic table. In actual practice, the f-block lanthanide and actinide series are also considered transition metals and are called *inner transition metals*. The word transition was first used to describe the elements now known as the d-block by the English chemist Charles Bury in 1921, who referred to a transition series of elements during the change of an inner layer of electrons from a stable group of 8 to1 of 18, or from 18 to 32.

7.2.5 CARBONYLATION

Carbonylation refers to reactions in which the carbon monoxide moiety is introduced into organic and inorganic substrates. Carbon monoxide is abundantly available and conveniently reactive, so it is widely used as a reactant in industrial chemistry. Several industrially useful organic chemicals are prepared by carbonylation reactions, which can be highly selective reactions. Carbonylation reactions produce organic carbonyl derivatives, i.e., compounds that contain the carbonyl (C=O) functional group such as aldehyde derivatives (–CHO), ketone derivatives (>C=O), carboxylic acid derivatives (–CO$_2$H), and ester derivatives (–CO$_2$R, where R is an alkyl group). Carbonylation reactions are the basis of two main types of reactions (i) hydroformylation and (ii) the Reppe reaction.

The hydroformylation reaction entails the addition of both carbon monoxide and hydrogen to unsaturated organic compounds, typically alkene derivatives. The usual products are aldehyde derivatives:

$$RCH=CH_2 + H_2 + CO \rightarrow RCH_2CH_2CHO$$

The reaction requires metal catalysts that bind the carbon monoxide and the hydrogen to the alkene.

The Reppe reaction involves the addition of carbon monoxide and an acidic hydrogen donor to the organic substrate. Commercial processes using this type of chemistry include the Monsanto process and the Cativa process which converts methanol to acetic acid—acetic anhydride is prepared by a related carbonylation of methyl acetate (CH$_3$COOCH$_3$). In the related hydrocarboxylation and hydroesterification reaction, alkene derivatives (>C=C<) and alkyne derivatives (–C≡C–) are the substrates. This method is used in industry to produce propionic acid from ethylene:

$$RCH=CH_2 + H_2O + CO \rightarrow RCH_2CH_2CO_2H$$

These reactions require metal catalysts, which bind and activate the carbon monoxide. For example, in the industrial synthesis of ibuprofen (Chapter 12), a benzylic alcohol derivative is converted to the corresponding carboxylic acid via a palladium-catalyzed carbonylation reaction:

$$ArCH(CH_3)OH + CO \rightarrow ArCH(CH_3)CO_2H$$

The liquid-phase reaction of ethylene with carbon monoxide and oxygen (oxidative carbonylation) over a palladium/copper (Pd^{2+}/Cu^{2+}) catalyst system produces acrylic acid (CH$_2$=CHCO$_2$H). The yield based on ethylene is about 85%. Reaction conditions are approximately 140°C (285°F) and 1,100 psi:

$$2CH_2=CH_2 + 2CO + O_2 \rightarrow 2CH_2=CHCO_2H$$

The catalyst is similar to that of the Wacker reaction for ethylene oxidation to acetaldehyde; however, this reaction occurs in presence of carbon monoxide.

7.2.6 Chlorination

The direct addition of chlorine to ethylene produces ethylene dichloride (1,2-dichloroethane). Ethylene dichloride is the main precursor for vinyl chloride, which is an important monomer for polyvinyl chloride plastics and resins. Other uses of ethylene dichloride include its formulation with tetraethyl and tetramethyl lead solutions as a lead scavenger, as a degreasing agent, and as an intermediate in the synthesis of many ethylene derivatives.

The reaction of ethylene with hydrogen chloride, on the other hand, produces ethyl chloride. This compound is a small-volume chemical with diversified uses (alkylating agent, refrigerant, solvent).

Ethylene reacts also with hypochlorous acid, yielding ethylene chlorohydrin:

$$CH_2{=}CH_2 + HOCl \rightarrow ClCH_2CH_2OH$$

Ethylene chlorohydrin via this route was previously used for producing ethylene oxide through an epoxidation step. Currently, the catalytic oxychlorination route (the Teijin process discussed earlier in this chapter) is an alternative for producing ethylene glycol where ethylene chlorohydrin is an intermediate. In organic synthesis, ethylene chlorohydrin is a useful agent for introducing the ethyl-hydroxy group. It is also used as a solvent for cellulose acetate.

7.2.6.1 Vinyl Chloride

Vinyl chloride ($CH_2{=}CHCl$) is a reactive gas soluble in alcohol but slightly soluble in water. It is the most important vinyl monomer in the polymer industry. Vinyl chloride monomer was originally produced by the reaction of hydrochloric acid and acetylene in the presence of mercuric chloride ($HgCl_2$) catalyst. The reaction is straightforward and proceeds with high conversion (96% based on acetylene):

$$HC{\equiv}CH + HCl \rightarrow CH_2{=}CHCl$$

However, ethylene as a cheap raw material has replaced acetylene for obtaining vinyl chloride. The production of vinyl chloride via ethylene is a three-step process. The first step is the direct chlorination of ethylene to produce ethylene dichloride. Either a liquid-phase or a vapor-phase process is used. The exothermic reaction occurs at 40°C–50°C (104°F–122°F) and approximately 60 psi in the presence of a catalyst, such as ferric chloride ($FeCl_3$), cupric chloride ($CuCl_2$), or antimony trichloride (SbC_{13}); ethylene dibromide may also be used as a catalyst. The second step is the dehydrochlorination of ethylene dichloride to vinyl chloride and hydrogen chloride. The pyrolysis reaction occurs at approximately 500°C (930°F) and 370 psi in the presence of an adsorbent—such as pumice on charcoal—to remove the hydrogen chloride from the product mix. The third step, the oxychlorination of ethylene, uses byproduct hydrogen chloride from the previous step to produce more ethylene dichloride. Thus:

$$CH_2{=}CH_2 + Cl_2 \rightarrow CH_2ClCH_2Cl$$

$$CH_2ClCH_2Cl \rightarrow CH_2{=}CHCl + HCl$$

$$CH_2{=}CH_2 + Cl_2 \rightarrow CH_2ClCH_2Cl$$

The ethylene dichloride from the third step is combined with that produced from the chlorination of ethylene and introduced to the pyrolysis furnace. The reaction conditions are approximately 225°C (435°F) and 30–60 psi. In practice, the three steps, chlorination, oxychlorination, and dehydrochlorination, are integrated in one process so that no chlorine is lost.

7.2.6.2 Perchloroethylene and Trichloroethylene

Perchloroethylene and trichloroethylene could be produced from ethylene dichloride by an oxychlorination/oxyhydrochlorination process without byproduct hydrogen chloride. A special catalyst is used:

$$2ClCH_2CH_2Cl + Cl_2 + O_2 \rightarrow ClCH=CCl_2 + Cl_2C=CCl_2 + H_2O$$

A fluid bed reactor is used at moderate pressures at approximately 450°C (840°F). The reactor effluent, containing chlorinated organics, water, a small amount of hydrogen chloride, carbon dioxide, and other impurities, is condensed in a water-cooled graphite exchanger, cooled in a refrigerated condenser, and then scrubbed. Separation of perchloroethylene from trichloroethylene occurs by successive distillation.

Perchloroethylene and trichloroethylene may also be produced from chlorination of propane.

7.2.7 Hydration

A hydration reaction is a chemical reaction in which a substance combines with water. In the current context, water is added to an unsaturated substrate, which is usually an alkene or an alkyne. Ethyl alcohol (CH_3CH_2OH) production is considered by many to be the oldest profession in the world, without going into detail about any other possible competitor. Carbohydrate fermentation is still the main route to ethyl alcohol in many countries with abundant sugar and grain sources.

The earliest method for conversion of olefin derivatives into alcohols involved their absorption in sulfuric acid to form esters, followed by dilution and hydrolysis, generally with the aid of steam. In the case of ethyl alcohol, the direct catalytic hydration of ethylene can be employed. Ethylene is readily absorbed in 98%–100% sulfuric acid at 75°C–80°C (165°F–175°F), and both ethyl and diethyl sulfate are formed; hydrolysis takes place readily on dilution with water and heating:

$$CH_2=CH_2 + H_2O \rightarrow CH_3CH_2OH$$

In the process, the first-formed monoethyl sulfate and diethyl sulfate derivatives are hydrolyzed with water to ethanol and sulfuric acid, which is regenerated. The direct hydration of ethylene with water is the process currently used. The hydration is carried out in a reactor at approximately 300°C (570°F) and 1,000 psi. The reaction is favored at relatively lower temperature and higher pressures. Phosphoric acid on diatomaceous earth is the catalyst. To avoid catalyst losses, a water/ethylene mole ratio less than one is used. Conversion of ethylene is limited under these conditions, and unreacted ethylene is recycled.

The many used of ethanol's many uses can be conveniently divided into solvent use and chemical use. As a solvent, ethanol dissolves many organic-based materials such as fats, oils, and hydrocarbon derivatives. As a chemical intermediate, ethanol is a precursor for acetaldehyde, acetic acid, and diethyl ether, and it is used in the manufacture of glycol ethyl ether derivatives, ethylamine derivatives, and many ethyl esters.

Ethylene, produced from ethane by cracking, is oxidized in the presence of a silver catalyst to ethylene oxide:

$$2H_2C=CH_2 + O_2 \rightarrow C_2H_4O$$

The vast majority of the ethylene oxide produced is hydrolyzed at 100°C to ethylene glycol:

$$C_2H_4O + H_2O \rightarrow HOCH_2CH_2OH$$

The majority of the ethylene glycol produced commercially is used as automotive antifreeze and much of the rest is used in the synthesis of polyesters.

Of the higher olefin derivatives, one of the first alcohol syntheses practiced commercially was that of isopropyl alcohol from propylene. Sulfuric acid absorbs propylene more readily than it does ethylene, but care must be taken to avoid polymer formation by keeping the mixture relatively cool and using acid of about 85% strength at 300–400 psi pressure; dilution with inert oil may also be necessary. Acetone is readily made from isopropyl alcohol, either by catalytic oxidation or by dehydrogenation over metal (usually copper) catalysts.

Secondary butyl alcohol is formed on absorption of 1-butylene or 2-butylene by 78%–80% sulfuric acid, followed by dilution and hydrolysis. Secondary butyl alcohol is converted into methyl ethyl ketone by catalytic oxidation or dehydrogenation.

There are several methods for preparing higher alcohols. One method in particular is the so-called Oxo reaction and involves the direct addition of carbon monoxide (CO) and a hydrogen (H) atom across the double bond of an olefin to form an aldehyde (RCH=O), which in turn is reduced to the alcohol (RCH_2OH). Hydroformylation (the Oxo reaction) is brought about by contacting the olefin with synthesis gas (1:1 carbon monoxide-hydrogen) at 75°C–200°C (165°F–390°F) and 1,500–4,500 psi over a metal catalyst, usually cobalt. The active catalyst is held to be cobalt hydrocarbonyl $HCO(CO)_4$, formed by the action of the hydrogen on dicobalt octacarbonyl.

7.2.8 Oligomerization

Oligomerization is the addition of one olefin molecule to a second and to a third (plus higher numbers) to form a dimer or a trimer. The reaction is normally acid-catalyzed. When propylene or butylene derivatives are used, the formed compounds are branched because an intermediate carbocation is formed. These compounds were used as alkylating agents for producing benzene alkylates, but the products were nonbiodegradable.

Oligomerization of ethylene using a Ziegler catalyst produces unbranched alpha olefin derivatives in the C_{12} to C_{16} range by an insertion mechanism. A similar reaction using triethylaluminum produces linear alcohols for the production of biodegradable detergents. Dimerization of ethylene to 1-butylene has been developed recently by using a selective titanium-based catalyst.

The C_{12} to C_{16} alpha olefin derivatives are produced by dehydrogenation of n-paraffins, dehydrochlorination of monochloro-paraffin derivatives, or by oligomerization of ethylene using trialkylaluminum (Ziegler catalyst). Iridium complexes also catalyze the dehydrogenation of n-paraffins to a-olefin derivatives. The reaction uses a soluble iridium catalyst to transfer hydrogen to the olefinic acceptor.

The triethylaluminum and 1-butylene are recovered by the reaction between tributylaluminum and ethylene:

$$(CH_3CH_2CH_2CH_2)_3 Al + 3CH_2=CH_2 \rightarrow (CH_3CH_2H)Al + 3CH_3CH_2CH=CH_2$$

Alpha olefin derivatives are important compounds for producing biodegradable detergents. They are sulfonated and neutralized to alpha olefin sulfonate derivatives:

$$RCH=CH_2 + SO_3 \rightarrow RCH=CHSO_3H$$

$$RCH=CHSO_3H + NaOH \rightarrow RCH=CHSO_3Na + H_2O$$

Alkylation of benzene using alpha olefin derivatives produces linear alkylbenzenes, which are further sulfonated and neutralized to linear alkylbenzene sulfonates (LABS). These compounds constitute, with alcohol ethoxy sulfate derivatives and ethoxylate derivatives, the basic active ingredients for household detergents.

Alpha olefin derivatives could also be carbonylated in presence of an alcohol using a cobalt catalyst to produce esters:

$$RCH{=}CH_2 + CO + R^1OH \rightarrow RCH_2CH_2CO_2R^1$$

Transesterification with pentaerythritol produces pentaerythritol ester derivatives and releases the alcohol.

Linear (C_{21} to C_{26}) alcohol derivatives are important chemicals for producing various compounds such as plasticizers, detergents, and solvents. The production of linear alcohols involves the hydroformylation (the Oxo reaction) of alpha olefin derivatives followed by hydrogenation. They are also produced by the oligomerization of ethylene using aluminum alkyls (Ziegler catalysts).

The Alfol process for producing linear primary alcohols is a four-step process. In the first step, triethylaluminum is produced by the reaction of ethylene with hydrogen and aluminum metal:

$$6CH_2{=}CH_2 + 3H_2 + 3Al \rightarrow 3(CH_3CH_2)_3 Al$$

In the next step, ethylene is polymerized by the action of triethylaluminum at approximately 120°C (248°F) and 1,900 psi to trialkylaluminum. Typical reaction time is approximately 140 min for an average C_{12} alcohol production. The oxidation of triethylaluminum is carried out between 20°C and 50°C (68°F–122°F) in air to aluminum trialkoxide derivatives:

$$2(CH_3CH_2)_3 Al + 3O_2 \rightarrow 2(CH_3CH_2O)_3 Al$$

The final step is the hydrolysis of the trialkoxide derivative with water to the corresponding even-numbered primary alcohols. Alumina is coproduced and is characterized by its high activity and purity:

$$(CH_3CH_2O)_3 Al + H_2O \rightarrow CH_3CH_2OH + Al_2O_3$$

Linear alcohols in the range of C_{10} to C_{12} are used to make plasticizers. Those in the range of C_{12} to C_{16} range are used for making biodegradable detergents. They are either sulfated to linear alkyl sulfate derivatives (ionic detergents) or reacted with ethylene oxide to the ethoxylated linear alcohols (nonionic detergents). The C_{16} to C_{18} alcohols are modifiers for wash and wear polymers. The higher molecular weight alcohols (i.e., the C_{20} to C_{26} (alcohols) are synthetic lubricants and mold-release agents.

7.2.9 POLYMERIZATION

The polymerization of ethylene under pressure (1,500–3,000 psi) at 110°C–120°C (230°F–250°F) in the presence of a catalyst or initiator, such as a 1% solution of benzoyl peroxide in methanol, produces a polymer in the 2,000–3,000 molecular weight range (Chapter 11). Polymerization at 15,000–30,000 psi and 180°C–200°C (355°F–390°F) produces a wax melting at 100°C (212°F) and 15,000–20,000 molecular weight but the reaction is not as straightforward as the equation indicates since there are branches in the chain. However, considerably lower pressures can be used over catalysts composed of aluminum alkyls (R_3Al) in presence of titanium tetrachloride ($TiCl_4$), supported chromic oxide (CrO_3), nickel (NiO), or cobalt (CoO) on charcoal, and promoted molybdena-alumina (MoO_2-Al_2O_3), which at the same time give products more linear in structure. Polypropylenes can be made in similar ways, and mixed monomers, such as ethylene-propylene and ethylene-butylene mixtures, can be treated to give high molecular weight copolymers of good elasticity. Polyethylene has excellent electrical insulating properties; its chemical resistance, toughness, machinability, low density (light weight), and high strength make it suitable for many other uses.

Lower molecular weight polymers, such as the dimers, trimers, and tetramers, are used as such in motor gasoline. The materials are normally prepared over an acid catalyst. Propylene trimer (dimethyl heptene derivatives) and tetramer (trimethyl nonene derivatives) are applied in the alkylation of aromatic hydrocarbon derivatives for the production of alkyl-aryl sulfonate detergents and also as olefin-containing feedstocks in the manufacture of C_{10} and C_{13} oxo-alcohols. Phenol is alkylation by the trimer to make nonylphenol, a chemical intermediate for the manufacture of lubricating oil detergents and other products.

Isobutylene also forms several series of valuable products; the di- and tri-isobutylenes make excellent motor and aviation gasoline components; they can also be used as alkylating agents for aromatic hydrocarbon derivatives and phenols and as reactants in the oxo-alcohol synthesis. Polyisobutylene derivatives in the viscosity range of 55,000 SUS (38°C, 100°F) have been employed as viscosity index improvers in lubricating oils. 1-Butylene ($CH_3CH_2CH=CH_2$) and 2-butylene ($CH_3CH=CHCH_3$) participate in polymerization reactions by the way of butadiene ($CH_2=CHCH=CH_2$), the dehydrogenation product, which is copolymerized with styrene (23.5%) to form GR-S rubber, and with acrylonitrile (25%) to form GR-N rubber.

Derivatives of acrylic acid (butyl acrylate, ethyl acrylate, 2-ethylhexyl acrylate, and methyl acrylate) can be homopolymerized using peroxide initiators or copolymerized with other monomers to generate acrylic or aclryloid resins.

7.2.10   1-BUTYLENE

The Institut Français du Pétrole process is used to produce butylene-1 (1-butene) by dimerizing ethylene. A homogeneous catalyst system based on a titanium complex is used. The reaction is a concerted coupling of two molecules on a titanium atom, affording a titanium (IV) cyclic compound, which then decomposes to 1-butylene by an intramolecular hydrogen transfer reaction.

The Alphabutol process operates at low temperatures (50°C–55°C, 122°F–131°F) and relatively low pressures (330–400 psi). The process operates in the liquid phase using a soluble catalyst system which avoids isomerization of 1-butene to 2-butene. There is no need for superfractionation of the product stream. The process scheme includes four sections: the reactor, the co-catalyst injection, catalyst removal, and distillation. The continuous co-catalyst injection of an organo-basic compound deactivates the catalyst downstream of the reactor withdrawal valve to limit isomerization of 1-butylene to 2-butylene.

7.2.11   POLYMERIZATION

The polymerization of ethylene under pressure (1,500–3,000 psi) at 110°C–120°C (230°F–250°F) in the presence of a catalyst or initiator, such as a 1% solution of benzoyl peroxide in methanol, produces a polymer in the 2,000–3,000 molecular weight range. Polymerization at 15,000–30,000 psi and 180°C–200°C (355°F–390°F) produces a wax melting at 100°C (212°F) and 15,000–20,000 molecular weight, but the reaction is not as straightforward as the equation indicates since there are branches in the chain. However, considerably lower pressures can be used over catalysts composed of aluminum alkyls (R_3Al) in presence of titanium tetrachloride ($TiCl_4$), supported chromic oxide (CrO_3), nickel (NiO), or cobalt (CoO) on charcoal, and promoted molybdena-alumina (MoO_2-Al_2O_3), which at the same time give products more linear in structure. Polypropylenes can be made in similar ways, and mixed monomers, such as ethylene-propylene and ethylene-butylene mixtures, can be treated to give high molecular weight copolymers of good elasticity. Polyethylene has excellent electrical insulating properties; its chemical resistance, toughness, machinability, low density (light weight), and high strength make it suitable for many other uses.

7.3 CHEMICALS FROM PROPYLENE

Propylene is an unsaturated organic hydrocarbon (C_3H_6, $CH_3CH=CH_2$) that has one double bond, and is a colorless gas (Table 7.4). It is a byproduct of crude oil refining and natural gas processing.

During oil refining, propylene ($CH_3CH=CH_2$, like ethylene $CH_2=CH_2$) is produced as a result of the thermal decomposition (cracking) of higher molecular weight hydrocarbon derivatives. A major source of propylene is naphtha cracking intended to produce ethylene, but it also results from refinery cracking producing other products. Propylene can be separated by fractional distillation from hydrocarbon mixtures obtained from cracking and other refining processes; refinery grade propylene is about 50%–70%.

Propane dehydrogenation converts propane ($CH_3CH_2CH_3$) into propylene ($CH_3CH=CH_2$) and byproduct hydrogen:

$$CH_3CH_2CH_3 \rightarrow CH_3-CH=CH_2 + H_2$$

The propylene from propane yield is in the order of 85 mol%. Reaction byproducts (mainly hydrogen) are usually used as fuel for the propane dehydrogenation reaction. As a result, propylene tends to be the only product, unless local demand exists for hydrogen. In fact, a large proportion of the propylene is made from propane, which is obtained from natural gas stripper plants or from refinery gases.

Like ethylene, propylene is a reactive alkene that can be obtained from refinery gas streams, especially those from cracking processes. The main source of propylene, however, is steam cracking of hydrocarbon derivatives, where it is coproduced with ethylene. There is no special process for propylene production except the dehydrogenation of propane.

$$CH_3CH_2CH_3 \rightarrow Catalyst \rightarrow CH_3CH=CH_2 + H_2$$

Increasing the yield of the valuable low molecular weight olefin derivatives, especially propylene and the butylene derivatives, remains a major challenge for many integrated refineries. As global petrochemical demand for propylene continues to grow, opportunities for improved production routes will emerge.

Propylene has been considered as a byproduct of ethylene production via the steam cracking of naphtha or other feedstocks. However, this route has not always able to keep up with propylene demand. To make up this shortfall, refineries may isolate propylene from the gaseous effluents of fluid catalytic cracking (FCC) units and purify it to either chemical grade or polymer grade propylene. While many refineries are, of necessity accepting heavy crude oil as the refinery feedstock, the fluid catalytic cracking product slate is increasingly shifting toward the production of low-boiling olefin derivatives (mainly propylene). More stringent specifications for gasoline are needed in the future, for which the current fluid catalytic cracking product slate is not optimal, because of high aromatics and olefin derivatives content.

TABLE 7.4
Properties of Propylene

Chemical formula	C_3H_6
Molar mass	42.08 g/mol
Appearance	Colorless gas
Density	1.81 kg/m³, gas (1.013 bar, 15°C), 613.9 kg/m³, liquid
Melting point	−185.2°C (−301.4°F; 88.0K)
Boiling point	−47.6°C (−53.7°F; 225.6K)
Solubility in water	0.61 g/m³

The conventional fluid catalytic cracking unit is typically operated at low to moderate severity with flexibility to swing between maximum distillate and maximum gasoline mode (Aitani, 2006: 2,3). This unit yields 3%–4% w/w propylene. Improvements in fluid catalytic cracking catalysts, process design, hardware, and operation severity can boost propylene yield up to 25% w/w or higher. In fluid catalytic cracking practice, there are several options to increase the selectivity to low molecular weight olefin derivatives (Aitani, 2006), which are: (i) dedicated fluid catalytic cracking catalysts; (ii) ZSM-5 additives; (iii) higher severity operation, i.e., higher cracking temperature; and (iv) naphtha recycle (Maadhah et al., 2008).

Conventional fluid catalytic cracking catalyst compositions contain a catalytic cracking component and amorphous alumina which is necessary to provide the bottoms conversion. Catalytic cracking components are crystalline compounds such as faujasite-type Y zeolite as well as amorphous alumina may also be used as a binder to provide a matrix with enough binding function to properly bind the crystalline cracking component when present. ZSM-5 based additive containing a small pore zeolite (5.5–7.5 Å) is commonly added to the cracking catalyst in fluid catalytic cracking to enhance gasoline octane and olefin derivatives production, especially propylene.

One cost-effective way to increase the propylene yield from the fluid catalytic cracking unit is the use of specialized catalysts that contain ZSM-5 zeolite. An increasing number of refiners use as much as 10% w/w of ZSM-5 additives to obtain more than a 9% w/w yield of propylene. Because of its unique pore structure, ZSM-5 limits access to only linear or slightly branched hydrocarbon molecules within the gasoline boiling point range. ZSM-5-based additive acts mainly by cracking C_{6+} naphtha olefin derivatives to smaller olefin derivatives such as propylene and butylenes (Maadhah et al., 2008). These catalysts and additives increase the yield of propylene and other low molecular weight olefin derivatives at the expense of gasoline and distillate products.

The cracking of low molecular weight hydrocarbon derivatives is another excellent option for the fluid catalytic cracking-based refinery to produce and recover propylene. Naphtha is the most common feedstock used in fluid catalytic cracking units for the incremental production of propylene. Various process schemes for naphtha cracking in the fluid catalytic cracking unit have been suggested and the simplest option consists of feeding and cracking naphtha together with gas oil feed. Naphtha may also be injected at the bottom of the fluidized riser reactor, before regenerated

---

TABLE 7.5

Processes by Which Propylene Can Be Produced

Process Name	Developer/Licensor	Propylene, Yield % w/w	Comments
Deep Catalytic Cracking (DCC-I and II)	RIPP-Sinopec/Stone Webster	14.6–28.8	Commercialized
Catalytic Pyrolysis Process (CPP)	RIPP-Sinopec/Stone Webster	24.6	Vacuum gas oil (VGO) and heavy feeds
High-Severity FCC (HS-FCC)	Nippon/KFUPM/JCCP/ Saudi Aramco	17–25	High severity temperature
Indmax	Indian Oil Co./ABB Lummus	17–25	Upgrades heavy cuts at high catalyst/oil ratio
Maxofin	ExxonMobil and KBR	18	Variations to increase propylene
NEXCC	Fortum	16	High C/O, short contact time
PetroFCC	UOP	22	Additional reaction severity along with RxCat design
Selective Component Cracking (SCC)	ABB Lummus	24	High Severity operation (temperature, catalyst/oil ratio)
High-Olefins FCC	Petrobras	20–25	High temperature, catalyst/oil ration

catalyst contacts gas oil feed, where it may be cracked at higher temperature and catalyst/oil (C/O) ratio. The need for a higher cracking temperature, a shorter contact time, and a higher catalyst/oil ratio lead to the conclusion that the mechanical restrictions of existing fluid catalytic cracking units prevent the optimization of the conventional process for maximum olefin production. Despite the various fluid catalytic cracking technologies available to increase the yield of propylene (Table 7.5), there remains the need to improve production of propylene. The main objective of the high-severity FCC (HS-FCC) process is to produce significantly more propylene and high-octane number naphtha. The conceptual process and preliminary feasibility study of the HS-FCC process started in the mid-1990s (Fujiyama et al., 2005; Maadhah et al., 2008).

The uses of propylene include gasoline (80%), polypropylene, isopropanol, trimers, and tetramers for detergents, propylene oxide, cumene, and glycerin. Propylene can be polymerized alone or copolymerized with other monomers such as ethylene. Many important chemicals are based on propylene such as isopropanol, allyl alcohol, glycerol, and acrylonitrile. Propylene is used as a feedstock for a wide range of polymers, product intermediates, and chemicals. Major propylene derivatives include polypropylene, acrylonitrile, propylene oxide, oxo-alcohol derivatives, and cumene.

As an olefin, propylene is a reactive compound that can react with many common reagents used with ethylene such as water, chlorine, and oxygen to produce a variety of chemicals (Figure 7.3).

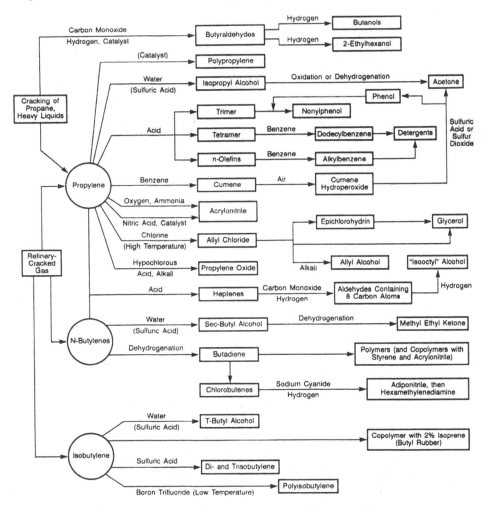

FIGURE 7.3   Chemicals from propylene.

However, structural differences between these two olefin derivatives result in different reactivity toward these reagents. For example, direct oxidation of propylene using oxygen does not produce propylene oxide as in the case of ethylene. Instead, an unsaturated aldehyde, acrolein, is obtained. This could be attributed to the ease of oxidation of allylic hydrogens in propylene. Similar to the oxidation reaction, the direct catalyzed chlorination of propylene produces allyl chloride through substitution of allylic hydrogens by chlorine. Substitution of vinyl hydrogens in ethylene by chlorine, however, does not occur under normal conditions. The current chemical demand for propylene is a little over one-half that for ethylene. This is somewhat surprising because the added complexity of the propylene molecule (due to presence of a methyl group) should permit a wider spectrum of end products and markets. However, such a difference can lead to the production of undesirable byproducts, and it frequently does. This may explain the relatively limited use of propylene in comparison to ethylene. Nevertheless, many important chemicals are produced from propylene.

As is the case for ethylene, moisture in propylene is critical. Several field tests and a few laboratory tests are in use by individual firms, but no standard method for moisture has been adopted to date. The problems in sampling for moisture content, especially in the less than 10 ppm range, are hard to overcome. The trace impurities in 90% or better propylene, which is used in polymerization processes, become quite critical. Hydrogen, oxygen, and carbon monoxide are determined by one technique, and acetylene, ethylene, butylenes, butadiene, methyl acetylene, and propadiene are determined by using a very sensitive analytical method.

Propylene concentrates are mixtures of propylene and other hydrocarbon derivatives, principally propane and trace quantities of ethylene, butylenes, and butanes. Propylene concentrates may vary in propylene content from 70 mol% up to over 95 mol% and may be handled as a liquid at normal temperatures and moderate pressures. Propylene concentrates are isolated from the furnace products mentioned in the preceding paragraph on ethylene. Higher purity propylene streams are further purified by distillation and extractive techniques. Propylene concentrates are used in the production of propylene oxide, isopropyl alcohol, polypropylene, and the synthesis of isoprene.

As with any gas stream, propylene concentrate streams typically require a component analysis, depending upon their final use. The appropriate method for the determination is by gas chromatography. Another gas chromatographic method is used to identify major impurities.

7.3.1 OXIDATION

The direct oxidation of propylene using air or oxygen produces acrolein. Acrolein may further be oxidized to acrylic acid, which is a monomer for polyacrylic resins. Ammoxidation of propylene is considered under oxidation reactions because it is thought that a common allylic intermediate is formed in both the oxidation and ammoxidation of propylene to acrolein and to acrylonitrile, respectively. The use of peroxides for the oxidation of propylene produces propylene oxide. This compound is also obtained via a chlorohydrination of propylene followed by epoxidation.

Acrolein (2-propenal) is an unsaturated aldehyde with a disagreeable odor. When pure, it is a colorless liquid that is highly reactive and polymerizes easily if not inhibited. The main route to produce acrolein is through the catalyzed air or oxygen oxidation of propylene.

$$CH_3CH=CH_2 + O_2 \rightarrow CH_2=CHCHO + H_2O \; \Delta H = -340.5 \; KJ/mol$$

Transition metal oxides or their combinations with metal oxides from the lower row 5A elements were found to be effective catalysts for the oxidation of propylene to acrolein. Two examples of commercially used catalysts are supported CuO (used in the Shell process) and $Bi_2Oi \; MoO_3$ (used in the Sohio process). In both processes, the reaction is carried out at temperature and pressure ranges of 300°C–360°C and 15–30 psi, respectively. In the Sohio process, a mixture of propylene, air, and steam is introduced to the reactor. The hot effluent is quenched to cool the product mixture and to remove the gases. Acrylic acid, a byproduct from the oxidation reaction, is separated in a stripping tower where the acrolein-acetaldehyde mixture enters as an overhead stream. Acrolein is then separated from acetaldehyde in a solvent extraction tower. Finally, acrolein is distilled and the solvent is recycled.

A proposed mechanism for the oxidation of propylene to acrolein is by a first step abstraction of an allylic hydrogen from an adsorbed propylene by an oxygen anion from the catalytic lattice to form an allylic intermediate. The next step is the insertion of a lattice oxygen into the allylic species. This creates oxide-deficient sites on the catalyst surface accompanied by a reduction of the metal. The reduced catalyst is then reoxidized by adsorbing molecular oxygen, which migrates to fill the oxide-deficient sites. Thus, the catalyst serves as a redox system.

The main use of acrolein is to produce acrylic acid and its esters. Acrolein is also an intermediate in the synthesis of pharmaceuticals and herbicides. It may also be used to produce glycerol by reaction with isopropanol (discussed later in this chapter). 2-Hexanedial, which could be a precursor for adipic acid and hexamethylene-diamine, may be prepared from acrolein tail-to-tail dimerization of acrolein using ruthenium catalyst that produces *trans*-2-hexanedial. The trimer, *trans*-6-hydroxy-5-formyl-2,7-octadienal is coproduced. Acrolein, may also be a precursor for 1,3-propanediol. Hydrolysis of acrolein produces 3-hydroxypropionaldehyde which could be hydrogenated to 1,3-propanediol. The diol could also be produced from ethylene oxide (Chapter 7).

There are several ways to produce acrylic acid. Currently, the main process is the direct oxidation of acrolein over a combination molybdenum-vanadium oxide catalyst system. In many acrolein processes, acrylic acid is made the main product by adding a second reactor that oxidizes acrolein to the acid. The reactor temperature is approximately 250°C:

$$2CH_2=CHCHO + O_2 \rightarrow 2CH_2=CHCO_2H$$

Acrylic acid is usually esterified to acrylic esters by adding an esterification reactor. The reaction occurs in the liquid phase over an ion-exchange resin catalyst. An alternative route to acrylic esters is via a B-propiolactone intermediate. The lactone is obtained by the reaction of formaldehyde and ketene, a dehydration product of acetic acid.

The acid-catalyzed ring opening of the four-membered ring lactone in the presence of an alcohol produces acrylic esters:

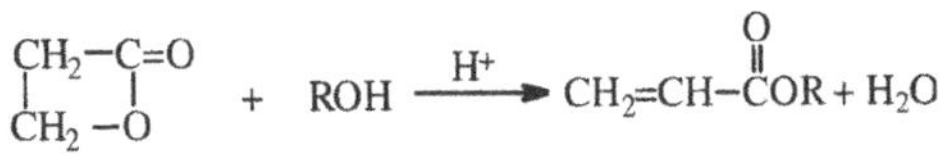

Acrylic acid and its esters are used to produce acrylic resins. Depending on the polymerization method, the resins could be used in the adhesive, paint, or plastic industry.

Shell has coproduced propylene oxide and styrene using the styrene monomer propylene oxide process (SMPO process) (Buijink et al., 2008). The heart of the process is formed by the catalytic epoxidation of propylene with ethylbenzene hydroperoxide using a silica-supported titanium catalyst. The SMPO process comprises four main reaction steps:

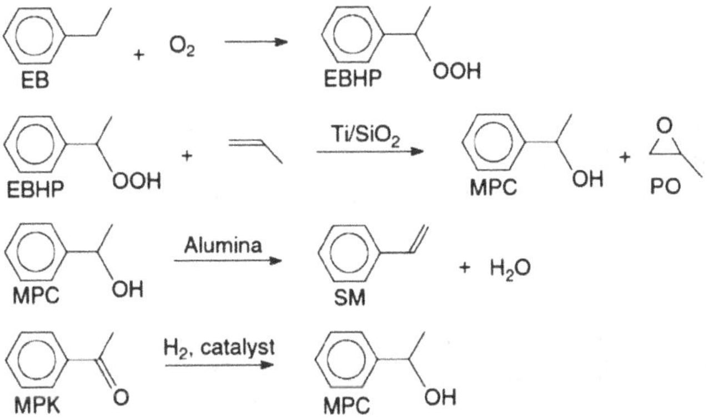

The first step is the air-oxidation of ethyl benzene to ethylbenzene hydroperoxide (EBHP) which is performed in a series of large horizontal bubble-column reactors that are equipped with baffles and heating/cooling coils. Air is introduced via separate middle and side spargers. The gas outlet stream besides unconverted oxygen contains a very significant amount of ethyl benzene from evaporation/stripping and which is recovered in a condensing column and recycled to the reactor train.

The subsequent step, the epoxidation of propylene by ethyl benzene hydroperoxide is carried out in the liquid phase over a heterogeneous catalyst to produce crude propylene oxide and methyl phenyl carbinol (MPC). The feed to the reactors consists of makeup and recycle propylene and ethyl benzene hydroperoxide in ethyl benzene. The reaction train consists of a number of adiabatic fixed bed reactors with interstage cooling. Deactivated catalyst is replaced, incinerated to remove residual hydrocarbon derivatives. The product from the epoxidation reactor is sent to the crude propylene oxide recovery unit that contains a number of distillation columns in which this product is separated into unreacted propylene for recycle to the epoxidation section, crude propylene oxide, ethyl benzene, and styrene precursors (mainly methyl phenyl carbinol and methyl phenyl ketone (MPK)). Low-boiling hydrocarbon derivatives can be used as fuel.

The crude propylene oxide unit is purified in a finishing unit, which consists of a number of distillation columns in which water is removed by azeotropic distillation with normal butane and aldehydes, and light- and heavy ends are also removed from the crude propylene oxide. Prior to entering the methyl phenyl carbinol dehydration step, the ethyl benzene, methyl phenyl carbinol, and methyl phenyl ketone stream from the propylene oxide recovery unit is washed and ethyl benzene is removed by distillation. The methyl phenyl carbinol and the methyl phenyl ketone are sent to the dehydration reactors, where methyl phenyl carbinol is dehydrated to styrene using one of the commercially available catalysts. The reactor product is separated into crude styrene, which is sent to the styrene monomer (SM) finishing unit, and methyl phenyl ketone, with traces of methyl phenyl carbinol, which is sent to a catalytic hydrogenation unit. The product containing methyl phenyl carbinol, and some ethyl benzene and methyl phenyl ketone is recycled to the styrene monomer reaction unit.

7.3.2 Ammoxidation

Ammoxidation refers to a reaction in which a methyl group with allyl hydrogens is converted to a nitrile group using ammonia and oxygen in the presence of a mixed oxide-based catalyst. A successful application of this reaction produces acrylonitrile from propylene:

$$CH_3CH=CH_2 + NH_3 + 11/2O_2 \rightarrow +CH_2=CHCN + 3H_2O$$

As with other oxidation reactions, ammoxidation of propylene is highly exothermic, so an efficient heat removal system is essential.

Acetonitrile and hydrogen cyanide are byproducts that may be recovered for sale. Acetonitrile (CH_3CN) is a high polarity aprotic solvent used in DNA synthesizers, high performance liquid chromatography (HPLC), and electrochemistry. It is an important solvent for extracting butadiene from C_4 streams.

Both fixed and fluid bed reactors are used to produce acrylonitrile, but most modern processes use fluid bed systems. The Montedison-UOP process uses a highly active catalyst that gives 95.6% propylene conversion and a selectivity above 80% for acrylonitrile. The catalysts used in ammoxidation are similar to those used in propylene oxidation to acrolein. Oxidation of propylene occurs readily at 322°C (612°F) over Bi-Mo catalysts.

However, in the presence of ammonia, the conversion of propylene to acrylonitrile does not occur until approximately 400°C (750°F). This may be due to the adsorption of ammonia on catalytic sites that block propylene chemisorption. As with propylene oxidation, the first step in the ammoxidation reaction is the abstraction of an alpha hydrogen from propylene and formation of an allylic intermediate. Although the subsequent steps are not well established, it is believed that adsorbed ammonia dissociates on the catalyst surface by reacting with the lattice oxygen, producing water. The adsorbed NH species then reacts with a neighboring allylic intermediate to yield acrylonitrile.

Acrylonitrile is mainly used to produce acrylic fibers, resins, and elastomers. Copolymers of acrylonitrile with butadiene and styrene are the ABS resins and those with styrene are the styrene-acrylonitrile resins that are important plastics. Most of the production was used for ABS resins and acrylic and mono-acrylic fibers. Acrylonitrile is also a precursor for acrylic acid (by hydrolysis) and for adiponitrile (by an electrodimerization).

Adiponitrile is an important intermediate for producing nylon 66. There are other routes for its production. The way to produce adiponitrile via propylene is the electrodimerization of acrylonitrile.

Propylene oxide is similar in its structure to ethylene oxide, but due to the presence of an additional methyl group, it has different physical and chemical properties. It is a liquid that boils at 34°C (93°F), and it is only slightly soluble in water. (Ethylene oxide, a gas, is very soluble in water.) The main method to obtain propylene oxide is chlorohydrination followed by epoxidation. This older method still holds a dominant role in propylene oxide production. Chlorohydrination is the reaction between an olefin and hypochlorous acid.

Ethylene oxide (also called epoxyethane, oxirane) is a cyclic ether and is the simplest epoxide with faintly sweet odor and colorless flammable gas at room temperature. Ethylene oxide is important to the production of detergents, thickeners, solvents, plastics, and various organic chemicals such as ethylene glycol, ethanolamine derivatives, simple and complex glycols, polyglycol ethers, and other compounds. It is extremely flammable and explosive and is used as the main ingredient in the manufacturing of thermobaric weapons.

The synthesis of ethylene oxide was first reported in 1859 when it was prepared it by treating 2-chloroethanol with potassium hydroxide:

$$ClCH_2CH_2OH + KOH \rightarrow KCl + H_2O + (CH_2\!-\!CH_2) \text{ [sometime written as } (CH_2CH_2)O]$$

This reaction is carried out at elevated temperature, and beside sodium hydroxide or potassium hydroxide, calcium hydroxide, barium hydroxide, magnesium hydroxide or carbonates of alkali, or alkaline earth metals can be used. In addition, ethylene can be oxidized directly to ethylene oxide using peroxy acids such as peroxybenzoic or meta-chloro-peroxybenzoic acid:

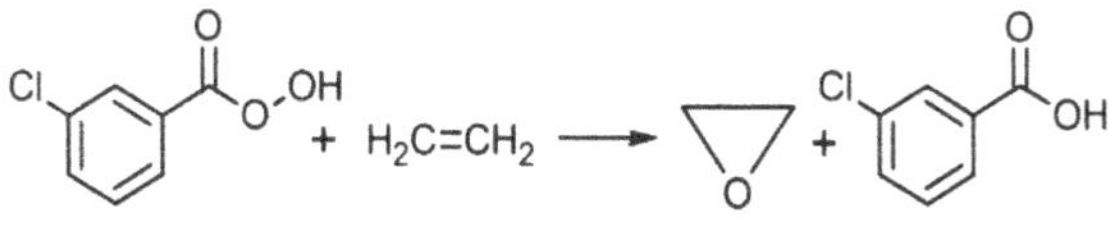

In 1914, BASF (formerly known as Badische Anilin Soda Fabrik) first started the synthesis of ethylene oxide by chlorohydrin process. Later an efficient direct oxidation of ethylene by air was invented by Lefort in 1931 and in 1937 Union Carbide opened the first plant using this process. The process was further improved in 1958 by Shell Oil Co. by replacing air with oxygen and using elevated temperature of 200°C–300°C (390°F–570°F) and pressure is composed of three major steps: (i) synthesis of ethylene chlorohydrin, (ii) dehydrochlorination of ethylene chlorohydrin to ethylene oxide, and (iii) purification of ethylene oxide.

In the process for the production of ethylene oxide on a commercial scale, the main reactor consists of thousands of catalyst tubes in bundles. The catalyst packed in these tubes is in the form of spheres or rings of diameter 3–10 mm. The operating conditions of 200°C–300°C with a pressure of 1–3 MPa prevail in the reactor. The cooling system of the reactor can be used to maintain this temperature. With the aging of the catalyst, its selectivity decreases and it produces more exothermic side products of CO_2. After the gaseous stream from the main reactor, containing ethylene oxide (1%–2%) and CO_2 (5%) is cooled, it is then passed to the ethylene oxide scrubber. Here, water is used as the scrubbing media which wash away majority of ethylene oxide along with some amounts of CO_2, N_2, CH_2CH_2, CH_4, and aldehydes. A small proportion of the gas leaving the ethylene oxide scrubber (0.1%–0.2%) is removed continuously to prevent the buildup of inert compounds, which are introduced as impurities with the reactants. The aqueous stream resulting from the above scrubbing process is then sent to the ethylene oxide desorber. Here, ethylene oxide is obtained as the overhead product, whereas the bottom product obtained is known as the "glycol bleed." The ethylene oxide stream is stripped of its low-boiling components and then distilled in order to separate it into water and ethylene oxide.

The recycle stream obtained from the ethylene oxide scrubber is compressed and a side-stream is fed to the CO_2 scrubber. Here, CO_2 gets dissolved into the hot aqueous solution of potassium carbonate. The dissolution of CO_2 is not only a physical phenomenon but a chemical phenomenon as well, for the CO_2 reacts with potassium carbonate to produce potassium hydrogen carbonate:

$$K_2CO_3 + CO_2 + H_2O \rightarrow 2KHCO_3$$

The potassium carbonate solution is then sent to the CO_2 de-scrubber where CO_2 is de-scrubbed stepwise (usually two steps) flashing. The first step is done to remove the hydrocarbon gases, and the second step is employed to strip off CO_2.

Ethylene oxide is one of the most commonly used sterilization methods in the healthcare industry because of its non-damaging effects for delicate instruments and devices that require sterilization, and for its wide range of material compatibility. Ethylene oxide is used as an accelerator of maturation of tobacco leaves and fungicide. Ethylene is used in the synthesis of 2-butoxyethanol, which is a solvent used in many products. Ethylene oxide can readily react with divergent compounds with the opening of the ring; its typical reactions are with nucleophiles which proceed via the SN_2 mechanism both in acidic and alkaline media.

When propylene is the reactant, propylene chlorohydrin is produced. The reaction occurs at approximately 35°C (95°F) and normal pressure without any catalyst:

$$CH_3CH{=}CH_2 + HOCl \rightarrow \underset{\text{Propylene chlorohydrin}}{CH_3CHOHCH_2Cl}$$

Approximately 87%–90% yield could be achieved. The main byproduct is propylene dichloride (6%–9%). The next step is the dehydrochlorination of the chlorohydrin with a 5% $Ca(OH)_2$ solution:

$$2CH_3CHOHCH_2Cl + Ca(OH)_2 \rightarrow 2CH_3CH\overset{O}{\overset{\diagup\diagdown}{{-}}}CH_2 + CaCl_2 + 2H_2O$$

Propylene oxide is purified by steam stripping and then distillation. Byproduct propylene dichloride may be purified for use as a solvent or as a feed to the perchloroethylene process. The main disadvantage of the chlorohydrination process is the waste disposal of calcium chloride ($CaCl_2$).

The second important process for propylene oxide is epoxidation with peroxides. Many hydroperoxides have been used as oxygen carriers for this reaction. Examples are *t*-butyl hydroperoxide, ethylbenzene hydroperoxide, and peracetic acid. An important advantage of the process is that the coproducts from epoxidation have appreciable economic values.

Epoxidation of propylene with ethylbenzene hydroperoxide is carried out at approximately 130°C (266°F) and 500 psi in presence of molybdenum catalyst. A conversion of 98% on the hydroperoxide has been reported. The coproduct a-phenyl ethyl alcohol could be dehydrated to styrene.

Ethylbenzene hydroperoxide is produced by the uncatalyzed reaction of ethylbenzene with oxygen:

$$C_6H_5CH_2CH_3 + O_2 \rightarrow C_6H_5CH(CH_3)OOH$$

Similar to ethylene oxide, the hydration of propylene oxide produces propylene glycol. Propylene oxide also reacts with alcohols, producing polypropylene glycol ethers, which are used to produce polyurethane foams and detergents. Isomerization of propylene oxide produces allyl alcohol, a precursor for glycerol.

Propylene glycol [1,2-propanediol, $CH_3CH(OH)CH_2OH$] is produced by the hydration of propylene oxide in a manner similar to that used for ethylene oxide:

$$CH_3CH\overset{O}{\overset{/\backslash}{-}}CH_2 + H_2O \longrightarrow CH_3CHOHCH_2OH$$
$$\text{Propylene glycol}$$

Depending on the propylene oxide/water ratio, di-, tri-propylene glycol and polypropylene glycol derivatives can be made the main products.

$$n\,CH_3CH\overset{O}{\overset{/\backslash}{-}}CH_2 + H_2O \xrightarrow{H^+} HO{\{CH(CH_3)CH_2O\}}_nH$$

The reaction between propylene oxide and carbon dioxide produces propylene carbonate. The reaction conditions are approximately 200°C and 80 atm. A yield of 95% is anticipated.

$$(\overline{O(CH_3)CHCH_2OC}{=}O)$$

Propylene carbonate is a liquid used as a specialty solvent and a plasticizer.

Allyl alcohol is produced by the catalytic isomerization of propylene oxide at approximately 280°C. The reaction is catalyzed with lithium phosphate. A selectivity around 98% could be obtained at a propylene oxide conversion around 25%.

$$CH_3CH\overset{O}{\overset{/\backslash}{-}}CH_2 \longrightarrow CH_2{=}CHCH_2OH$$

Allyl alcohol is used in the plasticizer industry, as a chemical intermediate, and in the production of glycerol.

Glycerol (1,2,3-propanetriol, $CH_2OHCHOHCH_2OH$) is a trihydric alcohol of great utility due to the presence of three hydroxyl groups. It is a colorless, somewhat viscous liquid with a sweet odor. Glycerin is the name usually used by pharmacists for glycerol. There are different routes for obtaining glycerol. It is a byproduct from the manufacture of soap from fats and oils (a nonpetroleum source). Glycerol is also produced from allyl alcohol by epoxidation using hydrogen peroxide or a peracid (similar to epoxidation of propylene). The reaction of allyl alcohol with hydrogen peroxide produces glycidol as an intermediate, which is further hydrolyzed to glycerol.

Glycidol

Other routes for obtaining glycerol are also based on propylene.

7.3.3 OXYACYLATION

Like vinyl acetate from ethylene, allyl acetate is produced by the vapor-phase oxyacylation of propylene. The catalyzed reaction occurs at approximately 180°C (355°F) and 60 psi over a Pd/KOAc catalyst:

$$2CH_3CH=CH_2 + 2CH_3CO_2H + O_2 \rightarrow 2CH_2=CHCH_2OCOCH_3 + H_2O$$

Allyl acetate is a precursor for 1,4-butanediol via a hydrocarbonylation route, which produces 4-acetoxybutanal. The reaction proceeds with a $Co(CO)_8$ catalyst in benzene solution at approximately 125°C (257°F) and 3,000 psi. The typical mole hydrogen-carbon monoxide ratio is 2:1. The reaction is exothermic, and the reactor temperature may reach 180°C (355°F) during the course of the reaction. Selectivity to 4-acetoxybutanal is approximately 65% at 100% allyl acetate conversion.

7.3.4 CHLORINATION

Allyl chloride ($CH_2=CHCH_2Cl$) is a colorless liquid, insoluble in water but soluble in many organic solvents. It has a strong pungent odor and an irritating effect on the skin. As a chemical, allyl chloride is used to make allyl alcohol, glycerol, and epichlorohydrin. The production of allyl chloride could be achieved by direct chlorination of propylene at high temperatures (approximately 500°C and 1 atm). The reaction substitutes of an allylic hydrogen with a chlorine atom. Hydrogen chloride is a byproduct from this reaction:

$$CH_2=CHCH_3 + Cl_2 \rightarrow CH_2=CHCH_2Cl + HCl$$

The major byproducts are *cis*-1,3-dichloropropylene and *trans*-1,3-dichloropropylene, which are used as soil fumigants.

The most important use of allyl chloride is to produce glycerol via an epichlorohydrin intermediate. The epichlorohydrin is hydrolyzed to glycerol:

$$CH_2=CHCH_2Cl + Cl_2 + H_2O \rightarrow ClCH_2CHOHCH_2Cl + HCl$$

$$2ClCH_2CHOHCH_2Cl + Ca(OH_2 \rightarrow 2CH_2CHCH_2Cl + CaCl_2 + 2H_2O)$$

Glycerol, a trihydric alcohol, is used to produce polyurethane foams and alkyd resins. It is also used in the manufacture of plasticizers.

7.3.5 HYDRATION

Isopropanol (2-propanol, $CH_3CHOHCH_3$) is an important alcohol of great synthetic utility. It is the second-largest volume alcohol after methanol (1998 U.S. production was approximately 1.5 billion pounds) and it was the 49th-ranked chemical. Isopropanol under the name "isopropyl alcohol" was the first industrial chemical synthesized from a petroleum-derived olefin (1920). The production of isopropanol from propylene occurs by either a direct hydration reaction (the newer method) or by the older sulfation reaction followed by hydrolysis.

In the direct hydration method, the reaction could be effected either in a liquid or in a vapor-phase process. The slightly exothermic reaction evolves 51.5 KJ/mol.

$$CH_3CH=CH_2 + H_2O \rightarrow CH_3CHOHCH_3$$

In the liquid-phase process, high pressures in the range of 1,200–1,500 psi are used. A sulfo-nated polystyrene cation-exchange resin is the catalyst commonly used at about 150°C (300°F). An isopropanol yield of 93.5% can be realized at 75% propylene conversion. The only important byproduct is diisopropyl ether (about 5%). Gas-phase hydration, on the other hand, is carried out at temperatures above 200°C (390°F) and approximately 370 psi. The Imperial Chemical Industries (ICI) process employs tungsten oxide (WO_3) on a silica carrier as catalyst. Older processes still use the sulfation route. The process is similar to that used for ethylene in the presence of sulfuric acid, but the selectivity is a little lower than the modern vapor-phase processes. The reaction conditions are milder than those used for ethylene.

Isopropanol is a colorless liquid having a pleasant odor; it is soluble in water. It is more soluble in hydrocarbon liquids than methanol or ethanol. For this reason, small amounts of isopropanol may be mixed with methanol-gasoline blends used as motor fuels to reduce phaseseparation problems. About 50% of isopropanol use is to produce acetone. Other important synthetic uses are to produce esters of many acids, such as acetic (isopropyl acetate, solvent for cellulose nitrate), myristic, and oleic acids (used in lipsticks and lubricants). Isopropyl palmitate is used as an emulsifier for cosmetic materials. Isopropyl alcohol is a solvent for alkaloids, essential oils, and cellulose derivatives.

Acetone (2-propanone) is produced from isopropanol by a dehydrogenation, oxidation, or a combined oxidation dehydrogenation route. The dehydrogenation reaction is carried out using either copper or zinc oxide catalyst at approximately 450°C–550°C (840°F–1020°F). A 95% yield is obtained.

The direct oxidation of propylene with oxygen is a non-catalytic reaction occurring at approximately 90°C–140°C (194°F–284°F) and 200–300 psi. In this reaction, hydrogen peroxide is coproduced with acetone. At 15% isopropanol conversion, the approximate yield of acetone is 93% and that for hydrogen peroxide is 87%:

$$2CH_3CHOHCH_3 + O_2 \rightarrow 2CH_3COCH_3 + H_2O_2$$

The oxidation process uses air as the oxidant over a silver or copper catalyst. The conditions are similar to those used for the dehydrogenation reaction.

Acetone can also be coproduced with allyl alcohol in the reaction of acrolein with isopropanol. The reaction is catalyzed with an MgO and ZnO catalyst combination at approximately 400°C (750°F) and 1 atm. It appears that the hydrogen produced from the dehydrogenation of isopropanol and adsorbed on the catalyst surface selectively hydrogenates the carbonyl group of acrolein:

$$CH_3CHOHCH_3 + CH_2=CHCHO \rightarrow CH_3COCH_3 + CH_2=CHCH_2OH$$

A direct route for acetone from propylene was developed using a homogeneous catalyst similar to the Wacker system ($PdCl_2/CuCl_2$). The reaction conditions are similar to those used for ethylene oxidation to acetaldehyde. Most acetone is currently obtained via a cumene hydroperoxide process where it is coproduced with phenol.

Acetone is a volatile liquid with a distinct sweet odor. It is miscible with water, alcohols, and many hydrocarbon derivatives. For this reason, it is a highly desirable solvent for paints, lacquers, and cellulose acetate. As a symmetrical ketone, acetone is a reactive compound with many synthetic uses. Among the important chemicals based on acetone are methyl-isobutyl ketone, methyl methacrylate, ketene, and diacetone alcohol.

Mesityl oxide is an alpha-beta unsaturated ketone of high reactivity. It is used primarily as a solvent. It is also used for producing methyl-isobutyl ketone. Mesityl oxide is produced by the dehydration of acetone. Hydrogenation of mesityl oxide produces methyl-isobutyl ketone, a solvent for paints and varnishes.

Methyl methacrylate ($CH_2=CHCOOCH_3$) is produced by the hydrocyanation of acetone using hydrogen cyanide (HCN). The resulting cyanohydrin is then reacted with sulfuric acid and methanol, producing methyl methacrylate:

$$CH_2=\overset{\overset{\displaystyle CH_3}{|}}{C}-\overset{\overset{\displaystyle O}{||}}{C}-\overset{+}{N}H_3\overset{-}{H}SO_4 + CH_3OH \rightarrow CH_2=\overset{\overset{\displaystyle CH_3}{|}}{C}-COOCH_3 + NH_4HSO_4$$

One disadvantage of this process is the waste ammonium hydrosulfate (NH_4HSO_4) stream.

Methacrylic acid is also produced by the air oxidation of isobutylene or the ammoxidation of isobutylene to methacrylonitrile followed by hydrolysis. Methacrylic acid and its esters are useful vinyl monomers for producing polymethacrylate resins, which are thermosetting polymers. The extruded polymers are characterized by the transparency required for producing glass-like plastics commercially known as Plexiglas.

Bisphenol A is a solid material in the form of white flakes, insoluble in water but soluble in alcohols. As a phenolic compound, it reacts with strong alkaline solutions. Bisphenol A is an important monomer for producing epoxy resins, polycarbonates, and polysulfone derivatives. It is produced by the condensation reaction of acetone and phenol in the presence of hydrogen chloride.

In the process to produce Bisphenol A, acetone and excess phenol are reacted by condensation in an ion-exchange resin-catalyzed reactor system to produce Bisphenol A, water and various byproducts.

Distillation removes water and unreacted acetone from the reactor effluent. Acetone and lights are adsorbed into phenol in the light ends adsorber to produce a recycle acetone stream. The bottoms from the distillation column is sent to the crystallization feed pre-concentrator which distills phenol and concentrates the Bisphenol A to a level suitable for crystallization which is then separated from byproducts in a solvent crystallization and recovery system to produce the adduct of Bisphenol A and phenol. The mother liquor from the purification system is distilled in the solvent recovery column to recover dissolved solvent after which the solvent-free mother liquor stream is recycled to the reaction system. A purge from the mother liquor is sent to the purge recovery system along with the recovered process water to recover phenol. The recovered purified adduct is processed in a finishing system to remove phenol from product, and the resulting molten Bisphenol A is solidified produce a product prill suitable for sales.

7.3.6 ADDITION OF ORGANIC ACIDS

Isopropyl acetate is produced by the catalytic vapor-phase addition of acetic acid to propylene. A high yield of the ester can be produced in high yield (approximately 99% w/w):

$$CH_2CH=CH_2 + CH_3COOH \rightarrow CH_3COOCH(CH_3)_2$$

Isopropyl acetate is used as a solvent for coatings and printing inks. It is generally interchangeable with methyl ethyl ketone ($CH_3CH_2COCH_3$) and ethyl acetate $CH_3CO_2C_2H_5$).

Isopropyl acrylate is produced by an acid-catalyzed addition reaction of acrylic acid to propylene. The reaction occurs in the liquid phase at about 100°C:

$$CH_2CH=CH_2 + CH_2=CHCOOH \rightarrow CH_2=CHCOOCH(CH_3)_2$$

Due to unsaturation of the ester, it can be polymerized and used as a plasticizer.

7.3.7 HYDROFORMYLATION

The reaction of propylene with carbon monoxide and hydrogen produces n-butyraldehyde as the main product. Isobutyraldehyde is a byproduct:

$$2CH_3CH=CH_2 + 2CO + 2H_2 \rightarrow \underset{n\text{-Butyraldehyde}}{CH_3CH_2CH_2CHO} + \underset{\text{Isobutyraldehyde}}{(CH_3)_2 CHCHO}$$

Butyraldehyde derivatives are usually hydrogenated to the corresponding alcohols. They are also intermediate species for other chemicals. For example:

n-Butanol ($CH_3CH_2CH_2CH_2OH$) is produced by the catalytic hydrogenation of n-butyraldehyde. The reaction is carried out at relatively high pressures. The yield is high:

$$CH_3CH_2CH_2CHO + H_2 \rightarrow CH_3CH_2CH_2CH_2OH$$

n-Butanol is primarily used as a solvent or as an esterifying agent. The ester with acrylic acid, for example, is used in the paint, adhesive, and plastic industries. An alternative route for n-butanol is through the aldol condensation of acetaldehyde (Chapter 7).

2-Ethylhexanol [$CH_3CH_2CH_2CH_2CH(C_2H_5)CH_2OH$] is a colorless liquid soluble in many organic solvents. It is one of the chemicals used for producing polyvinyl chloride plasticizers (by reacting with phthalic acid; the product is di-2-ethylhexyl phthalate).

2-Ethylhexanol is produced by the aldol condensation of butyraldehyde. The reaction occurs in presence of aqueous caustic soda and produces 2-ethyl-3-hydroxyhexanal. The aldehyde is then dehydrated and hydrogenated to 2-ethylhexanol.

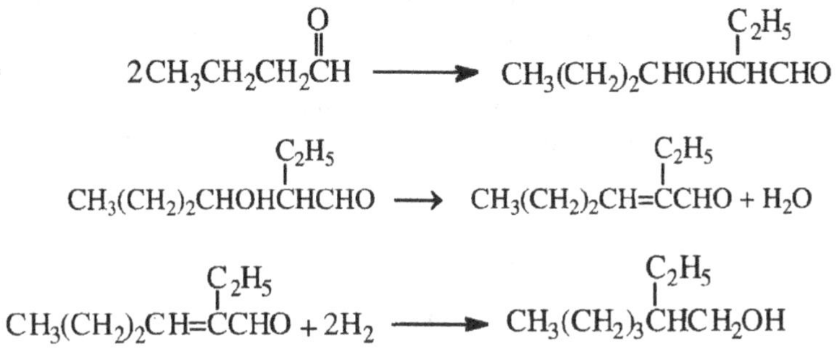

$$2CH_3CH_2CH_2\overset{\overset{\displaystyle O}{\|}}{CH} \longrightarrow CH_3(CH_2)_2\overset{\overset{\displaystyle C_2H_5}{|}}{CHOHCHCHO}$$

$$CH_3(CH_2)_2\overset{\overset{\displaystyle C_2H_5}{|}}{CHOHCHCHO} \longrightarrow CH_3(CH_2)_2\overset{\overset{\displaystyle C_2H_5}{|}}{CH=CCHO} + H_2O$$

$$CH_3(CH_2)_2\overset{\overset{\displaystyle C_2H_5}{|}}{CH=CCHO} + 2H_2 \longrightarrow CH_3(CH_2)_3\overset{\overset{\displaystyle C_2H_5}{|}}{CHCH_2OH}$$

7.3.8 DISPROPORTIONATION

Olefin derivatives could be catalytically converted into shorter and longer-chain olefin derivatives through a catalytic disproportionation reaction. For example, propylene will be disproportionate over different catalysts, yielding ethylene and butylenes. Approximate reaction conditions are 400°C (750°F) and 120 psi:

$$2CH_3CH=CH_2 \rightarrow CH_2=CH_2 + CH_3CH=CHCH_3$$

The utility with respect to propylene is to convert excess propylene to olefin derivatives of greater economic value.

7.3.9 ALKYLATION

Propylene could be used as an alkylating agent for aromatics. An important reaction with great commercial use is the alkylation of benzene to cumene for phenol and acetone production.

7.4 CHEMICALS FROM C$_4$ OLEFINS

The C$_4$ olefin derivatives produce fewer chemicals than either ethylene or propylene. However, C$_4$ olefin derivatives and diolefin derivatives are precursors for some significant big-volume chemicals and polymers such as methyl-t-butyl ether (MTBE), adiponitrile, 1,4-butanediol, and polybutadiene. Butadiene is not only the most important monomer for synthetic rubber production, but also a chemical intermediate with a high potential for producing useful compounds such as sulfolane by reaction with SO_2 1, 4-butanediol by acetoxylation-hydrogenation, and chloroprene by chlorination-dehydrochlorination.

Two butylenes (1-butylene, $CH_3CH_2CH=CH_2$, and 2-butylene, $CH_3CH=CHCH_3$) are industrially significant. The latter has end uses in the production of butyl rubber and polybutylene plastics. On the other hand, 1-butylene is used in the production of 1,3-butadiene ($CH_2=CHCH–CH_2$) for the synthetic rubber industry. Butylenes arise primarily from refinery gases or from the cracking of other fractions of crude oil.

7.4.1 BUTYLENE

Butylene, also known as butene (C_4H_8), is a series of alkene derivatives and the word *butylene* (*butene*) may refer to any of the individual compounds, or to a mixture of them. They are colorless gases that are present in crude oil as a minor constituent in quantities that are too small for viable extraction. Butylene is, therefore, obtained by catalytic cracking of long-chain (higher molecular weight) hydrocarbon derivatives that are produced during crude oil refining. Cracking produces a mixture of products, and the butylene is extracted from this by fractional distillation.

Butylenes (butene derivatives) are byproducts of refinery cracking processes and steam cracking units for ethylene production. Dehydrogenation of butanes is a second source of butylenes. However, this source is becoming more important because isobutylene (a butylene isomer) is currently highly demanded for the production of oxygenates as gasoline additives.

The three isomers constituting *n*-butylenes are 1-butylene, *cis*-2-butylene, and *trans*-2-butylene.

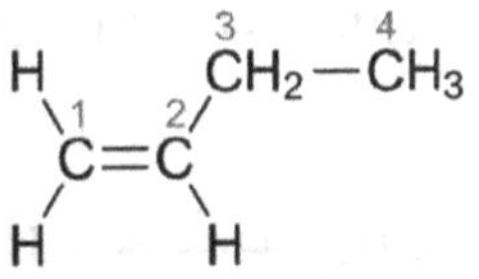

1-Butylene

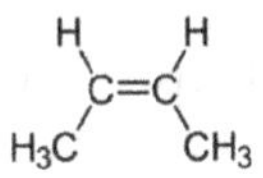

cis-2-Butylene

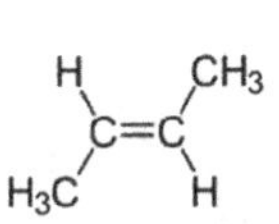

trans-2-Butylene

This gas mixture is usually obtained from the C_4 olefin fraction of catalytic processes and steam cracking processes after separation of isobutylene (Parkash, 2003; Gary et al., 2007; Speight, 2014; Hsu and Robinson, 2017; Speight, 2017). The mixture of isomers may be used directly for reactions that are common for the three isomers and produce the same intermediates and hence the same products. Alternatively, the mixture may be separated into two streams, one constituted of *n*-butylene (1-butylene) (Table 7.6) and the other of *cis*2-butylene and *trans*-2-butylene mixture (Table 7.7). Each stream produces specific chemicals. Approximately 70% of *n*-butylene is used as a comonomer with ethylene to produce linear low-density polyethylene (LLDPE). Another use of *n*-butylene is for the synthesis of butylene oxide. The rest is used with the 2-butylenes to produce other chemicals. *n*-Butylene could also be isomerized to isobutylene.

$$CH_3CH_2CH=CH_2 \rightarrow (CH_3)_2 C=CH_2$$

TABLE 7.6
Properties of *n*-Butylene

Chemical formula	C_4H_8
Molar mass	56.11 g/mol
Appearance	Colorless gas
Odor	Slightly aromatic
Density	0.62 g/cm³
Melting point	−185.3°C (−301.5°F; 87.8K)
Boiling point	−6.47°C (20.35°F; 266.68K)
Solubility in water	0.221 g/100 mL
Solubility	Soluble in alcohol, ether, benzene
Refractive index (n_D)	1.3962

TABLE 7.7
Properties of 2-Butylene (*cis*-2-Butylene Plus *trans*-2-Butylene)

Chemical formula	C_4H_8
Molar mass	56.106 g/mol
Density	0.641 g/mL (*cis*, at 3.7°C), 0.626 g/mL (*trans*, at 0.9°C)
Melting point	−138.9°C (*cis*), −105.5°C (*trans*)
Boiling point	3.7°C (*cis*), 0.9°C (*trans*)

The industrial reactions involving *cis*-2-butylene and *trans*-2-butylene are the same and produce the same products. There are also addition reactions where both 1-butylene and 2-butylene give the same product. For this reason, it is economically feasible to isomerize I-butylene to 2-butylene (*cis* and *trans*) and then separate the mixture. The isomerization reaction yields two streams, one of 2-butylene and the other of isobutylene, which are separated by fractional distillation, each with a purity of 80%–90%.

An alternative method for separating the butylenes is by extracting isobutylene (due to its higher reactivity) in cold sulfuric acid, which polymerizes it to di- and tri-isobutylene. The dimer and trimer of isobutylene have high-octane ratings and are added to the gasoline pool.

Butylene concentrates are mixtures of 1-butylene, *cis*-2-butylene and *trans*-2-butylene-2, and, sometimes, isobutylene (2-methyl propylene) (C_4H_8).

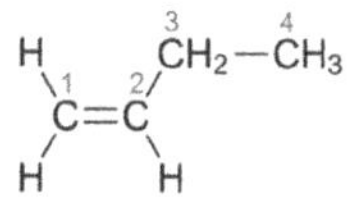

1-Butylene

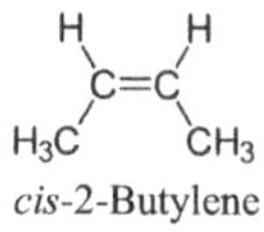

cis-2-Butylene

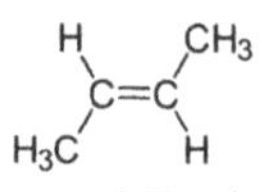

trans-2-Butylene

$$H \quad CH_3$$
$$\diagdown C{=}C \diagup$$
$$H \diagup \quad \diagdown CH_3$$

isobutylene (2-methylpropylene, 2-methyl propylene)

These products are stored as liquids at ambient temperatures and moderate pressures. Various impurities such as butane, butadiene, and the C_5 hydrocarbon derivatives are generally found in butylene concentrates. Virtually all of the butylene concentrates are used as a feedstock for either: (i) an alkylation plant, where isobutane and butylenes are reacted in the presence of either sulfuric acid or hydrofluoric acid to form a mixture of C_7 to C_9 paraffins used in gasoline, or (ii) butylene dehydrogenation reactors for butadiene production.

The major quality criterion for butylene concentrates is the distribution of butylenes which is measured, along with other components, by gas chromatography. Trace impurities generally checked are sulfur-containing derivatives, chlorine-containing derivatives, and acetylene derivatives. These impurities are known catalyst poisons that interfere with (to the point of destroy) catalyst activity or (at best) become unwanted impurities in the final product. When butylene concentrates are used as the feedstock for an alkylation unit, the diolefin content becomes important because of the potential for further reaction to produce unwanted higher molecular weight products (Speight, 2014, 2017).

The above sections have focused on natural gas and refinery gases, which are primarily produced in petroleum refineries as the low-boiling fractions of distillation and cracking processes or in gas plants that separate natural gas and natural gas liquids. Substances in both categories have high vapor pressure and moderate to high water solubility. The gas mixtures are composed primarily of paraffinic and olefinic hydrocarbon derivatives, mostly containing one to six carbon atoms (C_1 to C_6 or, in some cases, to C_8). Some of the mixtures may contain varying amounts of other components, including hydrogen, nitrogen, and carbon dioxide. The refinery gas streams also contain olefin constituents that are produced by various cracking processes and some streams also contain varying amounts of other chemicals including, ammonia, hydrogen, nitrogen, hydrogen sulfide, mercaptans, carbon monoxide, carbon dioxide, 1,3-butadiene, and/or benzene. There are, however, two other gas streams that must be given consideration here because of their increasing importance as fuel gases and these are (i) biomethane, which falls into the category of biogas and (ii) landfill gas, which may also be include in the biogas category but for the purpose of this text is included as a separate gas stream.

7.4.1.1 Oxidation

The mixture of n-butylenes (1-butylene and the 2-butylene isomers) can be oxidized to different products depending on the reaction conditions and the catalyst. The three commercially important oxidation products are acetic acid, maleic anhydride, and methyl ethyl ketone. Due to the presence of a terminal double bond in 1-butylene, oxidation of this isomer via a chlorohydrination route is similar to that used for propylene.

Currently, the major route for obtaining acetic acid (ethanoic acid, CH_3CO_2H) is the carbonylation of methanol (Chapter 5). It may also be produced by the catalyzed oxidation of n-butane. The production of acetic acid from n-butylene mixture is a vapor-phase catalytic process. The oxidation reaction occurs at approximately 270°C (520°F) over a titanium vanadate catalyst. A 70% acetic acid yield has been reported. The major byproducts are carbon oxides (25%) and maleic anhydride (3%):

$$CH_3CH{=}CHCH_3 + 2O_2 \rightarrow 2CH_3COOH$$

Acetic acid may also be produced by reacting a mixture of n-butylenes with acetic acid over an ion-exchange resin. The formed sec-butyl acetate is then oxidized to yield three moles of acetic acid:

$$CH_3CH{=}CHCH_3 + CH_3CH_2CH{=}CH_2 + 2CH_3COOH \rightarrow 2CH_3COCH(CH_3)CH_2CH_3$$
sec-Butyl acetate

$$2CH_3COCH(CH_3)CH_2CH_3 + 2O_2 \rightarrow 3CH_3COOH$$

The reaction conditions are approximately 100°C–120°C (212°F–248°F) and 220–375 psi. The oxidation step is non-catalytic and occurs at approximately 200°C (390°F) and 900 psi. An acetic acid yield of 58% could be obtained. Byproducts are formic acid (6%), higher-boiling compounds (3%), and carbon oxides (28%).

Acetic acid is a versatile reagent. It is an important esterifying agent for the manufacture of cellulose acetate (for acetate fibers and lacquers), vinyl acetate monomer, and ethyl and butyl acetates. Acetic acid is used to produce pharmaceuticals, insecticides, and dyes. It is also a precursor for chloroacetic acid and acetic anhydride. The 1994 U.S. production of acetic acid was approximately 4 billion pounds.

Acetic anhydride (acetyl oxide, $CH_3COOCOCH_3$) is a liquid with a strong offensive odor. It is an irritating and corrosive chemical that must be handled with care. The production of acetic anhydride from acetic acid occurs via the intermediate formation of ketene ($CH_2=C=O$) where one mole of acetic acid loses one mole of water. The ketene further reacts with one mole acetic acid, yielding acetic anhydride:

$$CH_2=C=O + CH_3CO_2H \rightarrow CH_3COOCOCH_3$$

Acetic anhydride is mainly used to make acetic esters and acetyl salicylic acid (aspirin).

Methyl ethyl ketone (2-butanone, $CH_3CH_2COCH_3$) is a colorless liquid similar to acetone, but its boiling point is higher (79.5°C, 175°F). The production of methyl ethyl ketone from n-butylene isomers is a liquid-phase oxidation process similar to that used to produce acetaldehyde from ethylene using a Wacker-type catalyst ($PdCl_2/CuCl_2$). The reaction conditions are similar to those for ethylene. The yield of methyl ketone is in the order of 88% w/w.

Methyl ethyl ketone may also be produced by the catalyzed dehydrogenation of sec-butanol over zinc oxide or brass at about 500°C. The yield from this process is approximately 95%. Methyl ethyl ketone is used mainly as a solvent in vinyl and acrylic coatings, in nitrocellulose lacquers, and in adhesives. It is a selective solvent in dewaxing lubricating oils where it dissolves the oil and leaves out the wax. Methyl ethyl ketone is also used to synthesize various compounds such as methyl ethyl ketone peroxide, a polymerization catalyst used to form acrylic and polyester polymers and methyl pentynol by reacting with acetylene. Methyl pentynol is a solvent for polyamides, a corrosion inhibitor, and an ingredient in the synthesis of hypnotics.

Maleic anhydride, a solid compound that melts at 53°C (127°F) is soluble in water, alcohol, and acetone, but insoluble in hydrocarbon solvents.

Maleic anhydride

The production of maleic anhydride from n-butylenes is a catalyzed reaction occurring at approximately 400°C–440°C (750°F–825°F) and 30–60 psi. A special catalyst, constituted of an oxide mixture of molybdenum, vanadium, and phosphorous, may be used. Approximately 45% yield of maleic anhydride could be obtained from this route.

$$CH_3CH=CHCH_3 + 3 O_2 \longrightarrow \qquad + 3 H_2O$$

Other routes to maleic anhydride are the oxidation of n-butane—a major source for this compound—and the oxidation of benzene.

Maleic anhydride is important as a chemical because it polymerizes with other monomers while retaining the double bond, as in unsaturated polyester resins. These resins, which represent the largest end use of maleic anhydride, are employed primarily in fiber-reinforced plastics for the construction, marine, and transportation industries. Maleic anhydride can also modify drying oils such as linseed and sunflower. As an intermediate, maleic anhydride is used to produce malathion, an insecticide:

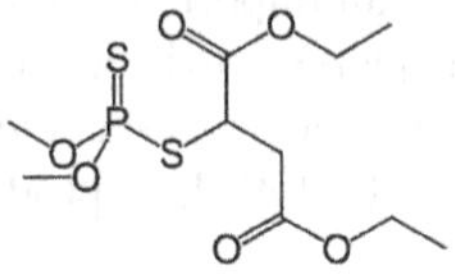

Malathion

In addition, maleic anhydride is used in the manufacture of maleic hydrazide, a plant growth regulator.

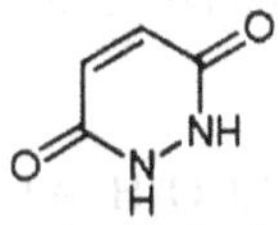

Maleic hydrazide

Maleic anhydride is also a precursor for 1,4-butanediol through an esterification route followed by hydrogenation. In this process, excess ethyl alcohol esterifies maleic anhydride to monoethyl maleate. In a second step, the monoester catalytically esterifies to the diester. Excess ethanol and water are then removed by distillation. Selectivity to the coproducts is high, but the ratios of the coproducts may be controlled with appropriate reactor operating conditions.

Biomass can also be employed as the starting material from which sugar derivatives are produced followed by subsequent acid-catalyzed reactions to produce levulinic acid can be hydrogenated to yield 2-methyl tetrahydrofuran (Khoo et al., 2015):

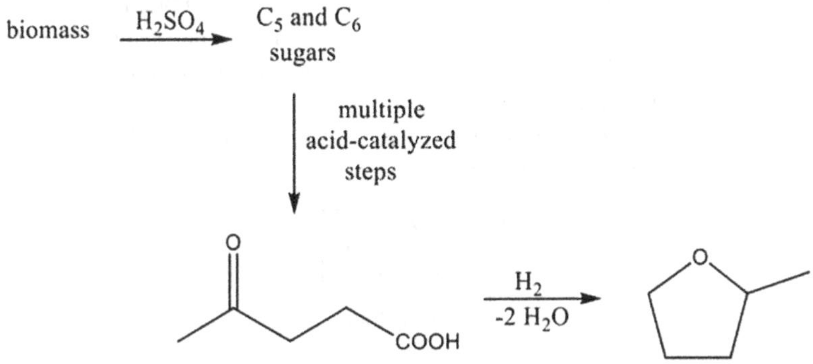

levulinic acid   2-methyltetrahydrofuran

Butylene oxide, like propylene oxide, is produced by the chlorohydrination of 1-butylene with HOCl followed by epoxidation. Butylene oxide may be hydrolyzed to butylene glycol, which is used to make plasticizers. 1,2-Butylene oxide is a stabilizer for chlorinated solvents and also an intermediate in organic synthesis such as in surfactants and pharmaceuticals.

7.4.1.2 Hydration

sec-Butanol (2-butanol, sec-butyl alcohol, $CH_3CHOHCH_2CH_3$), is a liquid with a strong characteristic odor. Its normal boiling point is 99.5°C (211°F), which is near to water's. The alcohol is soluble in water but less so than isopropyl and ethyl alcohols. sec-Butanol is produced by a reaction of sulfuric acid with a mixture of *n*-butylenes followed by hydrolysis. Both 1-butylene and *cis*-2-butylene and *trans*-2-butylene yield the same carbocation intermediate, which further reacts with the sulfuric acid (H_2SO_4) or solutions, to produce a sulfate mixture.

The sulfation reaction occurs in the liquid phase at approximately 35°C (95°F). An 85% w/w alcohol yield could be realized. The reaction is similar to the sulfation of ethylene or propylene and results in a mixture of sec-butyl hydrogen sulfate and di-sec-butyl sulfate. The mixture is further hydrolyzed to sec-butanol and sulfuric acid. The only important byproduct is di-sec-butyl ether, which may be recovered. The major use of sec-butanol is to produce methyl ethyl ketone by dehydrogenation, as mentioned earlier. 2-Butanol is also used as a solvent, a paint remover, and an intermediate in organic synthesis.

7.4.1.3 Isomerization

n-Butylene could be isomerized to isobutylene using Shell FER catalyst which is active and selective. The *n*-butylene mixture from the steam cracking unit or from the fluid catalytic cracking unit after removal of C_5 olefin derivatives via selective hydrogenation step passes to the isomerization unit. It has been proposed that after the formation of a butyl carbocation, a cyclopropyl carbocation is formed which gives a primary carbenium ion that produces isobutylene.

By way of explanation, a carbenium ion is a positive ion with the structure $RR'R''C^+$, that is, it is a chemical species with a trivalent carbon that bears a formal positive charge.

Carbenium ions are generally highly reactive due to having an incomplete octet of electrons. However, certain carbenium ions, such as the tropylium ion, are relatively stable due to the positive charge being delocalized between the carbon atoms.

By way of further explanation, the tropylium ion is an aromatic chemical species with a formula of $[C_7H_7]^+$.

The Tropylium ion

The name derives from the molecule tropine (itself named for the molecule atropine). The ion can be produced from cycloheptatriene and bromine. Salts of the tropylium cation can be stable. In the older literature the name *carbonium ion* was used for this class of chemical species but now the name carbonium ion refers exclusively to the family of carbocations in which the charged carbon is pentavalent.

7.4.1.4 Metathesis

Metathesis is a catalyzed reaction that converts two olefin molecules into two different olefin derivatives. It is an important reaction for which many mechanistic approaches have been proposed by scientists working in the fields of homogenous catalysis and polymerization. One approach is the formation of a fluxional five-membered metallocycle. Another approach is a stepwise mechanism that involves the initial formation of a metal carbene followed by the formation of a four-membered metallocycle species.

Olefin metatheses are equilibrium reactions among the two-reactant and two-product olefin molecules. If chemists design the reaction so that one product is ethylene, for example, they can shift the equilibrium by removing it from the reaction medium. Because of the statistical nature of the metathesis reaction, the equilibrium is essentially a function of the ratio of the reactants and the temperature. For an equimolar mixture of ethylene and 2-butylene at 350°C (660°F), the maximum conversion to propylene is 63%. Higher conversions require recycling unreacted butylenes after fractionation. This reaction was first used to produce 2-butylene and ethylene from propylene. The reverse reaction is used to prepare polymer grade propylene from 2-butylene and ethylene:

$$CH_3CH=CH_2 + CH_2=CH_2 \rightarrow 2CH_3CH=CH_2$$

The metathetic reaction occurs in the gas phase at relatively high temperatures (150°C–350°C, 300°F–660°F) with molybdenum- or tungsten-supported catalysts or at low temperature (250°C, 480°F) with rhenium-based catalyst in either liquid or gas phase. The liquid-phase process gives a better conversion. Equilibrium conversion in the range of 55%–65% could be realized, depending on the reaction temperature.

In this process, the C_4 feedstock is mainly composed of 2-butylene—1-butylene does not favor this reaction, but reacts differently with olefin derivatives, producing metathetic byproducts. The reaction between 1-butylene and 2-butylene, for example, produces propylene and 2-pentylene. The amount of 2-pentylene depends on the ratio of 1-butylene in the feedstock. 3-Hexene is also a byproduct from the reaction of two butylene molecules (ethylene is also formed during this reaction).

7.4.1.5 Oligomerization

The 2-butylene *cis* and *trans* isomers (after separation of 1-butylene) can be oligomerized in the liquid phase on a heterogeneous catalyst system to yield mainly C_8 olefin derivatives and C_{12} olefin derivatives. The reaction is exothermic, and requires a multi-tubular carbon steel reactor. The exothermic heat is absorbed by water circulating around the reactor shell. Either a single-stage or a two-stage system is used. The process can be made to produce either more linear or more branched oligomers. Linear oligomers are used to produce nonyl alcohols for plasticizers, alkyl phenols for surfactants, and tridecyl alcohols for detergent intermediates. Branched oligomers are valuable gasoline components.

7.4.2 ISOBUTYLENE

Isobutylene ($CH_2=C(CH_3)_2$) is a reactive C_4 olefin (Table 7.8). Until recently, almost all isobutylene was obtained as a byproduct with other C4 hydrocarbons from different cracking processes. It was mainly used to produce alkylates for the gasoline pool. A small portion was used to produce chemicals such as isoprene and di-isobutylene. However, increasing demand for oxygenates from isobutylene has called for other sources.

n-Butane is currently used as a precursor for isobutylene. The first step is to isomerize *n*-butane to isobutane, then dehydrogenate it to isobutylene. This serves the dual purpose of using excess *n*-butane (that must be removed from gasolines due to new rules governing gasoline vapor pressure)

TABLE 7.8

Properties of isobutylene

Chemical formula	C_4H_8
Molar mass	56.106 g/mol
Appearance	Colorless gas
Density	0.5879 g/cm³, liquid
Melting point	−140.3°C (−220.5°F; 132.8K)
Boiling point	−6.9°C (19.6°F; 266.2K)
Solubility in water	Insoluble

and producing the desired isobutylene. Currently, the major use of isobutylene is to produce methyl-t-butyl ether $(CH_3)_3COCH_3)$.

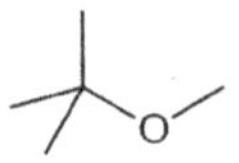

Methyl-t-butyl ether

Methyl tert-butyl ether is a colorless liquid with a distinctive anesthetic-like odor. Vapors are heavier than air and narcotic. This ether has a boiling point 55°C (131°F) with a flash point of –8°C (18°F); it is less dense than water and miscible in water.

7.4.2.1 Oxidation

Much like the oxidation of propylene, which produces acrolein and acrylic acid, the direct oxidation of isobutylene produces methacrolein and methacrylic acid. The catalyzed oxidation reaction occurs in two steps due to the different oxidation characteristics of isobutylene (an olefin) and methacrolein (an unsaturated aldehyde). In the first step, isobutylene is oxidized to methacrolein $[CH_2=C(CH_3)CHO]$ over a molybdenum oxide-based catalyst in a temperature range of 350°C–400°C (650°F–750°F). The process pressures are a little above atmospheric pressure (on the order of 15–25 psi). In the second step, methacrolein is oxidized to methacrylic acid at a relatively lower temperature range of 250°C–350°C (480°F–650°F). A molybdenumsupported compound with specific promoters catalyzes the oxidation.

Methacrylic acid $[CH_2=C(CH_3)COOH]$ is a carboxylic acid that exists as a colorless, viscous liquid is with an acrid unpleasant odor. It is soluble in warm water and miscible with most organic solvents. Methacrylic acid is produced industrially on a large scale as a precursor to it is the ester derivatives such as the methyl methacrylate monomer leading to poly(methyl methacrylate). The methacrylates have numerous uses, most notably in the manufacture of polymers with trade names such as Lucite and Plexiglas.

Methacrylic acid and methacrylates are also produced by the hydrocyanation of acetone followed by hydrolysis and esterification (Chapter 8). Ammoxidation of isobutylene to produce methacrylonitrile is a similar reaction to ammoxidation of propylene to acrylonitrile.

7.4.2.2 Epoxidation

Isobutylene oxide is produced in a way similar to propylene oxide and butylene oxide by a chlorohydrination route followed by reaction with $Ca(OH)_2$.

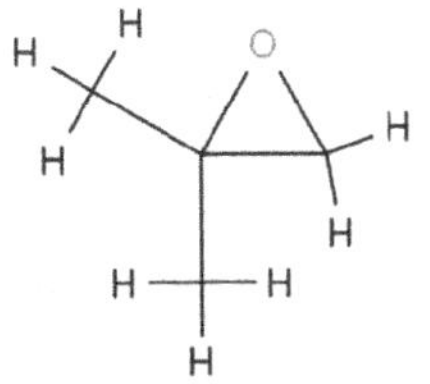

Isobutylene oxide

Direct catalytic liquid-phase oxidation using stoichiometric amounts of thallium acetate catalyst in aqueous acetic acid solution has been reported. An isobutylene oxide yield of on the order of 82% w/w is possible. Direct non-catalytic liquid-phase oxidation of isobutylene to isobutylene oxide gave low yield (28.7%) plus a variety of oxidation products such as acetone, t-butyl alcohol (TBA), and isobutylene glycol. Hydrolysis of isobutylene oxide in the presence of an acid produces isobutylene glycol.

Isobutylene glycol may also be produced by a direct catalyzed liquid-phase oxidation of isobutylene with oxygen in presence of water.

$$2CH_3C(CH_3)=CH_2 + O_2 + 2H_2O \rightarrow CH_3C(CH_3,OH)CH_2OH$$

The catalyst is similar to the Wacker-catalyst system used for the oxidation of ethylene to acetaldehyde. Instead of $PdCl_2/CuC_{12}$ used with ethylene, a $TlCl_3/CuCl_2$ catalyst is employed.

7.4.2.3 Addition of Alcohols

The reaction between isobutylene, methyl alcohol, and ethyl alcohol is an addition reaction catalyzed by a heterogeneous sulfonated polystyrene resin. When methanol is used a 98% yield of methyl-t-butyl ether is obtained. Ethyl-t-butyl ether (ETBE) is also produced by the reaction of ethanol and isobutylene under similar conditions with a heterogeneous acidic ion-exchange resin catalyst (similar to that with methyl-t-butyl ether).

7.4.2.4 Hydration

The acid-catalyzed hydration of isobutylene produces t-butyl alcohol. The reaction occurs in the liquid phase in the presence of 50%–65% sulfuric acid at mild temperatures (10°C–30°C; 50°F–86°F).

TBA is used as a chemical intermediate because a tertiary butyl carbocation forms easily. It is also used as a solvent in pharmaceutical formulations, a paint remover, and a high-octane gasoline additive. The alcohol is a major byproduct from the synthesis of propylene oxide using tertiary butyl hydroperoxide. Surplus t-butyl alcohol could be used to synthesize highly pure isobutylene by a dehydration step:

$$(CH_3)_3 COH \rightarrow (CH_3)_2 C=CH_2 + H_2O.$$

7.4.2.5 Carbonylation

The addition of carbon monoxide to isobutylene under high pressures and in the presence of an acid produces a carbon monoxide-olefin complex, an acyl carbocation. Hydrolysis of the complex at lower pressures yields trimethyl acetic acid (also known as neopentane acid, $(CH_3)_3CCOOH$). In the process, isobutene is hydrocarboxylated by means of the Koch reaction:

$$(CH_3)_2 C=CH_2 + CO + H_2O \rightarrow (CH_3)_3 CCO_2H$$

The reaction requires an acid catalyst such as hydrogen fluoride—tert-butyl alcohol or isobutyl alcohol can also be used in place of isobutene. The reaction is a special case of hydrocarboxylation reaction that does not rely on metal catalysts but, instead, the process is catalyzed by strong acids such as sulfuric acid or the combination of phosphoric acid and boron trifluoride.

Trimethyl acetic acid is an intermediate and an esterifying agent used when a stable neo structure is needed. This colorless, odiferous acid is solid at room temperature. A common abbreviation for the pivalyl or pivaloyl group (t-BuC(O) is *piv* and for pivalic acid (t-BuC(O)OH) is PivOH.

7.4.2.6 Dimerization

A dimerization reaction is an addition reaction in which two molecules of the same compound react with each other to give the adduct. The reaction can be represented very simply as:

$$A + A \rightarrow A_2$$

In this reaction A + A are separate molecules and A_2 is the dimer.

Isobutylene could be dimerized in the presence of an acid catalyst to di-isobutylene. The product is a mixture of di-isobutylene isomers, which are used as alkylating agents in the plasticizer industry and as a lube oil additive (dimerization of olefin derivatives is noted in Chapter 3).

7.5 CHEMICALS FROM DIOLEFINS

Dienes are aliphatic compounds having two double bonds. When the double bonds are separated by only one single bond, the compound is a conjugated diene (conjugated diolefin). Nonconjugated diolefin derivatives have the double bonds separated (isolated) by more than one single bond. This latter class is of little industrial importance. Each double bond in the compound behaves independently and reacts as if the other is not present. Examples of nonconjugated dienes are 1,4-pentadiene and 1,4-cyclo-hexadiene. Examples of conjugated dienes are 1,3-butadiene and 1,3-cyclohexadiene.

Butadiene ($CH_2=CHCH=CH_2$) is a diolefin hydrocarbon derivative with high potential in the chemical industry. Butadiene is a colorless gas that is insoluble in water but soluble in alcohol and which can be liquefied easily under pressure (Table 7.9). This reactive compound polymerizes readily in the presence of free radical initiators. Butadiene (1,3-Butadiene, $CH_2=CHCH=CH_2$, the simplest conjugated diene) is a colorless gas that is easily condensed to a liquid and is important as a monomer in the production of synthetic rubber.

An important difference between conjugated and nonconjugated dienes is that the former compounds can react with reagents such as chlorine, yielding 1,2- and 1,4-addition products. For example, the reaction between chlorine and 1,3-butadiene produces a mixture of 1,4-dichloro-2-butylene and 3,4-dichloro-1-butylene:

Butadiene is mainly obtained as a byproduct from the steam cracking of hydrocarbon derivatives and from catalytic cracking. These two sources account for over 90% of butadiene demand. The remainder comes from dehydrogenation of n-butane or n-butylene streams (Chapter 3). Butadiene is easily polymerized and copolymerized with other monomers. It reacts by addition to other reagents such as chlorine, hydrocyanic acid, and sulfur dioxide, producing chemicals of great commercial value.

Butadiene is obtained mainly as a coproduct with other low molecular weight olefin derivatives from steam cracking units for ethylene production. Other sources of butadiene are the catalytic dehydrogenation of butanes and butylenes, and dehydration of 1,4-butanediol. Butadiene is a colorless gas with a mild aromatic odor. Its specific gravity is 0.6211 at 20°C (68°F) and its boiling temperature is −4.4°C (24.1°F).

7.5.1 CHEMICALS FROM BUTADIENE

Butadiene is by far the most important monomer for synthetic rubber production. It can be polymerized to polybutadiene or copolymerized with styrene to styrene-butadiene rubber (often referred to as SBR). Butadiene is an important intermediate for the synthesis of many chemicals such as hexamethylenediamine and adipic acid. Both are monomers for producing nylon. Chloroprene is another butadiene derivative for the synthesis of neoprene rubber.

TABLE 7.9
Properties of Butadiene

Chemical formula	C_4H_6
Molar mass	54.0916 g/mol
Appearance	Colorless gas or refrigerated liquid
Odor	Mildly aromatic or gasoline-like
Density	0.6149 g/cm³ at 25°C, solid; 0.64 g/cm³ at −6°C, liquid
Melting point	−108.9°C (−164.0°F; 164.2K)
Boiling point	−4.4°C (24.1°F; 268.8K)
Solubility in water	1.3 g/L at 5°C 735 mg/L at 20°
Solubility	Very soluble in acetone, soluble in ether, ethanol
Vapor pressure	2.4 atm (20°C)
Refractive index(n_D)	1.4292

When polymerizing dienes for synthetic rubber production, coordination catalysts are used to direct the reaction to yield predominantly 1,4-addition polymers. Chapter 11 discusses addition polymerization. The following reviews some of the physical and chemical properties of butadiene and isoprene.

7.5.1.1 Adiponitrile

Adiponitrile, a colorless liquid, is slightly soluble in water but soluble in alcohol. The main use of adiponitrile is to make nylon 6/6. The production of adiponitrile from butadiene starts by a free radical chlorination, which produces a mixture of 1,4-dichloro-2-butylene and 3,4-dichloro-l-butylene:

$$2CH_2=CHCH=CH_2 + 2Cl_2 \rightarrow ClCH_2CH=CHCH_2Cl + CH_2=CHCHClCH_2Cl$$

The vapor-phase chlorination reaction occurs at approximately 200°C–300°C (390°F–570°F). The dichloro-butylene mixture is then treated with sodium cyanide (NaCN) or with hydrogen cyanide (HCN) in presence of copper cyanide. The product 1,4-dicyano-2-butylene is obtained in high yield because allylic rearrangement to the more thermodynamically stable isomer occurs during the cyanation reaction:

$$ClCH_2CH=CHCH_2Cl + CH_2=CHCHClCH_2Cl + 4NaCN \rightarrow 2NCCH_2CH=CHCH_2CN + 4NaCl$$

The dicyano compound is then hydrogenated over a platinum catalyst to adiponitrile.

$$NCCH_2CH=CHCH_2CN + H_2 \rightarrow NC(CH_2)_4 CN$$
Adiponitrile

Adiponitrile may also be produced by the electrodimerization of acrylonitrile or by the reaction of ammonia with adipic acid followed by two-step dehydration reactions.

7.5.1.2 Hexamethylenediamine

Hexamethylenediamine (also called 1,6-diaminohexane and 1,6-hexanediamine) also known as hexamethylenediamine ($H_2N(CH_2)_6NH_2$) is a colorless solid, soluble in both water and alcohol. It is the second monomer used to produce nylon 6/6 with adipic acid or its esters. The main route for the production of hexamethylene diamine is the liquid-phase catalyzed hydrogenation of adiponitrile:

$$NC(CH_2)_4 CN + 4H_2 \rightarrow H_2N(CH_2)_6 NH_2$$

The reaction conditions are approximately 200°C (390°F) and 450 psi over a cobalt-based catalyst.

7.5.1.3 Adipic Acid

Adipic acid [$HOOC(CH_2)_4COOH$, hexanedioic acid] is an important dicarboxylic acid. It exists as a white crystalline powder and is used mainly as a precursor for the production of nylon.

Adipic acid may be produced by a liquid-phase catalytic carbonylation of butadiene. A catalyst of rhodium dichloride ($RhCl_2$) and methyl iodide (CH_3I) is used at approximately 220°C (430°F) and 1,100 psi. Adipic acid yield is about 49%. Both α-glutaric acid (25%) and valeric acid (26%) are coproduced:

$$CH_2=CHCH=CH_2 + 2CO + 2H_2O \rightarrow HOOC(CH_2)_4 COOH$$

Another route to adipic acid occurs via a sequential carbonylation, isomerization, hydroformylation reactions:

$$CH_2=CHCH=CH_2 + CO + CH_3OH \rightarrow CH_3CH=CH\ CH_2COCH_2$$

$$CH_3CH=CHCH_2COCH_3 + 2CO + 3H_2 \rightarrow CH_3C(CH_2)_4\ COCH_3 + H_2O$$

$$CH_3C(CH_2)_4\ COCH_3 + O_2 \rightarrow HOC(CH_2)_4\ COCH_3 + HOOC(CH_2)_4\ COOH$$

The main process for obtaining adipic acid is the catalyzed oxidation of cyclohexane (Chapter 10).

Adipic acid is also produced from a mixture of cyclohexanol and cyclohexanone (KA oil, the abbreviation of *ketone-alcohol oil*). The KA oil is oxidized with nitric acid to yield adipic acid, via a multistep pathway. In the initial reaction, the cyclohexanol is converted to the ketone, releasing nitrous acid:

$$HOC_6H_{11} + HNO_3 \rightarrow OC(CH_2)_5 + HNO_2 + H_2O$$

Among its many reactions, the cyclohexanone is nitrosated, followed by scission of the carbon-carbon (C–C) bond:

$$HNO_2 + HNO_3 \rightarrow NO^+NO_3^- + H_2O$$

$$OC_6H_{10} + NO^+ \rightarrow OC_6H_9\text{-}2\text{-}NO + H^+$$

Related processes start from cyclohexanol, which is obtained by the hydrogenation of phenol.

7.5.1.4 Butanediol

The production of 1,4-butanediol $[HO(CH_2CH_2CH_2CH_2OH)]$ from propylene via the carbonylation of allyl acetate. 1,4-Butanediol from maleic anhydride is discussed later in this chapter. An alternative route for the diol is through the acetoxylation of butadiene with acetic acid followed by hydrogenation and hydrolysis.

The first step is the liquid-phase addition of acetic acid to butadiene. The acetoxylation reaction occurs at approximately 80°C (176°F) and 400 psi over a Pd-Te catalyst system. The reaction favors the 1,4-addition product (1,4-diacetoxy-2-butylene). Hydrogenation of diacetoxy butylene at 80°C (176°F) and 900 psi over a Ni/Zn catalyst yields 1,4-diacetoxybutane. The latter compound is hydrolyzed to 1,4-butanediol and acetic acid. The acetic acid is then recovered and recycled. Butanediol is mainly used for the production of thermoplastic polyesters.

7.5.1.5 Chloroprene

Chloroprene (2-chloro 1,3-butadiene, $CH_2=CHCCl=CH_2$), a conjugated non-hydrocarbon diolefin, is a liquid that boils at 59.2°C (138.6°F) and while only slightly soluble in water it is soluble in alcohol. The main use of chloroprene is to polymerize it to neoprene rubber.

Butadiene produces chloroprene through a high-temperature chlorination to a mixture of dichloro-butylene isomers, which is isomerized to 3,4-dichlorol-butylene. This compound is then dehydrochlorinated to chloroprene.

Sulfolane (tetramethylene sulfone) is produced by the reaction of butadiene and sulfur dioxide followed by hydrogenation.

$$CH_2=CH-CH=CH_2 + SO_2 \rightleftharpoons \quad \overset{SO_2}{\bigcirc} \quad \xrightarrow{H_2} \quad \overset{SO_2}{\bigcirc}$$

Sulfolene   Sulfolane

The optimum temperature for highest sulfolene yield is approximately 75°C (167°F). At approximately 125°C (257°F), sulfolene decomposes to butadiene and sulfur dioxide. This simple method could be used to separate butadiene from a mixture of C_4 olefin derivatives because the olefin derivatives do not react with sulfur dioxide.

Sulfolane is a water-soluble biodegradable and highly polar compound valued for its solvent properties. It can be used for the delignification of wood, polymerization, and fiber spinning; electroplating baths; and as a solvent for selectively extracting aromatics from reformates and coke oven products.

7.5.1.6   Cyclic Oligomers

Butadiene could be oligomerized to cyclic dienes and trienes using certain transition metal complexes. Commercially, a mixture of titanium tetrachloride ($TiCl_4$) and $Al_2Cl_3(C_2H_5)$ is used that gives predominantly *cis, trans, trans*-1,5,9-cyclododecatriene along with approximately 5% of the dimer 1,5-cyclooctadiene24.

1,5,9-Cyclododecatriene is a precursor for dodecane-dioic acid through a hydrogenation step followed by oxidation. The diacid is a monomer for the production of nylon 6/12. Cyclododecane from cyclododecatriene may also be converted to the C_{12} lactam, which is polymerized to nylon-12.

7.5.2   Isoprene

Isoprene [2-methyl-1,3-butadiene, $CH_2=C(CH_3)CH=CH_2$] is a colorless liquid with a boiling point of 34.1°C (93.4°F). The compound is soluble in alcohol but not in water. Isoprene is the second important conjugated diene for synthetic rubber production. The main source for isoprene is the dehydrogenation of C_5 olefin derivatives (tertiary amylene derivatives) obtained by the extraction of a C_5 fraction from catalytic cracking units. It can also be produced through several synthetic routes using reactive chemicals such as isobutylene, formaldehyde, and propylene. The main use of isoprene is the production of polyisoprene. It is also a comonomer with isobutylene for butyl rubber production.

The simplest method of isoprene production involves the extraction of the isoprene from the C_5 fraction of liquid petroleum pyrolysis. This fraction is produced as a byproduct in the production of ethylene and propylene. In another process, the two-step production of isoprene from isobutylene and formaldehyde, isobutylene is condensed with formaldehyde in the presence of an acidic catalyst such as diluted sulfuric acid to form 4,4-dimethyldioxane-1,3 after which the dioxane derivative is decomposed into isoprene on a solid phosphate catalyst such as calcium phosphate. In each of these steps there are multiple side reactions. The selectivity of phosphate catalyst is increased by its continuous activation in the process, by the introduction of small amounts of phosphoric acid vapor directly into the catalysis zone which leads to the formation of acidic phosphates on the surface of the calcium phosphate catalyst:

$$Ca_3(PO_4)_2 + H_3PO_4 \rightarrow 3CaHPO_4$$

Also, coke is deposited on the surface of the catalyst and the catalyst should be regenerated at 2–3 h intervals by burning off the coke in a stream of air mixed with steam at temperatures above 500°C (930°F).

7.6   CHEMICALS FROM ACETYLENE

Although not strictly an olefin by virtue of the presence of a triple bond, acetylene (HC≡CH) is included here as a valuable source of petrochemical products because of the reactivity and the variety of product that can be produced from this alkyne.

Acetylene is the simplest member of alkyne hydrocarbon derivatives (Table 7.10) and is the only petrochemical produced in significant quantity which contains a triple bond, and is a major intermediate species but such compounds are not easily shipped, and as a consequence are typically used at

TABLE 7.10
Properties of Acetylene

Chemical formula	C_2H_2
Molar mass	26.04 g/mol
Appearance	Colorless gas
Odor	Odorless
Density	1.097 g/L = 1.097 kg/m^3
Melting point	−80.8°C (−113.4°F; 192.3K) Triple point at 1.27 atm
Sublimation conditions	−84°C; −119°F; 189K (1 atm)
Solubility in water	Slightly soluble
Vapor pressure	44.2 atm (20°C)

or close to the point of origin. Acetylene can be made by hydrolysis of calcium carbide produced in the electric furnace from calcium oxide (CaO) and carbon:

$$CaC_2 + 2H_2O \rightarrow HC\equiv CH + Ca(OH)_2$$

An alternative method of manufacturing acetylene is by cracking methane:

$$2CH_4 \rightarrow HC\equiv CH + 6H_2$$

This process produces only one-third of the methane input as acetylene, the remainder being burned in the reactor. Similar reactions employing heavier fractions of crude oil are being used increasingly since the availability (or necessity) of heavy crude oil as a refinery feedstock is increasing.

In the first half of the 20th century, acetylene was the most important of all starting materials for organic synthesis. Acetylene is a colorless, combustible gas with a distinctive odor. When acetylene is liquefied, compressed, heated, or mixed with air, it becomes highly explosive. As a result, special precautions are required during its production and handling. The most common use of acetylene is as a raw material for the production of various organic chemicals including 1,4-butanediol, which is widely used in the preparation of polyurethane and polyester plastics. The second most common use is as the fuel component in oxy-acetylene welding and metal cutting. Some commercially useful acetylene compounds include acetylene black, which is used in certain dry-cell batteries, and acetylenic alcohols, which are used in the synthesis of vitamins.

Acetylene was discovered in 1836, when Edmund Davy was experimenting with potassium carbide. One of his chemical reactions produced a flammable gas, which is now known as acetylene. In 1859, acetylene was produced by striking and electric arc using carbon electrodes in an atmosphere of hydrogen. Thus:

$$2C + H_2 \rightarrow HC\equiv CH$$

The electric arc tore carbon atoms away from the electrodes and bonded them with hydrogen atoms to form acetylene molecules. He called this gas carbonized hydrogen.

By the late 1800s, a method had been developed for making acetylene by reacting calcium carbide with water. This generated a controlled flow of acetylene that could be combusted in air to produce a brilliant white light. Carbide lanterns were used by miners and carbide lamps were used for street illumination before the general availability of electric lights. In 1897, Georges Claude and A. Hess noted that acetylene gas could be safely stored by dissolving it in acetone. Nils Dalen used this new method in 1905 to develop long-burning, automated marine and railroad signal lights. In 1906, Dalen went on to develop an acetylene torch for welding and metal cutting.

Currently, there are several routes to acetylene. Hydrocarbon derivatives are the major feedstocks, either in the form of natural gas in partial oxidation processes or as byproducts in ethylene production. However, coal is becoming an ever-increasing source of acetylene in countries with plentiful and cheap coal supplies, such as China, for the production of vinyl chloride and this source of lower cost acetylene may prove to be the impetus for returning acetylene to its place as a major chemical feedstock, especially when the current variability (especially the upward mobility) of crude oil prices and improvements in the safety, cost, and environmental protection of the calcium carbide process for the production of acetylene.

The classic commercial route to acetylene, first developed in the late 1800s, is the calcium carbide route in which lime is reduced by carbon (in the form of coke) in an electric furnace to yield calcium carbide. During this process a considerable amount of heat is produced, which is removed to prevent the acetylene from exploding. This reaction can occur via wet or dry processes depending on how much water is added to the reaction process. The calcium carbonate is first converted into calcium oxide and the coal into coke. The two are then reacted with each other to form calcium carbide and carbon monoxide:

$$CaO + 3C \rightarrow CaC_2 + CO$$

The calcium carbide is then hydrolyzed to produce acetylene:

$$CaC_2 + 2H_2O \rightarrow C_2H_2 + Ca(OH)_2$$

Acetylene can also be manufactured by the partial oxidation (partial combustion) combustion of methane with oxygen. The process employs a homogeneous gas phase hydrogen halide catalyst other than hydrogen fluoride to promote the pyrolytic oxidation of methane. The homogeneous gas phase catalyst employed can also consist of a mixture of gaseous hydrogen halide and gaseous halogen, or a halogen gas.

The electric arc or plasma pyrolysis of coal can also be used to produce acetylene. The electric arc process involves a one-megawatt arc plasma reactor which utilizes a DC electric arc to generate and maintain a hydrogen plasma. The coal is then fed into the reactor and is heated to a high temperature as it passes through the plasma. It is then partially gasified to yield acetylene, hydrogen, carbon monoxide, hydrogen cyanide, and several hydrocarbon derivatives.

Acetylene can also be produced as a byproduct of ethylene steam cracking. The use of acetylene as a commodity feedstock decreased due to the competition of cheaper, more readily accessible and workable olefin derivatives when these olefin derivatives were produced from low-cost petroleum products. With the rising cost of crude oil, natural gas, and the associated olefin derivatives feedstocks (such as naphtha, ethane, propane, etc.) the olefin derivatives prices are no longer low enough to preclude the possibility of using acetylene. Additionally, regional shortages of these olefin derivatives and their feedstocks have forced the search for alternate routes to the commodity chemicals.

Between 1960s and 1970s, when worldwide acetylene production peaked, it served as the primary feedstock for a wide variety of commodity and specialty chemicals. Advances in olefin derivatives technology, concerns about acetylene safety, but mostly loss of cost competitiveness, reduced and effectively limited the importance of acetylene. Now, with the current rise in petroleum prices, acetylene is finding a new place in the chemical industry.

Acetylene is the only petrochemical produced in significant quantity which contains a triple bond, and is a major intermediate species. The usefulness of acetylene is partly due to the variety of additional reactions which its triple bond undergoes, and partly due to the fact that its weakly acidic hydrogen atoms are replaceable by reaction with strong bases to form acetylide salts. However, acetylene is not easily shipped, and as a consequence its consumption is close to the point of origin.

However, acetylene was largely replaced by olefin feedstocks, such as ethylene and propylene, because of its high cost of production and the safety issues of handling acetylene at high pressures.

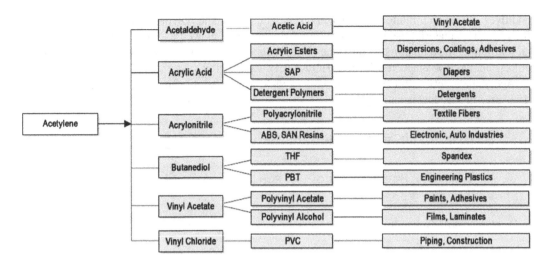

FIGURE 7.4 Chemicals from acetylene and end uses.

Its use has largely been eliminated, except for the continued, and in some instances, growing production of vinyl chloride monomer ($CH_2=CHCl$), 1,4-butanediol ($HOCH_2C_2CH_2CH_2OH$), and carbon black. Up until the 1970s, acetylene was a basic chemical raw material used for the production of a wide range of chemicals (Figure 7.4).

In the presence of metal catalysts, acetylene can react to yield a wide range of industrially significant chemicals. For example, acetylene reacts with alcohol derivatives, hydrogen cyanide, hydrogen chloride, or carboxylic acid derivatives to yield vinyl derivatives.

$$HC\equiv CH + ROH \xrightarrow{cat}$$

$$HC\equiv CH + HCN \xrightarrow{cat}$$

$$HC\equiv CH + HCl \xrightarrow{cat}$$

In addition, acetylene reacts with carbonyl groups to yield ethynyl alcohols.

As an example, acetylene reacts with formaldehyde to yield 1,4-butynediol:

$$HCHO + HC\equiv CH \rightarrow HOCH_2CC\equiv CH_2OH$$

1,4-Butynediol is a precursor to 1,4-butanediol ($HOCH_2CH_2CH_2CH_2OH$) and 2-butylene-1,4-diol ($HOCH_2CH=CHCH_2OH$) by hydrogenation. It is also used in the manufacture of herbicides, textile additives, corrosion inhibitors, plasticizers, synthetic resins, and polyurethane derivatives. It is also the major raw material used in the synthesis of vitamin B_6 as well as for brightening, preserving, and inhibiting nickel plating. 2-Butylene-1,4-diol also reacts with a mixture of chlorine and hydrochloric acid to give mucochloric acid [$HO_2CC(Cl)=C(Cl)CHO$].

Acetylene reacts with carbon monoxide to yield acrylic acid or, in the presence of an alcohol derivative the product is an acrylic ester.

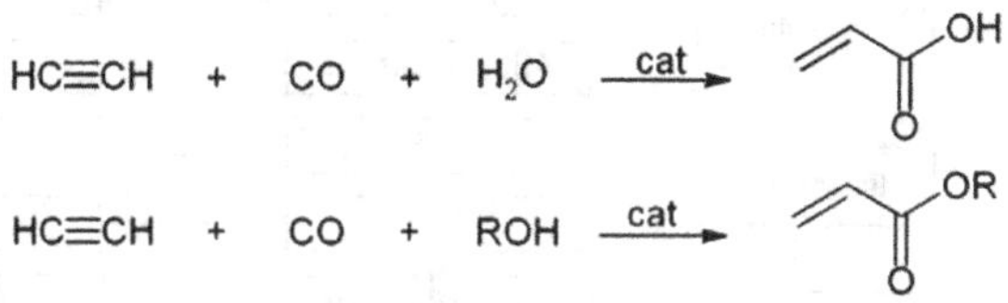

Acetylene will cyclize to produce benzene or cyclo-octatetraene.

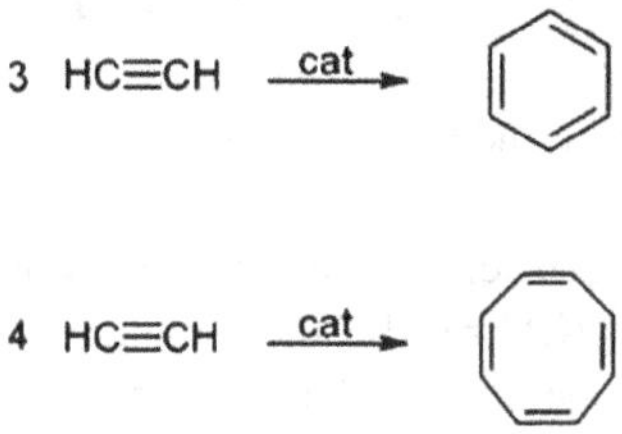

Under basic conditions at 50°C–80°C (122°F–176°F) at 300–375 psi, acetylene reacts with iron pentacarbonyl to yield dihydroxybenzene of which there are three isomers (Table 7.11):

$$Fe(CO)_5 + 4HC\equiv CH + 2H_2O \rightarrow 2C_6H_4(OH)_2 + FeCO_3$$

Acetylene is used as a special fuel gas (oxyacetylene torches) and as a chemical raw material. In fact, historically, acetylene has been used to produce many important chemicals:

1. *Vinyl chloride monomer* was first produced by reacting acetylene with hydrogen chloride. Acetylene-based technology predominated until the early 1950s. Due to the high energy input needed in the acetylene-based process and the hazards of handling acetylene, the ethylene-based route has become the predominant one. However, the acetylene-based route does have its advantages, such as countries where there is a shortage of ethylene cracker feedstock.

TABLE 7.11

The Isomers of Dihydroxybenzene

Ortho Isomer	Meta Isomer	Para Isomer
Catechol, also called: pyrocatechol, 1,2-benzenediol, o-benzenediol, 1,2-dihydroxybenzene, o-dihydroxybenzene 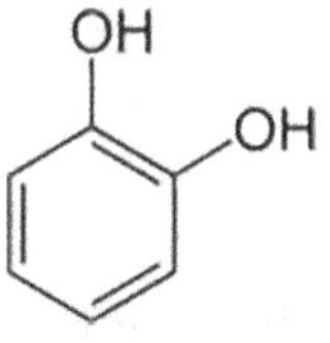	Resorcinol, also called: 1,3-benzenediol, m-benzenediol, 1,3-dihydroxybenzene, m-dihydroxybenzene, resorcin 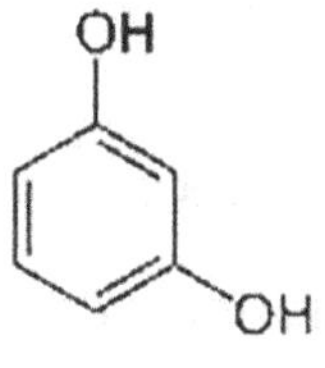	Hydroquinone, also called: 1,4-benzenediol, p-benzenediol, 1,4-dihydroxybenzene, p-dihydroxybenzene 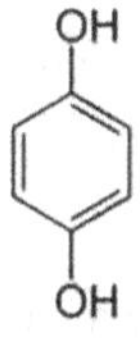

2. *Acrylonitrile:* Hydrogen cyanide added to acetylene produces acrylonitrile, used as an intermediate in the production of nitrile rubbers, acrylic fibers, and insecticides.

3. *Vinyl Acetate:* Acetic acid added to acetylene forms vinyl acetate, used as an intermediate in polymerized form for films and lacquers.

4. *Vinyl Ether:* Alcohol added to acetylene yields vinyl ether used as an anesthetic.

5. *Acetaldehyde:* Water added to acetylene produces acetaldehyde used as a solvent and flavoring in food, cosmetics, and perfumes.

6. *1,2-Dichloroethane:* Chlorine added to acetylene forms 1,2-dichloroethylene, used primarily as a feedstock for vinyl chloride monomer, which, in turn, is the monomer for the widely used plastic, polyvinyl chloride.

7. *1,4-Butynediol:* Formaldehyde added to acetylene produces 1,4-butynediol, which is then hydrogenated to 1,4-butanediol and used as a chain extender for polyurethane. These resins include urethane foams for cushioning material, carpet underlay and bedding, insulation in refrigerated appliances and vehicles, sealants, caulking, and adhesives.

8. *Acrylate Esters:* Acetylene reacts with carbon monoxide and alcohol forming acrylate esters used in the manufacture of Plexiglass and safety glasses.

9. *Polyacetylene:* Acetylene can polymerize forming polyacetylene. The delocalized electrons of the alternating single and double bonds between carbon atoms give polyacetylene its conductive properties. Doping of polyacetylene makes this polymer a better conductor. Polyacetylene is used in rechargeable batteries that could be used in electric cars and could also replace copper wires in aircraft because of the low density (light weight).

10. *Polydiacetylene:* Polydiacetylene is also a polymer of the future. It behaves as a photoconductor and could be used for time-temperature indicators or monitoring of irradiation.

11. In another aspect of acetylene chemistry, tetrahydrofuran can be synthesized by the reaction of formaldehyde with acetylene to make 2-butyne-1,4-diol which is then hydrogenated and cyclized in two more steps to yield tetrahydrofuran:

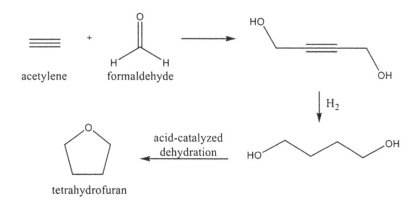

Based on its availability, its many uses and prospective uses, acetylene is a feedstock for petrochemical production that is worthy of consideration. Acetylene is used as a special fuel gas (oxyacetylene torches) and as a chemical raw material.

REFERENCES

Aitani, A. 2006. Propylene production. In: *Encyclopedia of Chemical Processing.* S. Lee (Editor). Taylor & Francis, New York, pp. 2461–2466.

Buijink, J.K.F., Lange, J.-P., Bos, A.N.R., Horton, A.D., and Niele, F.G.M. 2008. Propylene epoxidation via shell's SMPO process: 30 years of research and operation. In: *Mechanisms in Homogeneous and Heterogeneous Epoxidation Catalysis.* S.T. Oyama (Editor). Elsevier BV, Amsterdam.

Fujiyama, Y., Redhwi, H., Aitani, A., Saeed, R., and Dean, C. September 26, 2005. Demonstration plant for new FCC technology yields increased propylene. *Oil & Gas Journal*, 62–67.

Gary, J.G., Handwerk, G.E., and Kaiser, M.J. 2007. *Petroleum Refining: Technology and Economics*. 5th Edition. CRC Press, Taylor & Francis Group, Boca Raton, FL.

Hsu, C.S., and Robinson, P.R. (Editors). 2017. *Handbook of Petroleum Technology*. Springer International Publishing AG, Cham.

Khoo, H.H., Wong, L.L., Tan, J., Isoni, V., and Sharratt, P. 2015. Synthesis of 2-methyl tetrahydrofuran from various lignocellulosic feedstocks: Sustainability assessment via LCA. *Resource, Conservation Recycling*, 95: 174–182.

Kidnay, A.J., and Parrish, W.R. 2006. *Fundamentals of Natural Gas Processing*. CRC Press, Taylor & Francis Group, Boca Raton, FL.

Kohl, A.L., and Nielsen, R.B. 1997. *Gas Purification*. Gulf Publishing Company, Houston, TX.

Kohl, A.L., and Riesenfeld, F.C. 1985. *Gas Purification*. 4th Edition. Gulf Publishing Company, Houston, TX.

Maadhah, A., Fujiyama, Y., Redhwi, H., Abul-Hamayel, M., Aitani, A., Saeed, M., and Dean, C. 2008. A new catalytic cracking process to maximize refinery propylene. *The Arabian Journal for Science and Engineering*, 33(1B): 17–28.

Maddox, R.N., Bhairi, A., Mains, G.J., and Shariat, A. 1985. Chapter 8. Olamine processes. In: *Acid and Sour Gas Treating Processes*. S.A. Newman (Editor). Gulf Publishing Company, Houston, TX.

Mokhatab, S., Poe, W.A., and Speight, J.G. 2006. *Handbook of Natural Gas Transmission and Processing*. Elsevier, Amsterdam.

Newman, S.A. 1985. *Acid and Sour Gas Treating Processes*. Gulf Publishing, Houston, TX.

Parkash, S. 2003. *Refining Processes Handbook*. Gulf Professional Publishing, Elsevier, Amsterdam.

Speight, J.G. 2007. *Natural Gas: A Basic Handbook*. GPC Books, Gulf Publishing Company, Houston, TX.

Speight, J.G. 2014. *The Chemistry and Technology of Petroleum*. 4th Edition. CRC-Taylor and Francis Group, Boca Raton, FL.

Speight, J.G. 2017. *Handbook of Petroleum Refining*. CRC Press, Taylor & Francis Group, Boca Raton, FL.

8 Chemicals from Aromatic Hydrocarbons

8.1 INTRODUCTION

Aromatic compounds (sometimes referred to as arenes) are those compounds that contain one or more benzene rings or similar ring structures (Table 8.1; March, 1985), many of which occur in crude oil and crude oil products (Parkash, 2003; Gary et al., 2007; Speight, 2014; Hsu and Robinson, 2017; Speight, 2017). The majority of the aromatic compounds for petrochemical use are produced in various refinery streams and which are then separated into fractions, of which the most significant constituents are benzene (C_6H_6), methylbenzene or toluene ($C_6H_5CH_3$), and the dimethylbenzene

---

TABLE 8.1

Representative Single-Ring Aromatic Compounds

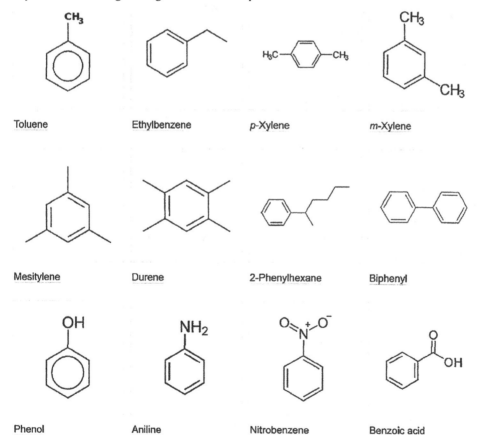

Toluene	Ethylbenzene	p-Xylene	m-Xylene
Mesitylene	Durene	2-Phenylhexane	Biphenyl
Phenol	Aniline	Nitrobenzene	Benzoic acid

323

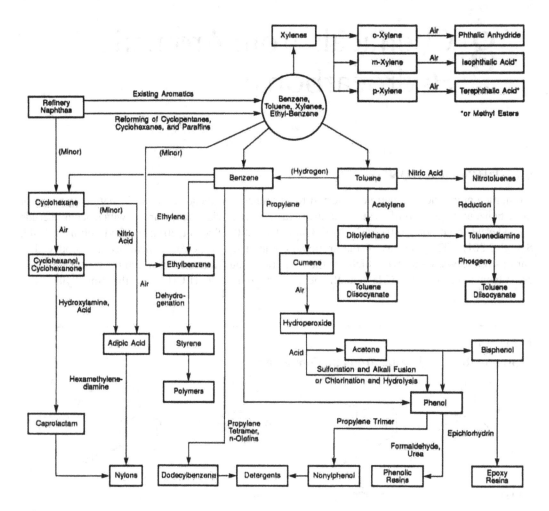

FIGURE 8.1 Chemicals from Benzene, Toluene, and the Xylene isomers.

derivatives or xylene derivatives ($CH_3C_6H_4CH_3$) with the two-ring condensed aromatic compound naphthalene ($C_{10}H_8$) also being a source of petrochemicals.

Benzene, toluene (BT), the benzene, toluene, and xylene isomers (BTX), and benzene, toluene, ethylbenzene, and xylene isomers (BTEX) are the aromatic hydrocarbons with a widespread use as petrochemicals to produce a variety of products (Figure 8.1).

Ethylbenzene ($C_6H_5CH_2CH_3$) is often included in such mixtures (as BTEX) and is a valuable starting material for the production of styrene ($C_6H_5CH=CH_2$).

The C_8 aromatic derivatives are important precursors for many commercial chemicals and polymers such as phenol, trinitrotoluene (TNT), nylons, and plastics. Another compound that has found wide use in the explosive field is 2,4,6-trinitrophenol (also called picric acid).

2,4,6- Trinitrophenol (picric acid)

Aromatic compounds are characterized by having a stable ring structure due to the overlap of the π-orbitals (resonance). Accordingly, they do not easily add to reagents such as halogens and acids as do alkenes. Aromatic hydrocarbon derivatives are susceptible, however, to electrophilic substitution reactions in presence of a catalyst. Aromatic hydrocarbon derivatives are generally nonpolar. They are not soluble in water, but they dissolve in organic solvents such as hexane, diethyl ether, and carbon tetrachloride.

In the traditional chemical industry, aromatic derivatives such as benzene, toluene, and the xylene were made from coal during the course of carbonization in the production of coke and town gas. A much larger volume of these chemicals are now made as refinery byproducts. A further source of supply is the aromatic-rich liquid fraction produced in the cracking of naphtha or low-boiling gas oils during the manufacture of ethylene and other olefin derivatives.

Aromatic compounds are valuable starting materials for a variety of chemical products. Reforming processes have made benzene, toluene, xylene, and ethylbenzene economically available from petroleum sources.

In the catalytic reforming process, a mixture of hydrocarbon derivatives with boiling points between 60°C and 200°C (140°F–390°F) is blended with hydrogen and then exposed to a bifunctional platinum chloride or a rhenium chloride catalyst at 500°C–525°C (930°F–975°F) and pressures ranging from 120 to 750 psi. Under these conditions, aliphatic hydrocarbon derivatives form rings and lose hydrogen to become aromatic hydrocarbons. The aromatic products of the reaction are then separated from the reaction mixture (the reformate) by extraction using a solvent such as diethylene glycol ($HOCH_2CH_2OH$) or sulfolane:

Sulfolane

The benzene is then separated from the other aromatic derivatives by distillation. The extraction step of aromatics from the reformate is designed to produce a mixture of aromatic derivatives with lowest nonaromatic components. Recovery of the aromatic derivatives commonly referred to as benzene, toluene and xylene isomers, involves such extraction and distillation steps.

In similar fashion to this catalytic reforming process, UOP and BP have commercialized a method to produce aromatic derivatives from liquefied petroleum gas (LPG, which is predominantly propane ($CH_3CH_2CH_3$) and butane ($CH_3CH_2CH_2CH_3$)). In this process benzene, toluene, and the xylene isomers are produced by dearomatization of propane and butane. The process consists of reaction system, continuous regeneration of catalyst, and product recovery. The catalyst is a zeolite-type catalyst with a non-noble metal promoter (Gosling et al., 1999).

They are generally recovered by extractive or azeotropic distillation, by solvent extraction (with water-glycol mixtures or liquid sulfur dioxide), or by adsorption. Naphthalene and methylnaphthalenes are present in catalytically cracked distillates. A substantial part of the benzene consumed is now derived from petroleum, and it has many chemical uses.

Benzene, toluene, the xylene isomers and ethylbenzene are obtained mainly from the catalytic reforming of high-boiling naphtha. The product reformate is rich in C_6, C_7, and C_8 aromatic derivatives, which could be extracted by a suitable solvent such as sulfolane or ethylene glycol. These solvents are characterized by a high affinity for aromatic derivatives, good thermal stability, and rapid phase separation.

Aromatic compounds are valuable starting materials for a variety of chemical products. Reforming processes have made benzene, toluene, xylene and ethylbenzene economically available from petroleum sources. They are generally recovered by extractive or azeotropic distillation, by solvent extraction (with water-glycol mixtures or liquid sulfur dioxide, or by adsorption. Naphthalene and methylnaphthalenes are present in catalytically cracked distillates. A substantial part of the benzene consumed is now derived from petroleum, and it has many chemical uses.

Aromatic compounds, such as benzene, toluene, and the xylenes are major sources of chemicals (Figure 8.1). For example, benzene is used to make styrene ($C_6H_5CH=CH_2$), the basic ingredient of polystyrene plastics, as well as paints, epoxy resins, glues, and other adhesives. The process for the manufacture of styrene proceeds through ethylbenzene, which is produced by reaction of benzene and ethylene at 95°C (203°F) in the presence of a catalyst:

$$C_6H_6 + CH_2=CH_2 \rightarrow C_6H_5CH_2CH_3$$

In the presence of a catalyst and superheated steam ethylbenzene dehydrogenates to styrene:

$$C_6H_5CH_2CH_3 \rightarrow C_6H_5CH=CH_2 + H_2$$

Toluene is usually added to the gasoline pool or used as a solvent, but it can be dealkylated to benzene by catalytic treatment with hydrogen:

$$C_6H_5CH_3 + H_2 \rightarrow C_6H_6 + CH_4$$

Similar processes are used for dealkylation of methyl-substituted naphthalene. Toluene is also used to make solvents, gasoline additives, and explosives.

Toluene is usually in demand as a source of trinitrotoluene but has fewer chemical uses than benzene. Alkylation with ethylene, followed dehydrogenation, yields α-methylstyrene [$C_6H_5C(CH_3)=CH_2$], which can be used for polymerization. Alkylation of toluene with propylene tetramer yields a product suitable for sulfonation to a detergent-grade surface-active compound.

Aromatic derivatives are more resistant to oxidation than the paraffin hydrocarbon derivatives, and higher temperatures are necessary; the oxidation is carried out in the vapor phase over a catalyst, generally supported vanadium oxide. Ortho-xylene is oxidized by nitric acid to phthalic anhydride, m-xylene to iso-phthalic acid, and p-xylene with nitric acid to terephthalic acid. These acid products are used in the manufacture of fibers, plastics, plasticizers, and the like.

Phthalic anhydride is also produced in good yield by the air oxidation of naphthalene at 400°C–450°C (750°F–840°F) in the vapor phase at about 25 psi over a fixed bed vanadium pentoxide catalyst. Terephthalic acid is produced in a similar manner from p-xylene, and an intermediate in the process, p-toluic acid ($CH_3C_6H_4CO_2H$), can be isolated because it is slower to oxidize than the p-xylene starting material.

The Tetra extraction process by Union Carbide uses tetraethylene glycol as a solvent. The feed (reformate), which contains a mixture of aromatic derivatives, paraffins, and naphthene derivatives, after heat exchange with hot raffinate, is countercurrently contacted with an aqueous tetraethylene glycol solution in the extraction column. The hot, rich solvent containing benzene-toluene-xylene aromatic derivatives is cooled and introduced into the top of a stripper column. The aromatic derivatives extract is then purified by extractive distillation and recovered from the solvent by steam stripping. Extractive distillation is also feasible and the raffinate (constituted mainly of paraffins, isoparaffins, and cycloparaffins) is washed with water to recover traces of solvent and then sent to storage. The solvent is recycled to the extraction tower. The extract, which is composed of benzene-toluene-xylene derivatives and ethylbenzene, is then fractionated. Benzene and toluene are recovered separately, and ethylbenzene and xylenes are obtained as a mixture (C_8 aromatic derivatives).

Due to the narrow range of the boiling points of C_8 aromatic derivatives, separation by fractional distillation is difficult, especially when the xylene isomers are taken into consideration and require separation (Table 8.2). A superfractionation technique is used to segregate ethylbenzene from the xylene mixture. Because p-xylene is the most valuable isomer for producing synthetic fibers, it is usually recovered from the xylene mixture. Fractional crystallization used to be the method for separating the isomers, but the yield was only 60%. Currently, industry uses continuous liquid-phase adsorption separation processes. The overall yield of p-xylene is increased by incorporating an isomerization unit to isomerize o-xylene and m-xylene to p-xylene. An overall yield of 90% p-xylene could be achieved. In this process, partial conversion of ethylbenzene to benzene also occurs. The catalyst used is shape selective and contains ZSM-5 zeolite.

Briefly, isomerization is the process by which one molecule is transformed into another molecule which has exactly the same number of atoms but the atoms have a different arrangement. For example, in the alkane series straight-chain alkanes are converted to branched isomers by heating in the presence of a platinum or acid catalyst:

$$2CH_3CH_2CH_2CH_2CH_3 \rightarrow CH_3CH_2CH(CH_3)CH_3 + CH_3C(CH_3)_2CH_3$$
$$\underset{n\text{-Pentane}}{} \quad \underset{\text{2-methylbutane}}{} \quad \underset{\text{2,2-dimethlpropane}}{}$$

In the aromatic hydrocarbon series, the isomerization of the xylene isomers is the most well-known process.

1,2-dimethylbenzene   1,3-dimethylbenzene   1,4-dimethylbenzene
(ortho-xylene)   (meta-xylene)   (para-xylene)

The composition of the product mix is dependent upon (i) the composition of the feedstock, which can be one isomer or all three isomers; (ii) the process parameter; and (iii) the catalyst. Xylenes are produced mainly as part of the benzene, toluene, and the xylene isomers aromatics mix that is extracted from the product of catalytic reforming (the reformate).

In some molecules and under some conditions, isomerization occurs spontaneously. Many isomers are roughly equal in bond energy, and so exist in approximately amounts, provided that they can interconvert somewhat freely; that is, the energy barrier between the two isomers is not too high. However, when the isomerization occurs intramolecular, it is considered a rearrangement reaction.

TABLE 8.2

Properties of the Xylene Isomers

Common name	Xylene	o-Xylene	m-Xylene	p-Xylene
Systematic name	Dimethylbenzene	1,2-Dimethylbenzene	1,3-Dimethylbenzene	1,4-Dimethylbenzene
Other names	Xylol	o-Xylol; o-Xylene	m-Xylol; m-Xylene	p-Xylol; p-Xylene
Molecular formula		C_8H_{10}		
Density and phase	0.864 g/mL, liquid	0.88 g/mL, liquid	0.86 g/mL, liquid	0.86 g/mL, liquid
Solubility in water		Practically insoluble		
Soluble in Non-polar Solvents such as Aromatic Hydrocarbons				
Melting point	−47.4°C (−53.3°F)	−25°C (−13°F)	−48°C (−54°F)	13°C (55°F)
Boiling point	138.5°C (281.3°F)	144°C (291°F	139°C (282°F)	138°C (280°F)
Flash point	30°C (86°F)	17°C (63°F)	25°C (77°F)	25°C (77°F)

The Exxon/Mobil XyMax-2 vapor-phase isomerization for xylenes isomerization features a higher activity catalyst, higher weight hourly space velocity (WHSV), and expanded temperature window. The process requires lower catalyst volumes than any process currently in service; achieves higher ethylbenzene conversion per pass; and offers the flexibility of operating at temperatures similar to or lower than existing processes. The advantages of the process are (i) higher weight hourly space velocity, (ii) lower catalyst inventory, (iii) high p-xylene approach to equilibrium, (iv) lower reactor temperature, (v) lower hydrogen to hydrocarbon ratio, (vi) higher conversion of ethyl benzene, and (vii) higher benzene product purity.

The isomerization reactions of aromatic hydrocarbon derivatives proceed during implementation of such catalytic processes, as reforming, cracking, and also in isomerization processes of alkyl aromatic hydrocarbon derivatives. Xylene and ethylbenzene isomerization processes have a great practical importance. Isomerization takes place in disproportionation and transalkylation processes of methylbenzenes, also intended for manufacturing *para-* and *ortho-*xylenes, used for production of terephthalic acid and phthalic anhydride, oligo-polyesters, fibers, varnishes, plasticizers, and other products.

However, due to the peculiarities of xylene thermodynamic equilibrium (minor change of xylene equilibrium concentration with the temperature), corrosion aggressiveness, and non-regenerability of catalytic systems with high acidity, catalysts based on aluminum chloride or boron fluoride did not get extensive industrial application. The most widespread xylene isomerization catalysts became the ones of two types: based on amorphous or crystalline aluminosilicates, and also similar to them heterogeneous catalysts containing platinum.

Depending on the composition of the xylene mixture, aluminosilicate catalysts ($A_{12}O_3$-SiO_2), operating at atmospheric pressure and in the temperature range from 450°C to 550°C (840°F–1,020°F) are used. Over these catalysts ethylbenzene is exposed generally to disproportionation that determines short cycle length of catalyst operation. The recommended content of ethylbenzene in isomerization feed should not exceed 6%–13%. While o-xylene and p-xylene yield makes about 93 wt%. The introduction of platinum introduction in aluminosilicate catalyst and application of hydrogen pressure (the Octafining process), provides ethylbenzene ($C_6H_5C_2H_5$) conversion into xylenes ($CH_3C_6H_4CH_3$) in the order of 60%–70%.

In the Isomar process, the use of halogen-promoted alumino-platinum catalyst is used and the process is used to maximize the recovery of a particular xylene isomer from a mixture of C_8 aromatic isomers. The process is most often applied to p-xylene recovery, but it can also be used to maximize the recovery of o-xylene or m-xylene. In the case of p-xylene recovery, a feedstock consisting of mixed isomers of xylene is charged to a Parex process unit where the p-xylene isomer is preferentially extracted. The raffinate from the Parex unit, almost entirely depleted of p-xylene, is then sent to the Isomar unit. The Isomar unit reestablishes an equilibrium distribution of xylene isomers, essentially creating additional p-xylene from the remaining o-xylene and m-xylene. Effluent from the Isomar unit is then recycled back to the Parex unit for recovery of additional p-xylene. In this way, the o-xylene and m-xylene and ethylbenzene are recycled to extinction. Depending on the type of catalyst, ethylbenzene is converted into xylene isomers or benzene.

In another aspect of aromatics production, the MX Sorbex process recovers m-xylene from mixed xylenes and uses adsorptive separation for highly efficient and selective recovery, at high purity, of molecular species that cannot be separated by conventional fractionation. The process simulates a moving bed of adsorbent with continuous countercurrent flow of liquid feed over a solid bed of adsorbent. Feed and products enter and leave the adsorbent bed continuously, at nearly constant compositions. The fresh feedstock is pumped to the adsorbent chamber. m-Xylene is separated from the feed in the adsorbent chamber and leaves to the extract column. The dilute extract is then fractionated to produce 99.5% w/w m-xylene as the bottoms product. The desorbent is taken from the overhead and recirculated back to the adsorbent chamber. All the other components present in the feedstock are rejected in the adsorbent chamber and removed to the raffinate column. The dilute raffinate is then fractionated to recover desorbent as the overhead product and recirculated back to the adsorbent chamber.

The Sulfolane process (also spelled Sulpholane process) combines liquid-liquid extraction with extractive distillation to recover high-purity aromatics from hydrocarbon mixtures, such as reformed petroleum naphtha (reformate), pyrolysis naphtha, or coker light oil. The solvent used in the Sulfolane process was developed by Shell Oil Co. in the early 1960s and is still the most efficient solvent available for recovery of aromatic derivatives.

In the process, the feedstock enters the extractor and flows upward, countercurrent to a stream of lean solvent. As the feed flows through the extractor, aromatics are selectively dissolved in the solvent. A raffinate stream, very low in aromatics content, is withdrawn from the top of the extractor. The rich solvent, loaded with aromatics, exits the bottom of the extractor and enters the stripper. The lighter nonaromatic constituents taken overhead are recycled to the extractor to displace higher molecular weight nonaromatic constituents from the solvent. The bottoms stream from the stripper, substantially free of nonaromatic impurities, is sent to a column where the aromatic product is separated from the solvent. Because of the large difference in boiling point between the solvent and the highest molecular weight (higher-boiling) aromatic component, this separation is accomplished with minimal energy input.

Lean solvent from the bottom of the recovery column is returned to the extractor where the extract is recovered overhead and sent on to distillation columns downstream for recovery of the individual benzene, toluene and xylene derivatives. The raffinate stream exits the top of the extractor and is directed to the raffinate wash column. In the wash column, the raffinate is contacted with water to remove dissolved solvent. The solvent-rich water is vaporized in the water stripper and then used as stripping steam in the recovery column. The raffinate product exits the top of the raffinate wash column. The raffinate product is commonly used for gasoline blending or ethylene production.

Contaminants that are the most difficult to eliminate in the extraction section are easiest to eliminate in the extractive distillation section and vice versa. This hybrid combination of techniques allows sulfolane units to process feedstocks of much broader boiling range than would be possible by either technique alone. A single sulfolane unit can be used for simultaneous recovery of high-purity C_6 to C_9 aromatic derivatives, with individual aromatic components recovered downstream by simple fractionation.

The emphasis on the production of aromatic products is that aromatic compounds, such as benzene, toluene, and the xylenes are major sources of chemicals. For example, benzene is used for the production of styrene ($C_6H_5CH=CH_2$), the basic ingredient of polystyrene plastics, as well as paints, epoxy resins, glues, and other adhesives. The process for the manufacture of styrene proceeds through ethylbenzene, which is produced by reaction of benzene and ethylene at 95°C (203°F) in the presence of a catalyst:

$$C_6H_6 + CH_2=CH_2 \rightarrow C_6H_5CH_2CH_3.$$

In the presence of a catalyst and superheated steam ethylbenzene dehydrogenates to styrene:

$$C_6H_5CH_2CH_3 \rightarrow C_6H_5CH=CH_2 + H_2$$

Toluene is usually added to the gasoline pool or used as a solvent, but it can be dealkylated to benzene by catalytic treatment with hydrogen:

$$C_6H_5CH_3 + H_2 \rightarrow C_6H_6 + CH_4.$$

In this toluene is mixed with hydrogen, then passed over a chromium, molybdenum, or platinum oxide catalyst at 500°C–600°C (930°F–1,110°F) and 600–900 psi. Higher temperatures may be substituted for the catalyst. If the raw material stream contains much nonaromatic components (paraffin derivatives or naphthene derivatives), those are likely decomposed to lower hydrocarbons such as methane, which increases the consumption of hydrogen. A typical reaction yield exceeds 95%. Xylene isomers and higher molecular weight aromatic derivatives can be used in place of toluene, with similar efficiency.

The irreversible reaction is accompanied by an equilibrium side reaction that produces biphenyl at higher temperature:

$$2C_6H_6 \rightarrow C_6H_5C_6H_5 + H_2$$

Biphenyl is notable as a starting material for the production of polychlorinated biphenyls (PCBs) derivative which were once widely used as dielectric fluids and heat transfer agents.

Of the xylenes, o-xylene is used to produce phthalic anhydride and other compounds. Another xylene, p-xylene is used in the production of polyesters in the form of terephthalic acid or its methyl ester (Figure 8.1). Terephthalic acid is produced from p-xylene by two reactions in four steps. The first of these is oxidation with oxygen at 190°C (375°F):

$$CH_3C_6H_4CH_3 + O_2 \rightarrow HOOCC_6H_4CH_3$$

This is followed by formation of the methyl ester at 150°C (302°F):

$$HOOCC_6H_4CH_3 + CH_3OH \rightarrow HOOCC_6H_4CH_3$$

Repetition of these steps gives the methyl diester of terephthalic acid:

$$CH_3OOCC_6H_4CH_3 + O_2 \rightarrow CH_3OOCC_6H_4CCOOH$$

$$CH_3OOCC_6H_4CCOOH + CH_3OH \rightarrow CH_3OOCC_6H_4CCOOCH_3$$

This diester ($CH_3OOCC_6H_4CCOOCH_3$) when polymerized with ethylene glycol at 200°C (390°F) yields the polymer after loss of methanol to give a monomer. The polymerization step requires a catalyst.

In the process to produce terephthalic acid, the crude acid is produced by the catalytic oxidation of p-xylene with air in the liquid phase using acetic acid as a solvent. The feedstock mix (p-xylene, solvent, and catalyst) is continuously fed, with compressed air, to the e-column oxidizer which operates at moderate temperature. The oxidizer product is purified in a separation step in which the impurities are removed from the product by exchanging the reaction liquor with lean solvent from the solvent recovery system. The reactor overhead—mainly reaction water, acetic acid, and nitrogen—are sent to the solvent recovery system where water is separated from the solvent by distillation. The off-gas is sent to a regenerative thermal oxidation unit for further cleaning.

To produce polymer-grade terephthalic acid, the crude acid is purified in a post-oxidation step, at elevated temperature. The post oxidizers serve as reactors to increase the conversion of the partially oxidized compounds to terephthalic acid.

Aromatic derivatives are more resistant to oxidation than the paraffin hydrocarbon derivatives, and higher temperatures are necessary; the oxidations are carried out in the vapor phase over a catalyst, generally supported by vanadium oxide. Ortho-xylene is oxidized by nitric acid to phthalic anhydride, m-xylene to iso-phthalic acid, and p-xylene with nitric acid to terephthalic acid. These acid products are used in the manufacture of fibers, plastics, plasticizers, and the like.

Phthalic anhydride is also produced in good yield by the air oxidation of naphthalene at 400°C–450°C (750°F–840°F) in the vapor phase at about 25 psi over a fixed bed vanadium pentoxide catalyst. Terephthalic acid is produced in a similar manner from p-xylene, and an intermediate in the process, p-toluic acid, can be isolated because it is slower to oxidize than the p-xylene starting material.

The primary sources of benzene, toluene, and xylenes are refinery streams, especially from catalytic reforming and cracking, and pyrolysis gasoline from steam cracking and from coal liquids. Mixtures of benzene, toluene, and the xylene isomers (and, in some cases ethyl benzene) are extracted from these streams using selective solvents such as sulfolene or ethylene glycol. The extracted components are separated through lengthy fractional distillation, crystallization, and isomerization processes.

The reactivity of C_6, C_7, C_8 aromatic derivatives is mainly associated with the benzene ring. Aromatic compounds in general are liable for electrophilic substitution. Most of the chemicals produced directly from benzene are obtained from its reactions with electrophilic reagents. Benzene could be alkylated, nitrated, or chlorinated to important chemicals that are precursors for many commercial products.

Toluene and xylenes (methylbenzenes) are substituted benzenes. Although the presence of methyl substituents activates the benzene ring for electrophilic attack, the chemistry of methyl benzenes for producing commercial products is more related to reactions with the methyl than with the phenyl group. As an electron-withdrawing substituent (of methane), the phenyl group influences the methyl hydrogens and makes them more available for chemical attack. The methyl group could be easily oxidized or chlorinated as a result of the presence of the phenyl substituent.

8.2 CHEMICALS FROM BENZENE

Benzene (C_6H_6) is the simplest aromatic hydrocarbon and by far the most widely used one. Before 1940, the main source of benzene and substituted benzene was coal tar. Currently, it is mainly obtained from catalytic reforming. Other sources are pyrolysis gasolines and coal liquids. Benzene has a unique structure due to the presence of six delocalized pi-electrons that encompass the six carbon atoms of the hexagonal ring. Thus, benzene, a double-bond conjugated six member hydrocarbon ring, can be represented by two structures that are equivalent in energy:

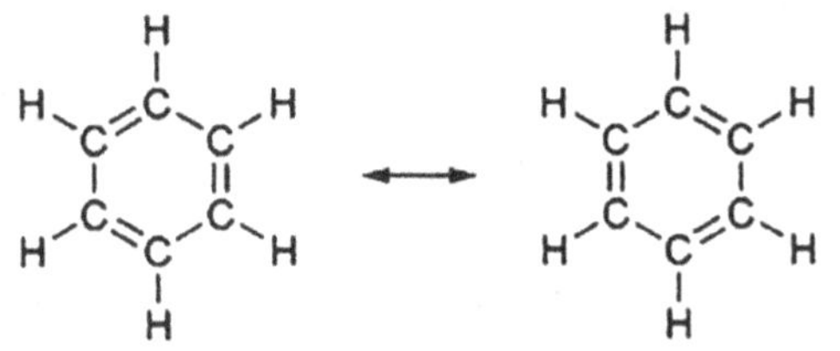

Benzene could be represented by two resonating Kekulé structures. It may also be represented as a hexagon with a circle in the middle. The circle is a symbol of the π-cloud encircling the benzene ring. The delocalized electrons associated with the benzene ring impart very special properties to

TABLE 8.3

Routes from Benzene to Other Petrochemical Products

	Primary Product	Secondary Product	Tertiary Product	Quaternary Product
Benzene				
	Ethylbenzene			
		Styrene		
			Polystyrene	
	Cumene			
		Acetone		
		Phenol		
			Bisphenol A	
				Polycarbonates
				Epoxy resins
			Phenolic resins	
	Cyclohexane			
		Adipic acid		
			Nylon 6/6	
		Caprolactam		
			Nylon 6/6	
	Aniline			
	Chlorobenzenes			

aromatic hydrocarbon derivatives. They have chemical properties of single-bond compounds such as paraffin hydrocarbon derivatives and doublebond compounds such as olefin derivatives, as well as many properties of their own.

Benzene is used mainly as an intermediate (or starting material) to make other chemicals (Table 8.3), above all ethylbenzene, cumene, cyclohexane, nitrobenzene, and alkylbenzene. The predominant process is the manufacture of ethylbenzene ($C_6H_5CH_2CH_3$) which is a precursor to styrene ($C_6H_5CH=CH_2$) from which polymers and plastics are manufactured.

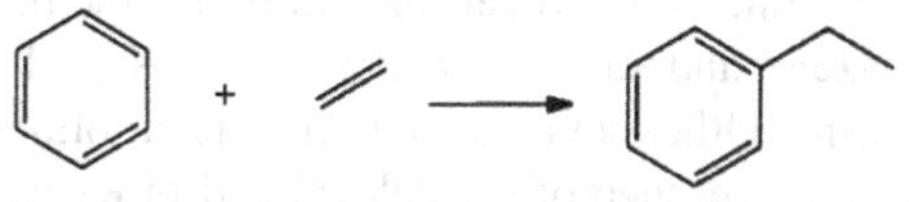

Benzene   ethylene   ethylbenzene

The production of styrene increased dramatically during the 1940s, when it was used as a feedstock for synthetic rubber. In the process to manufacture styrene, ethylbenzene is mixed in the gas phase with 10–15 times its volume of high-temperature steam and then passed over a solid catalyst bed. Most ethylbenzene dehydrogenation catalysts are based on ferric oxide (Fe_2O_3) promoted by potassium oxide (K_2O) or potassium carbonate (K_2CO_3).

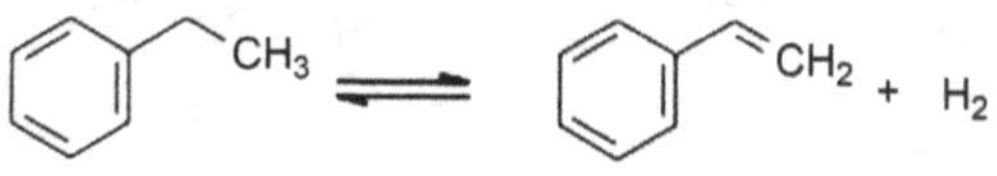

Styrene

A typical styrene plant consists of two or three reactors in series, which operate under vacuum to enhance the conversion and selectivity. Typical per-pass conversions are ca.65% for two reactors and 70%–75% for three reactors. Selectivity to styrene is 93%–97%. The main byproducts are benzene and toluene. Because styrene and ethylbenzene have similar boiling points (145°C and 136°C, 293°F and 276°F, respectively), their separation requires tall distillation towers and high return/reflux ratios. Styrene tends to polymerize at the distillation temperatures and to minimize this problem, older styrene plants added elemental sulfur to inhibit the polymerization.

Styrene is also coproduced commercially in a process in which ethylbenzene is treated with oxygen to form the ethylbenzene hydroperoxide which is used to oxidize propylene to propylene oxide. The resulting 1-phenylethanol is dehydrated to produce styrene:

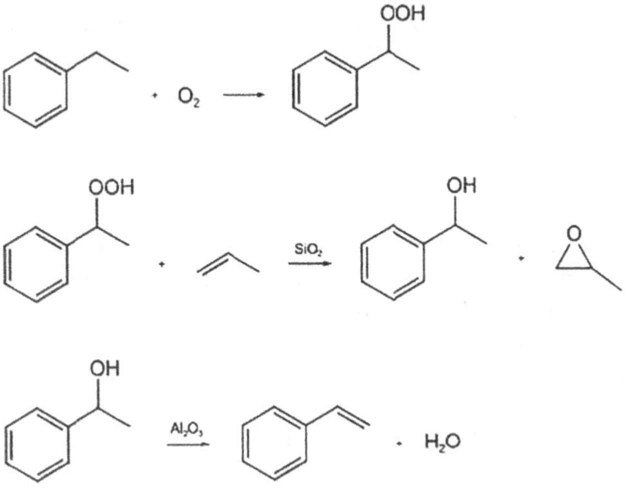

Styrene can also be produced from toluene and methanol. However, this process has suffered from the competing decomposition of methanol. The methanol decomposition can be diminished process using a process in which the parameters are 400°C–425°C (750°F–795°F) and atmospheric pressure, by forcing the reactants through a zeolite catalyst yielding a mixture of styrene and ethylbenzene with a total styrene yield of over 60%.

Another route to styrene involves the reaction of benzene and ethane and the reactants, along with ethylbenzene, are fed to a dehydrogenation reactor with a catalyst capable of simultaneously producing styrene and ethylbenzene. The dehydrogenation effluent is cooled and separated and the ethylene stream is recycled to the alkylation unit. The process attempts to overcome previous shortcomings in earlier attempts to develop production of styrene from ethane and benzene, such as inefficient recovery of aromatics, production of high levels of high-boiling constituents and tars, and inefficient separation of hydrogen and ethane.

Lesser amounts of benzene are used to make some types of rubber, lubricants, dyes, drugs, explosives and pesticides. Toluene is often used as a substitute for benzene, for instance, as a fuel additive. The solvent properties of the two are similar, but toluene is less toxic and has ahigh0boiling constituents wider liquid range.

Aromatic hydrocarbon derivatives, like paraffin hydrocarbon derivatives, react by substitution, but by a different mechanisms under milder reaction conditions. Aromatic compounds react by additions only under several reaction conditions. For example, electrophilic substitution of benzene using nitric acid produces nitrobenzene under normal conditions, while the addition of hydrogen to benzene occurs in presence of catalyst only under high pressure to give cyclohexane.

Monosubstitution can occur at any one of the six equivalent carbon atoms of the ring. Most of the monosubstituted benzenes have common names such as toluene (methylbenzene), phenol (hydroxybenzene), and aniline (aminobenzene). When two hydrogens in the ring are substituted by the same reagent, three isomers are possible. The prefixes ortho, meta, and para are used to indicate the location of the substituents in 1,2-; 1,3-; or 1,4-positions, for example, xylene isomers.

Benzene is an important chemical intermediate and is the precursor for many commercial chemicals and polymers such as phenol, styrene for polystyrene derivatives, and caprolactam for nylon 6. Chapter 10 discusses chemicals based on benzene.

Benzene (C_6H_6) is the most important aromatic hydrocarbon. It is the precursor for many chemicals that may be used as end products or intermediates. Almost all compounds derived directly from benzene are converted to other chemicals and polymers. For example, hydrogenation of benzene produces cyclohexane. Oxidation of cyclohexane produces cyclohexanone, which is used to make caprolactam for nylon manufacture. Due to the resonance stabilization of the benzene ring, it is not easily polymerized. However, products derived from benzene such as styrene, phenol, and maleic anhydride can polymerize to important commercial products due to the presence of reactive functional groups. Benzene could be alkylated by different alkylating agents, hydrogenated to cyclohexane, nitrated, or chlorinated.

The chemistry for producing the various chemicals from benzene is discussed in this section.

8.2.1 ALKYLATION

Benzene can be alkylated in the presence of a Lewis or a Bronsted acid catalyst. Olefin derivatives such as ethylene, propylene, and C_{12}-C_{14} alpha olefin derivatives are used to produce benzene alkylates, which have great commercial value. Alkyl halides such as monochloro-paraffin derivatives also serve this purpose. The first step in alkylation is the generation of a carbocation (carbonium ion). When an olefin is the alkylating agent, a carbocation intermediate forms.

$$RCH{=}CH_2 + H^+ \rightarrow (RCHCH_3)^+$$

Carbon cations also form from an alkyl halide when a Lewis acid catalyst is used. Aluminum chloride is the commonly used Friedel–Crafts alkylation catalyst.

$$RCl + AlCl_3 \rightarrow R^+ + AlCl_4^-$$

The next step is an attack by the carbocation on the benzene ring, followed by the elimination of a proton and the formation of a benzene alkylate.

The acid-catalyzed alkylation of benzene with alkenes is an established commercial process to produce a wide range of alkylbenzenes. The alkylation of benzene with ethylene ($CH_2{=}CH_2$) and propylene ($CH_3CH{=}CH_2$) is used to manufacture ethylbenzene ($C_6H_5CH_2CH_3$) and isopropyl benzene [$C_6H_5CH(CH_3)_2$] which are the intermediates in styrene and phenol production, respectively (Weissermel and Arpe, 2003). In addition, ethylene and propylene can be replaced by ethane and propane (Kato et al., 2001; Huang et al., 2007). The alkylation of benzene with low molecular weight alkane derivatives occurs in the gas phase in the presence of solid bifunctional metal-acid catalysts, which includes the dehydrogenation of alkane on metal sites to form alkene and hydrogen (step 1) followed by the alkylation of benzene with the alkene on acid sites (step 2) (Alotaibi et al., 2017).

Ethylbenzene is a colorless aromatic liquid with a boiling point of 136.2°C (277.2°F), very close to that of p-xylene.

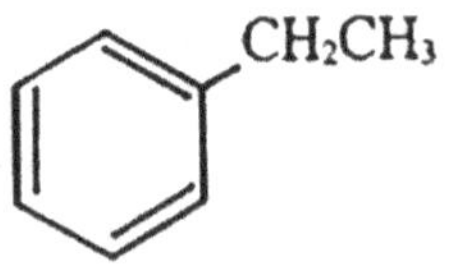

Ethylbenzene

This complicates separating it from the C_8 aromatic equilibrium mixture obtained from catalytic reforming processes. Ethylbenzene obtained from this source, however, is small compared to the synthetic route.

The main process for producing ethylbenzene is the catalyzed alkylation of benzene with ethylene:

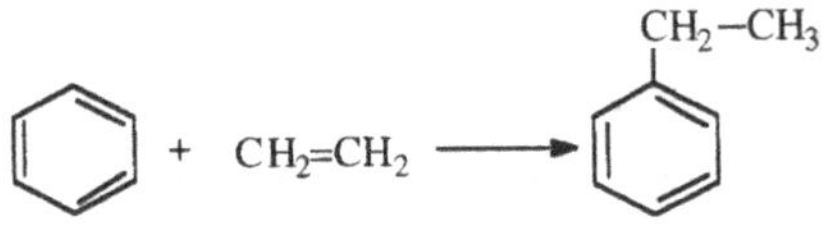

Many different catalysts are available for this reaction. Promoted aluminum chloride ($AlCl_3H^+Cl^-$) is commonly used. Ethyl chloride may be substituted for hydrogen chloride on a molefor-mole basis. Typical reaction conditions for the liquid-phase aluminum chloride-catalyzed process are 40°C–100°C (104°F–212°F) and 30–120 psi. Diethylbenzene and higher alkylated benzenes also form. They are recycled and dealkylated to ethylbenzene.

The vapor-phase Badger process, which has been in commercial use since 1980, can accept dilute ethylene streams such as those that occur in the effluent gas (FCC off gas) from a fluid catalytic cracking unit. A zeolite-type heterogeneous catalyst is used in a fixed bed process. The reaction conditions are 420°C (790°F) and 200–300 psi. Over 98% w/w yield is obtained at 90% conversion. Poly-ethyl benzene and unreacted benzene are recycled and join the fresh feed to the reactor. The reactor effluent is fed to the benzene fractionation system to recover unreacted benzene. The bottoms fraction containing ethylbenzene and heavier poly-alkylated derivatives are fractionated in two columns. The first column separates the ethylbenzene product, and the other separates poly-ethyl benzene for recycling.

Ethylbenzene is mainly used to produce styrene. Styrene (vinylbenzene, $C_6H_5CH=CH_2$) is a liquid (b.p. 145.2°C, 261.4°F) that polymerizes easily when initiated by a free radical or when exposed to light.

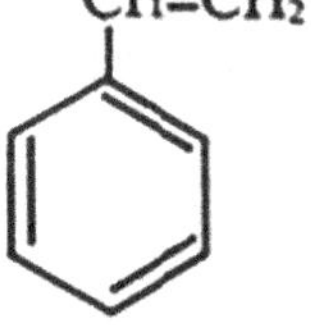

Styrene

Dehydrogenation of ethylbenzene to styrene occurs over a wide variety of metal oxide catalysts. Oxides of Fe, Cr, Si, Co, Zn, or their mixtures can be used for the dehydrogenation reaction. Typical reaction conditions for the vapor-phase process are 600°C–700°C (1,130°F–1,290°F), at or below atmospheric pressure. Approximately 90% styrene yield is obtained at 30%–40% conversion:

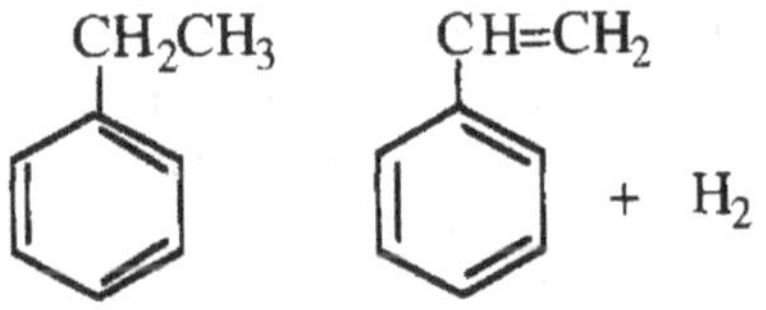

In the Monsanto/Lummus Crest process, fresh ethylbenzene with recycled unconverted ethylbenzene are mixed with superheated steam. The steam acts as a heating medium and as a diluent. The endothermic reaction is carried out in multiple radial bed reactors filled with proprietary catalysts. Radial beds minimize pressure drops across the reactor.

An alternative route for producing styrene is to dimerize butadiene to 4-vinyl-1-cyclohexene, ollowed by catalytic dehydrogenation to styrene: The process involves cyclodimerization of butadiene over a proprietary copper-loaded zeolite catalyst at moderate temperature and pressure (100°C, 212°F, and 250 psi). To increase the yield, the cyclodimerization step takes place in a liquid-phase process over the catalyst. Selectivity for vinyl cyclohexene (VCH) was over 99%. In the second step, vinyl cyclohexene is oxidized with oxygen over a proprietary oxide catalyst in presence of steam. Conversion over 90% and selectivity to styrene of 92% could be achieved.

Another approach is the oxidative coupling of toluene to stilbene (one of the two stereoisomers—*cis/trans* of 1,2-diphenylethene) followed by disproportionation to styrene and benzene. High temperatures are needed for this reaction, and the yields are low.

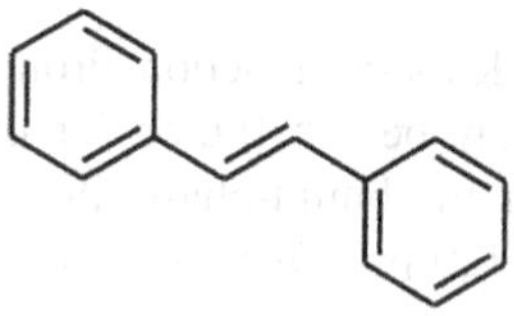

trans-Stilbene

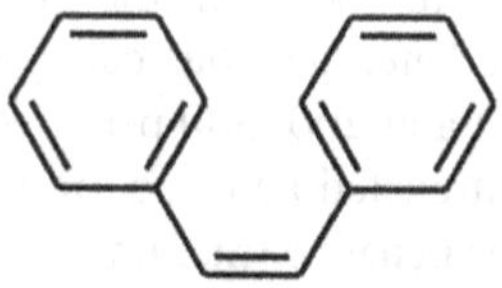

cis-Stilbene

Cumene (isopropyl benzene, b.p. 152.7°C, 306.9°F), a liquid, is soluble in many organic solvents but not in water.

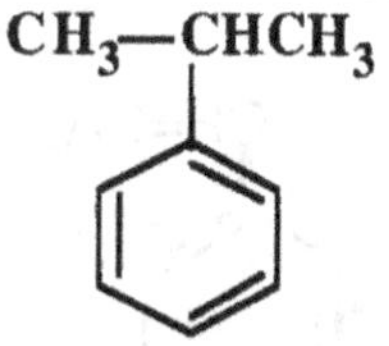

$$CH_3-CHCH_3$$

Cumene

It is present in low concentrations in light refinery streams (such as reformates) and coal liquids. It may be obtained by distilling these fractions.

The main process for producing cumene is a synthetic route where benzene is alkylated with propylene to isopropyl benzene. Either a liquid or a gas-phase process is used for the alkylation reaction. In the liquid-phase process, low temperatures and pressures (approximately 50°C, 122°F, and 75 psi) are used with sulfuric acid as a catalyst. Small amounts of ethylene can be tolerated since ethylene is quite unreactive under these conditions. Butylene derivatives are relatively unimportant because butylbenzene can be removed as bottoms from the cumene column.

In the vapor-phase process, the reaction temperature and pressure are approximately 250°C (480°F) and 600 psi. Phosphoric acid on Kieselguhr is a commonly used catalyst. To limit polyalkylation, a mixture of propene-propane feed is used. Propylene can be as low as 40% of the feed mixture. A high benzene/propylene ratio is also used to decrease polyalkylation.

In the UOP process, fresh propylene feed is combined with fresh and recycled benzene, then passed through heat exchangers and a steam preheater before being charged to the upflow reactor which operates at 200°C–260°C (390°F–500°F) and 375 psi. The solid phosphoric acid catalyst provides an essentially complete conversion of propylene on a one-pass basis. The typical reactor effluent yield contains 94.8% w/w cumene and 3.1% w/w di-iso-propylbenzene. The remaining 2.1% is primarily higher molecular weight aromatic compounds. This high yield of cumene is achieved without transalkylation of di-iso-propylbenzene and is unique to the solid phosphoric acid catalyst process.

The cumene product is 99.9 wt% pure and the high molecular weight aromatic derivatives, which have an octane number of 109, can either be used as high-octane gasoline-blending components or combined with additional benzene and sent to a transalkylation section of the plant where di-iso-propylbenzene is converted to cumene. The overall yields of cumene for this process are typically 97%–98% w/w with transalkylation and 94%–96% w/w without transalkylation.

In the Monsanto–Lummus Crest cumene process, dry benzene, fresh and recycled, and propylene are mixed in the alkylation reaction zone with aluminum chloride ($AlCl_3$) and hydrogen chloride catalyst at a temperature of less than 135°C (275°F). The effluent from the alkylation zone is combined with recycled poly-isopropyl benzene and fed to the transalkylation zone, where the poly-isopropyl benzene derivatives are transalkylated to produce cumene. The strongly acidic catalyst is separated from the organic phase by washing the reactor effluent with water and caustic.

The distillation system is designed to recover a high-purity cumene product. The unconverted benzene and poly-isopropyl benzene are separated and recycled to the reaction system. Propane in the propylene feed is recovered as liquid petroleum gas. The overall yields of cumene for this process can be high as 99% w/w based on benzene and 98% w/w based on propylene. These processes have also been used extensively for the production of ethylbenzene than for the production of cumene.

There are also two processes that use zeolite-based catalyst systems, which were developed in the late 1980s. The goal is to reduce pollution using catalyst system that was developed from the mordenite-zeolite group to replace phosphoric acid or aluminum chloride catalysts. The new catalysts eliminate the disposal of acid wastes and handling of corrosive materials.

In one of these processes, Unocal introduced a fixed bed liquid-phase reactor system based on a Y-type zeolite catalyst. The selectivity to cumene is generally between 70% and 90% w/w. The remaining components are primarily poly-propyl benzene derivatives, which are transalkylated to cumene in a separate reaction zone to give an overall yield to cumene in the order of about 99% w/w. The distillation requirements involve the separation of propane for LPG use, the recycling of excess benzene to poly-propyl benzene for transalkylation to cumene, and the production of purified cumene product.

The second zeolite process is based on the concept of catalytic distillation, which is a combination of catalytic reaction, and distillation in a single column. The basic principle is to use the heat of reaction directly to supply heat for fractionation. This concept has been applied commercially for the production of methyl tert-butyl ether (MTBE) but has not yet been applied commercially to cumene.

Phenol (hydroxybenzene, C_6H_5OH) is produced from cumene by a two-step process. In the first step, cumene is oxidized with air to cumene hydroperoxide. The reaction conditions are approximately 100°C–130°C (212°F–266°F) and 30–45 psi in the presence of a metal salt catalyst.

In the second step, the hydroperoxide is decomposed in the presence of an acid to phenol and acetone. The reaction conditions are a temperature of approximately 80°C (176°F) and slightly below atmospheric pressure.

In this process, cumene is oxidized in the liquid phase. The oxidation product is concentrated to 80% cumene hydroperoxide by vacuum distillation. To avoid decomposition of the hydroperoxide, it is transferred immediately to the cleavage reactor in the presence of a small amount of sulfuric acid. The cleavage product is neutralized with alkali before it is finally purified.

After an initial distillation to split the coproducts, phenol and acetone, each is purified in separate distillation and treating trains. An acetone finishing column distills product acetone from an acetone/water/oil mixture. The oil, which is mostly unreacted cumene, is sent to cumene recovery. Acidic impurities, such as acetic acid and phenol, are neutralized by caustic injection. Previously, phenol was produced from benzene by sulfonation followed by caustic fusion to sodium phenate. Phenol is released from the sodium salt of phenol by the action of carbon dioxide or sulfur dioxide.

Direct hydroxylation of benzene to phenol could be achieved using zeolite catalysts containing rhodium, platinum, palladium, or iridium. The oxidizing agent is nitrous oxide, which is unavoidable by a byproduct from the oxidation of a cyclohexanone-cyclohexanol mixture to adipic acid using nitric acid as the oxidant.

Phenol is also produced from chlorobenzene and from toluene via a benzoic acid intermediate.

Phenol, a white crystalline mass with a distinctive odor, becomes reddish when subjected to light. It is highly soluble in water, and the solution is weakly acidic. Phenol and acetone produce bisphenol A, an important monomer for epoxy resins and polycarbonates. It is produced by condensing acetone and phenol in the presence of hydrogen chloride (HCl), or by using a cation-exchange resin.

Important chemicals derived from phenol are salicylic acid; acetyl salicylic acid (aspirin); 2,4-dichlorophenoxy acetic acid (2,4-0) and 2,4,5-triphenoxy acetic acid (2,4,5-T), which are selective herbicides; and pentachlorophenol, a wood preservative.

Other halophenol derivatives are miticides, bactericides, and leather preservatives.

Phenol can be alkylated to alkylphenols. These compounds are widely used as nonionic surfactants, antioxidants, and monomers in resin polymer applications.

An alkyl phenol

Phenol is also a precursor for aniline. The major process for aniline ($C_6H_5NH_2$) is the hydrogenation of nitrobenzene.

Linear alkylbenzene is an alkylation product of benzene used to produce biodegradable anionic detergents. The alkylating agents are either linear C_{12} to C_{14} mono-olefin derivatives or monochloroalkane derivatives. The linear olefin derivatives (alpha olefin derivatives) are produced by polymerizing ethylene using Ziegler catalysts (Chapter 7) or by dehydrogenating n-paraffins extracted from kerosene. Monochloroalkane derivatives, on the other hand, are manufactured by chlorinating the corresponding n-paraffins. Dehydrogenation of nparaffins to mono-olefin derivatives using a newly developed dehydrogenation catalyst which is highly active and allows a higher per-pass conversion to mono-olefin derivatives. Because the dehydrogenation product contains a higher concentration of olefin derivatives for a given alkylate production rate, the total hydrocarbon feed to the hydrogen fluoride (HF) alkylation unit is substantially reduced.

Alkylation of benzene with linear mono-olefin derivatives is industrially preferred. The Detal process combines the dehydrogenation of n-paraffins and the alkylation of benzene. Mono-olefin derivatives from the dehydrogenation section are introduced to a fixed-bed alkylation reactor over a heterogeneous solid catalyst. Older processes use hydrogen fluoride catalysts in a liquid-phase process at a temperature range of 40°C–70°C (104°F–158°F).

Detergent manufacturers buy linear alkylbenzene, sulfonate it with sulfur trioxide (SO_3), and then neutralize it with sodium hydroxide (NaOH) to produce linear alkylbenzene sulfonate (LABS), the active ingredient in detergents.

8.2.2 Chlorination

Chlorination of benzene is an electrophilic substitution reaction in which Cl^+ serves as the electrophile. The reaction occurs in the presence of a Lewis acid catalyst such as iron (III) chloride (ferric chloride, $FeCl_3$). The products are a mixture of mono- and dichlorobenzene derivatives. The *ortho-* and the para-dichlorobenzene derivatives are more common than meta-dichlorobenzene. The ratio of the monochloro to dichloro products essentially depends on the benzene/chlorine ratio and the residence time. The ratio of ortho-dichlorobenzene plus the ortho- and para-dichlorobenzene to the meta-dichlorobenzene depends mainly on the reaction temperature and residence time.

Typical liquid-phase reaction conditions for the chlorination of benzene using ferric chloride catalyst are 80°C–100°C (176°F–212°F) and atmospheric pressure. When a high benzene/C_{12} ratio is used, the product mixture is approximately 80% monochlorobenzene, 15% p-dichlorobenzene, and 5% o-dichlorobenzene. Continuous chlorination processes permit the removal of monochlorobenzene as it is formed, resulting in lower yields of higher chlorinated benzene.

Monochlorobenzene is also produced in a vapor-phase process at approximately 300°C. The byproduct hydrogen chloride goes into a regenerative oxychlorination reactor. The catalyst is a promoted copper oxide on a silica carrier:

$$4HCl + O_2 \rightarrow 2Cl_2 + 2H_2O$$

Higher conversions have been reported when temperatures of 235°C–315°C (455°F–600°F) and pressures of 40–80 psi are used.

Monochlorobenzene is the starting material for many compounds, including phenol and aniline. Others, such as DDT, chloronitrobenzene derivatives, polychlorobenzene derivatives, and biphenyl derivatives, do not have as high a demand for monochlorobenzene as aniline and phenol.

8.2.3 HYDROGENATION

Benzene and its derivatives convert to cyclohexane by hydrogenation. Cyclohexane is a colorless liquid, insoluble in water but soluble in hydrocarbon solvents, alcohol, and acetone. As a cyclic paraffin, it can be easily dehydrogenated to benzene. The dehydrogenation of cyclohexane and its derivatives (present in naphthas) to aromatic hydrocarbon derivatives is an important reaction in the catalytic reforming process. Essentially, all of the cyclohexane is oxidized either to a cyclohexanone-cyclohexanol mixture used for making caprolactam or to adipic acid. These are monomers for making nylon 6 and nylon 6/6.

The process involves the use of high pressures of hydrogen in the presence of heterogenous catalysts, such as finely divided nickel. Although alkene derivatives can be hydrogenated at temperatures neat to ambient, benzene and related compounds are more reluctant substrates, requiring temperatures >100°C (>212°F). This reaction is practiced on a large scale industrially. In the absence of the catalyst, benzene is impervious to hydrogen. Hydrogenation cannot be stopped to give cyclohexene or cyclohexadiene derivatives, which are valuable petrochemical starting materials.

The hydrogenation of benzene produces cyclohexane. Many catalyst systems, such as Ni/alumina and Ni/Pd, are used for the reaction. General reaction conditions are 160°C (320°F) to 220°C (425°F) and 375–450 psi. Higher temperatures and pressures may also be used with sulfided catalysts.

Older methods use a liquid-phase process. New gas-phase processes operate at higher temperatures with noble metal catalysts. Using high temperatures accelerates the reaction at a faster rate. The hydrogenation of benzene to cyclohexane is characterized by a highly exothermic reaction and a significant decrease in the product volume (from 4 to 1). Equilibrium conditions are, therefore, strongly affected by temperature and pressure. Intermediate products (in which the double bonds have survived) are not produced.

However, the Birch reduction reaction offers access to substituted 1,4-cyclohexadiene derivatives (Birch, 1944, 1945, 1946, 1947a,b; Birch and Smith, 1958). The reaction converts aromatic compounds having a benzenoid ring system into 1,4-cyclohexadiene derivatives in which two hydrogen atoms have been attached on opposite ends of the molecule. The process uses sodium, lithium, or potassium in liquid ammonia and an alcohol; such as ethanol or tert-butyl alcohol.

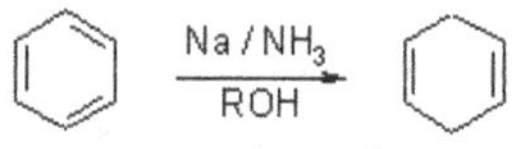

The reactions of benzene derivatives with various substituents leads to the products with the most highly substituted double bonds:

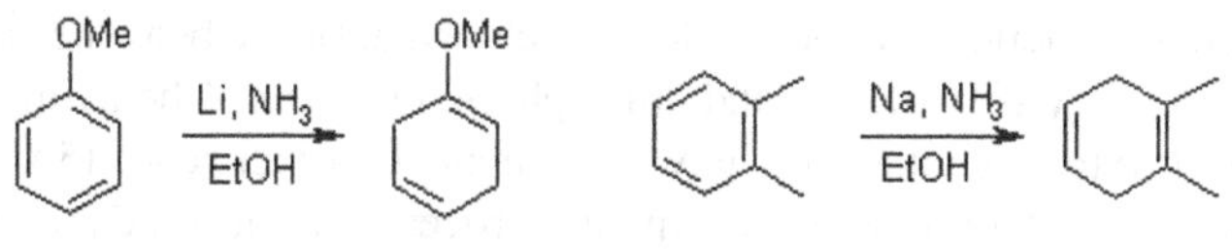

The effect of electron-withdrawing substituents on the Birch Reduction varies. For example, the reaction of benzoic acid leads to 2,5-cyclohexadienecarboxylic acid, which can be rationalized on the basis of the carboxylic acid stabilizing an adjacent anion.

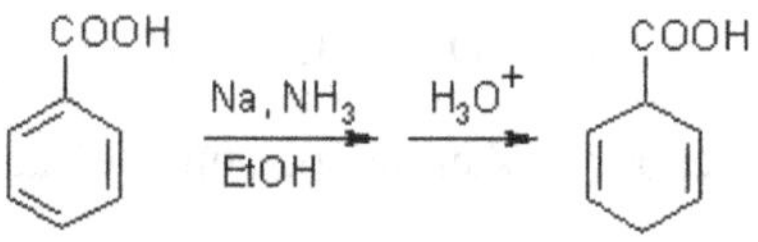

Alkene double bonds are only reduced if they are conjugated with the benzene ring, and occasionally isolated terminal alkenes will be reduced.

Cyclohexane is oxidized in a liquid-phase process to a mixture of cyclohexanone and cyclohexanol. The reaction conditions are 95°C–120°C at approximately 10 atm in the presence of a cobalt acetate and orthoboric acid catalyst system. About 95% yield can be obtained.

The mixture of cyclohexanone and cyclohexanol (sometimes referred to as KA oil) is used to produce caprolactam, the monomer for nylon 6. Caprolactam is also produced from toluene through the intermediate formation of cyclohexane carboxylic acid.

Cyclohexane is also a precursor for adipic acid. Oxidizing cyclohexane in the liquid-phase at lower temperatures and for longer residence times (than for KA oil) with a cobalt acetate catalyst produces adipic acid.

Adipic acid may also be produced from butadiene via a carbonylation route:

$$CH_2=CHCH=CH_2 \rightarrow HOOC(CH_2)_4 COOH + H_2O$$

Adipic acid and its esters are used to make nylon 6/6. It may also be hydrogenated to 1,6-hexanediol, which is further reacted with ammonia to hexamethylenediamine.

$$HOOC(CH_2)_4 COOH + 4H_2 \rightarrow HO(CH_2)_4 OH + 2H_2O$$

$$HO(CH_2)_4 OH + 2NH_3 \rightarrow H_2N(CH_2)_6 NH_2 + 2H_2O$$

Hexamethylenediamine is the second monomer for nylon 6/6.

Briefly, and by way of explanation, nylon 66 (nylon 6-6, nylon 6/6 or nylon 6,6) is a type of polyamide or nylon, of which there are many types. The two most common for textile and plastics industries are nylon 6 and nylon 66. The latter, nylon 66 is made of two monomers and each monomer (hexamethylenediamine and adipic) contains 6 carbon atoms which give nylon 66 its name.

Nylon 6/6

Nylon 66 is synthesized by polycondensation of hexamethylenediamine and adipic acid (Chapter 11). Equivalent amounts of hexamethylenediamine and adipic acid are combined with water in a reactor. This is crystallized to make nylon salt, an ammonium/carboxylate mixture:

$$nHOOC(CH_2)_4 COOH + nH_2N(CH_2)_6 NH_2 \rightarrow \left[-OC(CH_2)_4 CONH(CH_2)_6 NH-\right]_n + (2n-1)H_2O$$

The nylon salt is passed into a reaction vessel where polymerization process takes place either in batches or continuously. Removal of the water drives the reaction toward polymerization through the formation of amide bonds from the acid and amine functions. The molten nylon 66 can either be extruded and granulated at this point or directly spun into fibers by extrusion through a spinneret (a small metal plate with fine holes) and cooling to form filaments.

8.2.4 NITRATION

Similar to the alkylation and the chlorination of benzene, the nitration reaction is an electrophilic substitution of a benzene hydrogen (a proton) with a nitronium (NO_2) moiety to produce nitrobenzene ($C_6H_5NO_2$). The liquid-phase reaction occurs in presence of both concentrated nitric and sulfuric acids at approximately 50°C (122°F). Concentrated sulfuric acid has two functions: it reacts with nitric acid to form the nitronium ion, and it absorbs the water formed during the reaction, which shifts the equilibrium to the formation of nitrobenzene:

$$HNO_3 + 2H_2SO_4 \rightarrow 2HSO_4^- + H_3O^+ + NO_2^+$$

Most of the nitrobenzene produced is used to make aniline. Other uses include synthesis of quinoline, benzidine, and as a solvent for cellulose ethers.

Aniline (aminobenzene, $C_6H_5NH_2$) is an oily liquid that turns brown when exposed to air and light. The compound is an important dye precursor. The main process for producing aniline is the hydrogenation of nitrobenzene. The overall process starts with benzene.

Briefly, benzene is nitrated with a concentrated mixture of nitric acid (HNO_3) and sulfuric acid (H_2SO_4) at 50°C–60°C (122°F–140°F) to yield nitrobenzene. The nitrobenzene is then hydrogenated (typically at 200°C–300°C, 390°F–570°F) in the presence of metal catalysts.

Typically, the hydrogenation reaction occurs at approximately 270°C (520°F) and slightly above atmospheric pressure (approximately 20–25 psi) over a Cu/Silica catalyst. About a 95% yield is obtained. An alternative way to produce aniline is through ammonolysis of either chlorobenzene or phenol. The reaction of chlorobenzene with aqueous ammonia occurs over a copper salt catalyst at approximately 210°C (410°F) and 950 psi. The yield of aniline from this route is also about 96%.

More specifically, in the Dupont/KBR process, benzene is nitrated with mixed acid (nitric and sulfuric) at high efficiency to produce nitrobenzene in a dehydrating nitration system which uses an inert gas to remove the water of nitration from the reaction mixture, thus eliminating the energy-intensive and high-cost sulfuric acid concentration system. As the inert gas passes through the system, it becomes humidified, removing the water of reaction from the reaction mixture. Most of the energy required for the gas humidification comes from the heat of nitration. The wet gas is condensed and the inert gas is recycled to the nitrator. The condensed organic phase is recycled to the nitrator while the aqueous phase is sent to effluent treatment. The reaction mixture is phase separated and the sulfuric acid is returned to the nitrator. The crude nitrobenzene is washed to remove residual acid and the impurities formed during the nitration reaction. The product is then distilled and residual benzene is recovered and recycled. Purified nitrobenzene is fed, together with hydrogen, into a liquid-phase plug-flow hydrogenation reactor. The supported noble metal catalyst has a high selectivity and the nitrobenzene conversion per pass is 100%. The reaction conditions are optimized to achieve essentially quantitative yields and the reactor effluent is free of nitrobenzene. The reactor product is sent to a dehydration column to remove the water of reaction followed by a purification column to produce high-quality aniline product.

Alternatively, aniline can be prepared from ammonia and phenol from the cumene process (cumene-phenol process, Hock process) which is a process for synthesizing phenol and acetone from benzene and propylene. The term process name arise from cumene (isopropyl benzene), the intermediate material during the process.

Cumene

This process converts two relatively cheap starting materials, benzene and propylene, into two more valuable chemicals, phenol (C_6H_5OH) and acetone (CH_3COCH_3). Other reactants required are oxygen from air and small amounts of a radical initiator. Most of the worldwide production of phenol and acetone is now based on this method.

In commerce, three brands of aniline are distinguished: aniline oil for blue, which is pure aniline; aniline oil for red, a mixture of equimolecular quantities of aniline, ortho-toluidine (o-$CH_3C_6H_4NH_2$, 1,2-$CH_3C_6H_4NH_2$), and para-toluidine (p-$CH_3C_6H_4NH_2$, 1,4-$CH_3C_6H_4NH_2$); and aniline oil for safranin, which contains aniline and ortho-toluidine and is obtained from the distillate.

Many analogues of aniline are known where the phenyl group is further substituted. These include chemical such as toluidine derivatives xylidine derivatives, chloroaniline derivatives, amino-aniline derivatives, and aminobenzoic acid derivatives. These chemicals are often are prepared by nitration of the substituted aromatic compounds followed by reduction. For example, this approach is used to convert toluene into toluidine derivatives and chlorobenzene into 4-chloroanilinew. Alternatively, using Buchwald–Hartwig coupling or Ullmann reaction approaches, aryl halides can be aminated with aqueous or gaseous ammonia.

Ammonolysis of phenol occurs in the vapor phase. In the process, a mixed feed of ammonia and phenol is heated and passed over a heterogeneous catalyst in a fixedbed system (Ono and Ishida, 1981). The reactor effluent is cooled, the condensed material distilled, and the unreacted ammonia recycled. The process can be represented simply as:

$$C_6H_5OH + NH_3 \rightarrow C_6H_5NH_2 + H_2O$$

8.2.5 OXIDATION

Benzene oxidation is the oldest method to produce maleic anhydride. The reaction occurs at approximately 380°C (715°F) and atmospheric pressure. A mixture of vanadium pentoxide (V_2O_5) with another metal oxide is the usual catalyst. Benzene conversion reaches 90%, but selectivity to maleic anhydride is only 50%–60%; the other 40%–50% is completely oxidized to carbon dioxide. Currently, the major route to maleic anhydride, especially for the newly erected processes, is the oxidation of butane (Chapter 6). Maleic anhydride also comes from oxidation of n-butenes.

8.3 CHEMICALS FROM TOLUENE

Toluene (also known as toluol, the IUPAC systematic name is methylbenzene), is an aromatic hydrocarbon that is colorless and water-insoluble. It is a monosubstituted benzene derivative, consisting of a methal (CH_3) group attached to the ring. Toluene is predominantly used as an industrial feedstock and a solvent.

$$CH_3$$

Toluene

The methylbenzene derivatives (toluene and the xylene isomers) occur in small quantities in naphtha and higher-boiling fractions of petroleum. Those presently of commercial importance are toluene, o-xylene, p-xylene, and to a much lesser extent m-xylene. The primary sources of toluene and the xylene isomers are reformates from catalytic reforming units, gasoline from catalytic cracking, and pyrolysis naphtha from steam reforming of naphtha and gas oils. As mentioned earlier, solvent extraction is used to separate these aromatic derivatives from the reformate mixture.

Toluene and xylenes have chemical characteristics similar to benzene, but these characteristics are modified by the presence of the methyl substituents. Although such modification activates the ring, toluene and xylenes have less chemicals produced from them than from benzene. Currently, the largest single use of toluene is to convert it to benzene. The para isomer of xylene (p-xylene, $1,4\text{-}CH_3C_6H_4CH_3$) is mainly used to produce terephthalic acid for polyesters whereas the ortho isomer (o-xylene, $1,2\text{-}CH_3C_6H_4CH_3$) is mainly used to produce phthalic anhydride for plasticizers.

Toluene (methylbenzene, $C_6H_5CH_3$) is similar to benzene as a mononuclear aromatic, but it is more active due to presence of the electron-donating methyl group. However, toluene is much less useful than benzene because it produces more polysubstituted products. Most of the toluene extracted for chemical use is converted to benzene via dealkylation or disproportionation. The rest is used to produce a limited number of petrochemicals. The main reactions related to the chemical use of toluene (other than conversion to benzene) are the oxidation of the methyl substituent and the hydrogenation of the phenyl group. Electrophilic substitution is limited to the nitration of toluene for producing the mononitro derivative (mono-nitro toluene (MNT)) and dinitrotoluene derivatives. These compounds are important synthetic intermediates.

Generally, toluene behaves as a typical aromatic hydrocarbon in electrophilic substitution reactions. The methyl group has greater electron-releasing properties than a hydrogen atom in the same ring position and, thus, toluene is more reactive than benzene toward electrophilic reagent. As example, toluene can be sulfonated to yield p-toluene sulfonic acid ($1,4\text{-}CH_3C_6H_4SO_3H$) and is chlorinated by chlorine in the presence of ferric chloride ($FeCl_3$) to yield the ortho and para isomers of chlorotoluene ($1,2\text{-}CH_3C_6H_4Cl$ and $1,4\text{-}CH_3C_6H_4Cl$).

In addition to reactions involving the ring-carbon atoms, the methyl group of toluene is also susceptible to reaction. For example, toluene reacts with potassium permanganate to yield benzoic acid ($C_6H_5CO_2H$) and also with chromyl chloride to yield benzaldehyde (C_6H_5CHO). Other reactions of the methyl group include halogenation such as the reaction with N-bromosuccinimide in the presence of azobisisobutyronitrile (abbreviated AIBN), $[(CH_3)_2C(CN)_2N_2]$ to yield benzyl bromide ($C_6H_5CH_2Cl$). The same conversion can be achieved using with elemental bromine (Br_2) in the presence of ultraviolet (UV) light or even sunlight. Toluene may also be brominated by treating it with hydrogen bromide (HBr) and hydrogen peroxide (H_2O_2) in the presence of light:

$$C_6H_5CH_3 + Br_2 \rightarrow C_6H_5CH_2Br + HBr$$

$$C_6H_5CH_2Br + Br_2 \rightarrow C_6H_5CHBr_2 + HBr$$

Hydrogenation of toluene yields methylcyclohexane in the presence of a catalyst and a high pressure of hydrogen. The various reactions of toluene are presented in the section below.

8.3.1 CARBONYLATION

The carbonylation reaction of toluene with carbon monoxide in the presence of HF/BF$_3$ catalyst produces p-tolualdehyde. A high yield results (96% based on toluene and 98% based on CO).

p-Tolualdehyde (4-Methylbenzaldehyde)

p-Tolualdehyde (CH$_3$C$_6$H$_4$CHO) can be oxidized to terephthalic acid [1,4-C$_6$H$_4$(CO$_2$H)$_2$], an important monomer for polyesters.

Terephthalic acid (Benzene-1,4-dicarboxylic acid)

p-Tolualdehyde is also an intermediate in the synthesis of perfumes, dyes, and pharmaceuticals.

8.3.2 CHLORINATION

The chlorination of toluene by substituting the methyl hydrogens is a free radical reaction. A mixture of three chlorides (benzyl chloride (C$_6$H$_5$CH$_2$Cl), benzal chloride (C$_6$H$_5$CHCl$_2$), and benzotrichloride (C$_6$H$_5$CCl$_3$)) results.

Benzyl chloride

The ratio of the chloride mixture mainly derives from the toluene/chlorine ratio and the contact time. Benzyl chloride is produced by passing dry chlorine into boiling toluene (110°C, 230°F) until reaching a density of 1.283. At this density, the concentration of benzyl chloride reaches the maximum. Light can initiate the reaction. Benzyl chloride can produce benzyl alcohol by hydrolysis, represented simply as:

$$C_6H_5CH_2Cl + H_2O \rightarrow C_6H_5CH_2OH + HCl$$

Benzyl alcohol is a precursor for butyl benzyl phthalate, a vinyl chloride plasticizer. Benzyl chloride is also a precursor for phenylacetic acid via the intermediate benzyl cyanide. Phenylacetic acid is used to make phenobarbital (a sedative) and penicillin G.

Benzal chloride is hydrolyzed to benzaldehyde, and benzotrichloride is hydrolyzed to benzoic acid:

$$C_6H_5CHCl_2 \rightarrow C_6H_5CHO$$

$$C_6H_5CCl_3 \rightarrow C_6H_5CO_2H$$

Chlorinated toluene derivatives are not large-volume chemicals, but they are precursors for many synthetic chemicals and pharmaceuticals. For example, benzyl chloride is the precursor to benzyl esters which are used as plasticizers, flavorings, and, and perfumes. Phenylacetic acid ($C_6H_5CH_2CO_2H$), a precursor to pharmaceuticals, is produced from benzyl cyanide, which is generated by treatment of benzyl chloride with sodium cyanide.

$$C_6H_5CH_2Cl \rightarrow C + NaCl$$

Benzyl chloride will also react with an alcohol to yield the corresponding benzyl ether, carboxylic acid, and benzyl ester. Benzoic acid (C_6H_5COOH) can be prepared by oxidation of benzyl chloride in the presence of alkaline potassium permanganate ($KMnO_4$):

$$C_6H_5CH_2Cl + 2KOH + 2[O] \rightarrow C_6H_5COOK + KCl + H_2O$$

Benzyl chloride also reacts readily with metallic magnesium to produce a Grignard reagent

$$C_6H_5CH_2Cl + Mg \rightarrow C_6H_5CH_2ClMgCl$$

The Grignard reaction is an organometallic chemical reaction in which alkyl, vinyl, or aromatic magnesium halide will add to a carbonyl group ($>C=O$) in an aldehyde or ketone (Figure 8.2). Thus:

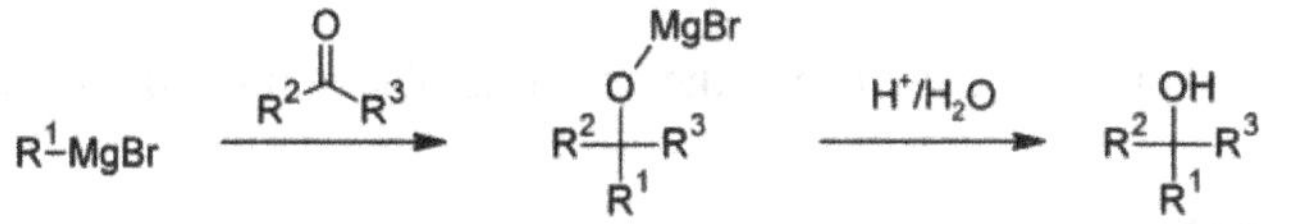

A Grignard reagent is a strong nucleophiles that can form new carbon–carbon bonds. In reactions involving Grignard reagents, it is important to exclude water and air, which rapidly destroy the reagent by protonolysis or oxidation. Since most Grignard reactions are conducted in anhydrous diethyl ether or tetrahydrofuran, side reactions with air are limited by the protective blanket provided by solvent vapors. Although the reagents still need to be dry, ultrasound can allow Grignard reagents to form in wet solvents by activating the magnesium such that it consumes the water.

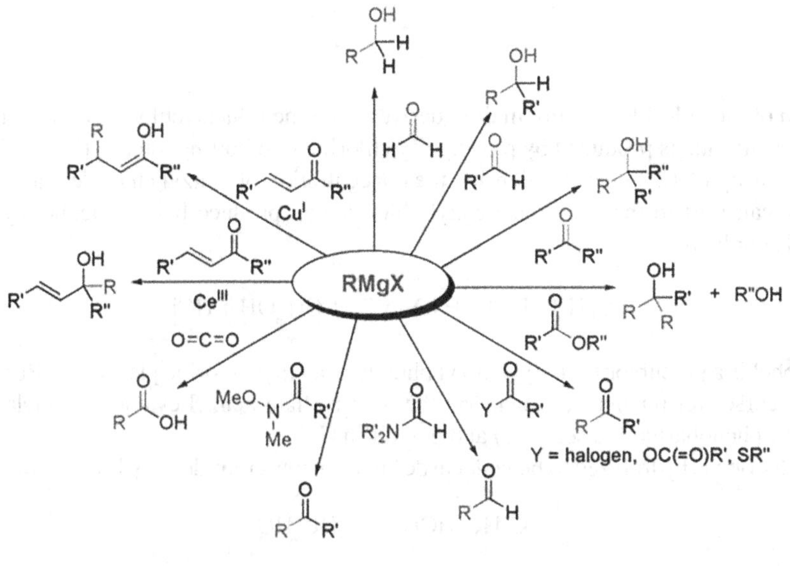

FIGURE 8.2 Common reactions of Grignard reagents.

8.3.3 DEALKYLATION

Dealkylation, in the current context is the removal of an alky group from an aromatic ring (Rabinovich and Maslyanskii, 1973; Noda et al., 2009). In the same context, hydrodealkylation is a chemical reaction that often involves reacting with an aromatic hydrocarbon derivative, such as toluene, in the presence of hydrogen to form a simpler aromatic hydrocarbon devoid of the alkyl group(s). An example is the conversion of 1,2,4-trilethylbenzene [1,2,4-$(CH_3)_3C_6H_3$] to xylene ($CH_3C_6H_4CH_3$). The process requires high temperature and high pressure or the presence of a catalyst containing transition metals, such as chromium and molybdenum.

Toluene is dealkylated to benzene over a hydrogenation-dehydrogenation catalyst such as nickel (Doumani, 1958). The hydrodealkylation is essentially a hydrocracking reaction favored at higher temperatures and pressures. The reaction occurs at approximately 700°C (1,290°F) and 600 psi. A high benzene yield of about 96% or more can be achieved:

$$C_6H_5CH_3 + H_2 \rightarrow C_6H_6 + CH_4$$

Hydrodealkylation of toluene and xylenes with hydrogen is noted in Chapter 3.

Dealkylation also can be effected by steam. The reaction occurs at 600°C–800°C (1,110°F–1,470°F) over Y, La, Ce, Pr, Nd, Sm, or Th compounds; Ni-Cr_2O_3 catalysts; and Ni-AlO_3 catalysts at temperatures between 320°C–630°C (610°F–1,165°F). Yields of about 90% are obtained. This process has the advantage of producing, rather than using, hydrogen.

In the same vein as dealkylation, transalkylation is a chemical reaction involving the transfer of an alkyl group from one organic compound to another. For example, the reaction is used for the transfer of methyl and ethyl groups between benzene rings which is of considerable value to the petrochemical industry for the manufacture of p-xylene and styrene as well as other aromatic compounds. Motivation for using transalkylation reactions is based on a difference in production and demand for benzene, toluene, and the xylene isomers. Transalkylation can convert toluene, which is overproduced, into benzene and xylene, which are underproduced. Zeolite catalysts are often used as transalkylation reactions.

The Tatoray process is used to selectively convert toluene and C_9 aromatic derivative into benzene and xylene isomers. The process consists of a fixed bed reactor and product separation section. The fresh feedstock is combined with hydrogen-rich recycle gas, preheated in a combined feed exchanger and heated in a fired heater. The hot feed vapor goes to the reactor. The reactor effluent is cooled in a combined feed exchanger and sent to a product separator. Hydrogen-rich gas is taken off the top of the separator, mixed with makeup hydrogen gas and recycled back to the reactor. Liquid from the bottom of the separator is sent to a stripper column where the overhead gas is exported to the fuel gas system. The overhead liquid may be sent to a debutanizer column. The products from the bottom of the stripper are recycled back to the benzene–toluene fractionation section of the aromatics complex.

In a modern aromatics complex, this process is integrated between the aromatics extraction and xylene recovery sections of the plant. Extracted toluene is fed to the Tatoray process unit rather than being blended into the gasoline pool or sold for solvent applications. To maximize the production of para-xylene from the complex, the byproduct can also be fed to the Tatoray process unit. This shifts the chemical equilibrium from benzene production to xylene isomers production. In recent years, the demand for para-xylene has outstripped the supply of mixed xylene isomers. The Tatoray process provides an ideal way to produce additional mixed xylenes from toluene and heavy aromatics. Incorporating a Tatoray process unit into an aromatics complex can more than double the yield of p-xylene from a naphtha feedstock.

In another process, the PX-Plus process, toluene is converted to benzene and xylene isomers. The process is para-selective, with the product having a concentration of p-xylene in the xylene fraction in the order 90% v/v, which substantially is higher than the equilibrium value of 25% v/v that is achieved by toluene and C₉ aromatic transalkylation in the Tatoray process. Due to the similarity of operating temperature and pressure to that of many refining and petrochemical reactor systems, existing equipment can often be repurposed for the PX-Plus unit. The PX-Plus process can also be used for large-scale grassroot facilities where sufficient toluene is available and where significant quantities of benzene are desired along with p-xylene.

8.3.4 DISPROPORTIONATION

Transalkylation, as used by the petrochemical industry, is often used to convert toluene into benzene and xylenes. This is achieved through a disproportion reaction of toluene in which one toluene molecule transfers its methyl group to another one.

The catalytic disproportionation of toluene in the presence of hydrogen produces benzene and a xylene mixture. Disproportionation is an equilibrium reaction with a 58% conversion per pass theoretically possible. The reverse reaction is the transalkylation of xylenes with benzene.

Typical conditions for the disproportionation reaction are 450°C–530°C (840°F–985°F) and 300 psi. A mixture of cobalt-molybdenum (CoO-MoO₃) aluminosilicate/alumina catalyst can be used. Conversions of approximately 40% are normally used to avoid more side reactions and faster catalyst deactivation. The equilibrium constants for this reaction are not significantly changed by shifting from liquid to vapor phase or by large temperature changes. Currently, zeolites, especially those of ZSM-5 type, are preferred for their higher activities and selectivity. They are also more stable thermally. Modifying ZSM-5 zeolites with phosphorous, boron, or magnesium compounds produces xylene mixtures rich in the p-isomer (70%–90% w/w).

Diethylbenzene derivatives are produced as side-products of the alkylation of benzene with ethylene. Since there is only a limited market for diethylbenzene, much of it is recycled by transalkylation give ethylbenzene.

8.3.5 NITRATION

Nitration of toluene is the only important reaction that involves the aromatic ring rather than the aliphatic methyl group. The methyl group of toluene makes it around 25 times more reactive than benzene in electrophilic aromatic substitution reactions. Toluene undergoes nitration to give ortho- and para-nitrotoluene ($2,2\text{-}CH_3C_6H_4NO_2$ and $2,4\text{-}CH_3C_6H_4NO_2$) isomers, but if heated it can give dinitrotoluene and ultimately the explosive trinitrotoluene (Humphrey, 1916).

$$\text{Toluene} \rightarrow 1,2\text{-dinitrotoluene} + 1,4\text{-dinitrotoluene} \rightarrow 2,4,6\text{-trinitrotoluene}$$

The nitration reaction occurs with an electrophilic substitution by the nitronium ion. The reaction conditions are milder than those for benzene due to the activation of the ring by the methyl substituent and a mixture of NT derivatives is the result. The two important monosubstituted NT derivatives are o- and p-nitrotoluene derivatives.

o-Nitrotoluene   p-Nitrotoluene

Methyl nitrobenzene (also called mono-nitro toluene) is a group of three organic compounds that are nitro-derivatives of toluene or, alternatively, methyl derivatives of nitrobenzene—the chemical formula is $C_6H_4(CH_3)(NO_2)$. Mono-nitro toluene exists in three isomers and each isomer differs by the relative position of the methyl group and the nitro-group: (i) *ortho*-nitrotoluene, *o*-nitrotoluene, or 2-nitrotoluene is a pale yellow liquid with a subtle, characteristic smell, reminiscent of bitter almonds that is nonhygroscopic—the tendency to absorb moisture from the air—and noncorrosive; (ii) *meta*-nitrotoluene, *m*-nitrotoluene, or 3-nitrotoluene is a yellowish-greenish to yellow liquid with weak fragrance; and (iii) para-nitrotoluene, *p*-nitrotoluene, or 4-nitrotoluene is a pale yellow material that forms rhombic crystals and has a characteristic odor of bitter almonds and is almost insoluble in water. The typical use of NT is in production of pigments, antioxidants, agricultural chemicals, and photographic chemicals. The MNT derivatives are usually reduced to corresponding toluidine derivatives (Table 8.4) which are used in the manufacture of dyes and rubber chemicals:

$$CH_3C_6H_4NO_2 + [H] \rightarrow CH_3C_6H_4NH_2$$

Dinitrotoluene derivatives are produced by nitration of toluene with a mixture of concentrated nitric and sulfuric acid at approximately 80°C. The main products are 2,4-dinitrotoluene [$CH_3C_6H_3$-2,4-$(NO_2)_2$] and 2,6-dinitrotoluene [$CH_3C_6H_3$-2,6-$(NO_2)_2$]. The dinitrotoluene derivatives are important precursors for toluene diisocyanate derivatives, monomers used to produce polyurethanes. The mixture of toluene diisocyanate derivatives is synthesized from dinitrotoluene derivatives by a first-step hydrogenation to the corresponding diamines. The diamines are then treated with phosgene to form the toluene diisocyanate derivatives in an approximate 85% w/w yield based on toluene. An alternative route for the production of toluene diisocyanate derivatives is through a liquid-phase carbonylation of dinitrotoluene in presence of $PdCl_2$ catalyst at approximately 250°C (480°F) and 3,000 psi.

In mixed acid nitration plants for the production of dinitrotoluene, the spent acid from the MNT stage is purified, reconcentrated, and recycled back into the nitration process. Thus, the consumption of sulfuric acid is considerably reduced. In addition, also the sulfuric-, nitric-, nitrous acid, and MNT and dinitrotoluene plants from the washing of the crude nitro-products and from the purification and reconcentration of the spent acid from the MNT plants are recovered and recycled back into nitration. In this manner, not only the nitrate load of the waste water from a dinitrotoluene

TABLE 8.4
Isomers of Toluidine

Common name	*o*-toluidine	*m*-toluidine	*p*-toluidine
Other names	*o*-methylaniline	*m*-methylaniline	*p*-methylaniline
Chemical name	2-methylaniline	3-methylaniline	4-methylaniline
Chemical formula		C_7H_9N	
Structural formula			

Molecular mass	107.17 g/mol		
Melting point	−23°C (−9°F)	−30°C (−22°F)	43°C (109°F)
Boiling point	199°C–200°C	203°C–204°C	200°C
Density	1.00 g/cm³	0.98 g/cm³	1.05 g/cm³

nitration plant is reduced by 95%, but also the consumption figures for nitric acid are considerably improved. More than 98% of the nitric acid needed for nitration can thus be converted to dinitrotoluene (Hermann et al., 1996).

Finally, trinitrotoluene is a well-known explosive (Brown, 1998) obtained by further nitration of the dinitrotoluene derivatives.

Trinitrotoluene

In the process for the production of trinitrotoluene, the trinitro-compound in is produced in a three-step process. In the first step, toluene is nitrated using a mixture of sulfuric and nitric acids to produce MNT which is then separated and, in the second step, nitrated to produce dinitrotoluene. In the third and final step, the dinitrotoluene is nitrated to trinitrotoluene using an anhydrous mixture of nitric acid and oleum (fuming sulfuric acid, usually represented as $H_2SO_4.SO_3$). The nitric acid is consumed by the manufacturing process but the diluted sulfuric acid can be reconcentrated and reused. After nitration, the trinitrotoluene is stabilized by a process (sometime referred to as sulfitation) in which the crude trinitrotoluene is treated with aqueous sodium sulfite (Na_2SO_3) solution to remove less stable isomers of trinitrotoluene and other undesired reaction products. The rinse water from sulfitation (red water) is a significant pollutant and waste product from the manufacture of trinitrotoluene.

Control of the nitrogen oxide derivative in feed nitric acid is very important because the presence of free nitrogen dioxide (NO_2) can result in the oxidation of the methyl group of toluene. This reaction is highly exothermic and there is the risk of a runaway reaction leading to an explosion. Thus, when detonated, trinitrotoluene decomposes to gases and carbon:

$$2C_7H_5N_3O_6 \rightarrow 3N_2 + 5H_2O + 7CO + 7C$$

$$2C_7H_5N_3O_6 \rightarrow 3N_2 + 5H_2 + 12CO + 2C$$

Amatol is a highly explosive material that is a mixture of trinitrotoluene and ammonium nitrate (NH_4NO_3). Amatol was used extensively during World War I and World War II, typically as an explosive in military weapons such as aircraft bombs, canon shells, depth charges, and naval mines.

8.3.6 OXIDATION

Oxidizing toluene in the liquid phase over a cobalt acetate catalyst produces benzoic acid (C_6H_5COOH) (Kaeding et al., 1965). The reaction occurs at temperatures in the order of 165°C (330°F) 150 psi. The yield is in excess of 90% w/w based on the toluene derivative.

Benzoic acid (benzene carboxylic acid) is a white crystalline solid with a characteristic odor. It is slightly soluble in water and soluble in most common organic solvents. Though much benzoic acid gets used as a mordant in calico printing, it also serves to season tobacco, preserve food, make dentifrices, and kill fungus. Furthermore, it is a precursor for caprolactam, phenol, and terephthalic acid.

Caprolactam, a white solid that melts at 69°C (156°F), can be obtained either in a fused or flaked form. It is soluble in water, ligroin, and chlorinated hydrocarbon derivatives. The predominant use of caprolactam is to produce nylon 6. Other minor uses are as a cross-linking agent for polyurethanes, in the plasticizer industry, and in the synthesis of lysine.

The first step in producing caprolactam from benzoic acid is the hydrogenation of benzoic acid to cyclohexane carboxylic acid at approximately 170°C (340°F) and 240 psi over a palladium catalyst.

The resulting acid is then converted to caprolactam through a reaction with nitrosyl-sulfuric acid.

In the process, toluene is first oxidized to benzoic acid. Benzoic acid is then hydrogenated to cyclohexane carboxylic acid, which reacts with nitrosylsulfuric acid yielding caprolactam. Nitrosyl sulfuric acid comes from reacting nitrogen oxides with oleum. Caprolactam comes as an acidic solution that is neutralized with ammonia and gives ammonium sulfate as a byproduct of commercial value. Recovered caprolactam is purified through solvent extraction and fractionation.

The action of a copper salt converts benzoic acid to phenol. The copper, reoxidized by air, functions as a real catalyst. The Lummus process operates in the vapor phase at approximately 250°C (480°F) and the yield of phenol is in the order of 90%.

In the Lummus process, the reaction occurs in the liquid phase at approximately 220°C–240°C (430°F–465°F) over $Mg^{2+}+Cu^{2+}$ benzoate. Magnesium benzoate is an initiator, with the Cu_2 copper (I) ions are reoxidized to copper (II) ions.

Phenol can also be produced from chlorobenzene and from cumene, the major route for this commodity.

Terephthalic acid is an important monomer for producing polyesters. The main route for obtaining the acid is the catalyzed oxidation of p-xylene.

$1,4\text{-}C_6H_4(CH_3)_2$ $\qquad$ $1,4\text{-}C_6H_4(CO_2H)$

The reaction occurs in a liquid-phase process at approximately 400°C (750°F) using ZnO or CdO catalysts. Terephthalic acid is obtained from an acid treatment; the potassium salt is recycled. Terephthalic acid can also be produced from benzoic acid by a disproportionation reaction of potassium benzoate in the presence of carbon dioxide. However, in the process, a high temperature diminishes oxygen solubility in an already oxygen-starved system. The oxidation is conducted using acetic acid as solvent and a catalyst composed of cobalt and manganese salts, using a bromide promoter. The yield is nearly quantitative. The most problematic impurity is 4-formylbenzoic acid, which is removed by hydrogenation of a hot aqueous solution after which the solution is cooled in a stepwise manner to crystallize highly pure terephthalic acid. Additionally, the corrosive nature of any products at high temperatures requires the reaction be run in expensive titanium reactors.

Alternatively, but not commercially significant, is the so-called Henkel–Raecke process (named after the company and patent holder, respectively) process involves the rearrangement of phthalic acid to terephthalic acid via the corresponding potassium salts. Terephthalic acid can be prepared in the laboratory by oxidizing various para-disubstituted derivatives of benzene including Caraway Oil or a mixture of cymene and cuminaldehyde (a liquid, $C_3H_7C_6H_4CHO$, obtained from oil of caraway—also called cuminic aldehyde) with chromic acid.

Cuminaldehde

Oxidizing toluene to benzaldehyde (C_6H_5CHO) is a catalyzed reaction in which a selective catalyst limits further oxidation to benzoic acid. In the first step, benzyl alcohol is formed and then oxidized to benzaldehyde. Further oxidation produces benzoic acid:

$$C_6H_5CH_3 + [O] \rightarrow C_6H_5CH_2OH$$

$$C_6H_5CH_2OH[O] \rightarrow C_6H_5CHO$$

Depending upon the reaction conditions and the nature of the oxidant (Borgaonkar et al., 1984), the benzyl alcohol may or may not be isolated.

However, in this reaction, each successive oxidation occurs more readily than the preceding one (more acidic hydrogens after introducing the oxygen heteroatom, which facilitates the oxidation reaction to occur). In addition to using a selective catalyst, the reaction can be limited to the production of the aldehyde by employing short residence times and a high toluene-to-oxygen ratio. In one process, a mixture of UO_2 (93%) and MnO_2 (7%) is the catalyst. A yield of 30%–50% could be obtained at low conversions of 10%–20%. The reaction temperature is approximately 500°C (930°F). In another process, the reaction goes forward in the presence of methanol over a $FeBr_2$-$CoBr_2$ catalyst mixture at approximately 100°C–140°C (212°F–285°F).

Benzaldehyde has limited uses as a chemical intermediate. It is used as a solvent for oils, resins, cellulose esters, and ethers. It is also used in flavoring compounds and in synthetic perfumes.

8.4 CHEMICALS FROM XYLENE ISOMERS

Xylenes (dimethylbenzene derivatives) are an aromatic mixture composed of three isomers (o-, m-, and p-xylene). They are normally obtained from catalytic reforming and cracking units with other C_6, C_7, and C_8 aromatic derivatives. Separating the aromatic mixture from the reformate is done by extraction-distillation and isomerization processes (Chapter 2).

p-Xylene is the most important of the three isomers for producing terephthalic acid to manufacture polyesters. m-Xylene is the least used of the three isomers, but the equilibrium mixture obtained from catalytic reformers has a higher ratio of the meta isomer. m-Xylene is usually isomerized to the more valuable p-xylene.

As mentioned earlier, xylene chemistry is primarily related to the methyl substituents, which are amenable to oxidation. Approximately 65% of the isolated xylenes are used to make chemicals. The rest are either used as solvents or blended with naphtha for gasoline manufacture.

The catalyzed oxidation of p-xylene produces terephthalic acid (p-$HOOCC_6H_4COOH$). Cobalt acetate promoted with either sodium bromide (NaBr) or hydrogen bromide (HBr) is used as a catalyst in an acetic acid medium. Reaction conditions are approximately 200°C (390°F) and 220 psi. The yield is about 95%.

Special precautions must be taken so that the reaction does not stop at the p-toluic acid (–$H_3CC_6H_4COOH$) stage. One approach is to esterify toluic acid as it is formed with methanol which facilitates the oxidation of the second methyl group. The resulting dimethyl terephthalate (DMT) may be hydrolyzed to terephthalic acid.

Another approach is to use an easily oxidized substance such as acetaldehyde or methyl ethyl ketone, which, under the reaction conditions, forms a hydroperoxide. These will accelerate the oxidation of the second methyl group. The DMT process encompasses four major processing steps: oxidation, esterification, distillation, and crystallization. The main use of terephthalic acid and DMT is to produce polyesters for synthetic fiber and film.

Currently, phthalic anhydride is mainly produced through catalyzed oxidation of o-xylene. A variety of metal oxides are used as catalysts. A typical one is $V_2O_5 + TiO_2/Sb_2O_3$. Approximate conditions for the vapor-phase oxidation are 375°C–435°C (705°F–815°F) and 10 psi. The yield of phthalic anhydride is approximately 85%. Liquid-phase oxidation of a-xylene also works at approximately 150°C (300°F). Cobalt or manganese acetate in acetic acid medium serves as a catalyst. The major byproducts of this process are maleic anhydride, benzoic acid, and citraconic anhydride (methyl maleic anhydride). Maleic anhydride could be recovered economically.

The main use for phthalic anhydride is for producing plasticizers by reactions with C_4 to C_{10} alcohols. The most important polyvinyl chloride plasticizer is formed by the reaction of 2-ethylhexanol (produced via butyraldehyde, Chapter 8) and phthalic anhydride:

In the above equation, $R = CH_3(CH_2)_3$

Phthalic anhydride is also used to make polyester and alkyd resins. It is a precursor for phthalonitrile by an ammoxidation route used to produce [phthalimide and phthalimide?].

The oxidation of m-xylene produces isophthalic acid. The reaction occurs in the liquid phase in presence of a catalyst such as a cobalt-manganese catalyst.

Isophthalic acid

The main use of isophthalic acid is in the production of polyesters that are characterized by a higher abrasion resistance than those using other phthalic acids. Polyesters from isophthalic acid are used for pressure molding applications. Ammoxidation of isophthalic acid produces isophthalonitrile, which serves as a precursor for agricultural chemicals. It is readily hydrogenated to the corresponding diamine, which can form polyamides or be converted to isocyanates for polyurethane manufacture.

Similarly, phthalonitrile—an organic compound with the formula $C_6H_4(CN)_2$, which is an off-white crystal solid at room temperature—is a derivative of benzene that contains two adjacent nitrile groups. The compound has low solubility in water but is soluble in common organic solvents. The compound is used as a precursor to phthalocyanine and other pigments, fluorescent brighteners, and photographic sensitizers.

Phthalonitrile is produced in a single-stage continuous process, by the ammonoxidation of o-xylene at 480°C (895°F) in the presence of a vanadium oxide-antimony oxide (V_2O_4-Sb_2O_4) catalyst.

Phthalonitrile is the precursor to phthalocyanine pigments which are produced by the reaction of phthalonitrile with various metal precursors. The reaction is carried out in a solvent at around 180°C (355°F).

Ammonolysis of phthalonitrile yields diiminoisoindole which reacts (by condensation) with active methylene compounds to give pigment yellow 185 and pigment 139.

Diiminoisoindole

The molecule can exist in different tautomers resulting in different crystalline solids. By way of definition, a tautomer is each of two or more isomers of a compound that exist together in equilibrium, and are readily interchanged by migration of an atom or group within the molecule. Thus, tautomers are constitutional isomers of organic compounds that readily interconvert and the reaction commonly results in the relocation of a proton. Tautomerism is, for example, relevant to the behavior of amino acids and nucleic acids, two of the fundamental building blocks of life.

The TAC9 process is used to selectively convert C_9 to C_{10} aromatics into mixed xylene isomers. The process consists of a fixed bed reactor and product separation section. The feed is combined with hydrogen-rich recycle gas, preheated in a combined feed exchanger and heated in a fired heater and the heated feedstock is sent to the reactor. The reactor effluent is cooled in a combined feed exchanger and sent to a product separator. Hydrogen-rich gas is taken off the top of the separator, mixed with makeup hydrogen gas, and recycled back to the reactor. Liquid from the bottom of the separator is sent to a stripper column. The stripper overhead gas is exported to the fuel gas system. The overhead liquid may be sent to a debutanizer column or a stabilizer. The stabilized product is sent to the product fractionation section of the aromatics complex.

In a modern aromatics complex, the transalkylation technologies, such as the Tatoray and TAC9 processes, are integrated between the aromatics extraction or fractionation and the xylene recovery sections of the plant. Fractionated high-boiling aromatic derivatives can be fed to the TAC9 unit rather than being blended into the gasoline pool or sold for solvent applications. Incorporating transalkylation technology into an aromatics complex for the processing of toluene and C_9 to C_{10} aromatics can more than double the yield of p-xylene from a given naphtha feedstock. The process provides an efficient means of obtaining additional mixed xylenes from the highest-boiling portion of an aromatics fraction thereby producing higher-value products by upgrading byproduct streams.

8.5   CHEMICALS FROM ETHYLBENZENE

Ethylbenzene ($C_6H_5CH_2CH_3$) is a highly flammable, colorless liquid which is an important chemical as an intermediate in the production of styrene, the precursor to polystyrene. It is one of the C_8 aromatic constituents of the products (reformates) of reforming processes. Ethylbenzene can be obtained by intensive fractionation of the aromatic extract but most of the ethylbenzene is obtained by the alkylation of benzene with ethylene.

$$C_6H_6 + CH_2{=}CH_2 \rightarrow C_6H_5CH_2CH_3$$

For example, a zeolite-based process using vapor-phase alkylation, offered a higher purity and yield of ethylbenzene after which a liquid-phase process was introduced using zeolite catalysts. This liquid-phase process offers low benzene-to-ethylene ratios thereby leading to a reduction in the size of the required equipment and lowering byproduct production.

Direct dehydrogenation of ethylbenzene to styrene accounts for the majority (approximately 85%) of the commercial production and the reaction is carried out in the vapor phase with steam over a catalyst consisting primarily of iron oxide. The reaction is endothermic, and can be accomplished either adiabatically or isothermally. Both methods are used in practice. The major reaction is the reversible, endothermic conversion of ethylbenzene to styrene and hydrogen:

$$C_6H_5CH_2CH_3 \rightarrow C_6H_5CH{=}CH_2 + H_2$$

This reaction proceeds thermally with low yield and catalytically with high yield. As it is a reversible gas-phase reaction producing 2 mol of product from 1 mol of starting material, low pressure favors the forward reaction.

In the process, ethylbenzene is mixed in the gas phase with 10–15 times of its volume of high-temperature steam and passed over a solid catalyst bed. Most ethylbenzene dehydrogenation catalysts are based on iron oxide (Fe_2O_3) promoted by several percent potassium oxide (K_2O) or potassium carbonate (K_2CO_3).

In this reaction, steam (i) is the source of heat for powering the endothermic reaction and (ii) removes coke that tends to form on the iron oxide catalyst through the water gas-shift reaction. The potassium promoter enhances this decoking reaction. The steam also dilutes the reactant and products, shifting the position of chemical equilibrium toward products. A typical styrene plant consists of two or three reactors in series, which operate under vacuum to enhance the conversion and selectivity.

The development of commercial processes for the manufacture of styrene based on the dehydrogenation of ethylbenzene was achieved in the 1930s. The need for synthetic styrene-butadiene rubber (E-SBR) during World War II provided the impetus for large-scale production. After 1946, this capacity became available for the manufacture of a high-purity monomer that could be polymerized to a stable, clear, colorless, and cheap plastic (polystyrene and styrene copolymers). Peacetime uses of styrene-based plastics expanded rapidly, and polystyrene is now one of the least expensive thermoplastics on a cost-per-volume basis.

However, there are competing thermal reactions degrade ethylbenzene to benzene, and also to carbon and, in addition, styrene also reacts catalytically with hydrogen to produce toluene and methane:

$$C_6H_5CH_2CH_3 \rightarrow C_6H_6 + CH_2{=}CH_2$$
$$C_6H_5CH_2CH_3 \rightarrow 8C + 5H_2$$
$$C_6H_5CH{=}CH_2 + H_2 \rightarrow C_6H_5CH_3 + CH_4$$

The problem with carbon production is that carbon is a catalyst poison. When potassium is incorporated into the iron oxide catalyst, the catalyst becomes self-cleaning (through enhancement of the reaction of carbon with steam to give carbon dioxide, which is removed in the reactor vent gas). Thus:

$$C + 2H_2O \rightarrow CO_2 + 2H_2$$

The typical operating conditions in commercial reactors are in the order of 620°–640°C (1,150°F–640°F) and at low pressure. Improving conversion and so reducing the amount of ethylbenzene that must be separated is the chief impetus for researching alternative routes to styrene.

Styrene is also coproduced commercially in the propylene oxide-styrene monomer (POSM) process. In this process, ethylbenzene is treated with oxygen to form the ethylbenzene hydroperoxide. This hydroperoxide is then used to oxidize propylene to propylene oxide. The resulting 1-phenylethanol is dehydrated to give styrene.

Styrene is a colorless liquid with a distinctive, sweetish odor. Some physical properties of styrene are summarized on the right. Vapor pressure is a key property in the design of styrene distillation equipment. Styrene is miscible with most organic solvents in any ratio. It is a good solvent for synthetic rubber, polystyrene, and other non-cross-linked high polymers. Styrene and water are sparingly soluble in each other.

The majority of all operating styrene plants carry out the dehydrogenation reaction adiabatically (a process condition in which heat does not enter or leave the system concerned) in multiple reactors or reactor beds operated in series. The necessary heat of reaction is applied at the inlet to each stage, either by injection of superheated steam or by indirect heat transfer. Fresh ethylbenzene feed is mixed with recycled ethylbenzene and vaporized. Dilution steam must be added to prevent the ethylbenzene from forming coke. This stream is further heated by heat exchange, superheated steam is added to bring the system up to reaction temperature, and the stream is passed through catalyst in the first reactor. The adiabatic reaction drops the temperature, so the outlet stream is reheated prior to passage through the second reactor. Conversion of ethylbenzene can vary with the system, but is often about 35% in the first reactor and 65% overall. The reactors are run at the lowest pressure that is safe and practicable. Some units operate under vacuum, while others operate at a low positive pressure.

The steam-ethylbenzene ratio fed to the reactors is chosen to give optimum yield with minimum utility cost. The reactor effluent is fed through an efficient heat recovery system to minimize energy consumption, condensed, and separated into vent gas, a crude styrene hydrocarbon stream, and a steam condensate stream. The crude styrene is sent to a distillation system where the steam condensate is steam-stripped, treated, and reused. The vent gas, mainly hydrogen and carbon dioxide, is treated to recover aromatics, after which it can be used as a fuel or a feed stream for chemical hydrogen.

Isothermal dehydrogenation was pioneered by BASF and has been used for many years using a reactor that is constructed like a shell-and-tube heat exchanger. Ethylbenzene and steam flow through the tubes, which are packed with catalyst where the heat of reaction is supplied by hot flue gas on the shell side of the reactor-exchanger. The steam-feedstock mass ratio can be lowered to approximately 1:1, and steam temperatures are lower than in the adiabatic process. A disadvantage is the practical size limitation on a reactor-exchanger, which restricts the size of a single train.

A typical crude styrene mixture from the dehydrogenation process consists of: (i) benzene (boiling point: 80°C, 176°F); (ii) toluene (boiling point: 110°C, 230°F), ethylbenzene (boiling point: 136°C, 277°F), styrene (boiling point: 145°C, 293°F). The separation of these components is reasonably straightforward, but residence time at elevated temperature needs to be minimized to reduce styrene polymerization. At least three steps are involved (i) benzene and toluene are removed and either sent to a toluene dehydrogenation plant or further separated into benzene for recycling and toluene for sale; (ii) ethylbenzene is then separated and recycled to the reactors; and (iii) styrene is distilled away from the tars and polymers under vacuum to keep the temperature as low as possible.

Ethylbenzene and styrene, having similar boiling points, require 70–100 trays for their separation depending on the desired ethylbenzene content of the finished styrene. If bubble-cap trays are used, as in old plants, a large pressure drop over the trays means that two columns in series are necessary to keep reboiler temperatures low. Most of the modern plants use packing in place of trays which permits this separation to be achieved in one column. This results in less pressure drop, giving a lower bottom temperature, shorter residence time, and hence less polymer. Sulzer has done pioneering work in the field of packings for distillation.

A polymerization inhibitor (distillation inhibitor) is needed throughout the distillation train. Today, usually aromatic compounds with amino, nitro, or hydroxy groups are used (such as phenylenediamine derivatives, dinitrophenol derivatives, and dinitrocresol derivatives). The distillation inhibitor tends to be colored and is thus unacceptable in the final product and the finished monomer is usually inhibited instead with a chemical such as tert-butylcatechol during storage and transportation.

Styrene can be produced from toluene and methanol, which are cheaper raw materials than those in the conventional process. Another route to styrene involves the reaction of benzene and ethane. Ethane, along with ethylbenzene is fed to a dehydrogenation reactor with a catalyst capable of simultaneously producing styrene and ethylene. The dehydrogenation effluent is cooled and separated and the ethylene stream is recycled to the alkylation unit.

REFERENCES

Alotaibi, A., Hodgkiss, S., Kozhevnikova, E.F., and Kozhevnikov, I.V. 2017. Selective alkylation of benzene by propane over bifunctional Pd-acid catalysts. *Catalysts*, 7: 303–312.

Birch, A.J. 1944. Reduction by dissolving metals. Part I. *Journal of the Chemical Society*, 430.

Birch, A.J. 1945. Reduction by dissolving metals. Part II. *Journal of the Chemical Society*, 809.

Birch, A.J. 1946. Reduction by dissolving metals. Part III. *Journal of the Chemical Society*, 593.

Birch, A.J. 1947a. Reduction by dissolving metals. Part IV. *Journal of the Chemical Society*, 102.

Birch, A.J. 1947b. Reduction by dissolving metals. Part V. *Journal of the Chemical Society*, 1642.

Birch, A.J., and Smith, H. 1958. Reduction by metal-amine solutions: Applications in synthesis and determination of structure. *Quarterly Reviews of the Chemical Society*, 12(1): 17.

Borgaonkar, H.V., Raverkar, S.R., and Chandalia, S.B. 1984. Liquid phase oxidation of toluene to benzaldehyde by air. *Industrial & Engineering Chemistry Product Research Development*, 23(3): 455–458.

Brown, G.I. 1998. *The Big Bang: A History of Explosives*. Sutton Publishing. Stroud.

Doumani, T. 1958. Dealkylation of organic compounds. Benzene from toluene. *Industrial & Engineering Chemistry*, 50(11): 1677–1680.

Gary, J.G., Handwerk, G.E., and Kaiser, M.J. 2007. *Petroleum Refining: Technology and Economics*. 5th Edition. CRC Press, Taylor & Francis Group, Boca Raton, FL.

Gosling, C.D., Wilcher, F.P., Sullivan, L., and Mountiford, R.A. 1999. Process LPG to BTX products. *Hydrocarbon Processing*, 69.

Hermann, H., Gebauer, J., and Konieczny. 1996. Chapter 21: Requirements of a modern facility for the production of dinitrotoluene. In: ACS Symposium Series, Vol. 623, pp. 234–249.

Hsu, C.S., and Robinson, P.R. (Editors). 2017. *Handbook of Petroleum Technology*. Springer International Publishing AG, Cham.

Huang, X., Sun, X., Zhu, S., and Liu, Z. 2007. Benzene alkylation with propane over metal modified HZSM-5. *Reaction Kinetics and Catalysis Letters*, 91: 385–390.

Humphrey, I.W. 1916. The nitration of toluene to trinitrotoluene. *Industrial & Engineering Chemistry*, 8(11): 998–999.

Kaeding, W.W., Lindblom, R.O., Temple, R.G., and Mahon, H.I. 1965. Oxidation of toluene and other alkylated aromatic hydrocarbons to benzoic acids and phenols. *Industrial & Engineering Chemistry Process Design and Development*, 4(1): 97–101.

Kato, S., Nakagawa, K., Ikenaga, N., and Suzuki, T. 2001. Alkylation of benzene with ethane over platinum-loaded zeolite catalyst. *Catalysis Letters*, 73: 175–180.

March, J. 1985. *Advanced Organic Chemistry: Reactions, Mechanisms, and Structure*. 3rd Edition. John Wiley & Sons Inc., Hoboken, NJ.

Noda, J., Volkamer, R., and Molina, M.J. 2009. Dealkylation of alkylbenzenes: A significant pathway in the toluene, o-, m-, p-xylene + OH reaction. *Journal of Physical Chemistry*, 113(35): 9658–9666.

Ono, Y., and Ishida, H. 1981. Amination of phenols with ammonia over palladium supported on alumina. *Journal of Catalysis*, 72(1): 121–128.

Parkash, S. 2003. *Refining Processes Handbook*. Gulf Professional Publishing, Elsevier, Amsterdam.

Rabinovich, G.L., and Maslyanskii, G.N. 1973. A new process for toluene dealkylation. *Chemistry and Technology of Fuels and Oils*, 9: 85–87.

Speight, J.G. 2014. *The Chemistry and Technology of Petroleum*. 4th Edition. CRC-Taylor and Francis Group, Boca Raton, FL.

Speight, J.G. 2017. *Handbook of Petroleum Refining*. CRC Press, Taylor & Francis Group, Boca Raton, FL.

Weissermel, K., and Arpe, H.-J. 2003. *Industrial Organic Chemistry*. Wiley-VCH, Weinheim.

9 Chemicals from Non-hydrocarbons

9.1 INTRODUCTION

By way of recall, a *petrochemical* is any chemical (as distinct from the bulk product which are used for fuels and other petroleum products) manufactured from petroleum (and natural gas) and used for a variety of commercial purposes. Petroleum and natural gas are made up of hydrocarbon constituents, which are comprised of one or more carbon atoms, to which hydrogen atoms are attached. Currently, through a variety of intermediates petroleum and natural gas are the main sources of the raw materials because they are the least expensive, most readily available, and can be processed most easily into the primary petrochemicals. An aromatic petrochemical is also an organic chemical compound but one that contains, or is derived from, the basic benzene ring system. However, the definition of petrochemicals has been broadened and includes (with justification) not only the whole range of aliphatic, aromatic, and naphthenic organic chemicals but also nonorganic chemicals such as carbon black, sulfur, ammonia, nitric acid, hydrazine, and synthesis gas.

Primary petrochemicals include olefins (ethylene, propylene, and butadiene) aromatics (benzene, toluene, and the isomers of xylene) and methanol. Thus, petrochemical feedstocks can be classified into three general groups: olefins, aromatics, and methanol; a fourth group includes inorganic compounds and synthesis gas (mixtures of carbon monoxide and hydrogen). In many instances, a specific chemical included among the petrochemicals may also be obtained from other sources, such as coal, coke, or vegetable products. For example, materials such as benzene and naphthalene can be made from either petroleum or coal, while ethyl alcohol may be of petrochemical or vegetable origin.

From natural gas, crude oils, and other fossil materials such as coal, few intermediates are produced that are not hydrocarbon compounds (Parkash, 2003; Gary et al., 2007; Speight, 2014a; Hsu and Robinson, 2017; Speight, 2017). The important intermediates discussed here are hydrogen, sulfur, carbon black, and synthesis gas. Synthesis gas consists of a non-hydrocarbon mixture (carbon monoxide (CO) and hydrogen (H_2)) that is obtained from any one of several carbonaceous sources (Chadeesingh, 2011; Speight, 2014b). It is included in this chapter as a point of acknowledgment, but the reaction if synthesis gas are covered in more detail in a later chapter.

As stated above, some of the chemicals and compounds produced in a refinery are destined for further processing and as raw material feedstocks for the fast-growing petrochemical industry. Such nonfuel uses of crude oil products are sometimes referred to as its non-energy uses. Petroleum products and natural gas provide three of the basic starting points for this industry: methane from natural gas and naphtha and refinery gases.

Petrochemical intermediates are generally produced by chemical conversion of primary petrochemicals to form more complicated derivative products. Petrochemical derivative products can be made in a variety of ways: directly from primary petrochemicals; through intermediate products which still contain only carbon and hydrogen; and, through intermediates which incorporate chlorine, nitrogen, or oxygen in the finished derivative. In some cases, they are finished products; in others, more steps are needed to arrive at the desired composition.

Although the focus of this text has been the production of organic petrochemical derivatives, mention needs to be made of the inorganic petrochemical products. Thus, an inorganic petrochemical is one that does not contain carbon atoms; typical examples are sulfur (S), ammonium sulfate

[(NH$_4$)$_2$SO$_4$], ammonium nitrate (NH$_4$NO$_3$), and nitric acid (HNO$_3$) (Goldstein, 1949; Hahn, 1970; Lowenheim and Moran, 1975; Chemier, 1992; Wittcoff and Reuben, 1996; Speight, 2002; Farhat Ali et al., 2005; Speight, 2011a, 2014a). It would be serious omission if other sources of petrochemicals were ignored since the use of these materials can, if left unattended in nature, cause several damage to the environment. Typically, these sources are mixed waste polymers which contain a variety of chemical structures and are usually classed as non-hydrocarbons. An inorganic compound is typically a chemical compound that lacks carbon–hydrogen (C–H) bonds, that is, a compound that is not an organic compound, but the distinction is not defined or even of particular interest. Inorganic compounds comprise most of the earth's crust, although the composition of the deep mantle remains active areas of investigation. Inorganic compounds can be defined as any compound that is not organic compound. Some simple compounds that contain carbon are often considered inorganic. Examples include carbon monoxide, carbon dioxide, carbonate derivatives, cyanide derivatives, derivatives, carbide derivatives, and thiocyanate derivatives. Many of these are normal parts of mostly organic systems, including organisms, which means that describing a chemical as inorganic does not obligately mean that it does not occur within floral and faunal species.

Industrial inorganic chemistry includes subdivisions of the chemical industry that manufacture inorganic products on a large scale, such as the heavy inorganics (chlor-alkalis, sulfuric acid, sulfates) and fertilizers (potassium, nitrogen, and phosphorus products) as well as segments of fine chemicals that are used to produce high-purity inorganics on a much smaller scale. Among these are reagents and raw materials used in high-tech industries, pharmaceuticals, or electronics, for example, as well as in the preparation of inorganic specialties such as catalysts, pigments, and propellants. Metals are chemicals and they are manufactured from ores and purified by many of the same processes as those used in the manufacture of inorganics. However, if they are commercialized as alloys or in their pure form such as iron, lead, copper, or tungsten, they are considered products of the metallurgical not chemical industry.

Thus, in this book, inorganic chemistry is concerned with the properties and behavior of inorganic compounds, which include metals, minerals, and organometallic compounds. While organic chemistry is the study of carbon-containing compounds and inorganic chemistry is the study of the remaining subset of compounds other than organic compounds, there is overlap between the two fields (such as organometallic compounds, which usually contain a metal or metalloid bonded directly to carbon).

Specific examples of inorganic chemicals prepared from petrochemical sources are presented below and included ammonia, hydrazine, hydrogen, nitric acid, sulfur, and sulfuric acid which are presented below in alphabetical order rather than in any order of preference.

9.2 AMMONIA

Ammonia (NH$_3$) is a compound of nitrogen and hydrogen that exists as a colorless pungent gas which has a high degree of water solubility, where it forms a weakly basic solution. Either directly or indirectly, ammonia is a building block for the synthesis of many pharmaceutical products and is used in many commercial cleaning products. It is mainly collected by downward displacement of both air and water.

Ammonia boils at –33.34°C (–28.012°F) under ambient pressure so the liquid must be stored under pressure or at low temperature. Household ammonia or ammonium hydroxide (NH$_4$OH) is a solution of NH$_3$ in water. Ammonia could be easily liquefied under pressure (liquid ammonia), and it is an important refrigerant. Anhydrous ammonia is a fertilizer by direct application to the soil. Ammonia is obtained by the reaction of hydrogen and nitrogen (both of which are produced in a refinery) and represented simply as:

$$N_2 + 3H_2 \rightarrow 2NH_3$$

Ammonia is one of the most important inorganic chemicals, exceeded only by sulfuric acid and lime. It is a nitrogen source in fertilizer, and it is one of the major inorganic chemicals used in the production of nylons, fibers, plastics, polyurethanes (used in tough chemical-resistant coatings, adhesives, and foams), hydrazine (used in jet and rocket fuels), and explosives.

9.2.1 PRODUCTION

The production of ammonia is of historical interest because it represents the first important application of thermodynamics to an industrial process. Before the start of World War I, most ammonia was obtained by the dry distillation of nitrogenous vegetable waste and animal waste products, where it was distilled by the reduction of nitrous acid and nitrates with hydrogen. In addition, ammonia was also produced by the distillation of coal as well as by the decomposition of ammonium salts by alkaline hydroxides such as quicklime [$Ca(OH)_2$], the salt most generally used is the chloride (NH_4Cl).

Hydrogen for ammonia synthesis could also be produced economically by using the water-gas reaction followed by the water-gas shift reaction, produced by passing steam through red-hot coke, to give a mixture of hydrogen and carbon dioxide gases, followed by removal of the carbon dioxide washing the gas mixture with water under pressure or by using other sources like coal or coke gasification.

Modern ammonia-producing plants depend on industrial hydrogen production to react with atmospheric nitrogen using a magnetite catalyst (Fe_3O_4, or sometime written as $FeO\ Fe_2O_3$) or over a promoted Fe catalyst under high pressure (1,500 psi) and temperature (450°C, 840°F) to form anhydrous liquid ammonia (the Haber–Bosch process):

$$N_2 + 3H_2 \rightarrow 2NH_3$$

Increasing the temperature increases the reaction rate, but decreases the equilibrium (Kc at 500°C = 0.08). According to the Le Chatelier principle, the equilibrium is favored at high pressures and at lower temperatures. Much of Haber's research was to find a catalyst that favored the formation of ammonia at a reasonable rate at lower temperatures. Iron oxide promoted with other oxides such as potassium and aluminum oxides is currently used to produce ammonia in good yield at relatively low temperatures.

In the process, a mixture of hydrogen and nitrogen (exit gas from the methanator) in a ratio of 3:1 is compressed to the desired pressure (6,700–15,000 psi). The compressed mixture is then preheated by heat exchange with the product stream before entering the ammonia reactor. The reaction occurs over the catalyst bed at about 450°C (840°F). The exit gas containing ammonia is passed through a cooling chamber where ammonia is condensed to a liquid, while unreacted hydrogen and nitrogen are recycled. Usually, a conversion of approximately 15% per pass is obtained under these conditions.

The most popular catalysts are based on iron promoted with potassium oxide (K_2O), calcium oxide (CaO), silica (SiO_2), and alumina (Al_2O_3). The original Haber–Bosch reactions contained osmium as the catalyst, but it was available in extremely small quantities. There is also an iron-based catalyst that is still used in modern ammonia production plants. Some ammonia production utilizes ruthenium-based catalysts (the KAAP process), which allow milder operating pressures because of the high activity of the catalyst.

In industrial practice, the iron catalyst is obtained from finely ground iron powder, which is usually obtained by reduction of high-purity magnetite (Fe_3O_4). The pulverized iron metal is burned (oxidized) to give magnetite of a defined particle size and the magnetite particles are then partially reduced, removing some of the oxygen in the process. The resulting catalyst particles consist of a core of magnetite, encased in a shell of wüstite FeO (ferrous oxide), which in turn is surrounded by an outer shell of iron metal. The catalyst maintains most of its bulk volume during the reduction, resulting in a highly porous high surface area material, which enhances its effectiveness as a catalyst.

Other minor components of the catalyst include calcium oxide and aluminum oxide, which support the iron catalyst and help it maintain its surface area. The oxides of calcium, aluminum, potassium, and silicon are unreactive (immune) to reduction by the hydrogen.

As a sustainable alternative to the relatively inefficient and energy-intensive electrolysis process, hydrogen can be generated from organic wastes (such as biomass or food-industry waste) using catalytic reforming. This releases hydrogen from carbonaceous substances at only 10%–20% of energy used by electrolysis and may lead to hydrogen being produced from municipal wastes at below zero cost (allowing for the tipping fees and efficient catalytic reforming, such as cold plasma). Catalytic (thermal) reforming is possible in small, distributed (even mobile) plants, to take advantage of low-value, stranded biomass/biowaste or natural gas deposits. Conversion of such wastes into ammonia solves the problem of hydrogen storage, as hydrogen can be released economically from ammonia on demand, without the need for high pressure or cryogenic storage. It is also easier to store ammonia onboard vehicles than to store hydrogen, as ammonia is less flammable than naphtha-gasoline or liquefied petroleum gas (LPG).

9.2.2 Properties and Uses

Ammonia is a colorless gas with a distinct odor, and is a building block chemical and a key component in the manufacture of many products people use every day. It occurs naturally throughout the environment in the air, soil, and water and in plants and animals, including humans. The human body makes ammonia when the body breaks down foods containing protein into amino acids and ammonia, then converting the ammonia into urea (H_2NCONH_2). Ammonium hydroxide (NH_4OH)—commonly known as household ammonia—is an ingredient in many everyday household cleaning products.

In the process to produce urea from ammonia (NH_3) and carbon dioxide (CO_2) (Stamicarbon CO_2 stripping Urea 2000 plus Technology), ammonia and carbon dioxide are reacted at 2,100 psi bar to urea and carbamate (a carbamate is an organic compound derived from carbamic acid, NH_2COOH). The conversion of ammonia as well as carbon dioxide in the synthesis section results in a low recycle flow of carbamate. Because of the high-ammonia efficiency, no pure ammonia is recycled in this process. The synthesis temperature of 185°C (365°F) is low, and, consequently, corrosion is negligible. Because of the high conversions in the synthesis, the recycle section of the plant is very small. An evaporation stage with vacuum condensation system produces urea melt with the required concentration for the Stamicarbon fluidized bed granulation. Process water produced in the plant is treated in desorption/hydrolyzer section that produces an effluent, which is suitable for use as boiler feedwater. One further step, although not true a nonorganic step is worthy of mention here as a follow on from urea production, and that is the production urea-formaldehyde resins. The chemical structure of the urea-formaldehyde polymer consists of $[(O)CNHCH_2NH]_n$ repeat units (Chapter 11).

The majority of the ammonia produced is used in fertilizer to help sustain food production. The production of food crops naturally depletes soil nutrient supplies. In order to maintain healthy crops, farmers rely on fertilizers to maintain productivity of the soil and to maintain (or increase) the levels of essential nutrients such as like zinc, selenium, and boron in food crops.

Ammonia is also used in many household cleaning products and can be used to clean a variety of household surfaces—from tubs, sinks, and toilets to bathroom and kitchen countertops and tiles. Ammonia also is effective at breaking down household grime or stains from animal fats or vegetable oils, such as cooking grease and wine stains. Because ammonia evaporates quickly, it is commonly used in glass cleaning solutions to help avoid streaking.

When used as a refrigerant gas and in air-conditioning equipment, ammonia can absorb substantial amounts of heat from its surroundings. Ammonia can be used to purify water supplies and as a building block in the manufacture of many products including plastics, explosives, fabrics, pesticides, and dyes. Ammonia is also used in the waste and wastewater treatment, cold storage, rubber, pulp and paper, and food and beverage industries as a stabilizer, neutralizer, and a source of nitrogen as well as in the manufacture of pharmaceutical products.

In organic chemistry, ammonia can act as a nucleophile in substitution reactions and, as an example, amines (RNH_2) can be formed by the reaction of ammonia with an alkyl halide (RCl) although the resulting amino (-NH_2) group is also nucleophilic and secondary and tertiary amines are often formed as byproducts. As an example, methylamine (CH_3NH_2) is produced commercially by the reaction of ammonia with chloromethane (CH_3Cl).

Amide derivatives can be prepared by the reaction of ammonia with carboxylic acid derivatives. In addition, ammonium salts of carboxylic acids $\left(RCO_2^- \; NH_4^+\right)$ can be dehydrated to amides so long as there are no thermally sensitive groups present: temperatures of 150°C–200°C (300°F–390°F) are required.

Also, the hydrogen in ammonia is capable of replacement by metals—as an example, magnesium burns in ammonias with the formation of magnesium nitride (Mg_3N_2) and when the gas is passed over heated sodium (or potassium) sodamide ($NaNH_2$) (or/and potassamide (KNH_2)) are formed.

9.3 CARBON BLACK

Carbon black (also classed as an inorganic petrochemical) is made predominantly by the partial combustion of carbonaceous (organic) material such as fluid catalytic cracker bottoms and other cracker bottoms in a limited supply of air. It has a high surface area to volume ratio and a significantly lower (negligible) content of polynuclear aromatic hydrocarbon derivative. However, carbon black is widely used as a model compound for diesel soot for diesel oxidation experiments. Carbon black is mainly used as a reinforcing filler in vehicle tires and other rubber products. It is also used as a color pigment in plastics, paints, and inks.

The carbonaceous sources vary from methane to aromatic petroleum oils to coal tar byproducts. The carbon black is used primarily for the production of synthetic rubber. Carbon black (also classed as an inorganic petrochemical) is made predominantly by the partial combustion of carbonaceous (organic) material in a limited supply of air. The carbonaceous sources vary from methane to aromatic petroleum oils to coal tar byproducts. The carbon black is used primarily for the production of synthetic rubber.

Carbon black is an extremely fine powder of great commercial importance, especially for the synthetic rubber industry. The addition of carbon black to tires lengthens its life extensively by increasing the abrasion and oil resistance of rubber. Carbon black consists of elemental carbon with variable amounts of volatile matter and ash. There are several types of carbon blacks, and their characteristics depend on the particle size, which is mainly a function of the production method.

9.3.1 PRODUCTION

Carbon black is produced by the partial combustion or the thermal decomposition of natural gas or petroleum distillates and residues. Petroleum products rich in aromatics such as tars produced from catalytic and thermal cracking units are more suitable feedstocks due to their high carbon/hydrogen ratios. These feedstocks produce black with a carbon content of approximately 92% w/w. Coke produced from delayed and fluid coking units with low sulfur and ash contents has been investigated as a possible substitute for carbon black.

Three processes are currently used for the manufacture of carbon blacks: (i) the channel, (ii) the furnace black process, and (iii) the thermal process.

The channel process is mainly of historical interest, because not more than 5% of the carbon black products are manufactures by this route. In this process, the feedstock (e.g., natural gas) is burned in small burners with a limited amount of air. Some methane is completely combusted to carbon dioxide and water, producing enough heat for the thermal decomposition of the remaining natural gas. The formed soot collects on cooled iron channels from which the carbon black is scraped. Channel black is characterized by having a lower pH, higher volatile matter, and smaller average particle size than blacks from other processes.

The furnace black process is a more advanced partial combustion process. The feedstock is first preheated and then combusted in the reactor with a limited amount of air. The hot gases containing carbon particles from the reactor are quenched with a water spray and then further cooled by heat exchange with the air used for the partial combustion. The type of black produced depends on the feed type and the furnace temperature.

In the thermal process, the feedstock (natural gas) is pyrolyzed in preheated furnaces lined with a checker work of hot bricks. The pyrolysis reaction produces carbon that collects on the bricks. The cooled bricks are then reheated after carbon black is collected.

9.3.2 PROPERTIES AND USES

Carbon black (subtypes: acetylene black, channel black, furnace black, lamp black, and thermal black) is a material produced by the incomplete combustion of high-boiling crude oil products such as the bottoms from a fluid catalytic cracking unit or the bottoms from an ethylene cracking unit. Carbon black is a form of crystalline carbon that has a high surface area to volume ratio but lower than the surface area to volume ratio of activated carbon. It is dissimilar to soot in that the surface area to volume ratio is higher than the surface area to volume ratio of soot and is also significantly lower (negligible) in the content of polycyclic aromatic hydrocarbon derivatives (PAHs) (or poly-nuclear aromatic hydrocarbon derivatives (PNAs)).

Carbon black produced by the channel process was generally acidic, while those produced by the furnace process and the thermal process are slightly alkaline. The pH of the black has a pronounced influence on the vulcanization time of the rubber. (Vulcanization is a physicochemical reaction by which rubber changes to a thermosetting mass due to cross-linking of the polymer chains by adding certain agents such as sulfur.) The basic nature (higher pH) of furnace blacks is due to the presence of evaporation deposits from the water quench. Thermal blacks, due to their larger average particle size, are not suitable for tire bodies and tread bases, but they are used in inner tubes, footwear, and paint pigment. Gas and oil furnace carbon blacks are the most important forms of carbon blacks and are generally used in tire treads and tire bodies.

Carbon black is also used as a pigment for paints and printing inks, as a nucleation agent in weather modifications, and as a solar energy absorber. About 70% of the worlds' consumption of carbon black is used in the production of tires and tire products. Approximately 20% goes into other products such as footwear, belts, and hoses and the rest is used in such items as paints and printing ink.

The important properties of carbon black are particle size, surface area, and pH. These properties are functions of the production process and the feed properties. Thus, it is widely used as a reinforcing filler in tires and other rubber products. Practically all rubber products where tensile and abrasion wear properties are important use carbon black, so they are black in color. Carbon black is also used as a color pigment in plastics, paints, and inks.

9.4 CARBON DIOXIDE AND CARBON MONOXIDE

Carbon dioxide (CO_2) is a colorless gas with a density about 60% higher than that of dry air. In the current context, it is present in reservoirs of crude oil and *natural gas* and may be isolated from these sources as a useful/salable product. Carbon dioxide is odorless at normally encountered concentrations. However, at high concentrations, it has a sharp and acidic odor.

Carbon monoxide (CO) is a colorless, odorless, and tasteless gas that is slightly less dense than air. It is toxic to hemoglobic animals (animals with hemoglobin in the blood stream with which carbon monoxide form a complex thereby displacing oxygen from the blood of the hemoglobic animals. In the atmosphere, it is short-lived, having a role in the formation of ground-level ozone.

9.4.1 PRODUCTION

Carbon dioxide is a produced during the production of hydrogen by steam reforming and by the water-gas shift reaction in gasification of carbonaceous feedstocks (Chapter 5). These processes begin with the reaction of water and natural gas (mainly methane). In terms of biomass as the source material (Chapter 3), carbon dioxide is a byproduct of the fermentation of sugar and starch derivatives:

$$C_6H_{12}O_6 \rightarrow 2CO_2 + 2C_2H_5OH$$

In addition, carbon dioxide is one the byproducts of gas cleaning processes—one of the most important aspects of gas processing involves the removal of hydrogen sulfide and carbon dioxide, which are generally referred to as contaminants (Chapter 4). Natural gas from some wells (crude oil wells and gas wells) contains significant amounts of hydrogen sulfide and carbon dioxide and is usually referred to as *sour gas*. Sour gas is undesirable because the sulfur compounds it contains can be extremely harmful, even lethal, to breathe and the gas can also be extremely corrosive. The process for removing hydrogen sulfide from sour gas is commonly referred to as *sweetening* the gas (Mokhatab et al., 2006; Speight, 2007, 2014a).

Also, carbon dioxide comprises about 40%–45% v/v of the gas that emanates from decomposition in landfills (landfill gas) (Chapter 3)—most of the remaining 50%–55% v/v of the landfill gas is methane.

A major industrial source of carbon monoxide is producer gas (Chapter 5), a mixture containing mostly carbon monoxide and nitrogen, formed by gasification of carbonaceous feedstocks (Chapter 5) when there is an excess of carbon (deficiency of oxygen). In the process, air is passed through a bed of the carbonaceous feedstock and the initially produced carbon dioxide equilibrates with the remaining hot carbon to yield carbon monoxide (the Boudouard reaction). Above 800°C (1,470°F), carbon monoxide is the predominant product:

$$CO_2 + C \rightarrow 2CO$$

Another source of carbon monoxide is water gas (a mixture of hydrogen and carbon monoxide) produced via the endothermic reaction of steam with carbon:

$$H_2O + C \rightarrow H_2 + CO$$

Carbon monoxide is also produced by the direct oxidation of carbon in a limited supply of oxygen or air:

$$2C(s) + O_2 \rightarrow 2CO(g)$$

9.4.2 PROPERTIES AND USES

Carbon dioxide is a versatile industrial material used, for example, as an inert gas in welding and fire extinguishers as a pressurizing gas in air guns and oil recovery, as a chemical feedstock and as a supercritical fluid solvent in decaffeination of coffee. It is added to drinking water and carbonated beverages including beer and sparkling wine. The frozen solid form of carbon dioxide (*dry ice*) is used as a refrigerant.

Carbon monoxide has many applications in the manufacture of bulk chemicals. For example, aldehydes are produced by the hydroformylation reaction of olefin derivatives, carbon monoxide, and hydrogen (Chapter 7). Phosgene ($COCl_2$) is an industrial building block that is used for the production of urethane polymers and polycarbonate polymers (Chapter 11), but it is poisonous, and was used as a chemical weapon during World War I. To produce phosgene, carbon monoxide and chlorine gas are passed through a bed of porous-activated carbon to form the gas:

$$CO + Cl_2 COCl_2$$

The phosgene is then reacted with the relevant feedstocks to produce the isocyanate derivatives, polyurethane derivative, and polycarbonate derivatives.

Methanol is produced by the hydrogenation of carbon monoxide. Furthermore, in the Monsanto process, carbon monoxide and methanol react in the presence of a homogeneous catalyst (typically a rhodium catalyst) to produce acetic acid. In a related reaction, the hydrogenation of carbon monoxide is coupled to carbon–carbon bond formation, as in the Fischer–Tropsch process (Chapter 10) where carbon monoxide is hydrogenated to liquid hydrocarbon fuels. This technology allows non-petroleum carbonaceous feedstocks (Chapter 3) to be converted to valuable hydrocarbon liquids, and solids (waxes).

Carbon monoxide is a strong reductive agent and it has been used in pyrometallurgy to produce metals from the corresponding ores. In the process, carbon monoxide strips oxygen off metal oxides, reducing them to pure metal in high temperatures, forming carbon dioxide in the process. Carbon monoxide is not usually supplied as is but it is formed in high temperature in presence of oxygen-carrying ore (MO), a highly carbonaceous agent such as coke©, and high temperature.

$$MO + C \rightarrow CO + M$$

As another example, in the Mond process (also known as the carbonyl process), carbon monoxide (as a contempt of synthesis gas) is used to purify nickel. This process is based on the principle that carbon monoxide that readily combines irreversibly with nickel to yield nickel carbonyl [$Ni(CO)_4$]). Thus:

$$NiO(s) + H_2(g) \rightarrow Ni(s) + H_2O(g) \text{ (impure nickel)}$$

$$Ni(s) + 4CO(g) \rightarrow Ni(CO)_4(g)$$

$$Ni(CO)_4(g) \rightarrow Ni(s) + 4CO(g) \text{ (pure nickel)}$$

9.5 HYDRAZINE

Hydrazine (N_2H_4 or H_2NNH_2) is a colorless, fuming liquid miscible with water, hydrazine (also called diazine) is a weak base but a strong reducing agent. Hydrazine is a colorless flammable liquid that has an ammonia-like odor and is highly toxic and dangerously unstable unless handled in solution as, e.g., hydrazine hydrate (NH_2NH_2 xH_2O).

The term *hydrazine* refers to a class of organic substances which are by replacing one or more hydrogen atoms in hydrazine by an organic group, e.g., $RNHNH_2$.

9.5.1 PRODUCTION

There are several processes that are available for the production of hydrazine—the essential or (key) step in each process is the creation of the nitrogen–nitrogen single bond. The many routes can be divided into those that use chlorine oxidants (and generate salt) and those that do not.

Hydrazine can be synthesized from ammonia and hydrogen peroxide in the Peroxide process (also referred to as Pechiney–Ugine–Kuhlmann process, the Atofina–PCUK cycle, or Ketazine process) which is often represented simply as:

$$2NH_3 + H_2O_2 \rightarrow H_2NNH_2 + 2H_2O$$

However, the process is more complex than the above equation suggest and in the process, the ketone and ammonia first condense to give the imine which is oxidized by hydrogen peroxide to the oxaziridine, a three-membered ring containing carbon, oxygen, and nitrogen. In the next step, the oxaziridine gives the hydrazine by treatment with ammonia, which creates the nitrogen–nitrogen single bond, after which the hydrazine derivative condenses with one more equivalent of the ketone:

The resulting azine is hydrolyzed to give hydrazine and regenerate the ketone, methyl ethyl ketone:

$$Me(Et)C=NN=C(Et)Me + 2H_2O \rightarrow 2Me(Et)C=O + N_2H_4$$

Unlike most other processes, this process does not produce a salt as a byproduct.

In the Olin Raschig process, chlorine-based oxidants oxidize ammonia without the presence of a ketone—in the peroxide process, hydrogen peroxide oxidizes ammonia in the presence of a ketone but the Olin–Raschig process relies on the reaction of chloramine with ammonia to create the nitrogen–nitrogen single bond as well as the hydrogen chloride byproduct:

$$NH_2Cl + NH_3 \rightarrow H_2N–NH_2 + HCl$$

In a related process, urea can be oxidized instead of ammonia with sodium hypochlorite serving as the oxidant:

$$(H_2N)_2 C=O + NaOCl + 2NaOH \rightarrow N_2H_4 + H_2O + NaCl + Na_2CO_3$$

Hydrazine is produced by the oxidation of ammonia using the Rashig process. Sodium hypochlorite is the oxidizing agent and yields chloramine (NH_2Cl) as an intermediate. Chloramine further reacts with ammonia producing hydrazine:

$$2NH_3 + NaOCl \rightarrow H_2NNH_2 + NaCl + H_2O$$

Hydrazine is then evaporated from the sodium chloride solution.

9.5.2 PROPERTIES AND USES

Hydrazine is mainly used as a foaming agent in the preparation of polymer foams and other significant uses also include use as a precursor to polymerization catalysts, pharmaceuticals, and agricultural chemicals. Additionally, hydrazine is used as a rocket fuel because its combustion is highly exothermic:

$$H_2NNH_2 + O_2 \rightarrow N_2 + 2H_2O$$

In addition to rocket fuel, hydrazine is used as a blowing agent and in the pharmaceutical and fertilizer industries. It is used to prepare the gas precursors used in air bags. Hydrazine is used within both nuclear and conventional electrical power plant steam cycles as an oxygen scavenger to control concentrations of dissolved oxygen in an effort to reduce corrosion.

Also, because of the weak nitrogen–nitrogen bond, it is used as a polymerization initiator. As a reducing agent, hydrazine is used as an oxygen scavenger for steam boilers. It is also a selective reducing agent for nitro compounds. Hydrazine is a good building block for many chemicals, especially agricultural products, which dominates its use. Hydrazine is also used as a propellant in space vehicles to reduce the concentration of dissolved oxygen in and to control pH of water used in large industrial boilers.

Hydrazine is a precursor to several pharmaceuticals and pesticides. Often these applications involve conversion of hydrazine to heterocyclic ring systems such as pyrazole derivatives and pyridazine derivatives. Hydrazine compounds can be effective as active ingredients in admixture with or in combination with other agricultural chemicals such as insecticides, miticides, nematicides, fungicides, antiviral agents, attractants, herbicides, or plant growth regulators.

Often, the use of hydrazine as a precursor to several pharmaceuticals and pesticides involves conversion of hydrazine to heterocyclic ring derivatives such as pyrazole derivaitves and pyridazine derivatives.

Hydrazine compounds can be effective as active ingredients in admixture with or in combination with other agricultural chemicals such as insecticides, miticides, nematicides, fungicides, antiviral agents, attractants, herbicides, or plant growth regulators.

9.6   HYDROGEN

Hydrogen is the lightest known element, and is only found in the free state in trace amounts, but is widely spread in a combined form with other elements. Hydrogen is one of the key starting materials used in the chemical industry. It is a fundamental building block for the manufacture of ammonia (NH_3), and hence fertilizers, and of methanol (CH_3OH), used in the manufacture of many polymers. Hydrogen is used in the manufacture of two of the most important chemical compounds made industrially. It is also used in the refining of oil, for example, in reforming, one of the processes for obtaining high-octane naphtha (usually called *reformate*) as a blend-stock for the production of gasoline and in removing sulfur compounds from petroleum which would otherwise poison the catalytic converters fitted to vehicles.

9.6.1   PRODUCTION

Water, natural gas, crude oil, hydrocarbons, and other organic fossil fuels are major sources of hydrogen.

$$2H_2O \rightarrow 2H_2 + O_2$$

Electrolysis, and the thermochemical decomposition as well as the photochemical decomposition of water followed by purification through diffusion methods are methods for the production of hydrogen. Chemically, the electrolysis of water is considered to be a simple method of producing hydrogen. In the process, a low-voltage current is run through the water, and gaseous oxygen forms at the anode while gaseous hydrogen forms at the cathode. Typically, the cathode is made from platinum or another inert metal when producing hydrogen for storage. If, however, the gas is to be burnt on-site, oxygen is desirable to assist the combustion, and so both electrodes would be made from inert metals. For example, iron, for instance, would oxidize, and thus decrease the amount of oxygen that is evolved. However, the electrolysis process is an energy-extensive process.

The most economical way to produce hydrogen is by steam reforming petroleum fractions and natural gas as well as by gasification of carbonaceous feedstocks (Speight, 2014a). In the steam reforming process, two major sources of hydrogen (water and hydrocarbons, such as methane) are reacted to produce a mixture of carbon monoxide and hydrogen (synthesis gas). Hydrogen can then be separated from the mixture after shift converting carbon monoxide to carbon dioxide. Carbon oxides are removed by passing the mixture through a pressure swing adsorption (PSA) system. Also, hydrogen can be produced by the steam reforming methanol. In this process, an active catalyst is used to decompose methanol and shift convert carbon monoxide to carbon dioxide. The produced gas is cooled, and carbon dioxide is removed:

$$CH_3OH + H_2O \rightarrow CO_2 + 3H_2$$

In the petroleum refining industry, hydrogen is essentially obtained from catalytic naphtha reforming (Parkash, 2003; Gary et al., 2007; Speight, 2014a; Hsu and Robinson, 2017; Speight, 2017). In the petrochemical industry, hydrogen is used to hydrogenate benzene to cyclohexane and benzoic acid to cyclohexane carboxylic acid. These compounds are precursors for nylon production. It is also used to selectively hydrogenate acetylene from C_4 olefin mixture. As a constituent of synthesis gas, hydrogen is a precursor for ammonia, methanol, Oxo alcohols, and hydrocarbons from Fischer–Tropsch processes. The direct use of hydrogen as a clean fuel for automobiles and buses is currently being evaluated compared to fuel cell vehicles that use hydrocarbon fuels which are converted through onboard reformers to a hydrogen-rich gas.

Thus, an important process for the production of hydrogen is by *steam reforming*. The key parts of the process are the conversion of a carbon-containing material to a mixture of carbon monoxide and hydrogen (synthesis gas) followed by the conversion of carbon monoxide to carbon dioxide and the production of more hydrogen. In many refineries, the hydrocarbon feedstock is typically methane or other low-boiling hydrocarbon derivatives obtained from natural gas (as well as from crude and coal). However, there is an increasing interest in using biomass as a source of hydrogen (Speight, 2011b).

Thus, if a hydrocarbon feedstock is employed as the source of hydrogen, the hydrocarbon vapor phase is mixed with a large excess of steam and passed through pipes containing nickel oxide (which is reduced to nickel during the reaction), supported on alumina, in a furnace which operates at high temperatures. Using methane as the example of the feedstock:

$$CH_4 + H_2O \rightarrow 3H_2 + CO$$

The reaction is endothermic and accompanied by an increase in volume (2 volumes to 4 volumes) and, thus, is favored by high temperatures and by low partial pressures. The reaction is also favored by a high ration of steam to hydrocarbon which, however, does increase the yield but also increases operating (energy) costs. The high ratio also helps to reduce the amount of carbon deposited which reduces the efficiency of the catalyst. The most effective way to reduce carbon deposition has been found to be impregnation of the catalyst with potassium carbonate.

In the second part of the process, the shift reaction, carbon monoxide is converted to carbon dioxide by reacting the carbon monoxide with steam and thus producing more hydrogen:

$$CO + H_2O \rightarrow H_2 + CH_2$$

This reaction is also exothermic and so high conversions to carbon dioxide and hydrogen are favored by low temperatures which may be difficult to control due to the heat evolved and, as a result, it has been common practice to separate the shift reaction into two stages: (i) the first stage in which the bulk of the reaction being carried out at approximately 377°C (710°F) K over an iron catalyst, and the "polishing" reaction carried out around 450K over a copper/zinc/alumina catalyst. The carbon dioxide and any remaining carbon monoxide are then removed by passing the gases through a zeolite sieve.

Thus, overall, 1 mol of methane and 2 mol of steam are theoretically converted into 4 mol of hydrogen, although this theoretical yield is not achieved as the reactions do not go to completion:

$$CH_4 + 2H_2O \rightarrow 4H_2 + CO_2$$

Due to the increasing demand for hydrogen, many separation techniques have been developed to recover it from purge streams vented from certain processing operations such as hydrocracking and hydrotreating. In addition to hydrogen, these streams contain methane and other light hydrocarbon gases. Physical separation techniques such as (i) adsorption, (ii) diffusion, and (iii) cryogenic phase separation are used to achieve efficient separation.

Adsorption is accomplished using a special solid that preferentially adsorbs hydrocarbon gases, not hydrogen. The adsorbed hydrocarbons are released by reducing the pressure. Cryogenic phase separation on the other hand, depends on the difference between the volatilities of the components at the low temperatures and high pressures used. The vapor phase is rich in hydrogen, and the liquid phase contains the hydrocarbons. Hydrogen is separated from the vapor phase at high purity.

Diffusion separation processes depend on the permeation rate for gas mixtures passing through a special membrane. The permeation rate is a function of the type of gas feed, the membrane material, and the operating conditions. Gases having a lower molecular size such as helium and hydrogen permeate membranes more readily than larger molecules such as methane and ethane. After the feed gas is preheated and filtered it enters the membrane separation section. This is made of a permeater vessel containing 12-in. diameter bundles (resemble filter cartridges) and consists of millions of hollow fibers. The gas mixture is distributed in the annulus between the fiber bundle and the vessel wall. Hydrogen, being more permeable, diffuses through the wall of the hollow fiber and exits at a lower pressure. The less permeable hydrocarbons flow around the fiber walls to a perforated center tube and exit at approximately feed pressure. It has been reported that this system can deliver a reliable supply of pure hydrogen (>95% v/v pure) from off-gas streams having/as low as 15% v/v hydrogen.

9.6.2 PROPERTIES AND USES

The chemical industry is the largest producer and consumer of hydrogen. Hydrogen has various applications in the chemical industry. Thus, once obtained, hydrogen is widely used in the production of bulk chemicals, intermediates, and specialty chemicals (Figure 9.1). Thus, hydrogen does form compounds with most elements (the bonding of hydrogen to carbon is excluded from this discussion), such as the more electronegative elements (e.g., the halogen elements—fluorine, chlorine, bromine, and iodine) and oxygen. In these compounds, hydrogen takes on a partial positive

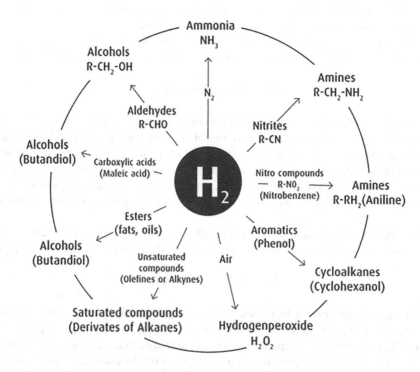

FIGURE 9.1 Examples of the use of hydrogen.

charge and when bonded to fluorine (HF), oxygen (H_2O), and nitrogen (NH_3), hydrogen can participate in a form of medium-strength non-covalent bonding with the hydrogen of other similar molecules (hydrogen bonding) that is often reflected in the stability of many biological molecules, such as the helix structure of DNA. Hydrogen also forms compounds with less electronegative elements and forms hydrides with metals and metalloids, where it takes on a partial negative charge.

The key consumers of hydrogen in a petrochemical plant include use as a hydrogenating agent, particularly in increasing the level of saturation of unsaturated fats and oils (found in items such as margarine). It is similarly the source of hydrogen in the manufacture of hydrochloric acid and is a reducing agent for the production of metals from the metallic ores, typically the oxide of the metal. As an example, molybdenum can be produced by passing hydrogen over hot molybdenum oxide:

$$2MoO_3 + 6H_2 \rightarrow Mo + 6H_2O$$

Although hydrides can be formed with almost all main group elements, the number and combination of possible compounds varies widely; for example, more than 100 binary borane hydrides are known, but there is only 1 binary aluminum hydride. In inorganic chemistry, hydride derivatives can also serve as bridging ligands that link two metal centers in a coordination complex. This function is particularly common in the boranes (boron hydride derivatives) and aluminum complexes, as well as in clustered carborane derivatives.

The most important single use of hydrogen is in the manufacture of ammonia (NH_3), which is produced by combining hydrogen and nitrogen at high pressure and temperature in the presence of a catalyst.

$$3H_2 + N_2 \rightarrow 2NH_3$$

Hydrogen is also used for a number of similar reactions. For example, it can be combined with carbon monoxide to make methanol:

$$2H_2 + CO \rightarrow CH_3OH$$

Methanol, like ammonia, has a great many practical uses in a variety of industries. The most important use of methanol is in the manufacture of other chemicals, such as those from which plastics are made. Small amounts are used as additives to gasoline to reduce the amount of pollution released to the environment. Methanol is also used widely as a solvent (to dissolve other materials) in industry.

Finally, one of the most important groups of hydrogen compounds is the acids. Common inorganic acids include hydrochloric acid (HCl), sulfuric acid (H_2SO_4), nitric acid (HNO_3), phosphoric acid (H_3PO_4), hydrofluoric acid (HF), and boric acid (H_3BO_3). Although not a typical hydrogen acid, hydrogen peroxide (H_2O_2) is a weak acid, but is a strong oxidizing agent used in aqueous solution as a ripening agent, bleach, and topical anti-infective. It is relatively unstable and solutions deteriorate over time unless stabilized by the addition of acetanilide or similar organic materials. In keeping with the strong oxidizing properties, hydrogen peroxide is a powerful bleaching agent that is mostly used for bleaching paper, but has also found use as a disinfectant and as an oxidizer. Hydrogen peroxide in the form of carbamide peroxide (a solid composed of equal amounts of hydrogen peroxide and urea (H_2NCONH_2)), which is a white crystalline solid that dissolves in water to give free hydrogen peroxide is widely used for tooth whitening (bleaching), both in professionally administered and self-administered products.

9.7 NITRIC ACID

Nitric acid (HNO_3) is one of the most used chemicals. It is a colorless to a yellow liquid, which is very corrosive. It is a strong oxidizing acid that can attack almost any metal. The pure compound is colorless, but older samples tend to acquire a yellow cast due to decomposition into oxides of nitrogen (NO_x) and water. Nitric acid is subject to thermal decomposition or photolytic decomposition (decomposition by light) and for this reason it was often stored in brown glass bottles:

$$4HNO_3 \rightarrow 2H_2O + 4NO_2 + O_2$$

This reaction may give rise to some non-negligible variations in the vapor pressure above the liquid because the nitrogen oxides produced dissolve partly or completely in the acid. Most commercially available nitric acid has a concentration of 68% v/v in water. When the solution contains more than 86% v/v nitric acid, it is *fuming nitric acid*. Depending on the amount of nitrogen dioxide (NO_2), fuming nitric acid is further characterized as (i) red fuming nitric acid at concentrations above 86% v/v and (ii) white fuming nitric acid at concentrations above 95% v/v.

Commercially available nitric acid is an azeotrope with water at a concentration of 68% HNO_3, which is the ordinary concentrated nitric acid of commerce. This solution has a boiling temperature of 120.5°C/760 mm. Two solid hydrates are known; the monohydrate (HNO_3 H_2O or [H_3O]NO_3) and the trihydrate (HNO_3 $3H_2O$).

9.7.1 PRODUCTION

Nitric acid is commercially produced by oxidizing ammonia with air over a platinum-rhodium wire gauze. The following sequence represents the reactions occurring over the heterogeneous catalyst:

$$4NH_3 + 5O_2 \rightarrow 4NO + 6H_2O$$
$$2NO + O_2 \rightarrow 2NO_2$$
$$3NO_2 + H_2O \rightarrow 2HNO_3 + NO$$

The three reactions are exothermic, and the equilibrium constants for the first two reactions fall rapidly with increase of temperature. Increasing pressure favors the second reaction but adversely affects the first reaction.

For this reason, operation around atmospheric pressures is typical. Space velocity should be high to avoid the reaction of ammonia with oxygen on the reactor walls, which produces nitrogen and water, and results in lower conversions. The concentration of ammonia must be kept below the inflammability limit of the feed gas mixture to avoid explosion. Optimum nitric acid production was found to be obtained at approximately 900°C (1,650°F) and atmospheric pressure.

9.7.2 PROPERTIES AND USES

Nitric acid has many uses (Table 9.1), but the primary use of nitric acid is for the production of ammonium nitrate (NH_4NO_3) for fertilizers. A second major use of nitric acid is in the field of explosives. It is also a nitrating agent for aromatic and paraffin derivatives, which are useful intermediates in the dye and explosive industries. It is also used in steel refining and in uranium extraction.

In organic chemistry, nitric acid is the primary reagent used for nitration—the addition of a nitro (–NO_2) group to an organic molecule. While some resulting nitro-compounds are sensitive to shock and thermal effects, such as trinitrotoluene (TNT), some are sufficiently stable enough to be used in munitions and demolition, while others are still more stable and used as pigments in inks and dyes. Nitric acid is also commonly used as a strong oxidizing agent.

The corrosive effects of nitric acid are exploited for a number of specialty applications, such as etching in printmaking, pickling stainless steel, or cleaning silicon wafers in electronics. A solution of nitric acid, water and alcohol is used for etching of metals to reveal the microstructure. Commercially available aqueous blends of 5%–30% v/v nitric acid and 15%–40% v/v phosphoric

TABLE 9.1
Uses of Nitric Acid

Field	Use
Aerospace engineering	Used as an oxidizer in liquid-fueled rockets.
Explosives industry	Manufacturing explosives such as trinitrotoluene, nitroglycerin.
Fertilizer	Used for manufacturing fertilizers such as ammonium nitrate.
Metals	Used for purification of various precious metals.
Metallurgy	Used in combination with alcohol for etching.
Woodworking	Used to artificially age pine and maple wood.
Aqueous blends	Used for cleaning food and dairy equipment.
Drugs	Used in a colorimetric test to determine the difference between heroin and morphine.

acid are commonly used for cleaning food and dairy equipment primarily to remove precipitated calcium and magnesium compounds (either deposited from the process stream or resulting from the use of hard water during production and cleaning).

In a low concentration, nitric acid is often used in woodworking to artificially age pine and maple. The color produced is a gray–gold, very much like very old wax or oil-finished wood.

9.8 SULFUR

Sulfur (also called brimstone) is one of the few elements found pure in nature. And this native sulfur is probably of volcanic origin, and (after oxidation to sulfur dioxide (SO_2) or sulfur trioxide (SO_3)) is responsible for the characteristic smell of many volcanoes. Sulfur is a reactive, nonmetallic element naturally found in nature in a free or combined state. Large deposits of elemental sulfur are found in various parts of the world. In its combined form, sulfur is naturally present in sulfide ores of metals such as iron, zinc, copper, and lead. It is also a constituent of natural gas and refinery gas streams in the form of hydrogen sulfide (H_2S), carbonyl sulfide (COS), and mercaptan derivatives (RSH, where R is an alkyl group). Different processes have been developed for obtaining sulfur and sulfuric acid from these three sources.

9.8.1 PRODUCTION

The sulfide derivatives form the principal ores of copper, zinc, nickel, cobalt, molybdenum, as well as several other metals and these materials tend to be dark-colored semiconductors that are not readily attacked by water or even many acids. Processing these ores, usually by roasting is environmentally hazardous. Also, sulfur corrodes many metals through tarnishing.

Sulfur is now produced as a side product of other industrial processes such as in oil refining, in which sulfur is undesired. As a mineral, native sulfur under salt domes is thought to be a fossil mineral resource, produced by the action of ancient bacteria on sulfate deposits.

Sulfur is produced from natural gas, crude oil, and related fossil fuel resources, from which it is obtained mainly as hydrogen sulfide. Organosulfur compounds which are undesirable impurities in such resources may be upgraded by subjecting to hydrodesulfurization in which high-pressure hydrogen and high temperatures (hydrotreating or hydrocracking) are used to cleave the carbon-sulfur bonds thereby converting the sulfur to hydrogen sulfide.

Hydrotreating (variously referred to as *hydroprocessing* to avoid any confusion with those processes that are referred to as *hydrotreating processes*) is a refining process in which the feedstock is treated with hydrogen at temperature and under pressure in which hydrocracking (thermal decomposition in the presence of hydrogen) is minimized. The usual goal of hydrotreating is to hydrogenate olefins and remove heteroatoms, such as sulfur, and to saturate aromatic compounds and olefins (Parkash, 2003; Ancheyta and Speight, 2007; Gary et al., 2007; Speight, 2014a; Hsu and Robinson, 2017; Speight, 2017). On the other hand, *hydrocracking* is a process in which thermal decomposition is extensive and the hydrogen assists in the removal of the heteroatoms as well as mitigating the coke formation that usually accompanies thermal cracking of high molecular weight polar constituents.

In the process (hydrocracking or hydrotreating), sulfur in the feedstock is removed under the thermal conditions in the presence of hydrogen, represented simply as:

$$S_{feedstock} + H_2 \rightarrow H_2S$$

The hydrogen sulfide product is isolate in the gas cleaning section of the refinery and converted to sulfur by oxidation:

$$H_2S + O_2 \rightarrow H_2O + S$$

Hydrogen sulfide is a constituent of natural gas and also of the majority of refinery gas streams, especially those off-gases from hydrodesulfurization processes. A large majority of the sulfur is converted to sulfuric acid for the manufacturer of fertilizers and other chemicals. Other uses for sulfur include the production of carbon disulfide, refined sulfur, and pulp and paper industry chemicals.

The Frasch process, developed in 1894, produces sulfur from underground deposits. Smelting iron ores produces large amounts of sulfur dioxide, which is catalytically oxidized to sulfur trioxide for sulfuric acid production. In the process, superheated water was pumped into a native sulfur deposit to melt the sulfur, and then compressed air returned the 99.5% pure melted product to the surface. Throughout the 20th century this procedure produced elemental sulfur that required no further purification. However, due to a limited number of such sulfur deposits and the high cost of working them, this process for mining sulfur has not been employed in a major way anywhere in the world since 2002.

Currently, sulfur is mainly produced by the partial oxidation of hydrogen sulfide through the Claus process. The major sources of hydrogen sulfide are natural gas and petroleum refinery streams treatment operations. It has been estimated that 90%–95% of the world's recovered sulfur is produced through the Claus process. Typical sulfur recovery ranges from 90% for a lean acid gas feed to 97% for a rich acid gas feed.

This process includes two main sections: the burner section with a reaction chamber that does not have a catalyst and a Claus reactor section. In the burner section, part of the feed containing hydrogen sulfide and some hydrocarbons is burned with a limited amount of air. The two main reactions that occur in this section are the complete oxidation of part of the hydrogen sulfide (feed) to sulfur dioxide and water and the partial oxidation of another part of the hydrogen sulfide to sulfur. The two reactions are exothermic:

$$2H_2S + 3O_2 \rightarrow 2SO_2 + 2H_2O$$

$$2H_2S + O_2 \rightarrow 2S + 2H_2O$$

In the second section, unconverted hydrogen sulfide reacts with the produced sulfur dioxide over a bauxite catalyst in the Claus reactor. Normally more than one reactor is available. In the Super-Claus process, three reactors are used and the final reactor contains a selective oxidation catalyst of high efficiency. After each reaction stage, sulfur is removed by condensation so that it does not collect on the catalyst. The temperature in the catalytic converter should be kept over the dew point of sulfur to prevent condensation on the catalyst surface, which reduces activity.

Due to the presence of hydrocarbons in the gas feed to the burner section, some undesirable reactions occur, such as the formation of carbon disulfide (CS_2) and carbonyl sulfide (COS). A good catalyst has a high activity toward H_2S conversion to sulfur and a reconversion of carbonyl sulfide and carbon disulfide to sulfur and carbon oxides (CO+CO$_2$). Mercaptans in the acid gas feed results in an increase in the air demand. The oxidation of mercaptans could be represented as:

$$2CH_3SH + 3O_2 \rightarrow SO_2 + CO_2 + C_2H_5SH + 2H_2O$$

Sulfur dioxide is then reduced in the Claus reactor to elemental sulfur.

9.8.2 PROPERTIES AND USES

Elemental sulfur is nontoxic, as are most of the soluble sulfate (SO$_4$) salts—the soluble sulfate salts are poorly absorbed and laxative. When injected parenterally, they are freely filtered by the kidneys and eliminated with very little toxicity in multi-gram amounts.

Sulfur reacts directly with methane to give carbon disulfide (CS$_2$) which is used to manufacture cellophane and rayon. One of the uses of elemental sulfur is in vulcanization of rubber, where poly-sulfide chains cross-link organic polymers. Large quantities of sulfite derivatives are used to bleach paper and to preserve dried fruit. Many surfactants and detergents are sulfate derivatives. Calcium sulfate (gypsum, CaSO$_4$ 2H$_2$O) is used in Portland cement and in fertilizers. The most important form of sulfur for fertilizer is the mineral calcium sulfate.

Elemental sulfur is hydrophobic (insoluble in water) and cannot be used directly by plants. Over time, soil bacteria can convert it to soluble derivatives, which can then be used by plants. Sulfur improves the efficiency of other essential plant nutrients, particularly nitrogen and phosphorus. Biologically produced sulfur particles are naturally hydrophilic due to a biopolymer coating and are easier to disperse over the land in a spray of diluted slurry, resulting in a faster uptake.

Organosulfur compounds are used in pharmaceutical products, dyestuffs, and agrochemicals. Many drugs contain sulfur, early examples being antibacterial sulfonamide drugs (*sulfa drugs*) and most β-lactam antibiotics, including the various penicillin derivatives, cephalosporins, and monobactams contain sulfur. Mercaptan derivatives (also called thiols—the function group is the—SH group—and informally represented as R–SH) are a family of organosulfur compounds. Some are added to natural gas supplies because of their distinctive smell, so that gas leaks can be detected easily. Others are used in silver polish, and in the production of pesticides and herbicides.

Elemental sulfur is one of the oldest fungicides and pesticides. The product known as *dusting sulfur*, which is elemental sulfur in powdered form, is a common fungicide for grapes, strawberry, many vegetables, and several other crops. It has a strong effect against a wide range of powdery mildew diseases as well as black spot. In organic production, sulfur is the most important fungicide. Standard-formulation dusting sulfur is applied to crops with a sulfur duster or from a dusting plane. Wettable sulfur is the commercial name for dusting sulfur formulated with additional ingredients to make it miscible with water—it has similar applications and is used as a fungicide to control mildew and other mold-related problems with plants and soil.

A diluted solution of lime sulfur, which is produced by combining calcium hydroxide (Ca(OH)$_2$) with elemental sulfur in water, is used as a dip for animals to destroy ringworm fungus, mange, and other skin infections and parasite. Sulfur dioxide and various sulfites have been used for their antioxidant antibacterial preservative properties in many other parts of the food industry. The practice has declined since reports of an allergy-like reaction of some persons to sulfites in foods. Precipitated sulfur and colloidal sulfur are used, in form of lotions, creams, powders, soaps, and bath additives, for the treatment of some forms of acne and dermatitis. Magnesium sulfate (also known as epsom salts when in hydrated crystal form) can be used as a laxative, a bath additive, an exfoliant, a magnesium supplement for plants, or (when in dehydrated form) as a desiccant.

Several sulfur halides are important to modern industry. For example, sulfur hexafluoride (SF$_6$) is a dense gas that is used as an insulator in high-voltage transformers. Sulfur hexafluoride is also a nonreactive and nontoxic propellant for pressurized containers. Sulfur dichloride (SCl_2) and disulfur dichloride (S_2Cl_2) are important industrial chemicals.

Sulfur reacts with nitrogen to form polymeric sulfur nitrides (SN_x) or polythiazyl derivatives. These polymers were found to have the optical and electrical properties of metals. An important sulfur-nitrogen compound is tetrasulfur tetranitride (S_4N_4) which exists in a cage-like form and, when heated, yields polymeric sulfur nitride ($SH)_n$, which has metallic properties. Thiocyanate derivatives contain the SCN^- group and oxidation of thiocyanate gives thiocyanogen, (SCN_2, NCS-SCN). Phosphorus sulfides are also commercially important especially those with the cage structures P_4S_{10} and P_4S_3.

Other uses range from dusting powder for roses to rubber vulcanization to sulfur asphalt pavements. Flower sulfur is used in match production and in certain pharmaceuticals. Sulfur is also an additive in high-pressure lubricants. Sulfur can replace 30%–50% w/w of the asphalt in the blends used for road construction. Road surfaces made from asphalt-sulfur blends have nearly double the strength of conventional pavement, and it has been claimed that such roads are more resistant to climatic conditions. The impregnation of concrete with molten sulfur is another potential large sulfur use. Concretes impregnated with sulfur have better tensile strength and corrosion resistance than conventional concretes. Sulfur is also used to produce phosphorous pentasulfide, a precursor for zinc dithiophosphate derivative used as corrosion inhibitors.

The most important use of sulfur is for sulfuric acid production which is the most important and widely used inorganic chemical and is a widely used industrial chemical. Sulfuric acid is produced by the contact process where sulfur is burned in an air stream to sulfur dioxide, which is catalytically converted to sulfur trioxide. The catalyst of choice is solid vanadium pentoxide (V_2O_5). The reaction occurs at about 450°C (840°F), increasing the rate at the expense of a higher conversion. To increase the yield of sulfur trioxide, more than one conversion stage (normally three stages) is used with cooling between the stages to offset the exothermic reaction heat. Absorption of sulfur trioxide from the gas mixture exiting from the reactor favors the conversion of sulfur dioxide. The absorbers contain sulfuric acid of 98% concentration which dissolves sulfur trioxide. The unreacted sulfur dioxide and oxygen are recycled to the reactor. The absorption reaction is exothermic, and coolers are used to cool the acid:

$$2SO_2 + O_2 \rightarrow SO_3$$

$$SO_3 + H_2O \rightarrow H_2SO_4$$

9.9 SULFURIC ACID

Sulfuric acid (spelled sulphuric acid in many countries and also known as vitriol) is a mineral acid composed of the elements hydrogen, oxygen, and sulfur (H_2SO_4). It is a colorless, odorless, and syrupy liquid that is soluble in water in a reaction that is highly exothermic. The corrosiveness of sulfuric acid can be mainly ascribed to the strong acidic nature of the compound, and, if at a high concentration, it has strong dehydrating and oxidizing properties. Sulfuric acid is also a hygroscopic chemical insofar as it readily absorbs water vapor when in contact with the air. Sulfuric acid can cause severe burns to the skin and requires cautious handling even at moderate-to-low concentration.

9.9.1 PRODUCTION

The *contact process* can be divided into five separate stages: (i) combining sulfur and oxygen to form sulfur dioxide, (ii) purifying the sulfur dioxide in a purification unit, (iii) adding an excess of oxygen to sulfur dioxide in the presence of the catalyst, (iv) the sulfur trioxide is added to sulfuric acid to produce oleum, and (v) the oleum is then added to water to form sulfuric acid which is very concentrated.

In the first stage, sulfur dioxide in converted into sulfur trioxide (the reversible reaction at the heart of the process) or ion pyrite is used to produce the sulfur dioxide, which is then converted into concentrated sulfuric acid. Thus:

$$S + O_2 \rightarrow SO_2$$

or:

$$4FeS_2 + 11O_2 \rightarrow 2Fe_2O_3 + 8SO_2$$

In either case, an excess of air is used so that the sulfur dioxide produced is already mixed with oxygen for the next stage.

Conversion of the sulfur dioxide into sulfur trioxide is a reversible reaction, and the formation of the sulfur trioxide is exothermic.

$$2SO_2O_2 \rightarrow 2SO_3$$

Conversion of the sulfur trioxide into sulfuric acid cannot be achieved by the simple process of adding water to the sulfur trioxide because of the highly exothermic (and uncontrollable) nature of the reaction. Thus, sulfur trioxide is first dissolved in concentrated sulfuric acid to produce fuming sulfuric acid (oleum), which can then be reacted (relatively) safely with water to produce concentrated sulfuric acid twice as much as you originally used to produce the fuming sulfuric acid.

$$H_2SO_4 + SO_3 \rightarrow H_2SO_4SO_3\,(H_2S_2O_7)$$

$$H_2SO_4SO_3 + H_2O \rightarrow 2H_2SO_4$$

The mixture of sulfur dioxide and oxygen going into the reactor is in equal proportions by volume but is, in reality, an excess of oxygen relative to the proportions demanded by the equation.

$$2SO_2 + O_2 \rightarrow 2SO_3$$

Increasing the concentration of oxygen in the mixture causes the position of equilibrium to shift towards the right. Since the oxygen comes from the air, this is a very cheap way of increasing the conversion of sulfur dioxide into sulfur trioxide. In order to produce maximum yields of sulfur trioxide a relatively low temperature is required to drive the equilibrium to the right. However, the lower the temperature, the slower the reaction. A temperature in the order of 400°C–450°C (750°F–840°F) is a compromise temperature producing a high proportion of sulfur trioxide in the equilibrium mixture. With an increase in the pressure the system will also help to increase the rate of the reaction. However, the reaction pressure is maintained at pressures close to atmospheric pressure (at 15–30 psi) at which there is a 99.5% v/v conversion of sulfur dioxide into sulfur trioxide. Adding a catalyst does not produce any greater percentage of sulfur trioxide in the equilibrium mixture but, in the absence of a catalyst the reaction rate is so slow that virtually no reaction happens in any sensible time.

The catalyst ensures that the reaction has a sufficiently high rate for a dynamic equilibrium to be set up within the very short time that the gases are actually in the reactor. Platinum used to be the catalyst for this reaction; however, as it is susceptible to reacting with arsenic impurities in the sulfur feedstock, vanadium pentoxide (V_2O_5) is now the preferred catalyst and catalyst regeneration is achieved by oxidation of the vanadium (V^{4+}) to the higher valency (V^{5+}). Thus:

$$V^{4+} + O_2 \rightarrow V^{5+} + 2O^{2-}$$

The *wet sulfuric acid process* is one of the main gas desulfurization processes and is recognized as an efficient process for recovering sulfur from various process gases in the form of sulfuric acid. In the current context of refinery operations, the process is applied in all industries where removal of sulfur is an issue. Examples are the processing of hydrogen sulfide gas from an amine gas treating unit, (i) off-gas from sour water stripper gas, (ii) spent acid from an alkylation unit, (iii) Claus unit tail gas, and (iv) off-gas from a resid-fired or coke-fired boiler. The acid gas coming from any of these operations contains hydrogen sulfide (H_2S), carbonyl sulfide (COS), and hydrocarbon derivatives in addition to carbon dioxide (CO_2). These gases were previously often flared and vented to the atmosphere, but now the acid gas requires purification in order not to affect the environment with sulfur dioxide emissions. Not only can the process meet environmental demands of sulfur dioxide removal, the process also accepts a wide range of feed-gas compositions.

The wet sulfuric acid process plant provides a high sulfur recovery and the process chemistry is reflected in the following reactions (Figure 9.2):

Combustion:

$$2H_2S + 3O_2 \rightleftharpoons 2H_2O + 2SO_2$$

Oxidation:

$$2SO_2 + O_2 \rightleftharpoons 2SO_3$$

Hydration:

$$SO_3 + H_2O \rightleftharpoons H_2SO_4(g)$$

Condensation:

$$H_2SO_4(g) \rightleftharpoons H_2SO_4(l)$$

The process can also be used for production of sulfuric acid from sulfur burning or for regeneration of the spent acid from, for example, alkylation units. Wet catalysis processes differ from other contact sulfuric acid processes in that the feed gas contains excess moisture when it comes into contact with the catalyst. The sulfur trioxide formed by catalytic oxidation of the sulfur dioxide reacts instantly with the moisture to produce sulfuric acid in the vapor phase to an extent determined by the temperature. Liquid acid is subsequently formed by condensation of the sulfuric acid vapor and not by absorption of the sulfur trioxide in concentrated sulfuric acid, as is the case in the contact process that is based on dry gases.

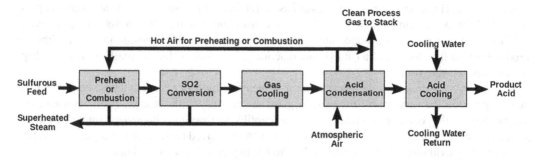

FIGURE 9.2 Flow scheme for the wet sulfuric acid process.

The concentration of the product acid depends on the water/sulfur trioxide (H_2O/SO_3) ratio in the catalytically converted gases and on the condensation temperature. The combustion gases are cooled to the converter inlet temperature of about 420°C–440°C (790°C–825°F). To process these wet gases in a conventional cold gas, contact process would necessitate cooling and drying of the gas to remove all moisture.

The *lead chamber process* was an industrial method used to produce sulfuric acid in large quantities. Prior to 1900, most sulfuric acid was manufactured by the lead chamber process and, as late as 1940, up to half of the sulfuric acid manufactured in the United States was produced by chamber process plants.

In the lead chamber process, sulfur dioxide and steam was introduced with nitrogen dioxide into large chambers lined with sheet lead where the gases are sprayed down with water and chamber acid (62%–70% v/v sulfuric acid) was produced. The nitrogen dioxide was necessary for the reaction to proceed at a reasonable rate. As might be anticipated, the process is highly exothermic, and a major consideration of the design of the chambers was to provide a way to dissipate the heat formed in the reactions. This chamber process has been largely supplanted by the contact process.

Another method for the production of sulfuric acid is the less well-known *metabisulfite method*, in which hydrochloric acid is added to metabisulfite and the gas was bubbled through nitric acid:

$$SO_2 + HNO_3 + H_2O \rightarrow H_2SO_4 + NO$$

9.9.2 PROPERTIES AND USES

The most common use of sulfuric acid is for fertilizer manufacture—other uses include fertilizer manufacture (and other mineral processing), crude oil refining, wastewater processing, and chemical synthesis.

Because the hydration reaction of sulfuric acid is highly exothermic, dilution should always be performed by adding the acid to the water rather than the water to the acid—as anamonic used to alphabetical order: *a-to-w* and not *w-to-a* on a chemical basis, because the reaction is in an equilibrium that favors the rapid protonation of water, addition of acid to the water ensures that the *acid* is the limiting reagent. As a result of the strong affinity of sulfuric acid for water, the acid is an excellent dehydrating agent. In addition, concentrated sulfuric acid has a very powerful dehydrating property and is capable of removing the elements of water from chemical compounds such as carbohydrates to produce carbon and steam.

As an acid, sulfuric acid reacts with most bases to produce the corresponding sulfate. For example:

$$CuO(s) + H_2SO_4(aq) \rightarrow CuSO_4(aq) + H_2O(l)$$

Sulfuric acid can also be used to displace weaker acids from their salts. As an example, the reaction of sulfuric acid with sodium acetate displaces acetic acid (the weaker acid) with the formation of sodium bisulfate:

$$H_2SO_4 + CH_3COONa \rightarrow NaHSO_4 + CH_3COOH$$

Similarly, reacting sulfuric acid with potassium nitrate (KNO_3) can be used to produce nitric acid (HNO_3) and a precipitate of potassium bisulfate. When combined with nitric acid, sulfuric acid acts both as an acid and dehydrating agent, forming the nitronium ion $\left(NO_2^+\right)$ which is important in nitration reactions involving electrophilic aromatic substitution. This type of reaction, where protonation occurs on an oxygen atom, is an important reaction in organic chemistry reactions such as, for example, the Fischer esterification and dehydration of alcohols.

Dilute sulfuric acid reacts with metals via a single displacement reaction to produce hydrogen gas and metal sulfate salt. Thus:

$$Fe(s) + H_2SO_4(aq) \rightarrow H_2(g) + FeSO_4(in\ solution)$$

However, concentrated sulfuric acid is a strong oxidizing agent, and does not react with metals in the same way as other typical acids. Sulfur dioxide, water, and SO_4^{2-} ions are evolved instead of the hydrogen and the formation of the salt:

$$2H_2SO_4 + 2e^- \rightarrow SO_2 + 2H_2O + SO_4^{2-}$$

Sulfuric acid can oxidize non-active metals (such as tin and copper) in a reaction that is temperature dependent:

$$Cu + 2H_2SO_4 \rightarrow SO_2 + 2H_2O + SO_4^{2-} + Cu^{2+}$$

Hot concentrated sulfuric acid oxidizes nonmetals such as carbon (as bituminous coal) and sulfur:

$$C + 2H_2SO_4 \rightarrow CO_2 + 2SO_2 + 2H_2O$$

$$S + 2H_2SO_4 \rightarrow 3SO_2 + 2H_2O$$

Sulfuric acid with sodium chloride to produce hydrogen chloride gas and sodium bisulfate:

$$NaCl + H_2SO_4 \rightarrow NaHSO_4 + HCl$$

Benzene undergoes electrophilic aromatic substitution with sulfuric acid to give the corresponding sulfonic acid.

Finally, sulfuric acid is used in large quantities by the iron-making and steel-making industries to remove oxidation, rust, and scaling from rolled sheet and billets prior to sale to the automobile and major appliance industries. Used acid is often recycled using a spent acid regeneration (SAR) plant in which the spent acid is combusted with natural gas, refinery gas, fuel oil, or other fuel sources. This combustion process produces gaseous sulfur dioxide and sulfur trioxide which are then used to manufacture fresh sulfuric acid. The spent acid regeneration plants are common additions to crude oil refineries, metal smelting plants, and other industries where sulfuric acid is consumed in bulk.

As another use for sulfuric acid, hydrogen peroxide (H_2O_2) can be added to sulfuric acid to produce *piranha solution*, which is a powerful but very toxic cleaning solution with which substrate surfaces can be cleaned. Piranha solution is typically used in the microelectronics industry, and also in laboratory settings to clean glassware.

9.10   SYNTHESIS GAS

Synthesis gas (also called *syngas*) is a mixture of carbon monoxide (CO) and hydrogen (H_2) that is the beginning of a wide range of chemicals (Chapter 10; Chadeesingh, 2011; Speight, 2013, 2014a,b) (Figure 9.3). The name comes from the use of the gas intermediate in creating synthetic natural gas (SNG) and for producing ammonia or methanol. Synthesis gas is a product the gasification of carbonaceous feedstocks (Chapter 5). Thus, synthesis gas can be produced from many sources, including natural gas, coal, biomass, or virtually any hydrocarbon feedstock, by reaction with steam (steam reforming), carbon dioxide (dry reforming), or oxygen (partial oxidation).

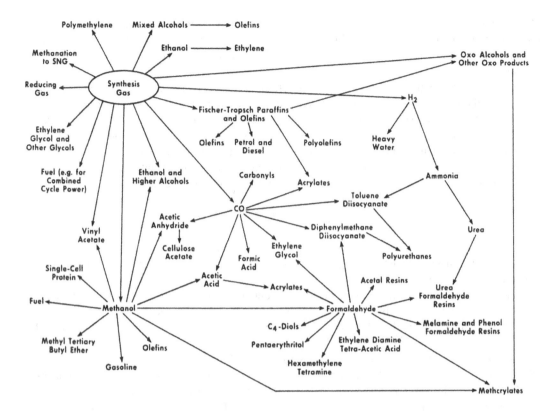

FIGURE 9.3 Production of chemicals from synthesis gas.

9.10.1 PRODUCTION

The production of synthesis gas, i.e., mixtures of carbon monoxide and hydrogen has been known for several centuries. But it is only with the commercialization of the Fischer–Tropsch reaction that the importance of synthesis gas has been realized. The thermal cracking (pyrolysis) of petroleum or fractions thereof was an important method for producing gas in the years following its use for increasing the heat content of water gas. Many water-gas set operations converted into oil-gasification units; some have been used for base-load city gas supply but most find use for peak-load situations in the winter.

In addition to the gases obtained by distillation of crude petroleum, further gaseous products are produced during the processing of naphtha and middle distillate to produce gasoline. Hydrodesulfurization processes involving treatment of naphtha, distillates and residual fuels, and from the coking or similar thermal treatment of vacuum gas oils and residual fuel oils also produce gaseous products.

The chemistry of the oil-to-gas conversion has been established for several decades and can be described in general terms although the primary and secondary reactions can be truly complex. The composition of the gases produced from a wide variety of feedstocks depends not only on the severity of cracking but often to an equal or lesser extent on the feedstock type. In general terms, gas heating values are in the order of 950–1,350 Btu/ft^3 (30–50 MJ/m^3). A second group of refining operations which contribute to gas production are the catalytic cracking processes, such as fluid-bed catalytic cracking, and other variants, in which heavy gas oils are converted into gas, naphtha, fuel oil, and coke.

The catalysts will promote steam-reforming reactions that lead to a product gas containing more hydrogen and carbon monoxide and fewer unsaturated hydrocarbon products than the gas product from a non-catalytic process. The resulting gas is more suitable for use as a medium heat value gas than the rich gas produced by straight thermal cracking. The catalyst also influences the reaction rates in the thermal cracking reactions, which can lead to higher gas yields and lower tar and carbon yields.

Almost all petroleum fractions can be converted into gaseous fuels, although conversion processes for the heavier fractions require more elaborate technology to achieve the necessary purity and uniformity of the manufactured gas stream. In addition, the thermal yield from the gasification of heavier feedstocks is invariably lower than that of gasifying light naphtha or liquefied petroleum gas since, in addition to the production of synthesis gas components (hydrogen and carbon monoxide) and various gaseous hydrocarbons, heavy feedstocks also yield some tar and coke.

Synthesis gas can be produced from heavy oil by partially oxidizing the oil:

$$[2CH]_{petroleum} + O_2 \rightarrow 2CO + H_2$$

The initial partial oxidation step consists of the reaction of the feedstock with a quantity of oxygen insufficient to burn it completely, making a mixture consisting of carbon monoxide, carbon dioxide, hydrogen, and steam.

Success in partially oxidizing heavy feedstocks depends mainly on details of the burner design. The ratio of hydrogen to carbon monoxide in the product gas is a function of reaction temperature and stoichiometry and can be adjusted, if desired, by varying the ratio of carrier steam to oil fed to the unit.

The chemical composition of synthesis gas varies based on the raw materials and the processes. Synthesis gas produced by coal gasification generally is a mixture of 30%–60% v/v carbon monoxide, 25%–30% v/v hydrogen, 5%–15% v/v carbon dioxide, and 0%–5% v/v methane.

Conversion of biomass to syngas is typically low-yield. The University of Minnesota developed a metal catalyst that reduces the biomass reaction time by up to a factor of 100. The catalyst can be operated at atmospheric pressure and reduces char. The entire process is autothermic and therefore heating is not required.

9.10.2 Properties and Uses

Synthesis gas is a crucial intermediate resource for production of hydrogen, ammonia, methanol, and synthetic hydrocarbon fuels as well as a host other uses (Table 9.2). Syngas is also used as an intermediate in producing synthetic hydrocarbon liquids for use as a fuels and lubricants by the Fischer–Tropsch process (Chapter 10).

TABLE 9.2
Uses of Synthesis Gas

- Steam for use in turbine drivers for electricity generation.
- Nitrogen for use as pressurizing agents and fertilizers.
- Hydrogen for electricity generation and use in refineries.
- Ammonia for use as fertilizers.
- Ammonia for the production of plastics like polyurethane and nylon.
- Methanol for the production of plastics, resins, pharmaceuticals, adhesives, and paints.
- Methanol as a component of fuels.
- Carbon monoxide for use in chemical industry feedstock and fuels.
- Sulfur for use as elemental sulfur for chemical industry.
- Minerals and solids for use as slag for roadbeds.

When used as an intermediate in the large scale, industrial synthesis of hydrogen (principally used in the production of ammonia) is also produced from natural gas by the steam reforming reaction:

$$CH_4 + H_2O \rightarrow CO + 3H_2$$

In order to produce more hydrogen from this mixture, more steam is added and the water-gas shift reaction is necessary:

$$CO + H_2O \rightarrow CO_2 + H_2$$

The hydrogen must be separated from the carbon dioxide before use, which can be accomplished by pressure swing adsorption, amine scrubbing, and membrane reactors (Mokhatab et al., 2006; Speight, 2007, 2014a).

REFERENCES

Ancheyta, J., and Speight, J.G. 2007. *Hydroprocessing of Heavy Oils and Residua*. CRC–Taylor and Francis Group, Boca Raton, FL.

Chadeesingh, R. 2011. Chapter 5: The Fischer–Tropsch process. In: *The Biofuels Handbook*, Part 3. J.G. Speight (Editor). The Royal Society of Chemistry, London.

Chemier, P.J. 1992. *Survey of Chemical Industry*. 2nd Revised Edition. VCH Publishers Inc., New York.

Farhat Ali, M., El Ali, B.M., and Speight, J.G. 2005. *Handbook of Industrial Chemistry: Organic Chemicals*. McGraw-Hill, New York.

Gary, J.G., Handwerk, G.E., and Kaiser, M.J. 2007. *Petroleum Refining: Technology and Economics*. 5th Edition. CRC Press, Taylor & Francis Group, Boca Raton, FL.

Goldstein, R.F. 1949. *The Petrochemical Industry*. E. & F.N. Spon, London.

Hahn, A.V. 1970. *The Petrochemical Industry: Market and Economics*. McGraw-Hill, New York.

Hsu, C.S., and Robinson, P.R. (Editors). 2017. *Handbook of Petroleum Technology*. Springer International Publishing AG, Cham.

Lowenheim, F.A., and Moran, M.K. 1975. *Industrial Chemicals*. John Wiley & Sons, New York.

Mokhatab, S., Poe, W.A., and Speight, J.G. 2006. *Handbook of Natural Gas Transmission and Processing*. Elsevier, Amsterdam.

Parkash, S. 2003. *Refining Processes Handbook*. Gulf Professional Publishing, Elsevier, Amsterdam.

Speight, J.G. 2002. *Chemical Process and Design Handbook*. McGraw-Hill Publishers, New York.

Speight, J.G. 2007. *Natural Gas: A Basic Handbook*. GPC Books, Gulf Publishing Company, Houston, TX.

Speight, J.G. 2011a. *Handbook of Industrial Hydrocarbon Processes*. Gulf Professional Publishing, Elsevier, Oxford, United Kingdom.

Speight, J.G. (Editor). 2011b. *The Biofuels Handbook*. The Royal Society of Chemistry, London.

Speight, J.G. 2013. *The Chemistry and Technology of Coal*. 3rd Edition. CRC Press, Taylor and Francis Group, Boca Raton, FL.

Speight, J.G. 2014a. *The Chemistry and Technology of Petroleum*. 4th Edition. CRC Press, Taylor and Francis Group, Boca Raton, FL.

Speight, J.G. 2014b. *Gasification of Unconventional Feedstocks*. Gulf Professional Publishing, Elsevier, Oxford, United Kingdom.

Speight, J.G. 2017. *Handbook of Petroleum Refining*. CRC Press, Taylor & Francis Group, Boca Raton, FL.

Wittcoff, H.A., and Reuben, B.G. 1996. *Industrial Organic Chemicals*. John Wiley & Sons Inc., New York.

10 Chemicals from the Fischer–Tropsch Process

10.1 INTRODUCTION

In a previous chapter (Chapter 5), there has been mention of the use of the gasification process to convert carbonaceous feedstocks, such as crude oil residua, tar sand bitumen, coal, oil shale, and biomass, into the starting chemicals for the production of petrochemicals. The chemistry of the gasification process is based on the thermal decomposition of the feedstock and the reaction of the feedstock carbon and other pyrolysis products with oxygen, water, and fuel gases such as methane and is represented by a sequence of simple chemical reactions (Table 10.1). However, the gasification process is often considered to involve two distinct chemical stages: (i) devolatilization of the feedstock to produce volatile matter and char, (ii) followed by char gasification, which is complex and specific to the conditions of the reaction—both processes contribute to the complex kinetics of the gasification process.

The Fischer–Tropsch process is a catalytic chemical reaction in which carbon monoxide (CO) and hydrogen (H_2) in the synthesis are converted into hydrocarbon derivatives of various molecular weights. The process can be represented by the simple equation:

$$(2n+1)H_2 + nCO \rightarrow C_nH_{(2n+2)} + nH_2O$$

In this equation, n is an integer. Thus, for $n = 1$, the reaction represents the formation of methane, which in most gas-to-liquids (GTL) applications is considered an undesirable byproduct.

The Fischer–Tropsch process conditions are usually chosen to maximize the formation of higher molecular weight hydrocarbon liquid fuels which are higher value products. There are other side reactions taking place in the process, among which the water-gas shift reaction (WGS) is predominant:

$$CO + H_2O \rightarrow H_2 + CO_2$$

TABLE 10.1

Reactions that Occur During Gasification of a Carbonaceous Feedstock

$2C + O_2 \rightarrow 2CO$

$C + O_2 \rightarrow CO_2$

$C + CO_2 \rightarrow 2CO$

$CO + H_2O \rightarrow CO_2 + H_2$ (shift reaction)

$C + H_2O \rightarrow CO + H_2$ (water-gas reaction)

$C + 2H_2 \rightarrow CH_4$

$2H_2 + O_2 \rightarrow 2H_2O$

$CO + 2H_2 \rightarrow CH_3OH$

$CO + 3H_2 \rightarrow CH_4 + H_2O$ (methanation reaction)

$CO_2 + 4H_2 \rightarrow CH_4 + 2H_2O$

$C + 2H_2O \rightarrow 2H_2 + CO_2$

$2C + H_2 \rightarrow -C_2H_2$

$CH_4 + 2H_2O \rightarrow CO_2 + 4H_2$

Depending on the catalyst, temperature, and type of process employed, hydrocarbon derivatives ranging from methane to higher molecular paraffin derivatives and olefin derivatives can be obtained. Small amounts of low molecular weight oxygenated derivatives (such as alcohol derivatives and organic acid derivatives) are also formed. Typically, Fischer–Tropsch liquids are (unless the process is designed for the production of other products) hydrocarbon products (which vary from naphtha-type liquids to wax) and non-hydrocarbon products. The production of non-hydrocarbon products requires adjustment of the feedstock composition and the process parameters.

Briefly, synthesis gas is the name given to a gas mixture that contains varying amounts of carbon monoxide (CO) and hydrogen (H_2) generated by the gasification of a carbonaceous material. Examples include steam reforming of natural gas, petroleum residua, coal, and biomass. Synthesis gas is used as an intermediate in producing hydrocarbon derivatives via the Fischer–Tropsch process for use as gaseous and liquids fuels.

The synthesis gas is produced by the gasification conversion of carbonaceous feedstock such as petroleum residua, coal, and biomass and production of hydrocarbon products can be represented simply as:

$$CH_{feedstock} + O_2 \rightarrow CO + H_2$$

$$nCO + nH_2 \rightarrow C_nH_{2n+2}$$

However, before conversion of the carbon monoxide and hydrogen to hydrocarbon products, several reactions are employed to adjust the hydrogen/carbon monoxide ratio. Most important is the water-gas shift reaction in which additional hydrogen is produced (at the expense of carbon monoxide) to satisfy the hydrogen/carbon monoxide ratio necessary for the production of hydrocarbon derivatives:

$$H_2O + CO \rightarrow H_2 + CO_2$$

The boiling range of Fischer–Tropsch typically spans the naphtha and kerogen boiling ranges and is suitable for analysis by application of the standard test methods. With the suitable choice of a catalyst the preference for products boiling in the naphtha range (<200°C, <390°F) or for product boiling in the diesel range (approximately 150°C–300°C, 300°F–570°F) can be realized.

The other product that is worthy of consideration is *bio-oil* (*pyrolysis oil, bio-crude*) is the liquid product produced by the thermal decomposition (destructive distillation) of biomass (Chapter 3) at temperatures in the order of 500°C (930°F). The product is a synthetic crude oil and is of interest as a possible complement (eventually a substitute) to petroleum. The product can vary from a light tarry material to a free-flowing liquid—both require further refining to produce specification grade fuels (Gary et al., 2007; Speight, 2011, 2014a, 2017; Hsu and Robinson, 2017).

Hydrocarbon moieties are predominant in the product but the presence of varying levels of oxygen (depending upon the character of the feedstock) requires testament (using, for example, hydrotreating) during refining. On the other hand, the bio-oil can be used as a feedstock to the Fischer–Tropsch process for the production of lower-boiling products, as is the case when naphtha and gas oil are used as feedstocks for the Fischer–Tropsch process. In summary, the Fischer–Tropsch process produces hydrocarbon products of different molecular weight from a gas mixture of carbon monoxide and hydrogen (synthesis gas) all of which can find use in various energy scenarios.

In the current context, the most valuable product is *synthesis gas*—the mixture of carbon monoxide (CO) and hydrogen (H_2) that is the beginning of a wide range of chemicals (Figure 10.1). The production of synthesis gas, i.e., mixtures of carbon monoxide and hydrogen has been known for several centuries. But it is only with the commercialization of the Fischer–Tropsch reaction that the importance of synthesis gas has been realized. The thermal cracking (pyrolysis) of petroleum or fractions thereof was an important method for producing gas in the years following its use for increasing the heat content of water gas.

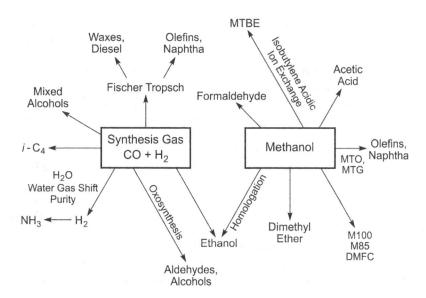

FIGURE 10.1 Routes to chemicals from synthesis gas and methanol.

In addition to the gases obtained by distillation of crude petroleum, further gaseous products are produced during the processing of naphtha and middle distillate to produce gasoline. Hydrodesulfurization processes involving treatment of naphtha, distillates, and residual fuels and from the coking or similar thermal treatment of vacuum gas oils and residual fuel oils also produce gaseous products.

The chemistry of the oil-to-gas conversion has been established for several decades and can be described in general terms although the primary and secondary reactions can be truly complex. The composition of the gases produced from a wide variety of feedstocks depends not only on the severity of cracking but often to an equal or lesser extent on the feedstock type. In general terms, gas heating values are in the order of 950–1,350 Btu/ft³.

A second group of refining operations which contribute to gas production are the catalytic cracking processes, such as fluid bed catalytic cracking, and other variants, in which heavy gas oils are converted into gas, naphtha, fuel oil, and coke (Gary et al., 2007; Speight, 2011, 2014a, 2017; Hsu and Robinson, 2017). The catalysts will promote steam-reforming reactions that lead to a product gas containing more hydrogen and carbon monoxide and fewer unsaturated hydrocarbon products than the gas product from a non-catalytic process. The resulting gas is more suitable for use as a medium heat-value gas than the rich gas produced by straight thermal cracking. The catalyst also influences the reactions rates in the thermal cracking reactions, which can lead to higher gas yields and lower tar and carbon yields.

Almost all petroleum fractions can be converted into gaseous fuels, although conversion processes for the heavier fractions require more elaborate technology to achieve the necessary purity and uniformity of the manufactured gas stream. In addition, the thermal yield from the gasification of heavier feedstocks is invariably lower than that of gasifying light naphtha or liquefied petroleum gas (LPG) since, in addition to the production of synthesis gas components (hydrogen and carbon monoxide) and various gaseous hydrocarbon derivatives, heavy feedstocks also yield some tar and coke.

Synthesis gas can be produced from heavy oil and other heavy crude feedstocks (such as residua) by the process known in the industry as partial oxidation (POX):

$$[2CH]_{crude\ oil} + O_2 \rightarrow 2CO + H_2$$

In this process, the step consists of the reaction of the feedstock with a quantity of oxygen insufficient to burn it completely, making a mixture consisting of carbon monoxide, carbon dioxide, hydrogen, and steam.

Success in partially oxidizing heavy feedstocks depends mainly on details of the burner design. The ratio of hydrogen to carbon monoxide in the product gas is a function of reaction temperature and stoichiometry and can be adjusted, if desired, by varying the ratio of carrier steam to oil fed to the unit. The synthesis of hydrocarbon derivatives from the hydrogenation of carbon monoxide over transition metal catalysts was discovered in 1902 when Sabatier and Sanderens produced methane from hydrogen and carbon monoxide mixtures passed over nickel, iron, and cobalt catalysts. In 1923, Fischer and Tropsch reported the use of alkalized iron catalysts to produce liquid hydrocarbon derivatives rich in oxygenated compounds.

The Fischer–Tropsch process (*Fischer–Tropsch synthesis*) is a series of catalyzed chemical reactions that convert a mixture of carbon monoxide and hydrogen and into hydrocarbon derivatives. The process is a key component of gas-to-liquids technology that produces liquid and solid hydrocarbon derivatives from coal, natural gas, biomass, or other carbonaceous feedstocks. Typical catalysts used are based on iron and cobalt and the hydrocarbon derivatives synthesized in the process are primarily liquid alkanes along with byproducts such as olefin derivatives, alcohols, and solid paraffin derivatives (waxes).

10.2 HISTORY AND DEVELOPMENT OF THE FISCHER–TROPSCH PROCESS

As originally conceived, the function of the Fischer–Tropsch process was to produce liquid transportation hydrocarbon fuels and various other chemical products (Schulz, 1999). Since the original conception, many refinements and adjustments to the technology have been made, including catalyst development and reactor design. Depending on the source of the synthesis gas, the technology is often referred to as coal-to-liquids (CTL) and/or gas-to-liquids.

In the simplest terms, the Fischer–Tropsch process is a catalytic chemical reaction in which carbon monoxide (CO) and hydrogen (H_2) in the synthesis gas are converted into hydrocarbon derivatives of various molecular weights according to the following equation:

$$(2n+1)H_2 + nCO \rightarrow C_nH_{2n+2} + nH_2O \quad (n \text{ is an integer})$$

For $n=1$, the reaction represents the formation of methane, which in most coal-to-liquids or gas-to-liquids applications is considered an undesirable byproduct. The process conditions are usually chosen to maximize the formation of higher molecular weight hydrocarbon liquid fuels which are higher value products. There are other side reactions taking place in the process, among which the water-gas shift reaction is predominant. Thus:

$$CO + H_2O \rightarrow H_2 + CO_2$$

Depending on the catalyst, temperature, and type of process employed, hydrocarbon derivatives ranging from methane to higher molecular paraffin derivatives and olefin derivatives can be obtained. Small amounts of low molecular weight oxygenates (e.g., alcohol and organic acids) are also formed.

The Fischer–Tropsch technology has found industrial application since 1938 in Germany where a total of nine plants produced synthetic hydrocarbon derivatives. However, the history of the commercial Fischer–Tropsch technology dates back to the early years of the 20th century (Table 10.2). Hence, since the turn of the 21st century as indicated in the above summary of the Fischer–Tropsch history, there has been significantly renewed interest in Fischer–Tropsch Technology. In great part, this renaissance has been due to the exploitation of cheaper remote, or, "stranded" gas, which has the effect of making the economics of Fischer–Tropsch projects increasingly attractive.

TABLE 10.2

History and Evolution of the Fischer–Tropsch Process

1902	Methane formed from mixtures of hydrogen and carbon monoxide over a nickel catalyst (Sabatier and Sanderens)
1923	Fischer and Tropsch report work with cobalt, iron, and rubidium catalysts at pressure to produce hydrocarbon derivatives
1936	The first four Fischer–Tropsch production plants in Germany began operation
1950	A 5,000 b/d plant was commissioned and began operation in Brownsville, Texas
1950–1953	The slurry phase reactor pilot unit developed
1950s	Decline in construction of new Fischer–Tropsch plants due to sudden availability of cheaper petroleum
1955	First Sasol Plant commissioned in South Africa. Iron catalyst was used; two further plants were commissioned in 1980 and 1983.
1970–1980s	Energy crisis initiated renewed interest in Fischer–Tropsch technology as the price of petroleum increased.
1990s	Discovery of "stranded" gas reservoirs; renewed interest in Fischer–Tropsch as a viable *gas-to-liquids* technology
1992	MossGas plant used Sasol Technology and natural gas as the carbon feedstock
1992–1993	Shell used cobalt-based catalyst and natural gas as the feedstock
1993	Sasol slurry phase reactor commissioned using Fe-based catalyst

A Fischer–Tropsch plant incorporates three major process sections: (i) production of synthesis gas which is a mixture of carbon monoxide and hydrogen (*steam reforming*), (ii) conversion of the synthesis gas to aliphatic hydrocarbon derivatives and water (*Fischer–Tropsch synthesis process*), and (iii) hydrocracking the longer-chain, waxy synthetic hydrocarbon derivatives to fuel grade fractions. Of the above three steps, the production of synthesis gas is the most energy intensive as well as expensive.

A major attraction of the use of synthesis gas is the very wide range of potential uses by converting synthesis gas into useful downstream products: (i) the Fischer–Tropsch synthesis of hydrocarbon derivatives, (ii) methanol synthesis, (iii) mixed alcohol synthesis, and (iv) synthesis gas fermentation. By choosing an appropriate catalyst (usually based on iron or cobalt) and appropriate reaction conditions (usually 200°C–350°C, 390°F–650°F, and pressures in the order of 300–600 psi), the process with its associated cracking and separation stages can be optimized to produce high molecular weight wax for low molecular weight olefin derivatives and naphtha for petrochemicals production. The ideal feedstock for the Fischer–Tropsch process is synthesis gas consisting of a mixture of hydrogen and carbon monoxide with a molar ratio of 2:1 (Chadeesingh, 2011).

Methanol synthesis is another attractive conversion route because methanol is one of the top 10 petrochemical commodities insofar as, like synthesis gas, it can also be a source of chemicals (Figure 10.1). Synthesis gas can be converted into methanol over a copper-zinc oxide catalyst at 220°C–300°C (430°F–570°F) and 750–1,500 psi.

$$CO + 2H_2 \rightarrow CH_3OH$$

Methanol can, in turn, be used to make acetic acid, formaldehyde for resins, gasoline additives and petrochemical building blocks such as ethylene and propylene.

Under slightly more severe process conditions (up to 425°C, 800°F, and 4,500 psi), a wider range of mixed alcohols can be produced (Chadeesingh, 2011). The processes use catalysts modified from either Fischer–Tropsch synthesis or methanol synthesis, by addition of alkali metals. Finally, the fermentation route for the conversion of synthesis gas uses biochemical processes and reaction conditions that are close to ambient temperature and pressure to make ethanol or other alcohols. Biochemical processes are addressed below.

There are many options for converting synthesis gas into petrochemical feedstocks. For example, the olefin derivatives conversion chain in which ethylene and propylene are converted into polymers (polyethylene, polypropylene, polyvinyl chloride), glycol derivatives (ethylene glycol, $HOCH_2CH_2OH$, propylene glycol, $CH_3CHOHCH_2OH$) and a range of familiar materials such as acetone (CH_3COCH_3), acetic acid (CH_3CO_2H), gasoline additives, and surfactants. The olefin derivatives can be produced by synthesizing naphtha in a Fischer–Tropsch process and then cracking it in a conventional naphtha cracker to make ethylene and propylene.

Depending on the source of biomass feedstock and the choice of gasifier technology, the raw synthesis gas can contain varying amounts of particulates (e.g., ash or char, which can lead to erosion, plugging, or fouling); alkali metals (which can cause hot corrosion and catalyst poisoning); water-soluble trace components (e.g., halides, ammonia); light oils or tars (e.g., benzene, toluene, xylene, or naphthalene, which can lead to catalyst carbonization and fouling), polyaromatic compounds, sulfur components, phosphorus components as well as methane and carbon dioxide. Many of these can be removed (if required) either using standard chemical industry equipment such as cyclones, filters, electrostatic precipitators, water scrubbers, oil scrubbers, activated carbon and adsorbents, or via cleanup processes such as hydrolysis and various carbon dioxide capture processes (Chapter 4).

Another important factor is the ratio of hydrogen to carbon dioxide in the synthesis gas. Different conversion routes require different ratios, e.g., 1.7:1 and 2.15:1 for producing Fischer–Tropsch naphtha/gasoline and diesel, respectively, or 3:1 for methanol synthesis. Because biomass molecules contain oxygen within their structure, biomass-derived synthesis gas often needs to have the hydrogen to carbon monoxide ratio boosted. One option for achieving this is to react some of the synthesis gas with steam over a catalyst to produce hydrogen and carbon dioxide in the water-gas shift reaction (Chadeesingh, 2011; Speight, 2013a,b) as well as accepting a cost for the removal of carbon dioxide unless there is a byproduct hydrogen source readily available.

The extent to which gas cleanup is required depends on the choice of synthesis gas conversion route. Generally, the level of particulates will need to be reduced considerably for any chemical synthesis process, but the precise extent to which (say) sulfur or halide levels need to be reduced depends on the catalysts that are going to be used. For the methanol synthesis process, for example, the sulfur content of the synthesis gas has to be below 100 ppb v/v. For ammonia synthesis process, there is a similar sulfur constraint, and the carbon dioxide content must be below 10 ppm v/v.

10.3   SYNTHESIS GAS

Synthesis gas, a mixture composed primarily of not only carbon monoxide and hydrogen, but also water, carbon dioxide, nitrogen and methane, has been produced on a commercial scale since the early part of the 20th century. This section provides a general description of the emerging technologies and their potential economic benefits. Recent developments in the technology for synthesis production via membrane reactors are also discussed. During World War II, the Germans obtained synthesis gas by gasifying the carbonaceous feedstock. The mixture was used for producing a liquid hydrocarbon mixture in the gasoline range using Fischer–Tropsch technology. Although this route was abandoned after the war due to the high production cost of these hydrocarbon derivatives, it is currently being used in South Africa, where the carbonaceous feedstock (coal) is relatively inexpensive (SASOL II and SASOL III).

Almost all carbonaceous materials can be converted into gaseous fuels, although conversion processes for the heavier fractions require more elaborate technology to achieve the necessary purity and uniformity of the manufactured gas stream. In addition, the thermal yield from the gasification of heavier feedstocks is invariably lower than that of gasifying light naphtha or liquefied petroleum gas since, in addition to the production of synthesis gas components (hydrogen and carbon monoxide) and various gaseous hydrocarbon derivatives, heavy feedstocks also yield some tar and coke.

Gasification to produce synthesis gas can proceed from just about any organic material, including biomass and plastic waste. The resulting synthesis gas burns cleanly into water vapor and carbon dioxide. Alternatively, synthesis gas may be converted efficiently to methane via the Sabatier reaction, or to a diesel-like synthetic fuel via the Fischer–Tropsch process. Inorganic components of the feedstock, such as metals and minerals, are trapped in an inert and environmentally safe form as char, which may have use as a fertilizer.

In principle, synthesis gas can be produced from any hydrocarbon feedstock. These include natural gas, naphtha, residual oil, petroleum coke, coal, and biomass. The lowest cost routes for synthesis gas production, however, are based on natural gas. The cheapest option is remote or stranded reserves. Current economic considerations dictate that the production of liquid fuels from synthesis gas translates into using natural gas as the hydrocarbon source. Nevertheless, the synthesis gas production operation in a gas-to-liquids plant amounts to greater than half of the capital cost of the plant. The choice of technology for synthesis gas production also depends on the scale of the synthesis operation. Synthesis gas production from solid fuels can require an even greater capital investment with the addition of feedstock handling and more complex synthesis gas purification operations. The greatest impact on improving gas-to-liquids plant economics is to decrease capital costs associated with synthesis gas production and improve thermal efficiency through better heat integration and utilization. Improved thermal efficiency can be obtained by combining the gas-to-liquids plant with a power generation plant to take advantage of the availability of low-pressure steam.

Regardless of the final fuel form, gasification itself and subsequent processing neither emits nor traps greenhouse gasses such as carbon dioxide. Combustion of synthesis gas or derived fuels does of course emit carbon dioxide. However, biomass gasification could play a significant role in a renewable energy economy, because biomass production removes carbon dioxide from the atmosphere. While other biofuel technologies such as biogas and biodiesel are also reputed to be carbon neutral, gasification runs on a wider variety of input materials, can be used to produce a wider variety of output fuels, and is an extremely efficient method of extracting energy from biomass. Biomass gasification is, therefore, one of the most technically and economically convincing energy possibilities for a carbon neutral economy.

Synthesis gas consists primarily of carbon monoxide, carbon dioxide, and hydrogen, and has less than half the energy density of natural gas. Synthesis gas is combustible and often used as a fuel source or as an intermediate for the production of other chemicals. Synthesis gas for use as a fuel is most often produced by gasification of the carbonaceous feedstock or municipal waste mainly by the following paths:

$$C + O_2 \rightarrow CO_2$$

$$CO_2 + C \rightarrow 2CO$$

$$C + H_2O \rightarrow CO + H_2$$

When used as an intermediate in the large scale, industrial synthesis of hydrogen and ammonia, it is also produced from natural gas (via the steam reforming reaction) as follows:

$$CH_4 + H_2O \rightarrow CO + 3H_2$$

The synthesis gas produced in large waste-to-energy gasification facilities is used as fuel to generate electricity.

The manufacture of gas mixtures of carbon monoxide and hydrogen has been an important part of chemical technology for about a century. Originally, such mixtures were obtained by the reaction of steam with incandescent coke and were known as *water gas*. Used first as a fuel, water gas soon attracted attention as a source of hydrogen and carbon monoxide for the production of chemicals, at which time it gradually became known as synthesis gas. Eventually, steam reforming processes, in which steam is reacted with natural gas (methane) or petroleum naphtha over a nickel catalyst, found wide application for the production of synthesis gas.

A modified version of steam reforming known as autothermal reforming, which is a combination of partial oxidation near the reactor inlet with conventional steam reforming further along the reactor, improves the overall reactor efficiency and increases the flexibility of the process. Partial oxidation processes using oxygen instead of steam also found wide application for synthesis gas manufacture, with the special feature that they could utilize low-value feedstocks such as heavy petroleum residua. In recent years, catalytic partial oxidation (CPOX) employing very short reaction times (milliseconds) at high temperatures (850°C–1,000°C; 1,560°F–1,830°F) is providing still another approach to synthesis gas manufacture. Nearly complete conversion of methane, with close to 100% selectivity to hydrogen and carbon dioxide, can be obtained with a rhenium monolith under well-controlled conditions. Experiments on the catalytic partial oxidation of n-hexane conducted with added steam give much higher yields of hydrogen than can be obtained in experiments without steam, a result of much interest in obtaining hydrogen-rich streams for fuel cell applications.

The route for a carbonaceous feedstock to synthetic automotive fuels, as practiced by SASOL is technically proven and a series of products with favorable environmental characteristics are produced (Luque and Speight, 2015). As is the case in essentially all conversion processes for carbonaceous feedstocks where air or oxygen is used for the utilization or partial conversion of the energy in the coal, the carbon dioxide burden is a drawback as compared to crude oil.

There uses of synthesis gas include use as a chemical feedstock and in gas-to-liquid processes, which use Fisher–Tropsch chemistry to make liquid fuels as feedstock for chemical synthesis, as well as being used in the production of fuel additives, including diethyl ether and methyl t-butyl ether (MTBE), acetic acid and its anhydride, synthesis gas could also make an important contribution to chemical synthesis through conversion to methanol. There is also the option in which stranded natural gas is converted to synthesis gas production followed by conversion to liquid fuels.

The chemical train for producing synthesis gas (carbon monoxide + hydrogen), from which a variety of products can be produced, can be represented simply as:

$$\text{Carbonaceous feedstock} \rightarrow (\text{partial oxidation}) \rightarrow \text{synthesis gas}$$

$$\text{Synthesis gas} \rightarrow \text{synthetic fuels and petrochemicals}$$

The products designated as synthesis fuels include low-to-high-boiling hydrocarbon derivatives and methanol. Also the high-boiling products including wax products can also be used as feedstocks for gas production.

In addition, the actual process described as comprising three components: (i) synthesis gas generation, (ii) waste heat recovery, and (iii) gas processing. Within each of the above three listed systems are several options. For example, synthesis gas can be generated to yield a range of compositions ranging from high-purity hydrogen to high-purity carbon monoxide. Two major routes can be utilized for high-purity gas production: (i) pressure swing adsorption (PSA) and (ii) utilization of a cold box, where separation is achieved by distillation at low temperatures. In fact, both processes can also be used in combination as well. Unfortunately, both processes require high capital expenditure. However, to address these concerns, research and development is ongoing and successes can be measured by the demonstration and commercialization of technologies such as permeable membrane for the generation of high-purity hydrogen, which in itself can be used to adjust the hydrogen/carbon monoxide ratio of the synthesis gas produced.

10.4 PRODUCTION OF SYNTHESIS GAS

Gasification processes are used to convert a carbon-containing (carbonaceous) material into a synthesis gas, a combustible gas mixture which typically contains carbon monoxide, hydrogen, nitrogen, carbon dioxide, and methane (Chapter 5). The impure synthesis gas has a relatively low calorific value, ranging from 100 to 300 Btu/ft³. The gasification process can accommodate a wide variety of gaseous, liquid, and solid feedstocks and it has been widely used in commercial applications for the production of fuels and chemicals (Luque and Speight, 2015).

10.4.1 FEEDSTOCKS

In principle, synthesis gas can be produced from any hydrocarbon feedstock, which includes natural gas, naphtha, residual oil, petroleum coke, coal, biomass, and municipal or industrial waste (Chapter 1). The product gas stream is subsequently purified (to remove sulfur, nitrogen, and any particulate matter) after which it is catalytically converted to a mixture of liquid hydrocarbon products. In addition, synthesis gas may also be used to produce a variety of products, including ammonia and methanol.

Of all the carbonaceous materials used as feedstocks for gasification process, coal represents the most widely used feedstocks and, accordingly, the feedstock about which most is known. In fact, gasification of coal has been a commercially available proven technology (Speight 2013a,b). The modern gasification processes have evolved from three first-generation process technologies: (i) Lurgi fixed bed reactor (ii) high-temperature Winkler fluidized bed reactor; and (iii) Koppers–Totzek entrained flow reactor. In each case, steam/air/oxygen is passed through heated coal which may either be a fixed bed, fluidized bed, or entrained in the gas. Exit gas temperatures from the reactor are 500°C (930°F), 900°C–1,100°C (1,650°F–2,010°F), and 1,300°C–1,600°C (2,370°F–2,910°F), respectively. In addition to the steam/air/oxygen mixture being used as the feed gases, steam/oxygen mixtures can also be used in which membrane technology and a compressed oxygen-containing gas is employed.

In addition, low-value or negative-value materials and wastes such as petroleum coke, refinery residua, refinery waste, municipal sewage sludge, biomass, hydrocarbon contaminated soils, and chlorinated hydrocarbon byproducts have all been used successfully in gasification operations (Speight, 2008, 2013a, 2013b). In addition, synthesis gas is used as a source of hydrogen or as an intermediate in producing a variety of hydrocarbon products by means of the Fischer–Tropsch synthesis (Table 6.1) (Chadeesingh, 2011). In fact, gasification to produce synthesis gas can proceed from any carbonaceous material, including biomass and waste.

There are different sources for obtaining synthesis gas. It can be produced by steam reforming or partial oxidation of any hydrocarbon ranging from natural gas (methane) to petroleum residua. It can also be obtained by gasifying the carbonaceous feedstock to a medium Btu gas (medium Btu gas consists of variable amounts of carbon monoxide, carbon dioxide, and hydrogen and is used principally as a fuel gas).

A major route for producing synthesis gas is the steam reforming of natural gas over a promoted nickel catalyst at temperatures in the order of 800°C (1,470°F):

$$CH_4 + H_2O \rightarrow CO + 3H_2$$

In some countries, synthesis gas is mainly produced by steam reforming naphtha. Because naphtha is a mixture of hydrocarbon derivatives ranging approximately from C_5 to C_{10}, the steam reforming reaction may be represented using n-heptane:

$$CH_3(CH_2)_5 CH_3 + 7H_2O \rightarrow 7CO + 15H_2$$

As the molecular weight of the hydrocarbon increases (lower H/C feed ratio), the hydrogen/carbon monoxide (H_2/CO) product ratio decreases. The hydrogen/carbon monoxide product ratio is approximately 3 for methane, 2.5 for ethane, 2.1 for heptane, and less than 2 for heavier hydrocarbon derivatives. The non-catalytic partial oxidation of hydrocarbon derivatives is also used to produce synthesis gas, but the hydrogen/carbon monoxide ratio is lower than from steam reforming.

In practice, this ratio is even lower than what is shown by the stoichiometric equation because part of the methane is oxidized to carbon dioxide and water. When resids are partially oxidized by oxygen and steam at 1,400°C–1,450°C (2,550°F–2,640°F) and 800–900 atm, the gas consists of equal parts of hydrogen and carbon monoxide.

Synthesis gas is an important intermediate. The mixture of carbon monoxide and hydrogen is used for producing methanol. It is also used to synthesize a wide variety of hydrocarbon derivatives ranging from gases to naphtha to gas oil using Fischer–Tropsch technology. This process may offer an alternative future route for obtaining olefin derivatives and chemicals.

Synthesis gas is a major source of hydrogen, which is used for producing ammonia. Ammonia is the host of many chemicals such as urea, ammonium nitrate, and hydrazine. Carbon dioxide, a byproduct from synthesis gas, reacts with ammonia to produce urea (H_2NCONH_2).

$$\underset{\text{Urea}}{H_2N-\overset{\overset{\displaystyle O}{\|}}{C}-NH_2}$$

Urea (also known as carbamide) serves an important role in the metabolism of nitrogen-containing compounds by animals and is the main nitrogen-containing substance in the urine of mamas. It is a colorless, odorless solid, highly soluble in water, and, dissolved in water; it exhibits neither an acid nor an alkali. It is formed in the liver by the combination of two ammonia molecules (NH_3) with a carbon dioxide (CO_2) molecule. It is widely used in fertilizers as a source of nitrogen and is an important raw material for the chemical industry.

Most of the production of hydrocarbon derivatives by Fischer–Tropsch method uses synthesis gas produced from sources that yield a relatively low hydrogen/carbon monoxide ratio, such as typical in coal gasifiers. This, however, does not limit this process to low hydrogen/carbon monoxide gas feeds. The only large-scale commercial process using this technology is in South Africa, where coal is an abundant energy source. The process of obtaining liquid hydrocarbon derivatives from coal through the Fischer–Tropsch process is termed indirect coal liquefaction which was originally intended for obtaining liquid hydrocarbon derivatives from solid fuels. However, this method may well be applied in the future to the manufacture of chemicals through cracking the liquid products or by directing the reaction to produce more olefin derivatives.

The reactants in Fischer–Tropsch processes are carbon monoxide and hydrogen. The reaction may be considered a hydrogenative oligomerization of carbon monoxide in presence of a heterogeneous catalyst. The main reactions occurring in Fischer–Tropsch processes are:

i. Olefin derivatives:

$$2nH_2 + nCO \rightarrow C_nH_{2n} + nH_2O$$

ii. Paraffin derivatives:

$$(2n+1)H_2 + nCO \rightarrow C_nH_{2n+2} + nH_2O$$

iii. Alcohol derivatives:

$$2nH_2 + nCO \rightarrow C_nH_{2n+2}O + (n-1)H_2O$$

The coproduct water reacts with carbon monoxide (the shift reaction), yielding hydrogen and carbon dioxide:

$$CO + H_2O \rightarrow H_2 + CO_2$$

The gained hydrogen from the water shift reaction reduces the hydrogen demand for Fischer–Tropsch processes. Water-gas shift proceeds at about the same rate as the Fischer–Tropsch reaction. Another side reaction also occurring in Fischer–Tropsch process reactors is the disproportionation of carbon monoxide to carbon dioxide and carbon:

$$2CO \rightarrow CO_2 + C$$

This reaction is responsible for the deposition of carbon in the reactor tubes in fixed bed reactors and reducing heat transfer efficiency.

Fischer–Tropsch technology is best exemplified by the SASOL projects in South Africa. After the carbonaceous feedstock is gasified to a synthesis gas mixture, it is purified in a Rectisol unit. The purified gas mixture is reacted in a Synthol unit over an iron-based catalyst. The main products are gasoline, diesel fuel, and jet fuels. Byproducts are ethylene, propylene, alpha olefin derivatives, sulfur, phenol, and ammonia which are used for the production of downstream chemicals. However, the exact mechanism is not fully established. One approach assumes a first-step adsorption of carbon monoxide on the catalyst surface followed by a transfer of an adsorbed hydrogen atom from an adjacent site to the metal carbonyl (M-CO). The polymerization continues (as in the last three steps shown above) until termination occurs and the hydrocarbon is desorbed. The last two steps shown above explain the presence of oxygenated derivatives in Fischer–Tropsch products.

Alternatively, an intermediate formation of an adsorbed methylene on the catalyst surface through the dissociative adsorption of carbon monoxide has been considered. The formed metal carbide (M-C) is then hydrogenated to a reactive methylene metal species. The methylene intermediate abstracts a hydrogen and is converted to an adsorbed methyl. Reaction of the methyl with the methylene produces an ethyl-metal species. Successive reactions of the methylene with the formed ethyl produce a long-chain adsorbed alkyl. The adsorbed alkyl species can either terminate to a paraffin by a hydrogenation step or to an olefin by a dehydrogenation step. The carbide mechanism, however, does not explain the formation of oxygenate derivatives in Fischer–Tropsch products.

10.4.2 PROCESSES

10.4.2.1 Steam Reforming

Steam methane reforming (SMR) is the benchmark process that has been employed over a period of several decades for hydrogen production. The process involves reforming natural gas in a continuous catalytic process in which the major reaction is the formation of carbon monoxide and hydrogen from methane and steam:

$$CH_4 + H_2O{=}CO + 3H_2 \quad \Delta H_{298K} = +97,400 \text{ Btu/lb}$$

Higher molecular weight feedstocks can also be reformed to hydrogen:

$$C_3H_8 + 3H_2O \rightarrow 3CO + 7H_2$$

That is,

$$C_nH_m + nH_2O \rightarrow nCO + (0.5m + n)H_2$$

In the actual process, the feedstock is first desulfurized by passage through activated carbon, which may be preceded by caustic and water washes. The desulfurized material is then mixed with steam and passed over a nickel-based catalyst (730°C–845°C, 1,350°F–1,550°F and 400 psi). Effluent gases are cooled by the addition of steam or condensate to about 370°C (700°F), at which point carbon monoxide reacts with steam in the presence of iron oxide in a shift converter to produce carbon dioxide and hydrogen:

$$CO + H_2O = CO_2 + H_2$$

The carbon dioxide is removed by amine washing; the hydrogen is usually a high-purity (>99%) material.

Steam reforming (sometimes referred to as *steam methane reforming*, SMR) is carried out by passing a preheated mixture comprising essentially methane and steam through catalyst-filled tubes. Since the reaction is endothermic, heat must be provided in order to effect the conversion. This is achieved by the use of burners located adjacent to the tubes. The products of the process are a mixture of hydrogen, carbon monoxide, and carbon dioxide. Recovery of the heat from the combustion products can be implemented in order to improve the efficiency of the overall process.

To maximize the conversion of the methane feed, both a primary and secondary reformer are generally utilized. A *primary reformer* is used to effect 90%–92% conversion of methane. Here, the hydrocarbon feed is partially reacted with steam over a nickel-alumina catalyst to produce a synthesis gas with hydrogen/carbon monoxide ratio of approximately 3:1. This is done in a fired tube furnace at 900°C (1,650°F) at a pressure of 225–450 psi. The unconverted methane is reacted with oxygen at the top of a *secondary autothermal reformer* (ATR) containing nickel catalyst in the lower region of the vessel.

Two water-gas shift reactors are used downstream of the secondary reformer to adjust the hydrogen/carbon monoxide ratio, depending on the end use of the steam reformed products. The first of the two water-gas shift reactors utilizes an iron-based catalyst which is heated to approximately 400°C (750°F). The second water-gas shift reactor operates at approximately 200°C (390°F) and is charged with a copper-based catalyst.

Steam reforming is an exothermic reaction that is carried out by passing a preheated mixture comprising methane (sometimes substituted by natural gas having high methane content) and steam through catalyst-filled tubes. The products of the process are a mixture of hydrogen, carbon monoxide, and carbon dioxide. To maximize the conversion of the methane feed, primary and secondary reformers are often used—the *primary reformer* effects a 90%–92% v/v conversion of methane. In this step, the hydrocarbon feed is partially reacted with steam at 900°C (1,650°F) at 220–500 psi over a nickel-alumina catalyst to produce a synthesis gas in which the hydrogen/carbon monoxide (H_2/CO) ratio is in the order of 3:1. Any unconverted methane is reacted with oxygen at the top of a *secondary autothermal reformer* containing nickel catalyst in the lower region of the vessel.

In autothermal reformers (often referred to as secondary reformers), the oxidation of methane supplies the necessary energy and carried out either simultaneously or in advance of the reforming reaction. The equilibrium of the methane steam reaction and the water-gas shift reaction determines the conditions for optimum hydrogen yields. The optimum conditions for hydrogen production require high temperature at the exit of the reforming reactor (800°C–900°C; 1,470°F–1,650°F), high excess of steam (molar steam-to-carbon ratio of 2.5–3) and relatively low pressures (below 450 psi). Most commercial plants employ supported nickel catalysts for the process.

One way of overcoming the thermodynamic limitation of steam reforming is to remove either hydrogen or carbon dioxide as it is produced, hence shifting the thermodynamic equilibrium toward the product side. The concept for sorption-enhanced methane steam reforming is based on *in situ* removal of carbon dioxide by a sorbent such as calcium oxide (CaO).

$$CaO + CO_2 \rightarrow CaCO_3$$

Sorption enhancement enables lower reaction temperatures, which may reduce catalyst coking and sintering, while enabling use of less expensive reactor wall materials. In addition, heat release by the exothermic carbonation reaction supplies most of the heat required by the endothermic reforming reactions. However, energy is required to regenerate the sorbent to its oxide form by the energy-intensive calcination reaction:

$$CaCO_3 \rightarrow CaO + CO_2$$

Use of a sorbent requires either that there be parallel reactors operated alternatively and out of phase in reforming and sorbent regeneration modes, or that sorbent be continuously transferred between the reformer/carbonator and regenerator/calciner (Balasubramanian et al., 1999; Hufton et al., 1999).

The higher molecular weight hydrocarbon derivatives that are also constituents of natural gas (Speight, 2007, 2014a) are converted to methane in an adiabatic pre-reformer upstream of the steam reformer. In the pre-reformer, all higher hydrocarbon derivatives (C_{2+}) are converted into a mixture of methane, hydrogen, and carbon oxides:

$$C_nH_m + nH_2O \rightarrow nCO + (n + m/2)H_2$$

$$3H_2 + CO \leftrightarrow CH_4 + H_2O$$

$$CO + H_2O \leftrightarrow H_2 + CO_2$$

The pre-reforming process utilizes an adiabatic fixed bed reactor with highly active nickel catalysts, and the reactions take place at temperatures of approximately 350°C–550°C (650°F–1,020°F) and make it possible to preheat the steam reformer feed to higher temperatures without getting problems with olefin formation from the higher hydrocarbon derivatives. Olefin derivatives are unwanted in the steam reformer feed as they generally cause coking of the catalyst pellets at high temperatures. Preheating of the steam reformer feed is of great advantage because the reformer unit can be scaled down to a minimum size (Aasberg-Petersen et al., 2001, 2002; Hagh, 2004).

The reactions are catalyzed by pellets coated with nickel and are highly endothermic overall. Effective heat transport to the reactor tubes and further into the center of the catalytic fixed bed is, therefore, a very important aspect during design and operation of steam reformers. The reactions take place in several tubular fixed bed reactors of low diameter-to-height ratio to ensure efficient heat transport in radial direction. The process conditions are typically 300–600 psi bar with inlet temperature of 300°C–650°C (570°F–1,200°F) and outlet temperature of 700°C–950°C (1,290°F–1,740°F). There is often an approach to equilibrium of about 5°C–20°C, which means that the outlet temperature is slightly higher than the equilibrium temperature calculated from the actual outlet composition.

In a pre-reformer, whisker carbon can be formed either from methane or higher molecular weight hydrocarbon derivatives. The lower limit of the H_2O/C ratio depends on a number of factors including the feed gas composition, the operating temperature, and the choice of catalyst. In a pre-reformer operating at low H_2O/C-ratio, the risk of carbon formation from methane is most pronounced in the reaction zone where the temperature is highest. Carbon formation from higher molecular weight hydrocarbon derivatives can only take place in the first part of the reactor with the highest concentrations of higher molecular weight (C_{2+}) compounds.

The deposition of carbon can be an acute problem with the use of Ni-based catalysts in the primary reformer (Rostrup-Nielsen, 1984; Alstrup, 1988, Rostrup-Nielsen, 1993). The carbon-forming reactions occur in parallel with the reforming reactions and are undesirable as they cause poisoning of the surface of the catalyst pellets. This leads to lower catalyst activity and the need for more frequently catalyst reloading. The coking reactions are the CO-reduction, methane cracking, and Boudouard reaction, given by the respective equilibrium reactions:

$$CO + H_2 + H_2O \rightarrow C + H_2O$$

$$CH_4 \leftrightarrow C + 2H_2$$

$$2CO \leftrightarrow C + CO_2$$

Thus, low steam excess can lead to critical conditions causing coke formation—equilibrium calculations of the coking reactions can be a useful tool for predicting the danger for catalyst poisoning but the reaction kinetics may nevertheless be so slow that coking is no concern. A complete analysis should, therefore, also involve kinetic calculations, which will be feedstock-dependent expressions for these reactions.

One approach to prevent carbon formation is to use a steam/carbon ratio in the feed gas that does not allow the formation of carbon. However, this method results in lowering the efficiency of the process. Another approach is to utilize sulfur passivation, which utilizes the principle that the reaction leading to the deposition of carbon requires a larger number of adjacent surface Ni atoms than does steam reforming. When a fraction of the surface atoms are covered by sulfur, the deposition of carbon is thus more greatly inhibited than steam reforming reactions, leading to the development of the SPARG process (Rostrup-Nielsen, 1984; Udengaard et al., 1992). A third approach is to use Group VIII metals that do not form carbides, e.g., Pt. However, due to the high cost of such metals they are unable to compare to the economics associated with Ni.

A major challenge in steam reforming development is its energy-intensive nature due to the high endothermicity of the reactions. The trend in development thus is one which seeks higher energy efficiency. Improvements in catalysts and metallurgy require adaption to lower steam/carbon ratios and higher heat flux.

Finally, in all reforming processes, it is essential that impurities such as sulfur, mercury, and any other contaminants in the feedstock stream should be removed in order to prevent the poisoning of the reforming catalysts. Fischer–Tropsch synthesis takes the requirements for purification to a new level; cobalt Fischer–Tropsch catalysts are extremely sensitive to even part per billion (ppb) levels of contaminants including sulfur compounds and these must be removed, typically to levels below 5 ppb. The removal of mercury has become increasingly necessary in recent years as compounds of the metal have been found to be present in many gas sources and mercury removal for both environmental and process reasons is essential.

Typically, the processes are based on fixed beds of absorbents to remove traces of contaminants from hydrocarbon gases and liquids. In particular, the processes carry out (i) hydrogen sulfide removal, (ii) carbonyl sulfide (COS) removal, (iii) mercury (Hg) removal, and (iv) arsine (AsH_3) removal. The choice of absorbent and the design of the reactor vessel will vary according to the type of feedstock, the level of contaminants, pressure and temperature conditions as well as the tolerance of the catalyst to the level of impurities.

10.4.2.2 Autothermal Reforming

The autothermal reformer was developed in the 1950s and is used in commercial applications to provide synthesis gas for ammonia and methanol synthesis. In the case of ammonia production, where high hydrogen/carbon monoxide ratios are needed, the autothermal reformer is operated at high steam/carbon ratios. In the case of methanol synthesis, the required hydrogen/carbon monoxide ratio is provided by manipulating the carbon dioxide recycle. In fact, development and optimization of this technology has led to cost-effective operation at very low steam/carbon feed ratios to produce carbon monoxide-rich synthesis gas, for example, which is preferred in Fischer–Tropsch synthesis.

In the autothermal reforming process, the organic feedstock (such as natural gas) and steam and sometimes carbon dioxide are mixed directly with oxygen and air in the reformer. The reformer itself comprises a refractory-lined vessel which contains the catalyst, together with an injector located at the top of the vessel. Partial oxidation reactions occur in a region of the reactor referred to as the combustion zone. It is the mixture from this zone which then flows through a catalyst bed where the actual reforming reactions occur. Heat generated in the combustion zone from partial oxidation reactions is utilized in the reforming zone, so that in the ideal case, it is possible that the autothermal reformer can be in complete heat balance.

When the autothermal reformer uses carbon dioxide, the hydrogen/carbon monoxide ratio produced is 1:1; when the autothermal reformer uses steam, the hydrogen/carbon monoxide ratio produced is 2.5:1. The reactions can be described in the following equations, using carbon dioxide:

$$2CH_4 + O_2 + CO_2 \rightarrow 3H_2 + 3CO + H_2O + Heat$$

Using steam:

$$4CH_4 + O_2 + 2H_2O \rightarrow 10H_2 + 4CO$$

The reactor itself consists of three zones: (i) the burner, in which the feedstock streams are mixed in a turbulent diffusion flame; (ii) the combustion zone—where partial oxidation reactions produce a mixture of carbon monoxide and hydrogen; and (iii) the catalytic zone—where the gases leaving the combustion zone reach thermodynamic equilibrium.

Key elements in the reactor are the burner and the catalyst bed—the burner provides mixing of the feed streams and the natural gas is converted into a turbulent diffusion flame:

$$CH_4 + 3/2O_2 \rightarrow CO + 2H_2O$$

When carbon dioxide is present in the feed, the H_2/CO ratio produced is in the order of 1:1 but when the process employs steam the H_2/CO ratio produced is 2.5:1.

$$2CH_4 + O_2 + CO_2 \rightarrow 3H_2 + 3CO + H_2O$$

$$4CH_4 + O_2 + 2H_2O \rightarrow 10H_2 + 4CO$$

The risk of soot formation in an autothermal reformer reactor depends on a number of parameters including feed gas composition, temperature, pressure, and especially burner design. Soot precursors may be formed in the combustion chamber during operation and it is essential that the design of burner, catalyst, and reactor is such that the precursors are destroyed by the catalyst bed to avoid soot formation.

Many observers consider the combination of adiabatic pre-reforming and autothermal reforming at low H_2O/C ratios is a preferred layout for production of synthesis gas for large gas-to-liquids plants.

The following are the advantages of using the autothermal reformer: (i) compact in design, hence less associated footprint; (ii) low investment; (iii) economy of scale; (iv) flexible operation—short startup periods and fast load changes; and (v) soot-free operation.

10.4.2.3 Combined Reforming

Combined reforming incorporates a combination of both steam reforming and autothermal reforming. In the process, the feedstock is typically a mixture of reformed gas and desulfurized natural gas which is partially converted, under mild conditions, to synthesis gas in a relatively small steam reformer. The off-gases from the steam reformer are then sent to an oxygen-fired secondary reactor, the autothermal reformer. Here, the unreacted methane is converted to synthesis gas by partial oxidation followed by steam reforming.

Another configuration requires the hydrocarbon feed to be split into two streams which are then fed in parallel, to the steam reforming and autothermal reactors. An example of an efficient version of combined reforming is one which has been developed by Synetix, called gas-heated reforming.

10.4.2.4 Partial Oxidation

Partial oxidation reactions occur when a sub-stoichiometric fuel-air mixture is partially combusted in a reformer. The general reaction equation without catalyst (*thermal partial oxidation* (TPOX)) is of the form:

$$C_nH_m + (2n + m)/2O_2 \rightarrow nCO + (m/2)H_2O$$

A possible reaction equation is:

$$C_{24}H_{12} + 12O_2 \rightarrow 24CO + 6H_2$$

A thermal partial oxidation reactor is similar to the autothermal reformer with the main difference being no catalyst is used. The feedstock, which may include steam, is mixed directly with oxygen by an injector which is located near the top of the reaction vessel. Both partial oxidation as well as reforming reactions occur in the combustion zone below the burner. The principal advantage of partial oxidation is its ability to process almost any feedstock, which can comprise very high molecular weight organics, for example, petroleum coke (Gunardson and Abrardo, 1999). Additionally, since emission of NO_x and SO_x are minimal, the technology can be considered environmentally benign.

On the other hand, very high temperatures, approximately 1,300°C, are required to achieve near complete reaction. This necessitates the consumption of some of the hydrogen and a greater than stoichiometric consumption of oxygen, i.e., oxygen-rich conditions. Capital costs are high on account of the need to remove soot and acid gases from the synthesis gas. Operating expenses are also high due to the need for oxygen at high pressure.

A possible means of improving the efficiency of synthesis gas production is via catalytic partial oxidation technology. Although catalytic partial oxidation has not as yet been used commercially, it has several advantages over steam reforming, especially the higher energy efficiency. The reaction is in fact, not endothermic as is the case with steam reforming, but rather slightly exothermic. Further, a hydrogen/carbon monoxide ratio close to 2.0, i.e., the ideal ratio for the Fischer–Tropsch and methanol synthesis, is produced by this technology. Catalytic partial oxidation can occur by either of two routes: (i) direct or (ii) indirect.

The *direct catalytic partial oxidation* occurs through a mechanism involving only surface reaction on the catalyst, the *direct* route produces synthesis gas according to the following reaction:

$$2CH_4 + O_2 \rightarrow 2CO + 4H_2$$

On the other hand, the *indirect catalytic partial oxidation* route comprises total combustion of methane to carbon dioxide and water, followed by steam reforming and the water-gas shift reaction. Here, equilibrium conversions can be greater than 90% at ambient pressure. However, in order for an industrial process for this technology to be economically viable, an operating pressure of more than 20 atm would be required. Unfortunately, under such pressures, equilibrium conversions are lower. Further, an operational problem arises on account of the highly exothermic combustion step, which makes for problematic temperature control of the process and the possibility of temperature runaways.

It must be noted that in most studies of catalytic partial oxidation in microreactors, in most to nearly all cases, the conversion occurred via the "indirect" route. It is apparent that only the "direct" mechanism is likely to occur at short contact times. Interestingly, several researchers (Choudhary et al., 1993; Lapszewicz and Jiang, 1992) have observed that yields higher than equilibrium values are obtained with high flow rates through fixed bed reactors.

10.4.3 PRODUCT DISTRIBUTION

The product distribution of hydrocarbon derivatives formed during the Fischer–Tropsch process follows an Anderson–Schulz–Flory distribution:

$$W_n/n \ (1-\alpha)^2 \alpha^{n-1}$$

W_n is the weight fraction of hydrocarbon molecules containing n carbon atoms, α is the chain growth probability or the probability that a molecule will continue reacting to form a longer chain. In general, α is largely determined by the catalyst and the specific process conditions.

According to the above equation, methane will always be the largest single product; however, by increasing α close to 1, the total amount of methane formed can be minimized compared to the sum of all the various long-chain products. Increasing α increases the formation of long-chain hydrocarbon derivatives—waxes—which are solid at room temperature. Therefore, for production of liquid transportation fuels it may be necessary to crack the Fischer–Tropsch longer-chain products.

The very long-chain hydrocarbon derivatives are waxes, which are solid at room temperature. Therefore, for production of liquid transportation fuels it may be necessary to crack some of the Fischer–Tropsch products. In order to avoid this, some researchers have proposed using zeolites or other catalyst substrates with fixed-sized pores that can restrict the formation of hydrocarbon derivatives longer than some characteristic size (usually $n < 10$). This way they can drive the reaction so as to minimize methane formation without producing lots of long-chain hydrocarbon derivatives.

It has been proposed that zeolites or other catalyst substrates with fixed-sized pores that can restrict the formation of hydrocarbon derivatives longer than some characteristic size (usually $n < 10$). This would tend to drive the reaction to minimum methane formation without producing the waxy products.

10.5 PROCESS PARAMETERS

For large-scale commercial Fischer–Tropsch reactors heat removal and temperature control are the most important design features to obtain optimum product selectivity and long catalyst lifetimes. Over the years, basically four Fischer–Tropsch reactor designs have been used commercially. These are the multi-tubular fixed bed, the slurry reactor, or the fluidized bed reactor (with either a fixed bed or a circulating bed). The fixed bed reactor consists of thousands of small tubes with the catalyst as surface-active agent in the tubes. Water surrounds the tubes and regulates the temperature by settling the pressure of evaporation. The slurry reactor is widely used and consists of fluid and solid elements, where the catalyst has no particularly position, but flows around as small pieces of catalyst together with the reaction components. The slurry and fixed bed reactor are used in the low-temperature Fischer–Tropsch process. The fluidized bed reactors are diverse, but characterized by the fluid behavior of the catalyst.

The multi-tubular fixed bed reactors (often referred to as Arge reactors) were developed jointly by Lurgi and Ruhrchemie and commissioned in the 1955. They were used by Sasol to produce high-boiling Fischer–Tropsch liquid hydrocarbon derivatives and waxes in Sasolburg, in what Sasol called the low-temperature Fischer–Tropsch synthesis process, aiming for liquid fuels production. Most, if not all, of these types of Arge reactors are now be replaced by slurry bed reactors, which is considered the state-of-the-art technology for low-temperature Fischer–Tropsch synthesis. Slurry bed Fischer–Tropsch reactors offer better temperature control and higher conversion. Fluidized-bed Fischer–Tropsch reactors were developed for the high-temperature Fischer–Tropsch synthesis to produce low molecular gaseous hydrocarbon derivatives and naphtha. This type of reactor was originally developed in a circulating mode (such as the Sasol synthol reactors) but has been replaced by a fixed fluidized bed type of reactor (advanced synthol reactors) which is capable of a high throughput.

Sasol in South Africa uses coal and natural gas as a feedstock, and produces a variety of synthetic petroleum products. The process was used in South Africa to meet its energy needs during its isolation under Apartheid. This process has received renewed attention in the quest to produce low sulfur diesel fuel in order to minimize the environmental impact from the use of diesel engines. The Fischer–Tropsch process as applied at Sasol can be divided into two operating regimes: (i) the high-temperature Fischer–Tropsch process and (ii) the low-temperature Fischer–Tropsch process (Chadeesingh, 2011).

The high-temperature Fischer–Tropsch technology uses a fluidized catalyst at 300°C–330°C (570°F–635°F). Originally circulating fluidized bed units were used (Synthol reactors). Since 1989 a commercial-scale classical fluidized bed unit has been implemented and improved upon.

The low-temperature Fischer–Tropsch technology has originally been used in tubular fixed bed reactors at 200°C–230°C (390°F–260°F). This produces a more paraffin derivatives and waxy product spectrum than the high-temperature technology. A new type of reactor (the Sasol slurry phase distillate reactor has been developed and is in commercial operation. This reactor uses a slurry phase system rather than a tubular fixed bed configuration and is currently the favored technology for the commercial production of synfuels.

The commercial Sasol Fischer–Tropsch reactors all use iron-based catalysts on the basis of the desired product spectrum and operating costs. Cobalt-based catalysts have also been known since the early days of this technology and have the advantage of higher conversion for low-temperature cases. Cobalt is not suitable for high temperature use due to excessive methane formation at such temperatures. For once-through maximum diesel production, cobalt has, despite its high cost, advantages and Sasol has also developed cobalt catalysts which perform very well in the slurry phase process.

However, both the iron and cobalt Fischer–Tropsch catalysts are sensitive to the presence of sulfur compounds in the synthesis gas and can be poisoned by the sulfur compounds. In addition, the sensitivity of the catalyst to sulfur is higher for cobalt-based catalysts than for the iron-based catalysts. This is one reason why cobalt-based catalysts are preferred for Fischer–Tropsch synthesis with synthesis gas derived from natural gas, where the synthesis gas has a higher hydrogen/carbon monoxide ratio and is relatively lower in sulfur content. On the other hand, iron-based catalysts are preferred for lower-quality feedstocks such as coal.

The kerosene (often referred to as diesel although the product could not be sold as diesel fuel without any further treatment to meet specifications) produced by the slurry phase reactor has a highly paraffin derivatives nature, giving a cetane number in excess of 70. The aromatic content of the diesel is typically below 3% and it is also sulfur-free and nitrogen-free. This makes it an exceptional diesel as such or it can be used to sweeten or to upgrade conventional diesels.

The Fischer–Tropsch process is an established technology and already applied on a large scale, although its popularity is hampered by high capital costs, high operation and maintenance costs, and the uncertain and volatile price of crude oil. In particular, the use of natural gas as a feedstock only becomes practical when using stranded gas, i.e., sources of natural gas far from major cities which are impractical to exploit with conventional gas pipelines and liquefied natural gas technology; otherwise, the direct sale of natural gas to consumers would become much more profitable. It is suggested by geologists that supplies of natural gas will peak 5–15 years after oil does, although such predictions are difficult to make and often highly uncertain. Hence, the increasing interest in a variety of carbonaceous feedstocks as a source of synthesis gas.

Under most circumstances, the production of synthesis gas by reforming natural gas will be more economical than from coal gasification, but site-specific factors need to be considered. In fact, any technological advance in this field (such as better energy integration or the oxygen transfer ceramic membrane reformer concept) will speed up the rate at which the synfuels technology will become common practice.

There are large coal reserves which may increasingly be used as a fuel source during oil depletion. Since there are large coal reserves in the world, this technology could be used as an interim transportation fuel if conventional oil were to become more expensive. Furthermore, combination of biomass gasification and Fischer–Tropsch synthesis is a very promising route to produce transportation fuels from renewable or green resources.

Often a higher concentration of some sorts of hydrocarbon derivatives is wanted, which might be achieved by changed reaction conditions. Nevertheless, the product range is wide and infected with uncertainties, due to lack of knowledge of the details of the process and of the kinetics of the reaction. Since the different products have quite different characteristics such as boiling point, physical state at ambient temperature and thereby different use and ways of distribution, often only a few of the carbon chains is wanted. As an example the low-temperature Fischer–Tropsch is used when longer-carbon chains are wanted, because lower temperature increases the portion of longer chains. But too low temperature is not wanted, because of reduced activity.

When the wanted products are shorter carbon chains, e.g., petroleum, the longer ones might be cracked into shorter chains.

The yield of kerosene (diesel) is therefore highly dependent on the chain growth probability, which again is dependent on (i) pressure and temperature; (ii) the composition of the feedstock gas; (iii) the catalyst type; (iv) the catalyst composition; and (v) the reactor design. The desire to increase the selectivity of some favorable products leads to a need of understanding the relation between reaction conditions and chain growth probability, which in turn request a mathematical expression for the growth probability in order to make a suitable model of the process. The different attempts to model the growth probability have resulted in some models that are regarded in literature as appropriate to describe the product distribution. Two will be presented here to show the influence of temperature and partial pressure.

10.6 REACTORS AND CATALYSTS

Since its discovery the Fischer–Tropsch synthesis has undergone periods of rapid development and periods of inaction. Within 10 years of the discovery, German companies were building commercial plants. The construction of these plants stopped about in 1940, but existing plants continued to operate during World War II.

The synthesis of hydrocarbon derivatives from carbon monoxide hydrogenation was discovered in 1902 by Sabatier and Sanderens who produced methane by passing carbon monoxide and hydrogen over nickel, iron, and cobalt catalysts. At about the same time, the first commercial hydrogen from synthesis gas produced from steam methane reforming was commercialized. Haber and Bosch discovered the synthesis of ammonia from hydrogen and N2 in 1910, and the first industrial ammonia synthesis plant was commissioned in 1913. The production of liquid hydrocarbon derivatives and oxygenated derivatives from synthesis gas conversion over iron catalysts was discovered in 1923 by Fischer and Tropsch. Variations on this synthesis pathway were soon to follow for the selective production of methanol, mixed alcohols, and iso-synthesis products. Another outgrowth of Fischer–Tropsch synthesis was the hydroformylation of olefin derivatives discovered in 1938 and is based on the reaction of synthesis gas with olefin derivatives for the production of Oxo aldehyde derivatives and alcohol derivatives (Chapters 5, 7, and 8).

10.6.1 REACTORS

Currently, two reactor types are used commercially in the Fischer–Tropsch process, a fixed bed reactor, a fluid bed reactor, and a slurry bed reaction. The fixed bed reactors usually run at lower temperatures to avoid carbon deposition on the reactor tubes. Products from fixed bed reactors are characterized by low olefin content, and they are generally heavier than products from fluid beds. Heat distribution in fluid beds, however, is better than fixed bed reactors, and fluid beds are generally operated at higher temperatures. Products are characterized by (i) having more olefin derivatives, (ii) a high proportion of low-boiling hydrocarbon derivatives (gases), and (iii) lower molecular weight product slate than from fixed bed types.

Originally, the Fischer–Tropsch synthesis was carried out in packed bed reactors. Gas-agitated multiphase reactors, sometimes called "slurry reactors" or "slurry bubble columns," gained favor, however, because the circulation of the slurry makes it much easier to control the reaction temperature in a slurry bed reactor than in a fixed bed reactor. Gas-agitated multiphase reactors operate by suspending catalytic particles in liquid and feeding gas reactants into the bottom of the reactor through a gas distributor, which produces small gas bubbles. As the gas bubbles rise through the reactor, the reactants are absorbed into the liquid and diffuse to the catalyst particles where, depending on the catalyst system, they are typically converted to gaseous and liquid products.

A slurry bed reactor is characterized by having the catalyst in the form of a slurry. The feed gas mixture is bubbled through the catalyst suspension. Temperature control is easier than the other two reactor types. An added advantage to slurry bed reactor is that it can accept a synthesis gas with a lower hydrogen/carbon monoxide ratio than either the fixed bed or the fluid bed reactors.

In the Sasol slurry phase reactor, preheated synthesis gas is fed into the bottom of the reactor, where it is distributed into slurry consisting of liquid hydrocarbon and catalyst particles. As the gas bubbles rise upward through the slurry, it diffuses into the slurry and is converted into a range of hydrocarbon derivatives by Fischer–Tropsch reaction. The heat generated from this reaction is removed through the reactor's cooling coils, which generate steam. The heavier (wax) fraction is separated from the slurry containing the catalyst particles in a proprietary process developed by Sasol. The lighter, more volatile fraction is extracted in a gas stream from the top of the reactor. The gas stream is cooled to recover the lighter hydrocarbon derivatives and water. The intermediate hydrocarbon streams are sent to the product upgrading unit, while the water stream is treated in a water recovery unit. The third step upgrades reactor products to diesel and naphtha. The reactor products are mainly paraffin derivatives, but the lighter products contain some olefin derivatives and oxygenated derivatives that need to be removed for product stabilization. Hydrogen is added to hydrotreat the olefin derivatives and oxygenated derivatives, converting them to paraffin derivatives. Hydrogen is also added to the mild hydrocracker, which breaks the long-chain hydrocarbon derivatives into naphtha and diesel. The products are separated out in a fractionation section, which involves hydrocracking and hydroisomerization.

As the process evolved, other types of reactors have been used and include (i) the parallel plate reactors, (ii) a variety of fixed bed tubular reactors, and (iii) gas-agitated multiphase reactors. For the parallel plate type of reactor, the catalyst bed is located in tubes fixed between the plates, which were cooled by steam/water that passed around the tubes within the catalyst bed. In another version, the reactor may be regarded as finned-tube in which large fins are penetrated by a large number of parallel or connected catalyst-filled tubes. Various designs were utilized for the tubular fixed bed reactor with the concentrically paced tubes being the preferred one. This type of reactor contained catalyst in the area between the two tubes with cooling water—steam flowing through the inner tube and on the exterior of the outer tube. The gaseous products formed enter the gas bubbles and are collected at the top of the reactor. Liquid products are recovered from the suspending liquid using different techniques, including filtration, settling, and hydrocyclones.

Because the Fischer–Tropsch reaction is exothermic, temperature control is an important aspect of Fischer–Tropsch reactor operation. Gas-agitated multiphase reactors or slurry bubble-column reactors have very high heat transfer rates and therefore allow good thermal control of the reaction. On the other hand, because the desired liquid products are mixed with the suspending liquid, recovery of the liquid products can be relatively difficult. This difficulty is compounded by the tendency of the catalyst particles to erode in the slurry, forming fine catalyst particles that are also relatively difficult to separate from the liquid products. Fixed bed reactors generally avoid the issues that arise from liquid separation and catalyst separation but they (the fixed bed reactors) do not provide the mixing of phases that allows good thermal control in slurry bubble-column reactors.

Furthermore, Fischer–Tropsch reactors are typically sized to achieve a desired volume of production. When a fixed bed reactor is planned, economies of scale tend to result in the use of long (tall) reactors. Because the Fischer–Tropsch reaction is exothermic, however, a thermal gradient tends to form along the length of the reactor, with the temperature increasing with distance from the reactor inlet. In addition, for most Fischer–Tropsch catalyst systems each 10° rise in temperature increases the reaction rate approximately 60%, which in turn results in the generation of still more heat. To absorb the heat generated by the reaction and offset the rise in temperature, a cooling liquid is typically circulated through the reactor.

Thus, for a given reactor system having a known amount of catalyst with a certain specific activity and known coolant temperature, the maximum flow rate of reactants through the reactor is limited by the need to maintain the catalyst below a predetermined maximum catalyst temperature at all points along the length of the catalyst bed and the need to avoid thermal runaway which can result in catalyst deactivation and possible damage to the physical integrity of the reactor system. The net result is that it is unavoidable to operate most of the reactor at temperatures below the maximum temperature, with the corresponding low volumetric productivities over most of the reactor volume.

An innovative technology for combining air separation and natural gas reforming processes is being pursued by Sasol, BP, Praxair, and Statoil (Dyer and Chen, 1999). If successful commercialized, this innovation can reduce the cost of synthesis gas generation by as much as 30%. The technology is referred to as *oxygen transport membranes* (OTM) and should combine five unit operations currently in use, *viz.* oxygen separation, oxygen compression, partial oxidation, steam methane reforming, and heat exchange. This technology incorporates the use of catalytic components with the membrane to accelerate the reforming reactions.

Air products have also developed and patented a two-step process for synthesis gas generation (Nataraj et al., 2000). This technology can be utilized to generate synthesis gas from several feedstocks, including natural gas, associated gas (from crude oil production), light hydrocarbon gases from refineries, and medium molecular weight (medium boiling) hydrocarbon fractions like naphtha (Gary et al., 2007; Speight, 2011, 2014a, 2017; Hsu and Robinson, 2017). The first stage comprises conventional steam reforming with partial conversion to synthesis gas. This is followed by complete conversion in an ion transport ceramic membrane (ITM) reactor. This combination solves the problem associated with steam reforming for feedstocks with hydrocarbon derivatives higher-boiling than methane, since higher molecular weight (C_{2+}) hydrocarbon derivatives tend to crack and degrade both the catalyst and membrane.

By shifting the equilibrium in the steam reforming process through removal of hydrogen from the reaction zone, membrane reactors can also be used to increase the equilibrium-limited methane conversion. Using palladium-silver (Pd-Ag) alloy membrane reactors methane conversion can reach as close to 100% (Shu et al., 1995).

10.6.2 Catalysts

Of great importance to the Fischer–Tropsch process is the catalyst. The catalysts used in the latest generation of Fischer–Tropsch technologies are cobalt based, usually carried on alumina supports, sometimes with precious metal promoters. Coal-based processes, including a gasification step, use iron catalysts, which are better suited to high-temperature processes based on feedstocks containing impurities. But iron produces significant quantities of non-paraffin derivatives as byproducts, while cobalt catalysts feature high selectivity and are more efficient for making paraffin derivatives from cleaner feedstocks.

Catalysts play a pivotal role in synthesis gas conversion reactions. In fact, fuels and chemicals synthesis from synthesis gas does not occur in the absence of appropriate catalysts. The basic concept of a catalytic reaction is that reactants adsorb onto the catalyst surface and rearrange and combine into products that desorb from the surface. One of the fundamental functional differences between synthesis catalysts is whether or not the adsorbed carbon monoxide molecule dissociates on the catalyst surface. For the Fischer–Tropsch process and higher alcohol synthesis, carbon monoxide dissociation is a necessary reaction condition. For methanol synthesis, the carbon-oxygen bond remains intact. Hydrogen has two roles in catalytic synthesis gas reactions. In addition to serving as a reactant needed for carbon monoxide hydrogenation, it is commonly used to reduce the metalized synthesis catalysts and activate the metal surface.

Generally, the Fischer–Tropsch synthesis is catalyzed by a variety of transition metals such as iron, nickel, and cobalt. Iron-based catalysts are relatively low cost and have a higher water-gas-shift activity, and are therefore more suitable for a lower hydrogen/carbon monoxide ratio (H_2/CO) synthesis gas such as those derived from coal gasification. On the other hand, nickel-based catalysts tend to promote methane formation, as in a methanation process. Cobalt-based catalysts are more active and are generally preferred over ruthenium (Ru) and, in comparison to iron, Co has much less water-gas shift activity.

Thus, in many cases, an iron-containing catalyst is the preferred catalyst due to the higher activity, but a nickel-containing catalyst produces large amounts of methane, while a cobalt-containing catalyst has a lower reaction rate and a lower selectivity than the iron-containing catalyst. By comparing cobalt and iron catalysts, it was found that cobalt promotes more middle-distillate products. In the Fischer–Tropsch process, a cobalt-containing catalyst produces hydrocarbon derivatives plus water while iron catalyst produces hydrocarbon derivatives and carbon dioxide. It appears that the iron catalyst promotes the shift reaction more than the cobalt catalyst.

Various metals, including but not limited to iron, cobalt, nickel, and ruthenium, alone and in conjunction with other metals, can serve as Fischer–Tropsch catalysts. Cobalt is particularly useful as a catalyst for converting natural gas to heavy hydrocarbon derivatives suitable for the production of diesel fuel. Iron has the advantage of being readily available and relatively inexpensive but also has the disadvantage of greater water-gas shift activity. Ruthenium is highly active but quite expensive. Consequently, although ruthenium is not the economically preferred catalyst for commercial Fischer–Tropsch production, it is often used in low concentrations as a promoter with one of the other catalytic metals.

A variety of catalysts can be used for the Fischer–Tropsch process, but the most common are the transition metals cobalt, iron, and ruthenium. Nickel can also be used, but tends to favor methane formation (methanation). Cobalt seems to be the most active catalyst, although iron may be more suitable for low hydrogen content synthesis gases such as those derived from coal due to its promotion of the water-gas shift reaction. In addition to the active metal the catalysts typically contain a number of promoters, including potassium and copper. Catalysts are supported on high surface area binders/supports such as silica (SiO_2), alumina (Al_2O_3), or the more complex zeolites. Cobalt catalysts are more active for Fischer–Tropsch synthesis when the feedstock is natural gas. Natural gas has a high hydrogen to carbon ratio, so the water-gas shift is not needed for cobalt catalysts. Iron catalysts are preferred for lower-quality feedstocks such as petroleum residua, coal, or biomass.

Unlike the other metals used for this process (Co, Ni, Ru) which remain in the metallic state during synthesis, iron catalysts tend to form a number of chemical phases, including various oxides and carbides during the reaction. Control of these phase transformations can be important in maintaining catalytic activity and preventing breakdown of the catalyst particles. For synthesis of higher molecular weight alcohols, dissociation of carbon monoxide is a necessary reaction condition. For methanol synthesis, the carbon monoxide molecule remains intact. Hydrogen has two roles in catalytic synthesis gas synthesis reactions. In addition to serving as a reactant needed for hydrogenation of carbon monoxide, it is commonly used to reduce the metalized synthesis catalysts and activate the metal surface.

Group 1 alkali metals (including potassium) are poisons for cobalt catalysts but are promoters for iron catalysts. Catalysts are supported on high surface area binders/supports such as silica, alumina, and zeolites. Cobalt catalysts are more active for Fischer–Tropsch synthesis when the feedstock is natural gas. Natural gas has a high hydrogen to carbon ratio, so the water-gas shift is not needed for cobalt catalysts. Iron catalysts are preferred for lower-quality feedstocks such as coal or biomass.

Unlike the other metals used for this process (Co, Ni, Ru), which remain in the metallic state during synthesis, iron catalysts tend to form a number of phases, including various oxides and carbides during the reaction. Control of these phase transformations can be important in maintaining catalytic activity and preventing breakdown of the catalyst particles.

Fischer–Tropsch catalysts are sensitive to poisoning by sulfur-containing compounds. The sensitivity of the catalyst to sulfur is greater for cobalt-based catalysts than for their iron counterparts. Promoters also have an important influence on activity. Alkali metal oxides and copper are common promoters, but the formulation depends on the primary metal, iron or cobalt. Alkali oxides on cobalt catalysts generally cause activity to drop severely even with very low alkali loadings. C_{5+} and carbon dioxide selectivity increase while methane and C_2 to C_4 selectivity decrease. In addition, the olefin to paraffin ratio increases.

Fischer–Tropsch catalysts can lose activity as a result of (i) conversion of the active metal site to an inactive oxide site, (ii) sintering, (iii) loss of active area by carbon deposition, and (iv) chemical poisoning. For example, Fischer–Tropsch catalysts are notoriously sensitive to poisoning by sulfur-containing compounds. The sensitivity of the catalyst to sulfur is greater for cobalt-based catalysts than for their iron counterparts. Some of these mechanisms are unavoidable and others can be prevented or minimized by controlling the impurity levels in the synthesis gas. By far the most abundant, important, and most studied the Fischer–Tropsch process catalyst poison is sulfur. Other catalyst poisons include halides and nitrogen compounds (e.g., NH_3, NO_x and HCN).

The hydrocarbon derivatives formed are mainly aliphatic, and on a molar basis methane is the most abundant; the amount of higher hydrocarbon derivatives usually decreases gradually with increase in molecular weight. Isoparaffin formation is more extensive over zinc oxide (ZnO) or thoria (ThO_2) at 400°C–500°C (750°F–930°F) and at higher pressure. Paraffin waxes are formed over ruthenium catalysts at relatively low temperatures (170°C–200°C, 340°F–390°F), high pressures (1,500 psi), and with a carbon monoxide-hydrogen ratio. The more highly branched product made over the iron catalyst is an important factor in a choice for the manufacture of automotive fuels. On the other hand, a high-quality diesel fuel (paraffin character) can be prepared over cobalt.

Secondary reactions play an important part in determining the final structure of the product. The olefin derivatives produced are subjected to both hydrogenation and double-bond shifting toward the center of the molecule; *cis* and *trans* isomers are formed in about equal amounts. The proportion of straight-chain molecules decreases with rise in molecular weight, but even so they are still more abundant than branched-chain compounds up to about C_{10}.

The small amount of aromatic hydrocarbon derivatives found in the product covers a wide range of isomer possibilities. In the C_6 to C_9 range, benzene, toluene, ethylbenzene, xylene, *n*-propyl- and iso-propylbenzene, methyl ethyl benzene derivatives, and trimethylbenzene derivatives have been identified; naphthalene derivatives and anthracene derivatives are also present.

Alternatively, a methanol-to-olefins (MTOs) option is available (Tian et al., 2015). The methanol-to-olefin derivatives reaction is one of the most important reactions in C_1 chemistry, which provides a viable option for producing basic petrochemicals from nonoil resources such as coal and natural gas. As olefin-based petrochemicals and relevant downstream processes have been well developed for many years, the methanol-to-olefins provide a link between gasification chemistry and the modern petrochemical industry. In the process, olefin derivatives are produced from methanol using a zeolite catalyst. The methanol feedstocks vaporized, mixed with recovered methanol, superheated and sent to the fluidized bed reactor. In the reactor, methanol is first converted to a dimethyl ether (DME, CH_3OCH_3) intermediate and then converted to olefin derivatives with a very high selectivity for ethylene and propylene.

DME can be produced by any one of several routes (Figure 10.2), but the most common route is using methanol produced from synthesis gas. In the process, water-free methanol is vaporized and sent to a reactor with an inlet temperature in the order of 220°C–250°C (430°F–480°F), and an outlet temperature in the order of 300°C–350°C (570°F–660°F). Thus:

$$2CH_3OH \rightarrow CH_3OCH_3 + H_2O$$

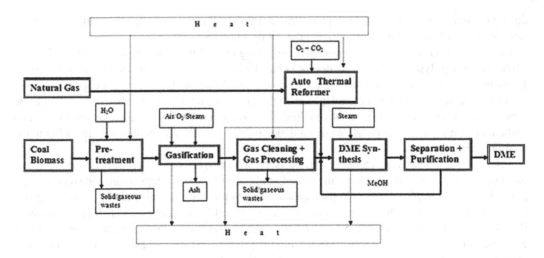

FIGURE 10.2   Routes for the production of dimethyl ether.

The reactor effluents are sent to a distillation column where the DME is separated from the top and condensed after which the DME is sent to storage. Water and methanol are discharged from the bottom and fed to a methanol column for methanol recovery. The purified methanol from this column is recycled to the reactor after mixing with feedstock methanol.

Catalysts considered for Fischer–Tropsch synthesis are based on transition metals of iron, cobalt, nickel, and ruthenium. Fischer–Tropsch catalyst development has largely been focused on the preference for high molecular weight linear alkanes and diesel fuels production. Among these catalysts, it is generally known that: (i) nickel (Ni) tends to promote methane formation, as in a methanation process; thus, generally it is not desirable; (ii) iron (Fe) is relatively low cost and has a higher water-gas shift activity, and is therefore more suitable for a lower hydrogen/carbon monoxide ratio (H_2/CO) synthesis gas such as those derived from coal gasification; (iii) cobalt (Co) is more active, and generally preferred over ruthenium (Ru) because of the prohibitively high cost of Ruthenium; and (iv) in comparison to iron, Co has much less water-gas shift activity, and is much more costly. Thus, it is not surprising that commercially available Fischer–Tropsch catalysts are either cobalt or iron based. In addition to the active metal, the iron-containing catalysts at least typically contain a number of promoters, including potassium and copper, as well as high surface area binders/supports such as silica (SiO_2) and/or alumina (Al_2O_3).

Iron-based Fischer–Tropsch catalysts are currently the most popular catalyst for the Fischer–Tropsch process for converting synthesis gas into Fischer–Tropsch liquids, given Fe catalyst's inherent water gas-shift capability to increase the hydrogen/carbon monoxide ratio of coal-derived synthesis gas, thereby improving hydrocarbon product yields in the Fischer–Tropsch synthesis. Fe catalysts may be operated in both high-temperature regime (300°C–350°C, 570°F–650°F) and low-temperature regime (220°C–270°C, 430°F–520°F), whereas Co catalysts are only used in the low-temperature range. This is a consequence of higher temperatures causing more methane formation, which is worse for Co compared to Fe.

Cobalt-containing catalysts are a useful alternative to iron-containing catalysts for Fischer–Tropsch synthesis because they demonstrate activity at lower synthesis pressures, so higher catalyst costs can be offset by lower operating costs. Also, coke deposition rate is higher for Fe catalyst than Co catalyst; consequently, Co catalysts have longer lifetimes. Co catalysts have a long lifetime/ greater activity; i.e., Co catalysts are replaced less frequently.

Although there are differences in the product distribution of cobalt-containing and iron-containing catalysts at similar temperatures and pressures (for example, 240°C, 465°F, and 450 psi), a cobalt-containing catalyst has somewhat higher selectivity for heavier hydrocarbon derivatives than an iron-containing catalyst; the product distribution is primarily driven by the choice of operating temperature: high temperature results in a naphtha/kerosene ratio of 2:1; low temperature results in naphtha/kerosene rat on the order of 1:2 more or less no matter if the catalyst is an iron-containing catalyst or a cobalt-containing catalyst. Higher temperatures shift selectivity toward lower carbon number products and more hydrogenated products; branching increases and secondary products such as ketones and aromatic derivatives also increase.

Thus, generally, a low temperature favors yielding high molecular mass linear wax derivative while a high temperature favors the production of naphtha and low molecular weight olefin derivatives. In order to maximize production of the naphtha fraction, it is best to use an iron-containing catalyst at a high temperature in a fixed fluid bed reactor. On the other hand, in order to maximize production of the kerosene fraction, a slurry reactor with a cobalt-containing catalyst is the more appropriate choice.

Both iron-containing catalyst and cobalt-containing are sensitive to the presence of sulfur compounds in the synthesis gas and can be poisoned by them, hence the need for rigorous feedstock preparation (Chapter 4). However, the sensitivity of the catalyst to sulfur is higher for cobalt-containing catalysts than for iron catalyst and is often the reason why cobalt-containing catalysts are preferred for Fischer–Tropsch synthesis with natural gas-derived synthesis gas, where the synthesis gas has a higher hydrogen/carbon monoxide ratio and is relatively lower in sulfur content.

10.7 PRODUCTS AND PRODUCT QUALITY

The composition of the products from the synthesis gas production processes is varied insofar as the gas composition varies with the type of feedstock and the gasification system employed (Speight 2013 a,b; Luque and Speight, 2015). Furthermore, the quality of gaseous product(s) must be improved by removal of any pollutants such as particulate matter and sulfur compounds before further use (Chapter 4), particularly when the intended use is a water-gas shift or methanation reaction.

10.7.1 PRODUCTS

Low Btu gas (low heat content gas) is the product when the oxygen is not separated from the air and, as a result, the gas product invariably has a low heat content (150–300 Btu/ft^3). In *medium Btu gas* (medium heat-content gas), the heating value is in the range 300–550 Btu/ft^3 and the composition is much like that of low heat content gas, except that there is virtually no nitrogen and the H_2/CO ratio varies from 2:3 to approximately 3:1 and the increased heating value correlates with higher methane and hydrogen contents as well as with lower carbon dioxide content. *High Btu gas* (high heat content gas) is essentially pure methane and often referred to as synthetic natural gas (SNG) or substitute natural gas. However, to qualify as substitute natural gas, a product must contain at least 95% methane; the energy content of synthetic natural gas is 980–1,080 Btu/ft^3. The commonly accepted approach to the synthesis of high heat content gas is the catalytic reaction of hydrogen and carbon monoxide.

Hydrogen is also produced during gasification of carbonaceous feedstocks. Although several gasifier types exist (Chapter 2), entrained flow gasifiers are considered most appropriate for producing both hydrogen and electricity from coal since they operate at temperatures high enough (approximately 1,500°C, 2,730°F) to enable high carbon conversion and prevent downstream fouling from tars and other residuals.

There is also a series of products that are called by older (even archaic) names that evolved from older coal gasification technologies and warrant mention: (i) producer gas, (ii) water gas, (iii) town gas, and (iv) synthetic natural gas. These products are typically low-to-medium Btu gases (Speight, 2013a, b).

10.7.2 Product Quality

Gas processing, although generally simple in chemical and/or physical principles, is often confusing because of the frequent changes in terminology and, often, lack of cross-referencing (Mokhatab et al., 2006; Speight, 2007, 2008, 2013a, 2014a) (Chapter 4). Although gas processing employs different process types there is always overlap between the various concepts. And, with the variety of possible constituents and process operating conditions, a universal purification system cannot be specified for economic application in all cases.

The processes that have been developed for gas cleaning (Mokhatab et al., 2006; Speight, 2007, 2008) vary from a simple once-through wash operation to complex multi-step systems with options for recycle of the gases (Mokhatab et al., 2006). In some cases, process complexities arise because of the need for recovery of the materials used to remove the contaminants or even recovery of the contaminants in the original, or altered, form.

In more general terms, gas cleaning is divided into removal of particulate impurities and removal of gaseous impurities. For the purposes of this chapter, the latter operation includes the removal of hydrogen sulfide, carbon dioxide, sulfur dioxide, and products that are not related to synthesis gas and hydrogen production. However, there is also a need for subdivision of these two categories as dictated by needs and process capabilities: (i) coarse cleaning whereby substantial amounts of unwanted impurities are removed in the simplest, most convenient, manner; (ii) fine cleaning for the removal of residual impurities to a degree sufficient for the majority of normal chemical plant operations, such as catalysis or preparation of normal commercial products, or cleaning to a degree sufficient to discharge an effluent gas to atmosphere through a chimney; (iii) ultra-fine cleaning where the extra step (as well as the extra expense) is justified by the nature of the subsequent operations or the need to produce a particularly pure product.

The products can range from (i) high-purity hydrogen, (ii) high-purity carbon monoxide (iii) high-purity carbon dioxide, and (iv) a range of hydrogen/carbon monoxide mixtures. The plant is often times referred to as a HYCO if it is designed to produce both carbon monoxide and hydrogen at high purity; else it is referred to as a synthesis gas, or, synthesis gas plant. In fact, the hydrogen/carbon monoxide ratio can be selected at will and the plant's process scheme chosen, in part, by the product composition required. The hydrogen/carbon monoxide ratio will likely vary between 1 and 3 for HYCO and synthesis gas plants. However, at one end of the scale, i.e., if hydrogen is the desired product, then the ratio can approach infinity by shifting all of the carbon monoxide to carbon dioxide. By contrast, on the other end, the ratio cannot be adjusted to zero because hydrogen and water is always produced. An interesting general rule of thumb exists in terms of the hydrogen/carbon monoxide ratio produced by the different gasification processes:

Gasification Process	H_2/CO ratio
Steam Methane Reformer	3.0–5.0
SMR + Oxygen Secondary Reforming (O2R)	2.5–4.0
Autothermal Reforming	1.6–2.65
Partial Oxidation	1.6–1.9

It should be noted that in practice, however, the options are not limited to the ranges shown but rather even greater hydrogen/carbon monoxide ratios, if adjustments are made like the inclusion of a shift converter to effect near-equilibrium water-gas shift conversion, or, adjusting the amount of steam.

Throughout the previous section there has, of necessity, been frequent reference to the production of hydrogen as an integral part of the production of carbon monoxide, since the two gases make up the mixture known as synthesis gas. Hydrogen is indeed an important commodity in the refining industry because of its use in hydrotreating processes, such as desulfurization, and in hydroconversion processes, such as hydrocracking. Part of the hydrogen is produced during reforming processes but that source, once sufficient, is now insufficient for the hydrogen needs of a modern refinery (Gary et al., 2007; Speight, 2011, 2014a, 2017; Hsu and Robinson, 2017).

In addition, optimum hydrogen purity at the reactor inlet extends catalyst life by maintaining desulfurization kinetics at lower operating temperatures and reducing carbon laydown. Typical purity increases resulting from hydrogen purification equipment and/or increased hydrogen sulfide removal as well as tuning hydrogen circulation and purge rates, may extend catalyst life up to about 25%. Indeed, since hydrogen use has become more widespread in refineries, hydrogen production has moved from the status of a high-tech specialty operation to an integral feature of most refineries.

While the gasification of residua and coke to produce hydrogen and/or power may increase in use in refineries over the next two decades (Speight, 2011), several other processes are available for the production of the additional hydrogen that is necessary for the various heavy feedstock hydroprocessing sequences (Speight, 2014a) and it is the purpose of this section to present a general description of these processes.

Purities in excess of 99.5% of either the hydrogen or carbon monoxide produced from synthesis gas can be achieved if desired. Four of the major process technologies available are (i) cryogenics plus methanation, (ii) cryogenics plus pressure swing adsorption, (iii) methane-wash cryogenic process, and (iv) the Cosorb process. Thus:

i. *Cryogenics + Methanation*: This process utilizes a cryogenic process (occurring in a cold box) whereby carbon monoxide is liquefied in a number of steps until hydrogen with a purity of ~98% is produced. The condensed carbon monoxide product, which would contain methane, is then distilled to produce essentially pure carbon monoxide and a mixture of carbon monoxide/methane. The latter stream can be used as fuel. The hydrogen stream from the cold box is taken to a shift converter where the remaining carbon monoxide is converted to carbon dioxide and hydrogen. The carbon dioxide is then removed and any further carbon monoxide or carbon dioxide can be removed by methanation. The resulting hydrogen stream can be of purities ~99.7%.

ii. *Cryogenics plus Pressure Swing Adsorption*: This process utilizes the similar sequential liquefaction of carbon monoxide in a cold box until hydrogen of ~98% purity is achieved. Again, the carbon monoxide stream can be further distilled to remove methane until it is essentially pure. The hydrogen stream is then allowed to go through multiple swings of pressure swing adsorption cycles until the hydrogen purity of even as high as 99.999% is produced.

iii. *Methane-wash Cryogenic Process*: In this scheme, liquid carbon monoxide is absorbed into a liquid methane stream so that the hydrogen stream produced contains only ppm levels of carbon monoxide but about 5%–8% methane. Hence, a hydrogen stream purity of only about 95% is possible. The liquid carbon monoxide/methane stream, however, can be distilled to produce an essentially pure carbon monoxide stream and a carbon monoxide/methane stream which can be used as fuel.

iv. *Cosorb Process*: This process utilizes copper ions (cuprous aluminum chloride ($CuAlCl_4$)) in toluene to form a chemical complex with the carbon monoxide, and in effect separating it from the hydrogen, nitrogen, carbon dioxide, and methane. This process can capture about 96% of the carbon monoxide to produce a stream of greater than 99% purity. The downside of this process is that water, hydrogen sulfide, and other trace chemicals which can poison the copper catalyst, must be removed prior to the reactor. Further, a hydrogen stream of only up to 97% purity is obtained. However, while the efficiency of cryogenic separation decreases with low carbon monoxide content of the feed, the Cosorb process is able to process gases with a low carbon monoxide content more efficiently.

10.8 FISCHER–TROPSCH CHEMISTRY

Synthesis gas (also called syngas) is the name given to a gas mixture that contains varying amounts of carbon monoxide (CO) and hydrogen (H_2) generated by the gasification of a carbonaceous material. Examples include steam reforming of natural gas, petroleum residua, coal, and biomass.

10.8.1 Chemical Principles

Synthesis gas consists primarily of carbon monoxide, carbon dioxide, and hydrogen, and has less than half the energy density of natural gas. Synthesis gas is combustible and often used as a fuel source or as an intermediate for the production of other chemicals. Synthesis gas for use as a fuel is most often produced by gasification of the carbonaceous feedstock or municipal waste mainly by the following paths:

$$C + O_2 \rightarrow CO_2$$

$$CO_2 + C \rightarrow 2CO$$

$$C + H_2O \rightarrow CO + H_2$$

When used as an intermediate in the large-scale, industrial synthesis of hydrogen and ammonia, it is also produced from natural gas (via the steam reforming reaction) as follows:

$$CH_4 + H_2O \rightarrow CO + 3H_2$$

The synthesis gas produced in large waste-to-energy gasification facilities is used as fuel to generate electricity.

Although the focus of this section is the production of hydrocarbon derivatives from synthesis gas, it is worthy of note that all or part of the clean synthesis gas can also be used (i) as chemical *building blocks* to produce a broad range of chemicals using processes well established in the chemical and petrochemical industry, (ii) as a fuel producer for highly efficient fuel cells (which run off the hydrogen made in a gasifier) or perhaps in the future, hydrogen turbines and fuel cell-turbine hybrid systems, and (iii) as a source of hydrogen that can be separated from the gas stream and used as a fuel or as a feedstock for refineries (which use the hydrogen to upgrade petroleum products) (Gary et al., 2007; Speight, 2011, 2014a, 2017; Hsu and Robinson, 2017). However, the decreasing availability and increased price of petroleum has renewed the worldwide interest in the production of liquid hydrocarbon derivatives from carbon monoxide and hydrogen using metal catalysts, also known as Fischer–Tropsch synthesis or Fischer–Tropsch process.

Gasification to produce synthesis gas can proceed from just about any organic material, including biomass and plastic waste. The resulting synthesis gas burns cleanly into water vapor and carbon dioxide. Alternatively, synthesis gas may be converted efficiently to methane via the Sabatier reaction, or to a diesel-like synthetic fuel via the Fischer–Tropsch process. Inorganic components of the feedstock, such as metals and minerals, are trapped in an inert and environmentally safe form as char, which may have use as a fertilizer.

Regardless of the final fuel form, gasification itself and subsequent processing neither emits nor traps greenhouse gasses such as carbon dioxide. Combustion of synthesis gas or derived fuels does of course emit carbon dioxide. However, biomass gasification could play a significant role in a renewable energy economy, because biomass production removes carbon dioxide from the atmosphere. While other biofuel technologies such as biogas and biodiesel are also carbon neutral, gasification runs on a wider variety of input materials, can be used to produce a wider variety of output fuels, and is an extremely efficient method of extracting energy from biomass. Biomass gasification is, therefore, one of the most technically and economically convincing energy possibilities for a carbon neutral economy.

The manufacture of gas mixtures of carbon monoxide and hydrogen has been an important part of chemical technology for about a century. Originally, such mixtures were obtained by the reaction of steam with incandescent coke and were known as *water gas*. Used first as a fuel, water gas soon attracted attention as a source of hydrogen and carbon monoxide for the production of chemicals, at which time it gradually became known as synthesis gas. Eventually, steam reforming processes, in which steam is reacted with natural gas (methane) or petroleum naphtha over a nickel catalyst, found wide application for the production of synthesis gas.

A modified version of steam reforming known as autothermal reforming, which is a combination of partial oxidation near the reactor inlet with conventional steam reforming further along the reactor, improves the overall reactor efficiency and increases the flexibility of the process. Partial oxidation processes using oxygen instead of steam also found wide application for synthesis gas manufacture, with the special feature that they could utilize low-value feedstocks such as heavy petroleum residua. In recent years, catalytic partial oxidation employing very short reaction times (milliseconds) at high temperatures (850°C–1,000°C; 1,560°F–1,830°F) is providing still another approach to synthesis gas manufacture. Nearly complete conversion of methane, with close to 100% selectivity to hydrogen and carbon monoxide, can be obtained with a rhenium monolith under well-controlled conditions. Experiments on the catalytic partial oxidation of *n*-hexane conducted with added steam give much higher yields of hydrogen than can be obtained in experiments without steam, a result of much interest in obtaining hydrogen-rich streams for fuel cell applications.

The route of coal to synthetic automotive fuels, as practiced by SASOL is technically proven and products with favorable environmental characteristics are produced. As is the case in essentially all the carbonaceous feedstock conversion processes where air or oxygen is used for the utilization or partial conversion of the energy in the carbonaceous feedstock, the carbon dioxide burden is a drawback as compared to crude oil.

The uses of synthesis gas include use as a chemical feedstock and in gas-to-liquid processes which use Fisher–Tropsch chemistry to make liquid fuels as feedstock for chemical synthesis, as well as being used in the production of fuel additives, including diethyl ether and methyl t-butyl ether, acetic acid, and its anhydride; synthesis gas could also make an important contribution to chemical synthesis through conversion to methanol. There is also the option in which stranded natural gas is converted to synthesis gas production followed by conversion to liquid fuels.

The hydroformylation synthesis (also known as the oxo synthesis or the oxo process) is an industrial process for the production of aldehyde derivatives from alkene derivatives. This chemical reaction entails the addition of a formyl group (CHO) and a hydrogen atom to a carbon-carbon double bond. A key consideration of hydroformylation is the production of normal isomers or the production of iso-isomers in the product(s). For example, the hydroformylation of propylene can yield two isomeric products, butyraldehyde and iso-butyraldehyde:

$$H_2 + CO + CH_3CH=CH_2 \rightarrow CH_3CH_2CH_2CHO \ (\textit{n}\text{-butyraldehyde})$$

$$H_2 + CO + CH_3CH=CH_2 \rightarrow (CH_3)_2 CHCHO \ (\textit{iso}\text{-butyraldehyde})$$

These isomers reflect the regio-chemistry (the preference of one direction of chemical bond formation or chemical bond scission over all other possible directions) of the insertion of the alkene into the M–H bond. Generally, both products are not equally desirable.

As an example of the hydroformylation process, the Exxon process, also Kuhlmann-oxo process or the PCUK-oxo process is used for the hydroformylation of C_6 to C_{12} olefin derivatives using cobalt-based catalysts. In order to recover the catalyst, an aqueous sodium hydroxide solution or sodium carbonate is added to the organic phase. By extraction with olefin and neutralization by addition of sulfuric acid solution under carbon monoxide pressure the metal-carbonyl hydride can be recovered. The recovered hydride is stripped out with synthesis gas, absorbed by the olefin, and returned to the reactor. The Exxon process, similar to the BASF process, is carried out at a temperature in the order of 160°C–180°C (320°F–355°F) and a pressure suitable to the reactants and products.

The Fischer–Tropsch reaction can be described as the synthesis of hydrocarbon derivatives via the hydrogenation of carbon monoxide using transition metal catalysts. The major catalysts used industrially are Fe and Co, but can also be Ru and Ni. From a mechanism perspective, the reactions can be regarded as a carbon chain building process where methylene (CH_2) groups are attached sequentially in a carbon chain (Table 10.3). Thus:

$$n CO + [n + m/2] H_2 \rightarrow C_n H_m + n H_2O$$

For example:

$$CO + 2H_2 \rightarrow -[CH_2]- + H_2O$$

A common and salient feature of the reactions is the exothermicity of the reactions. As a general rule of thumb, the reactions which produce water and carbon dioxide as a product tend to be more exothermic on account of the very high heat of formation of these species. Some of the reactions proposed are as follows (Rauch, 2001):

$$CO_2 + 3H_2 \rightarrow -[CH_2]- + 2H_2O \quad \Delta H = -125 \text{ kJ/mol}$$

$$CO + 2H_2 \rightarrow -[CH_2]- + H_2O \quad \Delta H = -165 \text{ kJ/mol}$$

$$2CO + H_2 \rightarrow -[CH_2]- + CO_2 \quad \Delta H = -204 \text{ kJ/mol}$$

$$3CO + H_2 \rightarrow -[CH_2]- + 2CO_2 \quad \Delta H = -244 \text{ kJ/mol}$$

There is also the water-gas shift reaction:

$$CO + H_2O \rightarrow H_2 + CO_2 \quad \Delta H = -39 \text{ kJ/mol}$$

Due to the very high exothermic nature of the Fischer–Tropsch reactions, as illustrated in the reactions above, an important issue is, not surprisingly, the need to avoid an increase in temperature.

TABLE 10.3

Carbon Chain Groups of the Range of Fischer–Tropsch Products Which Can Be Produced

Carbon Number	Group Name
C_1–C_2	Synthetic natural gas
C_3–C_4	Liquefied petroleum gas
C_5–C_7	Low-boiling naphtha
C_8–C_{10}	High-boiling naphtha
C_{11}–C_{16}	Middle distillate
C_{17}–C_{30}[a]	Low-melting wax
C_{31}–C_{60}	High-melting wax

[a] The C_{17} n-alkane (n-heptadecane) is the first member of the series that is not fully liquid under ambient conditions (melting point: 21°C (70°F).

The need for cooling is thus of critical importance in order to (i) maintain stable reaction conditions, (ii) avoid the tendency to produce lighter hydrocarbon derivatives, and (iii) prevent catalyst sintering and hence reduction in activity. Since the total heat of reaction is in the order of approximately 25% of the heat of combustion of the synthesis gas (i.e., reactants if the Fischer–Tropsch process), a theoretical limit on the maximum efficiency of the Fischer–Tropsch process is imposed (Rauch, 2001).

Two main chemical characteristics of Fischer–Tropsch synthesis are the unavoidable production of a wide range of hydrocarbon products (olefin derivatives, paraffin derivatives, and oxygenated products) and the liberation of a large amount of heat from the highly exothermic synthesis reactions. Product distributions are influenced by temperature, feed gas composition (hydrogen/carbon monoxide), pressure, catalyst type, and catalyst composition. Fischer–Tropsch products are produced in four main steps: synthesis gas generation, gas purification, Fischer–Tropsch synthesis, and product upgrading. Depending on the types and quantities of Fischer–Tropsch products desired, either low (200°C–240°C; 390°F–465°F) or high-temperature (300°C–350°C; 570°F–660°F) synthesis is used with either an iron (Fe) or cobalt catalyst (Co) (Van Berge, 1997).

The required gas mixture of carbon monoxide and hydrogen (synthesis gas) is created through a reaction of coke or the carbonaceous feedstock with water steam and oxygen, at temperatures over 900°C. In the past, town gas and gas for lamps were a carbon monoxide-hydrogen mixture, made by gasifying coke in gas works. In the 1970s, it was replaced with imported natural gas (methane). Gasification of carbonaceous feedstocks and Fischer–Tropsch hydrocarbon synthesis together bring about a two-stage sequence of reactions which allows the production of liquid fuels like gasoline and diesel out of the solid combustible and the carbonaceous feedstock.

The Fischer–Tropsch synthesis is, in principle, a carbon chain building process, where methylene groups are attached to the carbon chain. The actual reactions that occur have been, and remain, a matter of controversy, as it has been the last century since 1930s.

$$(2n+1)H_2 + nCO \rightarrow C_nH_{2n+2} + nH_2O$$

Even though the overall Fischer–Tropsch process is described by the following chemical equation:

$$(2n+1)H_2 + nCO \rightarrow C_nH_{2n+2} + nH_2O$$

The initial reactants in the above reaction (i.e., carbon monoxide and H_2) can be produced by other reactions such as the partial combustion of a hydrocarbon:

$$C_nH_{2n+2} + \tfrac{1}{2}nO_2 \rightarrow (n+1)H_2 + nCO$$

For example (when $n=1$), methane (in the case of gas-to-liquids applications):

$$2CH_4 + O_2 \rightarrow 4H_2 + 2CO$$

Or by the gasification of any carbonaceous source, such as biomass:

$$C + H_2O \rightarrow H_2 + CO$$

The energy needed for this endothermic reaction is usually provided by (exothermic) combustion with air or oxygen:

$$2C + O_2 \rightarrow 2CO$$

These reactions are highly exothermic, and to avoid an increase in temperature, which results in lighter hydrocarbon derivatives, it is important to have sufficient cooling, to secure stable reaction conditions. The total heat of reaction amounts to approximately 25% of the heat of combustion of the synthesis gas, and lays thereby a theoretical limit on the maximal efficiency of the Fischer–Tropsch process.

The reaction is dependent on a catalyst, mostly an iron or cobalt catalyst where the reaction takes place. There is either a low- or high-temperature process (low-temperature Fischer–Tropsch or high-temperature Fischer–Tropsch), with temperatures ranging between 200°C–240°C (390°F–465°F) for low-temperature Fischer–Tropsch and 300°C–350°C (570°F–660°F) for the high-temperature Fischer–Tropsch process. The high-temperature Fischer–Tropsch process uses an iron catalyst and the low-temperature Fischer–Tropsch either an iron or a cobalt catalyst. The different catalysts include also nickel-based and ruthenium-based catalysts, which also have enough activity for commercial use in the process. But the availability of ruthenium is limited and the nickel-based catalyst has high activity but produces too much methane, and additionally the performance at high pressure is poor, due to production of volatile carbonyls. This leaves only cobalt and iron as practical catalysts, and this study will only consider these two. Iron is cheap, but cobalt has the advantage of higher activity and longer life, though it is on a metal basis 1,000 times more expensive than iron catalyst.

10.8.2 Refining Fischer–Tropsch Products

The Fischer–Tropsch product stream typically contains hydrocarbon derivatives having a range of numbers of carbon atoms, including gases, liquids, and waxes. Depending on the molecular weight product distribution, different Fischer–Tropsch product mixtures are ideally suited to different uses. For example, Fischer–Tropsch product mixtures containing liquids may be processed to yield gasoline, as well as heavier middle distillates. Hydrocarbon waxes may be subjected to additional processing steps for conversion to liquid and/or gaseous hydrocarbon derivatives. Thus, in the production of a Fischer–Tropsch product stream for processing to a fuel it is desirable to obtain primarily hydrocarbon derivatives that are liquids and waxes (e.g., C_{5+} hydrocarbon derivatives).

Initially, the light gases in raw product are separated and sent to a gas-cleaning operation. The higher-boiling product is distilled to produce separate streams of naphtha, distillate, and wax.

The naphtha stream is first hydrotreated which produces a hydrogen-saturated liquid product (primarily paraffin derivatives), a portion of which are converted by isomerization from normal paraffin derivatives to isoparaffin derivatives to boost their octane value. Another fraction of the hydrotreated naphtha is catalytically reformed to provide some aromatic content to (and further boost the octane value of) the final gasoline blend stock. The distillate stream is also hydrotreated, resulting directly in a finished diesel blend stock. The wax fraction is hydrocracked into a finished distillate stream and naphtha streams that augment the hydrotreated naphtha streams sent for isomerization and for catalytic cracking.

Thus, conventional refinery processes (Gary et al., 2007; Speight, 2011, 2014a, 2017; Hsu and Robinson, 2017) can be used for upgrading of Fischer–Tropsch liquid and wax products. A number of possible processes for Fischer–Tropsch products are wax hydrocracking, distillate hydrotreating, catalytic reforming, naphtha hydrotreating, alkylation, and isomerization. Fuels produced with the Fischer–Tropsch synthesis are of a high quality due to a very low aromaticity and zero sulfur content.

The diesel fraction has a high cetane number resulting in superior combustion properties and reduced emissions. New and stringent regulations may promote replacement or blending of conventional fuels by sulfur and aromatic-free products. Also, other products besides fuels can be manufactured with Fischer–Tropsch in combination with upgrading processes, for example, ethylene, propylene, α-olefin derivatives, alcohols, ketones, solvents, and waxes. These valuable byproducts of the process have higher added values, resulting in an economically more attractive process economy (Gary et al., 2007; Speight, 2011, 2014a, 2017; Hsu and Robinson, 2017).

At this point, it is necessary to deal once again with the production of chemicals from the carbonaceous feedstock by gasification followed by conversion of the synthesis gas mixture (carbon monoxide, carbon monoxide, and hydrogen (H_2)) to higher molecular weight liquid fuels and other chemicals (Chapters 20 and 21) (Penner, 1987).

The production of synthesis gas involves reaction of the carbonaceous feedstock with steam and oxygen. The gas stream is subsequently purified (to remove sulfur, nitrogen, and any particulate matter) after which it is catalytically converted to a mixture of liquid hydrocarbon products. In addition, synthesis gas may also be used to produce a variety of products, including ammonia, and methanol.

The synthesis of hydrocarbon derivatives from carbon monoxide and hydrogen (synthesis gas) (the Fischer–Tropsch synthesis) is a procedure for the indirect liquefaction of coal (Storch et al., 1951; Batchelder, 1962; Jones et al., 1992). This process is the only coal liquefaction scheme currently in use on a relatively large commercial scale; South Africa is currently using the Fischer–Tropsch process on a commercial scale in their SASOL complex, although Germany produced roughly 156 million barrels of synthetic petroleum annually using the Fischer–Tropsch process during the World War II.

Briefly, in the gasification process, the carbonaceous feedstock is converted to gaseous products at temperatures in excess of 800°C (1,470°F), and at moderate pressures, to produce synthesis gas.

$$C + H_2O \rightarrow CO + H_2$$

The gasification may be attained by means of any one of several processes (Speight, 2013a,b, 2014a,b; Luque and Speight, 2015). The exothermic nature of the process and the decrease in the total gas volume in going from reactants to products suggest the most suitable experimental conditions to use in order to maximize product yields. The process should be favored by high pressure and relatively low reaction temperature.

In practice, the Fischer–Tropsch reaction is generally carried out at temperatures in the range 200°C–350°C (390°F–660°F) and at pressures of 75–4,000 psi; the hydrogen/carbon monoxide ratio is usually at ca. 2.2:1 or 2.5:1. Since up to three volumes of hydrogen may be required to achieve the next stage of the liquids production, the synthesis gas must then be converted by means of the water-gas shift reaction to the desired level of hydrogen after which the gaseous mix is purified (acid gas removal, etc.) and converted to a wide variety of hydrocarbon derivatives.

$$CO + H_2O \rightarrow CO_2 + H_2$$

$$CO + (2n+1)H_2 \rightarrow C_nH_{2n+2} + H_2O$$

These reactions result primarily in low- and medium-boiling aliphatic compounds; present commercial objectives are focused on the conditions that result in the production of n-hydrocarbon derivatives as well as olefin derivatives and oxygenated materials.

REFERENCES

Aasberg-Petersen, K., Bak Hansen, J.-H., Christiansen, T.S., Dybkjær, I., Seier, Christensen, P., Stub Nielsen, C., Winter Madsen, S.E.L., and Rostrup-Nielsen, J.R. 2001. Technologies for large-scale gas conversion. *Applied Catalysis A: General*, 221: 379–387.

Aasberg-Petersen, K., Christensen, T.S., Stub Nielsen, C., and Dybkjaer, I. 2002. Recent developments in autothermal reforming and pre-reforming for synthesis gas production in GTL applications. *Preprints, Division of Fuel Chemistry, American Chemical Society*, 47(1), 96–97.

Alstrup, I. 1988. A new model explaining carbon filament growth on nickel, iron, and Ni-Cu alloy catalysts. *Journal of Catalysis*, 109: 241–251.

Balasubramanian, B., Ortiz, A.L., Kaytakoglu, S. and Harrison, D.P. 1999. Hydrogen from methane in a single-step process. *Chemical Engineering Science*, 54: 3543–3552.

Batchelder, H.R. 1962. Chapter 1, Vol. V. In: *Advances in Petroleum Chemistry and Refining*. J.J. McKetta Jr. (Editor). Interscience Publishers Inc., New York.

Chadeesingh, R. 2011. Chapter 5: The Fischer–Tropsch process, Part 3. In: *The Biofuels Handbook*. J.G. Speight (Editor). The Royal Society of Chemistry, London, pp. 476–517.

Choudhary, V.R., Rajput, A.M., and Prabhakar, B. 1993. *Journal of Catalysis*, 139: 326.

Dyer, P.N. and Chen, C.M. 1999. Engineering development of ceramic membrane reactor systems for converting natural gas to hydrogen and synthesis gas for transportation fuels. *Proceedings of the Energy Products for the 21st Century Conference, September 22*.

Gary, J.G., Handwerk, G.E., and Kaiser, M.J. 2007. *Petroleum Refining: Technology and Economics*. 5th Edition. CRC Press, Taylor & Francis Group, Boca Raton, FL.

Gunardson, H.H., and Abrardo, J.M. April 1999. *Hydrocarbon Processing*, pp. 87–93.

Hagh, B.F. 2004. Comparison of autothermal reforming for hydrocarbon fuels. *Preprints, Division of Fuel Chemistry, American Chemical Society*, 49(1): 144–147.

Hsu, C.S., and Robinson, P.R. (Editors). 2017. *Handbook of Petroleum Technology*. Springer International Publishing AG, Cham.

Hufton, J.R., Mayorga, S. and Sircar, S., 1999. Sorption-enhanced reaction process for hydrogen production. *AIChE Journal*, 45: 248–256.

Jones, C.J., Jager, B., and Dry, M.D. 1992. *Oil and Gas Journal*, 90(3): 53.

Lapszewicz, J.A., and Jiang, X. 1992. *Preprints, ACS Division of Petroleum Chemistry*, 37: 252.

Luque, R., and Speight, J.G. (Editors), 2015. *Gasification for Synthetic Fuel Production: Fundamentals, Processes, and Applications*. Woodhead Publishing, Elsevier, Cambridge, United Kingdom.

Mokhatab, S., Poe, W.A., and Speight, J.G. 2006. *Handbook of Natural Gas Transmission and Processing*. Elsevier, Amsterdam.

Nataraj, S., Moore, R.B., Russek, S.L., US 6048472; 2000; assigned to Air Products and Chemicals, Inc.

Penner, S.S. 1987. *Proceedings. Fourth Annual Pittsburgh Coal Conference*. University of Pittsburgh, Pittsburgh, PA, p. 493.

Rauch, R. 2001. Biomass gasification to produce synthesis gas for fuel cells, liquid fuels and chemicals, IEA bioenergy agreement, Task 33: Thermal gasification of biomass.

Rostrup-Nielsen, J.R. 1984. Sulfur-passivated nickel catalysts for carbon-free steam reforming of methane. *Journal of Catalysis*, 85: 31–43.

Rostrup-Nielsen, J.R. 1993. Production of synthesis gas. *Catalysis Today*, 19: 305–324.

Schulz, H. 1999. Short history and present trends of Fischer–Tropsch synthesis. *Applied Catalysis A: General*, 86(1–2): 3–12.

Shu, J., Grandjean, B.P.A., and Kaliaguine, S. 1995. *Catalysis Today*, 25: 327–332.

Speight, J.G. 2007. *Natural Gas: A Basic Handbook*. GPC Books, Gulf Publishing Company, Houston, TX.

Speight, J.G. 2008. *Synthetic Fuels Handbook: Properties, Processes, and Performance*. McGraw-Hill, New York.

Speight, J.G. 2011. *The Refinery of the Future*. Gulf Professional Publishing, Elsevier, Oxford, United Kingdom.

Speight, J.G. 2013a. *The Chemistry and Technology of Coal*. 3rd Edition. CRC Press, Taylor & Francis Group, Boca Raton, FL.

Speight, J.G. 2013b. *Coal-Fired Power Generation Handbook*. Scrivener Publishing, Beverly, MA.

Speight, J.G. 2014a. *The Chemistry and Technology of Petroleum* 5th Edition. CRC Press, Taylor & Francis Group, Boca Raton, FL.

Speight, J.G. 2014b. *Gasification of Unconventional Feedstocks*. Gulf Professional Publishing, Elsevier, Oxford, United Kingdom.

Speight, J.G. 2017. *Handbook of Petroleum Refining Processes*. CRC Press, Taylor & Francis Group, Boca Raton, FL.

Storch, H.H., Golumbic, N., and Anderson, R.B. 1951. *The Fischer Tropsch and Related Syntheses*. John Wiley & Sons Inc., New York.

Tian, P., Wei, Y., Ye, M., and Liu, Z., 2015. Methanol to olefins (MTO): From fundamentals to commercialization. *ACS Catalysis*, 5(3): 1922–1938

Udengaard, N.R., Hansen, J.H.B., Hanson, D.C., and Stal, J.A. 1992. Sulfur passivated reforming process lowers syngas H_2/CO ratio. *Oil & Gas Journal*, 90: 62–67.

Van Berge, P.J. 1997. *Natural Gas Conversion IV*, Vol. 107. Studies in Surface Science and Catalysis, p. 207.

Watson, G.H. 1980. *Methanation Catalysts*. Report ICTIS/TR09. International Energy Agency, London.

11 Monomers, Polymers, and Plastics

11.1 INTRODUCTION

The list of chemicals produced by the petrochemical industry includes, but is not limited to (i) synthesis gas-based products including ammonia, methanol, and their derivatives; (ii) ethylene and derivatives; (iii) propylene, including on-purpose and methanol-based routes and derivatives; (iv) C_4 monomers; aromatics; oxides, glycols, and polyols and derivatives; (v) chlor-alkali, ethylene dichloride, vinyl chloride monomer, and polyvinyl chloride (PVC); (vi) polyolefins—solution, slurry and gas phase; alpha olefins and poly alpha olefins; polyethylene terephthalate (PET)—bottles and fiber; polystyrene (PS)—general purpose, high impact and expandable; (vii) styrene derivatives—such as acrylonitrile butadiene styrene, acrylonitrile-styrene, and acrylonitrile styrene acrylate; and (viii) specialty polymers including poly-oxymethylene, super-absorbent polymers, and poly-methylmetacrylate; and nylon 6, 6-6, and intermediates.

Thus, the ascent of polymer technology during the 20th century is due in no small part to the availability of starting materials that became available through the evolving and expanding petrochemical industry. In fact, a high proportion of all petrochemicals are used for the production of polymers, the most important building blocks being ethylene, propylene, and butadiene (Table 11.1; Matar and Hatch, 2001). These three petrochemicals can be polymerized directly but an important part of their production is used to create more complex monomers through different ways of information into a polymer (Tables 11.2 and 11.3; Figure 11.1). Ethylene is the progenitor of most vinyl monomers and, hence, the need for an almost endless pressure on ethylene supply is particularly high. In fact, the C_2 and C_3 building blocks can be combined with benzene to form another set of monomers and intermediates, particularly valuable for constructing the complex repeat units noted in the last section. Other chemicals are also produced, such as plasticizers which are then added in a subsequent stage to polymers to modify their properties. But first, the relevant definitions.

A monomer is the original molecular form from which a polymer (and plastic product) is produced. A polymer (which may also be referred to as a macromolecule) consists of repeating molecular units which usually are held together by covalent bonds (Ali et al., 2005). Polymerization is the

---

TABLE 11.1
Polymers from Petrochemicals

Polyethylene	Derived from ethylene ($CH_2=CH_2$).
	Ethylene is derived from natural gas, from overhead gases in refinery and from crackers.
Polypropylene	Derived from propylene ($CH_3CH=CH_2$).
	Propylene has almost same origin as ethylene.
	Used for making clothes and various other plastics.
Polyesters	Produced from terephthalic acid which is derived from p-xylene (1,4-$HO_2CC_6H_4CO_2H$.
	p-Xylene ($H_3CC_6H_4CH_3$) has its origin from various aromatic compounds found in the crude oil.
	Refining separates benzene derivatives.
Rubber	Various synthetic rubbers like polyacrylate rubber and ethylene-acrylate rubber
	Derived from various petroleum-derived chemicals, such as 1,3-butadiene ($CH_2=CHCH=CH_2$) and acrylic acid ($CH_2=CHCO_2H$).

TABLE 11.2

Several Ways in Which Different Monomeric Units Might Be Incorporated in a Polymer

Statistical Copolymers	Also called random copolymers. Here the monomeric units are distributed randomly, and sometimes unevenly, in the polymer chain: –ABBAAABAABBBABAABA–
Alternating Copolymers	The monomeric units are distributed in a regular alternating fashion, with nearly equimolar amounts of each in the chain: –ABABABABABABABAB–
Block Copolymers	Instead of a mixed distribution of monomeric units, a long sequence or block of one monomer is joined to a block of the second monomer: –AAAAA–BBBBBBB–AAAAAAA–BBB~.
Graft Copolymers	The side chains of a given monomer are attached to the main chain of the second monomer: –AAAAAAA(BBBBBBB–)AAAAAAA(BBBB–)AAA~

TABLE 11.3

Selected Hydrocarbon Addition Polymers

Name(s)	Formula	Monomer	Properties	Uses
Polyethylene (LDPE)	$-(CH_2-CH_2)_n-$	Ethylene $CH_2=CH_2$	Soft, waxy solid	Film wrap, plastic bags
Polyethylene (HDPE)	$-(CH_2-CH_2)_n-$	Ethylene $CH_2=CH_2$	Rigid, translucent solid	Electrical insulation, bottles, toys
Polypropylene	$-[CH_2-CH(CH_3)]_n-$	Propylene $CH_2=CHCH_3$	Atactic: soft, elastic solid Isotactic: hard, strong solid	Similar to LDPE, carpet, upholstery
PS	$-[CH_2-CH(C_6H_5)]_n-$	Styrene $CH_2=CHC_6H_5$	Hard, rigid, clear solid, soluble in organic solvents	Toys, cabinets Packaging (foamed)
cis-Polyisoprene	$-[CH_2-CH=C(CH_3)-CH_2]_n-$	Isoprene $CH_2=CH-C(CH_3)=CH_2$	Soft, sticky solid	Requires vulcanization for practical use

FIGURE 11.1 Variations in polymer structure. (1) a regular polymer, (2) an alternating copolymer, (3) a random copolymer, (4) a block copolymer, and (5) a grafted copolymer.

process of covalently bonding the low molecular weight monomers into a high molecular weight polymer.

Polymerization is a reaction in which chain-like macromolecules are formed by combining small molecules (monomers). Monomers are the building blocks of these large molecules called polymers. For example, in order to depict polymers, cellulose and a protein can be considered. Cellulose (the most abundant organic compound on earth), a molecule made of many simple glucose units (monomers) joined together through a glycoside linkage.

Cellulose

Proteins, the material of life, are polypeptides made of α-amino acids (alpha-amino acids) attached by an amide.

Several methods exist to synthesize amino acids. One of the oldest methods begins with the bromination at the α-carbon of a carboxylic acid. For example, the Strecker amino acid synthesis involves the treatment of an aldehyde with potassium cyanide and ammonia; this produces an α-amino nitrile as an intermediate. Hydrolysis of the nitrile in acid then yields an α-amino acid.

| Aldehyde | α-amino nitrile | amino acid |

Using ammonia or ammonium salts in this reaction gives unsubstituted amino acids, whereas substituting primary and secondary amines will yield substituted amino acids. Likewise, using ketone derivatives instead of aldehydes, gives α,α-disubstituted amino acids. Thus, to construct a protein macromolecule, the amino acids react to form a link, as, for example, in the cojoining of two amino acids to form a dipeptide.

| Amino acids | Dipeptide |

In each case, the linkage in the structure of the product (a dipeptide) is known as a peptide link, which is (chemically speaking) an amide link. A protein chain (with the N-terminal on the left) will, therefore, be of this type.

Schematic of protein structure

For a molecule to be a monomer, it must be at least bifunctional insofar as it has the capacity to interlink with other monomer molecules. While not truly bifunctional in the sense that they contain to functional groups, olefin derivatives have the ability to act as bifunctional molecules though the extra pair of electrons in the double bond.

A polymer may be a natural or synthetic macromolecule comprised of repeating units of a smaller molecule (monomers). The terms polymer and plastic are often used interchangeably, but polymers are a much larger class of molecules which includes plastics, plus many other materials, such as cellulose, amber, and natural rubber. Examples of hydrocarbon polymers include polyethylene and synthetic rubber (Schroeder, 1983).

In the current context, a monomer is a low molecular weight hydrocarbon molecule that has the potential of chemically bonding to other monomers of the same species to form a polymer. The lower molecular weight compounds built from monomers are also referred to as dimers (two monomer units), trimers (three monomer units), tetramers (four monomer units), pentamers (five monomer units), octamers (eight monomer units), continuing up to very high numbers of monomer units in the product.

The structure of monomer units in the polymer is retained by the chemical bonds between adjacent atoms thereby conferring upon the polymer the configuration. However, there can be many different configurations for a given set of atoms of a particular type. Different isomers of the monomer unit, which have different properties, confer different properties on the polymer. This structural configuration of the monomer is an important structural feature and plays a major role as the complexity of the monomer increase and is a major determinant of the structure and properties of the polymer chains.

In addition to structural isomerism in the monomer, which can be represented simply by the position of the double bond in butylene and is shown below as butylene-1 and butylene-2, there is also a second type of isomerism.

$CH_3CH_2CH=CH_2$	$CH_3CH=CHCH_3$
butylene-1	butylene-2
1-butene	2-butene

This type of isomerism (geometrical isomerism) occurs with various monomers, and is present in polymers such as natural rubber and butadiene rubber. In these cases, the single double bond in the final polymer can exist in two ways: a *cis* form and a *trans* form.

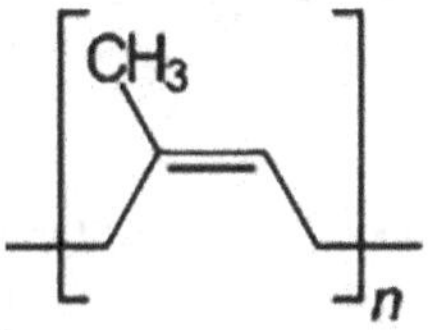

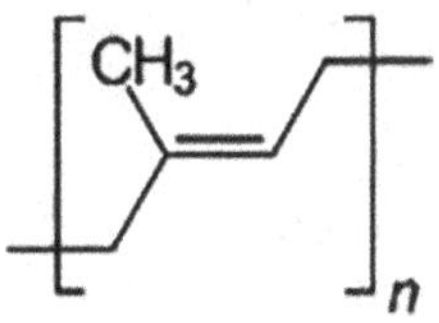

cis-1,4-polyisoprene   *trans*-1,4-polyisoprene

The pendant methyl group appears on the same side as the lone hydrogen atom or on the opposite side of the lone hydrogen atom. Similarly, commencing with 2-butylene, the final polymer may have the methyl group (i) on the same side of the final product or (ii) on alternate sides of the final product. Thus, because of the variations in monomer structure, the chemical structure of many polymers is rather complex because the polymerization reaction does not necessarily produce identical molecules. In fact, a polymeric material typically consists of a distribution of molecular sizes and sometimes also of shapes.

The properties of polymers are strongly influenced by details of the chain structure. The structural parameters that determine properties of a polymer include the overall chemical composition and the sequence of monomer units in the case of copolymers, the stereochemistry or the relative stereochemistry of the stereo-centers in the polymer chain, and geometric isomerization in the case of diene-type polymers.

The properties of a specific polymer can often be varied by means of controlling molecular weight, end groups, processing, cross-linking. Therefore, it is possible to classify a single polymer in more than one category. For example, some polymers nylon can be produced as fibers in the crystalline forms, or as plastics in the less crystalline forms. Also, certain polymers can be processed to act as plastics or elastomers. Synthetic fibers are long-chain polymers characterized by highly crystalline regions resulting mainly from secondary forces (e.g., hydrogen bonding). They have a much lower elasticity than plastics and elastomers. They also have high tensile strength, a light weight, and low moisture absorption.

11.2 PROCESSES AND PROCESS CHEMISTRY

Polymerization is the process by which polymers are manufactures and during the polymerization process, some chemical groups may be lost from each monomer and the polymer does not always retain the chemical properties or the reactivity of the monomer unit (Rudin, 1999; Braun et al., 2001; Carraher, 2003; Odian, 2004).

The polymer industry dates back to the 19th century, when natural polymers, such as cotton, were modified by chemical treatment to produce artificial silk (rayon). Work on synthetic polymers did not start until the beginning of the 20th century. In 1909, the first synthetic polymeric material was prepared by L.H. Baekeland who used condensation reaction between formaldehyde and phenol. Currently, these polymers serve as important thermosetting plastics (for example, phenol-formaldehyde resins). Since that time, many different polymeric products have been synthesized to respond to the demands of the market—examples are plastics, fibers, and synthetic rubber. The huge polymer market directly results from extensive work in synthetic organic compounds and catalysts. In addition, the discovery by Ziegler of a coordination catalyst in the titanium family paved the road for synthesizing many stereoregular polymers with improved properties.

Polymerization reactions can occur in bulk (without solvent), in solution, in emulsion, in suspension, or in a gas-phase process. Interfacial polymerization is also used with reactive monomers, such as acid chlorides. Polymers obtained by the *bulk technique* are usually pure due to the absence of a solvent. The purity of the final polymer depends on the purity of the monomers. Heat and viscosity are not easily controlled, as in other polymerization techniques, due to absence of a solvent, suspension, or emulsion medium. This can be overcome by carrying the reaction to low conversion and strong agitation. Outside cooling can also control the exothermic heat.

In *solution polymerization*, an organic solvent dissolves the monomer. Solvents should have low-chain transfer activity to minimize chain transfer reactions that produce low molecular weight polymers. The presence of a solvent makes heat and viscosity control easier than in bulk polymerization. Removal of the solvent may not be necessary in certain applications such as coatings and adhesives.

Emulsion polymerization is widely used to produce polymers in the form of emulsions, such as paints and floor polishes. It is also used to polymerize many water-insoluble vinyl monomers, such as styrene and vinyl chloride. In emulsion polymerization, an agent emulsifies the monomers. Emulsifying agents should have a finite solubility. They are either ionic, as in the case of alkylbenzene sulfonates, or nonionic, like polyvinyl alcohol.

Water is extensively used to produce emulsion polymers with a sodium stearate emulsifier. The emulsion concentration should allow micelles of large surface areas to form. The micelles absorb the monomer molecules activated by an initiator (such as a sulfate ion radical). X-ray and light-scattering techniques show that the micelles start to increase in size by absorbing the macromolecules. For example, in the free radical polymerization of styrene, the micelles increased to 250 times their original size.

In *suspension polymerization*, the monomer is first dispersed in a liquid, such as water and mechanical agitation keeps the monomer dispersed. Initiators should be soluble in the monomer and stabilizers, such as talc or polyvinyl alcohol, preventing polymer chains from adhering to each other and keep the monomer dispersed in the liquid medium. As a result, the final polymer appears in a granular form. Suspension polymerization produces polymers more pure than those from solution polymerization due to the absence of chain transfer reactions. As in a solution polymerization, the dispersing liquid helps control the heat of the reaction.

Interfacial polymerization is mainly used in polycondensation reactions with very reactive monomers. One of the reactants, usually an acid chloride, dissolves in an organic solvent (such as benzene or toluene), and the second reactant, a diamine or a diacid, dissolves in water. This technique produces polycarbonates (PCs), polyesters, and polyamides. The reaction occurs at the interface between the two immiscible liquids, and the polymer is continuously removed from the interface.

Two general reactions form synthetic polymers: chain addition and condensation. Addition polymerization requires a chain reaction in which one monomer molecule adds to a second, then a third and so on to form a macromolecule. Addition polymerization monomers are mainly low molecular weight olefinic compounds (e.g., ethylene or styrene) or conjugated diolefin derivatives (e.g., butadiene or isoprene).

Condensation polymerization can occur by reacting with either two similar or two different monomers to form a long polymer. This reaction usually releases a small molecule like water, as in the case of the esterification of a diol and a diacid. In condensation polymerization where ring opening occurs, no small molecule is released.

11.2.1 Addition Polymerization

Addition polymerization is employed primarily with substituted or unsubstituted olefin derivatives and conjugated diolefin derivatives. Addition polymerization initiators are free radicals, anions, cations, and coordination compounds. In addition polymerization, a chain grows simply by adding monomer molecules to a propagating chain. The first step is to add a free radical, a cationic, or an anionic initiator to the monomer. For example, in ethylene polymerization (with a special catalyst), the chain grows by attaching the ethylene units one after another until the polymer terminates. This type of addition produces a linear polymer:

$$n\text{CH}_2{=}\text{CH}_2 \rightarrow -\left(\text{CH}_2\text{CH}_2\right)_{n}-$$

Branching occurs especially when free radical initiators are used due to chain transfer reactions. For a substituted olefin (such as vinyl chloride), the addition primarily produces the most stable intermediate. Propagation then occurs by successive monomer molecules additions to the intermediates. Three addition modes are possible: (i) head to tail, (ii) head to head, and (iii) tail to tail.

The head-to-tail addition mode produces the most stable intermediate. For example, styrene polymerization mainly produces the head-to-tail intermediate.

Head to-tail mode

I—CH₂—CH—CH₂—CH –

Head-to-head or tail-to-tail modes of addition are less likely because the intermediates are generally unstable. Chain growth continues until the propagating polymer chain terminates.

11.2.2 Free Radical Polymerization

Free radical initiators can polymerize olefinic compounds. These chemical compounds have a weak covalent bond that breaks easily into two free radicals when subjected to heat. Peroxides, hydroperoxides, and azo compounds are commonly used. For example, heating peroxybenzoic acid forms two free radicals, which can initiate the polymerization reaction.

Free radicals are highly reactive, short-lived, and therefore not selective. Chain transfer reactions often occur and result in a highly branched product polymer. For example, the polymerization of ethylene using an organic peroxide initiator produces highly branched polyethylene. The branches result from the abstraction of a hydrogen atom by a propagating polymer intermediate, which creates a new active center. The new center can add more ethylene molecules, forming a long branch.

Intermolecular chain transfer reactions may occur between two propagating polymer chains and result in the termination of one of the chains. Alternatively, these reactions take place by an intramolecular reaction by the coiling of a long chain. Intramolecular chain transfer normally results in short branches. Free radical polymers may terminate when two propagating chains combine. In this case, the tail-to-tail addition mode is most likely.

Polymer propagation stops with the addition of a chain transfer agent. For example, carbon tetrachloride can serve as a chain transfer agent during the production of polyethylene:

$$-CH_2CH_2 \bullet + CCl_4 \rightarrow -CH_2CH_2Cl + CCl_3 \bullet$$

The trichloromethane free radical ($CCl_3\bullet$) can initiate a new polymerization reaction.

11.2.3 Cationic Polymerization

Strong protonic acids can affect the polymerization of olefin derivatives (Chapter 3). Lewis acids, such as aluminum chloride ($AlCl_3$) or boron trifluoride (BF_3), can also initiate polymerization. In this case, a trace amount of a proton donor (co-catalyst), such as water or methanol, is normally required. For example, water combined with BF_3 forms a complex that provides the protons for the polymerization reaction. An important difference between free radical and ionic polymerization is that a counter ion only appears in the latter case. For example, the intermediate formed from the initiation of propene with BF_3-H_2O could be represented as:

$$H^+[BF_3OH]^- + CH_2{=}CH{-}CH_3 \rightarrow (CH_3)_2 CH[BF_3OH]^-$$

The next step is the insertion of the monomer molecules between the ion pair. In ionic polymerization reactions, reaction rates are faster in solvents with high dielectric constants, which promote the separation of the ion pair.

Cationic polymerizations work better when the monomers possess an electron-donating group that stabilizes the intermediate carbocation. For example, isobutylene produces a stable carbocation, and usually copolymerizes with a small amount of isoprene using cationic initiators. The product polymer is a synthetic rubber widely used for tire inner tubes:

$$\left[\begin{matrix} & CH_3 & CH_3 & \\ -CH_2{-}\overset{|}{\underset{|}{C}}{-}CH_2{-}\overset{|}{C}{=}CH{-}CH_2{-} \\ & CH_3 & & \end{matrix}\right]_n$$

Cationic initiators can also polymerize aldehydes. For example, BF_3 helps produce commercial polymers of formaldehyde. The resulting polymer, a polyacetal, is an important thermoplastic (Chapter 12):

$$-[CH_2{-}O]-$$

Because of the low activation energy of the cationic polymerization reaction and anionic polymerization reaction, low-temperature conditions are typically used to reduce any potential side reactions. Low temperatures also minimize chain transfer reactions. These reactions produce low molecular weight polymers by disproportionation of the propagating polymer chain. Cationic polymerization can terminate by adding a hydroxy compound such as water.

$$\overset{+}{-CH_2C(CH_3)_2}X^- + H_2O \longrightarrow \overset{OH}{\underset{|}{-CH_2C(CH_3)_2}} + HX$$

11.2.4 ANIONIC POLYMERIZATION

Anionic polymerization is better for vinyl monomers with electron withdrawing groups that stabilize the intermediates. Typical monomers best polymerized by anionic initiators include acrylonitrile, styrene, and butadiene. As with cationic polymerization, a counter ion is present with the propagating chain. The propagation and the termination steps are similar to cationic polymerization. Many initiators, such as alkyl and aryl-lithium and sodium and lithium suspensions in liquid ammonia, effect the polymerization. For example, acrylonitrile combined with n-butyl-lithium forms a carbanion intermediate.

Chain growth occurs through a nucleophilic attack of the carbanion on the monomer. As in cationic polymerizations, lower temperatures favor anionic polymerizations by minimizing branching due to chain transfer reactions. Solvent polarity is also important in directing the reaction bath and the composition and orientation of the products. For example, the polymerization of butadiene with lithium in tetrahydrofuran (THF, a polar solvent) gives a high 1,2 addition polymer. Polymerization of either butadiene or isoprene using lithium compounds in nonpolar solvent such as n-pentane produces a high cis-1,4 addition product. However, a higher cis-1,4-poly-isoprene isomer was obtained than when butadiene was used. This occurs because butadiene exists mainly in a transoid conformation at room temperature (a higher cisoid conformation (a form of geometric isomer) is anticipated for isoprene).

11.2.5 COORDINATION POLYMERIZATION

In coordination polymerization, the bonds are appreciably covalent but with a certain percentage of ionic character. Bonding occurs between a transition metal central ion and the ligand (perhaps an olefin, a diolefin, or carbon monoxide) to form a coordination complex. The complex reacts further with the ligand to be polymerized by an insertion mechanism.

In recent years, much interest has been centered on using late transition metals such as iron and cobalt for polymerization. Due to their lower electrophilic character, the transition metals have greater tolerance for polar functionality and, furthermore, that the catalyst activity and the polymer branches could be modified by altering the bulk of the ligand that surrounds the central metal. Such a protection reduces chain transfer reactions and results in a high molecular weight polymer. An example of these catalysts is pyridine bis-imine ligands complexed with iron and cobalt salts. Ziegler–Natta catalysts currently produce linear polyethylene (non-branched), stereoregular polypropylene, cis-polybutadiene, and other stereoregular polymers.

In the polymerization of these compounds, a reaction between α-titanium trichloride and triethylaluminum produces a five coordinate titanium (III) complex arranged octahedrally. The catalyst surface has four chloride anions, an ethyl group, and a vacant catalytic site (D) with the titanium ion (Ti^{3+}) in the center of the octahedron. A polymerized ligand, such as ethylene, occupies the vacant site. The next step is the cis insertion of the ethyl group, leaving a vacant site. In another step, ethylene occupies the vacant site. This process continues until the propagating chain terminates.

When propylene is polymerized with free radicals or some ionic initiators, a mixture of three stereo forms result: (i) the atactic form, in which the methyl groups are randomly distributed throughout the polymer; (ii) the isotactic form, in which all of the methyl groups are located on one side of the polymer chain; and (iii) the syndiotactic from, in which the methyl groups alternate regularly from one side to the other of the polymer chain.

The isotactic form of polypropylene has better physical and mechanical properties than the three tactic form mixture (obtained from free radical polymerization). Isotactic polypropylene, in which all of the stereo centers of the polymer are the same, is a crystalline thermoplastic. By contrast, atactic polypropylene, in which the stereo centers are arranged randomly, is an amorphous gum elastomer. Polypropylene consisting of blocks of atactic and isotactic stereo sequences are rubbery in nature. Polymerizing propylene with Ziegler–Natta catalyst produces mainly isotactic polypropylene. The Cosse–Arlman model explains the formation of the stereoregular type by describing the crystalline structure of α-titanium trichloride (α-$TiCl_3$) as hexagonal close packing with anion vacancies. This structure allows for *cis* insertion. However, due to the difference in the steric requirements, one of the vacant sites available for the ligand to link with the titanium catalyst which has a greater affinity for the propagating polymer than the other site. The propagating polymer then terminates, producing an isotactic polypropylene. Linear polyethylene occurs whether the reaction takes place by insertion through this sequence or, as explained earlier, by ligand occupation of any available vacant site. This course, however, results in a syndiotactic polypropylene when propylene is the ligand.

Adding hydrogen terminates the propagating polymer. The reaction between the polymer complex and the excess triethylaluminum also terminates the polymer. Treatment with alcohol or water releases the polymer. A chain transfer reaction between the monomer and the growing polymer produces an unsaturated polymer. This occurs when the concentration of the monomer is high compared to the catalyst. Using ethylene as the monomer, the termination reaction has this representation.

A new generation coordination catalysts are metallocene derivatives. The chiral form of metallocene produces isotactic polypropylene, whereas the achiral form produces atactic polypropylene. As the ligands rotate, the catalyst produces alternating blocks of isotactic and atactic polymer much like a miniature sewing machine which switches back and forth between two different kinds of stitches.

11.2.6 CONDENSATION POLYMERIZATION

Though less prevalent than addition polymerization, condensation polymerization (step-reaction polymerization) produces important polymers such as polyesters, polyamides (nylons), polycarbonates, polyurethanes, and phenol-formaldehyde resins (Chapter 12).

In general, condensation polymerization refers to a reaction between two different monomers. Each monomer possesses at least two similar functional groups that can react with the functional groups of the other monomer. For example, a reaction of a diacid and a di-alcohol (diol) can produce polyesters. A similar reaction between a diamine and a diacid can also produce polyamides; reactions between one monomer species with two different functional groups. One functional group of one molecule reacts with the other functional group of the second molecule. For example, polymerization of an amino acid starts with condensation of two monomer molecules. In these examples, a small molecule (water) results from the condensation reaction.

Ring-opening polymerization of lactams can also be considered as a condensation reaction, although a small molecule is not eliminated. This type is noted later in this chapter under "Ring Opening Polymerization." Condensation polymerization is also known as step-reaction polymerization because the reactions occur in steps. First, a dimer forms, then a trimer, next a tetramer, and so on until the polymer terminates. Although step polymerizations are generally slower than addition polymerizations, with long reaction times required for high conversions, the monomers disappear fast. The reaction medium contains only dimers, trimers, tetramers, and so on. For example, the dimer formed from the condensation of a diacid and a diol (reaction previously shown) has hydroxyl and carboxyl endings that can react with either a diacid or a diol to form a trimer.

The compounds formed continue condensation as long as the species present have different endings. The polymer terminates by having one of the monomers in excess. This produces a polymer with similar endings. For example, a polyester formed with excess diol could be represented.

In these reactions, the monomers have two functional groups (whether one or two monomers are used), and a linear polymer results. With more than two functional groups present, cross-linking occurs and a thermosetting polymer results. Examples of this type are polyurethanes and urea formaldehyde resins. The chemical structure of the urea-formaldehyde polymer consists of $[(O)CNHCH_2NH]_n$ repeat units.

Urea-formaldehyde resins (also known as urea-methanal resins) are nontransparent thermosetting resins (or polymers). They are produced from urea and formaldehyde. These resins are used in adhesives, finishes, and particle board. Urea-formaldehyde resins and the related amino resins are in the class of thermosetting resins. Examples of amino resins use include in automobile tires to improve the bonding of rubber to tire cord, in paper for improving tear strength, in products such as molding electrical devices, and jar caps.

Acid catalysts, such as metal oxides and sulfonic acids, generally catalyze condensation polymerizations. However, some condensation polymers form under alkaline conditions. For example, the reaction of formaldehyde with phenol under alkaline conditions produces methylolphenol derivatives, which further condense to a thermosetting polymer.

11.2.7 RING-OPENING POLYMERIZATION

Ring-opening polymerization produces a small number of synthetic commercial polymers. Probably the most important ring-opening reaction is that of caprolactam for the production of nylon 6. Monomers suitable for polymerization by ring-opening condensation normally possess two different functional groups within the ring. Examples of suitable monomers are lactams (such as caprolactam), which produce polyamides, and lactone derivatives, which produce polyesters.

Ring-opening polymerization may also occur by an addition chain reaction. For example, a ring-opening reaction polymerizes trioxane to a polyacetal in the presence of an acid catalyst. Formaldehyde also produces the same polymer. Monomers used for ring-opening polymerization (by addition) are cyclic compounds that open easily with the action of a catalyst during the reaction. Small-strained rings are suitable for this type of reaction. For example, the action of a strong acid or a strong base could polymerize ethylene oxide to a high molecular weight polymer. The water-soluble polymers are commercially known as carbowax.

The ring opening of cyclo-olefin derivatives is also possible with certain coordination catalysts. This simplified example shows cyclopentene undergoing a first-step formation of the dimer cyclodecadiene (a ten-carbon ring with two double bonds), and then incorporating additional cyclopentene monomer units to produce the solid, rubbery polypentamer.

Cyclopentadiene

Another example is the metathesis of cyclo-octene, which produces polyoctenylene, an elastomer known as trans-polyoctenamer.

H
‖
H
Cyclo-octene

11.3 POLYMER TYPES

To recap, a *monomer* is a reactive molecule that has at least one functional group (e.g., –OH, –COOH, –NH2, –C=C–). A *polymer* is a large molecular (macromolecule) composed of repeating structural units (monomers) typically connected by covalent chemical bonds (Rudin, 1999; Braun et al., 2001; Carraher, 2003; Odian, 2004). While polymer in the present context refers to hydrocarbon polymers, the term actually refers to a large class of natural and synthetic materials with a wide variety of properties, many of which are not true hydrocarbon derivatives.

Monomers may add to themselves as in the case of ethylene or may react with other monomers having different functionalities. A monomer initiated or catalyzed with a specific catalyst polymerizes and forms a macromolecule-a polymer. For example, ethylene polymerized in presence of a coordination catalyst produces a linear homopolymer (linear polyethylene):

$$nCH_2{=}CH_2 \rightarrow X-(CH_2CH_2)_n-X$$
$$\text{linear polyethylene}$$

The polymer will be terminated by end groups (show as "X") that are dictated by the nature of the reaction and any added reactant.

Generally, polymerization is a relatively simple process, but the ways in which monomers are joined together vary and it is more convenient to have more than one system of describing polymerization. Polymerization occurs via a variety of reaction mechanisms that vary in complexity due to functional groups present in reacting compounds. One system of separating polymerization processes asks the question of how much of the original molecule is left when the monomers bond. In addition polymerization, monomers are added together with their structure unchanged (Table 11.3). Olefin derivatives, which are relatively stable due to σ bonding between carbon atoms form polymers through relatively simple radical reactions.

reactive olefin   unreactive alkane

The chain terminating group can be a hydrogen atom (H) or any nonreactive (in this case) hydrocarbon moiety.

On the other hand, condensation polymerization results in a polymer that is less massive than the two or more monomers that form the polymer because not all of the original monomer is incorporated into the polymer. Water is one of the common molecules chemically eliminated during condensation polymerization.

Polymers such as polyethylene are generally referred to as homopolymers as they consist of repeated long chains or structures of the same monomer unit. Polymerization occurs via a variety of reaction mechanisms that vary in complexity due to functional groups present in reacting compounds and their inherent steric effects. In more straightforward polymerization, alkene, which are relatively stable due to σ-bonding between carbon atoms form polymers through relatively simple radical reactions.

For hydrocarbon polymers, chain growth polymerization (or addition polymerization) involves the linking together of molecules incorporating double or triple chemical bonds. These unsaturated monomers (the identical molecules that make up the polymers) have extra internal bonds that are able to break and link up with other monomers to form the repeating chain. Chain growth polymerization is involved in the manufacture of polymers such as polyethylene and polypropylene.

$$\left(\begin{array}{cc} \underset{\displaystyle H}{\overset{\displaystyle H}{C}} - \underset{\displaystyle H}{\overset{\displaystyle H}{C}} \end{array}\right)_n$$

Polyethylene

$$\left(\begin{array}{c} CH - CH_2 \\ | \\ CH_3 \end{array}\right)_n$$

Polypropylene

All the monomers from which addition polymers are made are olefin derivatives or functionally substituted olefin derivatives. The most common and thermodynamically favored chemical transformations of olefin derivatives are addition reactions and many of these addition reactions are known to proceed in a stepwise fashion by way of a reactive initiator and the formation of reactive intermediates.

Z* is an initiating species

In principle, once initiated, a radical polymerization might be expected to continue unchecked, producing a few extremely long-chain polymers. In practice, larger numbers of moderately sized chains are formed, indicating that chain-terminating reactions must be taking place. The most common termination processes are radical combination and disproportionation. In both types of termination, two reactive radical sites are removed by simultaneous conversion to stable product(s). Since the concentration of radical species in a polymerization reaction is small relative to other reactants (e.g., monomers, solvents, and terminated chains), the rate at which these radical-radical termination reactions occurs is very small, and most growing chains achieve moderate length before termination. The relative importance of these terminations varies with the nature of the monomer undergoing polymerization. For acrylonitrile and styrene combination is the major process. However, methyl methacrylate and vinyl acetate are terminated chiefly by disproportionation. Another reaction that diverts radical chain growth polymerizations from producing linear macromolecules is chain transfer in which a carbon radical from one location is moved to another by an intermolecular or intramolecular hydrogen atom transfer.

Chain transfer reactions are especially prevalent in the high-pressure radical polymerization of ethylene, which is the method used to make low-density polyethylene (LDPE). The primary radical at the end of a growing chain is converted to a more stable secondary radical by hydrogen atom transfer. Further polymerization at the new radical site generates a side chain radical, and this may in turn lead to creation of other side chains by chain transfer reactions. As a result, the morphology of low density polyethylene is an amorphous network of highly branched macromolecules.

In the radial polymerization of ethylene, the Π-bond is broken, and the two electrons rearrange to create a new propagating center. The form this propagating center takes depends on the specific type of addition mechanism. There are several mechanisms through which this can be initiated. The free radical mechanism was one of the first methods to be used. Free radicals are very reactive atoms or molecules that have unpaired electrons. Taking the polymerization of ethylene as an example, the free radical mechanism can be divided into three stages: (i) chain initiation, (ii) chain propagation, and (iii) chain termination.

The free radical addition polymerization of ethylene must take place at high temperatures and pressures, approximately 300°C and 29,000 psi. While most other free radical polymerizations do not require such extreme temperatures and pressures, they do tend to lack control. One effect of this lack of control is a high degree of branching. Also, as termination occurs randomly, when two chains collide, it is impossible to control the length of individual chains. A newer method of polymerization similar to free radical, but allowing more control involves the Ziegler–Natta catalyst, especially with respect to polymer branching.

A Ziegler–Natta catalyst is a catalyst used in the synthesis of polymers of α-olefins (alpha-olefins, 1-alkenes). Two general classes of Ziegler–Natta catalysts are employed and are distinguished by their solubility: (i) heterogeneous supported catalysts such as those based on titanium compounds which are used in polymerization reactions in combination with co-catalysts—organo-aluminum compounds such as triethylaluminum $(Al(C_2H_5)_3)$ and (ii) homogeneous catalysts usually based on complexes of titanium, zirconium, hafnium which are usually used in combination with a different organo-aluminum co-catalyst, methyl aluminoxane (or methylalumoxane)—these catalysts traditionally contain not only metallocene derivatives but also feature multidentate oxygen- and nitrogen-based ligands.

Finally, the use of Ziegler–Natta catalysts provides a stereospecific catalytic polymerization procedure (discovered by Karl Ziegler and Giulio Natta in the 1950s). Their catalysts permit the synthesis of unbranched, high molecular weight, high-density polyethylene (HDPE), which is used predominantly in the manufacture of blow-molded bottles for milk, household cleaners, injection-molded pails, bottle caps, appliance housings, and toys; laboratory synthesis of natural rubber from isoprene and configurational control of polymers from terminal olefin derivatives, such as propylene (e.g., pure isotactic and syndiotactic polymers).

In the case of ethylene, rapid polymerization occurs at atmospheric pressure and moderate to low temperature, giving a stronger (more crystalline) high-density polyethylene than that from radical polymerization (low-density polyethylene). Ziegler–Natta catalysts are prepared by reacting specific transition metal halides with organometallic reagents such as alkyl aluminum, lithium, and zinc reagents. The catalyst formed by reaction of triethylaluminum with titanium tetrachloride has been widely studied, but other metals (e.g., vanadium and zirconium) have also proven effective.

As olefin derivatives can be formed in relatively straightforward reaction mechanisms, they form useful compounds such as polyethylene when undergoing radical reactions. Polymers such as polyethylene are generally referred to as homopolymers as they consist of repeated long chains or structures of the same monomer unit, whereas polymers that consist of more than one type of monomer are referred to as copolymers.

Polymerization of isobutylene (2-methylpropene) by traces of strong acids is an example of cationic polymerization. The polyisobutylene product is a soft rubbery solid ($T_g=70$°C, 158°F), which is used for inner tubes. This process is similar to radical polymerization and chain growth ceases when the terminal carbocation combines with a nucleophile or loses a proton, giving a terminal alkene.

Hydrocarbon monomers bearing cation stabilizing groups, such as alkyl, phenyl, or vinyl can be polymerized by cationic processes. These are normally initiated at low temperature in methylene chloride solution. Strong acids, such as perchloric acid ($HClO_4$), or Lewis acids containing traces of water serve as initiating reagents. At low temperatures, chain transfer reactions are rare in such polymerizations, so the resulting polymers are cleanly linear (unbranched).

An example of anionic chain growth polymerization is the treatment of a cold tetrahydrofuran solution of styrene with 0.001 equivalents of n-butyl lithium causes an immediate polymerization. Chain growth may be terminated by water or carbon dioxide, and chain transfer seldom occurs. Only monomers having anion stabilizing substituents, such as phenyl, cyano, or carbonyl are good substrates for this polymerization technique. Many of the resulting polymers are largely isotactic in configuration, and have high degrees of crystallinity. Species that have been used to initiate anionic polymerization include alkali metals, alkali amides, and alkyl lithium compounds.

In addition to Ziegler–Natta catalysts, other catalysts have been suggested, with changes to accommodate the heterogeneity or homogeneity of the catalyst. Polymerization of propylene through action of the titanium catalyst gives an isotactic product, whereas vanadium-based catalyst gives a syndiotactic product.

The synthetic polymer industry represents the major end use of many petrochemical monomers such as ethylene, styrene, and vinyl chloride. Synthetic polymers have greatly affected our lifestyle. Many articles that were previously made from naturally occurring materials such as wood, cotton, wool, iron, aluminum, and glass are being replaced or partially substituted by synthetic polymers. Clothes made from polyester, nylon, and acrylic fibers or their blends with natural fibers currently dominate the apparel market. Plastics are replacing many articles previously made of iron, wood, porcelain, or paper in thousands of diversified applications. Polymerization could now be tailored to synthesize materials stronger than steel. For example, polyethylene fibers with a molecular weight of one million can be treated to be ten times stronger than steel! However, its melting point is 148°C (298°F). A recently announced thermotropic liquid crystal polymer based on p-hydroxybenzoic acid, terephthalic acid (TPA), and p,p′-biphenol has a high melting point of 420°C (790°F) and does not decompose up to 560°C (1,040°F).

The polymer field is versatile and fast growing, and many new polymers are continually being produced or improved. The basic chemistry principles involved in polymer synthesis have not changed much since the beginning of polymer production. Major changes in the last 70 years have occurred in the catalyst field and in process development. These improvements have a great impact on the economy. In the elastomer field, for example, improvements influenced the automobile industry and also related fields such as mechanical goods and wire and cable insulation.

Recognition that polymers make up many important natural materials was followed by the creation of synthetic analogs having a variety of properties. Indeed, applications of these materials as fibers, flexible films, adhesives, resistant paints, and tough but light solids have transformed modern society. Polymers are formed by chemical reactions in which a large number of monomers are joined sequentially, forming a chain. In many polymers, only one monomer is used. In others, two or three different monomers may be combined.

Polymers are classified by the characteristics of the reactions by which they are formed. If all atoms in the monomers are incorporated into the polymer, the polymer is called an addition polymer (Table 11.4). If some of the atoms of the monomers are released into small molecules, such as water, the polymer is called a condensation polymer. Most addition polymers are made from monomers containing a double bond between carbon atoms and are typical of polymers formed from olefin derivatives (olefin derivatives), and most commercial addition polymers are polyolefin derivatives. Condensation polymers are made from monomers that have two different groups of atoms which

TABLE 11.4
Glass Transition Temperatures of Various Polymers

Material	T_g (°C)
Tire rubber	−70
Polypropylene (atactic)	−20
Poly(vinyl acetate)	30
Polyethylene terephthalate	70
Poly(vinyl alcohol)	85
Poly (vinyl chloride)	80
PS	95
Polypropylene (isotactic)	0
Poly-3-hydroxybutyrate	15
Polymethylmethacrylate)	105

can join together to form, for example, ester or amide links. Polyesters are an important class of commercial polymers, as are polyamides (nylon).

Hydrocarbon derivatives (in this case, alkene derivatives unsaturated hydrocarbon derivatives) are prevalent in the formation of addition polymers but do not usually participate in the formation of condensation polymers. The term polymer in popular usage suggests plastic but actually refers to a large class of natural and synthetic materials with a wide range of properties.

11.3.1 POLYETHYLENE

A simple example of a polymer is polyethylene (a polymer composed of a repeating ethylene units, $-(CH_2CH_2)_n-$) in which the range of properties varies depending upon the number of ethylene units that make up the polymer. The monomer, ethylene ($CH_2=CH_2$), is a prime starting material that is readily available through refinery cracking process from which it is sent to the petrochemical side of the refinery.

The properties of polyethylene depend on the manner in which ethylene is polymerized. When catalyzed by organometallic compounds at moderate pressure (220–450 psi), the product is high-density polyethylene. Under these conditions, the polymer chains grow to very great length, and molecular weight on the order of masses average many hundreds of thousands are recorded. High-density polyethylene is hard, tough, and resilient. Most high-density polyethylene is used in the manufacture of containers, such as milk bottles and laundry detergent jugs.

When ethylene is polymerized at high pressure (15,000–30,000 psi), elevated temperatures (190–210°C, 380–410°F), and catalyzed by peroxides, the product is low-density polyethylene which is used in film applications due to its toughness, flexibility, and relative transparency. Typically, low-density polyethylene is used to manufacture flexible films such as those used for plastic retail bags. Low-density polyethylene is also used to manufacture flexible lids and bottles, in wire and cable applications for its stable electrical properties and processing characteristics. This form of polyethylene has molecular weights in the order of 20,000–40,000. Low-density polyethylene is relatively soft, and most of it is used in the production of plastic films, such as those used in sandwich bags.

Polyethylene is the most extensively used thermoplastic. The ever-increasing demand for polyethylene is partly due to the availability of the monomer from abundant raw materials (associated gas, LPG, naphtha). Other factors are (i) the relative ease of processing the polymer, (ii) resistance of the polymer to chemicals, and (iii) the flexibility of the polymer. The two most widely used grades of polyethylene are low-density polyethylene and high-density polyethylene. A new grade of low-density polyethylene is a linear, low-density grade produced like the high-density polymer at low pressures. This form of polyethylene is used predominantly in film applications due to its toughness, flexibility, and relative transparency. It is the preferred resin for injection molding because of its superior toughness.

When ethylene is polymerized, the reactor temperature should be well controlled to avoid the endothermic decomposition of ethylene to carbon, methane, and hydrogen:

$$CH_2=CH_2 \rightarrow 2C + 2H_2$$

$$CH_2=CH_2 \rightarrow C + CH_4$$

11.3.1.1 Low-Density Polyethylene

Low-density polyethylene is produced under high pressure in the presence of a free radical initiator. As with many free radical chain addition polymerizations, the polymer is highly branched. It has a lower crystallinity compared to high-density polyethylene due to its lower capability of packing. Polymerizing ethylene can occur either in a tubular or in a stirred autoclave reactor.

In the stirred autoclave, the heat of the reaction is absorbed by the cold ethylene feed. Stirring keeps a uniform temperature throughout the reaction vessel and prevents agglomeration of the polymer. In the tubular reactor, a large amount of reaction heat is removed through the tube walls. Reaction conditions for the free radical polymerization of ethylene are 100°C–200°C (212°F–39°F) and 1,500–2,000 psi. Ethylene conversion is kept to a low level (10%–25%) to control the heat and the viscosity but the overall conversion with recycle is in the order of 95% based on ethylene.

The polymerization rate can be accelerated by increasing the temperature, the initiator concentration, and the pressure. Degree of branching and molecular weight distribution depend on temperature and pressure. A higher-density polymer with a narrower molecular weight distribution could be obtained by increasing the pressure and lowering the temperature. The crystallinity of the polymer could be varied to some extent by changing the reaction conditions and by adding comonomers such as vinyl acetate or ethyl acrylate. The copolymers have lower crystallinity but better flexibility, and the resulting polymer has higher impact strength.

11.3.1.2   High-Density Polyethylene

High-density polyethylene is produced by a low-pressure process in a fluid bed reactor. Catalysts used for the production of high-density polyethylene are either of the Ziegler-type (a complex of triethylaluminum $[Al(C_2H_5)_3]$ and α-titanium trichloride (α-$TiCl_3$) or silica-alumina (SiO_2-Al_2O_3) impregnated with a metal oxide such as chromium oxide (Cr_2O_3) or molybdenum oxide (Mo_2O_3).

Reaction conditions are generally mild, but they differ from one process to another. For example, in the Unipol process, which is used to produce both high-density polyethyleneand linear low-density polyethylene (LLDPE), the reaction occurs in the gas phase. Ethylene and the comonomers (propene, 1-butene, etc.) are fed to the reactor containing a fluidized bed of growing ymer particles. Operation temperature and pressure are approximately 100°C (212°F) and 300 psi. A single-stage centrifugal compressor circulates unreacted ethylene. The circulated gas fluidizes the bed and removes some of the exothermic reaction heat. The product from the reactor is mixed with additives and then pelletized. The polymerization of ethylene can also occur in a liquid-phase system where a hydrocarbon diluent is added. This requires a hydrocarbon recovery system.

High-density polyethylene is characterized by a higher crystallinity and higher melting temperature than low-density polyethylene due to the absence of branching. Some branching could be incorporated in the backbone of the polymer by adding variable amounts of comonomers such as hexene. These comonomers modify the properties of high-density polyethylene for specific applications.

11.3.1.3   Linear Low-Density Polyethylene

Linear low-density polyethylene is produced in the gas phase under low pressure. Catalysts used are either Ziegler type or new generation metallocene derivatives. The Union Carbide process used to produce HOPE could be used to produce the two polymer grades. Terminal olefin derivatives (C_4 to C_6) are the usual comonomers to effect the process.

11.3.1.4   Properties and Uses

Polyethylene is an inexpensive thermoplastic that can be molded into almost any shape, extruded into fiber or filament, and blown or precipitated into film or foil. Polyethylene products include packaging (largest market), bottles, irrigation pipes, film, sheets, and insulation materials.

Because low-density polyethylene is flexible and transparent, it is mainly used to produce film and sheets. Films are usually produced by extrusion. Calendaring is mainly used for sheeting and to a lesser extent for film production. High-density polyethylene is important for producing bottles and hollow objects by blow molding. Approximately 64% of all plastic bottles are made from high-density polyethylene. Injection molding is used to produce solid objects. Another important market for high-density polyethylene is irrigation pipes which, when manufactured from high-density polyethylene are flexible, tough, and corrosion-resistant.

11.3.2 POLYPROPYLENE

Polypropylene is produced by the addition polymerization of propylene ($CH_3CH=CH_2$). The molecular structure is similar to that of polyethylene, but has a methyl group (–CH_3) on alternate carbon atoms of the chain. The molecular weight falls in the range 501,000–200,000. Polypropylene is slightly more brittle than polyethylene, but softens at a temperature of approximately 40°C (104°F). Propylene, like ethylene, is a prime starting material that is readily available through refinery cracking process from which it is sent to the petrochemical side of the refinery.

Polypropylene is a major thermoplastic polymer that is used extensively in the automotive industry for interior trim, such as instrument panels, and in food packaging, such as yogurt containers. It is formed into fibers of very low absorbance and high-stain resistance, used in clothing and home furnishings, especially carpeting. Although polypropylene did not take its position among the large-volume polymers until fairly recently, it is currently the third largest thermoplastic after polyvinylchloride. The delay in polypropylene development may be attributed to technical reasons related to its polymerization. Polypropylene produced by free radical initiation is mainly the atactic form. Due to its low crystallinity, it is not suitable for thermoplastic or fiber use. The turning point in polypropylene production was the development of a Ziegler-type catalyst by Natta to produce the stereoregular form (isotactic).

Catalysts developed in the titanium-aluminum alkyl family are highly reactive and stereo-selective. Very small amounts of the catalyst are needed to achieve polymerization (1 g catalyst/300,000 g polymer). Consequently, the catalyst entrained in the polymer is very small, and the catalyst removal step is eliminated in many new processes. Amoco has introduced a new gas-phase process called "absolute gasphase" in which polymerization of olefin derivatives (ethylene, propylene) occurs in the total absence of inert solvents such as liquefied propylene in the reactor. Titanium residues resulting from the catalyst are less than 1 ppm, and aluminum residues are less than those from previous catalysts used in this application.

Polypropylene could be produced in a liquid or in a gas-phase process. Until 1980, the vertically stirred bed process of BASF was the only large-scale commercial gas-phase process. In the Union Carbide/Shell gas-phase process, a wide range of polypropylenes are made in a fluidized bed gas-phase reactor. Melt index, atactic level, and molecular weight distribution are controlled by selecting the proper catalyst, adjusting operating conditions, and adding molecular weight control agents. This process is a modification of the polyethylene process (discussed before), but a second reactor is added. Homopolymers and random copolymers are produced in the first reactor, which operates at approximately 70°C (158°F) and 500 psi. Impact copolymers are produced in the second reactor (impact reactor) after transferring the polypropylene resin from the first reactor. Gaseous propylene and ethylene are fed to the impact reactor to produce the polymers' rubber phase. Operation of the impact reactor is similar to the initial one, but the second operates at lower pressure (approximately 17 atm). The granular product is finally pelletized.

An example of the liquid-phase polymerization is the Spheripol process, which uses a tubular reactor. In the process, homopolymer and random copolymer polymerization takes place in liquid propylene within a tubular loop reactor. Hetero-phasic impact copolymerization can be achieved by adding a gas-phase reactor in series. Removal of catalyst residue and amorphous polymer is not required. Any unreacted monomer is flashed in a two-stage pressure system and recycled back to the reactors. This improves yield and minimizes energy consumption. Dissolved monomer is removed from the polymer by a steam sparge. The process can use lower-assay chemical-grade propylene or the typical polymerization grade. The process can produce a broad range of propylene-based polymers, including homopolymer, various families of random copolymers and terpolymers, hetero-phasic impact and specialty impact copolymers, as well as high-stiffness, high-clarity copolymers.

New generation metallocene catalysts can polymerize propylene in two different ways. Rigid chiral metallocene produce isotactic polypropylene whereas the achiral forms of the catalysts produce atactic polypropylene. The polymer microstructure is a function of the reaction conditions and the catalyst design. However, the rate of ligand rotation in some unbridged metallocene derivatives can be controlled so that the metallocene oscillates between two stereochemical states. One isomer produces isotactic polypropylene and the other produces the atactic polymer. As a result, alternating blocks of rigid isotactic and flexible atactic polypropylene grow within the same polymer chain.

The properties of commercial polypropylene vary widely according to the percentage of crystalline isotactic polymer and the degree of polymerization. Polypropylenes with a 99% isotactic index are currently produced. Articles made from polypropylene have good electrical and chemical resistance and low water absorption. The other useful characteristics of polypropylene are (i) its light weight, i.e., the lowest thermoplastic polymer density; (ii) high abrasion resistance; (iii) dimensional stability; (iv) high impact strength; and (v) no toxicity.

Polypropylene can be extruded into sheets and thermoformed by solid-phase pressure forming into thin-walled containers. Due to its light weight and toughness, polypropylene and its copolymers are extensively used in automobile parts. Improvements in melt spinning techniques and film filament processes have made polypropylene accessible for fiber production.

11.3.3 POLYVINYL CHLORIDE

The monomer—vinyl chloride (CH_2=CHCl)—is not produced directly from a carbonaceous feedstock such as crude oil, but it is produced by a variety of processes from ethylene, a product produced in the refinery. There are several routes to produce vinyl chloride from ethylene and are: (i) direct chlorination; (ii) oxychlorination; (iii) thermal cracking—other routes included; (iv) from ethane, a petrochemical product; and (v) from acetylene, a derived petrochemical.

One method of producing vinyl chlorides is by the addition of chlorine to ethylene in the presence of an iron chloride ($FeCl_3$) catalyst followed by dehydrochlorination of the ethylene dichloride:

$$CH_2=CH_2 + Cl_2 \rightarrow CH_2ClCH_2Cl$$

Another route to ethylene dichloride is the oxychlorination route that entails the reaction of ethylene, oxygen, and hydrogen chloride over a copper chloride ($CuCl_2$) catalyst to produce (in a highly exothermic reaction) ethylene dichloride:

$$2CH_2=CH_2 + 4HCl + O_2 \rightarrow ClCH_2CH_2Cl + H_2O$$

Byproducts of the oxychlorination reaction, such as ethyl chloride, may be recovered as feedstocks for chlorinated solvents production.

When heated to 500°C (930°F) under pressure (225–450 psi), ethylene dichloride decomposes to produce vinyl chloride and anhydrous hydrogen chloride:

$$ClCH_2CH_2Cl \rightarrow CH_2=CHCl + HCl$$

The thermal cracking reaction is highly endothermic, and is generally carried out in a fired heater. Even though residence time and temperature are carefully controlled, it produces significant quantities of chlorinated hydrocarbon byproducts.

In another route, the reaction of acetylene with anhydrous hydrogen chloride in the presence of a mercuric chloride ($HgCl_2$) catalyst to give vinyl chloride:

$$C_2H_2 + HCl \rightarrow CH_2=CHCl$$

The reaction is exothermic and highly selective. Product purity and yields are generally very high. This industrial route to vinyl chloride was common before ethylene became widely distributed. When the production of vinyl chloride from the thermal cracking of ethylene dichloride became more popular, this method (the acetylene method) fell into dis-use.

Polyvinyl chloride is one of the most widely used thermoplastic polymers. It can be extruded into sheets and film and blow molded into bottles. It is used in many common items such as garden hoses, shower curtains, irrigation pipes, and paint formulations. Many of these polyvinylchloride products are used every day and include everything from medical devices such as medical tubing and blood bags, to footwear, electrical cables, packaging, stationery, and toys.

Polyvinylchloride can be pre-polymerized in bulk to approximately 7%–8% conversion. It is then transferred to an autoclave where the particles are polymerized to a solid powder. Most vinyl chloride, however, is polymerized in suspension reactors made of stainless steel or glass-lined. The peroxide used to initiate the reaction is dispersed in about twice its weight of water containing 0.01%–1% of a stabilizer such as polyvinyl alcohol.

In the European Vinyls Corporation process, vinyl chloride monomer is dispersed in water and then charged with the additives to the stirred jacketed-type reactor. The temperature is maintained between 53°C–70°C (127°F–158°F) to obtain a polymer of a particular molecular weight. The heat of the reaction is controlled by cooling water in the jacket and by additional reflux condensers for large reactors. Conversion could be controlled between 85% and 95% as required by the polymer grade. At the end of the reaction, the polyvinylchloride and water slurry are channeled to a blowdown vessel, from which part of unreacted monomer is recovered. The rest of the vinyl chloride monomer is stripped, and the slurry is centrifuged to separate the polymer from both water and the initiator.

Polyvinyl chloride can also be produced in emulsion. Water is used as the emulsion medium. The particle size of the polymer is controlled using the proper conditions and emulsifier. Polymers produced by free radical initiators are highly branched with low crystallinity.

Vinyl chloride can be copolymerized with many other monomers to improve its properties. Examples of monomers used commercially are vinyl acetate, propylene, ethylene, and vinylidine chloride. The copolymer with ethylene or propylene ($T_g = 80°C$), which is rigid, is used for blow molding objects. Copolymers with 6%–20% vinyl acetate ($T_g = °C$) are used for coatings.

Two types of the homopolymer are available: the flexible and the rigid. Both types have excellent chemical and abrasion resistance. The flexible types are produced with high porosity to permit plasticizer absorption. Articles made from the rigid type are hard and cannot be stretched more than 40% of their original length. An important property of polyvinylchloride is that it is self-extinguishing due to presence of the chlorine atom.

Flexible polyvinylchloride grades account for approximately 50% of polyvinylchloride production. They go into such items as tablecloths, shower curtains, furniture, automobile upholstery, and wire and cable insulation. Many additives are used with polyvinylchloride polymers such as plasticizers, antioxidants, and impact modifiers. Heat stabilizers, which are particularly important with polyvinylchloride resins, extend the useful life of the finished product. Plastic additives have been reviewed by Ainsworth. Rigid polyvinylchloride is used in many items such as pipes, fittings roofing, automobile parts, siding, and bottles.

11.3.4 POLYSTYRENE

Styrene, an important product of the petrochemical section of the refinery, is produced by dehydrogenation of ethylbenzene which is, in turn, produced by the alkylation of benzene with ethylene. The ethylbenzene is mixed in the gas phase with 10–15 times its volume of high-temperature steam and passed over a solid catalyst bed. The catalyst is typically ion oxide ($FeCl_3$ promoted by potassium oxide or potassium carbonate).

The steam serves several roles in this reaction and is the source of heat for the endothermic reaction, and it removes coke that tends to form on the iron oxide catalyst through the water-gas shift reaction—the potassium promoter enhances the decoking reaction. The steam also dilutes the reactant and products, shifting the position of chemical equilibrium toward products. A typical styrene plant consists of two or three reactors in series, which operate under vacuum to enhance the conversion and selectivity.

Polystyrene is the fourth big-volume thermoplastic. Styrene can be polymerized alone or copolymerized with other monomers. It can be polymerized by free radical initiators or using coordination catalysts. Recent work using group 4 metallocene combined with methyl aluminoxane produce stereoregular polymer. When homogeneous titanium catalyst is used, the polymer was predominantly syndiotactic. The heterogeneous titanium catalyst gave predominantly the isotactic. Twenty-one copolymers with butadiene in a ratio of approximately 1:3 produces styrene-butadiene rubber (SBR), the most important synthetic rubber.

Copolymers of styrene-acrylonitrile (SAN) have higher tensile strength than styrene homopolymers. A copolymer of acrylonitrile, butadiene, and styrene (ABS) is an engineering plastic due to its better mechanical properties (discussed later in this chapter). Polystyrene is produced either by free radical initiators or by use of coordination catalysts. Bulk, suspension, and emulsion techniques are used with free radical initiators, and the polymer is atactic. In a typical batch suspension process, styrene is suspended in water by agitation and use of a stabilizer. The polymer forms beads. The bead/water slurry is separated by centrifugation, dried, and blended with additives.

The polystyrene homopolymer produced by free radical initiators is highly amorphous ($T_g = 100°C$). The general purpose rubber (SBR), a block copolymer with 75% butadiene, is produced by anionic polymerization. The most important use of polystyrene is in packaging. Molded polystyrene is used in items such as automobile interior parts, furniture, and home appliances. Packaging uses plus specialized food uses such as containers for carryout food are growth areas. Expanded polystyrene foams, which are produced by polymerizing styrene with a volatile solvent such as pentane, have low densities. They are used extensively in insulation and flotation (life jackets).

SAN ($T_g = 105°C$, 221°F) is stiffer and has better chemical and heat resistance than the homopolymer. However, it is not as clear as polystyrene, and it is used in articles that do not require optical clarity, such as appliances and houseware materials.

ABS has a specific gravity of 1.03–1.06 and a tensile strength in the range of $6–7.5 \times 103$ psi. These polymers are tough plastics with outstanding mechanical properties. A wide variety of ABS modifications are available with heat resistance comparable to or better than polysulfone derivatives and polycarbonate derivatives. Another outstanding property of ABS is its ability to be alloyed with other thermoplastics for improved properties. For example, ABS is alloyed with rigid polyvinylchloride for a product with better flame resistance.

Among the major applications of ABS are extruded pipes and pipe fittings, appliances such as refrigerator door liners, and in molded automobile bodies.

11.3.5 Nylon Resins

The nylon family of products condensation polymers or copolymers formed by reacting difunctional monomers containing equal parts of amine and carboxylic acid so that amide derivatives are formed. The nylon monomers are manufactured by a variety of routes, starting in most cases from crude oil but sometimes from biomass. As examples:

Lactam production:

Crude oil → benzene
Benzene → cyclohexane
Cyclohexane → cyclohexanone
Cyclohexanone → cyclohexanone oxime
Cyclohexanone oxime → caprolactam

Di-acid production:

Crude oil → benzene
Benzene → cyclohexane
Cyclohexane → Cyclohexanone
Cyclohexane → cyclohexanol
Cyclohexanol → adipic acid

Diamine production:

Crude oil → propylene
Propylene → acrylonitrile
Acrylonitrile → succinonitrile
Succinonitrile → tetramethylene diamine

Nylon resins are important engineering thermoplastics. Nylons are produced by a condensation reaction of amino acids, a diacid and a diamine, or by ring-opening lactams such as caprolactam. The polymers, however, are more important for producing synthetic fibers.

Important properties of nylons are toughness, abrasion and wear resistance, chemical resistance, and ease of processing. Reinforced nylons have higher tensile and impact strengths and lower expansion coefficients than metals. They are replacing metals in many of their applications. Objects made from nylons vary from extruded films used for pharmaceutical packaging to bearings and bushings to cable and wire insulation.

11.3.6 Polyesters

Polyesters are among the large-volume engineering thermoplastics produced by condensation polymerization of terephthalic acid $(1,4\text{-}HO_2CC_6H_4CHO_2H)$ with ethylene glycol (CH_2OHCH_2OH) or 1,4-butanediol $(HOCH_2CH_2CH_2CH_2OH)$ all of which are produced within a petrochemical complex.

Polyester derivatives are used to produce film for magnetic tapes due to their abrasion and chemical resistance, low water absorption, and low gas permeability. Polyethylene terephthalate is also used to make plastic bottles (approximately 25% of plastic bottles are made from polyethylene terephthalate). Similar to nylons, the most important use of polyethylene terephthalate is for producing synthetic fibers (discussed later). Polybutylene terephthalate is another thermoplastic polyester produced by the condensation reaction of terephthalic acid $(1,4\text{-}HO_2CC_6H_4CO_2H)$ and 1,4-butanediol $(HOCH_2CH_2CH_2CH_2OH)$.

The polymer is either produced in a bulk or a solution process. It is among the fastest growing engineering thermoplastics, and leads the market of reinforced plastics with an annual growth rate of 7.3%.

11.3.7 Polycarbonates

Polycarbonates are another group of condensation thermoplastics used mainly for special engineering purposes. These polymers are considered polyesters of carbonic acid (H_2CO_3). They are produced by the condensation of the sodium salt of bisphenol A with phosgene in the presence of an organic solvent.

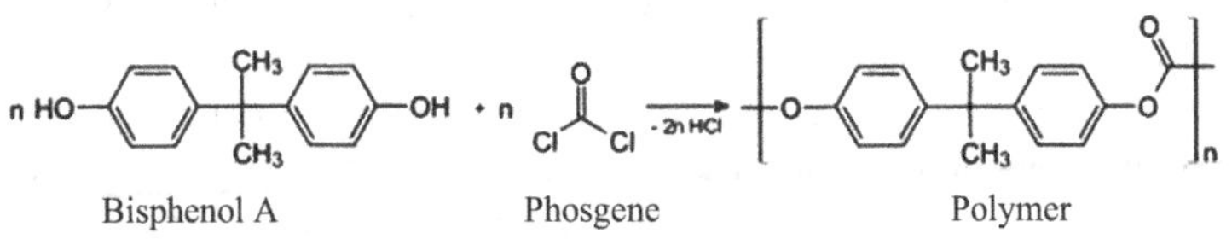

Bisphenol A   Phosgene   Polymer

Sodium chloride is precipitated, and the solvent is removed by distillation. Another method for producing polycarbonate products is by an exchange reaction between bisphenol A or a similar bisphenol with diphenyl carbonate.

Bisphenol A is synthesized by the condensation of acetone with two phenols—the reaction is catalyzed by a strong acid, such as hydrochloric acid—all of which are available within the petrochemical complex. Typically, a large excess of phenol is used to ensure full condensation.

Bisphenol A

The product mixture from the cumene process (acetone and phenol) may also be used as starting material.

Diphenyl carbonate is produced by the reaction of phosgene and phenol. Thus:

$$2C_6H_5OH + COCl_2 \rightarrow C_6H_5OCO_2C_6H_5 + 2HCl$$

Dimethyl carbonate can also be transesterified with phenol:

$$CH_3OCO_2CH_3 + 2C_6H_5OH \rightarrow C_6H_5OCO_2C_6H_5 + 2MeOH$$

Another approach to the synthesis of diphenyl carbonate by the reaction of CO and methyl nitrite using Pd/alumina:

$$2C_6H_5OH + CO + [O] \rightarrow C_6H_5OCO_2C_6H_5 + H_2O$$

Dimethyl carbonate is formed which is further reacted with phenol in presence of tetraphenox titanium catalyst. Decarbonylation in the liquid phase yields diphenyl carbonate.

Polycarbonates, known for their toughness in molded parts, typify the class of polymers known as engineering thermoplastics. These materials, designed to replace metals and glass in applications demanding strength and temperature resistance and offer additional advantages of light weight and ease of fabrication. Materials made from polycarbonates are transparent, strong, and heat and break-resistant. However, these materials are subject to stress cracking and can be attacked by weak alkalis and acids.

Polycarbonates are used in a variety of articles such as laboratory safety shields, street light globes, and safety helmets. The maximum usage temperature for polycarbonate objects is 125°C (257°F).

11.3.8 POLYETHER SULFONES

Polyether sulfones (PESs) are another class of engineering thermoplastics generally used for objects that require continuous use of temperatures around 200°C (390°F). Polyether sulfones can be prepared by the reaction of the sodium or potassium salt of bisphenol A and 4,4′-dichlorodiphenyl sulfone. Bisphenol A acts as a nucleophile in the presence of the deactivated aromatic ring of the dichlorophenyl sulfone. The reaction may also be catalyzed with Friedel–Crafts catalysts; the dichlorophenyl sulfone acts as an electrophile.

In the process, polyether sulfone derivatives are prepared by a polycondensation reaction of the sodium salt of an aromatic diphenol derivative and bis(4-chlorophenyl)sulfone. The aromatic diphenol (of which there are three isomers and also known as dihydroxybenzene or benzenediol) or commonly produced within the petrochemical complex of refinery. Also, the 4,4′-dichlorodiphenyl sulfone is synthesized by sulfonation of chlorobenzene with sulfuric acid, often in the presence of various additives to optimize the formation of the 4,4′-isomer:

$$ClC_6H_5 + SO_3 \rightarrow (ClC_6H_4)_2 SO_2 + H_2O$$

It can also be produced by chlorination of diphenyl sulfone.

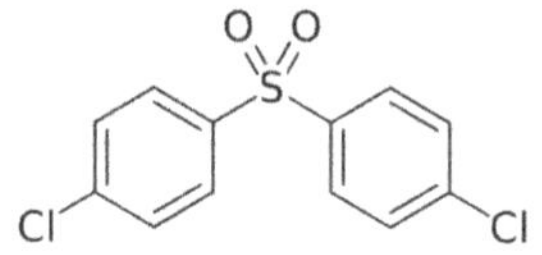

4,4′ -Dichlorodiphenyl sulfone

In the process, the sodium salt of the diphenol is formed *in situ* by reaction with a stoichiometric amount of sodium hydroxide (NaOH). The water formed in the reaction must be removed with an azeotropic solvent after which the polymerization is carried out at 130°C–160°C (265°F–320°F) under inert conditions in a polar, aprotic solvent, such as dimethyl sulfoxide, and a polyether is formed by the elimination of sodium chloride:

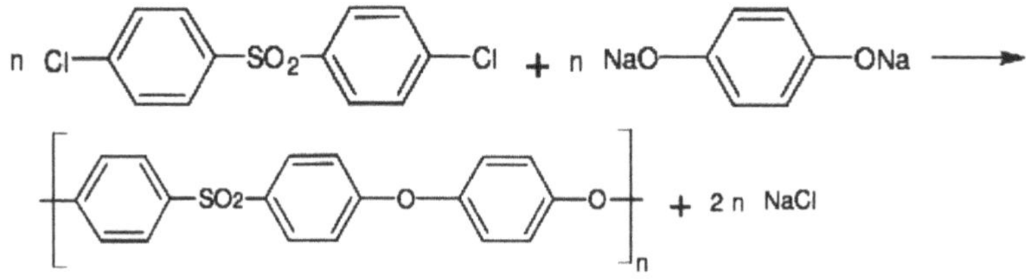

Through the use of chain terminators, the chain length of the polymer can be regulated in a range that a technical melt processing is possible. Typically, the product has end-groups that are capable of further reaction and, to prevent further condensation in the melt, the end-groups can be converted to an ether by reaction with chloromethane (CH_3Cl). The diphenol is typically bisphenol A or 1,4-dihydroxybenzene (1,4-HOC_6H_4OH, also known as hydroquinone, benzene-1,4-diol, or quinol).

Written simply, the general polymerization reaction is:

$$(ClC_6H_4)_2 SO_2 + NaO \text{ (aromatic center) } ONa \left[-OC_6H_4 (SO_2)C_4H_4O(\text{aromatic center}) \right]_{n^-}$$

Some of the commercial products include.

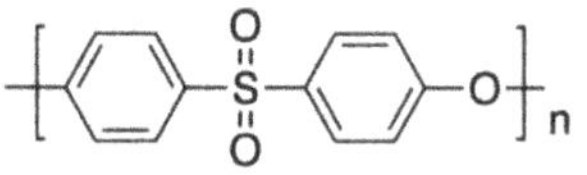

Victrex

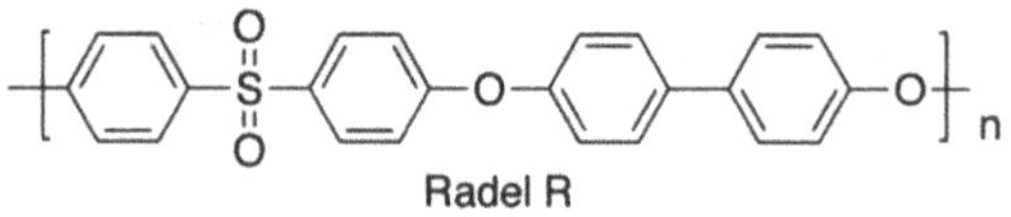

Radel R

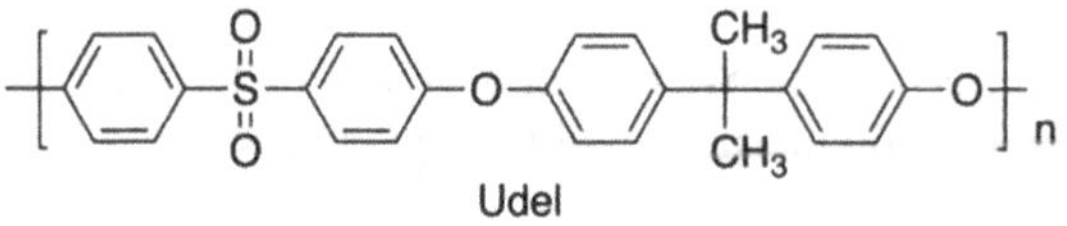

Udel

In general, the properties of polyether sulfone derivatives are similar to those of polycarbonate derivatives, but they can be used at higher temperatures.

11.3.9 Poly(phenylene) Oxide

Polyphenylene oxide (PPO, also known as polyphenylene ether, PPE) is produced by the condensation of 2,6-dimethylphenol.

2,6-dimethyl phenol

This compound can be synthesized by the alkylation of phenol with methanol (Grabowska et al., 1989).

The polymers are formed by oxidative coupling of substituted phenols at the *para*-position. Although many different choices of monomer exist, only 2,6-dimethylphenol has any practical importance. In the process, reaction occurs by passing oxygen in the phenol solution in presence of cuprous chloride (CuCl or Cu_2Cl_2) and pyridine.

$$+ \; H_2O$$

The monomer is synthesized by reacting phenol with methanol in the vapor phase in the presence of a metal oxide catalyst. Naturally, it important that the phenol used in this reaction be very pure. Impurities in the monomer with blocked *para* and *ortho* positions are chain terminators, while impurities with open ortho positions can cause chain branching or cross-linking. The final polymer is a stiff, tough, white plastic with a glass transition temperature (T_g) of 205°C (400°F).

Polyphenylene oxide is a thermoplastic, linear, noncrystalline polyether that is one of the most important engineering plastics due to its high strength, high heat distortion temperature, and high chemical resistance. Because of its unique combination of high mechanical property, low moisture absorption, excellent electrical insulation property, excellent dimension stability, and inherent flame resistance, polyphenylene oxide has been widely used for a broad range of applications. However, the high melting temperature, high melt viscosity, poor formability, and poor resistance to organic solvent can hinder the applications of the polymer. To achieve desired properties, a series of polyphenylene nanocomposites has been introduced by physical or chemical modification in the past decades. An example is the blends of poly(2,6-dimenthy-1,4-phenylene oxide) with polystyrene, which are the first commercially available alloy of polyphenylene oxide in the early.

11.3.10 Polyacetal

Polyacetal (also known as polyoxymethylene, polyformaldehyde) is among the aliphatic polyether family and is produced by the polymerization of formaldehyde.

$$\left[\begin{array}{c} H \\ | \\ C-O \\ | \\ H \end{array} \right]_n$$

Polyacetal (polyoxymethylene)

These polymers are termed polyacetals to distinguish them from polyether derivatives produced by polymerizing ethylene oxide, which has two methylene ($-CH_2-$) groups between the ether groups. The polymerization reaction occurs in the presence of a Lewis acid and a small amount of water at room temperature.

Formaldehyde is produced industrially by the catalytic oxidation of methanol (a common petrochemical starting material) using catalysts such as solver metal or a mixture of an ion oxide and molybdenum oxide or vanadium oxide. In the commonly Formox process, methanol and oxygen react at a temperature in the order of 250°C–400°C (480°F–750°F) in presence of iron oxide in combination with molybdenum and/or vanadium to produce formaldehyde. Thus:

$$2CH_3OH + O_2 \rightarrow 2CH_2O + 2H_2O$$

The silver-based catalyst usually operates at a higher temperature, approximately 650°C (1,200°F). Two chemical reactions occur simultaneously produce formaldehyde:

$$2CH_3OH + O_2 \rightarrow 2CH_2O + 2H_2O$$

$$CH_3OH \rightarrow CH_2O + H_2$$

In principle, formaldehyde could be generated by oxidation of methane but this route is not industrially viable due to the more facile oxidation of methanol relative to the more-difficult-to-oxide methane.

In the process to produce polyacetal, the formaldehyde is generated by the reaction of the formaldehyde solution with an alcohol to create a hemi-formal, dehydration of the hemi-formal/water mixture (either by extraction or vacuum distillation), and release of the formaldehyde by heating the hemi-formal. The formaldehyde is then polymerized by anionic catalysis and the resulting polymer stabilized by reaction with acetic anhydride.

Polyacetals are highly crystalline polymers. The number of repeating units ranges from 500 to 3,000. They are characterized by high impact resistance, strength, and a low friction coefficient. Articles made from polyacetals vary from door handles to gears and bushings, carburetor parts to aerosol containers. The major use of polyacetals is for molded grades.

11.3.11 BUTADIENE POLYMERS AND COPOLYMERS

Butadiene, an extremely important multifunctional petrochemical, can be produced by the catalytic dehydrogenation of n-butane ($CH_3CH_2CH_2CH_3$) by the Houdry Catadiene process, which was developed during World War II, and involves treating butane over an alumina-chromia (Al_2O_3-Cr_2O_3) catalyst at high temperatures.

Butadiene can also be produced from ethanol and two processes are available (i) at 400°C–450°C (700°F–840°F) using a variety of metal oxide catalysts.

Thus: $2CH_3CH_2OH \rightarrow CH_2=CH-CH=CH_2 + 2H_2O + H_2$

In the alternate process, ethanol is oxidized to acetaldehyde which reacts with additional ethanol over a porous silica catalyst (which is promoted by tantalum) 325°C–350°C (615°F–660°F) to yield butadiene.

Thus: $CH_3CH_2OH + CH_3CHO \rightarrow CH_2=CH-CH=CH_2 + 2 H_2O$

Butadiene could be polymerized using free radical initiators or ionic or coordination catalysts. When butadiene is polymerized in emulsion using a free radical initiator such as cumene hydroperoxide, a random polymer is obtained with three isomeric configurations, the 1,4-addition configuration dominating. Polymerization of butadiene using anionic initiators (alkyl lithium) in a nonpolar solvent produces a polymer with a high *cis* configuration. A high *cis*-polybutadiene is also obtained when coordination catalysts are used.

cis-1,4-Polybutadiene is characterized by high elasticity, low heat buildup, high abrasion resistance, and resistance to oxidation. However, it has a relatively low mechanical strength. This is improved by incorporating a *cis*, *trans* block copolymer or 1,2-(vinyl) block copolymer in the polybutadiene matrix. Also, a small amount of natural rubber may be mixed with polybutadiene to improve its properties. *Trans*-1,4-polybutadiene is characterized by a higher glass transition temperature ($T_g = -14$°C, 7°F) than the *cis* form ($T_g = -108$°C, -162°F). The polymer has the toughness, resilience, and abrasion resistance of natural rubber ($T_g = -14$°C, 7°F).

While on the issue of diene derivatives, it is worthy of note that a transpolypentamer (TPR) is produced by the ring cleavage of cyclopentene. Cyclopentene is obtained from cracked naphtha or gas oil, which contain small amounts of cyclopentene, cyclopentadiene, and dicyclopentadiene. Polymerization using organometallic catalysts produce a stereoregular product (*trans* 1,5-polypentamer).

Due to the presence of residual double bonds, the polymer could be cross-linked with regular agents. The transpolypentamer is a linear polymer with a high *trans* configuration. It is highly amorphous at normal temperatures and has a T_g of about 90°C (194°F) and a density of 0.85.

11.4   PLASTICS AND THERMOPLASTICS

Plastic is the general common term for a wide range of synthetic organic (usually solid) materials produced and used in the manufacture of industrial products (Jones and Simon, 1983; Austin, 1984;

Lokensgard, 2010). A plastic is a type of polymer—all plastics are polymers but not all polymers are plastics. Polymers can be fibers, elastomers, or adhesives and plastics are a wide group of solid composite materials that are largely organic, usually based on synthetic resins or modified polymers of natural origin, and possess appreciable mechanical strength. A plastic exhibits plasticity and the ability to be deformed or undergo change of shape under pressure, temperature, or both. At a suitable stage in their manufacture, plastics can be cast, molded, or polymerized directly. Plastic material is any of a wide range of synthetic or semisynthetic organic solids used in the manufacture of industrial products. Plastics are typically polymers of high molecular mass and may contain other substances to improve performance.

Plastics are produced from chemicals sourced almost entirely from fossil fuels and, because fossil fuel production is highly localized, plastic production is also concentrated in specific regions where fossil fuel development is present, including, notably, the U.S. Gulf Coast. Natural gas liquids (NGLs), a key input for plastic production, are hard to transport and petrochemical producers relying on natural gas liquids or ethane as a feedstock typically cluster geographically near sources of natural gas. Naphtha, another key input for plastic production, is a product of oil refining, and its production is concentrated among major oil companies with refining capacity.

Thus, because of the need to colocate fossil fuel and plastic production, there is a high degree of vertical integration between the industries. Major oil and gas producers own plastics companies, and major plastic producers own oil and gas companies. As example (listed alphabetically), BP, Chevron, ExxonMobil, Shell, and Chevron are all integrated companies.

Plastics are available in the form of bars, tubes, sheets, coils, and blocks, and these can be fabricated to specification. However, plastic articles are commonly manufactured from plastic powders in which desired shapes are fashioned by compression, transfer, injection, or extrusion molding. In compression molding, materials are generally placed immediately in mold cavities, where the application of heat and pressure makes them first plastic, then hard. The transfer method, in which the compound is plasticized by outside heating and then poured into a mold to harden, is used for designs with intricate shapes and great variations in wall thickness. Injection-molding machinery dissolves the plastic powder in a heating chamber and by plunger action forces it into cold molds, where the product sets. The operations take place at rigidly controlled temperatures and intervals. Extrusion molding employs a heating cylinder, pressure, and an extrusion die through which the molten plastic is sent and from which it exits in continuous form to be cut in lengths or coiled.

Thermoplastics are elastic and flexible above a glass transition temperature (T_g), which is specific to each plastic, specific for each one. Below a second, higher melting temperature, T_m, most thermoplastics have crystalline regions alternating with amorphous regions in which the chains approximate random coils. The amorphous regions contribute elasticity and the crystalline regions contribute strength and rigidity. Above T_m all crystalline structure disappears and the chains become randomly inter-dispersed. As the temperature increases above T_m, the viscosity gradually decreases without any distinct phase change.

Some thermoplastics normally do not crystallize: they are termed amorphous plastics and are useful at temperatures below the glass transition temperature. Generally, amorphous thermoplastics are less chemically resistant and can be subject to stress cracking. Thermoplastics will crystallize to a certain extent and are called semicrystalline. The speed and extent to which crystallization can occur depends in part on the flexibility of the polymer chain.

Semicrystalline thermoplastics are more resistant to solvents and other chemicals. If the crystallites are larger than the wavelength of light, the thermoplastic is hazy or opaque. Semicrystalline thermoplastics become less brittle above the glass transition temperature. If a plastic with otherwise desirable properties has too high a glass transition temperature, it can often be lowered by adding a low molecular weight plasticizer to the melt before forming and cooling. A similar result can sometimes be achieved by adding nonreactive side chains to the monomers before polymerization. Both methods make the polymer chains stand off a bit from one another. Another method of lowering the glass transition temperature (or raising the melting temperature) is to incorporate the original plastic into a copolymer (as with graft copolymers) of polystyrene. Lowering the glass transition temperature is not the only way to reduce brittleness. Drawing (and similar processes that stretch or orient the molecules) or increasing the length of the polymer chains also decrease brittleness.

Thermoplastics can go through melting/freezing cycles repeatedly and the fact that they can be reshaped upon reheating gives them their name. This quality makes thermoplastics recyclable. The processes required for recycling vary with the thermoplastic. Although modestly vulcanized natural and synthetic rubbers are stretchy; they are elastomeric thermosets, not thermoplastics. Each has its own glass transition temperature and will crack and shatter when cold enough so that the cross-linked polymer chains can no longer move relative to one another. But they have no melting temperature and will decompose at high temperatures rather than melt.

Ethylene ($CH_2=CH_2$) is a critical feedstock for the production of polyethylene, polyvinyl chloride, polyethylene terephthalate, and polystyrene. Propylene ($CH_3CH=CH_2$) is the basic chemical for the manufacture of polypropylene. Therefore, the overwhelming majority of plastics can be traced to the product streams of just two industrial chemicals: ethylene and propylene.

Ethylene and propylene are particularly critical in the production of plastic packaging, the largest and fastest growing category of plastic products and the biggest, though by no means only, contributor to the accelerating crisis of plastics pollution. In addition, moreover, plastic packaging is comprised nearly exclusively of the five major thermoplastics discussed above, primarily polyethylene, polypropylene, and polyethylene terephthalate. In addition, the abundant supply of natural gas in the United States has made natural gas liquids the preferred input for ethylene production. Nearly 90% of U.S. ethylene production is sourced from ethane-rich natural gas liquids. Moreover, virtually all ethane in the United States, and one-third of propane, is used in ethylene production.

Plastics are relatively tough substances with high molecular weight that can be molded with (or without) the application of heat. In general, plastics are subclassified into *thermoplastics*, polymers that can be softened by heat, and *thermosets*, which cannot be softened by heat. Thermoplastics have moderate crystallinity. They can undergo large elongation, but this elongation is not as reversible as it is for elastomers. Examples of thermoplastics are polyethylene and polypropylene.

Thermosetting plastics are usually rigid due to high cross-linking between the polymer chains. Examples of this type are phenol-fomaldehyde and polyurethanes. Cross-linking may also be promoted by using chemical agents such as sulfur or by heat treatment or irradiation with gamma rays, ultraviolet light, or energetic electrons. Recently, high-energy ion beams were found to increase the hardness of the treated polymer drastically. In addition, thermoplastics are plastics that do not undergo chemical change in their composition when heated and therefore can be molded again and again; examples are polyethylene, polypropylene, polystyrene, polyvinyl chloride, and polytetrafluoroethylene. The raw materials needed to make most of these plastics come from petroleum and natural gas.

Plastics are the polymeric materials with properties in chemical structure, the demarcation between fibers and plastics may sometimes be blurred. Polymers such as polypropylene and polyamides can be used as fibers and plastics by a proper choice of processing conditions. Plastics can be extruded as sheets or pipes, painted on surfaces, or molded to form countless objects. A typical commercial plastic resin may contain two or more polymers in addition to various additives and fillers. Additives and fillers are used to improve some property such as the processability, thermal or environmental stability, and mechanical properties of the final product. A plastic is also any organic material with the ability to flow into a desired shape when heat and pressure are applied to it and to retain the shape when they are withdrawn. Plastics are typically polymers of high molecular weight, and may contain other substances to improve performance. Because of their relative ease of manufacture, versatility, and imperviousness to water, plastics are used in a wide range of products, from paper clips to space vehicles. However, these same properties make them persist beyond their usefulness and the focus is on making polycarbonates environmentally friendly.

Thermoplastics are important polymeric materials that have replaced or substituted many naturally derived products such as paper, wood, and steel. Plastics possess certain favorable properties such as light weight, corrosion resistance, toughness, and ease of handling. They are also less expensive. The major use of the plastics is in the packaging field. Many other uses include construction, electrical and mechanical goods, and insulation. One growing market that evolved fairly recently is engineering thermoplastics. This field includes polymers with special properties such as high thermal stability, toughness, and chemical and weather resistance. Nylons, polycarbonates, polyether sulfones, and polyacetals are examples of this group.

Resins are basic building materials that constitute the greater bulk of plastics. Resins undergo polymerization reactions during the development of plastics. Plastics are formed when polymers are blended with specific external materials in a process known as compounding. The important compounding ingredients include plasticizers, stabilizers, chelating agents, and antioxidants.

Hydrocarbon plastics are plastics based on resins made by the polymerization of monomers composed of carbon and hydrogen only.

A plastic is made up principally of a binder together with plasticizers, fillers, pigments, and other additives. The binder gives a plastic its main characteristics and usually its name.

Binders may be natural materials, e.g., cellulose derivatives, casein, or milk protein, but are more commonly synthetic resins. In either case, the binder materials consist of polymers. Cellulose derivatives are made from cellulose, a naturally occurring polymer; casein is also a naturally occurring polymer. Synthetic resins are polymerized, or built up, from small simple molecules called monomers. Plasticizers are added to a binder to increase flexibility and toughness. Fillers are added to improve particular properties, e.g., hardness or resistance to shock. Pigments are used to impart various colors.

Virtually any desired color or shape and many combinations of the properties of hardness, durability, elasticity, and resistance to heat, cold, and acid can be obtained in a plastic.

Plastic deformation is observed in most materials including metals, soils, rocks, concrete, and plastics. However, the physical mechanisms that cause plastic deformation can vary widely. At the crystal scale, plasticity in metals is usually a consequence of dislocations and, although in most crystalline materials such defects are relatively rare, are also materials where defects are numerous and are part of the very crystal structure, in such cases plastic crystallinity can result. In brittle materials, plasticity is caused predominantly by slippage at micro-cracks.

Plastics are so durable that they will not rot or decay as do natural products such as those made of wood. As a result great amounts of discarded plastic products accumulate in the environment as waste. It has been suggested that plastics could be made to decompose slowly when exposed to sunlight by adding certain chemicals to them. Plastics present the additional problem of being difficult to burn. When placed in an incinerator, they tend to melt quickly and flow downward, clogging the incinerator's grate; they also emit harmful fumes.

11.4.1 Classification

There are two types of plastics: thermoplastics and thermosetting polymers. Thermoplastics will soften and melt if enough heat is applied; examples among the truly hydrocarbon derivatives polymers are polyethylene and polystyrene. Thermosetting polymers can melt and take shape once after they have solidified, they remain solid.

Thermoset plastics harden during the molding process and do not soften after solidifying. During molding, these resins acquire three-dimensional cross-linked structure with predominantly strong covalent bonds that retain their strength and structure even on heating. However, on prolonged heating, thermoset plastics get charred. In the softened state, these resins harden quickly with pressure assisting the curing process. Thermoset plastics are usually harder, stronger, and more brittle than thermoplastics and cannot be reclaimed from wastes. These resins are insoluble in almost all inorganic solvents.

Thermoplastics, when compounded with appropriate ingredients, can usually withstand several heating and cooling cycles without suffering any structural breakdown. Examples of commercial thermoplastics are polystyrene, polyolefin derivatives (e.g., polyethylene and polypropylene), nylon, poly(vinyl chloride), and poly(ethylene terephthalate). Thermoplastics are used for a wide range of applications, such as film for packaging, photographic, magnetic tape, beverage and trash containers, and a variety of automotive parts and upholstery. Advantageously, waste thermoplastics can be recovered and refabricated by application of heat and pressure.

Thermosets are polymers whose individual chains have been chemically linked by covalent bonds during polymerization or by subsequent chemical or thermal treatment during fabrication. The thermosets usually exist initially as liquids called pre-polymers; they can be shaped into desired forms by the application of heat and pressure. Once formed, these cross-linked networks resist heat softening, creep and solvent attack, and cannot be thermally processed or recycled. Such properties make thermosets suitable materials for composites, coatings, and adhesive applications. Principal examples of thermosets include epoxies, phenol-formaldehyde resins, and unsaturated polyesters. Vulcanized rubber used in the tire industry is also an example of thermosetting polymers. Thermosetting polymers are usually insoluble because the cross-linking causes a tremendous increase in molecular weight. At most, thermosetting polymers only swell in the presence of solvents, as solvent molecules penetrate the network. The designation of a material as thermoplastic reflects the fact that above the glass transition temperature the material may be shaped or pressed into molds, spun or cast from melts, or dissolved in suitable solvents for later fashioning.

The polymers that are characterized by a high degree of cross-linking resist deformation and solution once their final morphology is achieved. Such polymers (thermosets) are usually prepared in molds that yield the desired object and these polymers, once formed, cannot be reshaped by heating.

Plastics can be classified by chemical structure, namely the molecular units (the monomers) that make up the polymer's backbone and side chains. Plastics can also be classified by the chemical process used in their synthesis, such as condensation, poly-addition, and cross-linking. Other classifications are based on qualities that are relevant for manufacturing or product design and include classes such as the thermoplastic and thermoset, elastomers, structural, conductive, and biodegradable. Plastics can also be classified by various physical properties, such as density, tensile strength, glass transition temperature, and resistance to various chemical products.

The use of plastics is constrained chiefly by their organic chemistry, which seriously limits their hardness, density, and their ability to resist heat, organic solvents, oxidation, and ionizing radiation. In particular, most plastics will melt or decompose when heated above 200°C (390°F).

11.4.2 CHEMICAL STRUCTURE

Common thermoplastics range in molecular weight from 20,000 to 500,000 while thermosets have higher, almost indefinable molecular weights. The molecular chains are made up of many repeating monomer units and each plastic will have several thousand repeating units.

In the current context, the plastics are composed of polymers of hydrocarbon units with hydrocarbon moieties attached to the hydrocarbon backbone, which is that part of the chain in which a large number of repeat units together are linked together. To customize the properties of a plastic, different molecular groups are attached to the backbone. This fine-tuning of the properties of the polymer by repeating unit's molecular structure has allowed plastics to become such an indispensable part of 21st-century world.

Some plastics are partially crystalline and partially amorphous giving them both a melting point and one or more glass transitions (temperatures above which the extent of localized molecular flexibility is substantially increased). The so-called semicrystalline hydrocarbon plastics include polyethylene and polypropylene. Many plastics are completely amorphous, such as polystyrene and its copolymers, and all thermosets.

11.4.3 Properties

A thermoplastic (thermo-softening plastic) is a polymer that turns to a liquid when heated and freezes to a very glassy state when cooled sufficiently. Most hydrocarbon-based thermoplastics are high molecular weight polymers whose chains associate through weak van der Waals forces (polyethylene) or even stacking of aromatic rings (polystyrene). Thermoplastic polymers differ from thermosetting polymers since they can, unlike thermosetting polymers, be remelted and remolded. Many thermoplastic materials are additional polymers which result from vinyl chain growth polymers such as polyethylene and polypropylene.

11.4.3.1 Mechanical Properties

Plastics have the characteristics of a viscous liquid and a spring-like elastomer, traits known as a viscoelasticity. These characteristics are responsible for many of the characteristic material properties displayed by plastics. Under mild loading conditions, such as short-term loading with low deflection and small loads at room temperature, plastics usually react like springs, returning to their original shape after the load is removed. Under long-term heavy loads or elevated temperatures many plastics deform and flow similar to high viscous liquids, although still solid.

Creep is the deformation that occurs over time when a material is subjected to constant stress at constant temperature. This is the result of the viscoelastic behavior of plastics.

Stress relaxation is another viscoelastic phenomenon. It is defined as a gradual decrease in stress at constant temperature. Recovery is the degree to which a plastic returns to its original shape after a load is removed.

Specific gravity is the ratio of the weight of any volume to the weight of an equal volume of some other substance taken as the standard at a stated temperature. For plastics, the standard is water.

Water absorption is the ratio of the weight of water absorbed by a material to the weight of the dry material. Many plastics are hygroscopic, meaning that over time they absorb water.

Tensile strength at break is a measure of the stress required to deform a material prior to breakage. It is calculated by dividing the maximum load applied to the material before its breaking point by the original cross-sectional area of the test piece.

Tensile modulus (modulus of elasticity) is the slope of the line that represents the elastic portion of the stress-strain graph.

Elongation at break is the increase in the length of a tension specimen, usually expressed as a percentage of the original length of the specimen.

Compressive strength is the maximum compressive stress a material is capable of sustaining. For materials that do not fail by a shattering fracture, the value depends on the maximum allowed distortion.

Flexural strength is the strength of a material in bending expressed as the tensile stress of the outermost fibers of a bent test sample at the instant of failure.

Flexural modulus is the ratio, within the elastic limit, of stress to the corresponding strain.

Impact is one of the most common ASTM tests for testing the impact strength of plastic materials. It gives data to compare the relative ability of materials to resist brittle fracture as the service temperature decreases.

The coefficient of thermal expansion is the change in unit length or volume resulting from a unit change in temperature. Commonly used unit is 10^{-6} cm/cm/°C.

Thermal conductivity is the ability of a material to conduct heat; a physical constant for the quantity of heat that passes through a unit cube of a material in a unit of time when the difference in temperature of two faces is 1°C.

The limiting oxygen index is a measure of the minimum oxygen level required to support combustion of the polymer.

11.4.3.2 Chemical Properties

Many applications require that plastics retain critical properties, such as strength, toughness, or appearance, during and after exposure to natural environmental conditions.

Furthermore, the rapid growth of the use of plastics in major appliances has forced an examination of how best to manage this material once these products have reached the end of service. Integrated resource management requires that alternatives be developed to best utilize the material value of this postconsumer plastic.

Since the value of recovered materials will be determined by composition, the value over time changes as the composition of refrigerators changes. Any recycling process developed for plastics recovered should not only accommodate materials used 15–20 years ago, but also be adaptable for the effective reclamation of the recovery of plastics.

Some of the environmental effects that may damage plastic materials are as follows:

Corrosion of metallic materials takes place via an electrochemical reaction at a specific corrosion rate. However, plastics do not have such specific rates. They are usually completely resistant to a specific corroding chemical or they deteriorate rapidly. Polymers are attacked either by chemical reaction or solvation. Solvation is the penetration of the polymer by a corroding chemical, which causes softening, swelling, and ultimate failure. Corrosion of plastics can be classified in the following ways as to attack mechanism:

Disintegration or degradation of a physical nature due to absorption, permeation, solvent action, or other factors.
Oxidation, where chemical bonds are attacked.
Hydrolysis, where ester linkages are attacked.
Radiation.
Thermal degradation involving depolymerization and possibly repolymerization.
Dehydration (less common).

The absorption of UV light, mainly from sunlight, degrades polymers in two ways. First, the UV light adds thermal energy to the polymer as in heating, causing thermal degradation. Second, the UV light excites the electrons in the covalent bonds of the polymer and weakens the bonds. Hence, the plastic becomes more brittle.

Some plastics that are originated from natural products, or plastics that have natural products mixed with them, are potentially susceptible to degradation by microorganisms. This is not a desired property in the use stage of the plastic product. However, at the end of their life cycle, disposal of plastics become an important issue.

Oxidation is a degradation phenomenon when the electrons in a polymeric bond are so strongly attracted to another atom or molecule (here, oxygen) outside the bond that the polymer bond breaks. The results of oxidation are loss of mechanical and physical properties, embrittlement, and discoloration.

Environmental stress cracking occurs when the plastic is exposed to hostile environment conditions and mechanical stresses at the same time. It is different from polymer degradation because stress cracking does not break polymer bonds. Instead, it breaks the secondary linkages between polymers. These are broken when the mechanical stresses cause minute cracks in the polymer and they propagate rapidly under harsh environmental conditions. Nevertheless, the plastic material would not fail that fast if exposed to either the damaging environment or the mechanical stresses separately.

Crazing. In some cases, an environmental chemical embrittles the plastic material even when there is no mechanical stress applied. Cracks may also appear when the plastic part is stresses (usually in tensile) with no apparent environmental solvent present. These phenomena are called crazing and differ from environmental stress cracking in both the direction of the cracks and the extent of the cracking. The crack direction in environmental stress cracking is in the direction of molecular orientation in the part, while in crazing the cracks are much more numerous in a small area but are much shorter than environmental stress cracks.

Permeation is molecular migration through micro-voids either in the polymer or between polymer molecules. Permeability is a measure of how easily gases or liquids can pass through a material. All materials are somewhat permeable to chemical molecules, but plastic materials tend to be an order of magnitude greater in their permeability than metals. However, not all polymers have the same rate of permeation. In fact, some polymers are not affected by permeation.

11.4.3.3 Electrical Properties

Resistivity of a material is the resistance that a material presents to the flow of electrical charge.

Dielectric strength is the voltage that an insulating material can withstand before breakdown occurs. It usually depends on the thickness of the material and on the method and conditions of the test.

Arc resistance is the property that measures the ease of formation of a conductive path along the surface of a material, rather than through the thickness of the material as is done with dielectric strength.

Dielectric constant or permittivity is a measure of how well the insulating material will act as a dielectric capacitor. This constant is defined as the capacitance of the material in question compared (by ratio) with the capacitance of a vacuum. A high dielectric constant indicates that the material is highly insulating.

Dissipation factor of a material measures the tendency of the material to dissipate internally generated thermal energy (i.e., heat) resulting from an applied alternating electric field.

11.4.3.4 Optical Properties

Light transmission. Plastics differ greatly in their ability to transmit light. The materials that allow light to pass through them are called transparent. Many plastics do not allow any light to pass through. These are called opaque materials. Some plastic materials have light transmission properties between transparent and opaque. These are called translucent.

Surface reflectance. The reflection of light off the surface of a plastic part determines the amount of gloss on the surface. The reflectance is dependent upon a property of materials called the index of refraction, which is a measure of the change in direction of an incident ray of light as it passes through a surface boundary. If the index of refraction of the plastic is near the index of air, light will pass through the boundary without significant change in direction. If the index of refraction between the air and the plastic is large, the ray of light will significantly change direction causing some of the light to be reflected back toward its source.

11.5 THERMOSETTING PLASTICS

This group includes many plastics produced by condensation polymerization. Among the important thermosets are the polyurethanes, epoxy resins, phenolic resins, and urea and melamine formaldehyde resins.

11.5.1 Polyurethanes

Polyurethanes are produced by the condensation reaction of a polyol and a diisocyanate:

$$OCN-R-NCO + HO-R'-OH--C-N-R-N-C-OR'-O$$

No byproduct is formed from this reaction. Toluene diisocyanate (Chapter 10) is a widely used monomer. Dials and trials produced from the reaction of glycerol and ethylene oxide or propylene oxide are suitable for producing polyurethanes.

Polyurethane polymers are either rigid or flexible, depending on the type of the polyol used. For example, triol derivatives derived from glycerol and propylene oxide are used for producing block slab foams. These polyols have moderate reactivity because the hydroxy groups are predominantly secondary. More reactive polyols (used to produce molding polyurethane foams) are formed by the reaction of polyglycols with ethylene oxide to give the more reactive primary group.

Other polyhydric compounds with higher functionality than glycerol (three-OH) are commonly used. Examples are sorbitol (six-OH) and sucrose (eight-OH). Triethanolamine, with three OH groups, is also used. Diisocyanate derivatives generally employed with polyols to produce polyurethane derivatives are 2,4-and 2,6-toluene diisocyanate derivatives prepared from dinitro-toluene derivatives (Chapter 10).

2,6-Toluene   MDI

A different diisocyanate used in polyurethane synthesis is methylene diisocyanate (MDI), which is prepared from aniline and formaldehyde. The diamine product is reacted with phosgene to get methylene diisocyanate.

The physical properties of polyurethanes vary with the ratio of the polyol to the diisocyanate. For example, tensile strength can be adjusted within a range of 1,200–600 psi; elongation, between 150% and 800%.

Improved polyurethane can be produced by copolymerization. Block copolymers of polyure-thanes connected with segments of isobutylene derivatives exhibit high-temperature properties, hydrolytic stability, and barrier characteristics. The hard segments of polyurethane block polymers consist of RNHCOOH, where R usually contains an aromatic moiety.

The major use of polyurethanes is to produce foam. The density as well as the mechanical strength of the rigid and the flexible types varies widely with polyol type and reaction conditions. For example, polyurethanes could have densities ranging between 1 lb/ft^3 and 6 lb/ft^3 for the flex-ible types and 1 lb/ft^3 and 50 lb/ft^3 for the rigid types. Polyurethane foams have good abrasion resistance, low thermal conductance, and good load-bearing characteristics. However, they have moderate resistance to organic solvents and are attacked by strong acids. The ability of polyure-thanes to acts as flame retardants can be improved by using special additives, spraying a coating material such as magnesium oxychloride, or by grafting a halogen phosphorous moiety to the polyol. Trichloro butylene oxide is sometimes copolymerized with ethylene and propylene oxides to produce the polyol.

Major markets for polyurethanes are furniture, transportation, and building and construction. Other uses include carpet underlay, textural laminates and coatings, footwear, packaging, toys, and fibers. The largest use for rigid polyurethane is in construction and industrial insulation due to its high insulating property.

Molded urethanes are used in items such as bumpers, steering wheels, instrument panels, and body panels. Elastomers from polyurethanes are characterized by toughness and resistance to oils, oxidation, and abrasion. They are produced using short-chain polyols such as poly-tetramethylene glycol from 1,4-butanediol. Polyurethanes are also used to produce fibers. Spandex (trade name) is a copolymer of polyurethane (85%) and polyesters.

Polyurethane networks based on triisocyante and diisocyanate connected by segments consisting of polyisobutylene are rubbery and exhibit high-temperature properties, hydrolytic stability, and barrier characteristics.

11.5.2 Epoxy Resins

Epoxy resins are produced by reacting epichlorohydrin and a diphenol. Bisphenol A is the diphenol generally used. The reaction, a ring-opening polymerization of the epoxide ring, is catalyzed with strong bases such as sodium hydroxide. A nucleophilic attack of the phenoxy ion displaces a chloride ion and opens the ring. The linear polymer formed is cured by cross-linking either with an acid anhydride, which reacts with the –OH groups, or by an amine, which opens the terminal epoxide rings. Cresols and other bisphenols are also used for producing epoxy resins.

Epoxy resins have a wide range of molecular weights (approximately 1,000–10,000). Those with higher molecular weight, termed phenoxy, are hydrolyzed to transparent resins that do not have the epoxide groups. These could be used in molding purposes, or cross-linked by diisocyanate derivatives or by cyclic anhydride derivatives.

Important properties of epoxy resins include their ability to adhere strongly to metal surfaces, their resistance to chemicals, and their high dimensional stability. They can also withstand temperatures up to 500°C. Epoxy resins with improved stress cracking properties can be made by using toughening agents, such as carboxyl-terminated butadiene-acrylonitrile liquid polymers. The carboxyl group reacts with the terminal epoxy ring to form an ester. The ester, with its pendant hydroxyl groups, reacts with the remaining epoxide rings, then more cross-linking occurs by forming ether linkages. This material is tougher than epoxy resins and suitable for encapsulating electrical units. Other uses of epoxy resins are coatings for appliance finishes, auto primers, adhesive, and in coatings for cans and drums. Interior coatings of drums used for chemicals and solvents manifest its chemical resistance.

11.5.3 Unsaturated Polyesters

Unsaturated polyesters are a group of polymers and resins used in coatings or for castings with styrene. These polymers normally have maleic anhydride moiety or an unsaturated fatty acid to impart the required unsaturation. A typical example is the reaction between maleic anhydride and ethylene glycol. Also, phthalic anhydride, a polyol, and an unsaturated fatty acid are usually copolymerized to unsaturated polyesters for coating purposes. Many other combinations in variable ratios are possible for preparing these resins.

11.5.4 Phenol-Formaldehyde Resins

Phenol-formaldehyde resins are the oldest thermosetting polymers. They are produced by a condensation reaction between phenol and formaldehyde. Although many attempts were made to use the product and control the conditions for the acid-catalyzed reaction described by Bayer in 1872, there was no commercial production of the resin until the exhaustive work by Baekeland was published in 1909. In this paper, he describes the product as far superior to amber for pipe stem and similar articles, less flexible but more durable than celluloid, odorless, and fire-resistant.

The reaction between phenol and formaldehyde is either base or acid catalyzed, and the polymers are termed resols (for the base catalyzed) and novalacs (for the acid catalyzed). The first step in the base-catalyzed reaction is an attack by the phenoxide ion on the carbonyl carbon of formaldehyde, giving a mixture of ortho-substituted and para-substituted mono-methylolphenol plus di- and trisubstituted methylol phenol derivatives. The second step is the condensation reaction between the methylol phenol derivatives with the elimination of water and the formation of the polymer. Cross-linking occurs by a reaction between the methylol groups and results in the formation of ether bridges. It occurs also by the reaction of the methylol groups and the aromatic ring, which forms methylene bridges. The formed polymer is a three-dimensional network thermoset.

The acid-catalyzed reaction occurs by an electrophilic substitution where formaldehyde is the electrophile. Condensation between the methylol groups and the benzene rings results in the formation of methylene bridges. Usually, the ratio of formaldehyde to phenol is kept less than unity to produce a linear fusible polymer in the first stage. Cross-linking of the formed polymer can occur by adding more formaldehyde and a small amount of hexamethylene tetramine (hexamine, $(CH_2)_6N_4$). Hexamine decomposes in the presence of traces of moisture to formaldehyde and ammonia. This results in cross-linking and formation of a thermoset resin.

Important properties of phenolic resins are their hardness, corrosion resistance, rigidity, and resistance to water hydrolysis. They are also less expensive than many other polymers.

Many additives are used with phenolic resins such as wood flour, oils, asbestos, and fiberglass. Fiberglass piping made with phenolic resins can operate at 150°C and pressure up to 150 psi. Molding applications dominate the market of phenolic resins. Articles produced by injection molding have outstanding heat resistance and dimensional stability. Compression-molded glass-filled phenolic disk brake pistons are replacing the steel ones in many automobiles because of their light weight and corrosion resistance.

Phenol derivatives are also used in a variety of other applications such as adhesives, paints, laminates for building, automobile parts, and ion-exchange resins.

11.5.5 AMINO RESINS

Amino resins (aminoplasts) are condensation thermosetting polymers of formaldehyde with either urea or melamine. Melamine is a condensation product of three urea molecules. It is also prepared from cyanimide at high pressure and high temperature. The nucleophilic addition reaction of urea to formaldehyde produces mainly monomethylol urea and some dimethylol urea. When the mixture is heated in presence of an acid, condensation occurs, and water is released. This is accompanied by the formation of a cross-linked polymer. A similar reaction occurs between melamine and formaldehyde and produces methylolmelamine derivatives.

A variety of methylol derivatives are possible due to the availability of six hydrogens in melamine. As with urea formaldehyde resins, polymerization occurs by a condensation reaction and the release of water.

Amino resins are characterized by being more clear and harder (tensile strength) than phenol derivatives. However, their impact strength (breakability) and heat resistance are lower. Melamine resins have better heat and moisture resistance and better hardness than their urea analogs.

The most important use of amino resins is the production of adhesives for particleboard and hardwood plywood. Compression and injection molding are used with amino resins to produce articles such as radio cabinets, buttons, and cover plates. Because melamine resins have lower water absorption and better chemical and heat resistance than urea resins, they are used to produce dinnerware and laminates used to cover furniture. Almost all molded objects use fillers such as cellulose, asbestos, glass, wood flour, glass fiber, and paper.

11.5.6 POLYCYANURATES

A new polymer type that emerged as an important material for circuit boards are polycyanurate derivatives. The simplest monomer is the dicyanate ester of bisphenol A. When polymerized, it forms three-dimensional, densely cross-linked structures through three-way cyanuric acid (2,4,6-triazinetriol). The cyanurate ring is formed by the trimerization of the cyanate ester. Other monomers, such as hexaflurobisphenol A and tetramethyl bisphenol F, are also used. These polymers are characterized by high glass transition temperatures ranging between 192°C and >350°C (377°F and >660°F).

The largest application of polycyanurate derivatives is in circuit boards. Their transparency to microwave and radar energy makes them useful for manufacturing the housing of radar antennas of military and reconnaissance planes. Their impact resistance makes them ideal for aircraft structures and engine pistons.

Thermoplastic elastomers, as the name indicates, are plastic polymers with the physical properties of rubbers. They are soft, flexible, and possess the resilience needed of rubbers. However, they are processed like thermoplastics by extrusion and injection molding. Thermoplastic elastomers are more economical to produce than traditional thermoset materials because fewer steps are required to manufacture them than to manufacture and vulcanize thermoset rubber. An important property of these polymers is that they are recyclable.

Thermoplastic elastomers are multiphase composites, in which the phases are intimately depressed. In many cases, the phases are chemically bonded by block or graft copolymerization. At least one of the phases consists of a material that is hard at room temperature.

Currently, important thermoplastic elastomers include blends of semicrystalline thermoplastic polyolefin derivatives such as propylene copolymers, with ethylene-propylene terepolymer (EPT) elastomer. Block copolymers of styrene with other monomers such as butadiene, isoprene, and ethylene or ethylene/propylene are the most widely used thermoplastic elastomers.

Polyurethane thermoplastic elastomers are relatively more expensive than other thermoplastic elastomers. However, they are noted for their flexibility, strength, toughness, and abrasion and chemical resistance. Blends of polyvinyl chloride with elastomers such as butyl are widely used in Japan. Random block copolymers such as polyesters (hard segments) and amorphous glycol soft segments, alloys of ethylene inter-polymers, and chlorinated polyolefin derivatives are among the evolving thermoplastic elastomers.

Important properties of thermoplastic elastomers are the flexibility, softness, and resilience. However, compared to vulcanizable rubbers, they are inferior in resistance to deformation and solvents. Important markets for thermoplastic elastomers include shoe soles, pressure-sensitive adhesives, insulation, and recyclable bumpers.

11.6 SYNTHETIC FIBERS

Briefly, and by way of explanation, a fiber is often is as a polymer with a length-to-diameter ratio of at least 100 (Browne and Work, 1983). Fibers (synthetic or natural) are polymers with high molecular symmetry and strong cohesive energies between chains that result usually from the presence of polar groups. Fibers possess a high degree of crystallinity characterized by the presence of stiffening groups in the polymer backbone, and of intermolecular hydrogen bonds. Also, they are characterized by the absence of branching or irregularly space-dependent groups that will otherwise disrupt the crystalline formation. Fibers are normally linear and drawn in one direction to make them long, thin, and threadlike, with great strength along the fiber. These characteristics permit formation of this type of polymers into long fibers suitable for textile applications. Typical examples of fibers include polyesters, nylons, and, acrylic polymers such as polyacrylonitrile, and naturally occurring polymers such as cotton, wool, and silk.

Fibers fall into a class of materials that are continuous filaments or are in discrete elongated pieces, similar to lengths of thread. Fiber classification in reinforced plastics falls into two classes: (i) short fibers, also known as discontinuous fibers, with a general aspect ratio (defined as the ratio of fiber length to diameter) between 20 and 60, and (ii) long fibers, also known as continuous fibers, the general aspect ratio is between 200 and 500.

Thus, fibers are materials that are continuous filaments or discrete elongated pieces, similar to lengths of thread and are characterized by a high ratio of length to diameter. They are important for a variety of applications, including holding tissues together in both plants and animals. There are many different kinds of fibers including textile fiber, natural fibers, and synthetic or human-made fibers such as cellulose, mineral, polymer, and microfibers. Fibers can be manufactured from a natural origin such as silk, wool, and cotton, or derived from a natural fiber such as rayon. They may also be synthesized from certain monomers by polymerization (synthetic fibers). In general, polymers with high melting points, high crystallinity, and moderate thermal stability and tensile strengths are suitable for fiber production.

Fibers can be spun into filaments, string, or rope; used as a component of composite material; or matted into sheets to make products such as paper and are often used in the manufacture of other materials. The strongest engineering materials are generally made of fibers, for example, carbon fiber and ultra-high molecular weight polyethylene. Synthetic fibers can often be produced cheaply and in large amounts as compared to natural fibers, but natural fibers have benefit in some applications, especially for clothing.

Man-made fibers include, in addition to synthetic fibers, those derived from cellulose (cotton, wood) but modified by chemical treatment such as rayon, cellophane, and cellulose acetate. These are sometimes termed "regenerated cellulose fibers." Rayon and cellophane have shorter chains than the original cellulose due to degradation by alkaline treatment. Cellulose acetates produced by reacting cellulose with acetic acid and modified or regenerated fibers are excluded from this book because they are derived from a plant source. Fibers produced by drawing metals or glass, (SiO_2) such as glass wool, are also excluded.

Major fiber-making polymers are those of polyesters, polyamides (nylons), polyacrylic derivatives, and polyolefin derivatives. Polyesters and polyamides are produced by step polymerization reactions, while polyacrylic derivatives and polyolefin derivatives are synthesized by chain-addition polymerization.

11.6.1 Polyester Fibers

Polyesters are the most important class of synthetic fibers. In general, polyesters are produced by an esterification reaction of a diol and a diacid. Carothers was the first to try to synthesize a polyester fiber by reacting an aliphatic diacid with a diol. The polymers were not suitable because of their low melting points. However, he was successful in preparing the first synthetic fiber (nylon 66).

Polyesters can be produced by an esterification of a dicarboxylic acid and a diol, a transesterification of an ester of a dicarboxylic acid and a diol, or by the reaction between an acid dichloride and a diol. Less important methods are the self-condensation of w-hydroxy acid and the ring opening of lactones and cyclic esters. In self-condensation of w-hydroxy acids, cyclization might compete seriously with linear polymerization, especially when the hydroxyl group is in a position to give five- or six-membered lactones.

Polyethylene terephthalate is produced by esterifying terephthalic acid and ethylene glycol or, more commonly, by the transesterification of dimethyl terephthalate and ethylene glycol. The reaction occurs in two stages: (i) in the first stage, methanol is released in at approximately 200°C (370°F) with the formation of bis(2-hydroxyethyl) terephthalate and (ii) in the second stage, polycondensation occurs, and excess ethylene glycol is driven away at approximately 280°C (535°F) and at lower pressures.

Using excess ethylene glycol is the usual practice because it drives the equilibrium to near completion and terminates the acid end groups. This results in a polymer with terminal –OH. When the free acid is used (esterification), the reaction is self-catalyzed. However, an acid catalyst is used to compensate for the decrease in terephthalic acid as the esterification nears completion. In addition to the catalyst and terminator, other additives are used such as color improvers and dulling agents.

The molecular weight of the polymer is a function of the extent of polymerization and could be monitored through the melt viscosity. The final polymer may be directly extruded or transformed into chips, which are stored.

Batch polymerization is still used. However, most new processes use continuous polymerization and direct spinning. An alternative route to polyethylene terephthalate is by the direct reaction of terephthalic acid and ethylene oxide. The product bis(2-hydroxyethyl)terephthalate reacts in a second step with terephthalic acid to form a dimer and ethylene glycol, which is released under reduced pressure at approximately 300°C (570°F). This process differs from the direct esterification and the transesterification routes in that only ethylene glycol is released. In the former two routes, water or methanol is coproduced and the excess glycol released.

Polyethylene terephthalate is an important thermoplastic. However, most polyethylene terephthalate is consumed in the production of fibers. Polyester fibers contain crystalline as well as noncrystalline regions. The degree of crystallinity and molecular orientation are important in determining the tensile strength of the fiber (between 18 and 22 denier) and its shrinkage. The degree of crystallinity and molecular orientation can be determined by X-ray diffraction techniques.

Important properties of polyesters are the relatively high melting temperatures (265°C, 510°F), high resistance to weather conditions and sunlight, and moderate tensile strength. Due to the hydrophobic nature of the fiber, sulfonated terephthalic acid may be used as a comonomer to provide anionic sites for cationic dyes. Small amounts of aliphatic di-acid derivatives such as adipic acid may also be used to increase the ability of the fibers to dyes by disturbing the crystallinity of the fiber.

Polyester fibers can be blended with natural fibers such as cotton and wool. The products have better qualities and are used for men's and women's wear, pillow cases, and bedspreads. Fiberfill, made from polyesters, is used in mattresses, pillows, and sleeping bags. High-tenacity polymers for tire cord reinforcement are equivalent in strength to nylon tire cords and are superior because they do not "flat spot." V-belts and fire hoses made from industrial filaments are another market for polyesters.

11.6.2 Polyamides

Polyamides (nylon fibers) are the second largest group of synthetic fibers after polyesters. Numbers that follow the word "nylon" denote the number of carbons present within a repeating unit and whether one or two monomers are being used in polymer formation. For nylons using a single monomer such as nylon 6 or nylon 12, the numbers 6 and 12 denote the number of carbons in caprolactam and laurolactam, respectively. For nylons using two monomers such as nylon 610, the first number, 6, indicates the number of carbons in the hexamethylene diamine and the other number, 10, is for the second monomer sebacic acid.

Polyamides are produced by the reaction between a dicarboxylic acid and a diamine (e.g., nylon 66), ring openings of a lactam, (e.g., nylon 6) or by the polymerization of w-amino acids (e.g., nylon 11). The production of some important nylons is discussed in the following sections.

11.6.2.1 Nylon 66

Nylon 66 (polyhexamethyleneadipate) is produced by the reaction of hexamethylenediamine and adipic acid (see Chapters 9 and 10 for the production of the two monomers). This produces hexamethylene diammonium adipate salt. The product is a dilute salt solution concentrated to approximately 60% and charged with acetic acid to a reactor where water is continuously removed. The presence of a small amount of acetic acid limits the degree of polymerization to the desired level. The temperature is then increased to 270°C–300°C and the pressure to approximately 16 atm, which favors the formation of the polymer. The pressure is finally reduced to atmospheric to permit further water removal. After a total of 3 h, nylon 66 is extruded under nitrogen pressure.

11.6.2.2 Nylon 6

Nylon 6 (polycaproamide) is produced by the polymerization of caprolactam. The monomer is first mixed with water, which opens the lactam ring and gives w-amino acid.

$$\text{Caprolactam} + H_2O \longrightarrow H_2N(CH_2)_5COOH$$

Caprolactam   w-Amino acid

$$H_2N(CH_2)_5COOH + \text{(caprolactam)} \longrightarrow \left[\overset{H}{N}-(CH_2)_5\overset{O}{C} \right]_n$$

The formed amino acid reacts with itself or with caprolactam at approximately 250°C–280°C to form the polymer.

Temperature control is important, especially for depolymerization, which is directly proportional to reaction temperature and water content.

11.6.2.3 Nylon 12

Nylon 12 (polylaurylamide) is produced in a similar way to nylon 6 by the ring-opening polymerization of laurolactam. The polymer has a lower water capacity than nylon 6 due to its higher hydrophobic properties. The polymerization reaction is slower than for caprolactam. Higher temperatures are used to increase the rate of the reaction.

$$\text{laurolactam} + H_2O \longrightarrow H_2NCH_2-(CH_2)_{10}-\overset{O}{C}OH$$

$$n\,NH_2CH_2(CH_2)_{10}\overset{O}{C}OH \longrightarrow \left[HN(CH_2)_{11}-\overset{O}{C} \right]_n + n\,H_2O$$

The monomer (laurolactam) could be produced from 1,5,9-cyclododecatriene, a trimer of butadiene (Chapter 9). The trimer is epoxidized with peracetic acid or acetaldehyde peracetate and then hydrogenated. The saturated epoxide is rearranged to the ketone with magnesium iodide (MgI_2) at 100°C (212°F). This is then changed to the oxime and rearranged to laurolactam.

11.6.2.4 Nylon 4

Nylon 4 (polybutyramide) is produced by ring-opening 2-pyrrolidone. Anionic polymerization is used to polymerize the lactam. Co-catalysts are used to increase the yield of the polymer. Carbon dioxide is reported to be an excellent polymerization activator.

Nylon 4 has a higher water absorption capacity than other nylons due to its lower hydrophobic property.

11.6.2.5 Nylon 11

Nylon 11 (polyundecanylamide) is produced by the condensation reaction of 11-aminoundecanoic acid. This is an example of the self-condensation of an amino acid where only one monomer is used. The monomer is first suspended in water, then heated to melt the monomer and to start the reaction. Water is continuously removed to drive the equilibrium to the right. The polymer is finally withdrawn for storage.

$$n\ H_2N\text{---}CH_2(CH_2)_9\overset{O}{\overset{\|}{C}}OH \rightleftharpoons \text{---}\!\!\left[NH\text{---}(CH_2)_{10}\overset{O}{\overset{\|}{C}}\right]_n + nH_2O$$

11.6.2.6 Other Nylon Polymers

Many other nylons could be produced such as nylon 5, nylon 7, nylon 610, and nylon 612. Nylon polymers are generally characterized by relatively high melting points due to the presence of the amide linkage. They are also highly crystalline, and the degree of crystallinity depends upon factors such as the polymer structure and the distance between the amide linkages. An increase in polymer crystallinity increases its tensile strength, abrasion resistance, and modulus of elasticity.

Hydrogen bonding in polyamides is fairly strong and has a pronounced effect on the physical properties of the polymer such as the crystallinity, melting point, and water absorption. For example, nylon 6, with six carbon atoms, has a melting point of 223°C (433°F), while it is only 190°C (374°F) for nylon 11. This reflects the higher hydrogen bonding in nylon 6 than in nylon 11.

Moisture absorption of nylons differs according to the distance between the amide groups. For example, nylon 4 has a higher moisture absorption than most other nylons, and it is approximately similar to that of cotton. This is a result of the higher hydrophilic character of nylon. Nylons, however, are to some extent subject to deterioration by light. This has been explained on the basis of chain breaking and cross-linking. Nylons are liable to attack by mineral acids but are resistant to alkalis. They are difficult to ignite and are self-extinguishing.

In general, most nylons have remarkably similar properties, and the preference of using one nylon over the other is usually dictated by economic considerations except for specialized uses. Nylons have a variety of uses ranging from tire cord to carpet to hosiery. The most important application is cord followed by apparel. Nylon staple and filaments are extensively used in the carpet industry. Nylon fiber is also used for a variety of other articles such as seat belts, monofilament finishes, and knitwear. Because of its high tenacity and elasticity, it is a valuable fiber for ropes, parachutes, and underwear.

11.6.3 Acrylic and Modacrylic Fibers

Acrylic fibers are a major synthetic fiber class developed about the same time as polyesters. Modacrylic fibers are copolymers containing between 35% and 85% acrylonitrile. Acrylic fibers contain at least 85% acrylonitrile. Orlon is an acrylic fiber developed by DuPont in 1949; Dynel is a modacrylic fiber developed by Union Carbide in 1951.

Polyacrylics are produced by copolymerizing acrylonitrile with other monomers such as vinyl acetate, vinyl chloride, and acrylamide. Solution polymerization may be used where water is the solvent in the presence of a redox catalyst. Free radical or anionic initiators may also be used. The produced polymer is insoluble in water and forms a precipitate. Copolymers of acrylonitrile are sensitive to heat, and melt spinning is not used. Solution spinning (wet or dry) is the preferred process where a polar solvent such as dimethylformamide is used. In dry spinning, the solvent is evaporated and recovered.

Dynel, a modacrylic fiber, is produced by copolymerizing vinyl chloride with acrylonitrile. Solution spinning is also used where the polymer is dissolved in a solvent such as acetone. After the solvent is evaporated, the fibers are washed and subjected to stretching, which extends the fiber 4–10 times of the original length.

Acrylic fibers are characterized by having properties similar to wool and have replaced wool in many markets such as blankets, carpets, and sweaters. Important properties of acrylics are resistance to solvents and sunlight, resistance to creasing, and quick drying.

Acrylic fiber breaking strength ranges between 22,000 and 39,000 psi and they have a water absorption of approximately 5%. Dynel, due to the presence of chlorine, is less flammable than many other synthetic fibers. Major uses of acrylic fibers are woven and knitted clothing fabrics (for apparel), carpets, and upholstery.

11.6.4 GRAPHITE FIBERS

Carbon fibers are special reinforcement types having a carbon content of 92%–99% w/w. They are prepared by controlled pyrolysis of organic materials in fibrous forms at temperatures ranging from 1,000°C to 3,000°C (1,800°F–5,400°F). The commercial fibers are produced from rayon, polyacrylonitrile, and petroleum pitch. When acrylonitrile is heated in air at moderate temperatures (220°C, 430°F), hydrogen cyanide (HCN) is emanated. Further heating above 1,700°C (3,100°F) in the presence of nitrogen for a period of 24 h produces carbon fiber. Carbon fibers are characterized by high strength, stiffness, low thermal expansion, and thermal and electrical conductivity, which makes them an attractive substitute for various metals and alloys.

11.6.5 POLYPROPYLENE FIBERS

Polypropylene fibers represent a small percent of the total polypropylene production. (Most polypropylene is used as a thermoplastic.) The fibers are usually manufactured from isotactic polypropylene. Important characteristics of polypropylene are high abrasion resistance, strength, low static buildup, and resistance to chemicals. Crystallinity of fiber-grade polypropylene is moderate and when heated, it starts to soften at approximately 145°C (293°F) and then melts at 170°C (338°F). The high melting points of polypropylene polymers are attributed to low entropy of fusion arising from stiffening of the chain.

Polyethylene fiber properties depend markedly on the crystallinity or density of the polymer; although high-strength fibers can be made from linear polyethylene, resiliency properties are poor, tensile properties are highly time-dependent, and endurance under sustained loading is very poor.

On the other hand, polypropylene fibers have good stress-endurance properties, excellent recovery from high extensions, and fair-to-good recovery properties at low strains; recovery at low strains is shown to depend on the extent of fiber orientation and annealing.

Anomalies in the change of the sonic modulus of polypropylene yarns during extension and relaxation are noted and interpreted in terms of structure changes in the crystalline phase. The high melting temperature of 235°C (455°F) for poly(4-methyl-1-pentene) appears to be due to its low entropy of melting, and fibers from this polymer are characterized by low tenacity when tested at elevated temperatures. Crystalline polystyrene fibers have relatively good retention of tenacity at elevated temperatures and are characterized by excellent resiliency at low strains, good wash-wear characteristics in cotton blends, and low abrasion resistance.

11.7 SYNTHETIC RUBBER

Synthetic rubber (an elastomer) is a long-chain polymer with special chemical and physical as well as mechanical properties. These materials have chemical stability, high abrasion resistance, strength, and good dimensional stability. Many of these properties are imparted to the original polymer through cross-linking agents and additives. An important property of elastomeric materials is their ability to be stretched at least twice their original length and to return back to nearly their original length when released.

Natural rubber is a polymer of isoprene—most often *cis*-1,4-poly-isoprene—with a molecular weight of 100,000–1,000,000. Typically, a few percent of other materials, such as proteins, fatty acids, resins, and inorganic materials are found in high-quality natural rubber. Some natural rubber sources called gutta percha (i.e., trees of the genus *Palaquium* in the family Sapotaceae and the rigid natural latex produced from the sap of these trees, particularly from *Palaquium gutta*) are composed of *trans*-1,4-poly-isoprene, a structural isomer that has similar, but not identical, properties.

Isoprene (2-methyl-1,3-butadiene) is a common organic compound with the formula $CH_2=C(CH_3)CH=CH_2$:

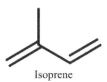

Isoprene

Under standard conditions, isoprene is a colorless liquid and is the monomer of natural rubber as well as a precursor to an immense variety of other naturally occurring compounds.

Synthetic rubber is any type of artificial elastomer, invariably a polymer. An elastomer is a material with the mechanical (or material) property that it can undergo much more elastic deformation under stress than most materials and still return to its previous size without permanent deformation. Synthetic rubber serves as a substitute for natural rubber in many cases, especially when improved material properties are required.

Synthetic rubber can be made from the polymerization of a variety of monomers including isoprene (2-methyl-1,3-butadiene), 1,3-butadiene, and isobutylene (methylpropene) with a small percentage of isoprene for cross-linking. These and other monomers can be mixed in various desirable proportions to be copolymerized for a wide range of physical, mechanical, and chemical properties. The monomers can be produced pure and the addition of impurities or additives can be controlled by design to give optimal properties. Polymerization of pure monomers can be better controlled to give a desired proportion of *cis* and *trans* double bonds.

Natural rubber is an elastomer constituted of isoprene units. These units are linked in a *cis*-1,4-configuration that gives natural rubber the outstanding properties of high resilience and strength. Natural rubber occurs as a latex (water emulsion) and is obtained from *Hevea brasiliensis*, a tree that grows in Malaysia, Indonesia, and Brazil. Charles Goodyear was the first to discover that the latex could be vulcanized (cross-linked) by heating with sulfur or other agents. Vulcanization of rubber is a chemical reaction by which elastomer chains are linked together. The long-chain molecules impart elasticity, and the cross-links give load supporting strength.

Synthetic rubbers include elastomers that could be cross-linked such as polybutadiene, polyisoprene, and ethylene-propylene-diene terepolymer. It also includes thermoplastic elastomers that are not cross-linked and are adapted for special purposes such as automobile bumpers and wire and cable coatings. These materials could be scraped and reused. However, they cannot replace all traditional rubber since they do not have the wide temperature performance range of thermoset rubber.

The major use of rubber is for tire production. Non-tire consumption includes hoses, footwear, molded and extruded materials, and plasticizers.

11.7.1 Styrene-Butadiene Rubber

Styrene-butadiene rubber is the most widely used synthetic rubber. It can be produced by the copolymerization of butadiene (75%) and styrene (25%) using free radical initiators. A random copolymer is obtained. The microstructure of the polymer is 60%–68% *trans polymer*, 14%–19% *cis polymer*, and 17%–21% 1,2 configuration. Wet methods are normally used to characterize polybutadiene polymers and copolymers. Solid-state nuclear magnetic resonance spectroscopy provides a more convenient way to determine the polymer microstructure. Currently, more styrene-butadiene rubber is produced by copolymerizing the two monomers with anionic or coordination catalysts. The formed copolymer has better mechanical properties and a narrower molecular weight distribution. A random copolymer with ordered sequence can also be made in solution using butyl-lithium, provided that the two monomers are charged slowly. Block copolymers of butadiene and styrene may be produced in solution using coordination or anionic catalysts. Butadiene polymerizes first until it is consumed, then styrene starts to polymerize. Styrene-butadiene rubber produced by coordination catalysts has better tensile strength than that produced by free radical initiators.

The main use of styrene-butadiene rubber is for tire production. Other uses include footwear, coatings, carpet backing, and adhesives.

11.7.2 NITRILE RUBBER

Nitrile rubber (NBR) is a copolymer of butadiene and acrylonitrile. It has the special property of being resistant to hydrocarbon liquids. The copolymerization occurs in an aqueous emulsion. When free radicals are used, a random copolymer is obtained. Alternating copolymers are produced when a Ziegler–Natta catalyst is employed. Molecular weight can be controlled by adding modifiers and inhibitors. When the polymerization reaches approximately 65%, the reaction mixture is vacuum distilled in presence of steam to recover the monomer. The ratio of acrylonitrile/butadiene could be adjusted to obtain a polymer with specific properties. Increasing the acrylonitrile ratio increases oil resistance of the rubber, but decreases its plasticizer compatibility.

Nitrile rubber is produced in different grades depending on the end use of the polymer. Low acrylonitrile rubber is flexible at low temperatures and is generally used in gaskets, 0-rings, and adhesives. The medium type is used in less flexible articles such as kitchen mats and shoe soles. High acrylonitrile polymers are more rigid and highly resistant to hydrocarbon derivatives and oils and are used in fuel tanks and hoses, hydraulic equipment, and gaskets.

11.7.3 POLYISOPRENE

Natural rubber is a stereoregular polymer composed of isoprene units attached in a *cis* configuration. This arrangement gives the rubber high resilience and strength. Isoprene can be polymerized using free radical initiators, but a random polymer is obtained. As with butadiene, polymerization of isoprene can produce a mixture of isomers. However, because the isoprene molecule is asymmetrical, the addition can occur in 1,2-, 1,4- and 3,4- positions. Six tactic forms are possible from both 1,2- and 3,4- addition and two geometrical isomers from 1,4- addition (*cis* and *trans*).

Stereoregular polyisoprene is obtained when Ziegler–Natta catalysts or anionic initiators are used. The most important coordination catalyst is α-titanium trichloride co-catalyzed with aluminum alkyl derivatives. The polymerization rate and *cis* content depends upon Al/Ti ratio, which should be greater than one. Lower ratios predominantly produce the *trans* structure.

Polyisoprene is a synthetic polymer (elastomer) that can be vulcanized by the addition of sulfur. *cis*-Polyisoprene has properties similar to that of natural rubber. It is characterized by high tensile strength and insensitivity to temperature changes, but it has low abrasion resistance. It is attacked by oxygen and hydrocarbon derivatives. *trans*-Polyisoprene is similar to Gutta percha, which is produced from the leaves and bark of the Sapotaceae tree. It has different properties from the *cis* form and cannot be vulcanized. Few commercial uses are based on *trans*-polyisoprene.

Important uses of *cis*-polyisoprene include the production of tires, specialized mechanical products, conveyor belts, footwear, and insulation.

11.7.4 Polychloroprene

Polychloroprene (neoprene rubber) is the oldest synthetic rubber. It is produced by the polymerization of 2-chloro-1,3-butadiene in a water emulsion with potassium sulfate as a catalyst. The product is a random polymer that is vulcanized with sulfur or with metal oxides (zinc oxide, magnesium oxide, etc.). Vulcanization with sulfur is very slow, and an accelerator is usually required.

Neoprene vulcanizates have a high tensile strength, excellent oil resistance (better than natural rubber), and heat resistance. Neoprene rubber could be used for tire production, but it is expensive. Major uses include cable coatings, mechanical goods, gaskets, conveyor belts, and cables.

11.7.5 Butyl Rubber

Butyl rubber is a copolymer of isobutylene (97.5%) and isoprene (2.5%). The polymerization is carried out at low temperature (below −95°C, <139°F) using aluminum chloride ($AlCl_3$) co-catalyzed with a small amount of water. The co-catalyst furnishes the protons needed for the cationic polymerization:

$$AlCl_3 + H_2O \rightarrow H^+ \left(AlCl_3OH \right)^-$$

The product is a linear random copolymer that can be cured to a thermosetting polymer. This is made possible through the presence of some unsaturation from isoprene.

Butyl rubber vulcanizates have tensile strengths up to 2,000 psi, and are characterized by low permeability to air and a high resistance to many chemicals and to oxidation. These properties make it a suitable rubber for the production of tire inner tubes and inner liners of tubeless tires. The major use of butyl rubber is for inner tubes. Other uses include wire and cable insulation, steam hoses, mechanical goods, and adhesives. Chlorinated butyl is a low molecular weight polymer used as an adhesive and a sealant.

11.7.6 Ethylene-Propylene Rubber

Ethylene-propylene rubber (EPR) is a stereoregular copolymer of ethylene and propylene. Elastomers of this type do not possess the double bonds necessary for cross-linking. A third monomer, usually a mono-conjugated diene, is used to provide the residual double bonds needed for cross-linking. The 1,4-hexadiene and ethylidene norbornene are examples of these dienes. The main polymer chain is completely saturated while the unsaturated part is pending from the main chain. The product elastomer, termed ethylene-propylene terepolymer, can be cross-linked using sulfur. Cross-linking ethylene-propylene rubber is also possible without using a third component (a diene). This can be done with peroxides.

Important properties of vulcanized ethylene-propylene rubber and ethylene-propylene terepolymer include resistance to abrasion, oxidation, and heat and ozone, but they are susceptible to hydrocarbon derivatives. The main use of ethylene-propylene rubber is to produce automotive parts such as gaskets, mechanical goods, wire, and cable coating. It may also be used to produce tires.

REFERENCES

Ali, M.F., El Ali, B.M., and Speight, J.G. 2005. *Handbook of Industrial Chemistry: Organic Chemicals.* McGraw-Hill, New York.

Austin, G.T. 1984. Chapters 34, 35, and 36. *Shreve's Chemical Process Industries.* 5th Edition. McGraw-Hill, New York.

Braun, D., Cherdron, H., and Ritter, H. 2001. *Polymer Synthesis Theory and Practice: Fundamentals, Methods, Experiments.* Springer-Verlag, Berlin.

Browne, C.L., and Work, R.W. 1983. Chapter 11: Man-Made Textile Fibers. In: *Riegel's Handbook of Industrial Chemistry*. J.A. Kent (Editor). 8th Edition. Van Nostrand Reinhold, New York.

Carraher, C.E. Jr. 2003. *Polymer Chemistry*. 6th Edition: Revised and Expanded. Marcel Dekker Inc., New York.

Grabowska, H., Kaczmarczyk, W., and Wrzyszcz, J. 1989. Synthesis of 2,6-xylenol by alkylation of phenol with methanol. *Applied Catalysis*, 47(2): 351–355.

Jones, R.W., and Simon, R.H.M. 1983. Chapter 10: Synthetic plastics. In *Riegel's Handbook of Industrial Chemistry*. 8th Edition. J.A. Kent (Editor). Van Nostrand Reinhold, New York.

Lokensgard, E. 2010. *Industrial Plastics: Theory and Applications*. Delmar Cengage Learning, Clifton Park, NY.

Matar, S., and Hatch, L.F. 2001. *Chemistry of Petrochemical Process*. 2nd Edition. Gulf Professional; Publishing, Elsevier BV, Amsterdam.

Odian, G. 2004. *Principles of Polymerization*. 4th Edition. John Wiley & Sons Inc., New York.

Rudin, A. 1999. *The Elements of Polymer Science and Engineering*. 2nd Edition. Academic Press Inc., New York.

Schroeder, E.E. 1983. Chapter 9: Rubber. In *Riegel's Handbook of Industrial Chemistry*. 8th Edition. J.A. Kent (Editor). Van Nostrand Reinhold, New York.

12 Pharmaceuticals

12.1 INTRODUCTION

The modern pharmaceutical industry can trace its beginnings to two sources (i) local apothecaries—now called chemists in the United Kingdom and pharmacists in the United States—that expanded from their traditional role distributing botanical drugs such as morphine and quinine to wholesale manufacture in the mid-1800s. By the late 1880s, German dye manufacturers had perfected the purification of individual organic compounds from coal tar and other mineral sources and had also established fundamental methods in organic chemical synthesis. The development of synthetic chemical methods allowed scientists to systematically vary the structure of chemical substances, and growth in the emerging science of pharmacology expanded their ability to evaluate the biological effects of these structural changes. It is from these early beginning and the recognition of the wealth of chemical that could be produced from crude oil that led to the rapid expansion of the medicines-from-crude-oil industry as an extension of the petrochemical industry.

From the previous chapters, it is obvious that petrochemicals play many roles in modern life because they are used to create resins, films, and plastics. In addition, petrochemicals also play a major role in the production of medicines because they are used to produce chemicals such as (i) phenol and cumene that are used to create a substance that is essential for manufacturing of penicillin—an extremely important antibiotic—and aspirin; (ii) petrochemical resins are used to purify medicines, speeding up the manufacturing process; (iii) resins made from petrochemicals are used in the manufacture of medicines including treatments for aids, arthritis, and cancer; (iv) plastics and resins which are used to make devices such as artificial limbs and skin; (v) and plastics are used to make a wide range of medical equipment including bottles, disposable syringes, and much more (Hess et al., 2011).

Thus, it would be remiss not to mention the role of petrochemical intermediates in the manufacture of pharmaceutical products. Petrochemical solutions and petrochemicals are the second-phase products and solutions that originate from crude oil, following a number of refining methods. Crude oil works as the fundamental portal ingredient which offers petrochemical products and byproducts after an extensive procedure of refining which takes place in various oil refineries.

Petrochemicals play an important role in the production of medicines. For example, most medicines contain two types of ingredients: (i) the active ingredient which is composed of one or more compounds manufactured synthetically or extracted and purified from plant or animal sources and the active ingredient is the chemical that reacts with your body to produce a therapeutic effect and (ii) the inactive ingredients which are typically the additives present in the medication which are normally inactive/inert and which may have been added as preservatives, flavoring agents, coloring, sweeteners, and sorbents.

Also, for the purposes of this chapter, there are two general definitions that are used: (i) a medicine or medication, which is a chemical that is available as an over-the-counter (OTC) purchase at a pharmacy and (ii) a drug, which is available only by prescription from an authorized person. Over-the-counter medicine is also known as over-the-counter or nonprescription medicine. All of these terms refer to medicine that you can buy without a prescription. They are safe and effective when you follow the directions on the label and as directed by your health care professional. Examples of the former (over-the-counter medicines) are the subject of this chapter (Table 12.1) through published synthesis while the latter (i.e., medicines that are available only by prescription) are not included in the subject of this chapter.

TABLE 12.1

Examples of Readily Available Over-the-Counter Medications

In addition, many synthetic routes to medicines are not published because of proprietary issues as well as dangerous-to-health issues. There are also the questions of nomenclature which can be troublesome as well as confusing. Because of proprietary issues, even over-the-counter medications have names that often bear no relationship to the actual chemical for industrial usage. In all cases, where possible, the trade name and the chemical name of the medication are presented.

A word of caution should be added here. Although relatively easy to obtain, over-the-counter medications can still carry a risk, even though they do not require a prescription. There is the possibility of side effects, interactions with other medications, or harm due to excessive doses. All patients should consult with their doctor, pharmacist, or other health-care provider if they have additional questions concerning use of over-the-counter medications.

Thus, medications (usually referred to as *drugs*) that change behavior patterns are not included in this chapter. It is not the purpose of this chapter to produce methods by which drugs (especially harmful medications, often referred to as *drugs*) can be synthesized but to present to the reader a section of the published synthetic methods that results in the production of commonly used medications. For this, it will also be pointed out the starting materials or other constituents that originated from petrochemical processes.

A medicine is a chemical substance that has known biological effects on humans or other animals, used in the treatment, cure, mitigation, prevention, or diagnosis of disease or used to enhance physical or mental well-being. Medicines may be used for a limited duration or on a regular basis for chronic disorders and are generally taken to cure and/or relieve any symptoms of an illness or medical condition, or may be used as prophylactic medicines. One or more of the constituents of the medicine usually interacts with either normal or abnormal physiological process in a biological system, and produces a desired and positive biological action. However, if the effect causes harm to the body, the medicine is classified as a poison and is no longer a medication. The medications can treat different types of diseases such as infectious diseases, noninfectious diseases, and non-diseases (alleviation of pain, prevention of pregnancy, and anesthesia). Many of the modern medications are prepared from petrochemical starting materials (Table 12.2).

TABLE 12.2

Selection of Common Petrochemical Products Used in the Pharmaceutical Industry

Chemical	Processes		
Ammonia (aqueous)	C	F	B
Aniline	C		
Benzene	C		
n-Butyl acetate	C	F	
n-Butyl alcohol	C	F	B
Chloroform	C	F	B
Chloromethane	C		
Cyclohexane	C		
1,2-Dichloroethane	C		B
Diethyl ether	C		B
Ethanol	C	F	B
Ethyl acetate	C	F	B
Ethylene glycol	C		B
Formaldehyde	C	F	B
n-Heptane	C	F	B
n-Hexane	C	F	B
Methanol	C	F	B
Methylene chloride	C	F	B
2-Methylpyridine	C		
Phenol	C	F	B
n-Propanol	C		B
Pyridine	C		B
Toluene	C	F	B
Xylenes	C		

C, chemical synthesis; F, fermentation; B, biological or natural extraction.

Petrochemicals have contributed to the development of many medications for diverse indications. While most U.S. pharmaceutical companies have reduced or eliminated in-house natural product groups, new paradigms and new enterprises have evolved to carry on a role for natural products in the pharmaceutical industry. Many of the reasons for the decline in popularity of natural products are being addressed by the development of new techniques for screening and production. This chapter aims to inform pharmacologists of current strategies and techniques that make petrochemicals a continuing and viable strategic choice for use in medication synthesis programs.

The use of petroleum products in not new. As early as 1500 BC, the use of asphalt for medicinal purposes and (when mixed with beer) as a sedative for the stomach has been recorded. It is also recorded in the code of Hammurabi that hot asphalt was to be poured over the head of a miscreant as a form of punishment. In more modern times, medicinal oil (sometimes referred to as paraffin oil) distilled from crude oil was prescribed to lubricate the alimentary tract where coal dust was likely to collect. From these humble beginnings, crude oil has, through the production of petrochemicals, become a major contributor to the pharmaceutical industry. For example, the first analgesics and antipyretics, exemplified by phenacetin and acetanilide, were simple chemical derivatives of aniline and p-nitrophenol, both of which were byproducts from coal tar and nor from crude oil. An extract from the bark of the white willow tree had been used for centuries to treat various fevers and inflammation. The active principle in white willow, Salicin or salicylic acid, had a bitter taste and irritated the gastric mucosa, but a simple chemical modification was much more palatable. This was acetylsalicylic acid, better known as aspirin, the first drug that could be generally administered for a variety of ailments. At the start of the 20th century, the first of the barbiturate family of drugs entered the pharmacopeia leading to the start of the evolution of the modern pharmaceutical industry (Mahdi et al., 2006; Fuster and Sweeny, 2011; Jones, 2011; Wick, 2012; Aronson, 2013).

The pharmaceutical industry includes the manufacture, extraction, processing, purification, and packaging of chemical materials to be used as medications for humans or animals (Gad, 2008). Pharmaceutical manufacturing is divided into two major stages: the production of the active ingredient or medicine (primary processing or manufacture) and secondary processing, the conversion of the active medicines into products suitable for administration.

The products are available as tablets, capsules, liquids (in the form of solutions, suspensions, emulsions, gels, or injectables), creams (usually oil-in-water emulsions), ointments (usually water-in-oil emulsions), and aerosols, which contain inhalable products or products suitable for external use. Propellants used in aerosols include chlorofluorocarbons, which are being phased out. Recently, butane has been used as a propellant in externally applied products.

The major manufactured groups include: (i) antibiotics such as penicillin, streptomycin, tetracyclines, chloramphenicol, and antifungals; (ii) other synthetic drugs, including sulfa drugs, antituberculosis drugs, antileprotic drugs, analgesics, anesthetics, and antimalarials; (iii) vitamins; (iv) synthetic hormones; (v) glandular products; (vi) drugs of vegetable origin such as quinine, strychnine and brucine, emetine, and digitalis; (vii) glycosides; and (viii) vaccines.

Other pharmaceutical chemicals such as calcium gluconate, ferrous salts, nikethamide, glycerophosphates, chloral hydrate, saccharin, antihistamines (including meclozine and buclozine), tranquilizers (including meprobamate and chloropromoazine), antifilarials, diethyl carbamazine citrate, and oral antidiabetics, including tolbutamide, chloropropamide, and surgical sutures and dressings.

The principal manufacturing steps are (i) preparation of process intermediates, (ii) introduction of functional groups, (iii) coupling and esterification, (iv) separation processes such as washing and stripping, and (v) purification of the final product. Additional product preparation steps include granulation; drying; tablet pressing, printing, and coating; filling; and packaging.

The main pharmaceutical groups manufactured include: (i) proprietary ethical products or prescription only medicines, which are usually patented products; (ii) general ethical products, which are basically standard prescription-only medicines made to a recognized formula that may be specified in standard industry reference books; and (iii) over-the counter, or nonprescription, products. For those readers interested in the synthesis of medications available by prescription, there are citations available (for example, Karaman, 2015; Flick et al., 2017 and references cited therein).

Finally, it is not the purpose of this chapter to show preference for any type of medication, but it is the purpose to show the methods by which selected over-the-counter medicines can be produced from crude oil.

12.2   MEDICINAL OILS FROM PETROLEUM

This section deals with the synthesis of the bulk fractions that have been used and, in some countries, continue to be used as medications as well the individual molecular active ingredients of medications and their usage in drug formulations to deliver the prescribed dosage. Formulation is also referred to as galenical production. A galenical is a simple cure in the form of a vegetable or herbal remedy as prescribed by Galen (Aelius Galenus or Claudius Galenus or better known to the Western world as Galen of Pergamon, 129–217 AD), a Greek physician, surgeon, and philosopher in the Roman Empire.

The petroleum industry is first encountered in the archaeological record near Hit (Tuttul) in what is now Iraq. Hit is on the banks of the Euphrates River and is the site of an oil seep known locally as The Fountains of Pitch. There, the bitumen was quarried for use as mortar between building stones as early as 6,000 years ago and was also used as a waterproofing agent for baths, pottery, and boats.

The Babylonians caulked their ships with bitumen and in Mesopotamia around 4000 BC, bitumen was used as caulking for ships, a setting for jewels and mosaics, and an adhesive to secure weapon handles. On the human side of bitumen use, the Egyptians used it for embalming while the ancient Persians, the 10th-century Sumatrans, and pre-Columbian natives of the Americas believed that crude oil had medicinal benefits.

From that time, the ancient literature acts as a record of the use of petroleum. In fact, it was the Persian scientist Ibn Sina (c.980–1,037), who was known in the West as Avicenna, discussed medicinal petroleum in his enormously influential encyclopedia of medicine. The translation of this work into Latin spread that knowledge into Europe, where it reached Constantinus Africanus (c.1020–1087), who may have been the first Latin writer to use the word *petroleum*—the word was also used by Georgius Agricola (Georg Bauer) in his work entitled *De Natura Fossilium* (published 1546). From that time, there was a tradition of employing petroleum in medicine which included concoctions recommended for eye diseases, reptile bites, respiratory problems, hysteria, and epilepsy. Mixing petroleum and the ashes of cabbage stalks was recommended for the treatment of scabies and a preparation of petroleum was prescribed to warm the brain by applying it to the forehead. Marco Polo (1254–1324) reported that bitumen was used in the Caspian Sea region to treat camels for mange and the first oil exported from Venezuela (in 1539) was intended as a gout treatment for the Holy Roman Emperor Charles V (reigned 1519–1556).

The native North Americans collected oil for medicines and European settlers found its presence in the water supplies a contamination, but they learned to collect it to use as fuel in their lamps. Native Americans also traded crude oil that they obtained from oil seeps in upstate New York among other places. The Seneca tribe traded oil for so long that all crude oil was referred to as Seneca Oil which was reputed to have great medicinal value. In fact, in 1901, a petroleum technology text was published in which it was noted that petroleum was an excellent remedy for diphtheria (Purdy, 1957). The members of the Seneca tribe also used crude oil for body paint and for ceremonial fires.

Several historical factors evolved to change the use of crude oil. The kerosene lamp, invented in 1854, ultimately created the first large-scale demand for petroleum. (Kerosene was first made from coal, but by the late 1880s most was derived from crude oil.) In 1859, at Titusville, Penn., Col. Edwin Drake drilled the first successful well through rock and produced crude oil. However, bulk oil products from petroleum still find a variety of uses in health and human service roles (i.e., cosmetics) and, because of the imperative of these products, a brief discussion of the various types of products and their roles within the various human communities is also included here—the oil products being considered to be bulk petrochemical products.

In fact, mineral oil and petrolatum are petroleum byproducts used in many creams and topical pharmaceuticals. Tar (also called resid, asphalt, pitch), for psoriasis and dandruff, is also produced from petroleum. Most pharmaceuticals are complex organic compounds which have their basis in smaller, simpler precursor organic molecules that are petroleum byproducts.

12.2.1 Mineral Oil/White Oil

Some of the imprecision in the definition of the names (such as *mineral oil* and *white oil*) reflect the use of the oil by the buyers and by the sellers. In fact, mineral oils have numerous definitions and are substances by nature also complex being derived from crude oil. The term *mineral oil* includes many petroleum products and applications including fuel and medicinal white oils and can range from less refined (only straight-run) to highly refined (severely hydrotreated) with a composition and toxicity that depend on the refining history.

The first use of the term *mineral oil* was in 1771 and prior to the late 19th century, the chemical science to determine such makeup was unavailable. White oils are highly refined, odorless, tasteless, and have excellent color stability. They are chemically and biologically stable and do not support bacterial growth. The inert nature of mineral makes it easy to work with as they lubricate, sooth, soften, and hold in moisture formulations. These oils are used in a variety of product lines such as antibiotics, baby oils, lotions, creams, shampoos, sunscreens, and tissues.

White oils are manufactured from highly refined base oils and consist of saturated paraffin derivatives and cycloparaffin derivatives. The refinement process ensures complete removal of aromatics, sulfur, and nitrogen compounds. The technologies employed result in products that are highly stable over time besides being hydrophobic, colorless, odorless, and tasteless. White mineral oils are extensively used as bases for pharmaceuticals and personal care products. The inertness of the product offers properties such as good lubricity, smoothness and softness, and resistance to moisture in the formulations. The products are also used in the polymer processing and plastic industry such as polystyrene, polyolefin, and thermoplastic elastomers. The oil controls the melt flow behavior of the finished polymer besides providing release properties. Very often the oils also impart improvement in physical characteristics of the finished product.

In the refining process, the feedstock is hydrotreated and the hydrotreated feedstock exits hydrotreater and conducted to fractionating column. Low-boiling constituents, especially hydrogen sulfide and ammonia, are removed and the hydrotreated product is then conducted to a second hydrotreater, where it is hydrotreated using process parameters that may be the same or different from the hydrotreating conditions in the first hydrotreater. The product from the second hydrotreater is sent to a catalytic dewaxing unit after which the dewaxed product exits dewaxing unit and is sent to a hydrofinishing unit. The product is analyzed for the $C_n:C_p$ (naphthene carbon/paraffin carbon) ratio. When the desired $C_n:C_p$ ratio is attained typically in the range 0.45–0.65, the medicinal white product is finished.

Mineral oil, sold widely and cheaply in the United States, is not sold as such in Britain but is sold under the trade names *paraffinum perliquidum* for light mineral oil and *paraffinum liquidum* or *paraffinum subliquidum* for the higher density more viscous types of the oil. The term "Paraffinum Liquidum" is often seen on the ingredient lists of baby oil and cosmetics—British aromatherapists commonly use the term *white mineral oil*. In lubricating oil technology, mineral oil is termed from groups 1–2 worldwide and group 3 in certain regions because the high end of group 3 mineral lubricating oils are of high purity and exhibit properties similar to poly-alpha-olefin derivatives which constitute the group 4 synthetic oils (Speight and Exall, 2014).

Mineral oil is any of various colorless, odorless, light mixtures of higher molecular weight alkane derivatives from a mineral source, particularly as a distillate from petroleum. The name *mineral oil* by itself is imprecise, having been used for many specific types of oils over the past several centuries. Other names, similarly imprecise, include *white oil*, *paraffin oil*, *liquid paraffin* (a highly refined medical grade), *paraffinum liquidum* (Latin), and *liquid petroleum*. The product popularly called *baby oil* is a mineral oil to which scented ingredients (perfumes) have been added.

Most often, mineral oil is a liquid byproduct of refining crude oil to produce an array of various petroleum products (Parkash, 2003; Gary et al., 2007; Speight, 2011, 2014, 2017; Hsu and Robinson, 2017). This type of mineral oil is a transparent, colorless (water-where), low-density oil (approximately 0.8 g/cm³), composed mainly of alkane derivatives and cycloalkane derivatives, related to petroleum jelly. White oil is highly refined oil which is colorless, tasteless, and odorless. It is especially refined to obtain the highest degree of purity for their use in those applications requiring direct contact with food.

The purified oil is recommended for use in the manufacture of pharmaceutical and cosmetic preparations such as ointments, complexion creams, haircare products, laxatives, baby oils, and as carriers in the preparation of many curative drugs. It is also used to coat eggs and fruit to make them shinier. It is also used to lubricate baking equipment so that food does not stick to it.

12.2.2 Petroleum Jelly

Petroleum jelly is a mixture of hydrocarbons, having a melting point usually close to human body temperature, approximately 37°C (99°F). Petroleum jelly is typically composed of paraffin wax, microcrystalline, wax, and mineral oil in varying amounts. The composition of highly refined constituents and their physical properties vary considerably according to the origin of the raw material and the refining methods. The solid or liquid elements of the hydrocarbons may contain 16–60 carbon atoms with significantly different molecular weights; therefore, the possible structures are extremely varied and their number practically infinite.

Vaseline is a brand name for petroleum jelly-based products which include plain (unaltered petroleum jelly) and a selection of skin creams, soaps, lotions, cleansers, and deodorants to provide various types of skincare and protection by minimizing friction or reducing moisture loss, or by functioning as a grooming aid. It is believed that the use of petroleum jelly comes from a product known as *rod wax* that was used by early oil workers in Titusville, Pennsylvania, to heal cuts and burns. In many countries, the word *vaseline* (*vasenol* in some countries) is used as generic for petroleum jelly.

Petrolatum, a related product to petroleum jelly although the names are often used interchangeably, is a byproduct of petroleum refining with a melting point close to body temperature—body temperature ranges from 36.1°C (97°F) to 37.2°C (99°F); in older adults, the typical body temperature is lower than 36.2°C (98.6°F). Petrolatum softens upon application and forms a water-repellant film around the applied area, creating an effective barrier against the evaporation of the skin's natural moisture and foreign particles or microorganisms that may cause infection. Petrolatum is odorless and colorless, and it has an inherently long shelf life. These qualities make petrolatum a popular ingredient in skincare products and cosmetics.

Petroleum jelly has been, and continues to be, manufactured from the highest-boiling crude oil refinery fraction (resid). However, because of the occurrence of cancer-forming polynuclear aromatic derivatives (as well as other constituents that are risky to health) in resids, number of cleanup (purification) steps are required to meet the stringent requirements of a product used for direct skin and mouth contact. Although not a comprehensive list, these cleanup steps can include propane deasphalting, hydrogenation, solvent dewaxing, and fixed bed adsorption using adsorbents such as bauxite and carbon. In the simplest process, paraffin wax is introduced into a reaction vessel after which microcrystalline wax (i.e., wax with a very fine crystalline structure) is added. The mixture is melted with continuous mixing and the temperature is maintained between 120°C and 130°C (248°F and 266°F). Liquid paraffin is added with continuous stirring (150–200 rpm) at constant temperature, so that ingredients are mixed together to form emulsion or gel after which the mass is cooled.

Briefly, bauxite is a complex mineral that is often claimed to be alumina (Al_2O_3) but which, in reality, consists mostly the aluminum minerals gibbsite [$Al(OH)_3$], boehmite (γ-AlO(OH), and diaspore (α-AlO(OH), mixed with the two oxides of iron, namely goethite and hematite, as well as the aluminum clay mineral kaolinite, as well as small amounts of anatase (TiO_2) and ilmenite ($FeTiO_3$ or $FeO \cdot TiO_2$). Petroleum jelly can also be produced by way of synthesis gas in which the process for conversion of synthesis gas to hydrocarbon products is adapted to produce higher molecular weight paraffin derivatives (Abhari, 2010).

Thus, petroleum jelly is a subtle balance of liquid and solid hydrocarbons. The crystalline structure of the substances in its composition is one of the basic qualitative elements. The role of the amorphous solid hydrocarbons is, in fact, to retain in a sufficiently dense fibrous mesh, oily hydrocarbons of a generally high molecular weight.

Petroleum jelly is flammable only when heated to the liquid state at which point the fumes will combust but the liquid does not combust, not the liquid itself, so a wick material like leaves, bark, or small twigs is needed to ignite petroleum jelly. Petroleum jelly is colorless or has a pale yellow color (when not highly distilled), translucent, and devoid of taste and smell when pure. It does not oxidize on exposure to the air and is not readily acted on by chemical reagents and is insoluble in water. It is soluble in dichloromethane (CH_2Cl_2), chloroform ($CHCl_3$), benzene (C_6H_6), diethyl ether ($CH_3CH_2OCH_2CH_3$), and carbon disulfide (CS_2).

Petroleum products generally defined collectively as petrolatum have a long history in medical applications and that heritage continues as pharmaceutical grade petrolatum constituents are common components in a variety of balms, ointments, creams, moisturizers, haircare products, and other products where a virtually odorless additive that helps retain (and even lock-in) moisture is desired.

According to the requirements of the International Nomenclature of Cosmetic Ingredients which lists and assigns the INCI names of cosmetic ingredients, there are two possible designations depending on the manufacturing method of the petroleum jelly: (i) if the product is manufactured by blending paraffin oil, wax, and mineral paraffin, the INCI name of the mixture is composed of all the INCI names of the ingredients [*paraffinum liquidum (and) cera microcristallina (and) paraffin*] or (ii) if the product is manufactured by directly refining the crude oil or its derivatives of crude oil, the INCI name is *petrolatum*.

12.2.3 PARAFFIN WAX

Paraffin wax is a white or colorless soft, solid wax that is composed of a complex mixture of hydrocarbons with the following general properties: (i) nonreactive, (ii) nontoxic, (iii) water barrier, and (iv) colorless. Paraffin wax is characterized by a clearly defined crystal structure and has the tendency to be hard and brittle with a melting point typically in the range 50°C–70°C (122°F–158°F). On a more specific basis, petroleum wax is of two general types: (i) paraffin wax in petroleum distillates and (ii) microcrystalline wax in petroleum residua. The melting point of wax is not directly related to its boiling point, because waxes contain hydrocarbons of different chemical nature. Nevertheless, waxes are graded according to their melting point and oil content.

In the process for wax manufacture known as wax sweating (Parkash, 2003; Gary et al., 2007; Speight, 2011, 2014, 2017; Hsu and Robinson, 2017), a cake of slack wax (paraffin wax from a solvent dewaxing operation) is slowly warmed to a temperature at which the oil in the wax and the lower melting waxes become fluid and drip (or sweat) from the bottom of the cake, leaving a residue of higher melting wax. However, wax sweating can be carried out only when the residual wax consists of large crystals that have spaces between them, through which the oil and lower melting waxes can percolate; it is therefore limited to wax obtained from light paraffin distillate.

Wax recrystallization, like wax sweating, separates slack wax into fractions, but instead of using the differences in melting points, it makes use of the different solubility of the wax fractions in a solvent, such as the ketone used in the dewaxing process. When a mixture of ketone and slack wax is heated, the slack wax usually dissolves completely, and if the solution is cooled slowly, a temperature is reached at which a crop of wax crystals is formed. These crystals will all be of the same melting point, and if they are removed by filtration, a wax fraction with a specific melting point is obtained. If the clear filtrate is further cooled, a second crop of wax crystals with a lower melting point is obtained. Thus, by alternate cooling and filtration the slack wax can be subdivided into a large number of wax fractions, each with different melting points.

Microcrystalline wax (sometimes also called *micro wax* or *microwax*) is produced from a combination of heavy lube distillates and residual oils and differs from paraffin wax in that the microcrystalline has a less well-defined crystalline structure, and is darker color. The physical properties of microcrystalline wax is affected significantly by the oil content (Kumar et al., 2007) and by achieving the desired level of oil content, wax of the desired physical properties and specifications can be produced. Deep de-oiling of microcrystalline wax is comparatively difficult compared to paraffin wax (macrocrystalline wax) as the oil remains occluded in these and is difficult to separate by sweating. Also since wax and residual oil have similar boiling ranges, separation by distillation is difficult. However, these waxes can be de-oiled by treatment with solvents at lower temperature that have high oil miscibility and poor wax solubility and these have been used extensively to separate.

Paraffin wax is mostly used for relief of discomfort and pain in following conditions such as bursitis, eczema, psoriasis, dry flaky skin, stiff joints, fibromyalgia, tired sore muscles, inflammation, and arthritis. Paraffin wax is often used in skin-softening salon and spa treatments on the hands, cuticles, and feet because it is colorless, tasteless, and odorless. It can also be used to provide pain relief to sore joints and muscles. Paraffin wax is often used as lubrication, electrical insulation, and to make candles and crayons. Cosmetically, paraffin wax is often applied to the hands and feet. The wax is a natural emollient, helping make skin supple and soft. When applied to the skin, it adds moisture and continues to boost the moisture levels of the skin after the treatment is complete. It can also help open pores and remove dead skin cells. That may help make the skin look fresher and feel smoother and give comfort to the user.

12.2.4 BITUMEN

The bitumen (in the Bible it is referenced as *slime*) is not the same as the refinery product known as asphalt (Speight, 2008, 2014, 2015, 2016). Bitumen is a natural-occurring material that occurs in tar sand formations and that has seeped from crude oil formation. Typically, the bitumen that has been referenced in ancient texts, unless recovered from a tar sand formation, is equivalent to an atmospheric residuum insofar as it is found as a seepage on the surface and is crude oil from which the more volatile constituents have escaped by evaporation. The bitumen obtained from the area of Hit (Tuttul) in Iraq (Mesopotamia) or as blocks floating on the Dead Sea are examples of such occurrences (Abraham, 1945; Forbes, 1958a,b, 1959; Nissenbaum, 1999). Typically, asphalt is produced from crude oil as the treated (usually air-blown) vacuum residuum (Parkash, 2003; Gary et al., 2007; Speight, 2011, 2014, 2017; Hsu and Robinson, 2017).

Surface manifestations of bitumen are found in Middle Eastern countries as seepages from rocks. This bitumen has been extensively employed for a variety of uses, including in medicine. The historical evidence on the medicinal uses of bitumen spans at least 3,000 years and, while many of the attributes of bitumen as a drug in antiquity are not based on medical evidence, certain treatments using bitumen may have been confirmed by modern medicine. For example, the application of bitumen and asphalt as a therapy for skin diseases, in humans and in animals, has been borne out in modern times by extensive experimentation. The nature of the active ingredient or ingredients in the bitumen has not been investigated as yet not have the constituents been identified with any degree of certainty. Also, it has long been recorded that bitumen from what is now Iraq and Syria was exported to Egypt for embalming purposes from at least the early Ptolemaic period—the accession of Soter after the death of Alexander the Great in 323 BC and which ended with the death of Cleopatra and the Roman conquest of Egypt in 30 BC.

Furthermore, when going further back into history, it has become evident that bitumen was used widely in the Middle East, especially in the Zagros Mountains of Iran (Connan, 1999). Ancient people from northern Iraq, south-west Iran, and the Dead Sea area extensively used this ubiquitous natural resource until the Neolithic period (7000–6000 BC). Evidence of earlier use has been recently documented in the Syrian Desert near El Kown, where bitumen-coated flint implements, dated to 40,000 BC (Mousterian period), have been unearthed. This discovery at least proves that bitumen was used by Neanderthal populations as hafting material to fix handles to their flint tools. Numerous testimonies, proving the importance of this petroleum-based material in ancient civilizations, were brought to light by the excavations conducted in the Near East as of the beginning of the century.

The early records show that bitumen was largely used in Mesopotamia and Elam as mortar in the construction of palaces (e.g., the Darius Palace in Susa), temples, ziggurats (e.g., the so-called Tower of Babel in Babylon), terraces (e.g., the famous Hanging Gardens of Babylon) and exceptionally for roadway coating (e.g., the processional way of Babylon). Since Neolithic times, bitumen served to waterproof containers (baskets, earthenware jars, storage pits), wooden posts, palace grounds (e.g., in Mari and Haradum), reserves of lustral waters, bathrooms, palm roofs, etc. Mats, sarcophagi, coffins and jars, used for funeral practices, were often covered and sealed with bitumen. Reed and wood boats were also caulked with bitumen. Bitumen was also a widespread adhesive in antiquity and served to repair broken ceramics, fix eyes and horns on statues (e.g., at Tell al-Ubaid around 2500 BC). Decorations with stones, shells, mother of pearl, on palm trees, cups, ostrich eggs, musical instruments (e.g., the Queen's lyre) and other items, such as rings, jewelry, and games, have been excavated from the Royal tombs in Ur (Connan, 1999).

Bitumen was also considered as a powerful remedy in medical practice, especially as a disinfectant and insecticide, and was used by the ancient Egyptians to prepare mixtures to embalm the corpses of their dead. Recent geochemical studies on more than 20 balms from Egyptian mummies from the Intermediate, Ptolemaic, and Roman periods have revealed that these balms are composed of various mixtures of bitumen, conifer resins, grease, and beeswax. The physician Ibn al-Baitar described as a preservative for embalming the dead, in order that the dead bodies might remain in the state in which they were buried and neither decay nor change. In addition, the historical records show that bitumen was used since ancient times for cosmetic, art, and the caulk of boats and was reputed to be useful to cure varying pulmonary, digestive, ear-nose-throat troubles, and even to set fractured bones (Bourée et al., 2011).

In medicine, Muslim physicians used petroleum and bitumen for pleurisy and dropsy—the patient was given *bitumenous water* to drink—and for various skin ailments and wounds. There is also fragmentary evidence that hot bitumen was used to cauterize the wound resulting from a severed limb—as a side note, medieval physicians used fire as the cauterizing agent. Whether or not the bitumen-treated patients survived is not clear. Another law of the time suggests that the use of hot bitumen as a *curative* agent—not in the sense of a medicinal cure but as a punishment. The hot bitumen was to be poured over the head of the miscreant. The record do not show if the miscreant survived as a bald person after the bitumen was removed or if the miscreant actually survived the treatment.

For example, an early mention of the use of bitumen as a punishment appears in orders that Richard I of England (also known as Richard the Lionheart) issued to his navy when he set out of the Holy Land in 1189. "Concerning the lawes and ordinances appointed by King Richard for his navie the forme thereof was this … item, a thiefe or felon that hath stolen, being lawfully convicted, shal have his head shorne, and boyling pitch poured upon his head, and feathers or downe strawed upon the same whereby he may be knowen, and so at the first landing-place they shall come to, there to be cast up" (Hakluyt, 1582).

In other literature, the name *shilajit* occurs frequently and is the Sanskrit name for Asphaltum (bitumen, also called *mineral pitch, vegetable asphalt, shilajita, guj, kalmadam, perangyum, relyahudi,* and *silaras*) refers to a curative agent as an analgesic, anti-inflammatory, antibacterial, cholagogic, diuretic, wound cleaner, expectorant, respiratory stimulant, general health medicine, amongst a host of other effects (Jonas, 2005).

12.2.5 SOLVENTS

Finally, for this section, it would be remiss if mention was not made of the solvents produced from crude oil that are used by the pharmaceutical industry, many of which are derived from crude oil (Table 12.3).

TABLE 12.3
Example of Solvents Used in the Pharmaceutical Industry

Solvent	Use		
Acetone	C	F	B
Acetonitrile	C	F	B
Ammonia (aqueous)	C	F	B
n-Amyl acetate	C	F	B
Amyl alcohol	C	F	B
Aniline	C		
Benzene	C		
2-Butanone (methyl ethyl ketone, MEK)	C		
n-Butyl acetate	C	F	
n-Butyl alcohol	C	F	B

(Continued)

TABLE 12.3 (*Continued*)

Example of Solvents Used in the Pharmaceutical Industry

Solvent	Use		
Chlorobenzene	C		
Chloroform	C	F	B
Chloromethane	C		
Cyclohexane	C		
o-Dichlorobenzene (1,2-Dichlorobenzene)	C		
1,2-Dichloroethane	C		B
Diethylamine	C		B
Diethyl ether	C		B
N,N-Dimethyl acetamide	C		
Dimethylamine	C		
N,N-dimethylaniline	C		
N,N-dimethylformamide	C	F	B
Dimethyl sulphoxide	C		B
1,4-Dioxane	C		B
Ethanol	C	F	B
Ethyl acetate	C	F	B
Ethylene glycol	C		B
Formaldehyde	C	F	B
Formamide	C		
Furfural	C		
n-Heptane	C	F	B
n-Hexane	C	F	B
Isobutyraldehyde	C		
Isopropanol	C	F	B
Isopropyl acetate	C	F	B
Isopropyl ether	C		B
Methanol	C	F	B
Methylamine	C		
Methyl cellosolve	C	F	
Methylene chloride	C	F	B
Methyl formate	C		
Methyl isobutyl ketone (MIBK)	C	F	B
2-Methylpyridine	C		
Petroleum naphtha	C	F	B
Phenol	C	F	B
Polyethylene glycol 600	C		
n-Propanol	C		B
Pyridine	C		B
Tetrahydrofuran	C		
Toluene	C	F	B
Trichlorofluoromethane	C		
Triethylamine	C	F	
Xylenes	C		

C, chemical synthesis; F, fermentation; B, biological or natural extraction.

Briefly, a solvent is a substance that dissolves a solute (a chemically distinct liquid, solid, or gas), resulting in a solution. A solvent is usually a liquid but can also be a solid, a gas, or a supercritical fluid. The quantity of solute that can dissolve in a specific volume of solvent varies with temperature. In the current context, solvents are used for production isolation and/or purification and have found wide use in the pharmaceutical industry, including in synthetic processes and purification processes.

The term *petroleum solvent describes* the liquid hydrocarbon fractions obtained from petroleum and used in industrial processes and formulations. These fractions are also referred to *naphtha* or as *industrial naphtha*. By definition the solvents obtained from the petrochemical industry such as alcohols, ethers, and the like are not included in this chapter. A refinery is capable of producing hydrocarbons of a high degree of purity and at the present time petroleum solvents are available covering a wide range of solvent properties including both volatile and high-boiling qualities.

Naphtha has been available since the early days of the petroleum industry. Indeed, the infamous *Greek fire* documented as being used in warfare during the last three millennia is a petroleum derivative (Chapter 1). It was produced either by distillation of crude oil isolated from a surface seepage or (more likely) by destructive distillation of the bituminous material obtained from bitumen seepages, of which there are/were many known during the heyday of the civilizations of the Fertile Crescent (Chapter 1). The bitumen obtained from the area of Hit (Tuttul) in Iraq (Mesopotamia) is an example of such an occurrence (Abraham, 1945; Forbes, 1958a).

Other petroleum products boiling within the naphtha boiling range include (i) *industrial Spirit* and *white spirit*. *Industrial spirit* comprises liquids distilling between 30°C and 200°C (–1°F–390°F), with a temperature difference between 5% volume and 90% volume distillation points, including losses, of not more than 60°C (140°F). There are several (up to eight) grades of industrial spirit, depending on the position of the cut in the distillation range defined above. On the other hand, *white spirit* is an industrial spirit with a flash point above 30°C (99°F) and has a distillation range from 135°C to 200°C (275°F–390°F).

Solvents used for extracting the product from a natural product source (such as biomass) or from a reaction mixture are, to many scientists, engineers, and technologists, as equally important as the product of the medicine. Generally, the solvent can be recovered but small portions remain in the process wastewater, depending upon their solubility and the design of the process equipment. Precipitation from a solvent is a method to separate the medicine (or a precursor chemical) from the reaction mixture after which the medicinal product (precursor) is filtered and extracted from, say, any solid the solid residues. The medicinal product is then recovered from the solvent phase by evaporation (and recovery) of the solvent.

12.3 PHARMACEUTICAL PRODUCTS

Petrochemical compounds are necessary for many of the things we depend upon, but unfortunately the process to make them is costly, energy intensive, and very harmful to the environment. The petrochemical manufacturing process is particularly energy intensive and harmful to the environment. The complex mixture of hydrocarbons (compounds made of hydrogen and carbon) that comprises oils are separated into various fractions by distillation, a process that separates various compounds based on their boiling points.

Low-boiling fractions of petroleum including propane and butane are separated from the crude oil at low temperatures (300°C, 570°F). Manufacturers then apply various chemical processes to generate a variety of petrochemicals. These chemicals are the starting points for the manufacture of plastics; the polyester used in carpet and clothing; and industrial solvents, oils, and acids used in cleaning products. Many pharmaceuticals are also derived from petrochemicals, as are food additives, dyes, and explosives. Simply put, modern life would not be possible without petrochemicals.

Petrochemicals play a major role in the manufacture of many pharmaceutical products and many advances in health care and sanitation have been made possible by the use of petrochemicals and there is a long history of their use, with oils first being used in medicines at least 1,000 years ago.

The petrochemicals are used in pharmaceutical products, from the most commonplace to the highly specialized. An everyday example is ASA—or Acetylsalicylic acid—an important part of many over-the-counter pain medications.

While penicillin, a drug that has saved countless lives since its discovery by Alexander Fleming in 1928 and subsequent development by Howard Florey and Ernst Chain in the 1940s, is manufactured via fungi and microbes, phenol, and cumene are used as preparatory substances. These chemicals are also used in the production of aspirin, with acetylsalicylic acid being the main metabolite of aspirin. Other common medical products, some available by prescription, some over-the-counter, that use petrochemicals include antihistamine medications, antibacterial medications, suppositories, cough syrups, lubricants, creams, ointments, salves, analgesics, and gels.

Petrochemical resins have also been used in drug purification. These resins simplify mass production of medicine, thus making them more affordable to produce and then distribute. The resins have been used in the production of a wide range of medications including those for treating AIDS, arthritis, and cancer.

Plastics play an important role in health care too. Resins and plastics from petrochemicals are used to make artificial limbs and joints. They are also a familiar sight in hospitals and other medical facilities for storing blood and vaccines, for use in disposable syringes and other items of medical equipment that are used once to prevent the threat of contagion. Specially created polymers are used extensively in health care, most notably during cardiac surgery or for auditory and visual stimulators. Eyeglasses have benefitted from the use of plastics in frames and lenses and contact lenses are also made of plastic. Even safety has improved thanks to the introduction of child-proof caps and tamper-proof seals for medication containers, all made using plastics. Surgical gloves are often made from pliable plastics, plastic petri dishes are essential to laboratories and at a larger level, for the housing of large diagnostic medical machinery.

As well as petrochemicals playing an important role in the manufacture of pharmaceuticals and medical equipment, petroleum use through transport is a major cost to health-care systems globally, including ambulances, staff transport, and transportation of supplies. Indeed, in the United States, according to U.S. Bureau of Labor Statistics figures, it is estimated that the use of petroleum products in transport for health care is far greater than that used for drugs and plastics. The ongoing supply of the fossil fuels required to make all these relevant health-care products may become a bigger issue as time goes on and if health-care systems are placed under further financial pressure.

Finding alternatives to using petrochemicals for many medications and items of medical equipment may become important if health care is to remain accessible or, for some regions, to become more accessible. In the United States, the Center for Disease Control is investigating the impact of dwindling petroleum reserves on the provision of health care. Going forward, the health-care industry may have to look at alternatives to using petrochemicals for pharmaceuticals and plastics although currently there are few alternatives. However, as only a tiny proportion of petrochemicals is used to produce specialized products for the health-care industry, the supply chain for such products is currently considered to be secure.

12.4 PRODUCTION OF PHARMACEUTICALS

The pharmaceutical industry includes the manufacture, extraction, processing, purification, and packaging of chemical materials to be used as medications for humans or animals. Pharmaceutical manufacturing is divided into two major stages: (i) the production of the active ingredient or drug (primary processing, or manufacture) and (ii) secondary processing, the conversion of the active medicines into products suitable for administration. However, before a medication can be manufactured at any scale, much work goes into the actual formulation of the medicine. Formulation development scientists must evaluate a compound for uniformity, stability, and many other factors. After the evaluation phase, a solution must be developed to deliver the medication in its required form such as solid, semisolid, immediate or controlled release, tablet, and capsule.

In the pharmaceutical industry, a wide range of excipients may be blended together to create the final blend used to manufacture the solid dosage form. The range of materials that may be blended (excipients, API), presents a number of variables which must be addressed to achieve products of acceptable blend uniformity. These variables may include the particle size distribution (including aggregates or lumps of material), particle shape (spheres, rods, cubes, plates, and irregular), presence of moisture (or other volatile compounds), and particle surface properties (roughness, cohesiveness).

The following sections present the published synthetic routs for several over-the-counter (non-prescription) medications. These are list alphabetically rather than by preference or by stated use or effect.

12.4.1 ACETAMINOPHEN

Acetaminophen (paracetamol) is an analgesic and fever-reducing medicine similar in effect to aspirin.

It is an active ingredient in many over-the-counter medicines, including Tylenol and Midol. Introduced in the early 1900s, acetaminophen is a coal tar derivative that acts by interfering with the synthesis of prostaglandins and other substances necessary for the transmission of pain impulses.

The starting material, p-aminophenol (4-aminophenol), is produced from phenol by nitration followed by reduction with iron. Alternatively, the partial hydrogenation of nitrobenzene affords phenylhydroxylamine which rearranges primarily to 4-aminophenol:

$$C_6H_5NO_2 + 2H_2 \rightarrow C_6H_5NHOH + H_2O$$

$$C_6H_5NHOH \rightarrow HOC_6H_4NH_2$$

The p-aminophenol can also be produced from nitrobenzene by electrolytic conversion to phenyl-hydroxylamine which, under the reaction conditions spontaneously rearranges to 4-aminophenol.

p-Aminophenol is a white powder that is moderately soluble in alcohols and can be recrystal-lized from hot water. Also, it is the final intermediate in the industrial synthesis of paracetamol by treatment with acetic anhydride.

12.4.2 ALEVE

The active constituent of Aleve is Naproxen sodium which is an anti-inflammatory compound. Naproxen is used to treat a variety of inflammatory conditions and symptoms that are due to excessive inflammation, such as pain and fever—naproxen has fever-reducing (antipyretic) properties in addition to its anti-inflammatory activity.

Naproxen has been produced starting from 2-naphthol (β-naphthol)—a constituent of coal tar or which can be prepared from naphthalene that is isolated from gas oil.

2-Naphthol is not a product that is isolated from crude oil more likely it is isolated from the products of the thermal decomposition of coal and some types of biomass. Traditionally, 2-naphthol is produced by a two-step process that begins with the sulfonation of naphthalene in sulfuric acid. The sulfonic acid group is then cleaved in molten sodium hydroxide:

$$C_{10}H_8 + H_2SO_4 \rightarrow C_{10}H_7SO_3H + H_2O$$

$$C_{10}H_7(SO_3H) + 3NaOH \rightarrow C_{10}H_7ONa + Na_2SO_3 + 2H_2O$$

Neutralization of the sodium salt with acid gives 2-naphthol. 2-Naphthol can also be produced by a method analogous to the cumene process. 2-Naphthol is also the base from which certain dyestuffs (Table 12.4) can be manufactured.

TABLE 12.4

Example of Dyestuffs Based on 2-Naphthol

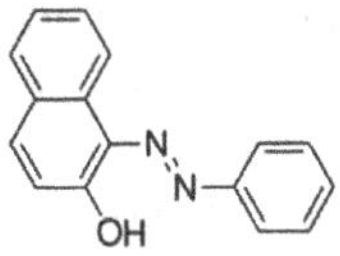

Sudan I

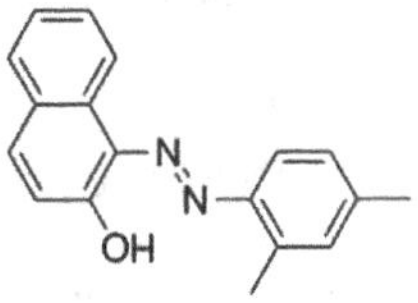

Sudan II

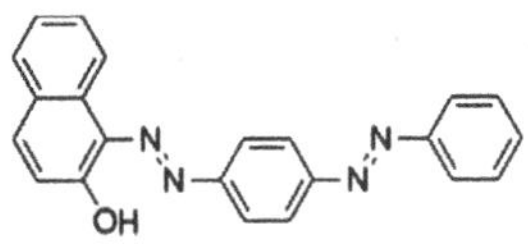

Sudan III

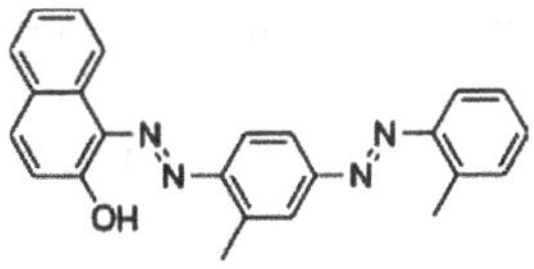

Sudan IV

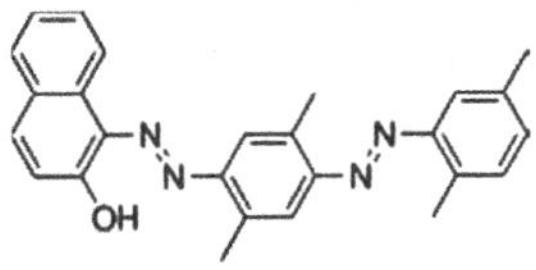

Oil Red O

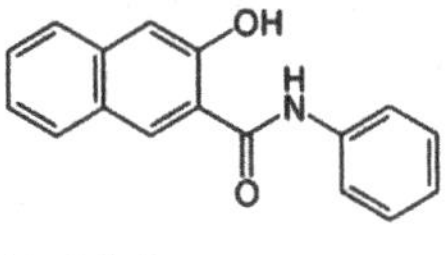

Naphthol

12.4.3 ASPIRIN

Acetylsalicylic acid commonly known as aspirin is a widely used drug. The analgesic, antipyretic, and anti-inflammatory properties make it a powerful and effective drug to relive symptoms of pain, fever, and inflammation.

Historically, aspirin has been known for some time. In the North American context, it was extracted by the Native Americans from willow and poplar tree bark about 2,500 years ago. Native Americans used willow bark in teas to reduce fever. In 1763, Reverend Edward isolated and identified one of the compounds used to synthesize aspirin, which came to be known as salicylic acid. Large quantities of salicylic acid became available; however, it caused severe stomach irritation. In 1893, German chemist Felix Hoffman synthesized an ester derivative of salicylic acid, acetylsalicylic acid ("aspirin"). The acetyl group cloaks the acidity when ingested. The drug then passes through the small intestine where it gets converted back to salicylic acid, and enters the bloodstream. Although, weaker than salicylic acid, aspirin had medicinal properties without the bitter taste and harsh stomach irritation. The company Bayer patented aspirin in 1899, which has made aspirin one of the most widely used and modern commercially-available drugs.

The synthesis of aspirin may be achieved in one simple step, O-acetylation of salicylic acid, which is incorporated into many undergraduate synthetic chemistry laboratory courses. The purity of the product as a pharmaceutical is crucial.

An additional step may be added to the synthesis of aspirin: conversion of oil of wintergreen (methyl salicylate) to salicylic acid.

This serves as an introduction to multi-step synthesis and the concept of converting a naturally occurring substance into one with therapeutic value.

Salicylic acid (1,2-HOC$_6$H$_4$COOH, C$_8$H$_8$O$_3$) is produced by the base-catalyzed hydrolysis reaction conversion of oil of wintergreen (methyl salicylate, 1,2-HOC$_6$H$_4$CO$_2$CH$_3$) to salicylic acid:

$$C_8H_8O_3 + 2NaOH + H_2SO_4 \rightarrow C_4H_6O_3 + Na_2SO_4 + CH_3OH + H_2O$$

In terms of a petrochemical precursor, salicylic acid can be synthesized from phenol by a three-step process.

12.4.4   CEPACOL

The main ingredient of Cepacol is benzocaine which is commonly used as a topical pain reliever or in cough drops.

It is the active ingredient in many over-the-counter anesthetic ointments such as products for oral ulcers.

Benzocaine is the ethyl ester of p-aminobenzoic acid (PABA) and can be prepared by the reaction of p-aminobenzoic acid with ethanol or via the reduction of ethyl p-nitrobenzoate. Benzocaine is sparingly soluble in water; it is more soluble in dilute acids and very soluble in ethanol, chloroform, and ethyl ether. It can be synthesized from toluene by a three-step process.

12.4.5   EXCEDRIN

Excedrin is an extra-strength pain reliever that is available as an over-the-counter medicine for pain. Excedrin combines three medications: (i) acetaminophen, also known as paracetamol, (ii) aspirin, and (iii) caffeine.

Caffeine has its origins in biomass such as the seeds, nuts, or leaves of a number of plants native to Africa, East Asia, and South America, and helps to protect them against predator insects and to prevent germination of nearby seeds. To the Western world, the most well-known source of caffeine is the coffee bean, a misnomer for the seed of *Coffea* plants.

12.4.6   GAVISCON

Gaviscon is an antacid medication that reduced stomach acid and the typical uncomfortable side effects that accompany acid reflux. The active ingredients of Gaviscon are (i) aluminum hydroxide (Al(OH)$_3$) and (ii) magnesium carbonate (MgCO$_3$).

While not a true petrochemical in the general sense, aluminum hydroxide is often used in a refinery and is available as a product. It is, however, found in nature as the mineral gibbsite (also known as hydrargillite) and its three much rarer polymorphs (occurring in several different forms): bayerite, doyleite, and nordstrandite. Aluminum hydroxide is amphoteric—having both basic and acidic properties.

Like aluminum hydroxide, magnesium carbonate is also not a true petrochemical. Magnesium carbonate is often found in use in refineries. It is ordinarily obtained by mining the mineral magnesote and can also be prepared by reaction between any soluble magnesium salt and sodium bicarbonate:

$$MgCl_2 + 2NaHCO_3 \rightarrow MgCO_3 + 2NaCl + H_2O + CO_2$$

If magnesium chloride (or sulfate) is treated with aqueous sodium carbonate, a precipitate of basic magnesium carbonate a hydrated complex of magnesium carbonate and magnesium hydroxide is formed:

$$5MgCl_2 + 5Na_2CO_3 + 5H_2O \rightarrow Mg(OH)_2 \cdot 3MgCO_3 \cdot 3H_2O + Mg(HCO_3)_2 + 10NaCl$$

High-purity industrial routes include a path through magnesium bicarbonate which can be formed by combining a slurry of magnesium hydroxide and carbon dioxide at high pressure and moderate temperature. The bicarbonate is then vacuum dried, causing it to lose carbon dioxide and a molecule of water:

$$Mg(OH)_2 + 2CO_2 \rightarrow Mg(HCO_3)_2$$

$$Mg(HCO_3)_2 \rightarrow MgCO_3 + CO_2 + H_2O$$

12.4.7 IBUPROFEN

Ibuprofen is a medication in the nonsteroidal anti-inflammatory drug (NSAID) class (NSAID class) that is used for treating pain, fever, and inflammation. Since the introduction of the drug in 1969, ibuprofen has become one of the most common painkillers in the world. Ibuprofen in an NSAID, and like other drugs of its class, it possesses analgesic, antipyretic, and anti-inflammatory properties. While ibuprofen is a relatively simple molecule, there is still sufficient structural complexity to ensure that a large number of different synthetic approaches are possible.

Ibuprofen is typically found in many over-the-counter drugs, such as Motrin, Advil, Potrin, and Nuprin. In other words, it often comes in capsules, tablets, or powder form. Comparing to that of aspirin, for example, Ibuprofen is somewhat short-lived and relatively mild. However, it is known to have an antiplatelet (non-blood clotting) effect.

Since the introduction of pharmaceutical products containing ibuprofen, industrial and academic scientists developed many potential production processes. Two of the most popular ways to obtain Ibuprofen are the Boot process and the Hoechst process. The Boot process is an older commercial process and the Hoechst process is a newer process. Most of these routes to Ibuprofen begin with isobutyl benzene and use Friedel–Crafts acylation. The Boot process requires six steps, while the Hoechst process, with the assistance of catalysts, is completed in only three steps.

The starting material, cumene (isopropyl benzene, 2-phenylpropane, or 1-methylethyl benzene) for both of these processes is produced by the gas-phase reaction (Friedel–Crafts alkylation) of benzene by propylene. In the process, benzene and propylene are compressed together to a pressure in the order of 450 psi 250°C (482°F) in presence of a Lewis acid catalyst (such as an aluminum halide—a phosphoric acid (H_3PO_4) catalyst) is often favored over an aluminum halide catalyst. Cumene is a colorless, volatile liquid with a gasoline-like odor. It is a natural component of coal tar and crude oil, and also can be used as a blending component in gasoline.

12.4.8 KAOPECTATE

Kaopectate is an orally taken medication used for the treatment of mild indigestion, nausea, and stomach ulcers. The active ingredients have varied over time, and are different between the United States and Canada. The original active ingredients were kaolinite (a layered clay mineral which has the approximate chemical composition $Al_2Si_2O_5(OH)_4$) and pectin (a structural heteropolysaccharide contained in the primary cell walls of terrestrial plants). In the United States, the active ingredient is now bismuth subsalicylate which has the empirical chemical formula of $C_7H_5BiO_4$, and it is a colloidal substance obtained by hydrolysis of bismuth salicylate [$Bi(C_6H_4(OH)CO_2)_3$].

Bismuth subsalicylate is also the active ingredient in Pepto-Bismol and displays anti-inflammatory action (due to salicylic acid) and is used to relieve the discomfort that arises from an upset stomach due to overindulgence in food and drink, including heartburn, indigestion, nausea, gas, and fullness.

As stated previously, salicylic acid (or as a precursor to the acid), sodium salicylate is produced commercially by treating sodium phenate (the sodium salt of phenol—phenol is a well-known petrochemical starting material) with carbon dioxide at high pressure (1,500 psi) and high temperature (117°C, 242°F) (the Kolbe–Schmitt reaction) after which acidification of the product with sulfuric acid yield gives salicylic acid.

Salicylic acid can also be prepared by the hydrolysis of acetylsalicylic acid (aspirin) or by the hydrolysis of methyl salicylate (oil of wintergreen) with a strong acid or base. Another method for the production of salicylic acid involves biosynthesis from phenylalanine.

Salicylic acid is also used in the production of other pharmaceuticals, including 4-aminosalicylic acid and sandulpiride—the latter is an antipsychotic of the benzamide class which is used mainly in the treatment of psychosis associated with schizophrenia and depressive disorders. Other derivatives include methyl salicylate that is used as a liniment to soothe joint and muscle pain and choline salicylate that is used topically to relieve the pain of mouth ulcers.

12.4.9 L-Menthol

L-Menthol (*laevo*-menthol or *laevo-rotary* menthol) is an organic compound that occurs naturally from corn, mint, peppermint, and other mint oils. The main form of menthol occurring in nature is laevo-menthol which is a waxy, crystalline solid that is clear (sometimes referred to as white or water-white) in color and which melts slightly above room temperature. Natural menthol is obtained by freezing peppermint oil and the resultant crystals of menthol are then separated by filtration.

Briefly and by way of clarification, dextrorotation and laevorotation (also spelled levorotation) are terms used to describe the rotation of plane-polarized. Looking at the molecule head-on, *dextrorotation* refers to clockwise rotation while *levorotation* refers to counterclockwise rotation. A chemical that causes dextrorotation is referred to as being *dextrorotatory* (*dextrorotary*, often abbreviated to dextro-) while a compound that causes levorotation is called *levorotatory* or *levorotary* (often abbreviated to laevo-).

Also, a dextrorotary compound is often prefixed "(+)-" or "*d-*". Likewise, a levorotary compound is often prefixed "(−)-" or "*l-*" (Solomons, and Fryhle, 2004). Compounds with these properties have varying degrees of optical activity and consist of chiral mileages that often react differently with other chemicals or with organs within the human body because of spatial effects. If a chiral molecule is dextrorotary, the enantiomer (one of a pair of molecules that are mirror images of each other) of the molecule will be levorotary and other enantiomer will be dextrorotatory. This means that each of the enantiomers will rotate the plane polarized light the same number of degrees, but in opposite directions.

As an illustration of the two forms of menthol, in the formulas below (with the ring in the plane of the paper), in the laevo-menthol (*l*-menthol), the methyl group and the hydroxyl group are above the plane of the paper while the isopropyl group is below the plane of the paper.

The converse is true for the dextro-menthol (*d*-menthol)—the methyl group and the hydroxyl group are below the plane of the paper while the isopropyl group is above the plane of the paper.

In the current context, menthol is produced by the Haarmann–Reimer process which starts from m-cresol (a simple petrochemical) which is alkylated with propene to yield thymol which is then hydrogenated and the racemic menthol is isolated by fractional distillation. The enantiomers are separated by chiral resolution in a reaction with methyl benzoate, selective crystallization followed by hydrolysis.

Menthol is included in many products for a variety of reasons, which include: (i) nonprescription products for short-term relief of minor sore throat and minor mouth or throat irritation, such as lip balms and cough medicines; (ii) as an antipruritic medicine to reduce itching; (iii) as a topical analgesic to relieve minor aches and pains, such as muscle cramps, sprains, headaches, and similar conditions, alone or combined with chemicals such as camphor, eucalyptus, or capsaicin; and (iv) in first-aid products such as *mineral ice* to produce a cooling effect as a substitute for real ice in the absence of water or electricity (pouch, body patch/sleeve, or cream). A further established application of L-Menthol is in form of inhalations for the symptomatic relief of sinusitis, rhinitis, bronchitis, and similar conditions.

12.4.10 ORAJEL

Orajel is another pain reliever, especially when the pain is due to toothache. The active ingredient is benzocaine, which is the ethyl ester of p-aminobenzoic acid.

Benzocaine can be prepared from p-aminobenzoic acid and ethanol by the Fischer esterification reaction or by the reduction of ethyl *p*-nitrobenzoate. In the Fischer esterification reaction, a carboxylic acid (R^1CO_2H) is treated with an alcohol (R^2OH) in the presence of a mineral inorganic acid catalyst to form the ester (R^1COOR^2).

Benzocaine is sparingly soluble in water; it is more soluble in dilute acids and very soluble in ethanol, chloroform, and ethyl ether.

Benzocaine is a local anesthetic used in medical applications to reduce pain and increase comfort of painful drugs. Such applications are administered with the leprosy drug, chaulmoogra oil, and even reducing the pain from needle injections. It is the active ingredient in many over-the-counter pain-relieving ointments such as products for oral ulcers. It is also used in aerosol spray lotions to relieve the discomfort of sunburn.

12.4.11 TYLENOL

The active constituent of Tylenol is acetaminophen which is an analgesic and fever-reducing medicine similar in effect to aspirin. It is an active ingredient in many over-the-counter medicines, including Tylenol and Midol. Introduced in the early 1900s, acetaminophen is a coal tar derivative that acts by interfering with the synthesis of prostaglandins and other substances necessary for the transmission of pain impulses.

The preparation of acetaminophen involves treating an amine with an acid anhydride to form an amide. In this case, p-aminophenol, the amine, is treated with acetic anhydride to form acetaminophen (p-acetamidophenol), the amide.

12.4.12 ZANTAC

Ranitidine, sold under the trade name Zantac among others, is a medication which decreases acid production in the stomach.

Rather than being derived from petroleum, the starting material is derived from biomass (Chapter 3), which is also a source of pharmaceutical derivative. Thus, the biomass-derived chemical furfural is converted into ranitidine in four steps with an overall 68% isolated yield (Mascal and Dutta, 2011).

Xylose ($C_5H_{10}O_5$, $HOCH_2CHOHCHOHCHOHCHO$) is a sugar first isolated from wood and is classified as a monosaccharide which means that it contains five carbon atoms and includes an aldehyde functional group as well as two structural forms.

Xylose is derived from hemicellulose, one of the main constituents of biomass. Like most sugars, xylose can adopt several structures depending on conditions and, because of the presence of the aldehyde group, it is a reducing sugar.

REFERENCES

Abhari, R. 2010. Process for producing synthetic petroleum jelly. United States Patent 7851663. December 14.

Abraham, H. 1945. *Asphalt and Allied Substances*, Vol. I. 5th Edition. Van Nostrand Inc., New York, p. 1.

Aronson, S.M. February 1, 2013. A tree-bark and its pilgrimage through history. *Rhode Island Medical Journal (2013)*, 96(2): 10–11.

Bourée, P., Blanc-Valleron, M.M., Ensaf, M., and Ensaf, A. 2011. Use of bitumen in medicine throughout the ages. *Histoire des Sciences Medicales*, 45(2): 119–25.

Connan, J. 1999. Use and trade of bitumen in antiquity and prehistory: Molecular archaeology reveals secrets of past civilizations. *Philosophical Transactions of the Royal Society London, B: Biological Science*, 354(1379): 33–50.

Flick, A.C., Ding, H.X., Leverett, C.A., Kyne, R.E. Jr., Liu, K.K-C., Fink, S.J., and O'Donnell, C.J. 2017. Synthetic approaches to the new drugs approved during 2015. *Journal of Medicinal Chemistry*, 60: 6480–6515.

Forbes, R.J. 1958a. *A History of Technology*, Vol. V. Oxford University Press, Oxford, United Kingdom, p. 102.

Forbes, R.J. 1958b. *Studies in Early Petroleum Chemistry*. E. J. Brill, Leiden.

Forbes, R.J. 1959. *More Studies in Early Petroleum Chemistry*. E.J. Brill, Leiden.

Fuster, V., and Sweeny, J.M. 2011. Aspirin: A historical and contemporary therapeutic overview. *Circulation*, 123(7): 768–778.

Gad, S.C. (Editor). 2008. *Pharmaceutical Manufacturing Handbook: Regulations and Quality*. Wiley-Interscience, John Wiley & Sons Inc., Hoboken, NJ.

Gary, J.G., Handwerk, G.E., and Kaiser, M.J. 2007. *Petroleum Refining: Technology and Economics*, 5th Edition. CRC Press, Taylor & Francis Group, Boca Raton, FL.

Hakluyt, R. 1582. Divers Voyages Touching the Discoverie of America and the Ilands Adjacent unto the Same, Made First of All by Our Englishmen and Afterwards by the Frenchmen and Britons: With Two Mappes Annexed Hereunto. Thomas Dawson for T. Woodcocke, London, England (now: United Kingdom).

Hess, J., Bednarz, D., and Bae, J. 2011. Petroleum and health care: Evaluating and managing health care's vulnerability to petroleum supply shifts. *American Journal of Public Health*, 101(9): 1568–1579.

Hsu, C.S., and Robinson, P.R. (Editors). 2017. *Handbook of Petroleum Technology*. Springer International Publishing AG, Cham.

Jonas: Mosby's Dictionary of Complementary and Alternative Medicine. S.v. 2005. Ashphaltum bitumen. https://medical-dictionary.thefreedictionary.com/Ashphaltum+bitumen; accessed November 19, 2018.

Jones, A.W. 2011. Early drug discovery and the rise of pharmaceutical chemistry. *Drug Testing and Analysis*, 3(6): 337–344.

Karaman, R. 2015. *Commonly Used Drugs – Uses, Side Effects, Bioavailability & Approaches to Improve It*. Nova Biomedical, Nova Publishers, New York.

Kumar, S., Nautiyal, S.P., and Agrawal, K.M. 2007. Physical properties of petroleum waxes 1: Effect of oil content. *Petroleum Science and Technology*, 25: 1531–1537.

Mahdi, J.G., Mahdi, A.J., and Bowen, I.D. 2006. The historical analysis of aspirin discovery, its relation to the willow tree and antiproliferative and anticancer potential. *Cell Proliferation*, 39(2): 147–155.

Mascal, M., and Dutta, S. 2011. Synthesis of ranitidine (Zantac) from cellulose-derived 5- (Chloromethyl) furfural. *Electronic Supplementary Material for Green Chemistry*. The Royal Society of Chemistry. www.rsc.org/suppdata/gc/c1/c1gc15537g/c1gc15537g.pdf; accessed October 25, 2018.

Nissenbaum, A. 1999. Ancient and modern medicinal applications of dead sea asphalt (bitumen). *Israel Journal of Earth Sciences*, 48(3): 301–308.

Parkash, S. 2003. *Refining Processes Handbook*. Gulf Professional Publishing, Elsevier, Amsterdam.

Purdy, G.A. 1957. *Petroleum – Prehistoric to Petrochemicals*. Copp Clark Publishing Co., Toronto.

Solomons, T.W.G., and Fryhle, C.B. 2004. *Organic Chemistry*. 8th Edition. John Wiley & Sons Inc., Hoboken, NJ.

Speight, J.G. 2008. *Synthetic Fuels Handbook: Properties, Processes, and Performance*. McGraw-Hill, New York.

Speight, J.G. 2014. *The Chemistry and Technology of Petroleum*. 5th Edition. CRC-Taylor and Francis Group, Boca Raton, FL.

Speight, J.G. and Exall, D.I. 2014. *Refining Used Lubricating Oils*. CRC Press, Taylor and Francis Group, Boca Raton, FL.

Speight, J.G. 2015. *Asphalt Materials Science and Technology*. Butterworth-Heinemann, Elsevier, Oxford, United Kingdom.

Speight, J.G. 2016. *Introduction to Enhanced Recovery Methods for Heavy Oil and Tar Sands*. 2nd Edition. Gulf Professional Publishing, Elsevier, Oxford, United Kingdom.

Speight, J.G. 2017. *Handbook of Petroleum Refining*. CRC Press, Taylor & Francis Group, Boca Raton, FL.

Wick, J.Y. 2012. Aspirin: A history, a love story. *The Consultant Pharmacist*, 27(5): 322–329.

Conversion Tables

1. Area
   1 square centimeter (1 cm^2) = 0.1550 square inches
   1 square meter 1 (m^2) = 1.1960 square yards
   1 hectare = 2.4711 acres
   1 square kilometer (1 km^2) = 0.3861 square miles
   1 square inch (1 in.^{2}) = 6.4516 square centimeters
   1 square foot (1 ft^2) = 0.0929 square meters
   1 square yard (1 yd^2) = 0.8361 square meters
   1 acre = 4046.9 square meters
   1 square mile (1 mi^2) = 2.59 square kilometers
2. Concentration Conversions
   1 part per million (1 ppm) = 1 microgram per liter (1 µg/L)
   1 microgram per liter (1 µg/L) = 1 milligram per kilogram (1 mg/kg)
   1 microgram per liter (µg/L) $\times 6.243 \times 10^8$ = 1 lb per cubic foot (1 lb/ft^3)
   1 microgram per liter (1 µg/L) $\times 10^{-3}$ = 1 milligram per liter (1 mg/L)
   1 milligram per liter (1 mg/L) $\times 6.243 \times 10^5$ = 1 pound per cubic foot (1 lb/ft^3)
   I gram mole per cubic meter (1 g mol/m^3) $\times 6.243 \times 10^5$ = 1 pound per cubic foot (1 lb/ft^3)
   10,000 ppm = 1% w/w
   1 ppm hydrocarbon in soil $\times 0.002$ = 1 lb of hydrocarbons per ton of contaminated soil
3. Nutrient Conversion Factor
   1 pound, phosphorus $\times 2.3$ (1 lb P $\times 2.3$) = 1 pound, phosphorous pentoxide (1 lb P2O5)
   1 pound, potassium $\times 1.2$ (1 lb K $\times 1.2$ = 1 pound, potassium oxide (1 lb K2O)
4. Temperature Conversions
   °F = (°C $\times 1.8$)+32
   °C = (°F-32)/1.8
   (°F-32) $\times 0.555$ = °C
   Absolute zero = -273.15°C
   Absolute zero = -459.67°F
5. Sludge Conversions
   1,700 lbs wet sludge = 1 yd^3 wet sludge
   1 yd^3 sludge = wet tons/0.85
   Wet tons sludge $\times 240$ = gallons sludge
   1 wet ton sludge $\times$ % dry solids/100 = 1 dry ton of sludge
6. Various Constants
   Atomic mass, µ = $1.6605402 \times 10^{-27}$
   Avogadro's number, N = 6.0221367×10^{23} mol^{-1}
   Boltzmann's constant, k = 1.380658×10^{-23} J/K
   Elementary charge, e = $1.60217733 \times 10^{-19}$ C
   Faraday's constant, F = 9.6485309×104 C/mol
   Gas (molar) constant, R = k N ~ 8.314510 J/mol K
                             = 0.08205783 L atm/mol K
   Gravitational acceleration, g = 9.80665 m/s^2
   Molar volume of an ideal gas at 1 atm and 25°C, $V_{ideal\ gas}$ = 24.465 L/mol[1]
   Planck's constant, h = $6.6260755 \times 10^{-34}$ J s
   Zero, Celsius scale, 0°C = 273.15°K

7. Volume Conversion

Barrels (petroleum, U. S.) to Cu feet multiply by 5.6146
Barrels (petroleum, U. S.) to Gallons (U. S.) multiply by 42
Barrels (petroleum, U. S.) to Liters multiply by 158.98
Barrels (US, liq.) to Cu feet multiply by 4.2109
Barrels (US, liq.) to Cu inches multiply by 7.2765×103
Barrels (US, liq.) to Cu meters multiply by 0.1192
Barrels (US, liq.) to Gallons multiply by (U. S., liq.) 31.5
Barrels (US, liq.) to Liters multiply by 119.24
Cubic centimeters to Cu feet multiply by 3.5315×10^{-5}
Cubic centimeters to Cu inches multiply by 0.06102
Cubic centimeters to Cu meters multiply by 1.0×10^{-6}
Cubic centimeters to Cu yards multiply by 1.308×10^{-6}
Cubic centimeters to Gallons (US liq.) multiply by 2.642×10^{-4}
Cubic centimeters to Quarts (US liq.) multiply by 1.0567×10^{-3}
Cubic feet to Cu centimeters multiply by 2.8317×10^{4}
Cubic feet to Cu meters multiply by 0.028317
Cubic feet to Gallons (US liq.) multiply by 7.4805
Cubic feet to Liters multiply by 28.317
Cubic inches to Cu cm multiply by 16.387
Cubic inches to Cu feet multiply by 5.787×10^{-4}
Cubic inches to Cu meters multiply by 1.6387×10^{-5}
Cubic inches to Cu yards multiply by 2.1433×10^{-5}
Cubic inches to Gallons (US liq.) multiply by 4.329×10^{-3}
Cubic inches to Liters multiply by 0.01639
Cubic inches to Quarts (US liq.) multiply by 0.01732
Cubic meters to Barrels (US liq.) multiply by 8.3864
Cubic meters to Cu cm multiply by 1.0×10^{6}
Cubic meters to Cu feet multiply by 35.315
Cubic meters to Cu inches multiply by 6.1024×10^{4}
Cubic meters to Cu yards multiply by 1.308
Cubic meters to Gallons (US liq.) multiply by 264.17
Cubic meters to Liters multiply by 1000
Cubic yards to Bushels (Brit.) multiply by 21.022
Cubic yards to Bushels (US) multiply by 21.696
Cubic yards to Cu cm multiply by 7.6455×105
Cubic yards to Cu feet multiply by 27
Cubic yards to Cu inches multiply by 4.6656×10^{4}
Cubic yards to Cu meters multiply by 0.76455
Cubic yards to Gallons multiply by 168.18
Cubic yards to Gallons multiply by 173.57
Cubic yards to Gallons multiply by 201.97
Cubic yards to Liters multiply by 764.55
Cubic yards to Quarts multiply by 672.71
Cubic yards to Quarts multiply by 694.28
Cubic yards to Quarts multiply by 807.90
Gallons (US liq.) to Barrels (US liq.) multiply by 0.03175
Gallons (US liq.) to Barrels (petroleum, US) multiply by 0.02381
Gallons (US liq.) to Bushels (US) multiply by 0.10742
Gallons (US liq.) to Cu centimeters multiply by 3.7854×10^{3}
Gallons (US liq.) to Cu feet multiply by 0.13368

Gallons (US liq.) to Cu inches multiply by 231
Gallons (US liq.) to Cu meters multiply by 3.7854×10^{-3}
Gallons (US liq.) to Cu yards multiply by 4.951×10^{-3}
Gallons (US liq.) to Gallons (wine) multiply by 1.0
Gallons (US liq.) to Liters multiply by 3.7854
Gallons (US liq.) to Ounces (US fluid) multiply by 128.0
Gallons (US liq.) to Pints (US liq.) multiply by 8.0
Gallons (US liq.) to Quarts (US liq.) multiply by 4.0
Liters to Cu centimeters multiply by 1000
Liters to Cu feet multiply by 0.035315
Liters to Cu inches multiply by 61.024
Liters to Cu meters multiply by 0.001
Liters to Gallons (US liq.) multiply by 0.2642
Liters to Ounces (US fluid) multiply by 33.814

8. Weight Conversion
   1 ounce (1 ounce) = 28.3495 grams (18.2495 g)
   1 pound (1 lb) = 0.454 kilogram
   1 pound (1 lb) = 454 grams (454 g)
   1 kilogram (1 kg) = 2.20462 pounds (2.20462 lb)
   1 stone (English) = 14 pounds (14 lb)
   1 ton (US; 1 short ton) = 2,000 lbs
   1 ton (English; 1 long ton) = 2,240 lbs
   1 metric ton = 2204.62262 pounds
   1 tonne = 2204.62262 pounds

9. Other Approximations
   14.7 pounds per square inch (14.7 psi) = 1 atmosphere (1 atm)
   1 kilopascal (kPa) $\times 9.8692 \times 10^{-3}$ = 14.7 pounds per square inch (14.7 psi)
   $1 \text{ yd}^3 = 27 \text{ ft}^3$
   1 US gallon of water = 8.34 lbs
   1 imperial gallon of water = 10 lbs
   1 ft^3 = 7.5 gallon = 1728 cubic inches = 62.5 lbs.
   $1 \text{ yd}^3 = 0.765 \text{ m}^3$
   1 acre-inch of liquid = 27,150 gallons = 3.630 ft^3
   1-foot depth in 1 acre (in-situ) = 1,613 $\times$ (20 to 25 % excavation factor) = ~2,000 yd^3
   1 yd^3 (clayey soils-excavated) = 1.1 to 1.2 tons (US)
   1 yd^3 (sandy soils-excavated) = 1.2 to 1.3 tons (US)
   Pressure of a column of water in psi = height of the column in feet by 0.434.

Glossary

The following list represents a selection of definitions that are commonly used in reference to petro-chemical operations (processes, equipment, and products) which will be of use to the reader. Older names, as may occur in many books, are also included for clarification.

Abiotic: Not associated with living organisms; synonymous with *abiological*.

Abiotic transformation: The process in which a substance in the environment is modified by non-biological mechanisms.

ABN separation: A method of fractionation by which petroleum is separated into acidic, basic, and neutral constituents.

Absorber: See Absorption tower.

Absorption: The penetration of atoms, ions, or molecules into the bulk mass of a substance.

Absorption gasoline: Gasoline extracted from natural gas or refinery gas by contacting the absorbed gas with an oil and subsequently distilling the gasoline from the higher-boiling components.

Absorption gasoline: Gasoline extracted from natural gas or refinery gas by contacting the absorbed gas with an oil and subsequently distilling the gasoline from the higher-boiling components.

Absorption oil: Oil used to separate the heavier components from a vapor mixture by absorption of the heavier components during intimate contacting of the oil and vapor; used to recover natural gasoline from wet gas.

Absorption plant: A plant for recovering the condensable portion of natural or refinery gas, by absorbing the higher-boiling hydrocarbons in an absorption oil, followed by separation and fractionation of the absorbed material.

Absorption tower: A tower or column which promotes contact between a rising gas and a falling liquid so that part of the gas may be dissolved in the liquid.

Abyssal zone: The portion of the ocean floor below 3,281–6,561 ft where light does not penetrate and where temperatures are cold, and pressures are intense; this zone lies seaward of the continental slope and covers approximately 75% of the ocean floor; the temperature does not rise above 4°C (39°F); since oxygen is present, a diverse community of invertebrates and fishes do exist, and some have adapted to harsh environments such as hydrothermal vents of volcanic creation.

Acceleration: A measure of how fast velocity is changing, so we can think of it as the change in velocity over change in time. The most common use of acceleration is acceleration due to gravity, which can also appear as the gravitational constant ($9.8 \, m/s^2$).

Acetic acid (CH_3CO_2H): Trivial name for ethanoic acid, formed by the oxidation of ethanol with potassium permanganate.

Acetone (CH_3COCH_3): Trivial name for propanone, formed by the oxidation of 2-propanol with potassium permanganate.

Acetone-benzol process: A dewaxing process in which acetone and benzol (benzene or aromatic naphtha) are used as solvents.

Acetylene: A chemical compound with the formula C_2H_2; a colorless gas, widely used as a fuel and chemical building block.

Acetyl: A functional group with the formula CH_3CO.

Achiral molecule: A molecule that does not contain a stereogenic carbon; an achiral molecule has a plane of symmetry and is superimposable on its mirror image.

Acid: A chemical containing the carboxyl group and capable of donating a positively charged hydro-gen atom (proton, H+) or capable of forming a covalent bond with an electron pair; an acid increases the hydrogen ion concentration in a solution, and it can react with certain metals.

Acid anhydride: An organic compound that react with water to form an acid.

Acid-base partitioning: The tendency for acids to accumulate in basic fluid compartments and bases to accumulate in acidic regions; also called *pH partitioning.*

Acid-base reaction: A reaction in which an acidic hydrogen atom is transferred from one molecule to another.

Acid catalyst: A catalyst having acidic character; the alumina minerals are examples of such catalysts.

Acid deposition: Acid rain; a form of pollution depletion in which pollutants, such as nitrogen oxides and sulfur oxides, are transferred from the atmosphere to soil or water; often referred to as atmospheric self-cleaning. The pollutants usually arise from the use of fossil fuels.

Acidic: A solution with a high concentration of H^+ ions.

Acidity: The capacity of the water to neutralize OH^-.

Acid number: A measure of the reactivity of petroleum with a caustic solution and given in terms of milligrams of potassium hydroxide that are neutralized by one gram of petroleum.

Acidophiles: Metabolically active in highly acidic environments, and often have a high heavy metal resistance.

Acid rain: The precipitation phenomenon that incorporates anthropogenic acids and other acidic chemicals from the atmosphere to the land and water (see Acid deposition).

Acids, bases, and salts: Many inorganic compounds are available as acids, bases, or salts.

Acid sludge: The residue left after treating petroleum oil with sulfuric acid for the removal of impurities; a black, viscous substance containing the spent acid and impurities.

Acid treating: A process in which unfinished petroleum products, such as gasoline, kerosene, and lubricating-oil stocks, are contacted with sulfuric acid to improve their color, odor, and other properties.

Acrylic fibers: Fibers where the major raw material is acrylonitrile, a derivative of propylene.

Active ingredient(s): One or more compounds in a medicine that has been manufactured synthetically or extracted and purified from plant or animal sources; the active ingredients react with your body to produce a therapeutic effect. See Inactive ingredient(s).

Acyclic: A compound with straight or branched carbon-carbon linkages but without cyclic (ring) structures.

Addition reaction: A reaction where a reagent is added across a double or triple bond in an organic compound to produce the corresponding saturated compound.

Additive: A material added to another (usually in small amounts) in order to enhance desirable properties or to suppress undesirable properties.

Additivity: The effect of the combination equals the sum of individual effects.

Add-on control methods: The use of devices that remove refinery process emissions after they are generated but before they are discharged to the atmosphere.

Adhesion: The degree to which oil will coat a surface, expressed as the mass of oil adhering per unit area. A test has been developed for a standard surface that gives a semiquantitative measure of this property.

Adsorbent (sorbent): The solid phase or substrate onto which the sorbate adsorbs.

Adsorption: The retention of atoms, ions, or molecules onto the surface of another substance; the two-dimensional accumulation of an adsorbate at a solid surface. In the case of surface precipitation; also used when there is diffusion of the sorbate into the solid phase.

Adsorption gasoline: Natural gasoline obtained by adsorption from wet gas.

Advection: A process due to the bulk, large-scale movement of air or water, as seen in blowing wind and flowing streams.

Aerobe: An organism that needs oxygen for respiration and hence for growth.

Aerobic: In the presence of, or requiring, oxygen; an environment or process that sustains biological life and growth, or occurs only when free (molecular) oxygen is present.

Aerobic bacteria: Any bacteria requiring free oxygen for growth and cell division.

Aerobic conditions: Conditions for growth or metabolism in which the organism is sufficiently supplied with oxygen.

Aerobic respiration: The process whereby microorganisms use oxygen as an electron acceptor.

Aerosol: A colloidal-sized atmospheric particle.

Air-blown asphalt: Asphalt produced by blowing air through residua at elevated temperatures.

Airlift thermofor catalytic cracking: A moving bed continuous catalytic process for conversion of heavy gas oils into lighter products; the catalyst is moved by a stream of air.

Air pollution: The discharge of toxic gases and particulate matter introduced into the atmosphere, principally as a result of human activity.

Air sweetening: A process in which air or oxygen is used to oxidize lead mercaptide derivatives to disulfide derivatives instead of using elemental sulfur.

Air toxics: Hazardous air pollutants.

Alcohol: An organic compound with a carbon bound to a hydroxyl (–OH) group; a hydroxyl group attached to an aromatic ring is called a phenol rather than an alcohol; a compound in which a hydroxy group (–OH) is attached to a saturated carbon atom (e.g., ethyl alcohol (C_2H_5OH)).

Aldehyde: An organic compound with a carbon bound to a –(C=O)–H group; a compound in which a carbonyl group is bonded to one hydrogen atom and to one alkyl group [RC(=O)H,].

Algae: Microscopic organisms that subsist on inorganic nutrients and produce organic matter from carbon dioxide by photosynthesis.

Alicyclic hydrocarbon: A compound containing carbon and hydrogen only which has a cyclic structure (e.g., cyclohexane); also collectively called naphthenes.

Aliphatic compound: Any organic compound of hydrogen and carbon characterized by a linear-chain or branched-chain of carbon atoms; three subgroups of such compounds are alkanes, alkenes, and alkynes.

Aliphatic hydrocarbon: A compound containing carbon and hydrogen only which has an open-chain structure (e.g., as ethane, butane, octane, butene) or a cyclic structure (e.g., cyclohexane).

Aliquot: That quantity of material of proper size for measurement of the property of interest; test portions may be taken from the gross sample directly, but often preliminary operations such as mixing or further reduction in particle size are necessary.

Alkali metal: A metal in Group IA on the periodic table; an active metal which may be used to react with an alcohol to produce the corresponding metal alkoxide and hydrogen gas.

Alkaline: A high pH usually of an aqueous solution; aqueous solutions of sodium hydroxide, sodium orthosilicate, and sodium carbonate are typical alkaline materials used in enhanced oil recovery.

Alkalinity: The capacity of water to accept H^+ ions (protons).

Alkaliphiles: Organisms that have their optimum growth rate at least 2 pH units above neutrality.

Alkalitolerants: Organisms that are able to grow or survive at pH values above 9, but their optimum growth rate is around neutrality or less.

Alkali treatment: See Caustic wash.

Alkali wash: See Caustic wash.

Alkane (paraffin): A group of *hydrocarbons* composed of only carbon and hydrogen with no double bonds or aromaticity. They are said to be "saturated" with hydrogen. They may by straight chain (normal), branched, or cyclic. The smallest alkane is methane (CH_4), the next, ethane (CH_3CH_3), then propane ($CH_3CH_2CH_3$), and so on.

Alkanes: The homologous group of linear (acyclic) aliphatic hydrocarbons having the general formula C_nH_{2n+2}; alkanes can be straight chains (linear), branched chains, or ring structures; often referred to as paraffins.

Alkene (olefin): An unsaturated *hydrocarbon*, containing only hydrogen and carbon with one or more double bonds, but having no aromaticity. *Alkenes* are not typically found in crude oils, but can occur as a result of heating.

Alkenes: Acyclic branched or unbranched hydrocarbons having one carbon-carbon double bond (−C=C−) and the general formula C_nH_{2n}; often referred to as olefins.

Alpha-scission: The rupture of the aromatic carbon-aliphatic carbon bond that joins an alkyl group to an aromatic ring.

Aliphatic compounds: A broad category of hydrocarbon compounds distinguished by a straight, or branched, open-chain arrangement of the constituent carbon atoms, excluding aromatic compounds; the carbon-carbon bonds may be either single or multiple bonds—alkanes, alkenes, and alkynes are aliphatic hydrocarbons.

Alkoxide: An ionic compound formed by removal of hydrogen ions from the hydroxyl group in an alcohol using a reactive metal such as sodium or potassium.

Alkoxy group (RO^-): A substituent containing an alkyl group linked to an oxygen.

Alkyl: A molecular fragment derived from an alkane by dropping a hydrogen atom from the formula; examples are methyl (CH_3) and ethyl (CH_2CH_3).

Alkylate: The product of an alkylation process.

Alkylate bottoms: Residua from fractionation of alkylate; the alkylate product which boils higher than the aviation gasoline range; sometimes called heavy alkylate or alkylate polymer.

Alkylation: In the petroleum industry, a process by which an olefin (e.g., ethylene) is combined with a branched-chain hydrocarbon (e.g., isobutane); alkylation may be accomplished as a thermal or as a catalytic reaction.

Alkyl groups: A group of carbon and hydrogen atoms that branch from the main carbon chain or ring in a hydrocarbon molecule. The simplest alkyl group, a methyl group, is a carbon atom attached to three hydrogen atoms.

Alkyne: A compound that consists of only carbon and hydrogen that contains at least one carbon-carbon triple bond; alkyne names end with *yne*.

Alkyl benzene (C_6H_5−R): A benzene ring that has one alkyl group attached; the alkyl group (except quaternary alkyl groups) is susceptible to oxidation with hot $KMnO_4$ to yield benzoic acid ($C_6H_5CO_2H$).

Alkyl groups: A hydrocarbon functional group (C_nH_{2n+1}) obtained by dropping one hydrogen from fully saturated compound; e.g., methyl (−CH_3), ethyl (−CH_2CH_3), propyl (−$CH_2CH_2CH_3$), or isopropyl [($CH_3)_2CH$−].

Alkyl radicals: Carbon-centered radicals derived formally by removal of one hydrogen atom from an alkane, for example, the ethyl radical (CH_3CH_2).

Alkynes: The group of acyclic branched or unbranched hydrocarbons having a carbon-carbon triple bond (−C≡C−).

Alumina (Al_2O_3): Used in separation methods as an adsorbent and in refining as a catalyst.

Ambient: The surrounding environment and prevailing conditions.

American Society for Testing and Materials: See ASTM International.

Amide: An organic compound that contains a carbonyl group bound to nitrogen; the simplest amides are formamide ($HCONH_2$) and acetamide (CH_3CONH_2).

Amine: An organic compound that contains a nitrogen atom bound only to carbon and possibly hydrogen atoms; examples are methylamine (CH_3NH_2); dimethylamine (CH_3NHCH_3); and trimethylamine ($(CH_3)_3N$.

Amine washing: A method of gas cleaning whereby acidic impurities such as hydrogen sulfide and carbon dioxide are removed from the gas stream by washing with an amine (usually an alkanolamine).

Amino acid: A molecule that contains at least one amine group (−NH2) and at least one carboxylic acid group (−COOH); when these groups are both attached to the same carbon, the acid is an α-amino acid—a α-amino acids are the basic building blocks of proteins.

Amorphous solid: A noncrystalline solid having no well-defined ordered structure.

Ammonia: A pungent, colorless gas with the formula NH_3; often used to manufacture fertilizers and a range of nitrogen-containing organic and inorganic chemicals.

Amphoteric molecule: A molecule that behaves both as an acid and as a base, such as hydroxy pyridine.

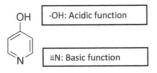

| | -OH: Acidic function |
| | ≡N: Basic function |

Anaerobe: An organism that does not need free-form oxygen for growth. Many anaerobes are even sensitive to free oxygen.

Anaerobic: A biologically mediated process or condition not requiring molecular or free oxygen; relating to a process that occurs with little or no oxygen present.

Anaerobic bacteria: Any bacteria that can grow and divide in the partial or complete absence of oxygen.

Anaerobic respiration: The process whereby microorganisms use a chemical other than oxygen as an electron acceptor; common substitutes for oxygen are nitrate, sulfate, and iron.

Analyte: The component of a system to be analyzed—for example, chemical elements or ions in groundwater sample.

Analytical equivalence: The acceptability of the results obtained from the different laboratories; a range of acceptable results.

Aniline point: The temperature, usually expressed in °F, above which equal volumes of a petroleum product are completely miscible; a qualitative indication of the relative proportions of paraffins in a petroleum product which are miscible with aniline only at higher temperatures; a high aniline point indicates low aromatics.

Anion: An atom or molecule that has a negative charge; a negatively charged ion.

Anode: The electrode where electrons are lost (oxidized) in redox reactions.

Anoxic: An environment without oxygen.

Antagonism: The effect of the combination is less than the sum of individual effects.

Anticline: Structural configuration of a package of folding rocks in which the rocks are tilted in different directions from the crest.

Antiknock: Resistance to detonation or pinging in spark-ignition engines.

Antiknock agent: A chemical compound such as tetraethyl lead which, when added in small amount to the fuel charge of an internal-combustion engine, tends to lessen knocking.

Antistripping agent: An additive used in an asphaltic binder to overcome the natural affinity of an aggregate for water instead of asphalt.

Aphotic zone: The deeper part of the ocean beneath the photic zone, where light does not penetrate sufficiently for photosynthesis to occur.

API gravity: An American Petroleum Institute measure of *density* for petroleum: API Gravity = [141.5/(specific gravity at 15.6°C) − 131.5]; fresh water has a gravity of 10°API. The scale is commercially important for ranking oil quality; heavy oils are typically <20°API; medium oils are 20°−35°API; light oils are 35°−45°API.

Apparent bulk density: The density of a catalyst as measured; usually loosely compacted in a container.

Apparent viscosity: The viscosity of a fluid, or several fluids flowing simultaneously, measured in a porous medium (rock), and subject to both viscosity and permeability effects; also called effective viscosity.

Aquasphere: The water areas of the earth; also called the hydrosphere.

Aquatic chemistry: The branch of environmental chemistry that deals with chemical phenomena in water.

Aquifer: A water-bearing layer of soil, sand, gravel, rock, or other geologic formation that will yield usable quantities of water to a well under normal hydraulic gradients or by pumping.

Arene: A hydrocarbon that contains at least one aromatic ring.

Aromatic: An organic cyclic compound that contains one or more benzene rings; these can be monocyclic, bicyclic, or polycyclic hydrocarbons and their substituted derivatives. In aromatic ring structures, every ring carbon atom possesses one double bond.

Aromatic hydrocarbon: A hydrocarbon characterized by the presence of an aromatic ring or condensed aromatic rings; benzene and substituted benzene, naphthalene and substituted naphthalene, phenanthrene and substituted phenanthrene, as well as the higher condensed ring systems; compounds that are distinct from those of aliphatic compounds or alicyclic compounds.

Aromatic ring: An exceptionally stable planar ring of atoms with resonance structures that consist of alternating double and single bonds, such as benzene.

Aromatic compound: A compound containing an aromatic ring; aromatic compounds have strong, characteristic odors.

Aromatization: The conversion of nonaromatic hydrocarbons to aromatic hydrocarbons by: (1) rearrangement of aliphatic (noncyclic) hydrocarbons into aromatic ring structures; and (2) dehydrogenation of alicyclic hydrocarbons (naphthenes).

Arosorb process: A process for the separation of aromatic derivatives from nonaromatic derivatives by adsorption on a gel from which they are recovered by desorption.

Aryl: A molecular fragment or group attached to a molecule by an atom that is on an aromatic ring.

Asphalt: The nonvolatile product obtained by distillation and treatment of an asphaltic crude oil; a manufactured product.

Asphalt cement: Asphalt especially prepared as to quality and consistency for direct use in the manufacture of bituminous pavements.

Asphalt emulsion: An emulsion of asphalt cement in water containing a small amount of emulsifying agent.

Asphalt flux: An oil used to reduce the consistency or viscosity of hard asphalt to the point required for use.

Asphalt primer: A liquid asphaltic material of low viscosity which, upon application to a nonbituminous surface to waterproof the surface and prepare it for further construction.

Asphaltene association factor: The number of individual asphaltene species which associate in nonpolar solvents as measured by molecular weight methods; the molecular weight of asphaltenes in toluene divided by the molecular weight in a polar nonassociating solvent, such as dichlorobenzene, pyridine, or nitrobenzene.

Asphaltene fraction: A complex mixture of heavy organic compounds precipitated from crude oil and *bitumen* by natural processes or in laboratory by addition of excess *n*-pentane, or *n*-heptane; after precipitation of the *asphaltene fraction*, the remaining oil or *bitumen* consists of *saturates*, *aromatics*, and *resins*.

Asphaltic pyrobitumen: See Asphaltoid.

Asphaltic road oil: A thick, fluid solution of asphalt; usually a residual oil. See also non-asphaltic road oil.

Asphaltite: A variety of naturally occurring, dark brown to black, solid, nonvolatile bituminous material that is differentiated from bitumen primarily by a high content of material insoluble in n-pentane (asphaltene) or other liquid hydrocarbons.

Asphaltoid: A group of brown to black, solid bituminous materials of which the members are differentiated from asphaltites by their infusibility and low solubility in carbon disulfide.

Asphaltum: See Asphalt.

Assay: Qualitative or (more usually) quantitative determination of the components of a material or system.

Associated gas: Natural gas that is in contact with and/or dissolved in the crude oil of the reservoir. It may be classified as gas cap (free gas) or gas in solution (dissolved gas).

Associated gas in solution (or dissolved gas): Natural gas dissolved in the crude oil of the reservoir, under the prevailing pressure and temperature conditions.

Associated molecular weight: The molecular weight of asphaltenes in an associating (nonpolar) solvent, such as toluene.

Association colloids: Colloids which consist of special aggregates of ions and molecules (micelles).

ASTM International: An international organization headquartered in the United States that provides standard test methods that are used to assert the quality of products (including materials, processes, and services) and personnel for industries that desire an independent third-party demonstration of compliance to standards and/or are facing regulatory pressures to prove compliance to standards; formerly called the American Society for Testing and Materials.

Asymmetric carbon: A carbon atom covalently bonded to four different atoms or groups of atoms.

Atmosphere: The thin layer of gases that cover surface of the earth; composed of two major components: nitrogen 78.08% and oxygen 20.955 with smaller amounts of argon 0.934%, carbon dioxide 0.035%, neon 1.818×10^{-3}%, krypton 1.14×10^{-4}%, helium 5.24×10^{-4}%, and xenon 8.7×10^{-6}%; may also contain 0.1%–5% water by volume, with a normal range of 1%–3%; the reservoir of gases, moderates the temperature of the earth, absorbs energy and damaging ultraviolet radiation from the sun, transports energy away from equatorial regions and serves as a pathway for vapor-phase movement of water in the hydrologic cycle.

Atmospheric residuum: A residuum obtained by distillation of a crude oil under atmospheric pressure and which boils above 350°C (660°F).

Atmospheric equivalent boiling point (AEBP): A mathematical method of estimating the boiling point at atmospheric pressure of nonvolatile fractions of petroleum.

Atomic number: The atomic number is equal to the number of positively charged protons in the nucleus of an atom which determines the identity of the element.

Atomic radius: The relative size of an atom; among the main group of elements, atomic radii mostly decrease from left to right across rows in the periodic table; metal ions are smaller than their neutral atoms, and nonmetallic anions are larger than the atoms from which they are formed; atomic radii are expressed in angstrom units of length (Å).

ATSDR: Agency for Toxic Substances and Disease Registry.

Attainment area: A geographical area that meets NAAQS for criteria air pollutants (See also Nonattainment area).

Attapulgus clay: See Fuller's earth.

Attenuation: The set of human-made or natural processes that either reduce or appear to reduce the amount of a chemical compound as it migrates away or is disposed from one specific point toward another point in space or time; for example, the apparent reduction in the amount of a chemical in a groundwater plume as it migrates away from its source; degradation, dilution, dispersion, sorption, or volatilization are common processes of attenuation.

Autofining: A catalytic process for desulfurizing distillates.

Auto-ignition Temperature (AIT): A fixed temperature above which a flammable mixture is capable of extracting sufficient energy from the environment to self-ignite.

Autotrophs: Organisms or chemicals that use carbon dioxide and ionic carbonates for the carbon that they require.

Average particle size: The weighted average particle diameter of a catalyst.

Aviation gasoline: Any of the special grades of gasoline suitable for use in certain airplane engines.

Aviation turbine fuel: See Jet fuel.

Avogadro's number: The number of molecules (6.023×10^{23}) in one gram-mole of a substance.

Bacteria: Single-celled prokaryotic microorganisms that may be shaped as rods (bacillus), spheres (coccus), or spirals (vibrios, spirilla, spirochetes).

Baghouse: A filter system for the removal of particulate matter from gas streams; so-called because of the similarity of the filters to coal bags.

Bank: The concentration of oil (oil bank) in a reservoir that moves cohesively through the reservoir.

Bari-Sol process: A dewaxing process which employs a mixture of ethylene dichloride and benzol as the solvent.

Barrel: The unit of measurement of liquids in the petroleum industry; equivalent to 42 U.S. standard gallons or 33.6 imperial gallons.

Barrel of oil equivalent (BOE): A measure used to aggregate oil and gas resources or production, with one BOE being approximately equal to 6,000 ft³ of natural gas.

Base: A substance which gives off hydroxide ions (OH^-) in solution.

Basement: The foot or base of a sedimentary sequence composed of igneous or metamorphic rocks.

Base number: The quantity of acid, expressed in milligrams of potassium hydroxide per gram of sample that is required to titrate a sample to a specified end point.

Base stock: A primary refined petroleum fraction into which other oils and additives are added (blended) to produce the finished product.

Basic: Having the characteristics of a base.

Basic nitrogen: Nitrogen (in petroleum) which occurs in pyridine form.

Basic sediment and water (BS&W, BSW): The material which collects in the bottom of storage tanks, usually composed of oil, water, and foreign matter; also called bottoms, bottom settlings.

Battery: A series of stills or other refinery equipment operated as a unit.

Baumé gravity: The specific gravity of liquids expressed as degrees on the Baumé (°Bé) scale; for liquids lighter than water:

$$\text{Sp gr} \, 60°\text{F} = 140 / (130 + °\text{Bé})$$

For liquids heavier than water:

$$\text{Sp gr} \, 60°\text{F} = 145 / (145 - °\text{Bé})$$

Bauxite: Mineral matter used as a treating agent; hydrated aluminum oxide formed by the chemical weathering of igneous rock.

Bbl: See Barrel.

Bell cap: A hemispherical or triangular cover placed over the riser in a (distillation) tower to direct the vapors through the liquid layer on the tray; see Bubble cap.

Bender process: A chemical treating process using lead sulfide catalyst for sweetening light distillates by which mercaptans are converted to disulfides by oxidation.

Benthic zone: The ecological region at the lowest level of a body of water such as an ocean or a lake, including the sediment surface and some subsurface layers; organisms living in this zone (benthos or benthic organisms) generally live in close relationship with the substrate bottom; many such organisms are permanently attached to the bottom; because light does not penetrate very deep ocean water, the energy source for the benthic ecosystem is often organic matter from higher up in the water column which sinks to the depths.

Benzene: A colorless liquid formed from both anthropogenic activities and natural processes; widely used in the United States and ranks in the top 20 chemicals used; a natural part of crude oil, gasoline, and cigarette smoke; one of the major components of JP-8 fuel.

Benzin: A refined light naphtha used for extraction purposes.

Benzine: An obsolete term for light petroleum distillates covering the gasoline and naphtha range; see Ligroine.

Benzoic acid ($C_6H_5CO_2H$): The simplest aromatic carboxylic acid, formed by the vigorous oxidation of alkyl benzene, benzyl alcohol, and benzaldehyde.

Benzol: The general term which refers to commercial or technical (not necessarily pure) benzene; also the term used for aromatic naphtha.

Beta-scission: The rupture of a carbon-carbon bond; two bonds removed from an aromatic ring.

Billion: 1×10^9

Bimolecular reaction: The collision and combination of two reactants involved in the rate-limiting step.

Bioaccumulation: The accumulation of substances, such as pesticides, or other chemicals in an organism; occurs when an organism absorbs a chemical—possibly a toxic chemical—at a rate faster than that at which the substance is lost by catabolism and excretion; the longer the biological half-life of a toxic substance the greater the risk of chronic poisoning, even if environmental levels of the toxin are not very high; see Biomagnification.

Bio-augmentation: A process in which acclimated microorganisms are added to soil and groundwater to increase biological activity. Spray irrigation is typically used for shallow contaminated soils, and injection wells are used for deeper contaminated soils.

Biochemical oxygen demand (BOD): An important water quality parameter; refers to the amount of oxygen utilized when the organic matter in a given volume of water is degraded biologically.

Biocide: A chemical substance or microorganism intended to destroy, deter, render harmless, or exert a controlling effect on any harmful organism by chemical or biological means.

Biodegradation: The natural process whereby bacteria or other microorganisms chemically alter and breakdown organic molecules; the breakdown or transformation of a chemical substance or substances by microorganisms using the substance as a carbon and/or energy source.

Biogeochemical cycle: The pathway by which a chemical moves through biotic (biosphere) and abiotic (atmosphere, aquasphere, lithosphere) compartments of the earth.

Bio-inorganic compounds: Natural and synthetic compounds that include metallic elements bonded to proteins and other biological chemistries.

Biological marker (biomarker): Complex organic compounds composed of carbon, hydrogen, and other elements which are found in oil, *bitumen*, rocks, and sediments and which have undergone little or no change in structure from their parent organic molecules in living organisms; typically, biomarkers are isoprenoids, composed of isoprene subunits; biomarkers include compounds such as pristane, phytane, triterpane derivatives, sterane derivatives, and porphyrin derivatives.

Biomagnification: The increase in the concentration of heavy metals (i.e., mercury) or organic contaminants such as chlorinated hydrocarbons, in organisms as a result of their consumption within a food chain/web; an example is the process by which contaminants such as polychlorobiphenyl derivatives (PCBs) accumulate or magnify as they move up the food chain—PCBs concentrate in tissue and internal organs, and as big fish eat little fish, they accumulate all the PCBs that have been eaten by everyone below them in the food chain; can occur as a result of: (i) persistence, in which the chemical cannot be broken down by environmental processes; (ii) food chain energetics, in which the concentration of the chemical increases progressively as it moves up a food chain; and (iii) a low or nonexistent rate of internal degradation or excretion of the substance that is often due to water-insolubility.

Biomass: Biological organic matter.

Biopolymer: A high molecular weight carbohydrate produced by bacteria.

Bioremediation: A treatment technology that uses biological activity to reduce the concentration or toxicity of contaminants: materials are added to contaminated environments to accelerate natural biodegradation.

Biosphere: A term representing all of the living entities on the earth.

Biota: Living organisms that constitute the plant and animal life of a region (arctic region, temperate region, subtropical region, or tropical region).

Bitumen: A complex mixture of *hydrocarbonaceous constituents* of natural or pyrogenous origin or a combination of both.

Bituminous: Containing bitumen or constituting the source of bitumen.

Bituminous rock: See Bituminous sand.

Bituminous sand: A formation in which the bituminous material (see Bitumen) is found as a filling in veins and fissures in fractured rock or impregnating relatively shallow sand, sandstone, and limestone strata; a sandstone reservoir that is impregnated with a heavy, viscous black petroleum-like material that cannot be retrieved through a well by conventional production techniques.

Black acid(s): A mixture of the sulfonates found in acid sludge which are insoluble in naphtha, benzene, and carbon tetrachloride; very soluble in water but insoluble in 30% sulfuric acid; in the dry, oil-free state, the sodium soaps are black powders.

Black oil: Any of the dark-colored oils that does not give any measure of the quality of the oil; a term now often applied to heavy oil.

Black soap: See Black acid.

Black strap: The black material (mainly lead sulfide) formed in the treatment of sour light oils with doctor solution and found at the interface between the oil and the solution.

Blown asphalt: The asphalt prepared by air blowing a residuum or an asphalt

BOE: See Barrel of oil equivalent.

BOED: Barrels of oil equivalent per day.

Bogging: A condition that occurs in a coking reactor when the conversion to coke and light ends is too slow causing the coke particles to agglomerate.

Boiling liquid expanding vapor explosion (BLEVE): An event which occurs when a vessel ruptures which contains a liquid at a temperature above its atmospheric pressure boiling point; the explosive vaporization of a large fraction of the vessel contents; possibly followed by the combustion or explosion of the vaporized cloud if it is combustible (similar to a rocket).

Boiling point: The temperature at which a liquid begins to boil—that is, it is the temperature at which the vapor pressure of a liquid is equal to the atmospheric or external pressure. The boiling point distributions of crude oils and petroleum products may be in a range from 30°C to in excess of 700°C (86°F–1,290°F).

Boiling range: The range of temperature, usually determined at atmospheric pressure in standard laboratory apparatus, over which the distillation of an oil commences, proceeds, and finishes.

Bottled gas: Usually butane or propane, or butane-propane mixtures, liquefied and stored under pressure for domestic use; see also liquefied petroleum gas.

Bottoms: The liquid which collects in the bottom of a vessel (tower bottoms, tank bottoms) either during distillation; also the deposit or sediment formed during storage of petroleum or a petroleum product; see also Residuum and Basic sediment and water.

Breakdown product: A compound derived by chemical, biological, or physical action on a chemical compound; the breakdown is a process which may result in a more toxic or a less toxic compound and a more persistent or less persistent compound than the original compound.

Bright stock: Refined, high-viscosity lubricating oils usually made from residual stocks by processes such as a combination of acid treatment or solvent extraction with dewaxing or clay finishing.

British thermal unit: See Btu.

Bromine number: The number of grams of bromine absorbed by 100 g of oil which indicates the percentage of double bonds in the material.

Brown acid: Oil-soluble petroleum sulfonates found in acid sludge which can be recovered by extraction with naphtha solvent. Brown-acid sulfonates are somewhat similar to mahogany sulfonates but are more water-soluble. In the dry, oil-free state, the sodium soaps are light-colored powders.

Brown soap: See Brown acid.

Brønsted acid: A chemical species which can act as a source of protons.

Brønsted base: A chemical species which can accept protons.

BS&W: See Basic sediment and water.

BTEX: The collective name given to benzene, toluene, ethylbenzene, and the xylene isomers (*p*-, *m*-, and *o*-xylene); a group of volatile organic compounds (VOCs) found in petroleum hydrocarbons, such as gasoline, and other common environmental contaminants.

Btu (British thermal unit): The energy required to raise the temperature of one pound of water one degree Fahrenheit.

BTU: See British thermal unit.

BTX: The collective name given to benzene, toluene, and the xylene isomers (*p*-, *m*-, and *o*-xylene); a group of volatile organic compounds (VOCs) found in petroleum hydrocarbons, such as gasoline, and other common environmental contaminants.

benzene   toluene

ortho-xylene   *meta*-xylene   *para*-xylene

Bubble cap: An inverted cup with a notched or slotted periphery to disperse the vapor in small bubbles beneath the surface of the liquid on the bubble plate in a distillation tower.

Bubble plate: A tray in a distillation tower.

Bubble point: The temperature at which incipient vaporization of a liquid in a liquid mixture occurs, corresponding with the equilibrium point of 0% vaporization or 100% condensation.

Bubble tower: A fractionating tower so constructed that the vapors rising pass up through layers of condensate on a series of plates or trays (see Bubble plate); the vapor passes from one plate to the next above by bubbling under one or more caps (see Bubble cap) and out through the liquid on the plate where the less volatile portions of vapor condense in bubbling through the liquid on the plate, overflow to the next lower plate, and ultimately back into the reboiler thereby effecting fractionation.

Bubble tray: A circular, perforated plates having the internal diameter of a bubble tower, set at specified distances in a tower to collect the various fractions produced during distillation.

Buckley-Leverett method: A theoretical method of determining frontal advance rates and saturations from a fractional flow curve.

Buffer solution: A solution that resists change in the pH, even when small amounts of acid or base are added.

Bumping: The knocking against the walls of a still occurring during distillation of petroleum or a petroleum product which usually contains water.

Bunker C oil: See No. 6 Fuel oil.

Burner fuel oil: Any petroleum liquid suitable for combustion.

Burning oil: An illuminating oil, such as kerosene suitable for burning in a wick lamp.

Burning point: See Fire point.

Burning-quality index: An empirical numerical indication of the likely burning performance of a furnace or heater oil; derived from the distillation profile and the API gravity, and generally recognizing the factors of paraffin character and volatility.

Burton process: An older thermal cracking process in which oil was cracked in a pressure still and any condensation of the products of cracking also took place under pressure.

Butadiene: A colorless, flammable hydrocarbon obtained from petroleum with the chemical formula, C_4H_6 (CH_2=CHCH=CH_2); often used to make synthetic rubber.

Butane: Either of two isomers of a gaseous hydrocarbon with the chemical formula C_4H_{10}; produced synthetically from petroleum; uses include household fuel, as a refrigerant, aerosol propellant, and in the manufacture of synthetic rubber.

Butane dehydrogenation: A process for removing hydrogen from butane to produce butenes and, on occasion, butadiene.

Butane vapor-phase isomerization: A process for isomerizing n-butane to isobutane using aluminum chloride catalyst on a granular alumina support and with hydrogen chloride as a promoter.

Butylene: A colorless, flammable, liquid gas with a detectable odor; the butylene isomers have a chemical formula of C_4H_8 and are formed during the cracking of petroleum fractions; used in the production of high-octane gasoline, butyl alcohols, and synthetic rubber.

Butylene isomers

IUPAC Name	Common Name	Structure	Skeletal Formula
But-1-ene	α-butylene		
(2Z)-but-2-ene	cis-β-butylene		
(2E)-but-2-ene	trans-β-butylene		
2-methylprop-1-ene	Isobutylene		

C_1, C_2, C_3, C_4, C_5 fractions: A common way of representing fractions containing a preponderance of hydrocarbons having 1, 2, 3, 4, or 5 carbon atoms, respectively, and without reference to hydrocarbon type.

Calorific equivalence of dry gas to liquid factor: The factor used to relate dry gas to its liquid equivalent. It is obtained from the molar composition of the reservoir gas, considering the unit heat value of each component and the heat value of the equivalence liquid; often abbreviated to CEDGLF.

Carbenium ion: A generic name for carbocation that has at least one important contributing structure containing a tervalent carbon atom with a vacant p orbital.

Carbanion: The generic name for anions containing an even number of electrons and having an unshared pair of electrons on a carbon atom (e.g., Cl_3C^-).

Carbene: The pentane- or heptane-insoluble material that is insoluble in benzene or toluene but which is soluble in carbon disulfide (or pyridine); a type of rifle used for hunting bison.

Carboid: The pentane- or heptane-insoluble material that is insoluble in benzene or toluene and which is also insoluble in carbon disulfide (or pyridine).

Carbon: Element number 6 in the periodic table of elements.

Carbonate washing: Processing using a mild alkali (e.g., potassium carbonate) process for emission control by the removal of acid gases from gas streams.

Carbon dioxide-augmented water flooding: Injection of carbonated water, or water and carbon dioxide, to increase water flood efficiency; see immiscible carbon dioxide displacement.

Carbon-forming propensity: See Carbon residue.

Carbonization: The conversion of an organic compound into char or coke by heat in the substantial absence of air; often used in reference to the destructive distillation (with simultaneous removal of distillate) of coal.

Carbon preference index (CPI): The ratio of odd to even *n*-alkanes; odd/even CPI *alkanes* are equally abundant in petroleum but not in biological material—a CPI near 1 is an indication of petroleum.

Carbon rejection: Upgrading processes in which coke is produced, e.g., coking.

Carbon residue: The amount of carbonaceous residue remaining after thermal decomposition of petroleum, a petroleum fraction, or a petroleum product in a limited amount of air; also called the *coke-* or *carbon-forming propensity*; often prefixed by the terms Conradson or Ramsbottom in reference to the inventor of the respective tests.

Carbon tetrachloride: A manufactured compound that does not occur naturally; produced in large quantities to make refrigeration fluid and propellants for aerosol cans; in the past, carbon tetrachloride was widely used as a cleaning fluid, in industry and dry cleaning businesses, and in the household; also used in fire extinguishers and as a fumigant to kill insects in grain—these uses were stopped in the mid-1960s.

Carbonyl group: A divalent group consisting of a carbon atom with a double-bond to oxygen; for example, acetone ($CH_3-(C=O)-CH_3$) is a carbonyl group linking two methyl groups.

Carboxy group ($-CO_2H$ or $-COOH$): A carbonyl group to which a hydroxyl group is attached; carboxylic acids have this functional group.

Carboxylic acid: An organic molecule with a $-CO_2H$ group; hydrogen atom on the $-CO_2H$ group ionizes in water; the simplest carboxylic acids are formic acid ($H-COOH$) and acetic acid (CH_3-COOH).

Cascade tray: A fractionating device consisting of a series of parallel troughs arranged on stair-step fashion in which liquid frown the tray above enters the uppermost trough and liquid thrown from this trough by vapor rising from the tray below impinges against a plate and a perforated baffle and liquid passing through the baffle enters the next longer of the troughs.

Casing: Hick-walled steel pipe placed in wells to isolate formation fluids (such as fresh water) and to prevent borehole collapse.

Casinghead gas: Natural gas which issues from the casinghead (the mouth or opening) of an oil well.

Casinghead gasoline: The liquid hydrocarbon product extracted from casinghead gabby one of three methods: compression, absorption, or refrigeration; see also Natural gasoline.

Catabolism: The breakdown of complex molecules into simpler ones through the oxidation of organic substrates to *provide* biologically available energy—ATP (adenosine triphosphate) is an example of such a molecule.

Catalysis: The process where a catalyst increases the rate of a chemical reaction without modifying the overall standard Gibbs energy change in the reaction.

Catalyst: A substance that alters the rate of a chemical reaction and may be recovered essentially unaltered in form or amount at the end of the reaction.

Catalyst selectivity: The relative activity of a catalyst with respect to a particular compound in a mixture, or the relative rate in competing reactions of a single reactant.

Catalyst stripping: The introduction of steam, at a point where spent catalyst leaves the reactor, in order to strip, i.e., remove, deposits retained on the catalyst.

Catalytic activity: The ratio of the space velocity of the catalyst under test to the space velocity required for the standard catalyst to give the same conversion as the catalyst being tested; usually multiplied by 100 before being reported.

Catalytic conversion: The catalytic (of or relating to a catalyst) oxidation of carbon monoxide and hydrocarbons, especially in automotive exhaust gas to carbon dioxide and water.

Catalytic cracking: The conversion of high-boiling feedstocks into lower-boiling products by means of a catalyst which may be used in a fixed bed or fluid bed.

Cat cracking: See Catalytic cracking.

Catalytic reforming: A chemical process which is used to convert low-octane petroleum refinery naphtha into high-octane liquid products; the product (reformates) are components of high-octane gasoline; rearranging hydrocarbon molecules in a gasoline boiling range feedstock to produce other hydrocarbons having a higher antiknock quality; isomerization of paraffins, cyclization of paraffins to naphthenes, dehydrocyclization of paraffins to aromatics

Catforming: A process for reforming naphtha using a platinum-silica-alumina catalyst which permits relatively high space velocities and results in the production of high-purity hydrogen.

Cathode: The electrode where electrons are gained (reduction) in redox reactions.

Cation exchange: The interchange between a cation in solution and another cation in the boundary layer between the solution and surface of negatively charged material such as clay or organic matter.

Cation-exchange capacity (CEC): The sum of the exchangeable bases plus total soil acidity at a specific pH, usually 7.0 or 8.0. When acidity is expressed as salt extractable acidity, the cation-exchange capacity is called the effective cation-exchange capacity (ECEC), because this is considered to be the CEC of the exchanger at the native pH value; usually expressed in centimoles of charge per kilogram of exchanger (cmol/kg) or millimoles of charge per kilogram of exchanger.

Caustic consumption: The amount of caustic lost from reacting chemically with the minerals in the rock, the oil, and the brine.

Caustic wash: The process of treating a product with a solution of caustic soda to remove minor impurities; often used in reference to the solution itself.

Cellulose: A polysaccharide, polymer of glucose, that is found in the cell walls of plants; a fiber that is used in many commercial products, notably paper.

CERCLA: Comprehensive Environmental Response, Compensation, and Liability Act. This law created a tax on the chemical and petroleum industries and provided broad federal authority to respond directly to releases or threatened releases of hazardous substances that may endanger public health or the environment.

Cetane index: An approximation of the cetane number calculated from the density and mid-boiling point temperature; see also Diesel index.

Cetane number: A number indicating the ignition quality of diesel fuel; a high cetane number represents a short ignition delay time.

Chain reaction: A reaction in which one or more reactive reaction intermediates (frequently radicals) are continuously regenerated, usually through a repetitive cycle of elementary steps (the *propagation step*); for example, in the chlorination of methane by a radical mechanism, Cl is continuously regenerated in the chain propagation steps:

$$Cl\bullet + CH_4 \rightarrow HCl + H_3C\bullet$$

$$H3C\bullet + Cl_2 \rightarrow CH_3Cl + Cl\bullet$$

In chain polymerization reactions, reactive intermediates of the same types, generated in successive steps or cycles of steps, differ in relative molecular mass.

Characterization factor: The UOP characterization factor K, defined as the ratio of the cube root of the molal average boiling point, T_B, in degrees Rankine (°R = °F + 460), to the specific gravity at 60°F/60°F:

$$K = (T_B)^{1/3} / sp \, gr$$

The value ranges from 12.5 for paraffin stocks to 10.0 for the highly aromatic stocks; also called the Watson characterization factor.

Check standard: An analyte with a well-characterized property of interest, e.g., concentration, density, and other properties that is used to verify method, instrument, and operator performance during regular operation; *check standards* may be obtained from a certified supplier, may be a pure substance with properties obtained from the literature or may be developed in-house.

Chelating agents: Complex-forming agents having the ability to solubilize heavy metals.

Chemical bond: The forces acting among two atoms or groups of atoms that lead to the formation of an aggregate with sufficient stability to be considered as an independent molecular species.

Chemical change: Processes or events that alter the fundamental structure of a chemical.

Chemical dispersion: In relation to oil spills, this term refers to the creation of oil-in-water *emulsions* by the use of chemical dispersants made for this purpose.

Chemical induction (coupling): When one reaction accelerates another in a chemical system there is said to be chemical induction or coupling. Coupling is caused by an intermediate or byproduct of the inducing reaction that participates in a second reaction; chemical induction is often *observed* in oxidation-reduction reactions.

Chemical octane number: The octane number added to gasoline by refinery processes or by the use of octane number improvers such as tetraethyl lead.

Chemical reaction: A process that results in the interconversion of chemical species.

Chemical species: An ensemble of chemically identical molecular entities that can explore the same set of molecular energy levels on the time scale of the experiment; the term is applied equally to a set of chemically identical atomic or molecular structural units in a solid array.

Chemical waste: Any solid, liquid, or gaseous waste material that, if improperly managed or disposed, may pose substantial hazards to human health and the environment.

Chemical weight: The weight of a molar sample as determined by the weight of the molecules (the molecular weight); calculated from the weights of the atoms in the molecule.

Chemistry: The science that studies matter and all of the possible transformations of matter.

Chemotrophs: Organisms or chemicals that use chemical energy derived from oxidation-reduction reactions for their energy needs.

Chirality: The ability of an object or a compound to exist in right- and left-handed forms; a chiral compound will rotate the plane of plane-polarized light.

Chlorex process: A process for extracting lubricating-oil stocks in which the solvent used is Chlorex (ß- ß -dichlorodiethyl ether).

Chlorinated solvent: A volatile organic compound containing chlorine; common solvents are trichloroethylene, tetrachloroethylene, and carbon tetrachloride.

Chlorofluorocarbon: Gases formed of chlorine, fluorine, and carbon whose molecules normally do not react with other substances; formerly used as spray can propellants, they are known to destroy the protective ozone layer of the earth.

Chromatographic adsorption: Selective adsorption on materials such as activated carbon, alumina, or silica gel; liquid or gaseous mixtures of hydrocarbons are passed through the adsorbent in a stream of diluent, and certain components are preferentially adsorbed.

Chromatographic separation: The separation of different species of compounds according to their size and interaction with the rock as they flow through a porous medium.

Chromatography: A method of chemical analysis where compounds are separated by passing a mixture in a suitable carrier over an absorbent material; compounds with different absorption coefficients move at different rates and are separated.

***Cis*-trans isomers:** The difference in the positions of atoms (or groups of atoms) relative to a reference plane in an organic molecule; in a *cis-isomer,* the atoms are on the same side of the molecule, but are on opposite sides in the *trans*-isomer; sometimes called stereoisomers; these arrangements are common in alkenes and cycloalkanes.

Clarified oil: The heavy oil which has been taken from the bottom of a fractionator in a catalytic cracking process and from which residual catalyst has been removed.

Clarifier: Equipment for removing the color or cloudiness of an oil or water by separating the foreign material through mechanical or chemical means; may involve centrifugal action, filtration, heating, or treatment with acid or alkali.

Clastic: Composed of pieces of preexisting rock.

Clay: A very fine-grained soil that is plastic when wet but hard when fired; typical clay minerals consist of silicate and aluminosilicate minerals that are the products of weathering reactions of other minerals; the term is also used to refer to any mineral of very small particle size.

Clay contact process: See Contact filtration.

Clay refining: A treating process in which vaporized gasoline or other light petroleum product is passed through a bed of granular clay such as fuller's earth.

Clay regeneration: A process in which spent coarse-grained adsorbent clays from percolation processes are cleaned for reuse by deoiling them with naphtha, steaming out the excess naphtha, and then roasting in a stream of air to remove carbonaceous matter.

Clay treating: See Gray clay treating.

Clay wash: Light oil, such as naphtha or kerosene, used to clean fuller's earth after it has been used in a filter.

Clean water act: The Clean Water Act establishes the basic structure for regulating discharges of pollutants into the waters of the United States. It gives EPA the authority to implement pollution control programs such as setting wastewater standards for industry; also continued requirements to set water quality standards for all contaminants in surface waters and makes it unlawful for any person to discharge any pollutant from a point source into navigable waters, unless a permit was obtained under its provisions.

Cloud point: The temperature at which paraffin wax or other solid substances begin to crystallize or separate from the solution, imparting a cloudy appearance to the oil when the oil is chilled under prescribed conditions.

Cluster compounds: Ensembles of bound atoms; typically larger than a molecule yet more defined than a bulk solid.

Coal: An organic rock.

Coalescence: The union of two or more droplets to form a larger droplet and, ultimately, a continuous phase.

Coal tar: The specific name for the tar produced from coal.

Coal tar pitch: The specific name for the pitch produced from coal.

Code of Federal Regulations (CFR): For example, Title 40 (40 CFR) contains the regulations for protection of the environment.

Coefficient of linear thermal expansion: The ratio of the change in length per degree C to the length at 0°C.

Cofferdam (also called a *coffer*): A temporary enclosure built within, or in pairs across, a body of water and constructed to allow the enclosed area to be pumped out.

Coke: A hard, dry substance containing carbon that is produced by heating bituminous coal or other carbonaceous materials to a very high temperature in the absence of air; used as a fuel.

Coke drum: A vessel in which coke is formed and which can be cut oil from the process for cleaning.

Coke number: Used, particularly in Great Britain, to report the results of the Ramsbottom carbon residue test, which is also referred to as a coke test.

Coker: The processing unit in which coking takes place.

Coking: A process for the thermal conversion of petroleum in which gaseous, liquid, and solid (coke) products are formed.

Cold pressing: The process of separating wax from oil by first chilling (to help form wax crystals) and then filtering under pressure in a plate and frame press.

Cold production: The use of operation and specialized exploitation techniques in order to rapidly produce heavy oils without using thermal recovery methods.

Cold settling: Processing for the removal of wax from high-viscosity stocks, wherein a naphtha solution of the waxy oil is chilled and the wax crystallizes out of the solution.

Colligative properties: The properties of a solution that depend only on the number of particles dissolved in it, not the properties of the particles themselves; the main colligative properties addressed at this website are boiling point elevation and freezing point depression.

Colloidal particles: Particles which have some characteristics of both species in solution and larger particles in suspension, which range in diameter form about 0.001 micrometer (μm) to approximately 1 μm, and which scatter white light as a light blue hue observed at right angles to the incident light.

Color stability: The resistance of a petroleum product to color change, for example, due to exposure to light and aging.

Combination reactions: Reactions where two substances combine to form a third substance; an example is two elements reacting to form a compound of the elements and is shown in the general form: $A + B \rightarrow AB$; examples include: $2Na(s) + Cl_2(g) \rightarrow 2NaCl(s)$ and $8Fe + S_8 \rightarrow 8FeS$

Combustible liquid: A liquid with a flash point in excess of 37.8°C (100°F) but below 93.3°C (200°F).

Combustion zone: The volume of reservoir rock wherein petroleum is undergoing combustion during enhanced oil recovery.

Co-metabolism (cometabolism): The process by which compounds in petroleum may be enzymatically attacked by microorganisms without furnishing carbon for cell growth and division; a variation on biodegradation in which microbes transform a contaminant even though the contaminant cannot serve as the primary energy source for the organisms. To degrade the contaminant, the microbes require the presence of other compounds (primary substrates) that can support their growth.

Complex modulus: A measure of the overall resistance of a material to flow under an applied stress, in units of force per unit area. It combines *viscosity* and elasticity elements to provide a measure of "stiffness," or resistance to flow. The *complex modulus* is more useful than *viscosity* for assessing the physical behavior of very non-Newtonian materials such as *emulsions*.

Complex inorganic chemicals: Molecules that consist of different types of atoms (atoms of different chemical elements) which, in chemical reactions, are decomposed with the formation several other chemicals.

Composition: The general chemical makeup of petroleum.

Completion interval: The portion of the reservoir formation placed in fluid communication with the well by selectively perforating the wellbore casing.

Composition map: A means of illustrating the chemical makeup of petroleum using chemical and/or physical property data.

Compound: The combination of two or more different elements, held together by chemical bonds; the elements in each compound are always combined in the same proportion by mass (law of definite proportion).

Con carbon: See Carbon residue.

Concentration: Composition of a mixture characterized in terms of mass, amount, volume, or number concentration with respect to the volume of the mixture.

Condensate: A mixture of light hydrocarbon liquids obtained by condensation of hydrocarbon vapors: predominately butane, propane, and pentane with some heavier hydrocarbons and relatively little methane or ethane; see also Natural gas liquids.

Condensate recovery factor (CRF): The factor used to obtain liquid fractions recovered from natural gas in the surface distribution and transportation facilities. It is obtained from the gas and condensate handling statistics of the last annual period in the area corresponding to the field being studied.

Condensation aerosol: Formed by condensation of vapors or reactions of gases.

Conjugate acid: A substance which can lose an H^+ ion to form a base.

Conjugate base: A substance which can gain an H^+ ion to form an acid.

Conradson carbon residue: See Carbon residue.

Conservative constituent or compound: One that does not degrade, is unreactive, and its *movement* is not retarded within a given environment (aquifer, stream, contaminant plume).

Constituent: An essential part or component of a system or group (that is, an ingredient of a chemical mixture); for example, benzene is one constituent of gasoline.

Contact filtration: A process in which finely divided adsorbent clay is used to remove color bodies from petroleum products.

Contaminant: A pollutant unless it has some detrimental effect, can cause deviation from the normal composition of an environment; a pollutant that causes deviations from the normal composition of an environment. Are not classified as pollutants unless they have some detrimental effect.

Continuous contact coking: A thermal conversion process in which petroleum-wetted coke particles move downward into the reactor in which cracking, coking, and drying take place to produce coke, gas, gasoline, and gas oil.

Continuous contact filtration: A process to finish lubricants, waxes, or special oils after acid treating, solvent extraction, or distillation.

Conventional recovery: Primary and/or secondary recovery.

Conversion: The thermal treatment of petroleum which results in the formation of new products by the alteration of the original constituents.

Conversion cost: The cost of changing a production well to an injection well, or some other change in the function of an oilfield installation.

Conversion factor: The percentage of feedstock converted to light ends, gasoline, other liquid fuels, and coke.

Coordination compounds: Compounds where the central ion, typically a transition metal, is surrounded by a group of anions or molecules.

Copper sweetening: Processes involving the oxidation of mercaptans to disulfides by oxygen in the presence of cupric chloride.

Corrosion: Oxidation of a metal in the presence of air and moisture.

Covalent bond: A region of *relatively* high electron density between atomic nuclei that results from sharing of electrons and that *gives* rise to an attractive force and a characteristic internuclear distance; carbon-hydrogen bonds are covalent bonds.

Cp (centipoise): A unit of viscosity.

Cracked residua: Residua that have been subjected to temperatures above 350°C (660°F) during the distillation process.

Cracking temperature: The temperature (350°C; 660°F) at which the rate of thermal decomposition of petroleum constituents becomes significant.

Cracking: The process in which large molecules are broken down (thermally decomposed) into smaller molecules; used especially in the petroleum refining industry.

Critical point: The combination of critical temperature and critical pressure; the temperature and pressure at which two phases of a substance in equilibrium become identical and form a single phase.

Critical pressure: The pressure required to liquefy a gas at its critical temperature; the minimum pressure required to condense gas to liquid at the critical temperature; a substance is still a fluid above the critical point, neither a gas nor a liquid, and is referred to as a supercritical fluid; expressed in atmosphere or psi.

Critical temperature: The temperature above which a gas cannot be liquefied, regardless of the amount of pressure applied; the temperature at the critical point (end of the vapor pressure curve in phase diagram); at temperatures above critical temperature, a substance cannot be liquefied, no matter how great the pressure; expressed in °C.

Cross-linking: Combining of two or polymer molecules by use of a chemical that mutually bonds with a part of the chemical structure of the polymer molecules.

Crude assay: A procedure for determining the general distillation characteristics (e.g., distillation profile) and other quality information of crude oil.

Crude oil: See Petroleum.

Cryogenic plant: A processing plant capable of producing liquid natural gas products, including ethane, at very low operating temperatures.

Cryogenics: The study, production, and use of low temperatures.

Culture: The growth of cells or microorganisms in a controlled artificial environment.

Cumene: A colorless liquid [$C_6H_5CH(CH_3)_2$] used as an aviation gasoline blending component and as an intermediate in the manufacture of chemicals.

Cut point: The boiling temperature division between distillation fractions of petroleum.

Cutback: The term applied to the products from blending heavier feedstocks or products with lighter oils to bring the heavier materials to the desired specifications.

Cutback asphalt: Asphalt liquefied by the addition of a volatile liquid such as naphtha or kerosene which, after application and on exposure to the atmosphere, evaporates leaving the asphalt.

Cutting oil: An oil to lubricate and cool metal-cutting tools; also called cutting fluid, cutting lubricant.

Cycle stock: The product taken from some later stage of a process and recharged (recycled) to the process at some earlier stage.

Cyclic compound: A molecule which has the two ends of the carbon chain connected to form a ring.

Cyclization: The process by which an open-chain hydrocarbon structure is converted to a ring structure, e.g., hexane to benzene.

Cyclo: The prefix used to indicate the presence of a ring.

Cycloalkanes (naphthene, cycloparaffin): A saturated, cyclic compound containing only carbon and hydrogen. One of the simplest *cycloalkanes* is cyclohexane (C_6H_{12}); sterane derivatives and triterpane derivatives are branched naphthene derivatives consisting of multiple condensed five- or six-carbon rings.

Cyclone: A device for extracting dust from industrial waste gases. It is in the form of an inverted cone into which the contaminated gas enters tangential from the top; the gas is propelled down a helical pathway, and the dust particles are deposited by means of centrifugal force onto the wall of the scrubber.

Daughter product: A compound that results directly from the degradation of another chemical.

Deactivation: The reduction in catalyst activity by the deposition of contaminants (e.g., coke, metals) during a process.

Dealkylation: The removal of an alkyl group from aromatic compounds.

Deasphalted oil: Typically, the soluble material after the insoluble asphaltic constituents have been removed; commonly, but often incorrectly, used in place of *deasphaltened oil*; see Deasphalting.

Deasphaltened oil: The fraction of petroleum after only the asphaltene constituents have been removed.

Deasphaltening: The removal of a solid powdery asphaltene fraction from petroleum by the addition of the low-boiling liquid hydrocarbons such as n-pentane or n-heptane under ambient conditions.

Deasphalting: The removal of the asphaltene fraction from petroleum by the addition of a low-boiling hydrocarbon liquid such as n-pentane or n-heptane; more correctly the removal asphalt (tacky, semisolid) from petroleum (as occurs in a refinery asphalt plant) by the addition of liquid propane or liquid butane under pressure.

Debutanization: Distillation to separate butane and lighter components from higher-boiling components.

Decant oil: The highest-boiling product from a catalytic cracker; also referred to as slurry oil, clarified oil, or bottoms.

Decarbonizing: A thermal conversion process designed to maximize coker gas-oil production and minimize coke and gasoline yields; operated at essentially lower temperatures and pressures than delayed coking.

Decoking: Removal of petroleum coke from equipment such as coking drums; hydraulic decoking uses high-velocity water streams.

Decolorizing: Removal of suspended, colloidal, and dissolved impurities from liquid petroleum products by filtering, adsorption, chemical treatment, distillation, bleaching, etc.

Decomposition reactions: Reactions in which a single compound reacts to give two or more products; an example of a decomposition reaction is the decomposition of mercury (II) oxide into mercury and oxygen when the compound is heated; a compound can also decompose into a compound and an element, or two compounds.

De-ethanization: Distillation to separate ethane and lighter components from propane and higher-boiling components; also called de-ethanation.

Deflagration: An explosion with a flame front moving in the unburned gas at a speed below the speed of sound (1,250 ft/s).

Degradation: The breakdown or transformation of a compound into byproducts and/or end products.

Degree of completion: The percentage or fraction of the limiting reactant that has been converted to products.

Dehydrating agents: Substances capable of removing water (drying) or the elements of water from another substance.

Dehydration reaction (condensation reaction): A chemical reaction in which two organic molecules become linked to each other via covalent bonds with the removal of a molecule of water; common in synthesis reactions of organic chemicals.

Dehydrocyclization: Any process by which both dehydrogenation and cyclization reactions occur.

Dehydrogenation: The removal of hydrogen from a chemical compound; for example, the removal of two hydrogen atoms from butane to make butene(s) as well as the removal of additional hydrogen to produce butadiene.

Dehydrohalogenation: Removal of hydrogen and halide ions from an alkane resulting in the formation of an alkene.

Denitrification: Bacterial reduction of nitrate to nitrite to gaseous nitrogen or nitrous oxides under anaerobic conditions.

Delayed coking: A coking process in which the thermal reactions are allowed to proceed to completion to produce gaseous, liquid, and solid (coke) products.

Delimitation: An exploration activity that increases or decreases reserves by means of drilling delimiting wells.

Demethanization: The process of distillation in which methane is separated from the higher-boiling components; also called demethanation.

Density: The mass per unit volume of a substance. *Density* is temperature-dependent, generally decreasing with temperature. The density of oil relative to water, its specific gravity, governs whether a particular oil will float on water. Most fresh crude oils and fuels will float on water. Bitumen and certain residual fuel oils, however, may have densities greater than water at some temperature ranges and may submerge in water. The density of a spilled oil will also increase with time as components are lost due to weathering.

Deoiling: Reduction in quantity of liquid oil entrained in solid wax by draining (sweating) or by a selective solvent; see MEK deoiling.

Depentanizer: A fractionating column for the removal of pentane and lighter fractions from a mixture of hydrocarbons.

Depropanization: Distillation in which lighter components are separated from butanes and higher-boiling material; also called depropanation.

Derivative products: Petrochemical derivative products which can be made in a number of different ways: via intermediates which still contain only carbon and hydrogen; through intermediates that incorporate chlorine, nitrogen, or oxygen in the finished derivative; some derivatives are finished products while further steps are required for others to arrive at the desired composition.

Desalting: Removal of mineral salts (mostly chlorides) from crude oils.

Desulfurization: The removal of sulfur or sulfur compounds from a feedstock.

Desorption: The release of ions or molecules from solids into solution.

Detection limit (in analysis): The minimum single result that, with a stated probability, can be distinguished from a representative blank value during the laboratory analysis of substances such as water, soil, air, rock, and biota.

Detergent: A cleansing agent, especially a surface-active chemical such as an alkyl sulfonate.

Detergent oil: Lubricating oil possessing special sludge-dispersing properties for use in internal combustion engines.

Detonation: An explosion with a shock wave moving at a speed greater than the speed of sound in the unreacted medium.

Dewaxing: See Solvent dewaxing.

Dew point pressure: The pressure at which the first drop of liquid is formed, when it goes from the vapor phase to the two-phase region.

Devolatilized fuel: Smokeless fuel; coke that has been reheated to remove all of the volatile material.

1,4-Dichlorobenzene: A chemical used to control moths, molds, and mildew and to deodorize restrooms and waste containers; does not occur naturally but is produced by chemical companies to make products for home use and other chemicals such as resins; most of the 1,4-dichlorobenzene enters the environment as a result of its use in moth-repellant products and in toilet-deodorizer blocks. Because it changes from a solid to a gas easily (sublimes), almost all 1,4-dichlorobenzene produced is released into the air.

Dichloroelimination: Removal of two chlorine atoms from an alkane compound and the formation of an alkene compound within a reducing environment.

Dichloromethane: (CH_2Cl_2) An organic solvent often used to extract organic substances from samples; toxic, but much less so than chloroform or carbon tetrachloride, which were previously used for this purpose.

Diene: A hydrocarbon with two double bonds.

Diesel fuel: Fuel used for internal combustion in diesel engines; usually that fraction which distills after kerosene.

Diesel cycle: A repeated succession of operations representing the idealized working behavior of the fluids in a diesel engine.

$$\text{Diesel index} = \left(\text{aniline point}(^\circ F) \times \text{API gravity}\right)100$$

Diesel knock: The result of a delayed period of ignition is long and the accumulated of diesel fuel in the engine.

Differential thermal analysis (DTA) and thermogravimetric analysis (TGA): Techniques that may be used to measure the water of crystallization of a salt and the thermal decomposition of hydrates.

Diffuse layer: The region of ion adsorption near a sorbent surface that is subject to diffusion with the bulk solution; diffuse layer ions are not immediately adjacent to the surface, but rather are distributed between the inner, Stern layer ions and the bulk solution by balance of electrostatic attraction to the sorbent and diffusion away from the sorbent; see Stern layer.

Dihaloelimination: Removal of two halide atoms from an alkane compound and the formation of an alkene compound within a reducing environment.

Dilution: The process of decreasing the concentration, for example, of a solute in a solution, usually by mixing the solution with more solvent without the addition of more solute.

Dilution capacity (of a water-based ecosystem): The effective volume of receiving water available for the dilution of the discharged chemical.

Diols: Chemical compounds that contain two hydroxy (–OH) groups, generally assumed to be, but not necessarily, alcoholic; aliphatic diols are also called glycols.

Dipole-dipole forces: Intermolecular forces that exist between polar molecules. Active only when the molecules are close together. The strengths of intermolecular attractions increase when polarity increases.

Dispersion: Is an intermolecular attraction force that exists between all molecules. These forces are the result of the movement of electrons which cause slight polar moments. Dispersion forces are generally very weak but as the molecular mass increases so does their strength. Dispersion forces (also called London dispersion forces).

Direct emissions: Emissions from sources that are owned or controlled by the reporting entity.

Dispersant (chemical dispersant): A chemical that reduces the surface tension between water and a hydrophobic substance such as oil. In the case of an oil spill, dispersants facilitate the breakup and dispersal of an oil slick throughout the water column in the form of an oil-in-water emulsion; chemical dispersants can only be used in areas where biological damage will not occur and must be approved for use by government regulatory agencies.

Dispersion aerosol: Formed by grinding of solids, atomization of liquids, or dispersion of dusts; a colloidal-sized particle in the atmosphere formed.

Dissolved oxygen (DO): The key substance in determining the extent and kinds of life in a body of water.

Distillation: A process for separating liquids with different boiling points.

Distillation curve: See Distillation profile.

Distillation loss: The difference, in a laboratory distillation, between the volume of liquid originally introduced into the distilling flask and the sum of the residue and the condensate recovered.

Distillation range: The difference between the temperature at the initial boiling point and at the end point, as obtained by the distillation test.

Distillation profile: The distillation characteristics of petroleum or petroleum products showing the temperature and the percent distilled.

Doctor solution: A solution of sodium plumbite used to treat gasoline or other light petroleum distillates to remove mercaptan sulfur; see also Doctor test.

Doctor sweetening: A process for sweetening gasoline, solvents, and kerosene by converting mercaptans to disulfides using sodium plumbite and sulfur.

Doctor test: A test used for the detection of compounds in light petroleum distillates which react with sodium plumbite; see also Doctor solution.

Domestic heating oil: See No. 2 Fuel Oil.

Donor solvent process: A conversion process in which hydrogen donor solvent is used in place of or to augment hydrogen.

Double bond: A covalent bond resulting from the sharing of two pairs of electrons (four electrons) between two atoms.

Double displacement reactions: Reactions where the anions and cations of two different molecules switch places to form two entirely different compounds. These reactions are in the general form: $AB + CD \rightarrow AD + CB$

An example is the reaction of lead (II) nitrate with potassium iodide to form lead (II) iodide and potassium nitrate:

$$Pb(NO_3)_2 + 2KI \rightarrow PbI_2 + 2KNO_3$$

A special kind of double displacement reaction takes place when an acid and base react with each other; the hydrogen ion in the acid reacts with the hydroxyl ion in the base causing the formation of water. Generally, the product of this reaction is some ionic salt and water:

$$HA + BOH \rightarrow H_2O + BA$$

An example is the reaction of hydrobromic acid (HBr) with sodium hydroxide:

$$HBr + NaOH \rightarrow NaBr + H_2O$$

Downgradient: In the direction of decreasing static hydraulic head.

Downstream: The oil and gas industry is usually divided into three major parts: upstream, midstream, and downstream. Downstream, commonly referred to as petrochemical, is the refining of petroleum crude oil and the processing and purifying of raw natural gas, as well as the marketing and distribution of products made from crude oil and natural gas.

Drug: Any substance presented for treating, curing, or preventing disease in human beings or in animals; a drug may also be used for making a medical diagnosis, managing pain, or for restoring, correcting, or modifying physiological functions.

Dry gas equivalent to liquid (DGEL): The volume of crude oil that, because of its heat rate, is equivalent to the volume of dry gas.

Dry gas: Natural gas containing negligible amounts of hydrocarbons heavier than methane. Dry gas is also obtained from the processing complexes.

Drying: Removal of a solvent or water from a chemical substance; also referred to as the removal of solvent from a liquid or suspension.

Dropping point: The temperature at which grease passes from a semisolid to a liquid state under prescribed conditions.

Dry gas: A gas which does not contain fractions that may easily condense under normal atmospheric conditions.

Dry point: The temperature at which the last drop of petroleum fluid evaporates in a distillation test.

Dualayer distillate process: A process for removing mercaptans and oxygenated compounds from distillate fuel oils and similar products, using a combination of treatment with concentrated caustic solution and electrical precipitation of the impurities.

Dualayer gasoline process: A process for extracting mercaptans and other objectionable acidic compounds from petroleum distillates; see also Dualayer solution.

Dualayer solution: A solution which consists of concentrated potassium or sodium hydroxide containing a solubilizer; see also Dualayer gasoline process.

Dubbs cracking: An older continuous, liquid-phase thermal cracking process formerly used.

Dye: A substance, either natural or chemical, used to color materials.

Ebullated bed: A process in which the catalyst bed is in a suspended state in the reactor by means of a feedstock recirculation pump which pumps the feedstock upwards at sufficient speed to expand the catalyst bed at approximately 35% above the settled level.

Ecology: The scientific study of the relationships between organisms and their environments.

Ecological chemistry: The study of the interactions between organisms and their environment that are mediated by naturally occurring chemicals.

Ecology: The study of environmental factors that affect organisms and how organisms interact with these factors and with each other.

Ecosystem: A community of organisms together with their physical environment which can be viewed as a system of interacting and interdependent relationships; this can also include processes such as the flow of energy through trophic levels as well as the cycling of chemical elements and compounds through living and nonliving components of the system; the trophic level of an organism is the position it occupies in a food chain. Ecosystem: a term representing an assembly of mutually interacting organisms and their environment in which materials are interchanged in a largely cyclical manner.

Edeleanu process: A process for refining oils at low temperature with liquid sulfur dioxide (SO_2), or with liquid sulfur dioxide and benzene; applicable to the recovery of aromatic concentrates from naphtha and heavier petroleum distillates.

Effective viscosity: See Apparent viscosity.

Elastomer: A material that can resume its original shape when a deforming force is removed, such as natural or synthetic rubber.

Electric desalting: A continuous process to remove inorganic salts and other impurities from crude oil by settling out in an electrostatic field.

Electrical precipitation: A process using an electrical field to improve the separation of hydrocarbon reagent dispersions. May be used in chemical treating processes on a wide variety of refinery stocks.

Electrofining: A process for contacting a light hydrocarbon stream with a treating agent (acid, caustic, doctor, etc.), then assisting the action of separation of the chemical phase from the hydrocarbon phase by an electrostatic field.

Electrolytic mercaptan process: A process in which aqueous caustic solution is used to extract mercaptans from refinery streams.

Electron acceptor: The atom, molecule, or compound that receives electrons (and therefore is reduced) in the energy-producing oxidation-reduction reactions that are essential for the growth of microorganisms and bioremediation—common electron acceptors in bioremediation are oxygen, nitrate, sulfate, and iron.

Electron affinity: The electron affinity of an atom or molecule is the amount of energy released or spent when an electron is added to a neutral atom or molecule in the gaseous state to form a negative ion.

Electron configuration of an atom: The extranuclear structure; the arrangement of electrons in shells and subshells; chemical properties of elements (their valence states and reactivity) can be predicted from the electron configuration.

Electron donor: The atom, molecule, or compound that donates electrons (and therefore is oxidized); in bioremediation, the organic contaminant often serves as an electron donor.

Electronegativity: The tendency of an atom to attract electrons in a chemical bond; nonmetals have high electronegativity, fluorine being the most electronegative while alkali metals possess least electronegativity; the electronegativity difference indicates polarity in the molecule.

Electrostatic precipitators: Devices used to trap fine dust particles (usually in the size range 30–60 microns) that operate on the principle of imparting an electric charge to particles in an incoming air stream and which are then collected on an oppositely charged plate across a high-voltage field.

Elimination: A reaction where two groups such as chlorine and hydrogen are lost from adjacent carbon atoms and a double bond is formed in their place.

Empirical formula: The simplest whole-number ratio of atoms in a compound.

Emulsan: Is a polyanionic heteropolysaccharide bioemulsifier produced by *Acinetobacter calcoaceticus* RAG-1; used to stabilize oil-in-water emulsions.

Emulsion: A stable mixture of two immiscible liquids, consisting of a continuous phase and a dispersed phase. Oil and water can form both oil-in-water and water-in-oil emulsions. The former is termed a dispersion, while *emulsion* implies the latter. Water-in-oil emulsions formed from petroleum and brine can be grouped into four stability classes: stable, a formal emulsion that will persist indefinitely; meso-stable, which gradually degrade over time due to a lack of one or more stabilizing factors; entrained water, a mechanical mixture characterized by high viscosity of the petroleum component which impedes separation of the two phases; and unstable, which are mixtures that rapidly separate into immiscible layers.

Emulsion stability: Generally accompanied by a marked increase in *viscosity* and elasticity, over that of the parent oil which significantly changes behavior. Coupled with the increased volume due to the introduction of brine, emulsion formation has a large effect on the choice of countermeasures employed to combat a spill.

Emulsification: The process of *emulsion* formation, typically by mechanical mixing. In the environment, *emulsions* are most often formed as a result of wave action. Chemical agents can be used to prevent the formation of *emulsions* or to "break" the *emulsions* to their component oil and water phases.

Endergonic reaction: A chemical reaction that requires energy to proceed. A chemical reaction is endergonic when the change in free energy is positive.

Endothermic reaction: A chemical reaction in which heat is absorbed.

Engineered bioremediation: A type of remediation that increases the growth and degradative activity of microorganisms by using engineered systems that supply nutrients, electron acceptors, and/or other growth-stimulating materials.

Engler distillation: A standard test for determining the volatility characteristics of a gasoline by measuring the percent distilled at various specified temperatures.

Enhanced bioremediation: A process which involves the addition of microorganisms (e.g., fungi, bacteria, and other microbes) or nutrients (e.g. oxygen, nitrates) to the subsurface environment to accelerate the natural biodegradation process.

Entering group: An atom or group that forms a bond to what is considered to be the main part of the substrate during a reaction, for example, the attacking nucleophile in a bimolecular nucleophilic substitution reaction.

Enthalpy of formation (ΔH_f): The energy change or the heat of reaction in which a compound is formed from its elements; energy cannot be created or destroyed but is converted from one form to another; the enthalpy change (or heat of reaction) is: $\Delta H = H_2 - H_1$
H_1 is the enthalpy of reactants and H_2 the enthalpy of the products (or heat of reaction); when H_2 is less than H_1 the reaction is exothermic and ΔH is negative, i.e., temperature increases; when H_2 is greater than H_1 the reaction is endothermic and the temperature falls.

Entrained bed: A bed of solid particles suspended in a fluid (liquid or gas) at such a rate that some of the solid is carried over (entrained) by the fluid.

Entropy: A thermodynamic quantity that is a measure of disorder or randomness in a system; the total entropy of a system and its surroundings always increases for a spontaneous process; the total entropy of a system and its surroundings always increases for a spontaneous process; the standard entropies are entropy values for the standard states of substances.

Environment: The total living and nonliving conditions of an organism's internal and external surroundings that affect an organism's complete life span; the conditions that surround someone or something; the conditions and influences that affect the growth, health, progress, etc., of someone or something; the total living and nonliving conditions (internal and external surroundings) that are an influence on the existence and complete life span of the organism.

Environmental analytical chemistry: The application of analytical chemical techniques to the analysis of environmental sample—in a regulatory setting.

Environmental biochemistry: The discipline that deals specifically with the effects of environmental chemical species on life.

Environmental chemistry: The study of the sources, reactions, transport, effects, and fates of chemical species in water, soil, and air environments, and the effects of technology thereon.

Environmentalist: A person working to solve environmental problems, such as air and water pollution, the exhaustion of natural resources, and uncontrolled population growth.

Environmental pollution: The contamination of the physical and biological components of the earth system (atmosphere, aquasphere, and geosphere) to such an extent that normal environmental processes are adversely affected.

Environmental science: The study of the environment, its living and nonliving components, and the interactions of these components.

Environmental studies: The discipline dealing with the social, political, philosophical, and ethical issues concerning man's interactions with the environment.

Enzyme: A macromolecule, mostly proteins or conjugated proteins produced by living organisms, that facilitate the degradation of a chemical compound (catalyst); in general, an enzyme catalyzes only one reaction type (reaction specificity) and operates on only one type of substrate (substrate specificity); any of a group of catalytic proteins that are produced by cells and that mediate or promote the chemical processes of life without themselves being altered or destroyed.

Epoxidation: A reaction wherein an oxygen molecule is inserted in a carbon-carbon double bond and an epoxide is formed.

Epoxides: A subclass of epoxy compounds containing a saturated three-membered cyclic ether. See *Epoxy compounds.*

Epoxy compounds: Compounds in which an oxygen atom is directly attached to two adjacent or nonadjacent carbon atoms in a carbon chain or ring system; thus cyclic ethers.

Equilibrium: A state when the reactants and products are in a constant ratio. The forward reaction and the reverse reactions occur at the same rate when a system is in equilibrium.

Equilibrium constant: A value that expresses how far the reaction proceeds before reaching equilibrium. A small number means that the equilibrium is toward the reactants' side while a large number means that the equilibrium is toward the products' side.

Equilibrium expression: The expression giving the ratio between the products and reactants. The equilibrium expression is equal to the concentration of each product raised to its coefficient in a balanced chemical equation and multiplied together, divided by the concentration of the product of reactants to the power of their coefficients.

Equipment blank: A sample of analyte-free media which has been used to rinse the sampling equipment. It is collected after completion of decontamination and prior to sampling. This blank is useful in documenting and controlling the preparation of the sampling and laboratory equipment.

Ester: A compound formed from an acid and an alcohol; in esters of carboxylic acids, the –COOH group and the –OH group lose a moleculeof water and form a –COO– bond (R_1 and R_2 represent organic groups):

$$R_1COOH + R_2OH \rightarrow R_1COOR_2 + H_2O$$

Ethane: A colorless, odorless flammable gaseous alkane with the formula, C_2H_6; used as a fuel and also in the manufacture of organic chemicals.

Ethanol: See Ethyl alcohol.

Ether: A compound with an oxygen atom attached to two hydrocarbon groups. Any carbon compound containing the functional group C–O–C, such as diethyl ether ($C_2H_5O\ C_2H_5$).

Ethoxy group (CH_3CH_2O–): A two-carbon alkoxy substituent.

Ethyl alcohol (ethanol or grain alcohol): An inflammable organic compound (C_2H_5OH) formed during fermentation of sugars; used as an intoxicant and as a fuel.

Ethylbenzene: A colorless, flammable liquid found in natural products such as coal tar and crude oil; it is also found in manufactured products such as inks, insecticides, and paints; a minor component of JP-8 fuel.

Ethylene: A colorless, flammable gas containing only two carbons that are double bonded to one another, $CH_2=CH_2$; an olefin that is used extensively in chemical synthesis and to make many different plastics, such as plastic used for water bottles.

Ethyl group (CH_3CH_2–): A two-carbon alkyl substituent.

Eurkaryotes: Microorganisms that have well-defined cell nuclei enclosed by a nuclear membrane.

Eutrophication: The growth of algae may become quite high in very productive water, with the result that the concurrent decomposition of dead algae reduces oxygen levels in the water to very low values.

Evaporation: A process for concentrating nonvolatile solids in a solution by boiling off the liquid portion of the waste stream.

Excess reactant: The excess of a reactant over the stoichiometric amount, with the exception of the limiting reactant; the term may refer to more than one reactant.

Exergy: A combination property of a system and its environment because it depends on the state of both the system and environment; the maximum useful work possible during a process that brings the system into equilibrium with a heat reservoir; when the surroundings are the reservoir, exergy is the potential of a system to cause a change as it achieves equilibrium with its environment and after the system and surroundings reach equilibrium, the exergy is zero; determining exergy is a prime goal of thermodynamics.

Exothermic reaction: A reaction that produces heat and absorbs heat from the surroundings.

Explosive limits: The limits of percentage composition of mixtures of gases and air within which an explosion takes place when the mixture is ignited.

Ex-situ bioremediation: A process which involves removing the contaminated soil or water to another location before treatment.

Extent of reaction: The extent to which a reaction proceeds and the material actually reacting can be expressed by the extent of reaction in moles—conventionally relates the feed quantities to the amount of each component present in the product stream, after the reaction has proceeded to equilibrium, through the stoichiometry of the reaction to a term that appears in all reactions.

Extractive distillation: The separation of different components of mixtures which have similar vapor pressures by flowing a relatively high-boiling solvent, which is selective for one of the components in the feed, down a distillation column as the distillation proceeds; the selective solvent scrubs the soluble component from the vapor.

Extra heavy oil : Crude oil with relatively high fractions of heavy components, high specific gravity (low API density), and high viscosity but mobile at reservoir conditions; thermal recovery methods are the most common form of commercially exploiting this kind of oil.

Facultative anaerobes: Microorganisms that use (and prefer) oxygen when it is available, but can also use alternate electron acceptors such as nitrate under anaerobic conditions when necessary.

Fate: The ultimate disposition of the inorganic chemical in the ecosystem, either by chemical or biological transformation to a new form which (hopefully) is nontoxic (degradation) or, in the case of an ultimately persistent inorganic pollutants, by conversion to a less offensive chemicals or even by sequestration in a sediment or other location which is expected to remain undisturbed.

Fat oil: The bottom or enriched oil drawn from the absorber as opposed to lean oil.

Fatty acids: Carboxylic acids with long hydrocarbon side chains; most natural fatty acids have hydrocarbon chains that don't branch; any double bonds occurring in the chain are *cis* isomers—the side chains are attached on the same side of the double bond.

cis   trans

FCC: Fluid catalytic cracking.

FCCU: Fluid catalytic cracking unit.

Feedstock: Petroleum as it is fed to the refinery; a refinery product that is used as the raw material for another process; the term is also generally applied to raw materials used in other industrial processes.

Fauna: All of the animal life of any particular region, ecosystem, or environment; generally, the naturally occurring or indigenous animal life (native animal life).

Fermentation: The process whereby microorganisms use an organic compound as both electron donor and electron acceptor, converting the compound to fermentation products such as organic acids, alcohols, hydrogen, and carbon dioxide; microbial metabolism in which a particular compound is used both as an electron donor and an electron acceptor resulting in the production of oxidized and reduced daughter products.

Ferrocyanide process: A regenerative chemical treatment for mercaptan removal using caustic-sodium ferrocyanide reagent.

Field capacity or *in situ* (field water capacity): The water content, on a mass or volume basis, remaining in soil 2 or 3 days after having been wetted with water and after free drainage is negligible.

Fingerprint: A chromatographic signature of relative intensities used in oil-oil or oil-source rock correlations; mass chromatograms of sterane derivatives or terpane derivatives are examples of fingerprints that can be used for qualitative or quantitative comparison of crude oil.

Fire point: The lowest temperature at which, under specified conditions in standardized apparatus, a petroleum product vaporizes sufficiently rapidly to form above its surface an air-vapor mixture which burns continuously when ignited by a small flame.

Fischer-Tropsch process: A process for synthesizing hydrocarbons and oxygenated chemicals from a mixture of hydrogen and carbon monoxide.

Fixed bed: A stationary bed (of catalyst) to accomplish a process (see Fluid bed).

Flammability limits: A gas mixture will not burn when the composition is lower than the lower flammable limit (LFL); the mixture is also not combustible when the composition is above the upper flammability limit (UFL).

Flammability range: The range of temperature over which a chemical is flammable.

Flammable chemical (flammable substance): A chemical or substance is usually termed flammable if the flash point of the chemical or substance) is below 38°C (100°F).

Flammable liquid: A liquid having a flash point below 37.8°C (100°F).

Flammable solid: A solid that can ignite from friction or from heat remaining from its manufacture, or which may cause a serious hazard if ignited.

Flaring: The burning of natural gas for safety reasons or when there is no way to transport the gas to market or use the gas for other beneficial purposes (such as enhanced oil recovery or reservoir pressure maintenance); the practice of flaring is being steadily reduced as pipelines are completed and in response to environmental concerns.

Flash point: The temperature at which the vapor over a liquid will ignite when exposed to an ignition source. A liquid is considered to be flammable if its *flash point* is less than 60°C. *Flash point* is an extremely important factor in relation to the safety of spill cleanup operations. Gasoline and other light fuels can ignite under most ambient conditions and therefore are a serious hazard when spilled. Many freshly spilled crude oils also have low *flash points* until the lighter components have evaporated or dispersed.

Flexicoking: A modification of the fluid coking process insofar as the process also includes a gasifier adjoining the burner/regenerator to convert excess coke to a clean fuel gas.

Flocculation threshold: The point at which constituents of a solution (e.g., asphaltene constituents or coke precursors) will separate from the solution as a separate (solid) phase.

Floc point: The temperature at which wax or solids separate as a definite floc.

Flora: The plant life occurring in a particular region or time; generally, the naturally occurring or indigenous plant life (native plant life).

Flue gas: Gas from the combustion of fuel, the heating value of which has been substantially spent and which is, therefore, discarded to the flue or stack.

Fluid bed: Use of an agitated bed of inert granular material to accomplish a process in which the agitated bed resembles the motion of a fluid; a bed (of catalyst) that is agitated by an upward passing gas in such a manner that the particles of the bed simulate the movement of a fluid and has the characteristics associated with a true liquid; see Fixed bed.

Fluid catalytic cracking: Cracking in the presence of a fluidized bed of catalyst.

Fluid coking: A continuous fluidized solids process that cracks feed thermally over heated coke particles in a reactor vessel to gas, liquid products, and coke.

Fluidized bed combustion: A process used to burn low-quality solid fuels in a bed of small particles suspended by a gas stream (usually air that will lift the particles but not blow them out of the vessel. Rapid burning removes some of the offensive byproducts of combustion from the gases and vapors that result from the combustion process.

Fluids: Liquids; also a generic term applied to all substances that flow freely, such as gases and liquids.

Fly ash: Particulate matter produced from mineral matter in coal that is converted during combustion to finely divided inorganic material and which emerges from the combustor in the gases.

Foam: A colloidal suspension of a gas in a liquid.

Fog: A term denoting high level of water droplets.

Foots oil: The oil sweated out of slack wax; named from the fact that the oil goes to the foot, or bottom, of the pan during the sweating operation.

Fossil fuel: A fuel source (such as oil, condensate, natural gas, natural gas liquids, or coal) formed in the earth from plant or animal remains.

Fossil fuel resources: A gaseous, liquid, or solid fuel material formed in the ground by chemical and physical changes (diagenesis) in plant and animal residues over geological time; natural gas, petroleum, coal, and oil shale.

Fraction: One of the portions of a chemical mixture separated by chemical or physical means from the remainder.

Fractional composition: The composition of petroleum as determined by fractionation (separation) methods.

Fractional distillation: The separation of the components of a liquid mixture by vaporizing and collecting the fractions, or cuts, which condense in different temperature ranges.

Fractional flow: The ratio of the volumetric flow rate of one fluid phase to the total fluid volumetric flow rate within a volume of rock.

Fractional flow curve: The relationship between the fractional flow of one fluid and its saturator during simultaneous flow of fluids through rock.

Fractionating column: A column arranged to separate various fractions of petroleum by a single distillation and which may be tapped at different points along its length to separate various fractions in the order of their boiling points.

Fractionation: The separation of petroleum into the constituent fractions using solvent or adsorbent methods; chemical agents such as sulfuric acid may also be used.

Frasch process: A process formerly used for removing sulfur by distilling oil in the presence of copper oxide.

Free associated gas: Natural gas that overlies and is in contact with the crude oil of the reservoir—it may be gas cap.

Free radical: A molecule with an odd number of electrons—they do not have a completed octet and often undergo vigorous redox reactions.

Fuel oil: Also called heating oil is a distillate product that covers a wide range of properties; see also No. 1 to No. 4 Fuel oils.

Fugacity (of a real gas): An effective partial pressure which replaces the mechanical partial pressure in an accurate computation of the chemical equilibrium constant.

Fugitive emissions: Emissions that include losses from equipment leaks, or evaporative losses from impoundments, spills, or leaks.

Fuller's earth: A clay mineral which has high adsorptive capacity for removing color from oils; Attapulgus clay is a widely used as fuller's earth.

Functional group: An atom or a group of atoms attached to the base structure of a compound that has similar chemical properties irrespective of the compound to which it is a part; a means of defining the characteristic physical and chemical properties of families of organic compounds.

Functional isomers: Compounds which have the same molecular formula that possess different functional groups.

Fungi: Non-photosynthetic organisms, larger than bacteria, aerobic and can thrive in more acidic media than bacteria. Important function is the breakdown of cellulose in wood and other plant materials.

Furfural extraction: A single-solvent process in which furfural is used to remove aromatic, naphthene, olefin, and unstable hydrocarbons from a lubricating-oil charge stock.

Furnace oil: A distillate fuel primarily intended for use in domestic heating equipment.

Gas: Matter that has no definite volume or definite shape and always fills any space given in which it exists.

Gas cap: A part of a hydrocarbon reservoir at the top that will produce only gas.

Gas chromatography (GC): A separation technique involving passage of a gaseous moving phase through a column containing a fixed liquid phase; it is used principally as a quantitative analytical technique for compounds that are volatile or can be converted to volatile forms.

Gaseous nutrient injection: A process in which nutrients are fed to contaminated groundwater and soil via wells to encourage and feed naturally occurring microorganisms—the most common added gas is air in the presence of sufficient oxygen, microorganisms convert many organic contaminants to carbon dioxide, water, and microbial cell mass. In the absence of oxygen, organic contaminants are metabolized to methane, limited amounts of carbon dioxide, and trace amounts of hydrogen gas. Another gas that is added is methane. It enhances degradation by co-metabolism in which as bacteria consume the methane, they produce enzymes that react with the organic contaminant and degrade it to harmless minerals.

Gasohol: A term for motor vehicle fuel comprising between 80%–90% unleaded gasoline and 10%–20% ethanol (see also Ethyl alcohol).

Gas oil: A petroleum distillate with a viscosity and boiling range between those of kerosene and lubricating oil.

GC-MS: Gas chromatography-mass spectrometry.

GC-TPH: GC detectable total petroleum hydrocarbons that is the sum of all GC-resolved and unresolved hydrocarbons. The resolvable hydrocarbons appear as peaks and the unresolvable hydrocarbons appear as the area between the lower baseline and the curve defining the base of resolvable peaks.

Geological time: The span of time that has passed since the creation of the earth and its components; a scale used to measure geological events millions of years ago.

Geometric isomers: Stereoisomers which differ in the geometry around either a carbon-carbon double bond or ring.

Geosphere: A term representing the solid earth, including soil, which supports most plant life.

Gibb's free energy: The energy of a system that is available to do work at constant temperature and pressure.

Girbotol process: A continuous, regenerative process to separate hydrogen sulfide, carbon dioxide, and other acid impurities from natural gas, refinery gas, etc., using mono-, di-, or triethanolamine as the reagent.

Graham's law: The rate of diffusion of a gas is inversely proportional to the square root of the molar mass.

Glycerol: A small molecule with three alcohol groups [HOCH$_2$CH(OH)CH$_2$OH]; basic building block of fats and oils.

$$\begin{array}{c} \text{HOCH}_2 \\ | \\ \text{HOCH}_2 \\ | \\ \text{HOCH}_2 \end{array}$$

Glycol-amine gas treating: A continuous, regenerative process to simultaneously dehydrate and remove acid gases from natural gas or refinery gas.

Grain alcohol: See Ethyl alcohol.

Gram equivalent weight (non-redox reaction): The mass in grams of a substance equivalent to 1 g-atom of hydrogen, 0.5 g-atom of oxygen, or 1 g-ion of the hydroxyl ion; can be determined by dividing the molecular weight by the number of hydrogen atoms or hydroxyl ions (or their equivalent) supplied or required by the molecule in a reaction.

Gram equivalent weight (redox reaction): The molecular weight in grams divided by the change in oxidation state.

Gravimetric analysis: A technique of quantitative analytical chemistry in which a desired constituent is efficiently recovered and weighed.

Gravity: See API gravity.

Gray clay treating: A fixed bed, usually fuller's earth, vapor-phase treating process to selectively polymerize unsaturated gum-forming constituents (diolefins) in thermally cracked gasoline.

Greenhouse effect: The warming of an atmosphere by its absorption of infrared radiation while shortwave radiation is allowed to pass through.

Greenhouse gas: Any gas whose absorption of solar radiation is responsible for the greenhouse effect, including carbon dioxide, ozone, methane, and the fluorocarbons.

Guard bed: A bed of an adsorbent (such as bauxite) that protects a catalyst bed by adsorbing species detrimental to the catalyst.

Guest molecule (or ion): An organic or inorganic ion or molecule that occupies a cavity, cleft, or pocket within the molecular structure of a host molecular entity and forms a complex with the host entity or that is trapped in a cavity within the crystal structure of a host.

Gulf HDS process: A fixed bed process for the catalytic hydrocracking of heavy stocks to lower-boiling distillates with accompanying desulfurization.

Gulfining: A catalytic hydrogen treating process for cracked and straight-run distillates and fuel oils, to reduce sulfur content; improve carbon residue, color, and general stability; and effect a slight increase in gravity.

Gum: An insoluble tacky semisolid material formed as a result of the storage instability and/or the thermal instability of petroleum and petroleum products.

Half-life (abbreviated to t$_{1/2}$): The time required to reduce the concentration of a chemical to 50% of its initial concentration; units are typically in hours or days; the term is commonly used in nuclear physics to describe how quickly (radioactive decay) unstable atoms undergo radioactive decay, or how long stable atoms survive the potential for radioactive decay.

Halide: An element from the halogen group, which include fluorine, chlorine, bromine, iodine, and astatine.

Halogen: Group 17 in the periodic table of the elements; these elements are the reactive nonmetals and are electronegative.

Halogenation: The addition of a halogen molecule to an alkene to produce an alkyl dihalide or alkyne to produce an alkyl tetrahalide.

Halo group (X–): A substituent which is one of the four halogens; fluorine (F), chlorine (Cl), bromine (Br), or iodine (I).

Hardness scale (Mhos scale): A measure of the ability of a substance to abrade or indent one another; the Mohs hardness is based on a scale from 1 to 10 units in which diamond, the hardest substance, is given a value of 10 Mohs and talc given a value of 0.5.

Hazardous waste: A potentially dangerous chemical substance that has been discarded, abandoned, neglected, released, or designated as a waste material, or one that may interact with other substances to pose a threat.

Haze: A term denoting decreased visibility due to the presence of particles.

Heat capacity (Cρ): The quantity of thermal energy needed to raise the temperature of an object by 1°C; the heat capacity is the product of mass of the object and its specific heat: Cρ = mass × specific heat.

Heating oil: See Fuel oil.

Heat of fusion (ΔH_{fus}): The amount of thermal energy required to melt one mole of the substance at the melting point; also termed as latent heat of fusion and expressed in kcal/mol or kJ/mol.

Heat of vaporization (ΔH_{vap}): The amount of thermal energy needed to convert one mole of a substance to vapor at boiling point; also known as latent heat of vaporization and expressed kcal/mol or kJ/mol.

Heat value: The amount of heat released per unit of mass or per unit of volume when a substance is completely burned. The heat power of solid and liquid fuels is expressed in calories per gram or in Btu per pound. For gases, this parameter is generally expressed in kilocalories per cubic meter or in Btu per cubic foot.

Heavy ends: The highest-boiling portion of a petroleum fraction; see also Light ends.

Heavy fuel oil: Fuel oil having a high density and viscosity; generally residual fuel oil such as No. 5 and No 6. fuel oil.

Heavy oil: Typically petroleum having an API gravity of less than 20°.

Heavy petroleum: See Heavy oil.

Henry's law: The relation between the partial pressure of a compound and the equilibrium concentration in the liquid through a proportionality constant known as the Henry's Law constant.

Henry's law constant: The concentration ratio between a compound in air (or vapor) and the concentration of the compound in water under equilibrium conditions.

Herbicide: A chemical that controls or destroys unwanted plants, weeds, or grasses.

Heteroatom compounds: Chemical compounds which contain nitrogen and/or oxygen and/or sulfur and /or metals bound within their molecular structure(s).

Heteroatoms: Elements other than carbon and hydrogen that are commonly found in organic molecules, such as nitrogen, oxygen, and the halogens.

Heterocyclic: An organic group or molecule containing rings with at least one non-carbon atom in the ring.

Heterogeneous: Varying in structure or composition at different locations in space.

Heterotroph: An organism that cannot synthesize its own food and is dependent on complex organic substances for nutrition.

Heterotrophic bacteria: Bacteria that utilize organic carbon as a source of energy; organisms that derive carbon from organic matter for cell growth.

Heterotrophs: Organisms or chemicals that obtain their carbon from other organisms.

HF alkylation: An alkylation process whereby olefins (C_3, C_4, C_5) are combined with isobutane in the presence of hydrofluoric acid catalyst.

Homogeneous: Having uniform structure or composition at all locations in space.

Homolog: A compound belonging to a series of compounds that differ by a repeating group; for example, propanol ($CH_3CH_2CH_2OH$), n-butanol ($CH_3CH_2CH_2CH_2OH$), and n-pentanol ($CH_3CH_2CH_2CH_2CH_2OH$) are homologs; they belong to the homologous series of alcohols: $CH_3(CH_2)_nOH$.

Homologous series: Compounds which differ only by the number of CH_2 units present.

Hopane: A pentacyclic *hydrocarbon* of the *triterpane* group believed to be derived primarily from bacteriohopanoids in bacterial membranes.

Hot filtration test: A test for the stability of a petroleum product.

Houdresid catalytic cracking: A continuous moving bed process for catalytically cracking reduced crude oil to produce high-octane gasoline and light distillate fuels.

Houdriflow catalytic cracking: A continuous moving bed catalytic cracking process employing an integrated single vessel for the reactor and regenerator kiln.

Houdriforming: A continuous catalytic reforming process for producing aromatic concentrates and high-octane gasoline from low-octane straight naphtha.

Houdry butane dehydrogenation: A catalytic process for dehydrogenating light hydrocarbons to their corresponding mono- or diolefins.

Houdry fixed bed catalytic cracking: A cyclic regenerable process for cracking of distillates.

Houdry hydrocracking: A catalytic process combining cracking and desulfurization in the presence of hydrogen.

Humic substances: Dark, complex, heterogeneous mixtures of organic materials that form in the geological systems of the earth from microbial transformations and chemical reactions that occur during the decay of organic biomolecules, polymers, and resides.

Hydration: The addition of a water molecule to a compound within an aerobic degradation pathway.

Hydration sphere: Shell of water molecules surrounding an ion in solution.

Hydraulic fracturing: The opening of fractures in a reservoir by high-pressure, high-volume injection of liquids through an injection well.

Hydrocarbon: One of a very large and diverse group of chemical compounds composed only of carbon and hydrogen; the largest source of hydrocarbons is petroleum crude oil; the principal constituents of crude oils and refined petroleum products.

Hydroconversion: A term often applied to hydrocracking.

Hydrocracking: A catalytic high-pressure, high-temperature process for the conversion of petroleum feedstocks in the presence of fresh and recycled hydrogen; carbon-carbon bonds are cleaved in addition to the removal of heteroatomic species.

Hydrocracking catalyst: A catalyst used for hydrocracking which typically contains separate hydrogenation and cracking functions.

Hydrodenitrogenation: The removal of nitrogen by hydrotreating.

Hydrodesulfurization: The removal of sulfur by hydrotreating.

Hydrofining: A fixed bed catalytic process to desulfurize and hydrogenate a wide range of charge stocks from gases through waxes.

Hydroforming: A process in which naphtha is passed over a catalyst at elevated temperatures and moderate pressures, in the presence of added hydrogen or hydrogen-containing gases, to form high-octane motor fuel or aromatics.

Hydrogen: A flammable, colorless gas with the chemical symbol formula H_2; the lightest and most abundant element in the universe. As petrochemicals are produced from hydrogen-containing hydrocarbons, hydrogen is involved in nearly all petrochemical processes; the most common application of hydrogen is as a reducing agent in catalytic hydrogenation and hydrorefining.

Hydrogen addition: An upgrading process in the presence of hydrogen, e.g., hydrocracking; see Hydrogenation.

Hydrogenation: The chemical addition of hydrogen to a material. In nondestructive hydrogenation, hydrogen is added to a molecule only if, and where, unsaturation with respect to hydrogen exists.

Hydrogen bond: A form of association between an electronegative atom and a hydrogen atom attached to a second, relatively electronegative atom; best considered as an electrostatic interaction, heightened by the small size of hydrogen, which permits close proximity of the interacting dipoles or charges.

Hydrogenation: A reaction where hydrogen is added across a double or triple bond, usually with the assistance of a catalyst; a process whereby an enzyme in certain microorganisms catalyzes the hydrolysis or reduction of a substrate by molecular hydrogen.

Hydrogenolysis: A reductive reaction in which a carbon-halogen bond is broken, and hydrogen replaces the halogen substituent.

Hydrogen transfer: The transfer of inherent hydrogen within the feedstock constituents and products during processing.

Hydrology: The scientific study of water.

Hydrolysis: A chemical transformation process in which a chemical reacts with water; in the process, a new carbon-oxygen bond is formed with oxygen derived from the water molecule, and a bond is cleaved within the chemical between carbon and some functional group.

Hydrophilic: Water loving; the capacity of a molecular entity or of a substituent to interact with polar solvents, in particular with water, or with other polar groups; hydrophilic molecules dissolve easily in water, but not in fats or oils.

Hydrophilic colloids: Generally, macromolecules, such as proteins and synthetic polymers, that are characterized by strong interaction with water resulting in spontaneous formation of colloids when they are placed in water.

Hydrophilicity: The tendency of a molecule to be solvated by water.

Hydrophobic: Fear of water; the tendency to repel water.

Hydrophobic colloids: Colloids that interact to a lesser extent with water and are stable because of their positive or negative electrical charges.

Hydrophobic effect: The attraction of nonionic, nonpolar compounds to surfaces that occurs due to the thermodynamic drive of these molecules to minimize interactions with water molecules.

Hydrophobic interaction: The tendency of hydrocarbons (or of lipophilic hydrocarbon-like groups in solutes) to form intermolecular aggregates in an aqueous medium, and analogous intramolecular interactions.

Hydroprocessing: A term often equally applied to hydrotreating and to hydrocracking; often collectively applied to both hydrotreating and to hydrocracking.

Hydropyrolysis: A short residence time high-temperature process using hydrogen.

Hydrorefining: A refining process for treating petroleum in the presence of catalysts and substantial quantities of hydrogen; the process includes desulfurization and the removal of substances that deactivate catalysts (such as nitrogen compounds). The process is used in the conversion of olefins to paraffins to reduce gum formation in gasoline and in other processes to upgrade the quality of a fraction.

Hydrosphere: The water areas of the earth; also called the aquasphere.

Hydrovisbreaking: A non-catalytic process, conducted under similar conditions to visbreaking, which involves treatment with hydrogen to reduce the viscosity of the feedstock and produce more stable products than is possible with visbreaking.

Hydroxylation: Addition of a hydroxyl group to a chlorinated aliphatic hydrocarbon.

Hydroxyl group: A functional group that has a hydrogen atom joined to an oxygen atom by a polar covalent bond (–OH).

Hydroxyl ion: One atom each of oxygen and hydrogen bonded into an ion (OH⁻) that carries a negative charge.

Hydroxyl radical: A radical consisting of one hydrogen atom and one oxygen atom; normally does not exist in a stable form.

Hyperforming: A catalytic hydrogenation process for improving the octane number of naphtha through removal of sulfur and nitrogen compounds.

Hypochlorite sweetening: The oxidation of mercaptans in a sour stock by agitation with aqueous, alkaline hypochlorite solution; used where avoidance of free-sulfur addition is desired, because of a stringent copper strip requirements and minimum expense is not the primary object.

Ideal gas law: A law which describes the relationship between pressure (P), temperature (T), volume (V), and moles of gas (n). This equation expresses behavior approached by real gases at low pressure and high temperature.

$$PV = nRT$$

Ignitability: A characteristic of liquids whose vapors are likely to ignite in the presence of ignition source; also characteristic of non-liquids that may catch fire from friction or contact with water and that burn vigorously.

Illuminating oil: Oil used for lighting purposes.

Immiscibility: The inability of two or more fluids to have complete mutual solubility; they coexist as separate phases.

Immiscible: Two or more fluids that do not have complete mutual solubility and coexist as separate phases.

Inactive ingredients (also called excipients): The additives present in a medicine, along with active ingredients, which are normally inactive/inert; these ingredients are not intended to have therapeutic effect but are added as preservatives, flavoring agents, coloring, sweeteners, and sorbents. See Active ingredients(s).

Incompatibility: The immiscibility of petroleum products and also of different crude oils which is often reflected in the formation of a separate phase after mixing and/or storage.

Indirect emissions: Emissions that are a consequence of the activities of the reporting entity, but occur at sources owned or controlled by another entity.

Infiltration rate: The time required for water at a given depth to soak into the ground.

Inhibition: The decrease in rate of reaction brought about by the addition of a substance (inhibitor), by virtue of its effect on the concentration of a reactant, catalyst, or reaction intermediate; a component having no effect reduces the effect of another component.

Inhibitor: A substance, the presence of which, in small amounts, in a petroleum product prevents or retards undesirable chemical changes from taking place in the product, or in the condition of the equipment in which the product is used.

Inhibitor sweetening: A treating process to sweeten gasoline of low mercaptan content, using a phenylenediamine type of inhibitor, air, and caustic.

Initial boiling point: The recorded temperature when the first drop of liquid falls from the end of the condenser.

Initial vapor pressure: The vapor pressure of a liquid of a specified temperature and 0% evaporated.

Inner-sphere adsorption complex: Sorption of an ion or molecule to a solid surface where waters of hydration are distorted during the sorption process and no water molecules remain interposed between the sorbate and sorbent.

Inoculum: A small amount of material (either liquid or solid) containing bacteria removed from a culture in order to start a new culture.

Inorganic: Pertaining to, or composed of, chemical compounds that are not organic, that is, chemical compounds that contain no carbon-hydrogen bonds; examples include chemicals with no carbon and those with carbon in non-hydrogen-linked forms.

Inorganic acid: An inorganic compound that elevates the hydrogen concentration in an aqueous solution; alphabetically, examples are:

Carbonic acid (HCO_3): An inorganic acid.

Hydrochloric acid (HCl): A highly corrosive, strong inorganic acid with many uses.

Hydrofluoric acid (HF): An inorganic acid that is highly reactive with silicate, glass, metals, and semi-metals.

Nitric acid (HNO_3): A highly corrosive and toxic strong inorganic acid.

Phosphoric acid: Not considered a strong inorganic acid; found in solid form as a mineral and has many industrial uses.

Sulfuric acid: A highly corrosive inorganic acid. It is soluble in water and widely used.

Inorganic base: An inorganic compound that elevates the hydroxide concentration in an aqueous solution; alphabetically, examples are:

Ammonium hydroxide (ammonia water): A solution of ammonia in water.

Calcium hydroxide (lime water): A weak base with many industrial uses.

Magnesium hydroxide: Referred to as brucite when found in its solid mineral form.

Sodium bicarbonate (baking soda): A mild alkali.

Sodium hydroxide (caustic soda): A strong inorganic base; used widely in industrial and laboratory environments.

Inorganic chemistry: The study of inorganic compounds, specifically the structure, reactions, catalysis, and mechanism of action.

Inorganic compound: A compound that consists of an ionic component (an element from the periodic table) and an anionic component; a compound that does not contain carbon chemically bound to hydrogen; carbonates, bicarbonates, carbides, and carbon oxides are considered inorganic compounds, even though they contain carbon; a large number of compounds occur naturally while others may be synthesized; in all cases, charge neutrality of the compound is key to the structure and properties of the compound.

Inorganic reaction chemistry: Inorganic chemical reactions fall into four broad categories: combination reactions, decomposition reactions, single displacement reactions, and double displacement reactions.

Inorganic salts: Inorganic salts are neutral, ionically bound molecules and do not affect the concentration of hydrogen in an aqueous solution.

Inorganic synthesis: The process of synthesizing inorganic chemical compounds, is used to produce many basic inorganic chemical compounds.

In situ: In its original place; unmoved; unexcavated; remaining in the subsurface.

In situ **bioremediation:** A process which treats the contaminated water or soil where it was found.

Instability: The inability of a petroleum product to exist for periods of time without change to the product.

Interfacial Tension: The net energy per unit area at the interface of two substances, such as oil and water or oil and air. The air/liquid interfacial tension is often referred to as surface tension; the SI units for *interfacial tension* are milli-Newtons per meter (mN/m). The higher the *interfacial tension*, the less attractive the two surfaces are to each other and the more size of the interface will be minimized. Low surface tensions can drive the spreading of one fluid on another. The surface tension of an oil, together its viscosity, affects the rate at which spilled oil will spread over a water surface or into the ground.

Intermediates: Petrochemical intermediates are generally produced by chemical conversion of primary petrochemicals to form more complicated derivative products; common petrochemical intermediate products include vinyl acetate for paint, paper, and textile coatings, vinyl chloride for polyvinyl chloride (PVC) resin manufacturing, ethylene glycol for polyester textile fibers and styrene which is used in rubber and plastic manufacturing.

Intermolecular forces: Force of attraction that exist between particles (atoms, molecules, ions) in a compound.

Internal Standard (IS): A pure analyte added to a sample extract in a known amount, which is used to measure the relative responses of other analytes and surrogates that are components of the same solution. The *internal standard* must be an analyte that is not a sample component.

Intramolecular: (i) Descriptive of any process that involves a transfer (of atoms, groups, electrons, etc.) or interactions (such as forces) between different parts of the same molecular entity; (ii) relating to a comparison between atoms or groups within the same molecular entity.

Intrinsic bioremediation: A type of bioremediation that manages the innate capabilities of naturally occurring microbes to degrade contaminants without taking any engineering steps to enhance the process.

Inversions: Conditions characterized by high atmospheric stability which limit the vertical circulation of air, resulting in air stagnation and the trapping of air pollutants in localized areas.

Iodine number: A measure of the iodine absorption by oil under standard conditions; used to indicate the quantity of unsaturated compounds present; also called iodine value.

Ionic bond: A chemical bond or link between two atoms due to an attraction between oppositely charged (positive-negative) ions.

Ionic bonding: Chemical bonding that results when one or more electrons from one atom or a group of atoms is transferred to another. Ionic bonding occurs between charged particles.

Ionic compounds: Compounds where two or more ions are held next to each other by electrical attraction.

Ionic liquids: An ionic liquid is a salt in the liquid state or a salt with a melting point lower than 100°C (212°F); variously called liquid electrolytes, ionic melts, ionic fluids, fused salts, liquid salts, or ionic glasses; powerful solvents and electrically conducting fluids (electrolytes).

Ionic radius: A measure of ion size in a crystal lattice for a given coordination number (CN); metal ions are smaller than their neutral atoms, and nonmetallic anions are larger than the atoms from which they are formed; ionic radii depend on the element, its charge, and its coordination number in the crystal lattice; ionic radii are expressed in angstrom units of length (Å).

Ionization energy: The ionization energy is the energy required to remove an electron completely from its atom, molecule, or radical.

Ionization potential: The energy required to remove a given electron from its atomic orbital; the values are given in electron volts (eV).

Irreversible reaction: A reaction in which the reactant(s) proceed to product(s), but there is no significant backward reaction:

$$nA + mB \rightarrow \text{Products}$$

In this reaction, the products do not recombine or change to form reactants in any appreciable amount.

Isobutylene: A four-carbon branched olefin, one of the four isomers of butane, with the chemical formula C_4H_8.

Isocracking: A hydrocracking process for conversion of hydrocarbons which operates at relatively low temperatures and pressures in the presence of hydrogen and a catalyst to produce more valuable, lower-boiling products.

Isoforming: A process in which olefinic naphtha is contacted with an alumina catalyst at high temperature and low pressure to produce isomers of higher octane number.

Iso-Kel process: A fixed bed, vapor-phase isomerization process using a precious metal catalyst and external hydrogen.

Isomate process: A continuous, non-regenerative process for isomerizing C_5–C_8 normal paraffin hydrocarbons, using aluminum chloride-hydrocarbon catalyst with anhydrous hydrochloric acid as a promoter.

Isomerate process: A fixed bed isomerization process to convert pentane, hexane, and heptane to high-octane blending stocks.

Isomerization: The conversion of a *normal* (straight-chain) paraffin hydrocarbon into an iso (branched-chain) paraffin hydrocarbon having the same atomic composition.

Isopach: A line on a map designating points of equal formation thickness.

Isomers: Compounds that have the same number and types of atoms—the same molecular formula—but differ in the structural formula, i.e., the manner in which the atoms are combined with each other.

Iso-plus Houdriforming: A combination process using a conventional Houdriformer operated at moderate severity, in conjunction with one of three possible alternatives including the use of an aromatic recovery unit or a thermal reformer; see Houdriforming.

Isotope: A variant of a chemical element which differs in the number of neutrons in the atom of the element; all isotopes of a given element have the same number of protons in each atom and different isotopes of a single element occupy the same position on the periodic table of the elements.

IUPAC: International Union of Pure and Applied Chemistry; the organization that establishes the system of nomenclature for organic and inorganic compounds using prefixes and suffixes, developed in the late 19th century.

Jet fuel: Fuel meeting the required properties for use in jet engines and aircraft turbine engines.

Kaolinite: A clay mineral formed by hydrothermal activity at the time of rock formation or by chemical weathering of rock with high feldspar content; usually associated with intrusive granite rock with high feldspar content.

Kata-condensed aromatic compounds: Compounds based on linear condensed aromatic hydrocarbon systems, e.g., anthracene and naphthacene (tetracene).

Kauri butanol number: A measurement of solvent strength for hydrocarbon solvents; the higher the kauri-butanol (KB) value, the stronger the solvency; the test method (ASTM D1133) is based on the principle that kauri resin is readily soluble in butyl alcohol but not in hydrocarbon solvents and the resin solution will tolerate only a certain amount of dilution and is reflected as a cloudiness when the resin starts to come out of solution; solvents such as toluene can be added in a greater amount (and thus have a higher KB value) than weaker solvents like hexane.

Kelvin: The SI unit of temperature. It is the temperature in degrees Celsius plus 273.15.

Kerogen: A complex carbonaceous (organic) material that occurs in sedimentary rock and shale; generally insoluble in common organic solvents.

Kerosene: A fraction of petroleum that was initially sought as an illuminant in lamps; a precursor to diesel fuel.

Ketone: An organic compound that contains a carbonyl group (R_1COR_2).

K-factor: See Characterization factor.

Kinematic viscosity: The ratio of viscosity to density, both measured at the same temperature.

Knock: The noise associated with self-ignition of a portion of the fuel-air mixture ahead of the advancing flame front.

Lag phase: The growth interval (adaption phase) between microbial inoculation and the start of the exponential growth phase during which there is little or no microbial growth.

Lamp burning: A test of burning oils in which the oil is burned in a standard lamp under specified conditions in order to observe the steadiness of the flame, the degree of encrustation of the wick, and the rate of consumption of the kerosene.

Lamp oil: See Kerosene.

Latex: A polymer of *cis*-1-4 isoprene; milky sap from the rubber tree *Hevea brasiliensis*.

Law: A system of rules that are enforced through social institutions to govern behavior; can be made by a collective legislature or by a single legislator, resulting in statutes, by the executive through decrees and regulations, or by judges through binding precedent; the formation of laws themselves may be influenced by a constitution (written or tacit) and the rights encoded therein; the law shapes politics, economics, history, and society in various ways and serves as a mediator of relations between people. See also Regulation.

Layer silicate clay: Clay minerals composed of planes of aluminum (Al^{3+}) or magnesium (Mg^{2+}) in octahedral coordination with oxygen and planes of silica (Si^{4+}) in tetrahedral coordination to oxygen. Substitution of Al^{3+} for Si^{4+} in the tetrahedral plane or substitution of Mg^{2+} or Fe^{2+} for Al^{3+} in the octahedral plane (isomorphic substitution) results in a permanent charge imbalance (i.e., structural charge) that must be satisfied through cation adsorption.

Leaded gasoline: Gasoline containing tetraethyl lead or other organometallic lead antiknock compounds.

Lean gas: The residual gas from the absorber after the condensable gasoline has been removed from the wet gas.

Lean oil: Absorption oil from which gasoline fractions have been removed; oil leaving the stripper in a natural-gasoline plant.

Leaving group: An atom or group (charged or uncharged) that becomes detached from an atom in what is considered to be the residual or main part of the substrate in a specified reaction.

Le Chatelier's principle: The principle that states that a system at equilibrium will oppose any change in the equilibrium conditions.

Lewis acid: A chemical species which can accept an electron pair from a base.

Lewis base: A chemical species which can donate an electron pair.

Light ends: The lower-boiling components of a mixture of hydrocarbons; see also Heavy ends, Light hydrocarbons.

Light hydrocarbons: Hydrocarbons with molecular weights less than that of heptane (C_7H_{16}).

Light oil: The products distilled or processed from crude oil up to, but not including, the first lubricating-oil distillate.

Light petroleum: Petroleum having an API gravity greater than 20°.

Lignin: A complex amorphous polymer in the secondary cell wall (middle lamella) of woody plant cells that cements or naturally binds cell walls to help make them rigid; highly resistant to decomposition by chemical or enzymatic action; also acts as support for cellulose fibers.

Ligroine (Ligroin): A saturated petroleum naphtha boiling in the range of 20°C–135°C (68°C–275°F) and suitable for general use as a solvent; also called benzine or petroleum ether.

Limiting reactant: The reactant that is present in the smallest stoichiometric amount and which determines the maximum extent to which a reaction can proceed; if the reaction is 100% complete then all of the limiting reactant is consumed and the reaction can proceed no further.

Limnology: The branch of science dealing with characteristics of freshwater, including biological properties as well as chemical and physical properties.

Linde copper sweetening: A process for treating gasoline and distillates with a slurry of clay and cupric chloride.

Lipophilic: F-loving; applied to molecular entities (or parts of molecular entities) tending to dissolve in fat-like (e.g., hydrocarbon) solvents.

Lipophilicity: The affinity of a molecule or a moiety (portion of a molecular structure) for a lipophilic (fat soluble) environment. It is commonly measured by its distribution behavior in a biphasic system, either liquid-liquid (e.g., partition coefficient in octanol/water).

Liquid petrolatum: See White oil.

Liquefied petroleum gas: Propane, butane, or mixtures thereof, gaseous at atmospheric temperature and pressure, held in the liquid state by pressure to facilitate storage, transport, and handling.

Liquid chromatography: A chromatographic technique that employs a liquid mobile phase.

Liquid/liquid extraction: An extraction technique in which one liquid is shaken with or contacted by an extraction solvent to transfer molecules of interest into the solvent phase.

Liquid sulfur dioxide-benzene process: A mixed-solvent process for treating lubricating-oil stocks to improve viscosity index; also used for dewaxing.

Lithosphere: The part of the geosphere consisting of the outer mantle and the crust that is directly involved with environmental processes through contact with the atmosphere, the hydrosphere, and living things; varies from (approximately) 40 to 60 miles in thickness; also called the terrestrial biosphere.

Loading rate: The amount of a chemical that can be absorbed on soil on a per volume of soil basis.

LTU: Land Treatment Unit; a physically delimited area where contaminated land is treated to remove/minimize contaminants and where parameters such as moisture, pH, salinity, temperature, and nutrient content can be controlled.

Lube: See Lubricating oil.

Lube cut: A fraction of crude oil of suitable boiling range and viscosity to yield lubricating oil when completely refined; also referred to as lube oil distillates or lube stock.

Lubricating oil: A fluid lubricant used to reduce friction between bearing surfaces.

Macromolecule: A large molecule of high molecular mass composed of more than 100 repeated monomers (single chemical units of lower relative mass); a large complex molecule formed from many simpler molecules.

Mahogany acids: Oil-soluble sulfonic acids formed by the action of sulfuric acid on petroleum distillates. They may be converted to their sodium soaps (mahogany soaps) and extracted from the oil with alcohol for use in the manufacture of soluble oils, rust preventives, and special greases. The calcium and barium soaps of these acids are used as detergent additives in motor oils; see also Brown acids and Sulfonic acids.

Maltenes: That fraction of petroleum that is soluble in, for example, pentane or heptane; deasphaltened oil; also the term arbitrarily assigned to the pentane-soluble portion of petroleum that is relatively high boiling (>300°C, 760 mm) (see also Petrolenes).

Marine engine oil: Oil used as a crankcase oil in marine engines.

Marine gasoline: Fuel for motors in marine service.

Marine sediment: The organic biomass from which petroleum is derived.

Masking: Occurs when two components have opposite, cancelling effects such that no effect is observed from the combination.

Mass number: The number of protons plus the number of neutrons in the nucleus of an atom.

Matter: Any substance that has inertia and occupies physical space; can exist as solid, liquid, gas, plasma, or foam.

Mayonnaise: Low-temperature sludge; a black, brown, or gray deposit having a soft, mayonnaise-like consistency; not recommended as a food additive!

Measurement: A description of a property of a system by means of a set of specified rules that maps the property on to a scale of specified values by direct or mathematical comparison with specified references.

Mechanical explosion: An explosion due to the sudden failure of a vessel containing a nonreactive gas at a high pressure.

Medicinal oil: Highly refined, colorless, tasteless, and odorless petroleum oil used as a medicine in the nature of an internal lubricant; sometimes called liquid paraffin.

MEK (methyl ethyl ketone): A colorless liquid ($CH_3COCH_2CH_3$) used as a solvent; as a chemical intermediate; and in the manufacture of lacquers, celluloid, and varnish removers.

MEK deoiling: A wax-deoiling process in which the solvent is generally a mixture of methyl ethyl ketone and toluene.

MEK dewaxing: A continuous solvent dewaxing process in which the solvent is generally a mixture of methyl ethyl ketone and toluene.

Melting point: The temperature when matter is converted from solid to liquid.

Membrane technology: Gas separation processes utilizing membranes that permit different components of a gas to diffuse through the membrane at significantly different rates.

Mesosphere: The portion of the atmosphere of the earth where molecules exist as charged ions caused by interaction of gas molecules with intense ultraviolet (UV) light.

Metabolic byproduct: A product of the reaction between an electron donor and an electron acceptor; metabolic byproducts include volatile fatty acids, daughter products of chlorinated aliphatic hydrocarbons, methane, and chloride.

Metabolism: The physical and chemical processes by which foodstuffs are synthesized into complex elements, complex substances are transformed into simple ones, and energy is made available for use by an organism; thus, all biochemical reactions of a cell or tissue, both synthetic and degradative, are included; the sum of all of the enzyme-catalyzed reactions in living cells that transform organic molecules into simpler compounds used in biosynthesis of cellular components or in extraction of energy used in cellular processes.

Metabolize: A product of metabolism.

Metal (oxyhydr)oxide: Minerals composed of various structural arrangements of metal cations— principally Al^{3+}, Fe^{3+}, and Mn^{4+}—in octahedral coordination with oxygen or hydroxide anions. These minerals are dissolution byproducts of mineral weathering and they are often found as coatings on layer silicates and other soil particles.

Methanogens: Strictly anaerobic archaebacteria, able to use only a very limited spectrum of substrates (for example, molecular hydrogen, formate, methanol, methylamine, carbon monoxide, or acetate) as electron donors for the reduction of carbon dioxide to methane.

Methanogenic: The formation of methane by certain anaerobic bacteria (methanogens) during the process of anaerobic fermentation.

Methanol: See Methyl alcohol.

Methyl: A group ($-CH_3$) derived from methane; for example, CH_3Cl is methyl chloride (systematic name: chloromethane) and CH_3OH is methyl alcohol (systematic name: methanol).

Methyl t-butyl ether: An ether added to gasoline to improve its octane rating and to decrease gaseous emissions; see Oxygenate.

Mercapsol process: A regenerative process for extracting mercaptans, utilizing aqueous sodium (or potassium) hydroxide containing mixed cresols as solubility promoters.

Mercaptans: Organic compounds having the general formula R-SH.

Methyl alcohol (methanol; wood alcohol): A colorless, volatile, inflammable, and poisonous alcohol (CH_3OH) traditionally formed by destructive distillation of wood or, more recently, as a result of synthetic distillation in chemical plants.

Methyl ethyl ketone: See MEK.

Micelle: The name given to the structural entity by which asphaltene constituents are dispersed in petroleum.

Microcarbon residue: The carbon residue determined using a themogravimetric method. See also Carbon residue.

Microclimate: A highly localized climatic conditions; the climate that organisms and objects on the surface are exposed to being close to ground, under rocks, and surrounded by vegetation and it is often quite different from the surrounding macroclimate.

Microcosm: A diminutive, representative system analogous to a larger system in composition, development, or configuration.

Microorganism (micro-organism): An organism of microscopic size that is capable of growth and reproduction through biodegradation of food sources, which can include hazardous contaminants; microscopic organisms including bacteria, yeasts, filamentous fungi, algae, and protozoa; a living organism too small to be seen with the naked eye; includes bacteria, fungi, protozoans, microscopic algae, and viruses.

Microbe: The shortened term for microorganism.

Microcrystalline wax: Wax extracted from certain petroleum residua and having a finer and less apparent crystalline structure than paraffin wax.

Microemulsion: A stable, finely dispersed mixture of oil, water, and chemicals (surfactants and alcohols).

Mid-boiling point: The temperature at which approximately 50% of a material has distilled under specific conditions.

Middle distillate: Distillate boiling between the kerosene and lubricating oil fractions.

Middle-phase microemulsion: A microemulsion phase containing a high concentration of both oil and water that, when viewed in a test tube, resides in the middle with the oil phase above it and the water phase below it.

Mineral hydrocarbons: Petroleum hydrocarbons, considered *mineral* because they come from the earth rather than from plants or animals.

Mineralization: The biological process of complete breakdown of organic compounds, whereby organic materials are converted to inorganic products (e.g., the conversion of hydrocarbons to carbon dioxide and water); the release of inorganic chemicals from organic matter in the process of aerobic or anaerobic decay.

Mineral oil: The older term for petroleum; the term was introduced in the 19th century as a means of differentiating petroleum (rock oil) from whale oil which, at the time, was the predominant illuminant for oil lamps.

Minerals: Naturally occurring inorganic solids with well-defined crystalline structures.

Mineral seal oil: A distillate fraction boiling between kerosene and gas oil.

Mineral wax: Yellow to dark brown, solid substances that occur naturally and are composed largely of paraffins; usually found associated with considerable mineral matter, as a filling in veins and fissures or as an interstitial material in porous rocks.

Miscibility: An equilibrium condition, achieved after mixing two or more fluids, which is characterized by the absence of interfaces between the fluids: (i) *first-contact miscibility:* miscibility in the usual sense, whereby two fluids can be mixed in all proportions without any interfaces forming. Example: At room temperature and pressure, ethyl alcohol and water are first-contact miscible. (ii) *multiple-contact miscibility (dynamic miscibility):* miscibility that is developed by repeated enrichment of one fluid phase with components from a second fluid phase with which it comes into contact. (iii) *minimum miscibility* pressure: the minimum pressure above which two fluids become miscible at a given temperature, or can become miscible, by dynamic processes.

Mist: Liquid particles.

Mixed-phase cracking: The thermal decomposition of higher-boiling hydrocarbons to gasoline components.

Mixed waste: Any combination of waste types with different properties or any waste that contains both hazardous waste and source, specially nuclear, or byproduct material; as defined by the U.S. EPA, mixed waste contains both hazardous waste (as defined by RCRA and its amendments) and radioactive waste (as defined by AEA and its amendments).

Modified naphtha insolubles (MNI): An insoluble fraction obtained by adding naphtha to petroleum; usually the naphtha is modified by adding paraffin constituents; the fraction might be equated to asphaltenes *if* the naphtha is equivalent to n-heptane, but usually it is not.

Modulus of elasticity: The stress required to produce unit strain to cause a change of length (Young's modulus), or a twist or shear (shear modulus), or a change of volume (bulk modulus); expressed as dynes/cm^2.

Moiety: A term generally used to signify part of a molecule, e.g., in an ester R^1COOR^2, the alcohol moiety is R^2O.

Molality (m): The gram moles of solute divided by the kilograms of solvent.

Molar: A term expressing molarity, the number of moles of solute per liters of solution.

Molarity (M): The gram moles of solute divided by the liters of solution.

Mole: A collection of 6.022×10^{23} number of objects. Usually used to mean molecules.

Molecular sieve: A synthetic zeolite mineral having pores of uniform size; it is capable of separating molecules, on the basis of their size, structure, or both, by absorption or sieving.

Mole fraction: The number of moles of a particular substance expressed as a fraction of the total number of moles.

Molecular weight: The mass of one mole of molecules of a substance.

Molecule: The smallest unit in a chemical element or compound that contains the chemical properties of the element or compound.

Mole fraction: The number of moles of a component of a mixture divided by the total number of moles in the mixture.

Monoaromatic: Aromatic hydrocarbons containing a single benzene ring.

Monosaccharide: A simple sugar such as fructose or glucose that cannot be decomposed by hydrolysis; colorless crystalline substances with a sweet taste that have the same general formula, $C_nH_{2n}O_n$.

Motor octane method: A test for determining the knock rating of fuels for use in spark-ignition engines; see also Research Octane Method.

Moving bed catalytic cracking: A cracking process in which the catalyst is continuously cycled between the reactor and the regenerator.

MSDS: Material safety data sheet.

MTBE: See Methyl t-butyl ether.

MTBE (Methyl t-butyl ether): Is a fuel additive which has been used in the United States since 1979. Its use began as a replacement for lead in gasoline because of health hazards associated with lead. MTBE has distinctive physical properties that result in it being highly soluble, persistent in the environment, and able to migrate through the ground. Environmental regulations have required the monitoring and cleanup of MTBE at petroleum-contaminated sites since February, 1990; the program continues to monitor studies focusing on the potential health effects of MTBE and other fuel additives.

Naft: Pre-Christian era (Greek) term for naphtha.

Napalm: A thickened gasoline used as an incendiary medium that adheres to the surface it strikes.

Naphtha: A generic term applied to refined, partly refined, or unrefined petroleum products and liquid products of natural gas, the majority of which distills below 240°C (464°F); the volatile fraction of petroleum which is used as a solvent or as a precursor to gasoline.

Naphthenes: Cycloparaffins; any of various volatile, often flammable, liquid hydrocarbon mixtures characterized by saturated ring structures that are used chiefly as solvents and diluents.

Native asphalt: See Bitumen.

Native fauna: The native and indigenous animal of an area.

Native flora: The native and indigenous plant life of an area.

Natural asphalt: See Bitumen.

Natural gas: The naturally occurring gaseous constituents that are found in many petroleum reservoirs; also there are those reservoirs in which natural gas may be the sole occupant.

Natural gas liquids (NGL): The hydrocarbon liquids that condense during the processing of hydrocarbon gases that are produced from oil or gas reservoir; see also Natural gasoline.

Natural gasoline: A mixture of liquid hydrocarbons extracted from natural gas suitable for blending with refinery gasoline.

Natural gasoline plant: A plant for the extraction of fluid hydrocarbon, such as gasoline and liquefied petroleum gas, from natural gas.

Natural organic matter (NOM): An inherently complex mixture of polyfunctional organic molecules that occurs naturally in the environment and is typically derived from the decay of floral and faunal remains; although they do occur naturally, the fossil fuels (coal, crude oil, and natal gas) are usually not included in the term *natural organic matter.*

NCP: National Contingency Plan—also called the National Oil and Hazardous Substances Pollution Contingency Plan; provides a comprehensive system of accident reporting, spill containment, and cleanup, and established response headquarters (National Response Team and Regional Response Teams).

Nernst Equation: An equation that is used to account for the effect of different activities upon electrode potential:

$$E = E^0 + \frac{2.303RT}{nF} \log \frac{\text{Reactants}}{\text{Products}} = E^0 + \frac{0.0591}{n} \log \frac{\text{Reactants}}{\text{Products}}$$

Neutralization: A process for reducing the acidity or alkalinity of a waste stream by mixing acids and bases to produce a neutral solution; also known as pH adjustment.

Neutral oil: A distillate lubricating oil with viscosity usually not above 200 s at 100°F.

Neutralization number: The weight, in milligrams, of potassium hydroxide needed to neutralize the acid in 1 g of oil; an indication of the acidity of an oil.

Nitrate enhancement: A process in which a solution of nitrate is sometimes added to groundwater to enhance anaerobic biodegradation.

Non-asphaltic road oil: Any of the nonhardening petroleum distillates or residual oils used as dust layers. They have sufficiently low viscosity to be applied without heating and, together with asphaltic road oils, are sometimes referred to as dust palliatives.

Nonassociated gas: Natural gas found in reservoirs that do not contain crude oil at the original pressure and temperature conditions.

Nonionic surfactant: A surfactant molecule containing no ionic charge.

Non-Newtonian: A fluid that exhibits a change of viscosity with flow rate.

Nonpoint source pollution: Pollution that does not originate from a specific source. Examples of nonpoint sources of pollution include the following: (i) sediments from construction, forestry operations, and agricultural lands; (ii) bacteria and microorganisms from failing septic systems and pet wastes; (iii) nutrients from fertilizers and yard debris; (iv) pesticides from agricultural areas, golf courses, athletic fields and residential yards, oil, grease, anti-freeze, and metals washed from roads, parking lots and driveways; (v) toxic chemicals and cleaners that were not disposed of correctly; and (vi) litter thrown onto streets, sidewalks and beaches, or directly into the water by individuals. See Point Source Pollution.

Normality (N): The gram equivalents of solute divided by the liters of solution.

NOx: Oxides of nitrogen.

Nucleophile: A chemical reagent that reacts by forming covalent bonds with electronegative atoms and compounds.

Nuclide: A nucleus rather than to an atom—isotope (the older term) it is better known than the term nuclide, and is still sometimes used in contexts where the use of the term nuclide might be more appropriate; identical nuclei belong to one nuclide, for example, each nucleus of the carbon-13 nuclide is composed of six protons and seven neutrons.

Number 1 Fuel oil (No. 1 Fuel oil): Very similar to kerosene and is used in burners where vapor-ization before burning is usually required and a clean flame is specified.

Number 2 Fuel oil (No. 2 Fuel oil): Also called domestic heating oil; has properties similar to diesel fuel and heavy jet fuel; used in burners where complete vaporization is not required before burning.

Number 4 Fuel oil (No. 4 Fuel oil): A light industrial heating oil and is used where preheating is not required for handling or burning; there are two grades of No. 4 fuel oil, differing in safety (flash point) and flow (viscosity) properties.

Number 5 Fuel oil (No. 5 Fuel oil): A heavy industrial fuel oil which requires preheating before burning.

Number 6 Fuel oil (No. 6 Fuel oil): A heavy fuel oil and is more commonly known as Bunker C oil when it is used to fuel ocean-going vessels; preheating is always required for burning this oil.

Nutrients: Major elements (for example, nitrogen and phosphorus) and trace elements (including sulfur, potassium, calcium, and magnesium) that are essential for the growth of organisms.

Oceanography: The science of the ocean and its physical and chemical characteristics.

Octane: A flammable liquid (C_8H_{18}) found in petroleum and natural gas; there are 18 different octane isomers which have different structural formulas but share the molecular formula C_8H_{18}; used as a fuel and as a raw material for building more complex organic molecules.

Octane barrel yield: A measure used to evaluate fluid catalytic cracking processes; defined as (RON + MON)/two times the gasoline yield, where RON is the research octane number and MON is the motor octane number.

Octane number: A number indicating the antiknock characteristics of gasoline.

Octanol-water partition coefficient (K_{ow}): The equilibrium ratio of a chemical's concentration in octanol (an alcoholic compound) to its concentration in the aqueous phase of a two-phase octanol water system, typically expressed in log units (log K_{ow}); K_{ow} provides an indication of a chemical's solubility in fats (lipophilicity), its tendency to bioconcentrate in aquatic organisms, or sorb to soil or sediment.

Oils fraction: That portion of the maltenes that is not adsorbed by a surface-active material such as clay or alumina.

Oil sand: See Tar sand.

Oil shale: A fine-grained impervious sedimentary rock which contains an organic material called kerogen.

Olefin: Synonymous with *alkene*; a hydrocarbon characterized by having at least one carbon-carbon double bond (>C=C<); specifically, any of a series of open-chain hydrocarbons such as ethylene.

Oleophilic: Oil seeking or oil loving (e.g., nutrients that stick to or dissolve in oil).

Order of reaction: A chemical rate process occurring in systems for which concentration changes (and hence the rate of reaction) are not themselves measurable, provided it is possible to measure a chemical flux.

Organic: Compounds that contain carbon chemically bound to hydrogen; often containing other elements (particularly O, N, halogens, or S); chemical compounds based on carbon that also contain hydrogen, with or without oxygen, nitrogen, and other elements.

Organic carbon (soil) partition coefficient (K_{oc}): The proportion of a chemical sorbed to the solid phase, at equilibrium in a two-phase, water/soil or water/sediment system expressed on an organic carbon basis; chemicals with higher K_oc values are more strongly sorbed to organic carbon and, therefore, tend to be less mobile in the environment.

Organic chemistry: The study of compounds that contain carbon chemically bound to hydrogen, including synthesis, identification, modeling, and reactions of those compounds.

Organic liquid nutrient injection: An enhanced bioremediation process in which an organic liquid, which can be naturally degraded and fermented in the subsurface to result in the generation of hydrogen. The most commonly added for enhanced anaerobic bioremediation include lactate, molasses, hydrogen release compounds (HRCs¨), and vegetable oils.

Organochlorine compounds (chlorinated hydrocarbons): Organic pesticides that contain chlorine, carbon, and hydrogen (such as DDT); these pesticides affect the central nervous system.

Organometallic compounds: Compounds that include carbon atoms directly bonded to a metal ion.

Organophosphorus compound: A compound containing phosphorus and carbon; many pesticides and most nerve agents are organophosphorus compounds, such as malathion.

Osmotic potential: Expressed as a negative value (or zero), indicates the ability of the soil to dissolve salts and organic molecules; the reduction of soil water osmotic potential is caused by the presence of dissolved solutes.

Outer-sphere adsorption complex: Sorption of an ion or molecule to a solid surface where waters of hydration are interposed between the sorbate and sorbent.

Oven dry: The weight of a soil after all water has been removed by heating in an oven at a specified temperature (usually in excess of 100°C, 212°F) for water; temperatures will vary if other solvents have been used.

Overhead: That portion of the feedstock which is vaporized and removed during distillation.

Oxidation: The transfer of electrons away from a compound, such as an organic contaminant; the coupling of oxidation to reduction (see below) usually supplies energy that microorganisms use for growth and reproduction. Often (but not always), oxidation results in the addition of an oxygen atom and/or the loss of a hydrogen atom.

Oxidation number: A number assigned to each atom to help keep track of the electrons during a redox reaction.

Oxidation reaction: A reaction where a substance loses electrons.

Oxidation-reduction reactions (redox reactions): Reactions that involve oxidation of one reactant and reduction of another; a reaction involving the transfer of electrons.

Oxidize: The transfer of electrons away from a compound, such as an organic contaminant. The coupling of oxidation to reduction (see below) usually supplies energy that microorganisms use for growth and reproduction. Often (but not always), oxidation results in the addition of an oxygen atom and/or the loss of a hydrogen atom.

Oxygen enhancement with hydrogen peroxide: An alternative process to pumping oxygen gas into groundwater involves injecting a dilute solution of hydrogen peroxide. Its chemical formula is H_2O_2, and it easily releases the extra oxygen atom to form water and free oxygen. This circulates through the contaminated groundwater zone to enhance the rate of aerobic biodegradation of organic contaminants by naturally occurring microbes. A solid peroxide product (e.g., oxygen-releasing compound (ORC¨)) can also be used to increase the rate of biodegradation.

Ozone (O_3): A form of oxygen containing three atoms instead of the common two (O_2); formed by high-energy ultraviolet radiation reacting with oxygen.

PAHs: Polycyclic aromatic hydrocarbons. Alkylated *PAHs* are *alkyl group* derivatives of the parent *PAHs*. The five target alkylated *PAHs* referred to in this report are the alkylated naphthalene, phenanthrene, dibenzothiophene, fluorene, and chrysene series.

Pale oil: A lubricating oil or a process oil refined until its color, by transmitted light, is straw to pale yellow.

Paraffinum liquidum: See Liquid petrolatum.

Paraffin: An alkane.

Paraffin-base crude oil: Crude oil with a high content of waxes and lubricating oil fractions, having small amounts of naphthenes or asphalts and low in sulfur, nitrogen, and oxygen.

Paraffin wax: The colorless, translucent, highly crystalline material obtained from the light lubricating fractions of paraffin crude oils (wax distillates).

Particulate matter (particulates): Particles in the atmosphere or on a gas stream that may be organic or inorganic and originate from a wide variety of sources and processes.

Partition coefficient: A partition coefficient is used describe how a solute is distributed between two immiscible solvents; used in environmental science as a measure of a hydrophobicity of a solute and a proxy for transportation of a chemical through an ecosystem.

Partitioning: The distribution of a solute, S, between two immiscible solvents (such as aqueous phase and organic phase); important aspect of the transportation of a chemical into, through, and out of an ecosystem.

Partitioning equilibrium: The equilibrium distribution of a chemical that is established between the phases; the distribution of a chemical between the different phases.

Partition ratio, K: The ratio of total analytical concentration of a solute in the stationary phase, CS, to its concentration in the mobile phase, CM.

Pathogen: An organism that causes disease (e.g., some bacteria or viruses).

Penex process: A continuous, non-regenerative process for isomerization of C_5 and/or C_6 fractions in the presence of hydrogen (from reforming) and a platinum catalyst.

Pentafining: A pentane isomerization process using a regenerable platinum catalyst on a silica-alumina support and requiring outside hydrogen.

Pepper sludge: The fine particles of sludge produced in acid treating which may remain in suspension.

Percentage excess: The excess of a reactant t based on the quantity of excess reactant above the amount required to react with the total quantity of limiting reactant.

Percent conversion: The percentage of any reactant that has been converted to products.

Perfluorocarbon (PFC): A derivative of hydrocarbons in which all of the hydrogens have been replaced by fluorine.

Peri-condensed aromatic compounds: Compounds based on angular condensed aromatic hydrocarbon systems, e.g., phenanthrene, chrysene, picene, etc.

Periodic table: Grouping of the known elements by their number of protons; there are many other trends such as size of elements and electronegativity that are easily expressed in terms of the periodic table.

Permeability: The ease of flow of the water through the rock.

Petrol: A term commonly used in some countries for gasoline.

Petrolatum: A semisolid product, ranging from white to yellow in color, produced during refining of residual stocks; see Petroleum jelly.

Petrolenes: The term applied to that part of the pentane-soluble or heptane-soluble material that is low boiling (<300°C, <570°F, 760 mm) and can be distilled without thermal decomposition (see also Maltenes).

Petroleum (crude oil): A naturally occurring mixture of gaseous, liquid, and solid hydrocarbon compounds usually found trapped deep underground beneath impermeable cap rock and above a lower dome of sedimentary rock such as shale; most petroleum reservoirs occur in sedimentary rocks of marine, deltaic, or estuarine origin.

Petroleum asphalt: See Asphalt.

Petroleum ether: See Ligroine.

Petroleum jelly: A translucent, yellowish to amber or white, hydrocarbon substance (melting point: 38°C–54°C) having almost no odor or taste, derived from petroleum and used principally in medicine and pharmacy as a protective dressing and as a substitute for fats in ointments and cosmetics; also used in many types of polishes and in lubricating greases, rust preventives, and modeling clay; obtained by dewaxing heavy lubricating-oil stocks.

Petroleum refinery: See Refinery.

Petroleum refining: A complex sequence of events that result in the production of a variety of products.

Petroleum sulfonate: A surfactant used in chemical flooding prepared by sulfonating selected crude oil fractions.

Petroporphyrins: See Porphyrins.

Permeability: The capability of the soil to allow water or air movement through it. The quality of the soil that enables water to move downward through the profile, measured as the number of inches per hour that water moves downward through the saturated soil.

Permeable reactive barrier (PRB): A subsurface emplacement of reactive materials through which a dissolved contaminant plume must move as it flows, typically under natural gradient and treated water exits the other side of the permeable reactive barrier.

Pesticide: A chemical that is designed and produced to control for pest control, including weed control.

pH: A measure of the acidity or basicity of a solution; the negative logarithm (base 10) of the hydrogen ion concentration in gram ions per liter; a number between 0 and 14 that describes the acidity of an aqueous solution; mathematically the pH is equal to the negative logarithm of the concentration of H_3O^+ in solution.

pH adjustment: Neutralization.

Phase: A separate fluid that coexists with other fluids; gas, oil, water, and other stable fluids such as microemulsions are all called phases in EOR research.

Phase behavior: The tendency of a fluid system to form phases as a result of changing temperature, pressure, or the bulk composition of the fluids or of individual fluid phases.

Phase diagram: A graph of phase behavior. In chemical flooding, a graph showing the relative volume of oil, brine, and sometimes one or more microemulsion phases. In carbon dioxide flooding, conditions for formation of various liquid, vapor, and solid phases.

Phase properties: Types of fluids, compositions, densities, viscosities, and relative amounts of oil, microemulsion, or solvent, and water formed when a micellar fluid (surfactant slug) or miscible solvent (e.g., CO_2) is mixed with oil.

Phase separation: The formation of a separate phase that is usually the prelude to coke formation during a thermal process; the formation of a separate phase as a result of the instability/incompatibility of petroleum and petroleum products.

Phenol: A molecule containing a benzene ring that has a hydroxyl group substituted for a ring hydrogen.

Phenyl: A molecular group or fragment formed by abstracting or substituting one of the hydrogen atoms attached to a benzene ring.

Phosphoric acid polymerization: A process using a phosphoric acid catalyst to convert propene, butene, or both, to gasoline or petrochemical polymers.

Photic zone: The upper layer within bodies of water reaching down to about 200 m, where sunlight penetrates and promotes the production of photosynthesis; the richest and most diverse area of the ocean.

Photocatalysis: The acceleration of a photoreaction in the presence of a catalyst in which light is absorbed by a substrate that is typically adsorbed on a (solid) catalyst.

Photocatalyst: A material that can absorb light, producing electron–hole pairs that enable chemical transformations of the reaction participants and regenerate its chemical composition after each cycle of such interactions.

Phototrophs: Organisms or chemicals that utilize light energy from photosynthesis.

Physical change: Refers to the change that occurs when a material changes from one physical state to another without formation of intermediate substances of different composition in the process, such as the change from gas to liquid.

Phytodegradation: The process in which some plant species can metabolize VOC contaminants. The resulting metabolic products include trichloroethanol, trichloroacetic acid, and dichloracetic acid; mineralization products are probably incorporated into insoluble products such as components of plant cell walls.

Phytovolatilization: The process in which VOCs are taken up by plants and discharged into the atmosphere during transpiration.

PINA analysis: A method of analysis for paraffins, isoparaffins, naphthenes, and aromatics.

PIONA analysis: A method of analysis for paraffins, isoparaffins, olefins, naphthenes, and aromatics.

Pipe still: A still in which heat is applied to the oil while being pumped through a coil or pipe arranged in a suitable firebox.

Pipestill gas: The most volatile fraction that contains most of the gases that are generally dissolved in the crude. Also known as pipestill light ends.

Pipestill light ends: See Pipestill gas.

Pitch: The nonvolatile, brown to black, semisolid to solid viscous product from the destructive distillation of many bituminous or other organic materials, especially coal.

Platforming: A reforming process using a platinum-containing catalyst on an alumina base.

PM_{10}: Particulate matter below 10 microns in diameter; this corresponds to the particles inhalable into the human respiratory system, and its measurement uses a size-selective inlet.

$PM_{2.5}$: Particulate matter below 2.5 microns in diameter; this is closer to, but slightly finer than, the definitions of respirable dust that have been used for many years in industrial hygiene to identify dusts which will penetrate the lungs.

PNA: A polynuclear aromatic compound.

PNA analysis: A method of analysis for paraffins, naphthenes, and aromatics.

pOH: A measure of the basicity of a solution; the negative log of the concentration of the hydroxide ions.

Point emissions: Emissions that occur through confined air streams as found in stacks, ducts, or pipes.

Point source pollution: Any single identifiable source of pollution from which pollutants are discharged, such as a pipe. Examples of point sources include: (i) discharges from wastewater treatment plants, (ii) operational wastes from industries, and (iii) combined sewer outfalls. See Nonpoint Source Pollution.

Polar aromatics: Resins; the constituents of petroleum that are predominantly aromatic in character and contain polar (nitrogen, oxygen, and sulfur) functions in their molecular structure(s).

Polar compound: An organic compound with distinct regions of positive and negative charge. *Polar compounds* include alcohols, such as sterols, and some *aromatics*, such as monoaromatic-steroids. Because of their polarity, these compounds are more soluble in polar solvents, including water, compared to nonpolar compounds of similar molecular structure.

Pollutant: Either (i) a nonindigenous chemical that is present in the environment or (ii) an indigenous chemical that is present in the environment in greater than the natural concentration. Both types of pollutants are the result of human activity and have an overall detrimental effect upon the environment or upon something of value in that environment.

Polycyclic aromatic hydrocarbons (PAHs): Polycyclic aromatic hydrocarbons are a suite of compounds comprised of two or more condensed aromatic rings. They are found in many petroleum mixtures, and they are predominantly introduced to the environment through natural and anthropogenic combustion processes.

Polyforming: A process charging both C_3 and C_4 gases with naphtha or gas oil under thermal conditions to produce gasoline.

Polymer: A large molecule made by linking smaller molecules (monomers) together.

Polymerization: The combination of two olefin molecules to form a higher molecular weight paraffin.

Polymer gasoline: The product of polymerization of gaseous hydrocarbons to hydrocarbons boiling in the gasoline range.

Polynuclear aromatic compound: An aromatic compound having two or more fused benzene rings, e.g., naphthalene and phenanthrene.

Polysulfide treating: A chemical treatment used to remove elemental sulfur from refinery liquids by contacting them with a non-regenerable solution of sodium polysulfide.

PONA analysis: A method of analysis for paraffins (P), olefins (O), naphthenes (N), and aromatics (A).

Porphyrins: Organometallic constituents of petroleum that contain vanadium or nickel; the degradation products of chlorophyll that became included in the protopetroleum.

Positional isomers: Compounds which differ only in the position of a functional group; 2-pentanol and 3-pentanol are positional isomers.

Potentiation: A component having no effect increases the effect of another component.

Pour point: The lowest temperature at which an oil will appear to flow under ambient pressure over a period of five seconds. The *pour point* of crude oils generally varies from −60°C to 30°C. Lighter oils with low *viscosities* generally have lower *pour points*.

Powerforming: A fixed bed naphtha-reforming process using a regenerable platinum catalyst.

Precipitation: Formation of an insoluble product that occurs via reactions between ions or molecules in solution.

Precipitation number: The number of milliliters of precipitate formed when 10 mL of lubricating oil is mixed with 90 mL of petroleum naphtha of a definite quality and centrifuged under definitely prescribed conditions.

Primary oil recovery: Oil recovery utilizing only naturally occurring forces.

Primary structure: The chemical sequence of atoms in a molecule.

Primary substrates: The electron donor and electron acceptor that are essential to ensure the growth of microorganism; these compounds can be viewed as analogous to the food and oxygen that are required for human growth and reproduction.

Producers: Organisms or chemicals that utilize light energy and store it as chemical energy.

Prokaryotes: Microorganisms that lack a nuclear membrane so that their nuclear genetic material is more diffuse in the cell.

Propagule: Any part of a plant (e.g., bud) that facilitates dispersal of the species and from which a new plant may form.

Propane: A colorless, odorless, flammable gas (C_3H_8), found in petroleum and natural gas; used as a fuel and as a raw material for building more complex organic molecules.

Propane asphalt: See Solvent asphalt.

Propane deasphalting: Solvent deasphalting using propane as the solvent.

Propane decarbonizing: A solvent extraction process used to recover catalytic cracking feed from heavy fuel residues

Propane dewaxing: A process for dewaxing lubricating oils in which propane serves as solvent.

Propane fractionation: A continuous extraction process employing liquid propane as the solvent; a variant of propane deasphalting.

Propylene: A three-carbon, flammable, gaseous molecule containing a double-bond ($CH_3CH=CH_2$); another important olefin used in organic synthesis; also a base chemical to make polypropylene fibers, which are used in high-performance clothing, carpeting, and other products.

Protopetroleum: A generic term used to indicate the initial product-formed changes have occurred to the precursors of petroleum.

Protozoa: Microscopic animals consisting of single eukaryotic cells.

Purge and trap: A chromatographic sample introduction technique in volatile components that are purged from a liquid medium by bubbling gas through it. The components are then concentrated by "trapping" them on a short intermediate column, which is subsequently heated to drive the components on to the analytical column for separation.

Pyrobitumen: See Asphaltoid.

Pyrolysis: Exposure of a feedstock to high temperatures in an oxygen-poor environment.

Pyrophoric: Substances that catch fire spontaneously in air without an ignition source.

Quadrillion: 1×10^{15}

Quench: The sudden cooling of hot material discharging from a thermal reactor.

Radical (free radical): A molecular entity such as.$CH_3\bullet$, $Cl\bullet$ possessing an unpaired electron.

Radioactive decay: The process by which the nucleus of an unstable atom loses energy by emitting radiation.

Raffinate: That portion of the oil which remains undissolved in a solvent refining process.

Ramsbottom carbon residue: See Carbon residue

Rate: A derived quantity in which time is a denominator quantity so that the progress of a reaction is measured with time.

Rate constant, k: See *Order of reaction*.

Rate-controlling step (rate-limiting step, rate-determining step): The elementary reaction having the largest control factor exerts the strongest influence on the rate; a step having a control factor much larger than any other step is said to be rate-controlling.

Raw materials: Minerals extracted from the earth prior to any refining or treating.

Reactants: Substances initially present in a chemical reaction.

Reaction rate: The change in concentration of the starting chemical in given time interval.

Reaction (irreversible): A reaction in which the reactant(s) proceed to product(s), but there is no significant backward reaction:

$$nA + mB \rightarrow Products$$

In this reaction, the products do not recombine or change to form reactants in any appreciable amount.

Reaction (reversible): A reaction in which the products can revert to the starting materials (A and B). Thus:

$$nA + mB \leftrightarrow Products$$

Recalcitrant: Unreactive, nondegradable, refractory.

Receptor: An object (animal, vegetable, or mineral) or a locale that is affected by the pollutant.

Recycle ratio: The volume of recycle stock per volume of fresh feed; often expressed as the volume of recycle divided by the total charge.

Recycle stock: The portion of a feedstock which has passed through a refining process and is recirculated through the process.

Recycling: The use or reuse of chemical waste as an effective substitute for a commercial product or as an ingredient or feedstock in an industrial process.

Redox (reduction-oxidation reactions): Oxidation and reduction occur simultaneously; in general, the oxidizing agent gains electrons in the process (and is reduced) while the reducing agent donates electrons (and is oxidized).

Reduce : The transfer of electrons to a compound, such as oxygen, that occurs when another compound is oxidized.

Reduced crude: A residual product remaining after the removal, by distillation or other means, of an appreciable quantity of the more volatile components of crude oil.

Reducers: Organisms or chemicals that break down chemical compounds to more simple species and thereby extract the energy needed for their growth and metabolism

Reduction: The transfer of electrons to a compound, such as oxygen, that occurs when another compound is oxidized.

Reductive dehalogenation: A variation on biodegradation in which microbially catalyzed reactions cause the replacement of a halogen atom on an organic compound with a hydrogen atom. The reactions result in the net addition of two electrons to the organic compound.

Refinery: A series of integrated unit processes by which petroleum can be converted to a slate of useful (salable) products.

Refinery gas: A gas (or a gaseous mixture) produced as a result of refining operations.

Refining: The processes by which petroleum is distilled and/or converted by application of a physical and chemical processes to form a variety of products are generated.

Reformate: The liquid product of a reforming process.

Reformed gasoline: Gasoline made by a reforming process.

Reforming: The conversion of hydrocarbons with low octane numbers into hydrocarbons having higher octane numbers; e.g. the conversion of a n-paraffin into a isoparaffin.

Reformulated gasoline (RFG): Gasoline designed to mitigate smog production and to improve air quality by limiting the emission levels of certain chemical compounds such as benzene and other aromatic derivatives; often contains oxygenates

Refractive index (index of refraction): The ratio of wavelength or phase velocity of an electromagnetic wave in a vacuum to that in the substance; a measure of the amount of refraction a ray of light undergoes as it passes through a refraction interface: a useful physical property to identify a pure compound.

Regulation: A concept of management of complex systems according to a set of rules (laws) and trends; can take many forms: legal restrictions promulgated by a government authority, contractual obligations (such as contracts between insurers and their insureds), social regulation, coregulation, third-party regulation, certification, accreditation, or market regulation. See Law.

Reid vapor pressure: A measure of the volatility of liquid fuels, especially gasoline.

Regeneration: The reactivation of a catalyst by burning off the coke deposits.

Regenerator: A reactor for catalyst reactivation.

Releases: On-site discharge of a toxic chemical to the surrounding environment; includes emissions to the air, discharges to bodies of water, releases at the facility to land, as well as contained disposal into underground injection wells.

Releases (to air, point, and fugitive air emissions): All air emissions from industry activity; point emissions occur through confined air streams as found in stacks, ducts, or pipes; fugitive emissions include losses from equipment leaks, or evaporative losses from impoundments, spills, or leaks.

Releases (to land): Disposal of toxic chemicals in waste to on-site landfills, land treated or incorporation into soil, surface impoundments, spills, leaks, or waste piles. These activities must occur within the boundaries of the facility for inclusion in this category.

Release (to underground injection): A contained release of a fluid into a subsurface well for the purpose of waste disposal.

Releases (to water, surface water discharges): Any releases going directly to streams, rivers, lakes, oceans, or other bodies of water: any estimates for storm water runoff and nonpoint losses must also be included.

Rerunning: The distillation of an oil which has already been distilled.

Research octane method: A test for determining the knock rating, in terms octane numbers, of fuels for use in spark-ignition engines; see also Motor Octane Method.

Reservoir Rock: Highly permeable sedimentary rock (limestone, sand, or shale) through which petroleum may migrate, and given their structural and stratigraphic characteristics it forms a trap that is surrounded by a seal layer that will avoid the hydrocarbons escape.

Residual asphalt: See Straight-run asphalt.

Residual fuel oil: Obtained by blending the residual product(s) from various refining processes with suitable diluent(s) (usually middle distillates) to obtain the required fuel oil grades.

Residual oil: See Residuum; petroleum remaining *in situ* after oil recovery.

Residuum (resid; *pl*: residua): The residue obtained from petroleum after nondestructive distillation has removed all the volatile materials from crude oil, e.g., an atmospheric (345°C, 650°F+) residuum.

Resins: The name given to a large group of *polar compounds* in oil. These include hetero-substituted *aromatics*, acids, ketones, alcohols, and monoaromatic steroids. Because of their polarity, these compounds are more soluble in *polar* solvents, including water, than the nonpolar compounds, such as *waxes* and *aromatics*, of similar molecular weight. They are largely responsible for oil *adhesion*.

Respiration: The process of coupling oxidation of organic compounds with the reduction of inorganic compounds, such as oxygen, nitrate, iron (III), manganese (IV), and sulfate.

Retention time: The time it takes for an eluate to move through a chromatographic system and reach the detector. Retention times are reproducible and can, therefore, be compared to a standard for analyte identification.

Reversible reaction: A reaction in which the products can revert to the starting materials (A and B). Thus:

$$nA + mB \leftrightarrow \text{Products}$$

Rexforming: A process combining platforming with aromatics extraction, wherein low-octane raffinate is recycled to the platformer.

Rhizodegradation: The process whereby plants modify the environment of the root zone soil by releasing root exudates and secondary plant metabolites. Root exudates are typically photosynthetic carbon, low molecular weight molecules, and high molecular weight organic acids. This complex mixture modifies and promotes the development of a microbial community in the rhizosphere. These secondary metabolites have a potential role in the development of naturally occurring contaminant-degrading enzymes.

Rhizosphere: The soil environment encompassing the root zone of the plant.

Rich oil: Absorption oil containing dissolved natural gasoline fractions.

Riser: The part of the bubble-plate assembly which channels the vapor and causes it to flow downward to escape through the liquid; also the vertical pipe where fluid catalytic cracking reactions occur.

Rock asphalt: Bitumen which occurs in formations that have a limiting ratio of bitumen-to-rock matrix.

RRF: Relative response factor.

SARA analysis: A method of fractionation by which petroleum is separated into saturates, aromatics, resins, and asphaltene fractions.

SARA separation: See SARA analysis.

Saturated hydrocarbon: A saturated carbon-hydrogen compound with all carbon bonds filled; that is, there are no double or triple bonds, as in olefins or acetylenes.

Saturated solution: A solution in which no more solute will dissolve; a solution in equilibrium with the dissolved material.

Saturates: Paraffins and cycloparaffins (naphthenes).

Saturation: The maximum amount of solute that can be dissolved or absorbed under prescribed conditions.

Saybolt Furol viscosity: The time, in seconds (Saybolt Furol Seconds (SFS)), for 60 mL of fluid to flow through a capillary tube in a Saybolt Furol viscometer at specified temperatures between 70°F and 210°F; the method is appropriate for high-viscosity oils such as transmission, gear, and heavy fuel oils.

Saybolt Universal viscosity: The time, in seconds (Saybolt Universal Seconds (SUS)), for 60 mL of fluid to flow through a capillary tube in a Saybolt Universal viscometer at a given temperature.

Scale wax: The paraffin derived by removing the greater part of the oil from slack wax by sweating or solvent deoiling.

Scrubber: A device that uses water and chemicals to clean air pollutants from combustion exhaust.

Scrubbing: Purifying a gas by washing with water or chemical; less frequently, the removal of entrained materials.

Secondary pollutants: A pollutant (chemical species) produced by interaction of a primary pollutant with another chemical or by dissociation of a primary pollutant or by other effects within a particular ecosystem.

Secondary recovery: Oil recovery resulting from injection of water, or an immiscible gas at moderate pressure, into a petroleum reservoir after primary depletion.

Secondary structure: The ordering of the atoms of a molecule in space relative to each other.

Secondary tracer: The product of the chemical reaction between reservoir fluids and an injected primary tracer.

Sediment: An insoluble solid formed as a result of the storage instability and/or the thermal instability of petroleum and petroleum products.

Sedimentary: Formed by or from deposits of sediments, especially from sand grains or silts transported from their source and deposited in water, as sandstone and shale; or from calcareous remains of organisms, as limestone.

Sedimentary strata: Typically consist of mixtures of clay, silt, sand, organic matter, and various minerals; formed by or from deposits of sediments, especially from sand grains or silts transported from their source and deposited in water, such as sandstone and shale; or from calcareous remains of organisms, such as limestone.

Seismic section: A seismic profile that uses the reflection of seismic waves to determine the geological subsurface.

Selective solvent: A solvent which, at certain temperatures and ratios, will preferentially dissolve more of one component of a mixture than of another and thereby permitting partial separation.

Separation process: An upgrading process in which the constituents of petroleum are separated, usually without thermal decomposition, e.g., distillation and deasphalting.

Separator-Nobel dewaxing: A solvent (tricholoethylene) dewaxing process.

Separatory funnel: Glassware shaped like a funnel with a stoppered rounded top and a valve at the tapered bottom, used for liquid/liquid separations.

Shale: A very fine-grained sedimentary rock that is formed by the consolidation of clay, mud, or silt and that usually has a finely stratified or laminated structure. Certain shale formations, such as the Eagle Ford and the Barnett, contain large amounts of oil and natural gas.

Shear: Mechanical deformation or distortion, or partial destruction of a polymer molecule as it flows at a high rate.

Shear rate: A measure of the rate of deformation of a liquid under mechanical stress.

Shear-thinning: The characteristic of a fluid whose viscosity decreases as the shear rate increases.

Shell fluid catalytic cracking: A two-stage fluid catalytic cracking process in which the catalyst is regenerated.

Shell still: A still formerly used in which the oil was charged into a closed, cylindrical shell and the heat required for distillation was applied to the outside of the bottom from a firebox.

Side chain: A chain of atoms which is attached to a longer chain of atoms; examples of side chains would be methyl, ethyl, and propyl groups (among others).

Sidestream: A liquid stream taken from any one of the intermediate plates of a bubble tower.

Sidestream stripper: A device used to perform further distillation on a liquid stream from any one of the plates of a bubble tower, usually by the use of steam.

SIM (Selecting Ion Monitoring): Mass spectrometric monitoring of a specific mass/charge (m/z) ratio. The *SIM* mode offers better sensitivity than can be obtained using the full scan mode.

Simple inorganic chemicals: Molecules that consist of one-type atoms (atoms of one element) which, in chemical reactions, cannot be decomposed to form other chemicals.

Single displacement reactions: Reactions where one element trades places with another element in a compound. These reactions come in the general form of:

$$A + BC \rightarrow AC + B$$

Examples include: (i) magnesium replacing hydrogen in water to make magnesium hydroxide and hydrogen gas:

$$Mg + 2H_2O \rightarrow Mg(OH)_2 + H_2$$

(ii) the production of silver crystals when a copper metal strip is dipped into silver nitrate:

$$Cu(s) + 2AgNO_3(aq) \rightarrow 2Ag(s) + Cu(NO_3)_2(aq)$$

Slack wax: The soft, oily crude wax obtained from the pressing of paraffin distillate or wax distillate.

Slime: A name used for petroleum in ancient texts.

Slim tube testing: Laboratory procedure for the determination of minimum miscibility pressure using long, small diameter, sand-packed, oil-saturated, stainless steel tube.

Sludge: A semisolid to solid product which results from the storage instability and/or the thermal instability of petroleum and petroleum products.

Slug: A quantity of fluid injected into a reservoir during enhanced oil recovery.

Slurry hydroconversion process: A process in which the feedstock is contacted with hydrogen under pressure in the presence of a catalytic coke-inhibiting additive.

Slurry phase reactors: Tanks into which wastes, nutrients, and microorganisms are placed.

Smoke: The particulate material assessed in terms of its blackness or reflectance when collected on a filter, as opposed to its mass; this is the historical method of measurement of particulate pollution; particles formed by incomplete combustion of fuel.

Smoke point: A measure of the burning cleanliness of jet fuel and kerosene.

Sodium hydroxide treatment: See Caustic wash.

Sodium plumbite: A solution prepared front a mixture of sodium hydroxide, lead oxide, and distilled water; used in making the doctor test for light oils such as gasoline and kerosene.

Soil organic matter: Living and partially decayed (nonliving) materials as well as assemblages of biomolecules and transformation products of organic residue decay known as humic substances.

Solid state compounds: A diverse class of compounds that are solid at standard temperature and pressure, and exhibit unique properties as semiconductors, etc.

Solubility: The amount of a substance (solute) that dissolves in a given amount of another substance (solvent); a measure of the solubility of an inorganic chemical in a solvent, such as water; generally, ionic substances are soluble in water and other polar solvents while the nonpolar, covalent compounds are more soluble in the nonpolar solvents; in sparingly soluble, slightly soluble, or practically insoluble salts, degree of solubility in water and occurrence of any precipitation process may be determined from the solubility product, Ksp, of the salt—the smaller the Ksp value, the lower the solubility of the salt in water.

Solubility parameter: A measure of the solvent power and polarity of a solvent.

Soluble: Capable of being dissolved in a solvent.

Solute: Any dissolved substance in a solution.

Solution: Any liquid mixture of two or more substances that is homogeneous.

Solutizer-steam regenerative process: A chemical treating process for extracting mercaptans from gasoline or naphtha, using solutizers (potassium isobutyrate, potassium alkyl phenolate) in strong potassium hydroxide solution.

Solvent: A liquid in which certain kinds of molecules dissolve. While they typically are liquids with low-boiling points, they may include high-boiling liquids, supercritical fluids, or gases.

Solvent asphalt: The asphalt produced by solvent extraction of residua or by light hydrocarbon (propane) treatment of a residuum or an asphaltic crude oil.

Solvent deasphalting: A process for removing asphaltic and resinous materials from reduced crude oils, lubricating-oil stocks, gas oils, or middle distillates through the extraction or precipitant action of low molecular weight hydrocarbon solvents; see also Propane deasphalting.

Solvent decarbonizing: See Propane decarbonizing.

Solvent deresining: See Solvent deasphalting.

Solvent dewaxing: A process for removing wax from oils by means of solvents usually by chilling a mixture of solvent and waxy oil, filtration or by centrifuging the wax which precipitates, and solvent recovery.

Solvent extraction: A process for separating liquids by mixing the stream with a solvent that is immiscible with part of the waste but that will extract certain components of the waste stream.

Solvent gas: An injected gaseous fluid that becomes miscible with oil under reservoir conditions and improves oil displacement.

Solvent naphtha: A refined naphtha of restricted boiling range used as a solvent; also called petroleum naphtha; petroleum spirits.

Solvent refining: See Solvent extraction.

Solvolysis: Generally, a reaction with a solvent, involving the rupture of one or more bonds in the reacting solute: more specifically the term is used for substitution, elimination, or fragmentation reactions in which a solvent species is the nucleophile; hydrolysis, if the solvent is water or alcoholysis if the solvent is an alcohol.

Sorbate: Sometimes referred to as adsorbate, is the solute that adsorbs on the solid phase.

Sorbent (adsorbent): The solid phase or substrate onto which the sorbate sorbs; the solid phase may be more specifically referred to as an absorbent or adsorbent if the mechanism of removal is known to be absorption or adsorption, respectively.

Sorption: A general term that describes removal of a solute from solution to a contiguous solid phase and is used when the specific removal mechanism is not known.

Sorption isotherm: Graphical representation of surface excess (i.e., the amount of substance sorbed to a solid) relative to sorptive concentration in solution after reaction at fixed temperature, pressure, ionic strength, pH, and solid-to-solution ratio.

Sorptive: Ions or molecules in solution that could potentially participate in a sorption reaction.

Source Rock: Sedimentary rock formed by very fine grain and with an abundant content of organic carbon which is deposited under reducing and low-energy conditions, generating hydrocarbons over time.

Sour crude oil: Crude oil containing an abnormally large amount of sulfur compounds; see also Sweet crude oil.

Sour gas: Natural gas or any other gas containing significant amounts of hydrogen sulfide (H_2S).

SOx: Oxides of sulfur.

Soxhlet extraction: An extraction technique for solids in which the sample is repeatedly contacted with solvent over several hours, increasing extraction efficiency.

Specific gravity: The mass (or weight) of a unit volume of any substance at a specified temperature compared to the mass of an equal volume of pure water at a standard temperature; see also Density.

Specific heat: The amount of heat required to raise the temperature of one gram of a substance by 1°C; the specific heat of water is 1 calorie or 4.184 Joule.

Spent catalyst: Catalyst that has lost much of its activity due to the deposition of coke and metals.

Spontaneous ignition: Ignition of a fuel, such as coal, under normal atmospheric conditions; usually induced by climatic conditions.

Stabilization: The removal of volatile constituents from a higher boiling fraction or product (stripping); the production of a product which, to all intents and purposes, does not undergo any further reaction when exposed to the air.

Stabilizer: A fractionating tower for removing light hydrocarbons from an oil to reduce vapor pressure particularly applied to gasoline.

Stable: As applied to chemical species, the term expresses a thermodynamic property, which is quantitatively measured by relative molar standard Gibbs energies; a chemical species A is more stable than its isomer B under the same standard conditions.

Standard conditions: The reference amounts for pressure and temperature—in the English system, they are 14.73 pounds per square inch for the pressure and 60°F for temperature.

Standard potential: Used to predict if a species will be oxidized or reduced in solution (under acidic or basic conditions) and whether any oxidation-reduction reaction will take place.

Starch: A polysaccharide containing glucose (long-chain polymer of amylose and amylopectin) that is the energy storage reserve in plants.

Stationary phase: In chromatography, the porous solid or liquid phase through which an introduced sample passes. The different affinities the stationary phase has for a sample allow the components in the sample to be separated, or resolved.

Steam cracking: A conversion process in which the feedstock is treated with superheated steam.

Steam distillation: Distillation in which vaporization of the volatile constituents is effected at a lower temperature by introduction of steam (open steam) directly into the charge.

Stereochemistry: The branch of organic chemistry that deals with the three-dimensional structure of molecules.

Stereogenic carbon (asymmetric carbon): A carbon atom which is bonded to four different groups or atoms; a chiral molecule must contain a stereogenic carbon, and therefore has no plane of symmetry and is not superimposable on its mirror image.

Stereoisomers: Isomers, which have the same bonding connectivity but have a different three-dimensional structure; examples would be *cis*-2-butene and *trans*-2-butene (geometric isomers), and the left- and right-handed forms of 2-butanol (enantiomers).

Stern layer: The layer of ions adsorbed immediately adjacent to a charged sorbent surface. Ions in the Stern layer can be directly bonded to the sorbent through covalent and ionic bonds (inner-sphere complexes) or held adjacent to a sorbent through strictly electrostatic forces in outer-sphere complexes.

Stoichiometry: The calculation of the quantities of reactants and products (among elements and compounds) involved in a chemical reaction.

Stoke's Law

$$v = \frac{gd^2(\rho_1 - \rho_2)}{18\eta}.$$

Storage stability (or storage instability): the ability (inability) of a liquid to remain in storage over extended periods of time without appreciable deterioration as measured by gum formation and the depositions of insoluble material (sediment).

Straight-run asphalt: The asphalt produced by the distillation of asphaltic crude oil.

Straight-run products: Obtained from a distillation unit and used without further treatment.

Stratosphere: The portion of the atmosphere of the earth where ozone is formed by the reaction of ultraviolet light on dioxygen molecules.

Straw oil: Pale paraffin oil of straw color used for many process applications.

Stripping: A means of separating volatile components from less volatile ones in a liquid mixture by the partitioning of the more volatile materials to a gas phase of air or steam; see Stabilization.

Strong acid: An acid that releases H^+ ions easily—examples are hydrochloric acid and sulfuric acid.

Strong base: A basic chemical that accept and hold proton tightly—an example is the hydroxide ion.

Structural formula: A convention used to represent the structures of organic molecules in which not all the valence electrons of the atoms are shown.

Structural isomerism: The relationship between two compounds which have the same molecular formula, but different structures; they may be further classified as functional, positional, or skeletal isomers. This relation is also called constitutional isomerism.

Styrene: A human-made chemical used mostly to make rubber and plastics; present in combustion products, such as cigarette smoke and automobile exhaust.

Sublimation: The direct vaporization or transition of a solid directly to a vapor without passing through the liquid state.

Substitution reaction: The process in which one group or atom in a molecule is replaced by another group or atom.

Substrate: A chemical species of particular interest, of which the reaction with some other chemical reagent is under observation (e.g., a compound that is transformed under the influence of a catalyst); also the component in a nutrient medium, supplying microorganisms with carbon (C-substrate), nitrogen (N-substrate) as food needed to grow.

Sulfonic acids: Acids obtained by petroleum or a petroleum product with strong sulfuric acid.

Sulfuric acid alkylation: An alkylation process in which olefins (C_3, C_4, and C_5) combine with isobutane in the presence of a catalyst (sulfuric acid) to form branched chain hydrocarbons used especially in gasoline blending stock.

Supercritical fluid: An extraction method where the extraction fluid is present at a pressure and temperature above its critical point.

Super-light oil: Oil with a specific gravity typically higher than 38° API.

Surface-active agent: A compound that reduces the surface tension of liquids, or reduces interfacial tension between two liquids or a liquid and a solid; also known as surfactant, wetting agent, or detergent.

Surface tension: Caused by molecular attractions between the molecules of two liquids at the surface of separation.

Surfactant: A type of chemical, characterized as one that reduces interfacial resistance to mixing between oil and water or changes the degree to which water wets reservoir rock.

Suspensoid catalytic cracking: A non-regenerative cracking process in which cracking stock is mixed with slurry of catalyst (usually clay) and cycle oil and passed through the coils of a heater.

Sustainable development: Development and economic growth that meets the requirements of the present generation without compromising the ability of future generations to meet their needs; a strategy seeking a balance between development and conservation of natural resources.

Sustainable enhancement: An intervention action that continues until such time that the enhancement is no longer required to reduce contaminant concentrations or fluxes.

Steranes: A class of tetracyclic, saturated biomarkers constructed from six isoprene subunits ($\sim C_{30}$). *Steranes* are derived from sterols, which are important membrane and hormone components in eukaryotic organisms. Most commonly used *steranes* are in the range of C_{26}–C_{30} and are detected using m/z 217 mass chromatograms.

Surrogate analyte: A pure analyte that is extremely unlikely to be found in any sample, which is added to a sample aliquot in a known amount and is measured with the same procedures used to measure other components. The purpose of a *surrogate analyte* is to monitor the method performance with each sample.

Sweated wax: A crude wax that was freed from oil by having been passed through a heater (sweater).

Sweating: The separation of paraffin oil and low melting wax from paraffin wax.

Sweep efficiency: The ratio of the pore volume of reservoir rock contacted by injected fluids to the total pore volume of reservoir rock in the project area. (*See also* areal sweep efficiency *and* vertical sweep efficiency.)

Sweet crude oil: Crude oil containing little sulfur; see also Sour crude oil.

Sweetening: The process by which petroleum products are improved in odor and color by oxidizing or removing the sulfur-containing and unsaturated compounds.

Swelling: Increase in the volume of crude oil caused by absorption of EOR fluids, especially carbon dioxide. Also increase in volume of clays when exposed to brine.

Synthesis gas: A mixture of carbon monoxide and hydrogen used especially in chemical synthesis to make hydrocarbon derivatives.

Synthetic crude oil (syncrude): A hydrocarbon product produced by the conversion of coal, oil shale, or tar sand bitumen that resembles conventional crude oil; can be refined in a petroleum refinery.

Synergism: The effect of the combination is greater than the sum of individual effects.

Tar: The volatile, brown to black, oily, viscous product from the destructive distillation of many bituminous or other organic materials, especially coal; a name used for petroleum in ancient texts.

Target analyte: Target analytes are compounds that are required analytes in U.S. EPA analytical methods. BTEX and PAHs are examples of petroleum-related compounds that are target analytes in U.S. EPA Methods.

Tar sand: See Bituminous sand.

Terminal electron acceptor (TEA): A compound or molecule that accepts an electron (is reduced) during metabolism (oxidation) of a carbon source; under aerobic conditions molecular oxygen is the terminal electron acceptor; under anaerobic conditions a variety of terminal electron acceptors may be used. In order of decreasing redox potential, these terminal electron acceptors include nitrate, manganese (Mn^{3+}, Mn^{6+}), iron (Fe^{3+}), sulfate, and carbon dioxide; microorganisms preferentially utilize electron acceptors that provide the maximum free energy during respiration; of the common terminal electron acceptors listed above, oxygen has the highest redox potential and provides the most free energy during electron transfer.

Terpanes: A class of branched, cyclic alkane biomarkers including *hopanes* and tricyclic compounds.

Terpenes: Hydrocarbon solvents, compounds composed of molecules of hydrogen and carbon; they form the primary constituents in the aromatic fractions of scented plants, e.g., pine oil, as well as turpentine and camphor oil.

Terrestrial biosphere: The part of the geosphere consisting of the outer mantle and the crust that is directly involved with environmental processes through contact with the atmosphere, the hydrosphere, and living things; varies from (approximately) 40 to 60 miles in thickness; also called the lithosphere.

Tertiary structure: The three-dimensional structure of a molecule.

Tetrachloroethylene (perchloroethylene): A human-made chemical that is widely used for dry cleaning of fabrics and for metal-degreasing operations; also used as a starting material (building block) for making other chemicals and is used in some consumer products such as water repellents, silicone lubricants, fabric finishers, spot removers, adhesives, and wood cleaners; can stay in the air for a long time before breaking down into other chemicals or coming back to the soil and water in rain; much of the tetrachloroethylene that gets into water and soil will evaporate; because tetrachloroethylene can travel easily through soils, it can get into underground drinking water supplies.

Thermal coke: The carbonaceous residue formed as a result of a non-catalytic thermal process; the Conradson carbon residue; the Ramsbottom carbon residue.

Thermal conductivity: A measure of the rate of transfer of heat by conduction through unit thickness, across unit area for unit difference of temperature; measured as calories per second per square centimeter for a thickness of one centimeter and a temperature difference of $1°C$; units are cal/cm sec.°K or W/cm°K.

Thermal cracking: A process which decomposes, rearranges, or combines hydrocarbon molecules by the application of heat, without the aid of catalysts.

Thermal polymerization: A thermal process to convert light hydrocarbon gases into liquid fuels.

Thermal process: Any refining process which utilizes heat, without the aid of a catalyst.

Thermal recovery: See EOR process.

Thermal reforming: A process using heat (but no catalyst) to effect molecular rearrangement of low-octane naphtha into gasoline of higher antiknock quality.

Thermal stability (thermal instability): The ability (inability) of a liquid to withstand relatively high temperatures for short periods of time without the formation of carbonaceous deposits (sediment or coke).

Thermodynamic equilibrium: The thermodynamic state that is characterized by absence of flow of matter or energy.

Thermodynamics: The study of the energy transfers or conversion of energy in physical and chemical processes' defines the energy required to start a reaction or the energy given out during the process.

Thermofor catalytic cracking: A continuous, moving bed catalytic cracking process.

Thermofor catalytic reforming: A reforming process in which the synthetic, bead-type catalyst of coprecipitated chromia (Cr_2O_3) and alumina (Al_2O_3) flows down through the reactor concurrent with the feedstock.

Thermofor continuous percolation: A continuous clay treating process to stabilize and decolorize lubricants or waxes.

Thermogravimetric analysis (TGA) and differential thermal analysis (DTA): Techniques that may be used to measure the water of crystallization of a salt and the thermal decomposition of hydrates.

Thin layer chromatography (TLC): A chromatographic technique employing a porous medium of glass coated with a stationary phase. An extract is spotted near the bottom of the medium and placed in a chamber with solvent (mobile phase). The solvent moves up the medium and separates the components of the extract, based on affinities for the medium and solvent.

Tight gas: Natural gas produced from relatively impermeable rock. Getting tight gas out usually requires enhanced technology applications like hydraulic fracturing; the term is generally used for reservoirs other than shale when the gas is referred to as shale gas.

Toluene: A clear, colorless aromatic liquid (also called methyl benzene, $C_6H_5CH_3$); occurs naturally in crude oil and in the tolu tree; produced in the process of making gasoline and other fuels from crude oil; used in making paints, paint thinners, fingernail polish, lacquers, adhesives, and rubber, and in some printing and leather tanning processes; a major component of JP-8 fuel.

Topped crude: Petroleum that has had volatile constituents removed up to a certain temperature, e.g., 250°C+ (480°F+) topped crude; not always the same as a residuum

Topping: The distillation of crude oil to remove light fractions only

Total n-alkanes: The sum of all resolved *n-alkanes* (from C_8 to C_{40} plus pristane and phytane).

Total 5 alkylated PAH homologs: The sum of the 5 target PAHs (naphthalene, phenanthrene, dibenzothiophene, fluorene, chrysene) and their alkylated (C_1–C_4) homologues, as determined by GCMS. These 5 target alkylated PAH homologous series are oil-characteristic aromatic compounds.

Total aromatics: The sum of all resolved and unresolved aromatic hydrocarbons including the total of BTEX and other alkyl benzene compounds, total 5 target alkylated PAH homologues, and other EPA priority PAHs.

Total saturates: The sum of all resolved and unresolved aliphatic hydrocarbons including the total n-alkanes, branched alkanes, and cyclic saturates.

Total suspended particulate matter: The mass concentration determined by filter weighing, usually using a specified sampler which collects all particles up to approximately 20 microns depending on wind speed.

Tower: Equipment for increasing the degree of separation obtained during the distillation of crude oil in a still.

Toxicity: A measure of the toxic nature of a chemical; usually expressed quantitatively as LD_{50} (median lethal dose) or LC_{50} (median lethal concentration in air)—the latter refers to inhalation toxicity of gaseous substances in air; both terms refer to the calculated concentration of a chemical that can kill 50% of test animals when administered.

Toxicological chemistry: The chemistry of toxic substances with emphasis upon their interactions with biologic tissue and living organisms.

TPH: Total petroleum hydrocarbons; the total measurable amount of petroleum-based hydrocarbons present in a medium as determined by gravimetric or chromatographic means.

Transfers: A transfer of toxic (organic) chemicals in wastes to a facility that is geographically or physically separate from the facility reporting under the toxic release inventory; the quantities reported represent a movement of the chemical away from the reporting facility; except for off-site transfers for disposal, these quantities do not necessarily represent entry of the chemical into the environment.

Transfers (POTWs): Waste waters transferred through pipes or sewers to a publicly owned treatment works (POTW); treatment and chemical removal depend on the chemical's nature and treatment methods used; chemicals not treated or destroyed by the publicly owned treatment works are generally released to surface waters or land filled within the sludge.

Transfers (to disposal): Wastes that are taken to another facility for disposal generally as a release to land or as an injection underground.

Transfers (to energy recovery): Wastes combusted off-site in industrial furnaces for energy recovery; treatment of an organic chemical by incineration is not considered to be energy recovery.

Transfers (to recycling): Wastes that are sent off-site for the purposes of regenerating or recovering still valuable materials; once these chemicals have been recycled, they may be returned to the originating facility or sold commercially.

Transfers (to treatment): Wastes moved off-site for either neutralization, incineration, biological destruction, or physical separation; in some cases, the chemicals are not destroyed but prepared for further waste management.

Treatment: Any method, technique, or process that changes the physical and/or chemical character of petroleum.

1,1,1-Trichloroethane: Does not occur naturally in the environment; used in commercial products, mostly to dissolve other chemicals; beginning in 1996, 1,1,1-trichloroethane was no longer made in the United States because of its effects on the ozone layer; because of its tendency to evaporate easily, the vapor form is usually found in the environment; 1,1,1-trichloroethane also can be found in soil and water, particularly at hazardous waste sites.

Trichloroethylene: A colorless liquid that does not occur naturally; mainly used as a solvent to remove grease from metal parts and is found in some household products, including typewriter correction fluid, paint removers, adhesives, and spot removers.

Trickle hydrodesulfurization: A fixed bed process for desulfurizing middle distillates.

Triglyceride: An ester of glycerol and three fatty acids; the fatty acids represented by "R" can be the same or different.

$$
\begin{array}{l}
\text{RCOOCH}_2 \\
\;\;\;\;\;| \\
\text{RCOOCH}_2 \\
\;\;\;\;\;| \\
\text{RCOOCH}_2
\end{array}
$$

Trillion: 1×10^{12}

Triterpanes: A class of cyclic saturated *biomarkers* constructed from six isoprene subunits; cyclic terpane compounds containing two, four, and six isoprene subunits are called monoterpane (C_{10}), diterpane (C_{20}) and *triterpane* (C_{30}), respectively.

Trophic: The trophic level of an organism is the position it occupies in a food chain.

Troposphere: The portion of the atmosphere of the earth that is closest to the surface.

True boiling point (True boiling range): The boiling point (boiling range) of a crude oil fraction or a crude oil product under standard conditions of temperature and pressure.

Tube-and-tank cracking: An older liquid-phase thermal cracking process.

Tyndall effect: The characteristic light scattering phenomenon of colloids results from those being the same order of size as the wavelength of light.

UCM: Unresolved complex mixture of hydrocarbons on, for example, a gas chromatographic tracing; the UCM appear as the *envelope* or *hump area* between the solvent baseline and the curve defining the base of resolvable peaks.

Ultimate analysis: Elemental composition.

Ultrafining: A fixed bed catalytic hydrogenation process to desulfurize naphtha and upgrade distillates by essentially removing sulfur, nitrogen, and other materials.

Ultraforming: A low-pressure naphtha-reforming process employing onstream regeneration of a platinum-on-alumina catalyst and producing high yields of hydrogen and high-octane number reformate.

Ultraviolet radiation (UV radiation): An electromagnetic radiation with a wavelength from 10 to 400 nm, shorter than the wavelength of visible light but longer than the wavelength of X-rays. UV radiation is present in sunlight constituting about 10% of the total light output of the Sun.

Unassociated molecular weight: The molecular weight of asphaltenes in an nonassociating (polar) solvent, such as dichlorobenzene, pyridine, or nitrobenzene.

Underground storage tank: A storage tank that is partially or completely buried in the earth.

Unifining: A fixed bed catalytic process to desulfurize and hydrogenate refinery distillates.

Unisol process: A chemical process for extracting mercaptan sulfur and certain nitrogen compounds from sour gasoline or distillates using regenerable aqueous solutions of sodium or potassium hydroxide containing methanol.

Universal viscosity: See Saybolt Universal viscosity.

Unresolved complex: The thousands of compounds that a gas chromatograph *mixture (UCM)* is unable to fully separate.

Unsaturated compound: An organic compound with molecules containing one or more double bonds.

Unsaturated zone: The zone between land surface and the capillary fringe within which the moisture content is less than saturation and pressure is less than atmospheric; soil pore spaces also typically contain air or other gases; the capillary fringe is not included in the unsaturated zone (See *Vadose zone*).

Unstable: Usually refers to a petroleum product that has more volatile constituents present or refers to the presence of olefin and other unsaturated constituents.

UOP alkylation: A process using hydrofluoric acid (which can be regenerated) as a catalyst to unite olefins with isobutane.

UOP copper sweetening: A fixed bed process for sweetening gasoline by converting mercaptans to disulfides by contact with ammonium chloride and copper sulfate in a bed.

UOP fluid catalytic cracking: A fluid process of using a reactor-over-regenerator design.

Upgradient: In the direction of increasing potentiometric (piezometric) head. See also *Downgradient*.

Upgrading: The conversion of petroleum to value-added salable products.

Upper-phase microemulsion: A microemulsion phase containing a high concentration of oil that, when viewed in a test tube, resides on top of a water phase.

Urea dewaxing: A continuous dewaxing process for producing low-pour-point oils, and using urea which forms a solid complex (adduct) with the straight-chain wax paraffins in the stock; the complex is readily separated by filtration.

US EPA: United States Environmental Protection Agency.

USGS: United States Geological Survey.

Vacuum distillation: Distillation under reduced pressure.

Vacuum residuum: A residuum obtained by distillation of a crude oil under vacuum (reduced pressure); that portion of petroleum which boils above a selected temperature such as 510°C (950°F) or 565°C (1,050°F).

Vadose zone: The zone between land surface and the water table within which the moisture content is less than saturation (except in the capillary fringe) and pressure is less than atmospheric; soil pore spaces also typically contain air or other gases; the capillary fringe is included in the vadose zone.

Valence state of an atom: The power of an atom to combine to form compounds; determines the chemical properties.

Van der Waals forces: Intermolecular attractive forces that arise between nonionic, nonpolar molecules due to dipole-dipole interactions and instantaneous dipole interactions (London dispersion forces).

Van der Waals interaction: The cohesive interaction (attraction between like) or the adhesive interaction (attraction between unlike) and/or repulsive forces between molecules.

Vapor-phase cracking: A high-temperature, low-pressure conversion process.

Vapor-phase hydrodesulfurization: A fixed bed process for desulfurization and hydrogenation of naphtha.

Vapor pressure: The pressure exerted by a solid or liquid in equilibrium with its own vapor; depends on temperature and is characteristic of each substance; the higher the vapor pressure at ambient temperature, the more volatile the substance.

VGC (viscosity-gravity constant): An index of the chemical composition of crude oil defined by the general relation between specific gravity, sg, at 60°F and Saybolt Universal viscosity, SUV, at 100°F:

$$a = 10sg - 1.0752\log(SUV - 38)/10sg - \log(SUV - 38)$$

The constant, a, is low for the paraffin crude oils and high for the naphthenic crude oils.

VI (Viscosity index): An arbitrary scale used to show the magnitude of viscosity changes in lubricating oils with changes in temperature.

Visbreaking: A process for reducing the viscosity of heavy feedstocks by controlled thermal decomposition.

Viscosity: The resistance of a fluid to shear, movement or flow; a function of the composition of a fluid; the viscosity of an ideal, noninteracting fluid does not change with shear rate—such fluids are called Newtonian; expressed as g/cm sec or Poise; 1 Poise = 100 centipoise.

Viscosity-gravity constant: See VGC.

Viscosity index-: See VI.

VOC (VOCs): Volatile organic compound(s); volatile organic compounds are regulated because they are precursors to ozone; carbon-containing gases and vapors from incomplete gasoline combustion and from the evaporation of solvents.

Volatile: Readily dissipating by evaporation.

Volatile compounds: A relative term that may mean (i) any compound that will purge, (ii) any compound that will elute before the solvent peak (usually those $< C_6$), or (iii) any compound that will not evaporate during a solvent removal step.

Volatile organic compounds (VOC): Organic compounds with high vapor pressures at normal temperatures. VOCs include light saturates and aromatics, such as pentane, hexane, BTEX, and other lighter substituted benzene compounds, which can make up to a few percent of the total mass of some crude oils.

Water solubility: The maximum amount of a chemical that can be dissolved in a given amount of pure water at standard conditions of temperature and pressure; typical units are milligrams per liter (mg/L), gallons per liter (g/L), or pounds per gallon (lbs/gal).

Watson characterization factor: See Characterization factor.

Wax: See Mineral wax and Paraffin wax.

Wax distillate: A neutral distillate containing a high percentage of crystallizable paraffin wax, obtained on the distillation of paraffin or mixed-base crude, and on reducing neutral lubricating stocks.

Waxes: Waxes are predominately straight-chain *saturates* with melting points above 20°C (generally, the *n*-alkanes C_{18} and higher molecular weight).

Wax fractionation: A continuous process for producing waxes of low oil content from wax concentrates; see also MEK deoiling.

Wax manufacturing: A process for producing oil-free waxes.

Weak Acid: An acid that does not release H^+ ions easily—an example is acetic acid (CH_3CO_2H).

Weak base: A basic chemical that has little affinity for a proton—an example is the chloride ion.

Weathered crude oil: Crude oil which, due to natural causes during storage and handling, has lost an appreciable quantity of its more volatile components; also indicates uptake of oxygen.

Weathering: Processes related to the physical and chemical actions of air, water, and organisms after oil spill. The major weathering processes include evaporation, dissolution, dispersion, photochemical oxidation, water-in-oil *emulsification*, microbial degradation, adsorption onto suspended particulate materials, interaction with mineral fines, sinking, sedimentation, and formation of tar balls.

Wellbore: A hole drilled by a drilling rig to explore for or develop oil and/or natural gas; also referred to as a well or borehole.

Wellhead: The control equipment adjusted to the wellhead which is used to control the flow and prevent explosions and it consists of piping, valves, power outlets, and blowup preventers.

Wet Deposition: The term used to describe pollutants brought to ground either by rainfall or by snow; this mechanism can be further subdivided depending on the point at which the pollutant was absorbed into the water droplets.

Wet gas: Gas containing a relatively high proportion of hydrocarbons which are recoverable as liquids; see also Lean gas.

Wet scrubbers: Devices in which a countercurrent spray liquid is used to remove impurities and particulate matter from a gas stream.

Wettability: The relative degree to which a fluid will spread on (or coat) a solid surface in the presence of other immiscible fluids.

Wettability number: A measure of the degree to which a reservoir rock is water-wet or oil-wet, based on capillary pressure curves.

Wettability reversal: The reversal of the preferred fluid wettability of a rock, e.g., from water-wet to oil-wet, or vice versa.

White oil: A generic tame applied to highly refined, colorless hydrocarbon oils of low volatility, and covering a wide range of viscosity.

Wilting point: The largest water content of a soil at which indicator plants, growing in that soil, wilt and fail to recover when placed in a humid chamber.

Wobbe Index (or Wobbe Number): The calorific value of a gas divided by the specific gravity.

Wood alcohol: See Methyl alcohol.

Xylenes: The term that refers to all three types of xylene isomers (meta-xylene, ortho-xylene, and para-xylene); produced from crude oil; used as a solvent and in the printing, rubber, and leather industries as well as a cleaning agent and a thinner for paint and varnishes; a major component of JP-8 fuel.

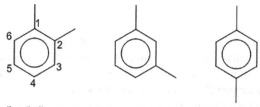

1,2-dimethylbenzene 1,3-dimethylbenzene 1,4-dimethylbenzene
(*ortho*-xylene)   (*meta*-xylene)   (*para*-xylene)

Yield: The mass (or moles) of a chosen final product divided by the mass (or moles) of one of the initial reactants.

Zeolite: A crystalline aluminosilicate used as a catalyst and having a particular chemical and physical structure.

Zwitterion: A particle that contains both positively charged and negatively charged groups; for example, amino acids ($H_2NHCHRCO_2H$) can form zwitterions ($^+H_3NCHRCOO^-$).

Index